Fundamentals of Organic Synthesis
The Retrosynthetic Analysis

Volume II

For Undergraduate (Honours) and Postgraduate Students

Fundamentals of Organic Synthesis
The Retrosynthetic Analysis

Volume II

Dr Ratan Kumar Kar, MSc, PhD
Reader in Chemistry
Vivekananda College (Thakurpukur)
The University of Calcutta

New Central Book Agency (P) Ltd

Co-Published by

Bijay Publications

E-mail : bijaypublications@gmail.com

Exclusively Distributed by

Bijay Pustak Bhandar (P) Ltd

90/6A M. G. Road, Bhawani Dutta Lane, Kolkata-700007

✆ : (033) 4605-7856, 8777898576, 8274916104

E-mail : bijaypustakbhandar@gmail.com

NCBA

REGD OFFICE
8/1 Chintamoni Das Lane, Kolkata 700 009, India
email: ncbakolkata@gmail.com

OVERSEAS
NCBA (UK) Ltd, 149 Park Avenue North, Northampton, NN3 2HY, UK
email: ncbauk@yahoo.com

CORPORATE OFFICE
212 Shahpur Jat, New Delhi 110 049
email: ncbaexports@ncbapvtltd.com

BIJAY PUBLICATIONS
90/6A M. G. Road, Bhawani Dutta Lane, Kolkata-7
E-mail : bijaypublications@gmail.com

BIJAY PUSTAK BHANDAR (P) LTD
90/6A M. G. Road, Bhawani Dutta Lane, Kolkata-7
E-mail : bijaypustakbhandar@gmail.com

FUNDAMENTALS OF ORGANIC SYNTHESIS THE RETROSYNTHETIC ANALYSIS (Vol II) ● Dr R K Kar

First Published: January 2007
Reprinted: 2008, 2018, **2023**

PUBLISHER • TYPESETTER • PRINTER
New Central Book Agency (P) Ltd
8/1 Chintamoni Das Lane, Kolkata 700 009

CO-PUBLISHED BY
Bijay Publications
90/6A M G Road, Bhawani Dutta Lane, Kolkata 700007

PROOF CHECKING
Author

PROJECT SUPERVISOR
Sukumar Patra

PROJECT TEAM
Joydeep Ghosh, Prabhat Jas and Ananta Bhowmik

ISBN: 81 7381 532 1

Price ₹ 995.00

Dedicated to my Children
Shramana and Shilanjan

Contents

Preface

While teaching, I have often felt that students of Chemistry Honours need a handy and concise book on Organic Synthesis—a volume comprising of the latest advances in the field including Synthetic Methodology, concept, synthon, synthetic equivalent, retro-synthetic analysis etc. Classified synthesis, synthesis of medium and large ring compounds, umpolung synthesis, synthesis of polymers, chiral synthesis, and synthesis of Dyes and Drugs should ideally be incorporated in the same text. The idea struck me as the students have often expressed their helplessness in absence of such a book in the market. I, for one, have never found out a single such book on the subject, as could cater to the needs of the students in every possible respect. Naturally, there has always remained a dearth of such a concise text with details of the synthetic procedure as included in the Honours and Postgraduate level syllabi of all major Indian universities.

It goes without saying that my present effort is primarily aimed at meeting that long-felt need. In the present volume special emphasis has been given to retrosynthetic analysis including ABC knowledge of synthesis, planning of synthesis, technique of synthesis, and synthetic methodology coupled with the knowledge of chemical instruments with which the synthesis is to be performed.

Of course, a few other titles on Organic Synthesis are very much available in the market penned by certain foreign authors. But I dare say such books are not tailor made for the students preparing for the degree level examinations in the Indian subcontinent. It holds true of other competitive examinations as well. I have written the present text according to the new course structure of BSc Honours and Postgraduation studies of Indian universities to alleviate the difficulties experienced by the students in collecting the scattered information from different books, journals, and periodicals while preparing the answers for the examinations.

My sincerest gratitude is offered to Mr Amitabha Sen, the Director of New Central Book Agency (P) Ltd, for successful completion of the project. Highly indebted I am to Sri Sukumar Patra and Sri Joydeep Ghosh for carrying out the extremely tedious job of typesetting with utmost dedication for months together.

However, I shall welcome any constructive criticism and suggestion in order to improve upon the existing text. I appeal to all my learned friends to point out the shortcomings of the volume, which shall be accepted in good humour for the betterment of the text in the best interest of the students.

January 2007

Ratan Kumar Kar

Reader in Chemistry
Vivekananda College (Thakurpukur)
Kolkata 700 063
West Bengal, India

Preface

While teaching, I have often felt that students of Chemistry Honours need a handy and concise book on Organic Synthesis—a volume comprising of the latest advances in the field including Synthetic Methodology, concept, synthon, synthetic equivalent, retrosynthetic analysis etc. Classified synthesis, synthesis of medium and large ring compounds, Umpolung synthons, synthesis of polymers, chiral synthesis, and synthesis of Dyes and Drugs should ideally be incorporated in the same text. The idea struck me as the students have often expressed their helplessness in absence of such a book in the market. I, for one, have never found out a single such book on the subject as could cater to the needs of the students in every possible respect. Naturally, there has always remained a dearth of such a concise text with details of the synthetic procedure as included in the Honours and Postgraduate level syllabi of all major Indian universities.

It goes without saying that my present effort is primarily aimed at meeting that long-felt need. In the present volume special emphasis has been given to retrosynthetic analysis including FGI, knowledge of synthesis, planning of synthesis, methods of synthesis and synthetic methodology coupled with the knowledge of chemical reagents with which the synthesis is to be performed.

Of course, a few other books on Organic Synthesis are very much available in the market penned by certain foreign authors. But I dare say such books are not rather made for the students preparing for the degree level examinations in the Indian subcontinent. It holds true of other competitive examinations as well. I have written the present text according to the new course structure of B.Sc. Honours and Postgraduation studies of Indian universities to alleviate the difficulties experienced by the students in collecting the scattered information from different books, journals, and periodicals while preparing the answers for the examinations.

My sincerest gratitude is offered to Mr. Amalendu Sen, the Director of New Central Book Agency (P) Ltd. for successful completion of the project. I am highly indebted to Sri Sukumar Paul and Sri Joydeep Ghosh for carrying out the extremely tedious job of typesetting with utmost dedication for months together.

However, I shall welcome any constructive criticism and suggestion in order to improve upon the existing text. I appeal to all my learned friends to point out the shortcomings of the volume, which shall be accepted in good humour for the betterment of the text in the best interest of the students.

[illegible] — [illegible]

Reader in Chemistry

Vivekananda College (Thakurpukur)

Kolkata 700 063

West Bengal, India

Abbreviations

aq.	aqueous	**Me**	methyl
Ac	acetyl	**Et**	ethyl
Boc	tertiary-butoxy carbonyl	**Py**	pyridine
Cond.	condensation	**LDA**	lithium diisopropyl amide
Conc.	concentrated	**TMS**	trimethyl silyl
DCC	dicyclohexyl carbodiimide	**Ts**	tosyl
DDQ	2, 3-dichloro-5, 6-dicyano -1, 4-benzoquinone	**r.t**	room temperature
de	diastereomeric excess	**SET**	single electron transfer
ee	enantiomeric excess	**M.A**	mixed acid
DMF	dimethyl formamide	**SN**	synthon
DMSO	dimethyl sulphoxide	**SE**	synthetic equivalent
THF	tetrahydrofuran	**FGI**	functional group inter conversion
Pr	n-propyl	**FGR**	functional group removal
iPr	isopropyl	**FGA**	functional group addition
tBu or Bu^t	tertiary butyl	**IllE**	illogical electrophile
mCPBA	metachloro per benzoic acid	**IllNu**	illogical nucleophile
HOMO	highest occupied molecular orbital	**dix**	disconnection
LUMO	lowest unoccupied molecular orbital	**RA**	retro-analysis
SOMO	singly occupied molecular orbital	**TM**	target molecule
EAA	ethyl aceto acetate	**PG**	protecting group
DEM	diethyl malonate	**TS**	transition state
ECA	ethyl cyano acetate	**Ea**	energy of activation
DA	Diels-Alder reaction	**X's**	excess
liq.	liquid	**DIBAL**	diisobutylaluminium hydride
hv	UV or visible light	**eq.**	equivalent
		Δ	solvent heated under reflux

Introduction

Synthesis is the central area of organic chemistry where the chemist's art and imagination are involved. In any synthesis there are two vital parts, one is proper planning which comes from disconnection of target molecule and the other is knowledge of chemical instruments which are starting materials or synthetic equivalents, name reactions and rearrangements, ring-closure and ring-opening, ring-contraction and ring-expansion, protection and deprotection of functional groups or position, activation and deactivation of functional groups or position, activation with protection, condensing agent, dehydrating agent, functionalisation, defunctionalisation and functional group interconversion with which the synthesis is to be performed.

The first part is discussed in "Disconnection Design: The Retrosynthetic Analysis" (chapter 2) and the second part is discussed in "Synthetic Methodology" (chapter 1).

Two kinds of reactions are involved in any organic synthesis and therefore synthesis may be divided into two categories for separate considerations.

(i) Skeleton construction: The actual carbon-skeleton must be constructed from smaller market available starting materials by reactions which form C—C bond. The process is central consideration in synthesis design.

(ii) Functional alteration: The functionality resulting from one C—C bond formation must be transformed into that required for next or ultimately into the final functional groups of the target molecule.

In designing synthesis of an organic molecule the most difficult job is to find out a logical precursor i.e. starting material which has to be a readily available organic compound. The retrosynthesis—the exact reverse of the actual synthesis starting from the target molecule is highly efficient to provide such logical precursor. In this imaginary process bond disconnection is done in the target molecule to give synthons (charged ions) and then a new but simple structure (synthetic equivalent) from which the target can be made is involved.

The disconnection approach (retrosynthetic analysis) may be illustrated by the following target molecule (TM).

Retro-analysis

(TM) —α, β unsat. C=O⟹ [1,5-dicarbonyl: Me, CHO (5), Me, OEt, positions 1, 2, 3, 4] —1, 5 dico; 3, 4 dix.; Michael cond.⟹ [CHO (5), Me, OEt, positions 4, 3, 2, 1] + Methyl vinyl ketone (SE)

[CHO (5), Me, OEt] —1, 5 dico; 3, 4 dix.; Michael cond.⟹ Propanal (SE) + Ethyl acrylate (SE)

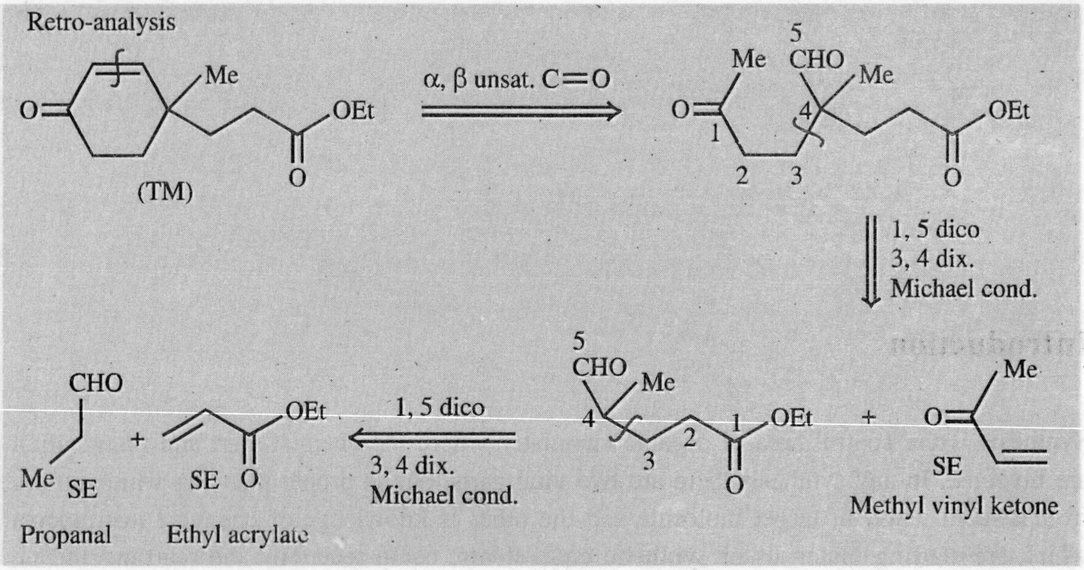

The retrosynthetic analysis reveals that the TM may be synthesized from three commercially available materials propanal, methyl vinyl ketone and ethyl acrylate which are synthetic equivalents or starting materials.

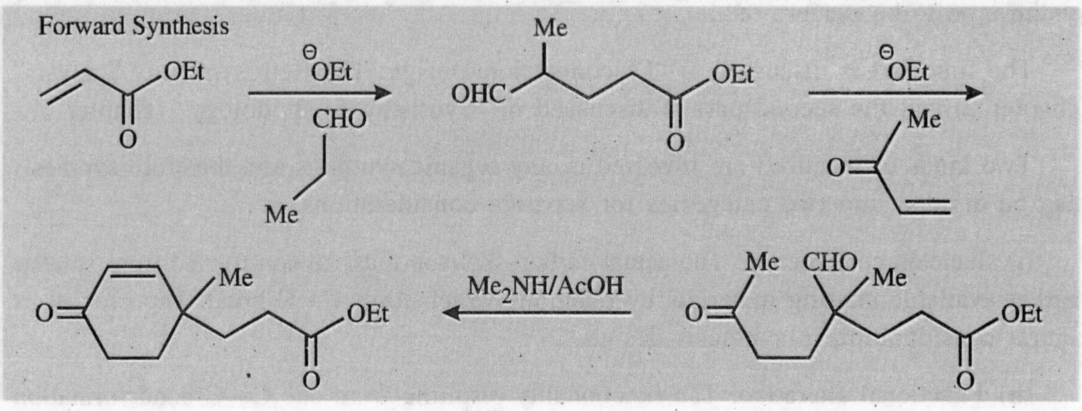

Forward Synthesis

Ethyl acrylate —$\overset{\ominus}{O}Et$, propanal (CHO, Me)→ OHC–CH(Me)–CH₂CH₂–CO–OEt —$\overset{\ominus}{O}Et$, methyl vinyl ketone→ [Me, CHO, Me, OEt keto-aldehyde] —$Me_2NH/AcOH$→ TM

An organic molecule can be synthesized by several, often many, different routes. A given synthetic route is a sequence of reactions designed to convert commercially available starting materials into desired target molecule.

The best practice usually is to devise a number of different routes by dissection of a given target molecule and compare them to select the best one, since the time spent in synthesis design on the paper is far less than that spent in the actual laboratory. This can be clarified by dissection of the following target molecule (TM) of which dissection (1) is the most suitable since it is one step process.

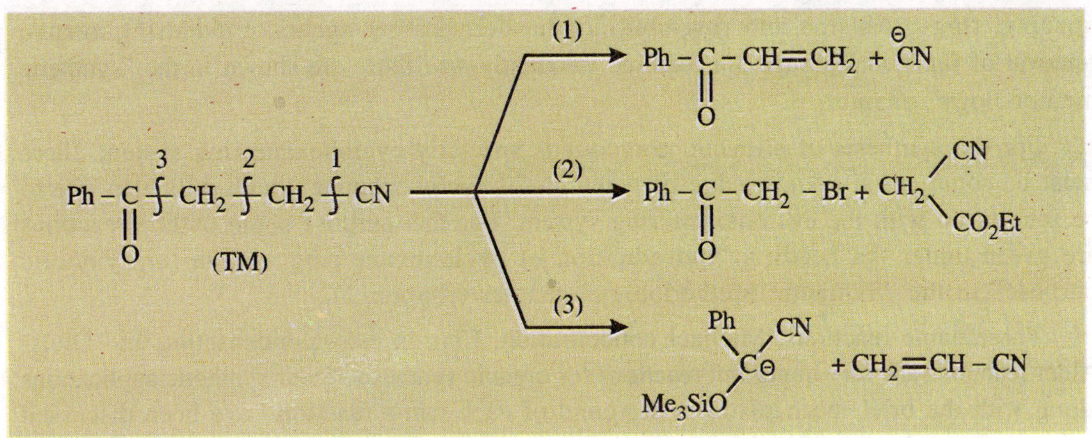

The retrosynthetic analysis not only give informations about the starting materials but also give informations of chemical instruments with which the synthesis is to be performed. During retro-analysis, the synthons developed which are intermediates in the real synthesis indicates how and what type of reactions should be performed in the forward synthesis.

The basic disconnections of various target molecules containing one to multiple functional groups, aliphatic, carboalicyclic, heteroalicyclic, carboaromatic, heteroaromatic and the compounds of various architecture and pattern have been shown under the heading "Guidelines for disconnection" in The Retrosynthetic Analysis chapter 2.

The C—C bond formation reactions are one of the most fundamental processes in synthetic organic chemistry. This is shown in "Synthetic Methodology" chapter (chapter 1) under the heading "Method of C—C bond formation".

For the synthesis of nitrogen containing target molecules, C—N and N—N bond formations are essential and for this purpose a short flow-sheet is given under the heading "Method of C—N and N—N bond formation" in the "Synthetic Methodology" chapter.

During organic synthesis alkylation, arylation, acylation, formylation, vinylation, benzoylation, carbethoxylation and methylenylation are frequently required. These are discussed under appropriate heading in the "Synthetic Methodology" chapter.

During the disconnection of target molecules, various nucleophilic (donor) and electrophilic (acceptor) synthons arises and the knowledge of corresponding nucleophilic and electrophilic synthetic equivalents are very important which will help one to take decision regarding starting materials. All these have been shown under the heading "Synthons and Synthetic equivalents" in The Retrosynthetic Analysis chapter (chapter 2). In addition to the knowledge of various synthons, synthetic equivalents there must be some knowledge of carbanion sources including acylanion equivalents—these are also shown in the retrosynthetic analysis chapter.

During the organic synthesis protection—deprotection of functional groups, position and ring, activation and deactivation of functional groups or position, ring-closure—ring-opening, ring-expansion and ring-contraction, dehydrating agents, condensing agents, removal of functional groups are required frequently—all these are shown in the "synthetic methodology" chapter.

For the synthesis of alicyclic compounds specially cyclohexane ring system, there must be some definite knowledge of reactions of cyclohexanone which will help one to be introduced with the cyclohexane ring system. For this purpose some outline reactions are given under the heading "Indroduction of cyclohexane ring system for synthetic purpose" in the "Synthetic Methodology" chapter (chapter 1).

Three name reactions, Michael condensation, Claisen ester condensation and Diels-Alder reactions are very important reactions for organic synthesis. Their synthetic applications along with the brief mechanistic background of each name reaction have been discussed in a separate chapter "Synthetic Application of Michael Condensation, Claisen Ester Condensation and Diels-Alder Reaction" (chapter 3).

Three compounds Ethyl acetoacetate, Diethylmalonate and Ethyl cyanoacetate are very important and these are versatile starting materials for various classes of compounds aliphatic, alicyclic, carboaromatic and heteroaromatic. Their synthetic applications as synthetic equivalents along with the brief introduction of these molecules are discussed in a separate chapter "Synthetic Application of Ethyl Aceto Acetate, Diethyl Malonate and Ethyl Cyano Acetate" (chapter 4).

Method of alkylation of cycloalkanone and synthesis of fused bicyclic and spirocyclic ring systems and angular alkylation are discussed in a separate chapter "Synthesis of Alkylated Cycloalkanone, Fused Bicyclic and Spirocyclic Ring System" (chapter 5).

The more classified synthesis mainly "Umpolung Synthesis" (chapter 6), "Synthesis of Medium and Large Ring Compounds" (chapter 7), "Chiral Synthesis" (chapter 8), "Synthesis of Dyes" (chapter 9), "Synthetic Drugs and Pharmaceuticals" (chapter 10), and "Synthesis of polymers" (chapter 11) are discussed in the respective chapters.

Elaborate discussions with huge examples and with different functional groups and also with different design of molecules with disconnection approach are discussed in the retrosynthetic analysis chapter (chapter 2).

Lastly one question must arise, what is the best synthesis?

In short, the best synthesis of an organic molecule involves conversion of the most available and cheapest starting material (which is a most logical precursor in a given sequence of reactions) into the desired target molecule by the least number of steps and in the highest possible overall yield with a purest grade of product.

The Criterion of a good synthesis then becomes one of minimising the number of steps, the number of starting material units and the number of operations adjusting functionality must be kept as small as possible.

The overall yield of the synthesis is dramatically reduced by poor yields in the individual steps in the multistep synthesis. The overall yield of the product depends on the individual steps. That is why convergent synthesis is better than linear synthesis in the multistep processes.

During the disconnection of a target molecule (TM) in the retrosynthetic way one should have to try to divide the molecule into approximately equal halves, and therefore aim to do the same with each subsequent disconnections. This will lead to a synthesis which is convergent i.e. several "mini-syntheses" leading to the target molecule. The alternative linear approach, in which small units are added to a growing chain, is less efficient.

A schematic representation of a convergent versus linear synthesis is shown below. The generalised molecule (TM) consists of four different groups linked together in a chain. If we assume a 70% yield for each synthetic manipulation on the pathway then the linear pathway would give a 34% overall yield for the target molecule whereas the convergent approach would lead to a 49% yield (based on the longest pathway).

Linear disconnection	Convergent disconnection
A–B–C–D (TM)	A–B–C–D (TM)
⇓ 1	⇓ 1
A + B–C–D	A–B C–D
⇓ 2	⇓ ⇓ 2
B + C–D	A + B C + D
⇓ 3	
C + D	
Longest pathway is three steps	Longest pathway is two steps

The following target molecule (TM) synthesis may be taken as for illustration of the two processes.

Disconnection for linear synthesis

$$\text{PhC(OH)(CH}_3\text{)CH}_2\text{CH}_2\text{NMe}_2 \text{ (TM)} \overset{1}{\Longrightarrow} \text{PhCOCH}_2\text{CH}_2\text{NMe}_2 + \text{CH}_3\text{MgBr}$$

$$\Downarrow 2$$

$$\text{PhCOCH}_3 + \text{CH}_2{=}\overset{\oplus}{\text{N}}\text{Me}_2 \Longrightarrow \text{CH}_2\text{O} + \text{Me}_2\text{NH}$$

$$\Downarrow 3$$

$$\text{C}_6\text{H}_6 + \text{CH}_3\text{COCl}$$

Forward Synthesis

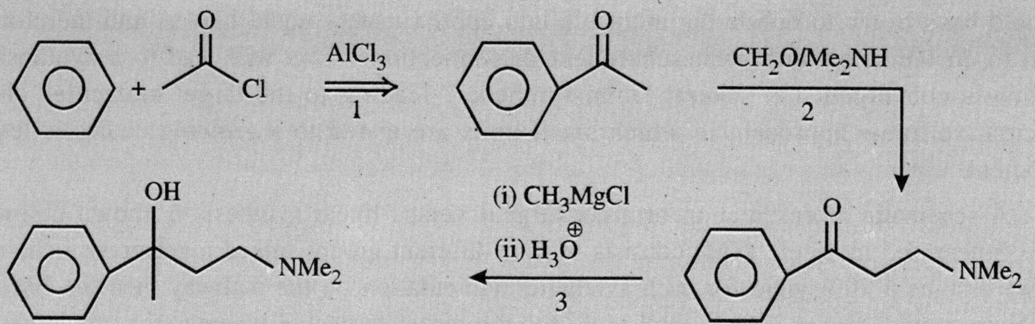

Disconnection for convergent synthesis

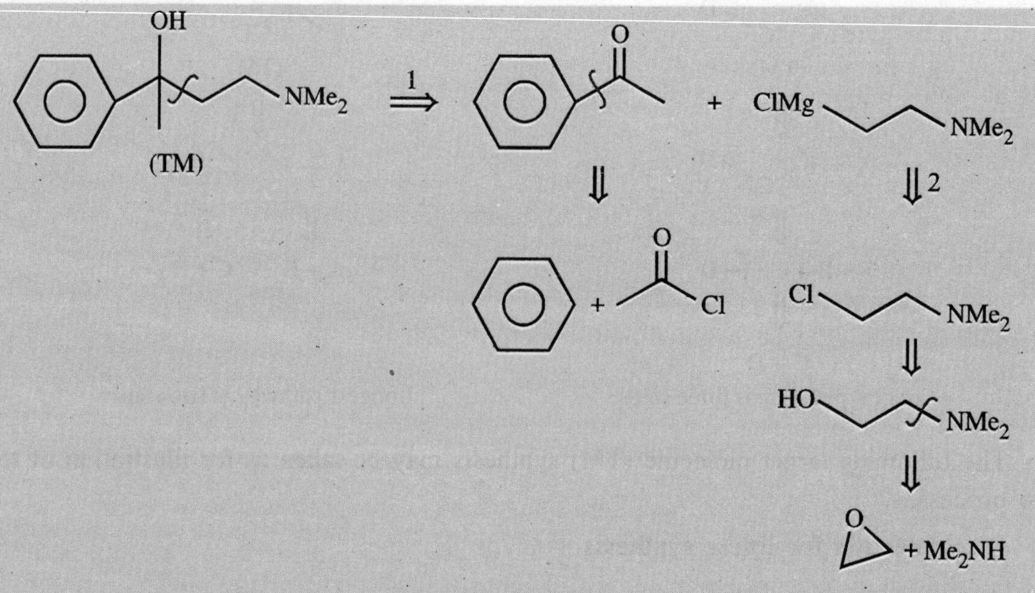

Forward Synthesis

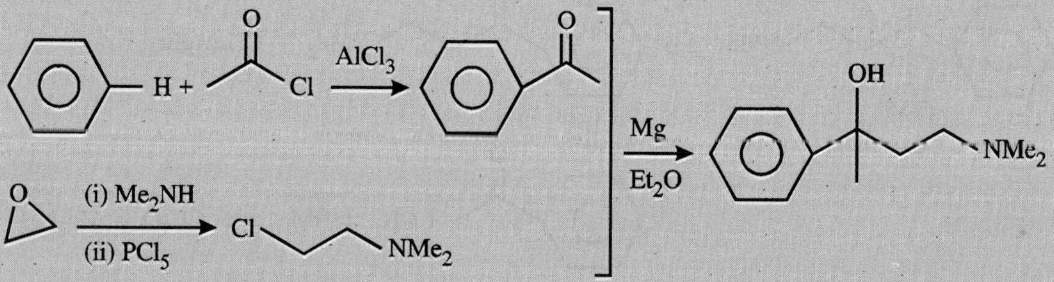

Synthesis of an organic molecule is of two types: Synthesis of academic interest (academic pleasure), synthesis of economic interest—economic route is the most prized one and most preferred due to obvious reason.

Every synthesis has its own story i.e. sequence of reactions. Solution of synthetic problems is a knowledge of "structure and reactions". That's why the best way to learn organic synthesis is by solving synthetic problems. Expertise in synthetic design, as in many aspects of organic chemistry, comes largely with practice. For this purpose solution of synthesis of hundred plus target molecules are discussed wih retrosynthetic analysis and hundred plus target molecules (synthetic problems) are given to work out retrosynthetically almost in every chapter and in every section and subsection of each chapter.

1

Synthetic Methodology

1.1 The background knowledge to organic synthesis

Organic synthesis is a creative science and is the ultimate target for learning organic chemistry. There are three main branches in organic chemistry: Reaction mechanism, Stereo chemistry and Organic synthesis. Of these three, organic synthesis is the applied branch as to synthesize organic molecules, knowledge of reaction mechanism (which will guide to disconnect any target molecule in the retrosynthetic analysis) and stereo chemistry (for desired stereo structure in the target molecule) are essential.

In the context to organic synthesis one should know about two basic questions:

1. What is organic synthesis?
2. How organic synthesis is performed?

The answer of both the questions demands long discussion. However a short description is given below.

1.2 What is organic synthesis?

The ideal synthesis of an organic molecule involves conversion of the most available and cheapest starting material which is the most logical precursor in a given sequence of reactions, into the desired target molecule by the least number of steps and in the highest possible overall yield with purest grade of products.

1.3 How organic synthesis is performed?

1. The vital part of an organic synthesis is the construction of C—C bond or Carbon-skeleton either by condensation or rearrangement with the help of Name reactions or any reliable reaction with simplified reasonable mechanism.

2. The Carbon skeleton must be constructed from smaller market available starting materials. This process is central consideration in the synthesis design.

3. The functionality resulting from one C—C bond formation must be transformed into the desired one as in the target molecule or any functional alteration may be done by the following techniques.

(a) Insertion of functional group—functionalisation.

(b) Removal of undesired functional group—defunctionalisation.

(c) Inter conversion of functional group—remainder of the molecule must be unaffected.

(d) Some functional groups cannot be introduced directly into a molecule and in that situation first functional group addition and then functional group inter conversion are necessary.

(e) A multistage organic synthesis usually involves steps in which the oxidation or reduction of a particular group is required. Particularly highly selective methods (redox) which make it possible to oxidise or reduce at a required stage in a molecule without affecting other oxidisable or reducible groups.

1.4 The chemical instruments required during organic synthesis

If one wants to build up a house, one should have a knowledge of architecture and also some knowledge of materials and instruments with which the construction is to be made. Likewise, during the build up (synthesis) of an organic molecule one should have a knowledge of proper planning (which comes from disconnection of target molecule—discussed in the retrosynthetic analysis chapter) and the knowledge of chemical instruments with which the synthesis is to be performed.

The various chemical instruments are needed during organic synthesis. The followings are the chart of chemical instruments which are very important steps in the synthetic methodology part.

1. Name reactions and Rearrangements applied during organic synthesis—Method of C—C bond formation or Carbon skeleton.
2. Method of C—N and N—N bond formation.
3. Protection and deprotection of functional groups / position / ring.
4. Activation and deactivation of functional groups / position—activation with protection.
5. Ring-closure—ring-opening.
6. Ring-expansion—ring-contraction.

7. Dehydrating agent.
8. Condensing agent.
9. Functional groups removal—defunctionalisation.

1.4.1 Method of C—C bond formation or carbon skeleton construction

C—C bond formation reaction is one of the most fundamental processes in synthetic organic chemistry. The basic reactions so far used to form C—C bond is only limited to those processes that involved carbanion and carbonium ion intermediates. The reactions are usually carried out in basic medium. The only important acid catalysed C—C bond formation is the Friedel-Crafts alkylation or acylation of aromatic ring and Mannich reaction.

The pericyclic reactions (especially Diels-Alder reaction) is the only class for C—C bond formation without giving intermediates.

Certain free radical C—C bond formation reactions are also known as for example, pinacolic reduction and acyloin condensation.

Certain rearrangements make (and break) C—C bond in a manner useful for synthesis, usually these reactions are used for expanding or contracting rings or chain.

The following name reactions and rearrangements are very frequently used for C—C bond formation (within the brackets indicates main target) which are applied during organic synthesis.

1. Michael condensation (C—C bond and 1, 5-dicarbonyl compound)

It is a condensation reaction between stabilised carbanion and α, β unsaturated carbonyl group
or α, β unsaturated C = C bond with a — M group (— M = — CN, — CO_2Et, — CHO, —C(=O)— etc). This reaction is utilised for generation of 1, 5-dicarbonyl compound or any 1, 5-dioxygen system (with functional group interconversion) during the synthesis. It is a basic medium reaction.

O, CO_2Et, (i) $\overset{\ominus}{O}Et$, (ii), O, O, O, CO_2Et, dil HCl, Δ, O, O

2. Aldol condensation (C—C bond and **α, β** unsaturated carbonyl)

This is a reaction between enolate and carbonyl group. This reaction is utilised mainly for generation of α, β unsaturated carbonyl or 1, 3-dioxygen system and also for cyclisation with C—C bond formation. It is a basic medium reaction.

(i) $\overset{\ominus}{O}H/H_2O$; CH_3CHO ; Δ ; CrO_3/Py

(ii) $\overset{\ominus}{O}H/H_2O$

(iii) $\overset{\ominus}{O}H/H_2O$

Fate of enolate in the process of synthesis

(a) For aldol condensation,

(b) For Claisen ester condensation,

(c) For Dieckmann reaction,

(d) For alkylation, acylation etc.

(e) For haloform reaction,

(f) For oxidative dimerisation of stabilised enolate eg EAA and DEM gives dimer with I_2,

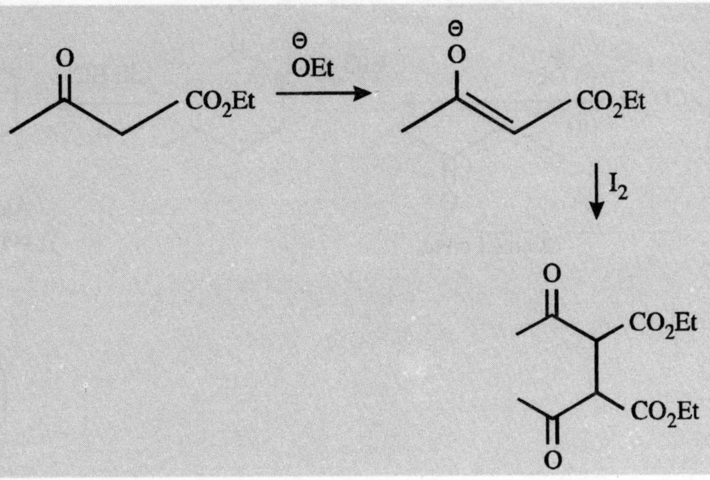

(g) As a Michael donor,

(h) In base catalysed deuterium exchange of α-H,

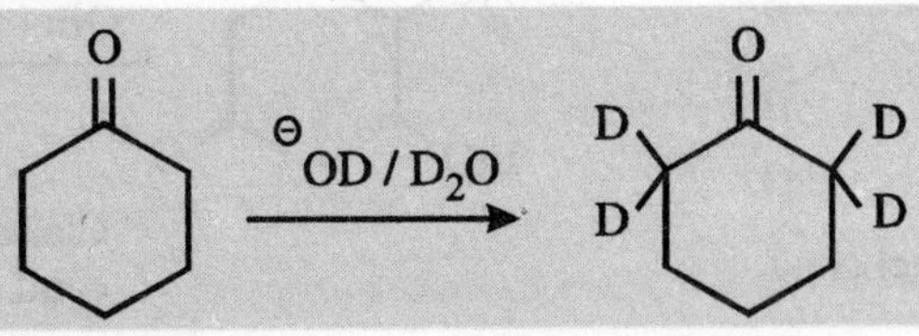

(i) In base catalysed racemisation of aldehyde, keto and acids.

3. Claisen ester condensation (C—C bond and 1, 3-dioxygen system)

This is a condensation reaction between carbanion (source α H containing ester or carbonyl) and ester carbonyl.

This reaction is utilised for generation of 1, 3-dicarbonyl or any 1, 3-dioxygen system. It is a basic medium reaction.

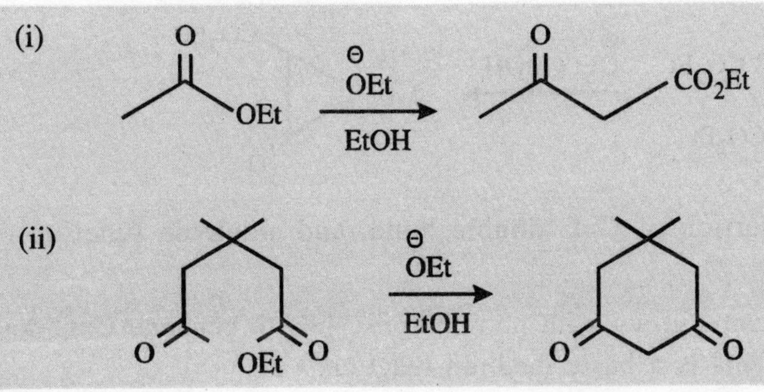

(i)

(ii)

4. Michael followed by Aldol condensation (C—C bond and for cyclisation)

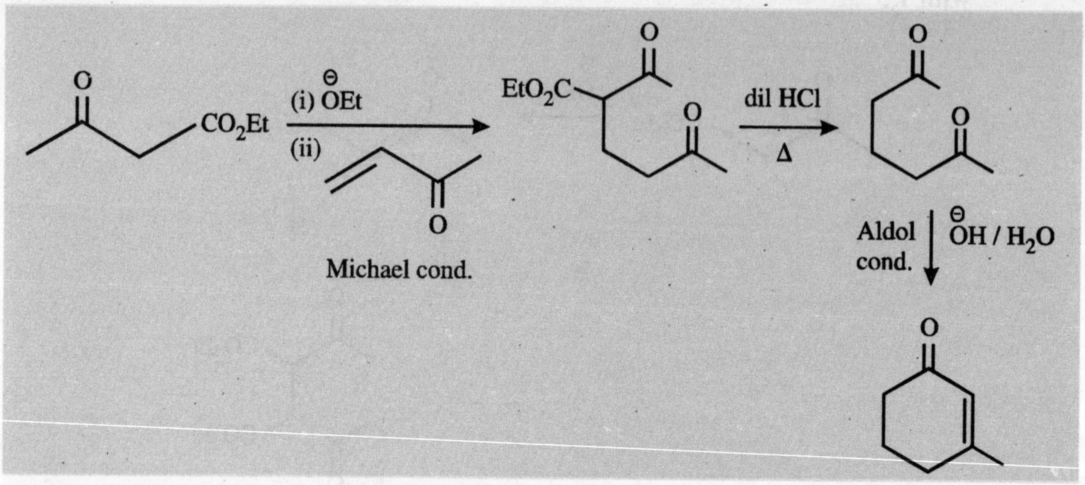

5. Michael followed by Claisen ester condensation (C—C bond and cyclisation)

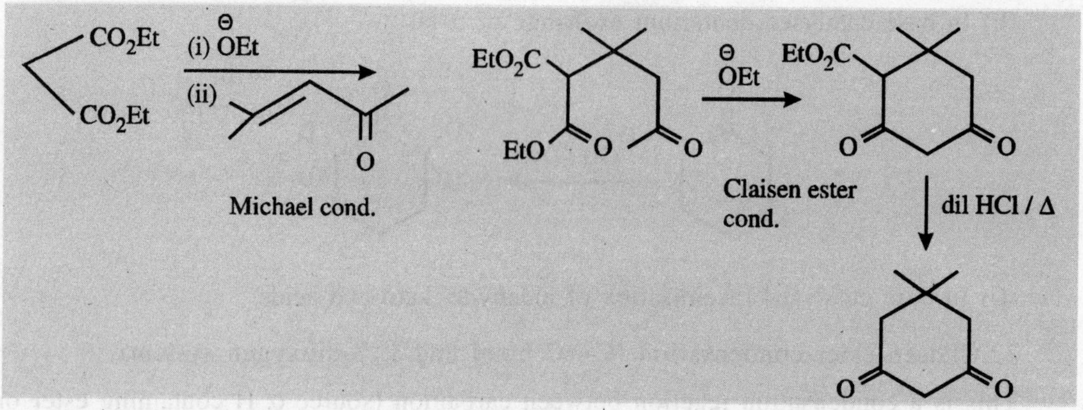

6. Dieckmann reaction (C—C bond and for cyclisation, specially to synthesize medium to large ring compounds).

This is a basic medium reaction.

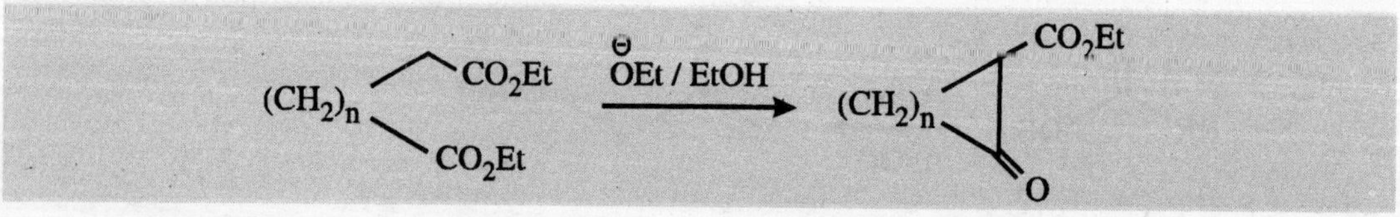

7. Wittig reaction (insertion of C=C double bond, and aldehyde function over carbonyl)

It is the reaction of carbonyl group with phosphorous ylide to generate C=C double bond during the synthesis. This is a basic medium reaction.

With the wittig reaction the followimg types of C = C double bond may be generated.

(i) Isolated

$(C_6H_5)_3P=CH_2$

(ii) α, β unsaturated (with stabilised ylide)

Br

(i) $(EtO)_3P$

(ii) $\overset{\ominus}{O}Et$

(ii) CH_2O, work up

(iii) β, γ unsaturated

$P(C_6H_5)_3$

CH_2O

This reaction can also be utilised to convert lower aldehyde to higher homologue of aldehyde i.e. insertion of aldehyde function over carbonyl function.

(i)

CHO

(i) $MeO-CH=P(C_6H_5)_3$

(ii) $H_3\overset{\oplus}{O}$

CHO

(ii)

(i) $MeO-CH=P(C_6H_5)_3$

(ii) $H_3\overset{\oplus}{O}$

CHO

8. Mannich reaction (C—C bond, potential source of **α, β** unsaturated carbonyl and for insertion of methylene group to the enolisable carbon)

This is a condensation reaction between formaldehyde, dimethyl amine and enolisable ketone to give Mannich base. This is an acid medium reaction.

H_5C_6

$CH_2=\overset{\oplus}{N}Me_2$

$CH_2O + Me_2NH{\cdot}HCl$

H_5C_6 — NMe_2

Mannich base

$\equiv H_5C_6$

Mannich base is used as a potential source for α, β unsaturated ketone in situ (since α, β unsaturated carbonyl readily polymerise) during the organic synthesis as in Robinson annulation (Michael condensation followed by aldol with cyclisation). The α, β unsaturated carbonyl obtained from Mannich base is a milder process.

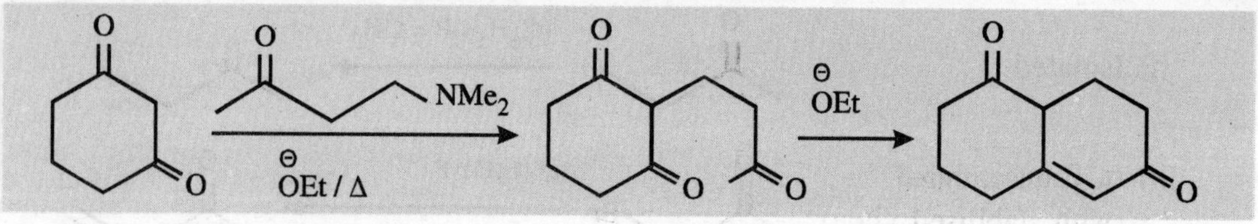

This reaction is also used for insertion of methylene group to enolisable carbon.

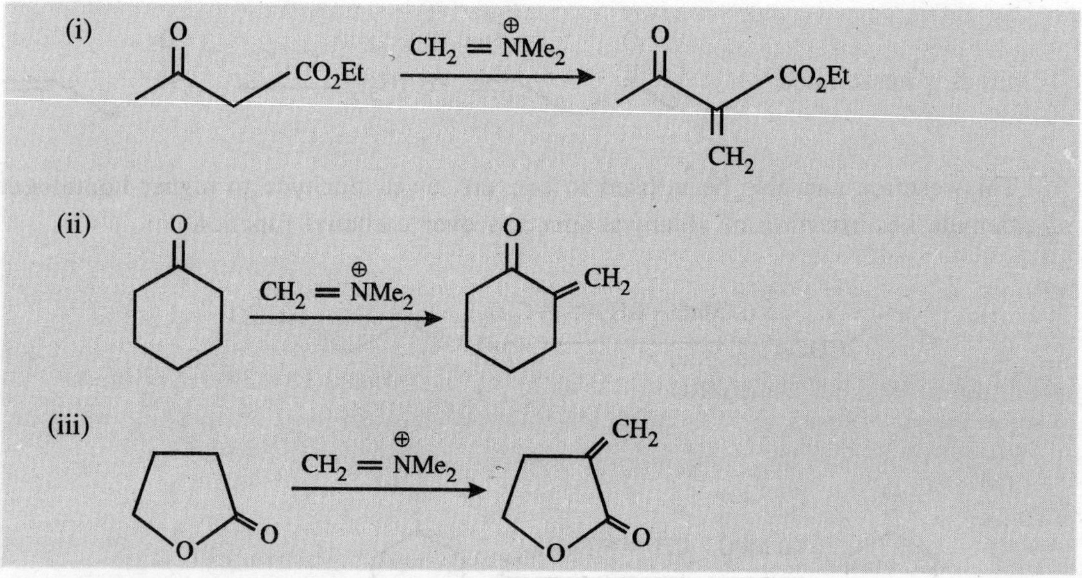

(i) $CH_2 = \overset{\oplus}{N}Me_2$

(ii) $CH_2 = \overset{\oplus}{N}Me_2$

(iii) $CH_2 = \overset{\oplus}{N}Me_2$

All the products are good Michael acceptors.

9. Benzoin condensation (C—C bond and aromatic α-Ketol)

This is a condensation reaction of benzaldehyde in presence of KCN in ethanolic solution for the preparation of aromatic α-Ketol or its modified form eg benzil.

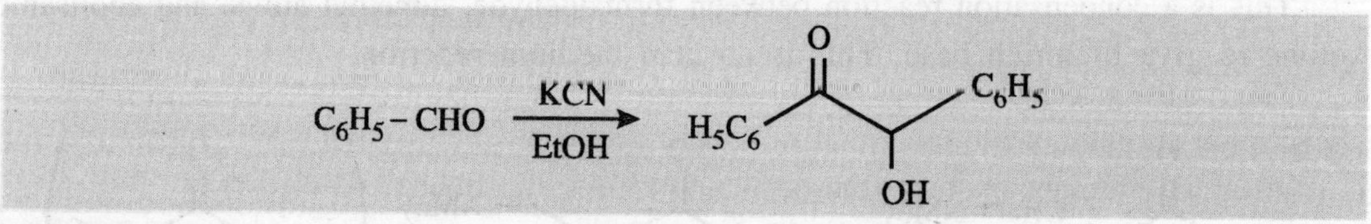

$C_6H_5-CHO \xrightarrow[EtOH]{KCN} H_5C_6-CO-CH(OH)-C_6H_5$

10. Acyloin condensation [C—C bond, to generate acyloin- (α-ketol) and also for cyclisation]

This is a reaction between two ester functions either intermolecular or intramolecular. This condensation is utilised for generation of acyloin or its modified form eg a diketo, mono keto, glycol etc. and also for cyclisation during synthesis.

(i) CH_3COOEt —(i) Na / Xylene; (ii) $H_3O^{\oplus}$→ $CH_3COCH(OH)CH_3$

(ii) $(CH_2)_4(CO_2Et)_2$ —(i) Na / Xylene; (ii) $H_3O^{\oplus}$→ 2-hydroxycyclohexanone —Ni/H_2→ cyclohexane-1,2-diol (OH, OH); —Zn/AcOH→ cyclohexanone (O); —CrO_3/Py→ cyclohexane-1,2-dione (O, O)

11. Pinacolic reduction (C—C bond, to generate glycol with tertiary carbon and also for cyclisation purpose)

This is a condensation reaction between two ketones to form Pinacol (usually tertiary glycol) and which is the synthetic equivalent of Pinacol Pinacolone rearrangement. This reaction is also utilised for cyclisation to give cyclic glycol.

(i) $(CH_3)_2C{=}O + O{=}C(CH_3)_2$ —(i) Mg / C_6H_6; (ii) $H_3O^{\oplus}$→ $(CH_3)_2C(OH)-C(OH)(CH_3)_2$ (OH HO)

(ii) diketone (O, O) —(i) Mg / C_6H_6; (ii) $H_3O^{\oplus}$→ 1,2-dimethylcyclohexane-1,2-diol (OH, OH)

12. Knoevenagel reaction (C = C) bond and for the preparation of α, β unsaturated acid, ester, cyanide and carbonyl usually used as a Michael acceptor)

This is the condensation reaction between the carbonyl group of either aliphatic or alicyclic or aromatic with an active methylene group in presence of a base. This reaction is utilised to generate α, β unsaturated acid, ester, cyanide or carbonyl.

$$Ph-CHO + CH_2(CO_2Et)_2 \xrightarrow[\Delta]{\text{pyridine}} Ph-CH{=}C(CO_2Et)_2$$

13. Reformatsky reaction: (C—C or C═C bond for the preparation of **β** or **γ**-hydroxy ester and **α**, **β** or **β**-**γ** unsaturated ester)

This is the reaction between a carbonyl function and ester enolate to give β or γ hydroxy ester and α, β or β-γ unsaturated ester.

(i)

Br–CH_2–CO_2Et —(Zn, ether)→ CH_2=C(O⁻ ZnBr⁺)(OEt)

↓ CH_3CHO

CH_3CH(OZnBr)CH_2CO_2Et —($H_3O^{\oplus}$)→ CH_3CH(OH)CH_2CO_2Et (β - hydroxy ester) —(Δ)→ CH_3CH=CHCO_2Et (α, β unsaturated ester)

(ii)

CH_2=CH–CO–OEt —(HBr)→ Br–CH_2CH_2–CO–OEt

↓ Zn/ether

Br Zn–CH_2CH_2–CO–OEt

↓ (i) CH_2O (ii) $H_3O^{\oplus}$

HO–$CH_2CH_2CH_2$–CO–OEt (γ - hydroxy ester) —(Δ)→ CH_2=CH–CH_2–CO–OEt (β, γ unsaturated ester)

14. Diels-Alder reaction : (C—C bond, to synthesize 6-membered cyclic carbo or hetero alicyclic ring and benzenoid carboaromatic and heterocyclic ring including polynuclear hydrocarbons)

This is a cyclisation reaction of diene and dienophile under thermal condition to give cyclic six-membered ring and after that their modified form to give various carbo and heteroalicyclic ring and carbo and heteroaromatic compound.

(i) OMe + CO2Et →(Δ) → ($H_3O^{\oplus}$) CO2Et

(ii) + O → → ($H_3O^{\oplus}$) OH

(iii) O + OEt → O OEt

(iv) N—CO2Et, N—CO2Et → (H_2/Pd) → (KOH/H_2O/Δ) N—H, N—H

15. Retro Diels-Alder (For the synthesis of various benzenoid and furan derivative and new diene and dienophile)

(i) S + CO2Et ≡ CO2Et →(Δ) S, CO2Et, CO2Et →(Retro, – S) CO2Et, CO2Et

(ii)

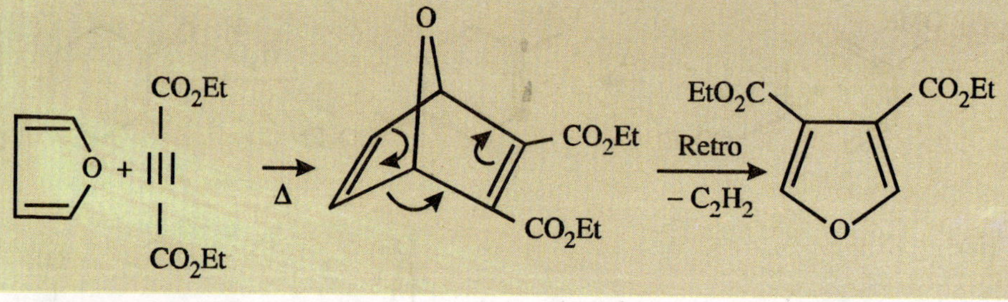

16. Enamine reaction (C—C bond, for alkylation or acylation of **α** position of a carbonyl compound)

When enamine is reacted with alkyl halide or acyl halide or any other appropriate electrophilic carbon then α-position is alkylated or acylated

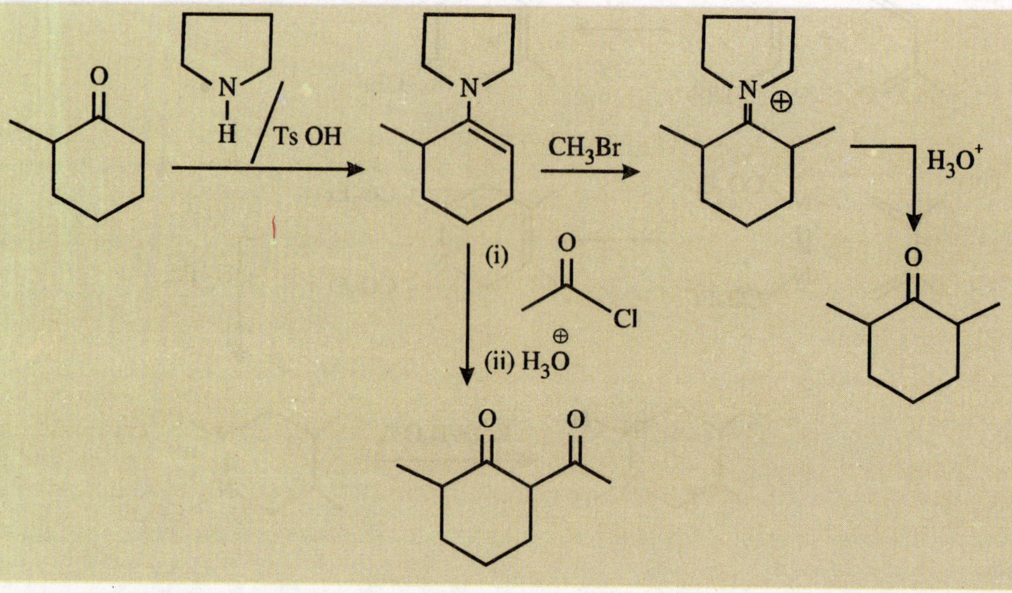

Enamine is activation of 'ene' and it may be considered as N-enolate.

Fate of enamine during organic synthesis.

(i) For alkylation, acylation, fomylation or arylation at the α-position of a carbonyl

(ii) For selective alkylation when there is a regio selective choice

(iii) Acts as a Michael donor

(iv) To synthesize 1, 3 and 1, 5 dicarbonyl and various derivatives

(v) For chiral synthesis

(vi) For blocking and deblocking purpose

(vii) For cyclisation purpose to synthesize fused bicyclic ring.

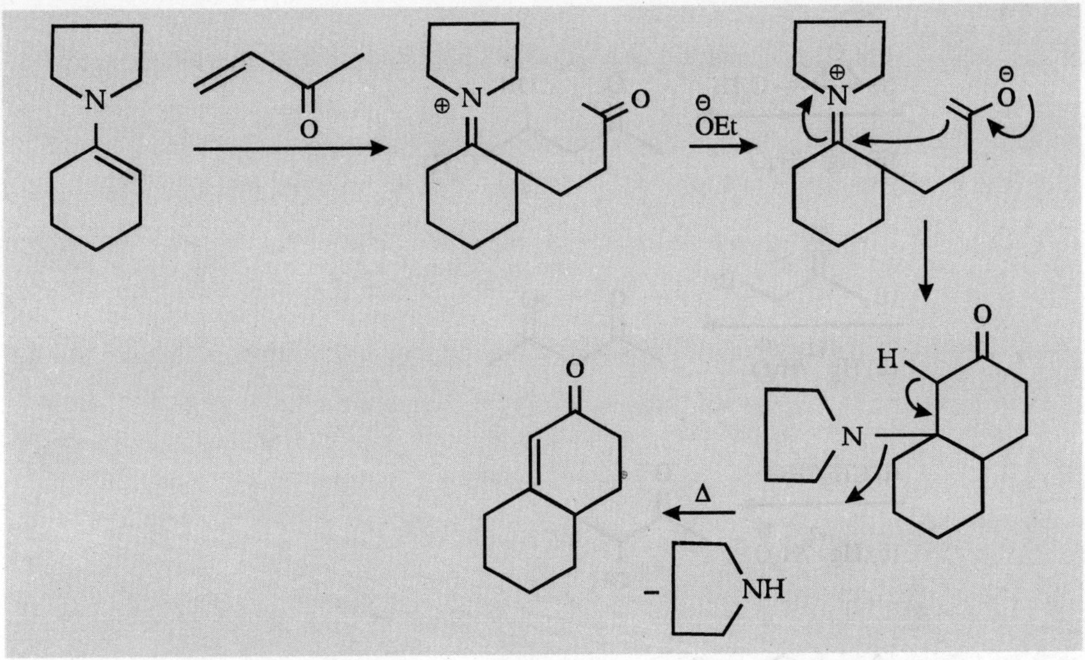

N O ⊕N O ⊖OEt ⊕N ⊖O O H N O Δ - NH

17. Umpolung synthesis (C—C bond, for deuteration, alkylation or acylation of acyl carbon, to synthesize 1, 2, 1, 3 and 1, 4 dioxygen system, to synthesize β γ unsaturated carbonyl and also to synthesize 3-7 membered ring)

The following examples will illustrate the facts:

(i) SH SH O H BF_3 S S H n BuLi S S ⊖ (i) D_2O (ii) $Hg^{+2}/H_3O^{\oplus}$ O D (i) CH_3I (ii) $Hg^{+2}/H_3O^{\oplus}$ O (i) CH_3COCl (ii) $Hg^{+2}/H_3O^{\oplus}$ O O

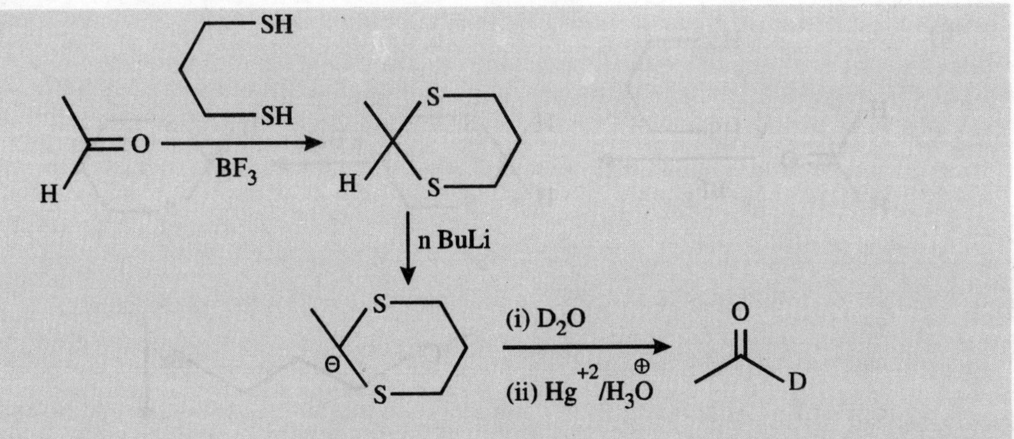

(i) epoxide–C_6H_5; (ii) $Hg^{+2}/H_3O^{\oplus}$ → O, OH, C_6H_5

(i) bromoacetone (O, Br); (ii) $Hg^{+2}/H_3O^{\oplus}$ → O, O

(i) CH_3CHO; (ii) $Hg^{+2}/H_3O^{\oplus}$ → O, OH

(i) prenyl bromide (Br); (ii) $Hg^{+2}/H_3O^{\oplus}$ → O

(ii)

H, H, =O; HS, HS; BF_3 → H, H, S, S; n BuLi → H, ⊖, S, S

Cl, Br ↓

Cl, H, S, S; n BuLi ← S, S; Hg^{+2}, H_3O^{+} ← =O

18. Oxidative dimerisation reaction (C—C bond, to synthesize 1, 4-dicarbonyl and 1, 4-dicarboxylic compound and to develop two alkyne units in the adjacent position)

(i) Stabilised enolate when refluxed with iodine dimerisation occurs.

(a)

$CH_3COCH_2CO_2Et \xrightarrow{\ominus OEt} CH_3C(O^{\ominus})=CHCO_2Et \xrightarrow{I_2} EtO_2C\text{-}CH(COCH_3)\text{-}CH(COCH_3)\text{-}CO_2Et$

$CH_3C(O^{\ominus})=CHCO_2Et \xrightarrow{I_2} CH_3COCH(I)CO_2Et$; $CH_3COCH(I)CO_2Et + CH_3C(O^{\ominus})=CHCO_2Et \xrightarrow{SN^2} EtO_2C\text{-}CH(COCH_3)\text{-}CH(COCH_3)\text{-}CO_2Et$

$\xrightarrow{\text{dil HCl}/\Delta} CH_3COCH_2CH_2COCH_3$

(b)

$CH_2(CO_2Et)_2 \xrightarrow{\ominus OEt} \overset{\ominus}{C}H(CO_2Et)_2 \xrightarrow{I_2} (EtO_2C)_2CH\text{-}CH(CO_2Et)_2$

$\xrightarrow{\text{(i) KOH, (ii) } H_3O^{\oplus}\text{, (iii) } 250^{\circ}C} HO_2C\text{-}CH_2CH_2\text{-}CO_2H$

(ii) When an alkyne where SP hybridized carbon is linked with hydrogen is treated with a base in presence of cupric ion dimerisation occurs.

$CH_3-C\equiv C-H \xrightarrow[Cu^{+2}]{Py} CH_3-C\equiv C-C\equiv C-CH_3$

$CH_3-C\equiv C-H \xrightarrow{Py} CH_3-C\equiv \overset{\ominus}{C} \xrightarrow{Cu^{+2}} CH_3-C\equiv C\bullet + Cu^{+} \xrightarrow{\bullet C\equiv C-CH_3} CH_3-C\equiv C-C\equiv C-CH_3$

19. Claisen rearrangement (C—C bond, to synthesize allyl derivative of phenol and **γ-δ** unsaturated carbonyl compound)

This is a rearrangement reaction of allyl vinyl ether under thermal condition applicable to aromatic, aliphatic or alicyclic system. This is a 3, 3-sigmatropic rearrangement.

This rearrangement is utilised in the synthesis for the following two cases mainly.

(i) For the preparation of ortho allyl derivative of phenol which is otherwise difficult to synthesize eg by Friedel crafts reaction.

Δ

OH

OH

K_2CO_3 / CH_3COCH_3

Br

O

(ii) For the preparation of γ-δ unsaturated aldehyde, ketone and ester.

(a)

Δ

≡

δ β O

H

γ α

OEt

$CH_3CH(OEt)_2$

−EtOH

OH

− EtOH

EtO

O

H

OEt

(b)

Δ

O

dilHCl

Δ

CO_2Et

CO_2Et

O

Py

CO_2Et

CO_2Et

Br

(c)

Δ

≡

O

OEt

OEt

OEt

$CH_3C(OEt)_3$ / CH_3CH_2COOH → ; –EtOH →

(d) Δ →

Δ → ; –EtOH ←

20. Cope-rearrangement (C—C bond, mainly to synthesize 1, 5-diene and sometimes with proper stereo chemistry. Unsaturated carbonyl compound may also be prepared via oxy-Cope. Para allyl derivative from allyl phenyl ether)

The thermal rearrangement of 1, 5-diene by 3, 3-sigmatropic shift is called Cope-rearrangment.

(i) (E, E) Δ → 1, 5 - diene

(ii) Δ → 1, 5 - diene

(iii) 1, 5 - Pentadiene - 3 - ol Δ → (Oxy-Cope) ⇌

(iv)

O OH

Δ Δ

Claisen Cope

21. Fries rearrangement (C—C bond, to introduce alkyl or acyl group into the benzene ring indirectly when Friedel-Crafts process is not possible)

This is a rearrangement reaction of alkyl or acyl phenyl ether in presence of $AlCl_3$ to give ortho/para, alkyl or acyl derivative of phenol

O – R OH OH

$AlCl_3$ R

Δ OR

R

R = alkyl, allyl or acyl group

22. Pinacol-Pina colone rearrangement (C—C bond and skeleton change, specially useful for the preparation of a carbonyl adjacent to a tertiary carbon and for ring-expansion and ring-contraction)

This is a rearrangement reaction of glycol in presence of acid (protic or Lewis) to give ketone adjacent to a tertiary carbon.

(i) PCl_5 O

OH HO

(ii) H_2SO_4

HO OH O

(iii) OH O

H_2SO_4

OH

23. Wolff-rearrangement (C—C bond and skeleton change. To prepare acyclic lower homologue to higher homologous acid and for cyclic ring-contraction)

This is a rearrangement reaction of diazoketone for conversion of lower homologue to higher homologous acid for acyclic system and for ring-contraction for alicyclic system.

(i)

$$R-CO_2H \xrightarrow{SOCl_2} R-\overset{\overset{\displaystyle O}{\|}}{C}-Cl \xrightarrow{CH_2N_2} R-\overset{\overset{\displaystyle O}{\|}}{C}-\overset{\ominus}{C}H-\overset{\oplus}{N}\equiv N \text{ (diazoketone)}$$

$$\xrightarrow{Ag_2O/\Delta} [R-CH=C=O] \xrightarrow{H_2O} R-CH_2-CO_2H$$

(ii)

Cyclohexanone $\xrightarrow{HNO_2}$ 2-nitrosocyclohexanone (NO) $\xrightarrow{H_2/Pd}$ 2-aminocyclohexanone (NH_2) $\xrightarrow{HNO_2}$ diazoketone ($\overset{\oplus}{N}\equiv N$) $\xrightarrow{h\nu \text{ or } \Delta}$ [carbene intermediate] $\longrightarrow$ cyclopentylidene $C=O$ $\xrightarrow{H_2O}$ cyclopentane-CO_2H

24. Favorskii-rearrangement (C—C bond and skeleton change. This reaction is utilised for ring-contraction and also to prepare tertiary acid)

This is a rearrangement reaction of α-haloketone in presence of a base to prepare tertiary acid for aliphatic system and tertiary or secondary acid in addition to ring-contraction for alicyclic system.

(i)

α-chloroketone (Cl) $\xrightarrow{\overset{\ominus}{O}H}$ [cyclopropanone] $\xrightarrow{\overset{\ominus}{O}H}$ $\overset{\ominus}{O}$ OH adduct $\longrightarrow$ $\succ$—COOH

(ii) 2-chlorocyclohexanone $\xrightarrow{\overset{\ominus}{O}H}$ [bicyclic cyclopropanone] $\xrightarrow[\text{(ii) } H_3\overset{\oplus}{O}]{\text{(i) } \overset{\ominus}{O}H}$ cyclopentanecarboxylic acid (COOH)

25. Benzilic acid rearrangement (C—C bond and skeleton change. To prepare α-hydroxy acid and also for ring-contraction)

This is a rearrangement reaction of α-diketo in presence of a base to give α-hydroxyl acid for aromatic α-diketo and α-hydroxy acid in addition to ring-contraction for alicyclic α-diketo.

(i) $Ph-\underset{\|}{\overset{}{C}}(O)-\underset{\|}{C}(O)-Ph \xrightarrow[\text{(ii) } H_3\overset{\oplus}{O}]{\text{(i) } \overset{\ominus}{O}H} (Ph)_2C(OH)CO_2H$

(ii) 3,3,6,6-tetramethylcyclohexane-1,2-dione $\xrightarrow[\text{(ii) } H_3\overset{\oplus}{O}]{\text{(i) } \overset{\ominus}{O}H}$ 2,2,5,5-tetramethyl-1-hydroxycyclopentanecarboxylic acid (OH, CO_2H)

1.4.2 Method of C—N and N—N bond formation

A large number of target molecules involve C—N and N—N bond as for example in aliphatic, heteroalicyclic, heterocyclic, benzofused heterocyclic, benzenoid and also in dyes, drugs and in pharmaceutical products. Therefore, C—N and N—N bond formation are very important part in the organic synthesis.

For the formation of C—N and N—N bond there must be some knowledge of electrophilic and nucleophilic nitrogen. For this purpose the following outlines are given.

Electrophilic Nitrogen

Synthon	Synthetic equivalent
$-\overset{\oplus}{N}O_2$	HNO_3, $HNO_3 + H_2SO_4$, $EtONO_2 + \overset{\ominus}{O}Et$, $AcONO_2$, $HNO_3 + Ac_2O$, $AgNO_3 + BF_3 + MeCN$
$\overset{\oplus}{N}O$	HNO_2, $EtONO + HCl$
$-\overset{\oplus}{N}\equiv N$	$H_5C_6-\overset{\oplus}{N}\equiv N$, $\overset{\ominus}{C}H_2-\overset{\oplus}{N}\equiv N$
$-\overset{\oplus}{N}H_2$	H_2NOCH_3, NH_2-Cl

$H–\ddot{N}$, $R–\ddot{N}$ N_3H, $CH_3–N =\ \ =CO$, $_5H_6–C =N =CO$, $_5H_6CNO_2$ $R–N_3$

$R–C(=O)–\ddot{N}$ $R–C(=O)–\overset{\ominus}{N}–\overset{\oplus}{N} \equiv N + h\nu$ or Cu/Δ

$(CH_3)_2C=\overset{\oplus}{N}$ $(CH_3)_2C=N–OH$

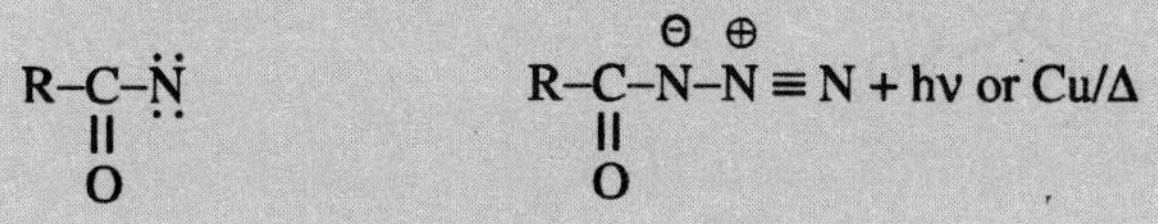

Nucleophilic nitrogen containing synthetic equivalent

NH_3, CH_3NH_2, Me_2NH, Me_3N, pyrrolidine (NH), $NH_2–NH_2$

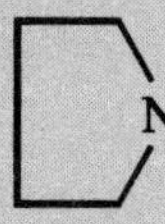

$TsNHNH_2$, $H_5C_6NH_2$, $H_5C_6NHNH_2$, H_5C_6NHMe, $H_5C_6–NH–NH–C_6H_5$,

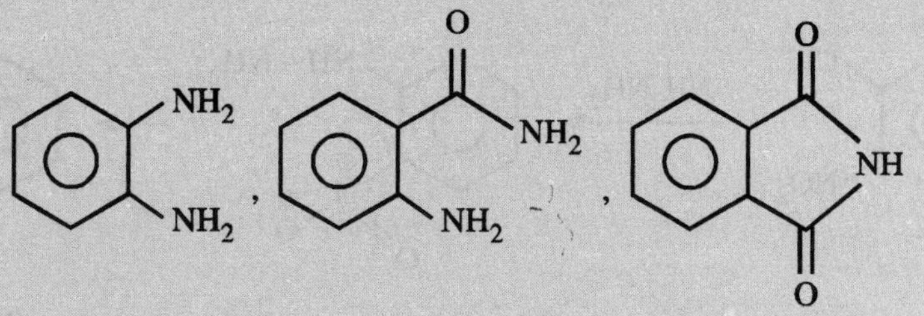

o-phenylenediamine (NH_2, NH_2), 2-aminobenzamide ($CONH_2$, NH_2), phthalimide (NH)

NH_2CONH_2, $NH_2CONHNH_2$, KCN, KNO_2, KN_3, NH_2OH, $CH_3C \equiv N$, $(CH_3)_3C–NH_2$,

piperidine (N–H), $(CH_3)_2CH–NH_2$, pyrrolidine (N–H)

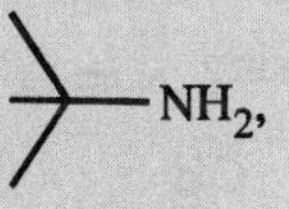

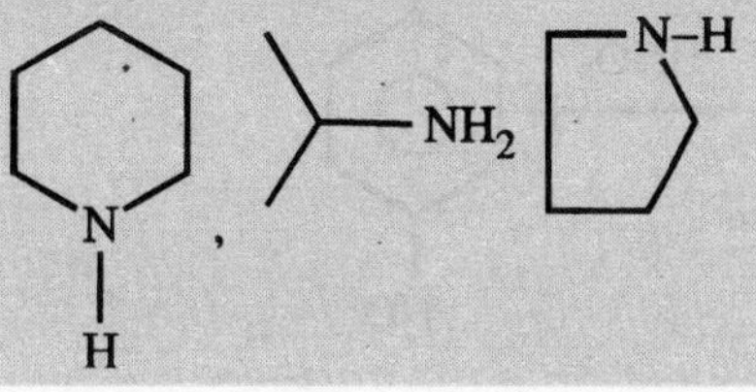

The following reactions (outlines) are some illustrations of C—N and N—N bond formation using electrophilic nitrogen.

1. Using $–\overset{\oplus}{NO_2}$

(i) 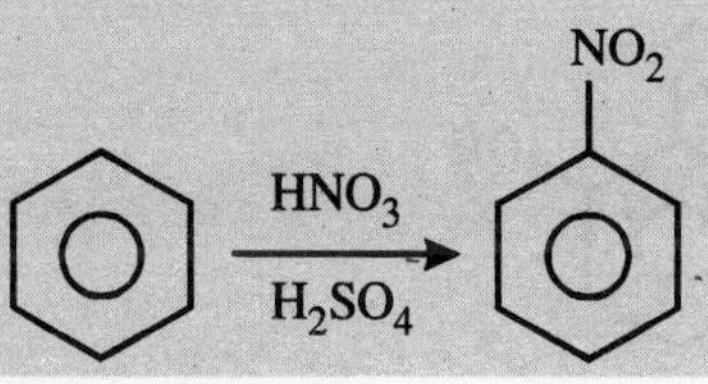 $C_6H_6 \xrightarrow[H_2SO_4]{HNO_3} C_6H_5NO_2$

(ii) 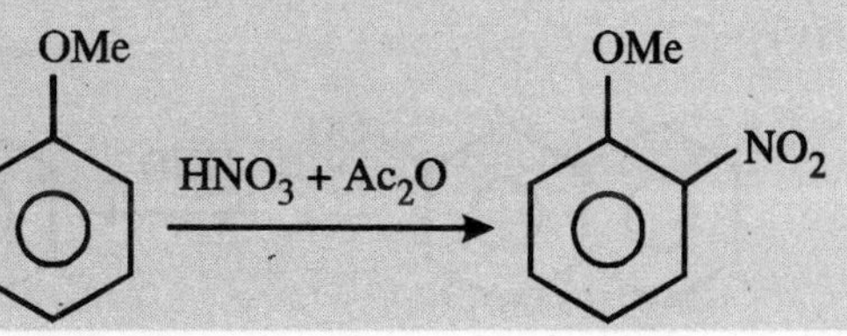 $C_6H_5OMe \xrightarrow{HNO_3 + Ac_2O}$ o-$NO_2C_6H_4OMe$

(iii) Furan $\xrightarrow{AcONO_2}$ 2-nitrofuran (NO_2)

(iv) $C_6H_5-CH_2CN \xrightarrow{\overset{\ominus}{O}Et\ /\ EtONO_2} C_6H_5-CH(NO_2)(CN) \xrightarrow[\Delta]{H_3\overset{\oplus}{O}} C_6H_5CH_2NO_2$

(v) $C_6H_5NO_2 \xrightarrow{Zn\ /\ NaOH} C_6H_5-NH-NH-C_6H_5$ (N – N bond)

(vi) Benzene $\xrightarrow[MeCN]{AgNO_3\ /\ BF_3}$ Nitrobenzene (NO_2)

(vii) 1-Cl-2-NO_2-benzene $\xrightarrow{NH_2NH_2}$ 2-($NH-NH_2$)-nitrobenzene ($\overset{\oplus}{N}(=O)\overset{\ominus}{O}$) $\rightarrow$ 1-hydroxybenzotriazole (N, N, N–OH)

(N–N bond)

2. Using $\overset{\oplus}{N}=O$

(i) Phenol (OH) $\xrightarrow{HNO_2}$ p-nitrosophenol (OH, NO)

(ii) NMe_2-benzene $\xrightarrow{HNO_2}$ p-nitroso-NMe_2-benzene (NMe_2, NO)

(iii) Cyclohexanone (O) $\xrightarrow{HNO_2}$ 2-nitrosocyclohexanone (O, NO) $\rightleftharpoons$ 2-oximinocyclohexanone (O, =NOH) $\xrightarrow{H_2\ /\ Pd}$ 2-aminocyclohexanone (O, NH_2)

(iv) 2-Naphthol (OH) $\xrightarrow{HNO_2}$ 1-nitroso-2-naphthol (NO, OH)

(v) OH → aq. NH_4HSO_3/Δ, Bucherer reaction → NH_2

(vi) O, CO_2Et → HNO_2 → O, CO_2Et, NO ⇌ O, CO_2Et, NOH → Zn / AcOH or H_2 / Pd → O, CO_2Et, NH_2

(vii) CO_2Et, CO_2Et → HNO_2 → HO–N = (CO_2Et)(CO_2Et) → H_2 / Pd → H_2N – (CO_2Et)(CO_2Et)

(viii) O or O O → $\overset{\ominus}{O}Et$ / EtONO → O, NO or O O, NO

(ix) NH_2 → HNO_2 → $\overset{\oplus}{N} \equiv N$ (N–N bond)

(x) NHMe → HNO_2 → N – Me, NO (N – N bond)

3. Using $-\overset{\oplus}{N}\equiv N$

(i)

$H_5C_6-\overset{\oplus}{N}\equiv N+$ $CH_3COCH_2CO_2Et$ $\longrightarrow$ $H_5C_6-N=N-CH(CO_2Et)(COCH_3)$ $\rightleftharpoons$ $C_6H_5-NH-N=C(CO_2Et)(COCH_3)$

(ii)

$\overset{\ominus}{C}H_2-\overset{\oplus}{N}\equiv N$ $+ Me-C\equiv C-H$ $\longrightarrow$ 3-methylpyrazole (N–H)

(iii)

$C_6H_5-\overset{\oplus}{N}\equiv N + H_2N-C_6H_4-$ $\rightarrow$ $C_6H_5-N=N-NH-C_6H_4-$

(N – N bond)

(iv)

o-phenylenediamine (NH_2, NH_2) $\xrightarrow{HNO_2}$ (NH_2, $\overset{\oplus}{N}\equiv N$) $\rightarrow$ benzotriazole (NH, N, N) (N–N bond)

(v)

2-aminobenzamide ($CONH_2$, NH_2) $\xrightarrow{HNO_2}$ ($CONH_2$, $\overset{\oplus}{N}\equiv N$) $\longrightarrow$ 4-hydroxy-1,2,3-benzotriazine (OH, N, N, N)

(N – N bond)

(vi)

2-aminoacetophenone ($COCH_3$, NH_2) $\xrightarrow{HNO_2}$ ($COCH_3$, $\overset{\oplus}{N}\equiv N$) $\longrightarrow$ 4-hydroxycinnoline (OH, N, N)

4. Using $\overset{\oplus}{N}H_2$

$$\text{(t-Bu)}\!-\!MgCl + NH_2OCH_3 \longrightarrow \text{(t-Bu)}\!-\!NH_2$$

5. Using $-\ddot{N}$

(i) $H\!-\!\overset{\ominus}{N}\!-\!\overset{\oplus}{N}\equiv N \xrightarrow{UV} H\!-\!\ddot{N} \xrightarrow{CH_2=CH_2}$ aziridine (N–H)

(ii) $R\!-\!N=C=O \xrightarrow{UV} R\!-\!\ddot{N} \xrightarrow{CH_2=CH_2}$ aziridine (N–R) $\quad R =$ alkyl / aryl

(iii) 2-nitrobiphenyl (NO_2) $\xrightarrow{(EtO)_3P}$ biphenyl nitrene ($\ddot{N}$) $\longrightarrow$ carbazole (N–H)

6. Using $R-\underset{\underset{O}{\|}}{C}-\ddot{N}$

(i) Hofmann, Lossen, Schmidt and Curtius reactions involve acyl nitrene ($R-\underset{\underset{O}{\|}}{C}-\ddot{N}$) intermediate and hence new C—N bond may be formed through these rearrangement.

Examples

(a) $RCO_2H \xrightarrow{N_3H} R-\underset{\underset{O}{\|}}{C}-\overset{\ominus}{N}-\overset{\oplus}{N}\equiv N \longrightarrow R-\underset{\underset{O}{\|}}{C}-\ddot{N} \longrightarrow R-N=C=O$

$R-N=C=O \xrightarrow{H_2O} R-NH_2$

$R-N=C=O \xrightarrow{EtOH} R-NHCO_2Et$

(b) anthranilic acid (NH_2, CO_2H) $\xrightarrow{N_3H}$ 2-aminobenzoyl azide (NH_2, $C(=O)N_3$) $\xrightarrow{\Delta}$ acyl nitrene (NH_2, $C(=O)\ddot{N}$) $\longrightarrow$ 2-aminophenyl isocyanate (NH_2, $N=C=O$) $\longrightarrow$ benzimidazolone (NH, NH, C=O)

7. Using $=\overset{\oplus}{N}$ (Nitrenium Cation)

During Beckmann rearrangement, new C—N bond is formed through nitrenium cation

$C_6H_5COCH_3 \xrightarrow{NH_2OH} C_6H_5(CH_3)C{=}N{-}OH \xrightarrow{PCl_5} [C_6H_5(CH_3)C{=}\overset{\oplus}{N}] \rightarrow CH_3C{=}\overset{\oplus}{N}{-}C_6H_5 \xrightarrow{H_2O} C_6H_5-NHCOCH_3$

The following reactions (outlines) are some illustrations for C—N bond formation using nucleophilic nitrogen synthetic equivalent.

1. Using NH_3

$NH_3 + RCHO \longrightarrow [RCH{=}NH] \xrightarrow{H_2/Pt} R-CH_2-NH_2$

$+ CH_3COCH_2CO_2Et \longrightarrow CH_3C(NH_2){=}CHCO_2Et$

$+ CH_3COCH_2CH_2COCH_3 \longrightarrow CH_3C(OH)(NH_2)CH_2CH_2COCH_3 \longrightarrow$ 2,5-dimethylpyrrole (N–H)

$+ (CH_3)_2C{=}CH{-}CO{-}CH{=}C(CH_3)_2 \longrightarrow$ HN (2,2,6,6-tetramethylpiperidin-4-one) =O

+ phthalic anhydride $\longrightarrow$ phthalimide (NH) $\xrightarrow[\text{(ii) R - Br}]{\text{(i) KOH}}$ N–R phthalimide $\xrightarrow[\text{(ii) } H_3\overset{\oplus}{O}]{\text{(i) KOH}}$ $R-NH_2$

+ 1-chloro-2,4-dinitrobenzene → 2,4-dinitroaniline

+ CH_3COCl → CH_3CONH_2

2. Using Me_2NH

$Me_2NH + CH_2O + HCl \longrightarrow CH_2{=}\overset{\oplus}{N}Me_2$

$+ CH_2O +$ acetone $+ HCl \longrightarrow$ $CH_3COCH_2CH_2NMe_2$

+ ethylene oxide ⟶ $HOCH_2CH_2NMe_2$

+ CH_3COCl ⟶ CH_3CONMe_2

+ $CH_2{=}CHCOCH_3$ ⟶ $CH_3COCH_2CH_2NMe_2$

3. Using $N_3^{\ominus}$

$N_3^{\ominus} + CH_3CH(Br)CO_2^{\ominus} \longrightarrow CH_3CH(N_3)CO_2^{\ominus} \xrightarrow{H_2/Pd} CH_3CH(NH_2)CO_2H$

4. Using CH_3-CN

$CH_3CN + CH_3MgBr \longrightarrow (CH_3)_2C{=}NMgBr \xrightarrow{Li/liqNH_3} (CH_3)_2CH{-}NH_2$

$+ (CH_3)_3C{-}OH + H_2SO_4 \longrightarrow (CH_3)_3C{-}NH_2$ (Ritter reaction)

5. Using NH_2OH

$NH_2OH + CH_3CHO \longrightarrow (H_3C)(H)C{=}NOH \xrightarrow{Ac_2O} CH_3-C\equiv N$

$+ (CH_3)_2C{=}O \longrightarrow (CH_3)_2C{=}NOH \xrightarrow[\text{or } H_2/Pd]{LiAlH_4} (CH_3)_2CH-NH_2$

6. Using pyrrolidine (N–H)

N–H + cyclohexanone $\xrightarrow{TsOH}$ 1-(cyclohex-1-enyl)pyrrolidine $\xrightarrow{CH_3I}$ iminium ($=\overset{\oplus}{N}$) $\xrightarrow{NaBH_4}$ 1-(2-methylcyclohexyl)pyrrolidine

+ $CH_3COCl \longrightarrow CH_3CO$–N(pyrrolidine)

7. Using NH_2CONH_2

$NH_2CONH_2 + CH_2(CO_2Et)_2 \xrightarrow[EtOH]{\overset{\ominus}{O}Et}$ barbituric acid (O, NH, O, NH, O)

8. Using $C_6H_5NH_2$

$C_6H_5NH_2 + Ac_2O \longrightarrow C_6H_5-NH-COCH_3$

$+ C_6H_5CHO \longrightarrow C_6H_5N = CHC_6H_5$

$$+ C_6H_5-N=C=O \longrightarrow C_6H_5-NH-\overset{O}{\overset{\|}{C}}-NHC_6H_5$$

$$+ CHCl_3 + \overset{\ominus}{O}H \longrightarrow C_6H_5-N=C:$$

$$+ CH_3COCl \xrightarrow{Et_3N} C_6H_5-NH\,COCH_3$$

9. Using $C_6H_5NHNH—C_6H_5$

$$C_6H_5NH\,NHC_6H_5 + CH_2(CO_2Et)_2 \xrightarrow[EtOH]{\overset{\ominus}{O}Et} \text{1,2-diphenylpyrazolidine-3,5-dione } (H_5C_6-N-N-C_6H_5)$$

1.4.3 Protection and deprotection of functional groups, position and ring

Protection (blocking) and deprotection (deblocking) of functional groups, position or ring is an important part during organic synthesis.

This is required for polyfunctional molecules and in multistep synthesis. This is also required when the choice of selective reagent is not sufficient to stop the attack of other functional groups present in the polyfunctional molecules. So, during the functional group inter conversion or C—C bond formation in the process of synthesis, protection is necessary. During the synthesis, if more than one choice of alkylation or cyclisation then one of the two positions must be protected.

Nature of protecting groups

1. The reagent for protecting group must be easily available and stable.
2. The reagent should not possess or introduce a chiral centre in the molecule.
3. The protected group must be stable throughout the required reaction sequence.
4. The protected group should be stable either in acid or base medium and also stable with redox system.
5. After the desired reaction, the protecting group should be easily removed (deprotected) from the functional groups, position or ring and in quantitative yield.
6. Sometimes protection and deprotection of a functional group is the process which is reversible functional group interconversion as for example in the enamine synthesis.
7. Protection is actually deactivation (completely inert) of a functional group, position or a ring in a given sequence of reaction.

1. Protection of functional groups

(a) Protection of Carbonyl

Protecting group (PG)	Protecting reagent	Deprotection	(PG) Stable with	(PG) Unstable or sensitive with
Acetal or ketal	ROH + HCl (BF_3) $HOCH_2CH_2OH$ / HCl gas	$H_3O^{\oplus}$	bases, nucleophiles, reducing agent, grignard reagent, $LiAlH_4$, $NaBH_4$ Me_2CuLi, $KMnO_4$ / $\overset{\ominus}{O}H$	electrophile oxidising agents
Cyclic thioketal	$HSCH_2CH_2CH_2SH$ / BF_3	(i) Hg^{+2} / $H_3O^{\oplus}$ (ii) if Hg^{+2} sensitive group present then CH_3I – H_2O/CH_3CN / $CaCO_3$	bases, acids $LiAlH_4$ Me_2CuLi $KMnO_4$ / $\overset{\ominus}{O}H$	electrophile
mono thioketal or hemithioketal	$HSCH_2CH_2CH_2OH$ / $ZnCl_2$	Raney Ni/ H_2/CH_3COCH_3	bases, $LiAlH_4$	
enol ether	CH_3OH / BF_3 CH_2N_2 $\overset{\ominus}{O}H$ / Me_2SO_4	$H_3O^{\oplus}$	bases $LiAlH_4$ CH_3MgBr	acids
trimethyl Silyl enol ether	Me_3SiCl / Et_3N	$nBu_4\overset{\oplus}{N}\overset{\ominus}{F}$ CH_3Li	bases weak acids	strong acid strong nucleophile
enamine	pyrrolidine (N–H) / TsOH	$H_3O^{\oplus}$	bases	acids
Metalloenamine	cyclohexyl–NH_2	$H_3O^{\oplus}$	bases	acids
hydrazone	$H_2N — NMe_2$	$NaIO_4/CH_3OH/H_2O$	bases	acids

Some illustrations are given below showing applications of carbonyl protection-deprotection using conversion used in the organic synthesis.

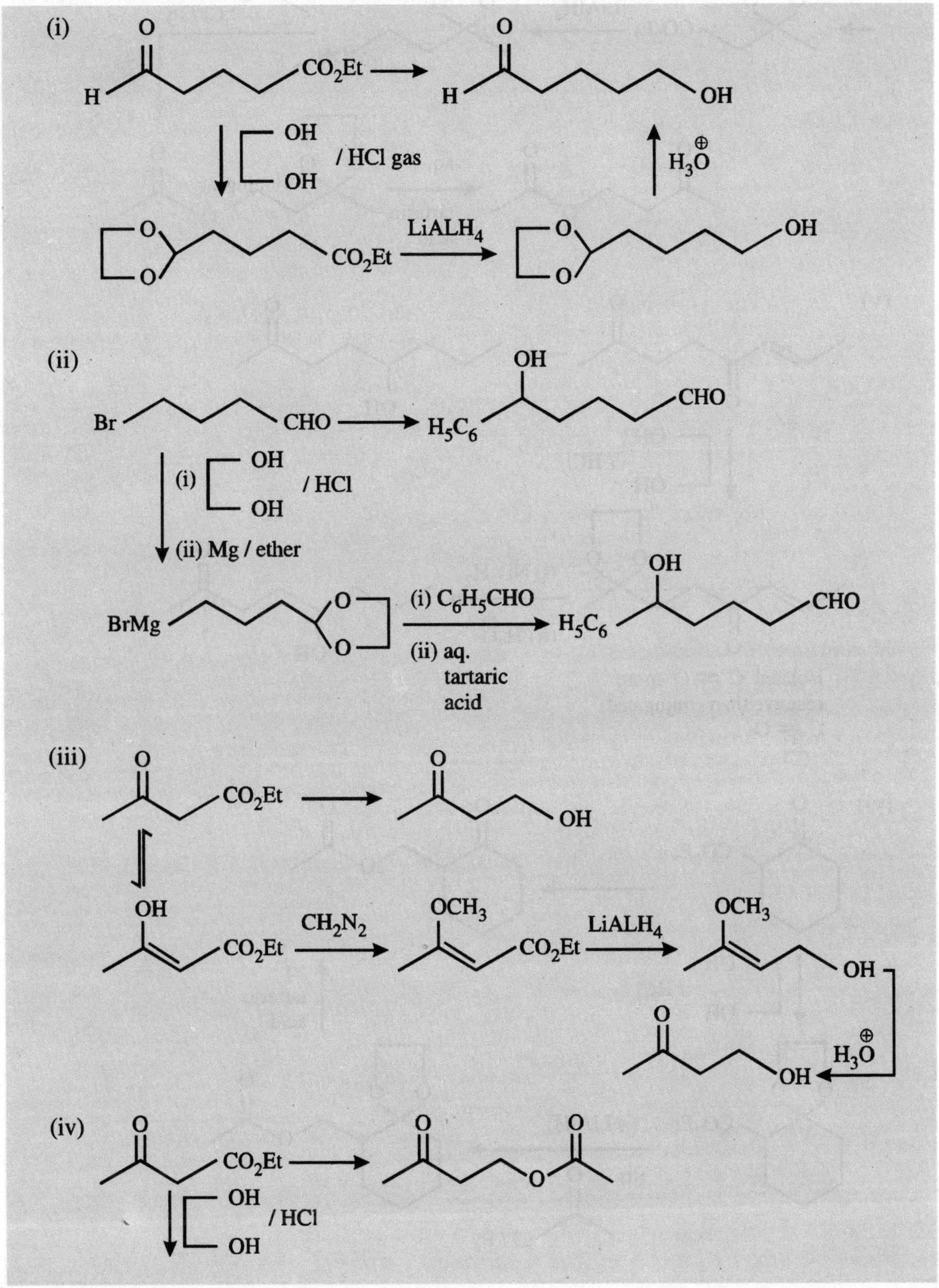

(i)

O / H / CO_2Et → O / H / OH

↓ OH / OH / HCl gas ↑ $H_3\overset{\oplus}{O}$

O / O / CO_2Et —LiALH$_4$→ O / O / OH

(ii)

Br / CHO → OH / H_5C_6 / CHO

↓ (i) OH / OH / HCl (ii) Mg / ether

BrMg / O / O —(i) C_6H_5CHO (ii) aq. tartaric acid→ OH / H_5C_6 / CHO

(iii)

O / CO_2Et → O / OH

⇅

OH / CO_2Et —CH_2N_2→ OCH_3 / CO_2Et —LiALH$_4$→ OCH_3 / OH

$H_3\overset{\oplus}{O}$ → O / OH

(iv)

O / CO_2Et → O / O / O

↓ OH / OH / HCl

LiAlH$_4$

Cl / Py

aq. tartaric acid

(v)

OH / HCl

OH

(i) Ni / H$_2$

(ii) H$_3$O$^{\oplus}$

isolated C = O more reactive than conjugated C = O

(vi)

CO$_2$Et

OH / HCl

OH

aq. tartaric acid

(i) LiAlH$_4$

(ii) Cl / Py

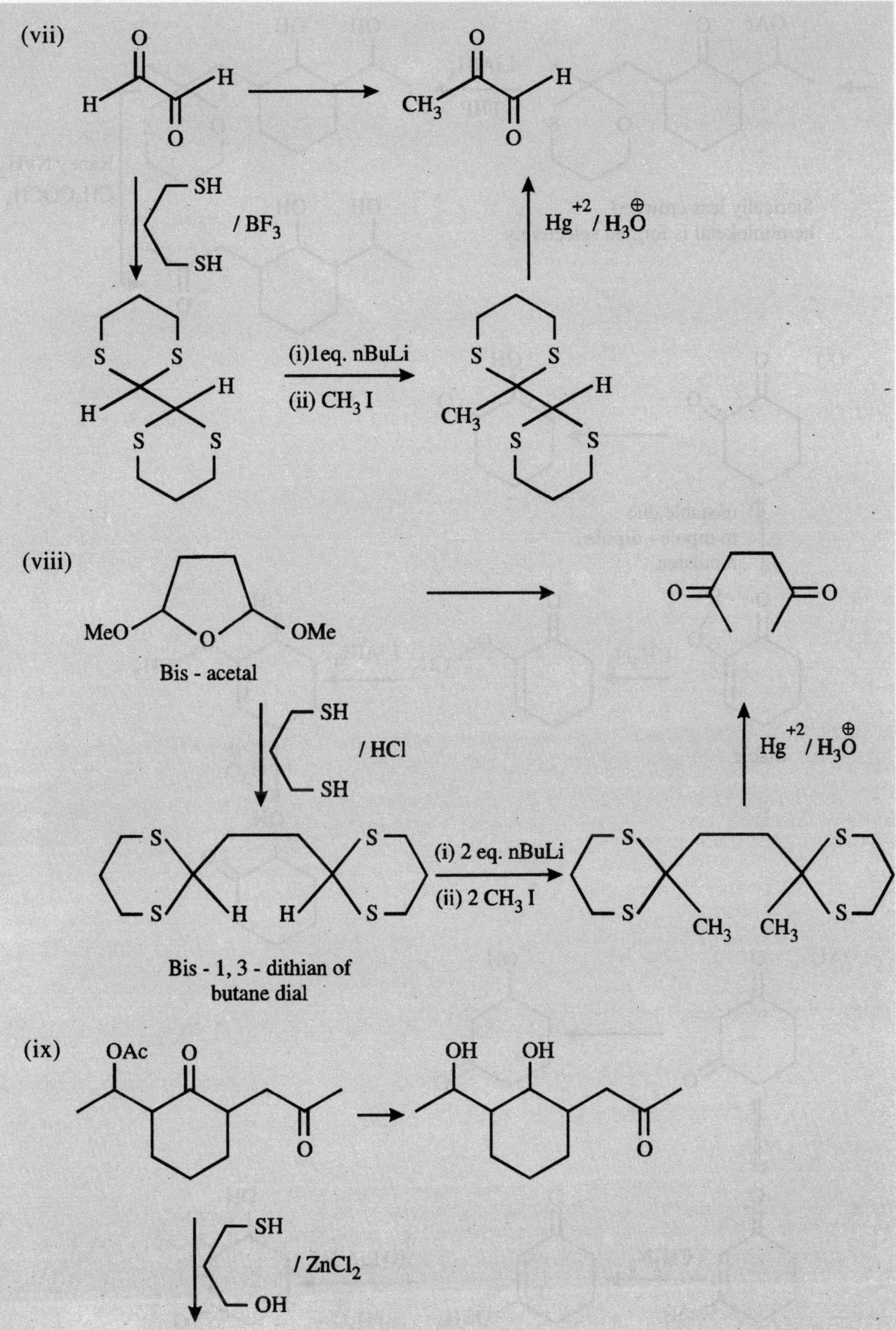
(vii)
O
H
H
O
O
H
CH3
O
SH
/ BF3
SH
Hg+2 / H3O⊕
S S
H
H
S S
(i)1eq. nBuLi
(ii) CH3 I
S S
H
CH3
S S
(viii)
MeO
O
OMe
Bis - acetal
O
O
Hg+2 / H3O⊕
SH
/ HCl
SH
S S
S S
H H
(i) 2 eq. nBuLi
(ii) 2 CH3 I
S S
S S
CH3 CH3
Bis - 1, 3 - dithian of
butane dial
(ix)
OAc
O
O
OH
OH
O
SH
/ ZnCl2
OH

OAc O

LiAlH$_4$
THF

OH OH

O S

Raney Ni/H_2
CH_3COCH_3

Sterically less crowded hemithioketal is formed selectively

OH OH

O

(x) O O OH O

unstable due to dipole - dipole repulsion

O······H O

Stable

CH_2N_2

O O CH_3

LiAlH$_4$

OH O CH_3

$H_3\overset{\oplus}{O}$

OH O

(xi) O O OH O

O

CH_2N_2

OH

O OCH$_3$

(i) LiAlH$_4$
(ii) $H_3\overset{\oplus}{O}$

OH O

(xii)

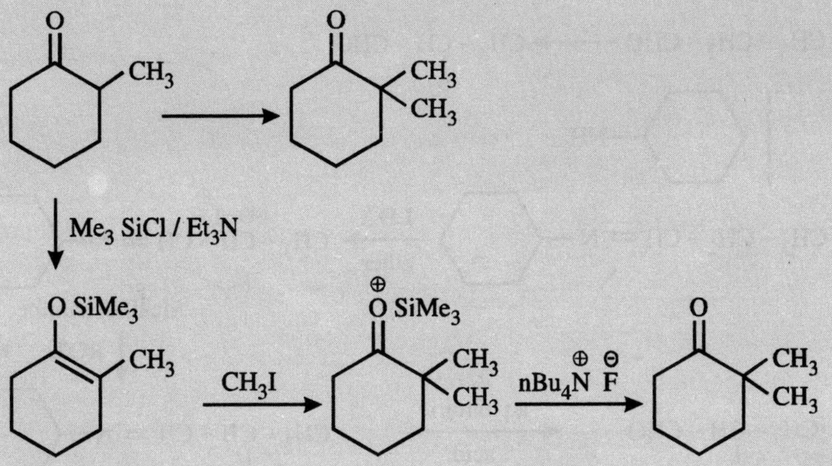

Me$_3$ SiCl / Et$_3$N

O SiMe$_3$ CH$_3$ $\xrightarrow{CH_3I}$ $\overset{\oplus}{O}$ SiMe$_3$ CH$_3$ CH$_3$ $\xrightarrow{nBu_4\overset{\oplus}{N}\ \overset{\ominus}{F}}$ O CH$_3$ CH$_3$

specific enolate
(other regio enol ether is
less stable)

(xiii)

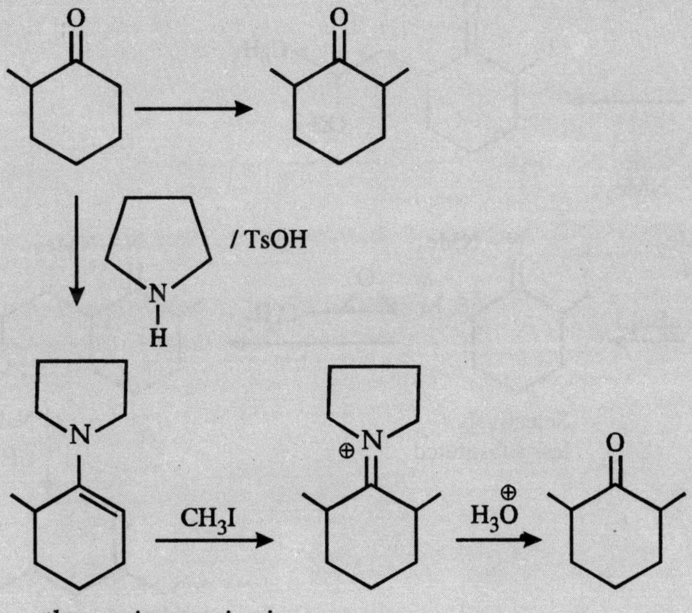

/ TsOH

$\xrightarrow{CH_3I}$ $\xrightarrow{H_3\overset{\oplus}{O}}$

other regio enamine is
less stable due to steric reason

(xiv) $CH_3-CH_2-CHO \longrightarrow CH_3-CH(R)-CHO$

$\downarrow$ cyclohexyl–NH_2

$CH_3-CH_2-CH=N$–cyclohexyl $\xrightarrow[\text{ether}]{\text{LDA}}$ $CH_3-\overset{\ominus}{C}H-CH=N$–cyclohexyl ($Li^{\oplus}$) $\longleftrightarrow$

Metalloenamine

$\downarrow$ ROTs

$CH_3-CH(R)-CHO \xleftarrow[\text{acid}]{\text{aq. tartaric}} CH_3-CH(R)-CH=N$–cyclohexyl

(xv) 2-methylcyclohexanone $\longrightarrow$ 2-methyl-6-(2-hydroxy-2-phenylethyl)cyclohexanone (C_6H_5, OH)

$\downarrow$ NH_2-NMe_2

$N-NMe_2$ (hydrazone) $\xrightarrow{\text{nBuLi}}$ $N-NMe_2$, $\ominus$ $Li^{\oplus}$ $\xrightarrow{\text{epoxide–}C_6H_5}$ $N-NMe_2$ ... C_6H_5, O^-

Selectively less substituted

$\downarrow$ $NaIO_4$, $CH_3OH + H_2O$

O ... C_6H_5, OH

(b) Protection of alcohol

Protecting group (PG)	Protecting reagent	Deprotection	PG Stable with	PG Un stable with
ethers				
R — O — Me	CH_2N_2/BF_3 $^{\ominus}OH/Me_2SO_4$	HI or HBr	bases, electrophiles,	HI/HBr
R — O — $CH_2C_6H_5$	$C_6H_5CH_2Br$/Base	H_2/Pd	Oxidising agents	HI/HBr
R — O — $SiMe_3$	Me_3 SiCl/Et_3N or Py	$nBu_4 \overset{\oplus}{N}\overset{\ominus}{F}$		
R — O — C $(C_6H_5)_3$	$(C_6H_5)_3$ CCl	HBr or H_2/Pd		acids
acetals				
$CH_3-CH(OR)_2$	CH_3CHO/HCl gas	$H_3\overset{\oplus}{O}$	bases	acids
	(dihydropyran) /HCl gas	$H_3\overset{\oplus}{O}$	bases	acids
esters				
R — O — $COCH_3$	CH_3COCl/Py	MeOH/EtOH	weak base	Nucleo phile
R — O — Ts	TsCl/Py		electrophiles	
R — O — Ms	MsCl/Py	$H_3\overset{\oplus}{O}$	oxidising agent	
tertiary butyl				
R — O — $C(CH_3)_3$	$(CH_3)_3C$—OH / HCl	aq.H_2SO_4	grignard reagent	acids
Phenol				
Ar — OMe	CH_2N_2	HI, HBr	bases,	acids
Ar — $OSiMe_3$	Me_3SiCl/Py	$\overset{\ominus}{F}$	oxidising agent, grignard reagent	attack by electro philes on the ring
Diol				
carbonate (for 1, 2-diol)	$COCl_2$/Py	$LiOH + H_2O$	bases,	acids
acetal or ketal (for 1, 2-diol)	CH_3CHO or CH_3COCH_3/HCl	$H_3\overset{\oplus}{O}$	$KMnO_4/\overset{\ominus}{O}H$ CrO_3/Py	
	$CH_2I_2/\overset{\ominus}{O}H$	PCl_5/H_2O	grignard	
	$(C_6H_5)_2CCl_2/\overset{\ominus}{O}H$	$H_3\overset{\oplus}{O}$	$LiAlH_4$	
for 1, 3 - diol	C_6H_5CHO/HCl gas CH_3CHO/HCl gas	$H_3\overset{\oplus}{O}$		

Some illustrations are given below showing application of alcohol protection-deprotection using conversion used in the organic synthesis.

(i) $HOCH_2C\equiv C-H \longrightarrow HOCH_2C\equiv C-CO_2H$

Protection: dihydropyran/HCl

THP–O–$CH_2C\equiv C-H$ $\xrightarrow[\text{(ii) } CO_2]{\text{(i) } CH_3MgBr/THF}$ THP–O–$CH_2C\equiv C-CO_2^{\ominus}$ $\xrightarrow{H_3O^{\oplus}}$ $HOCH_2C\equiv C-CO_2H$

(ii) $Br(CH_2)_4OH \longrightarrow D(CH_2)_4OH$

Protection: $(CH_3)_3C$–OH/HCl

$Br(CH_2)_4O$–$C(CH_3)_3$ $\xrightarrow{\text{(i) Mg/THF}}$ $BrMg(CH_2)_4O$–$C(CH_3)_3$ $\xrightarrow{D_2O}$ $D(CH_2)_4O$–$C(CH_3)_3$ $\xrightarrow{\text{aq. } H_2SO_4}$ $D(CH_2)_4OH$

(iii) p-$HOC_6H_4CH_3 \longrightarrow p$-$HOC_6H_4COOH$

Protection: TsCl/Py

p-$TsOC_6H_4CH_3$ $\xrightarrow{KMnO_4/\overset{\ominus}{O}H}$ p-$TsOC_6H_4CO_2^{\ominus}$ $\xrightarrow{H_3O^{\oplus}}$ p-HOC_6H_4COOH

(iv)

OH → OH, NO_2

↓ $\overset{\ominus}{O}H/Me_2SO_4$

OMe —HNO_3 / Ac_2O→ OMe, NO_2 —HI→ OH, NO_2

(v)

OH, Br → OH (allyl)

↓ Me_3SiCl/Py

$OSiMe_3$, Br —Mg/THF→ $OSiMe_3$, MgBr —(allyl bromide, Br)→ $OSiMe_3$ (allyl) —$\overset{\ominus}{F}, H_3\overset{\oplus}{O}$→ OH (allyl)

(vi)

OH, OH, OH → OH, OH, O

↓ CH_3COCH_3/HCl

O, O, OH —CrO_3/Py→ O, O, O —$H_3\overset{\oplus}{O}$→ OH, OH, O

(vii)

OH OH OH → O OH OH

↓ CH_3CHO/HCl

OH O O —(i) CrO_3/Py, (ii) $H_3\overset{\oplus}{O}$→ O OH OH

(viii)

OH OH OH → OMe OH OH

↓ $(C_6H_5)_2CCl_2/\overset{\ominus}{O}H$

$\overset{\ominus}{O}$ O O C_6H_5 C_6H_5 —Me_2SO_4→ OMe O O C_6H_5 C_6H_5 —$H_3\overset{\oplus}{O}$→ OMe OH OH

(ix)

HO HO → HO HO CO_2H

↓ $\overset{\ominus}{O}H/CHCl_3$

$\overset{\ominus}{O}$ $\overset{\ominus}{O}$ CHO —CH_2I_2→ CH_2 O O CHO

↓ $KMnO_4$

CH_2 O O COOH —PCl_5→ Cl Cl C O O COCl

↓ H_2O

HO HO COOH

(x) Glycerol (OH, OH, OH) → (OH, OH, OR)

R = alkyl or acyl
For alkyl introduction
$\overset{\ominus}{O}H/Me_2SO_4$

For acyl introduction
CH_3COCl/Py

↓ CH_3COCH_3/HCl

Acetonide (O, O, OH) $\xrightarrow{\overset{\ominus}{O}H/R_2SO_4}$ Acetonide (O, O, OR) $\xrightarrow{H_3\overset{\oplus}{O}}$ (OH, OH, OR)

(xi) (OH, OH, OH) → (OH, $OCOCH_3$, OH)

2eq. ↓ $(C_6H_5)_3CCl$

($OC(C_6H_5)_3$, OH, $OC(C_6H_5)_3$) $\xrightarrow{CH_3COCl/Py}$ ($OC(C_6H_5)_3$, $OCOCH_3$, $OC(C_6H_5)_3$) $\xrightarrow[H_2/Pd]{HBr\ or}$ (OH, $OCOCH_3$, OH)

Primary alcohol reacts faster with trityl chloride

This type of blocking-deblocking may also be done with TsCl/Py

(xii) (OH, OH, OH) → (OH, OCH_3, OH)

↓ C_6H_5CHO/HCl

Benzylidene acetal (O, OH, O)$-C_6H_5$ $\xrightarrow{\overset{\ominus}{O}H/Me_2SO_4}$ (O, OMe, O)$-C_6H_5$ $\xrightarrow{H_3\overset{\oplus}{O}}$ (OH, OCH_3, OH)

(xiii) 2,3-bis(MeO_2C)-1,4-benzoquinone → 2,3-bis(MeO_2C)-1,4,5-trihydroxybenzene (OH, OH, OH)

↓ Ac_2O/BF_3

ester exchanged deprotection of acyl group since OAc is an ester function - here MeOH is chosen, otherwise aryl - CO_2Me will give other ester (eg - CO_2Et if EtOH is chosen) by ester exchange

(c) Protection of amino

Protecting group (PG)	Protecting reagent	Deprotection	PG Stable with	PG Unstable with
acyl	Ac_2O/conc. H_2SO_4	$H_3O^{\oplus}$	electrophiles	acids
	CH_3COCl/base	$\overset{\ominus}{O}H/H_2O$		bases
Tosyl	TsCl/Py	$H_3O^{\oplus}$	bases	acids
Mesyl	$MeSO_2Cl$ (MsCl)/Py			
Trimethyl Silyl	Me_3 SiCl/Py	$\overset{\ominus}{F}$	bases	
Trityl	$(C_6H_5)_3$ CCl/Et_3N	H_2/Pd	bases	acids
Benzyloxy Carbonyl	$C_6H_5CH_2OCOCl$	H_2/Pd Na/liqNH_3	bases	acids
Phthalimide	Phthalic anhydride	NH_2NH_2		

Illustrations

(i) NH₂-C₆H₄-CH₃ → NH₂-C₆H₄-COOH

TsCl/Py

NHTs-C₆H₄-CH₃ $\xrightarrow{KMnO_4/\overset{\ominus}{O}H}$ NHTs-C₆H₄-$CO_2^{\ominus}$ $\xrightarrow{H_3\overset{\oplus}{O}}$ NH_2-C₆H₄-COOH

(ii) $H_2N-CH_2-COOH \longrightarrow H_2N-CH_2-\overset{\overset{O}{\|}}{C}-NH-\underset{CH_3}{\underset{|}{CH}}-COOH$

glycine — glycinyl alanine

$\downarrow C_6H_5-CH_2-O-COCl$

$C_6H_5-CH_2-O-\underset{\underset{O}{\|}}{C}-NH-CH_2-COOH$

amino group is blocked

$\xrightarrow{PCl_5}$ $C_6H_5-CH_2-O-\underset{\underset{O}{\|}}{C}-NH-CH_2-\underset{\underset{O}{\|}}{C}-Cl$

carboxyl group is activated

$\downarrow H_2N-\underset{CH_3}{\underset{|}{CH}}-COOCMe_3$

Carboxyl protected alanine

$C_6H_5-CH_2-O-\underset{\underset{O}{\|}}{C}-NH-CH_2-\underset{\underset{O}{\|}}{C}-NH-\underset{CH_3}{\underset{|}{CH}}-COOCMe_3$

(i) H_2/Pd
(ii) CF_3CO_2H

$H_2N-CH_2-\underset{\underset{O}{\|}}{C}-NH-\underset{CH_3}{\underset{|}{CH}}-COOH$

amino and carboxyl groups are deblocked

(iii) $H_2N-\underset{CH_3}{\underset{|}{CH}}-COOH \longrightarrow H_2N-\underset{CH_3}{\underset{|}{CH}}-\underset{\underset{O}{\|}}{C}-NH-CH_2-COOH$

alanine — alanyl glycine (Ala - Gly)

Phthalimide (NH) $\xrightarrow{\overset{\ominus}{O}H/ClCO_2Et}$ N-ethoxycarbonylphthalimide ($N-CO_2Et$)

alanine $\downarrow$

Phthalimido–$N-\underset{CH_3}{\underset{|}{CH}}-COOH$ $\xrightarrow{PCl_5}$ Phthalimido–$N-\underset{CH_3}{\underset{|}{CH}}-\underset{\underset{O}{\|}}{C}-Cl$

amino group is blocked — carboxyl group is activated

$\downarrow H_2N-CH_2-COOH$

→ N-phthaloyl-Ala-Gly: (phthalimido)N – CH(CH_3) – C(=O) – NH – CH_2 – COOH

$\xrightarrow{NH_2NH_2}$ $H_2N – CH(CH_3) – C(=O) – NH – CH_2 – COOH$ (Ala - Gly) – phthalhydrazide

(d) Protection of Carboxylic group

Protecting group (PG)	Protecting reagent	Deprotection	PG Stable with	PG Un-stable with
tertiary butyl	Trace H_2SO_4 / isobutylene	dry HCl anhydrous CF_3CO_2H		
ester	CH_2N_2 $EtOH/H^{\oplus}$	$\overset{\ominus}{O}H/H_2O$	weak bases	strong bases
benzyl	$C_6H_5CH_2$-Cl/Py	H_2/Pd		
trimethyl silyl	Me_3SiCl/Et_3N	CH_3OH	oxidising agents	
anion ($RC\bar{O}_2$, R-$\bar{O}$)	base	acid	nucleophiles	electrophiles

Illustrations

$H_2N – CH_2 – COOH$ → H_2N–CH(CH_3)–C(=O)–NH–CH_2–CO_2H

glycine — alanyl glycine (Ala Gly)

↓ isobutylene /Trace H_2SO_4

$H_2N – CH_2 – C(=O) – O – C(CH_3)_3$

+ C_6H_5–CH_2–O–C(=O)–NH–CH(CH_3)–C(=O)–Cl

alanine carboxyl group is activated and $–NH_2$ group is blocked

→ C_6H_5–CH_2–O–C(=O)–NH–CH(CH_3)–C(=O)–NH–CH_2–C(=O)–O–C(CH_3)$_3$

$\xrightarrow[\text{(ii) dry HCl or anhydrous } CF_3CO_2H]{\text{(i) } H_2/Pd}$ $H_2N-CH(CH_3)-CO-NH-CH_2-COOH$ Ala Gly

Amino and carboxyl groups are deblocked

(e) Protection of halogen

$Br(CH_2)_4Br \longrightarrow Br(CH_2)_4NO_2$ or $NC(CH_2)_4NO_2$

$Br(CH_2)_4Br \xrightarrow{\overset{\ominus}{O}Et/EOH} EtO(CH_2)_4Br \xrightarrow[CH_3COCH_3]{KNO_2} EtO(CH_2)_4NO_2 \xrightarrow{HBr} Br(CH_2)_4NO_2 \xrightarrow[CH_3COCH_3]{KCN} NC(CH_2)_4NO_2$

$\overset{\ominus}{O}Et/EtOH$ is added slowly to the dibromo compound in 1 : 1 proportion

(f) Protection of double bond

(i) $CH_2 = CH-CH_2OH \longrightarrow CH_2 = CH-COOH$

$CH_2 = CH-CH_2OH \xrightarrow{Br_2} CH_2Br-CHBr-CH_2OH \xrightarrow{HNO_3} CH_2Br-CHBr-COOH \xrightarrow[\Delta]{Zn/CH_3OH} CH_2 = CH-COOH$

(ii) p-Benzoquinone $\longrightarrow$ 2,3-epoxy-p-benzoquinone

p-Benzoquinone $\xrightarrow{\text{cyclopentadiene (1 : 1)}}$ Diels–Alder adduct $\xrightarrow{\overset{\ominus}{O}H/H_2O_2}$ epoxide of adduct $\xrightarrow{-\text{cyclopentadiene, Retro DA}}$ 2,3-epoxy-p-benzoquinone

(g) Protection of triple bond

(i) $CH_3-C\equiv C-C\equiv C-H \longrightarrow CH_3-CH=CH-C\equiv C-H$

$\downarrow CH_3MgBr$

$CH_3-C\equiv C-C\equiv CMgBr \xrightarrow{Me_3SiCl} CH_3-C\equiv C-C\equiv CSiMe_3$

$\downarrow H_2/Pd/CaCO_3/Pb(OAc)_2$ Lindlar catalyst

$CH_3-CH=CH-C\equiv C-H \xleftarrow[THF]{nBu_4\overset{\oplus\ominus}{N}F} CH_3-CH=CH-C\equiv C\,SiMe_3$

bulky - $SiMe_3$ group
protected reduction of adjacent yne

(ii) $H-C\equiv C-H \longrightarrow H-C\equiv C-C\equiv C-H/H-(C\equiv C)_n-H$

$\downarrow CH_3MgBr$ 1 : 1

$H-C\equiv CMgBr \xrightarrow[THF]{Me_3SiCl} H-C\equiv C-SiMe_3 \xrightarrow{Cu(OAc)_2/Py/O_2}$

$H-C\equiv C-C\equiv C-H \xleftarrow[THF]{nBu_4\overset{\oplus\ominus}{N}F} Me_3Si-C\equiv C-C\equiv C-SiMe_3$

(iii) $CH_3-\underset{Cl}{CH}-CH_2-C\equiv C-CH_3 \longrightarrow CH_3-\underset{CHO}{CH}-CH_2-C\equiv C-CH_3$

$\downarrow$ 1,3-dithiane /nBuLi

$CH_3-CH(\text{1,3-dithian-2-yl})-CH_2-C\equiv C-CH_3 \xrightarrow[CH_3CN/CaCO_3]{CH_3I-H_2O} CH_3-\underset{CHO}{CH}-CH_2-C\equiv C-CH_3$

1, 3-dithiane is usually deblocked by Hg^{+2}/H_3O^+ but this reagent is very sensitive with triple bond and converted to $-\underset{\underset{O}{\|}}{C}-CH_2-$

CH_3I — $H_2O/CH_3CN/CaCO_3$ is a neutral deblocking reagent without attack on the triple bond and the mechanism is given below and $CaCO_3$ is used to absorb HI

Dithiane (S:) $\xrightarrow{CH_3-I}$ S-methylsulfonium ($\overset{\oplus}{S}-CH_3$, I^-) $\xrightarrow{H_2\ddot{O}}$ ring-opened hemithioacetal (O–H, $S-CH_3$) + HI $\longrightarrow$ CHO + SH–(CH₂)₃–S–CH₃

(2) Protection of position

(a) Unsymmetrical cyclohexanone ring system

(i)

$\overset{\ominus}{O}Et/HCO_2Et$

CHO

LDA

2 eq

CHO

CH_3I

1 eq

CHO

H_2O

CHO

$\overset{\ominus}{O}H/H_2O$

$-HCO_2^{\ominus}$

(ii)

$\overset{\ominus}{O}Et/(CO_2Et)_2$

$COCO_2Et$

Borax/Δ

CO_2Et

$\overset{\ominus}{O}Et$

CH_3I

CO_2Et

dil HCl/Δ

(iii)

$\overset{\ominus}{O}Et/HCO_2Et$

CHO

CH – OH

C_4H_9SH

TsOH

$CHSC_4H_9$

(i) $\overset{\ominus\oplus}{OK}$ / OH

(ii) CH_3I

$CHSC_4H_9$

KOH/H_2O

OH

OH /Δ

(b) Benzene ring system

(i) Me

Me Me Me

$/AlCl_3$

Me

$CH_3Cl/AlCl_3$

Me Me Me

HF/Δ

Me Me Me

p - position is blocked

deblocked

(ii) Me → Me, NO_2 (meta)

$HNO_3 + H_2SO_4$

Me, NO_2 —H_2/Pd→ Me, NH_2 —$HNO_3 + H_2SO_4$→ Me, NO_2, NH_2 —(i) $NaNO_2$ / HCl (ii) H_3PO_2/Δ→ Me, NO_2

blocked

deblocked

(iii) NH_2 → NH_2, Br

Conc-H_2SO_4

NH_2, SO_3H → $\overset{\oplus}{N}H_3$, $SO_3^{\ominus}$ —Ac_2O→ $NHCOCH_3$, SO_3H —Br_2→ $NHCOCH_3$, Br, SO_3H —dil. H_2SO_4 100°C→ NH_2, Br

blocked

remains as zwitter ion

deblocked

Since NH_2 group is highly activating and hence on reaction with Br_2 it gives substituted aniline. Therefore, p-position is blocked by-SO_3H. This is now in zwitterionic form. The $-NH_2$ group is acetylated because of two reasons (a) to remove the zwitter ion, (b) to deactivate the ring for monobromination.

(iv) NH_2 → Br, Br (NH_2)

Conc-H_2SO_4

NH_2 / SO_3H → $NH_3^{\oplus}$ / $SO_3^{\ominus}$ remains as Zwitter ion →(PCl_5) NH_2 / SO_2Cl →(liq NH_3) NH_2 / SO_2NH_2 Sulphonamide blocked

$Br_2 + H_2O$ ↑ $HBr + H_2O_2$

NH_2, Br, Br, SO_2NH_2 →(dil. H_2SO_4, 100°C) NH_2, Br, Br deblocked

Sulphonamide is used due to two reasons (a) to stop the zwitter ion formation, (b) to keep $-NH_2$ group free to activate the ring for dibromination.

3. Protection of ring

(i) Glucose shows mutarotation and gives Fehlings' or Tollens' reagent test in aqueous medium. These two observations are due to the presence of cyclic ring structures which are hemiacetals equilibrium with open chain structure.

H–C–OH, H–C–OH, HO–C–H, H–C–OH, H–C (ring O), CH_2OH	⇌	CHO, H–C–OH, HO–C–H, H–C–OH, H–C–OH, CH_2OH	⇌	HO–C–H, H–C–OH, HO–C–H, H–C–OH, H–C (ring O), CH_2OH
α - D - glucose hemiacetal		open chain		β - D - glucose hemiacetal

The equilibrium mixture accounts for mutarotation and Fehlings' solution test. These two observations may be withdrawn by protection of anomeric (hemiacetal carbon is known as anomeric carbon) hydroxyl group by selective methylation which is acetal.

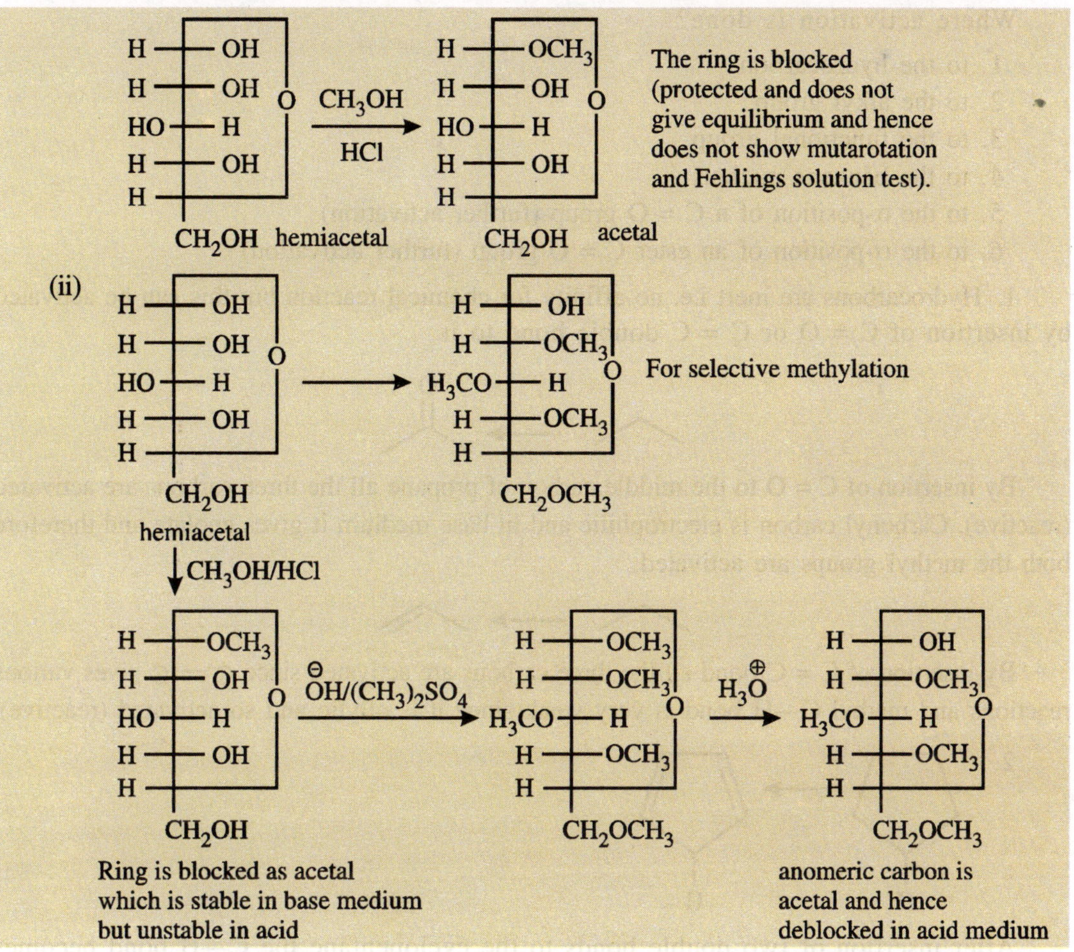

1.4.4 Activation and deactivation of functional groups,position—activation with protection

Sometimes rather than protection of one part of a molecule, it is better to activate another.

Any functional group can act as a protecting group provided it can easily be added and removed and provided, of course, that it does not react with the reagent.

Why activation is required?

1. to control the reaction
2. to stop the self condensation
3. to stop the further (repeated) reaction
4. to increase the acidity
5. to introduce the aromaticity within a ring

6. Hydrocarbons are compound having no functional group (deactivated) and therefore by addition of functional group (activation) to the hydrocarbon, disconnection is possible

Where activation is done?

1. to the hydrocarbon
2. to the alkyl group
3. to the functional group
4. to the benzene nucleus
5. to the α-position of a C = O group (further activation)
6. to the α-position of an ester C = O group (further activation)

1. Hydrocarbons are inert i.e. no affinity for chemical reaction but this can be activated by insertion of C = O or C = C double bond to it.

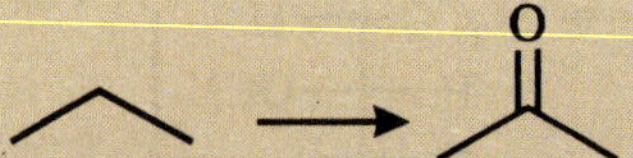

By insertion of C = O to the middle carbon of propane all the three carbons are activated (reactive). Carbonyl carbon is electrophilic and in base medium it gives enolate and therefore both the methyl groups are activated.

By insertion of C = C bond all the three carbons are activated since C = C gives various reactions and methyl C—H bond is very weak since it is allylic and so activated (reactive).

2.

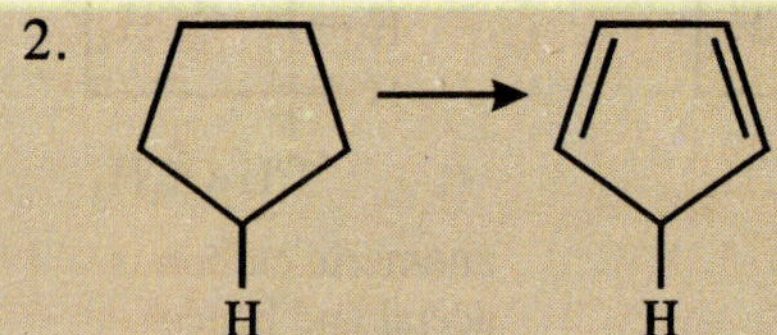

After insertion of two double bonds to the cyclopentane the C—H bond becomes highly activated because of its double allylic nature and after abstraction of proton which is highly acidic—the ring becomes aromatic.

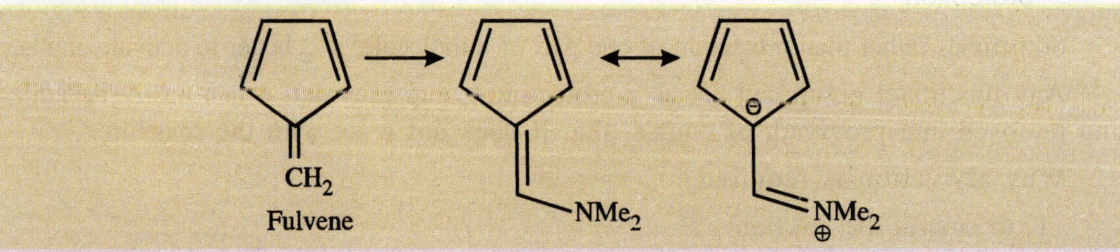

Fulvene is a very unstable molecule but insertion of—NMe_2 group—the ring becomes activated i,e aromatized.

3. (a) 'ene' (C = C) is usually nucleophilic and its nucleophilicity may be activated further by adding a +M group adjacent to it.

Enolate, dienolate and diene-diolate-dianion or silylated enol ether are nucleophilic activation of ene with carbonyl protection in base medium.

enolate:

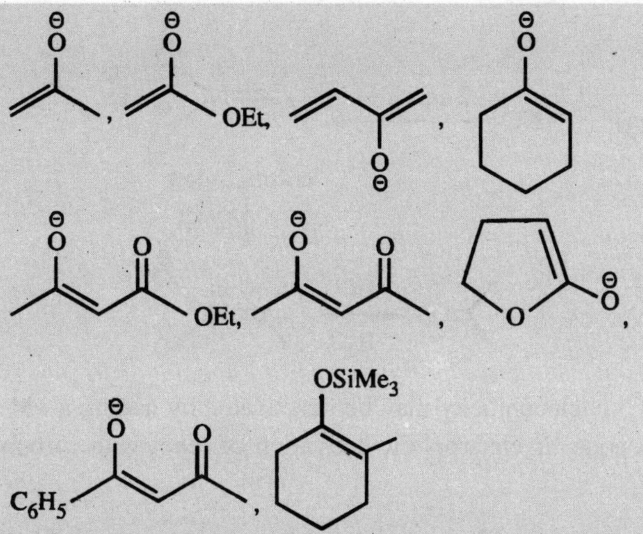

OEt, OEt, OSiMe$_3$ C_6H_5

dienolate:

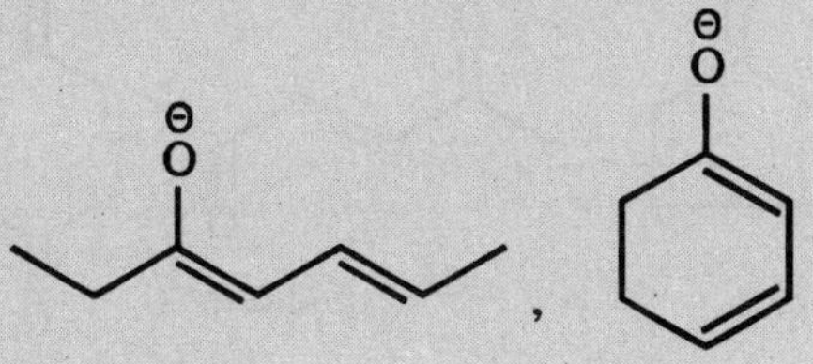

diene-diolate dianion:

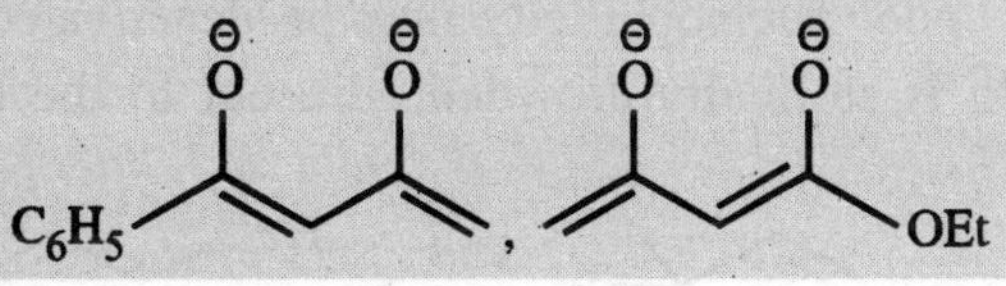

C_6H_5 OEt

(b) 'Enamine' is another class of intermediate which is nucleophilic activation of 'ene' with carbonyl protection in base medium.

NMe_2

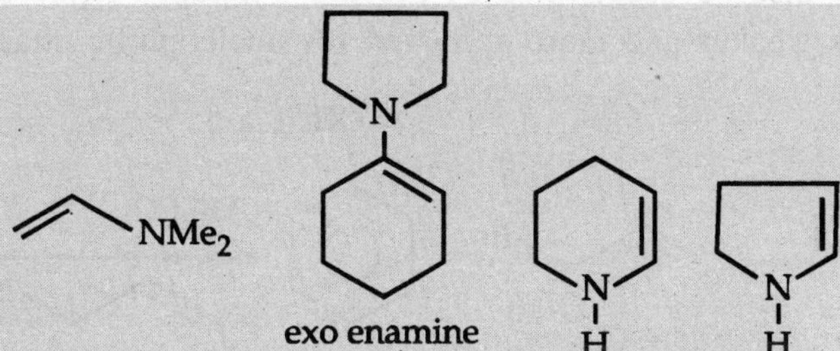

exo enamine

endo enamine

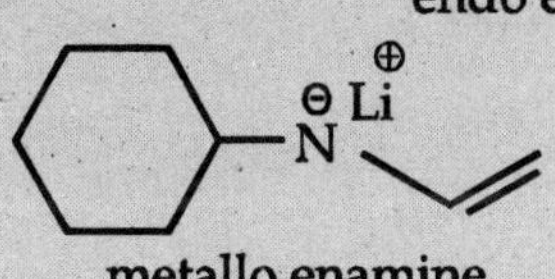

metallo enamine

(c) Nitro alkenium anion is potentially enolate with carbonyl protection in base medium.

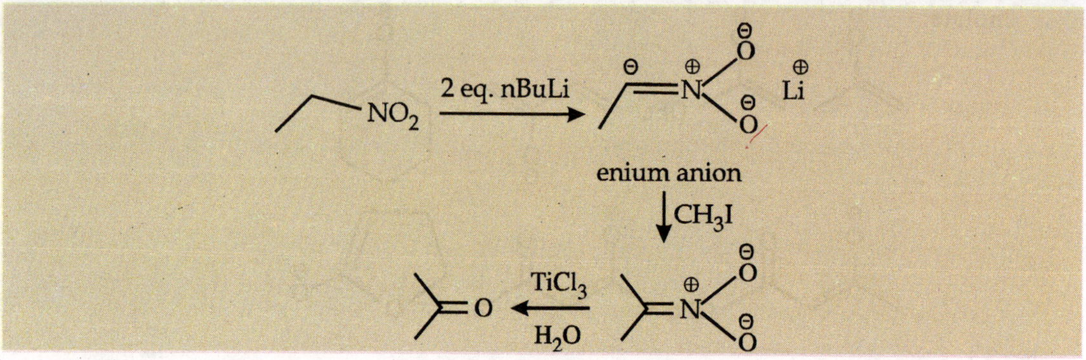

2 eq. nBuLi
enium anion
CH_3I
$TiCl_3$
H_2O

4. (a) 'ene' (C = C) nucleophilicity may be deactivated by adding a –M group adjacent to it. As for example 'enone' is electrophilic activation of 'ene' with carbonyl deactivation in base medium.

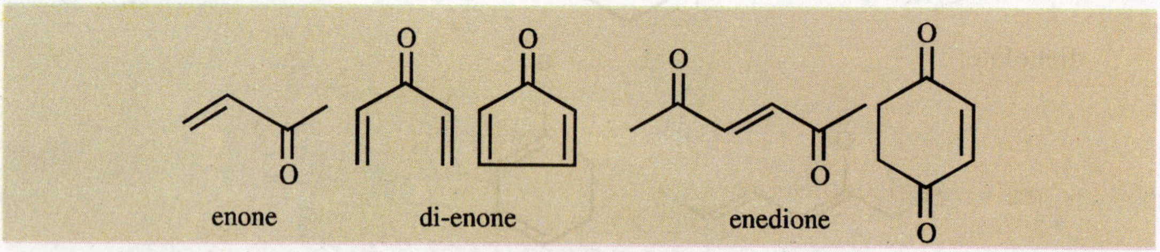

enone di-enone enedione

(b) Sometimes 'ene' nucleophilicity may be deactivated by placing a leaving group to the carbon which is sterically crowded adjacent to the ene.

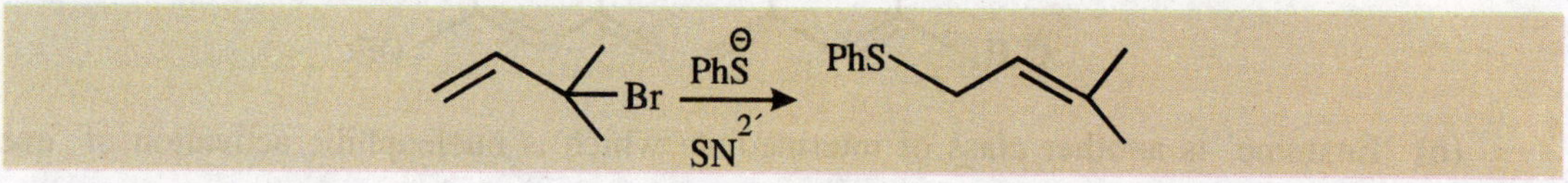

$PhS^{\ominus}$
$SN^{2'}$

(c) 'ene' may also be deactivated by placing a halogen adjacent to it since allylic C-halogen bond is weaker and more activated for nucleophilic attack in base medium.

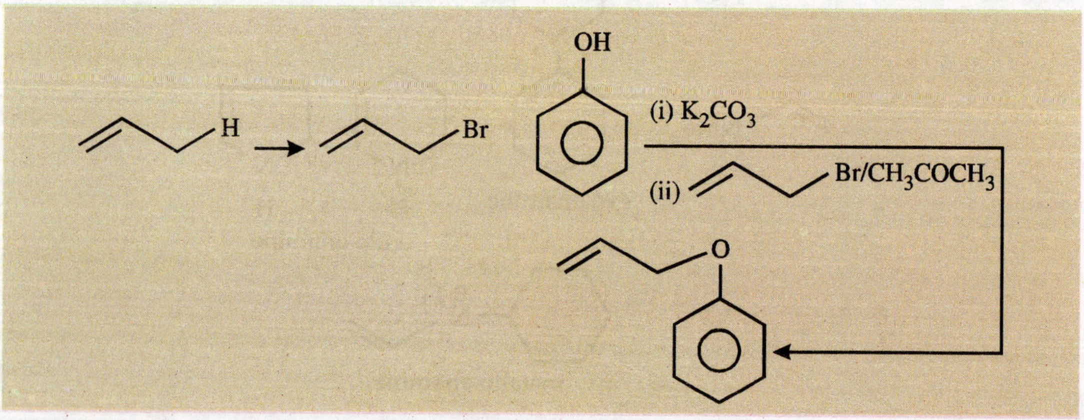

(i) K_2CO_3
(ii) Br/CH_3COCH_3

(d) 'ene' nucleophilicity may be completely deactivated i.e. polarity is completely reversed when it is converted to epoxy function.

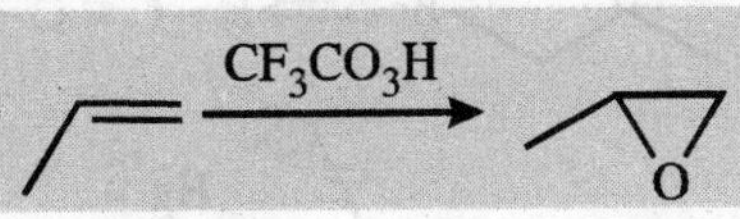

CF_3CO_3H

(e) Epichlorohydrin is a compound in which epoxy carbon is more activated for nucleophilic attack.

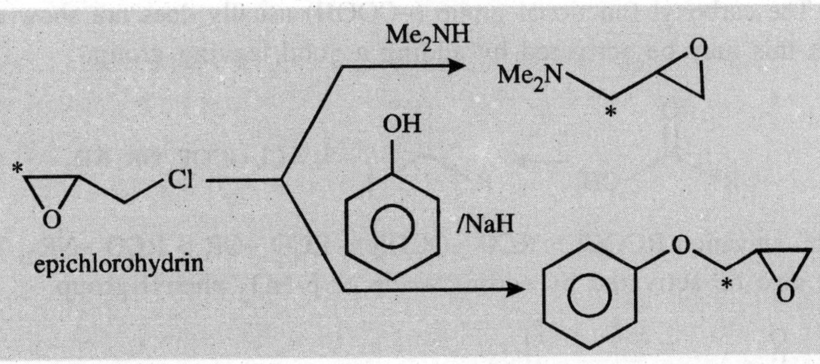

Me_2NH Me_2N * OH * Cl epichlorohydrin /NaH O *

(f) Enol-ether, enol-acetate are protection of enol (and hence carbonyl) with activation of 'ene' as nucleophile in base medium and inert with some redox agent e.g. $LiAlH_4$, CrO_3/Py.

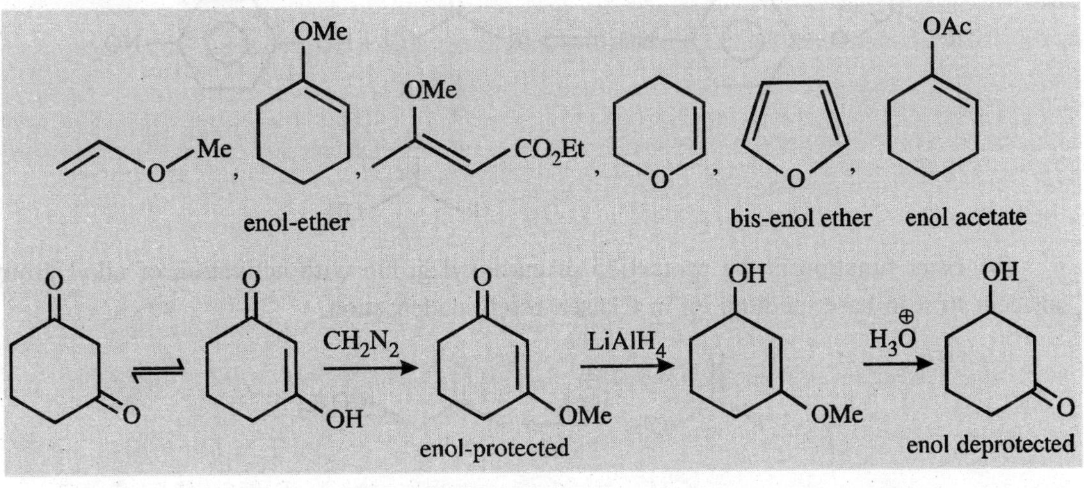

OMe, OMe, Me, CO_2Et, OAc

enol-ether bis-enol ether enol acetate

O O O OH OH CH_2N_2 $LiAlH_4$ $H_3\overset{\oplus}{O}$ O OH OMe OMe O

enol-protected enol deprotected

5. 'yne' ($C \equiv C$) is the protection of keto-methylene group ($-COCH_2-$) in base medium. The 'yne' is deactivating group in base medium.

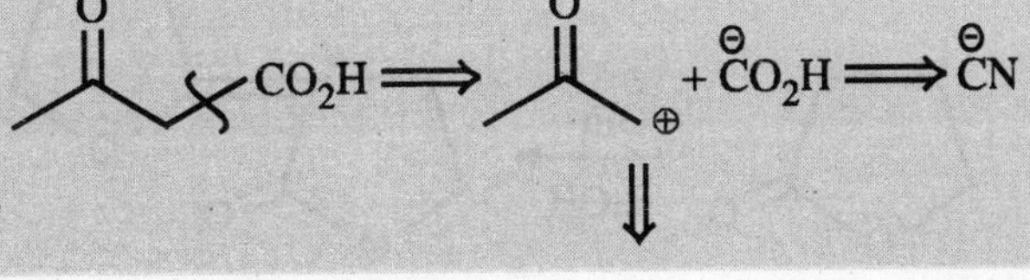

O $CO_2H \Longrightarrow$ O $\oplus$ $+ \overset{\ominus}{C}O_2H \Longrightarrow \overset{\ominus}{C}N$

$\Downarrow$

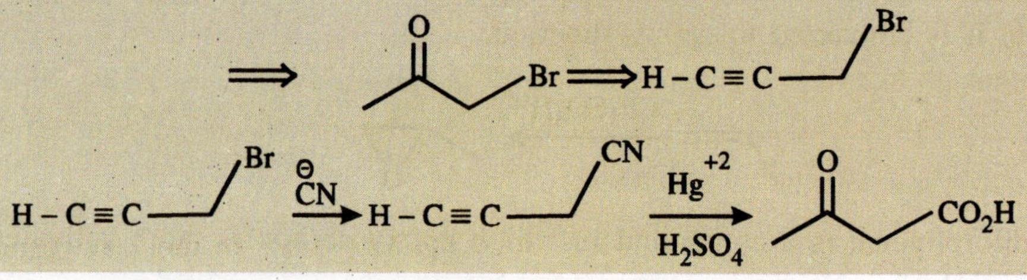

6. (a) The carboxyl functional group (–COOH) usually does not show any carbonyl activity but this may be activated by adding a good leaving group.

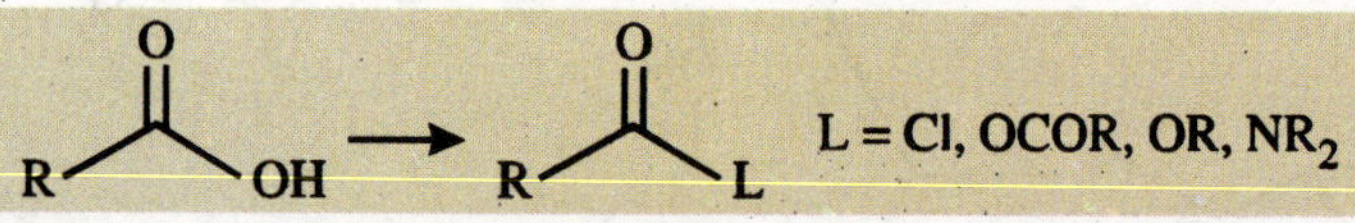

L = Cl, OCOR, OR, NR_2

order of activation RCOCl > RCO –OCOR > RCO –OR > RCO $–NR_2$. The carboxyl group may also be activated by adding azide or p-NO_2 phenyl group.

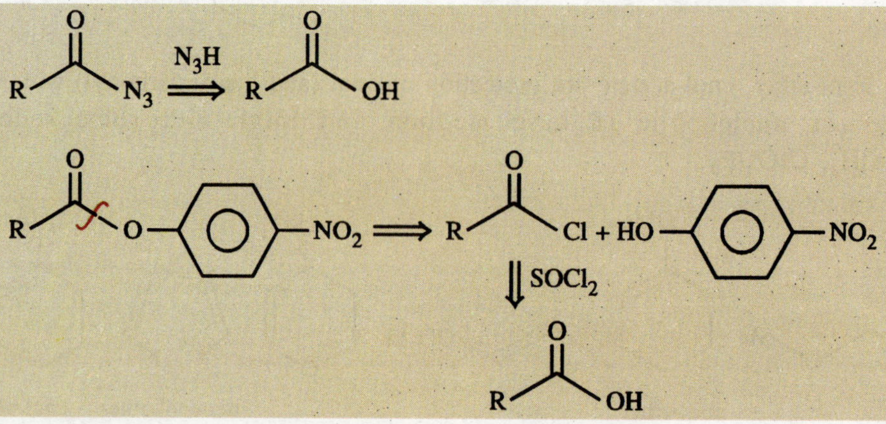

(b) Ester function is the protection of carboxyl group with activation of alkyl group adjacent to it in base medium eg in Claisen ester condensation.

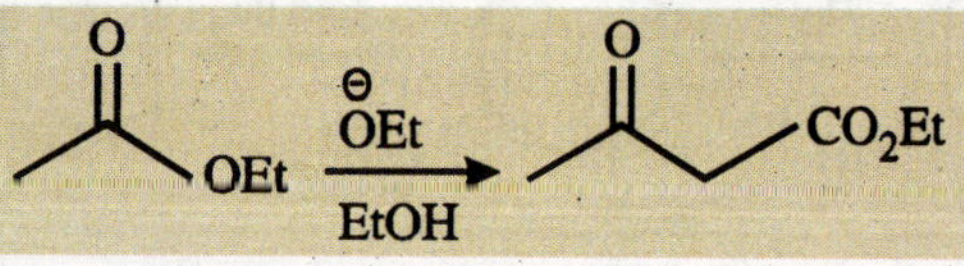

(c) Lactone is the protection of ester carbonyl in base medium e.g. γ-lactone gives Claisen ester condensation.

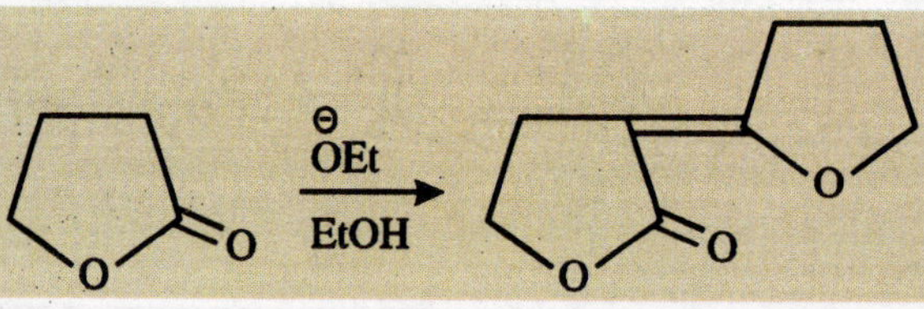

(d) Saturated anhydride (e.g. succinic anhydride) is the activation of carboxyl carbonyl group as in the Friedel-Crafts reaction.

(e) Saturated anhydride is the activation (acidity) of adjacent methylene group with the protection of vicinal dicarboxylic group in base medium e.g. in the Claisen ester condensation and in the Perkin reaction.

(f) Unsaturated anhydride is the activation of 'ene' as dienophile in the Diels-Alder reaction with protection of vicinal dicarboxylic group.

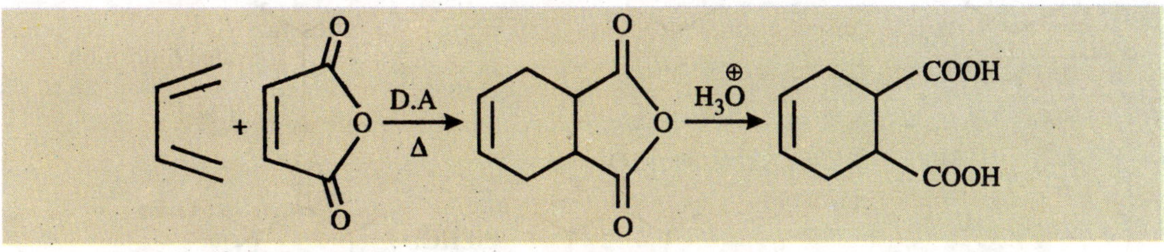

7. In the DA reaction diene and dienophile may be activated by placing ED and Ew group respectively.

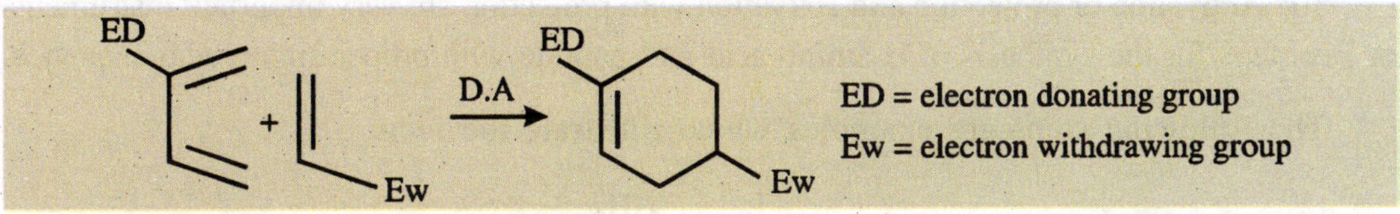

8. Alkyl group adjacent to carbonyl is pro-nucleophilic in base medium. It shows nucleophilicity in base medium via enolate but this nucleophilicity may be completely deactivated by placing halogen adjacent to carbonyl in base medium. This halocarbon is illogical electrophile.

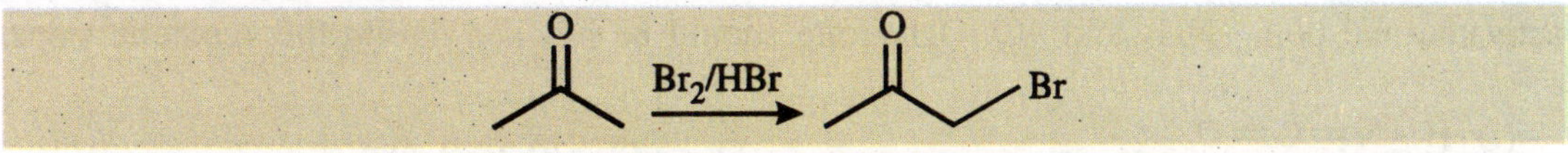

Carbonyl deactivation may also be done by placing 3-membered ring adjacent to carbonyl

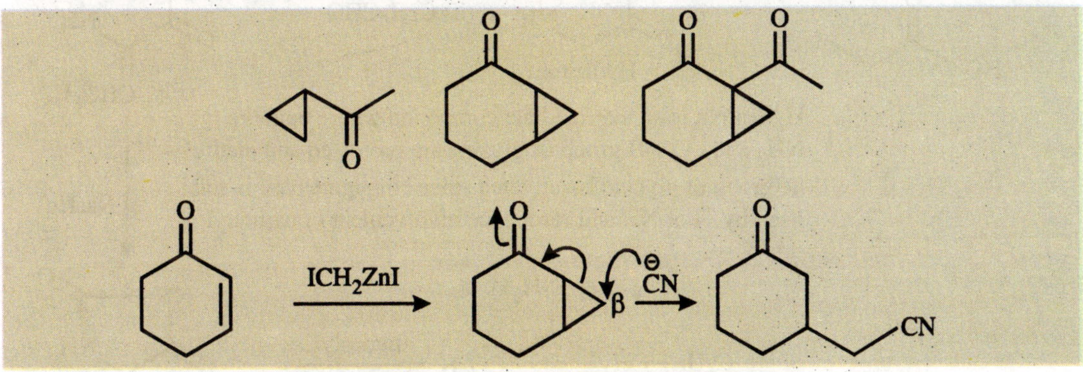

The ring βC is more electrophilic than carbonyl (which behaves as a Michael acceptor) because of steric strain relieves.

In the 'enone' ene may be considered as 3-membered ring.

9. Activation of position: The conversion of 2-methyl cyclohexanone to 2, 6-dimethyl cyclohexanone is difficult since isomeric gem dialkylated product parallel the reaction. Therefore, activation of 6-position is necessary.

$\overset{\ominus}{O}Et/(CO_2Et)_2$ → $COCO_2Et$ —Δ→ CO_2Et

activated 6-position
$\overset{\ominus}{O}Et/CH_3I$

CO_2Et —dil.HCl, Δ→

10. Activation or protection and activation with protection are very important instruments or key steps for the synthesis of α-amino acid and peptide with proper amino acid sequence.

The following steps are examples which illustrate the facts

(a)

$H_2N-CH_2-CO_2H \longrightarrow C_6H_5-CH_2-CH(NH_2)-CO_2H$

Glycine Phenyl alanine

For the conversion of glycine to phenyl alanine, methylene group of glycine should be activated and both $-NH_2$ and $-COOH$ group should be blocked during the synthetic steps.

(i) $H-N=C=O$ + $HO-C(=O)-CH_2-NH_2$ ⟶ Hydantoin

(i) CH_3CO_2Na/Ac_2O
(ii) C_6H_5CHO

CHC_6H_5

Hydantoin is an intermediate compound where both the NH_2 and $-CO_2H$ group of glycine are protected and methylene group of glycine is activated since conjugate base is stabilised by –I of NH and resonance involvement of carbonyl

Na(Hg)

C_6H_5

hydrolysis

$C_6H_5-CH_2-CH(NH_2)-CO_2H$

Phenyl alanine

(ii)

$H_2N\text{-}CH_2\text{-}CO_2H$ (Glycine) $\xrightarrow{C_6H_5COCl}$ $C_6H_5CO\text{-}NH\text{-}CH_2\text{-}COOH$ $\xrightarrow[-H_2O]{\overset{\ominus}{O}Ac/Ac_2O}$ Azlactone

benzoyl glycine (hippuric acid) $-NH_2$ is blocked

Azlactone activation of methylene with double protection of $-NH_2$ and $-COOH$

$\downarrow C_6H_5-CHO$

$\xrightarrow{Na(Hg)}$ $\xrightarrow[\text{deblocked}]{\text{hydrolysis}}$ $C_6H_5-CH_2-CH(NH_2)-CO_2H$

Phenyl alanine

(iii)

$H_2N\text{-}CH_2\text{-}CO_2H$ + $HO_2C\text{-}CH_2\text{-}NH_2$ $\xrightarrow[-2H_2O]{\Delta}$ Diketopiperazine $\xrightarrow[Ac_2O]{2C_6H_5CHO}$ $\xrightarrow[\text{deblocked}]{P/HI}$ $2\ C_6H_5-CH_2-CH(NH_2)-COOH$

α - amino acid on being heated dimerises

Diketopiperazine both-NH_2 and $-COOH$ groups are protected with activation of methylene

Phenyl alanine

(b) Activation or protection and activation with protection are very necessary caution during the synthesis of peptide with definite amino acid sequences. This may be explained with the synthesis of dipeptide glycinylalanine.

The synthesis of glycinyl alanine may be done by protection of amino group of glycine with activation of carboxylic group and protection of carboxylic group of alanine. The whole sequence may be shown as follows.

$H_2N\text{–}CH_2\text{–}CO_2H$ (Glycine) $\xrightarrow{C_6H_5CH_2OCOCl}$ $C_6H_5-CH_2-O-CO-NH-CH_2-CO_2H$

$-NH_2$ group is protected

$\downarrow PCl_5$

$C_6H_5-CH_2-O-CO-NH-CH_2-COCl$

$-COOH$ group is activated

$H_2N\text{–}CH(CH_3)\text{–}CO_2H$ (Alanine) $\xrightarrow{\text{isobutylene / Trace } H_2SO_4}$ $H_2N-CH(CH_3)-CO-O-C(CH_3)_3$

$-CO_2H$ group is protected

$C_6H_5-CH_2-O-CO-NH-CH_2-COCl$ $+ H_2N-CH(CH_3)-CO-O-C(CH_3)_3$

$\downarrow$

$C_6H_5-CH_2-O-CO-NH-CH_2-CO-NH-CH(CH_3)-CO-O-C(CH_3)_3$

(i) H_2/Pd (amino group is deblocked)

(ii) anhydrous CF_3CO_2H (Carboxyl group is deblocked)

$\downarrow$

$H_2N-CH_2-CO-NH-CH(CH_3)-CO_2H$

Glycinyl alanine (glyala)

11. Neighbouring group participation (NGP) is a protection which gives retention of configuration at the chiral centre with activation since NGP involves reaction always becomes faster.

$\overset{\oplus}{H_3N}-C(H)(Me)-CO_2^{\ominus}$ (S-Alanine) $\xrightarrow{HNO_2}$ $HO-C(H)(Me)-CO_2H$ (S-Lactic acid)

(S)-Alanine $\xrightarrow{HNO_2}$ $-\overset{\oplus}{N}\equiv N$ $\xrightarrow[S_N2]{NGP}$ (R) – $\xrightarrow[S_N2]{H_2\ddot{O}}$ (S) –

S-Lactic acid

12. Acetone $\longrightarrow$ EAA

(i) Acetone is activated because it can give enolate in base medium.

(ii) By addition of–CO_2Et in place of hydrogen it becomes further activated, doubly activated.

(iii) The utility of further activation is as follows

(a) It will stop self condensation of CH_3COCH_3 in base medium.

(b) Enolate of EAA will stop the self condensation (aldol or Claisen ester condensation) of it since it gives stabilised enolate-controlled.

(c) Very weak base can abstract methylene proton from EAA because of its double activation and the resulting enolate can act as a Michael donor which required milder condition.

(d) Because of double activation mono and dialkylation can be done in EAA since mono alkylation requires normal condition and dialkylation occurs in drastic condition.

(iv) By addition of – CO_2Et to acetone it is nothing but activation with protection because – CO_2Et may be removed easily.

$\xrightarrow[MeI]{\overset{\ominus}{O}Et}$ $\xrightarrow[\Delta]{\text{dil HCl}}$

13. Ethyl acetate $\longrightarrow$ DEM

(i) Ethyl acetate is activated since it can give ester enolate in base medium.

(ii) By addition of $-CO_2Et$ in place of hydrogen it becomes further activated, doubly activated.

(iii) The utility of further activation is as follows:

(a) It will stop the self-condensation (Claisen ester condensation) of ethyl acetate.

(b) Enolate of DEM does not take part in Claisen ester condensation because it is stabilised enolate — controlled.

(c) Very weak base can abstract proton from methylene of DEM because it is double activated and the resulting enolate can acts as a Michael donor which required milder condition.

(d) Because of double activation mono and dialkylated product may be isolated since dialkylation requires more vigorous condition

(e) By addition of $-CO_2Et$ to ethyl acetate it is nothing but double activation with protection since $-CO_2Et$ may be removed.

$$CH_2(CO_2Et)_2 \xrightarrow[\text{(ii) } CH_3-Br]{\text{(i) } \overset{\ominus}{O}Et} H_3C-CH(CO_2Et)_2 \xrightarrow[\text{(ii) } H_3\overset{\oplus}{O}/\Delta]{\text{(i) KOH}} CH_3-CH_2-COOH$$

14. $CH_2(CO_2Et)_2$ (DEM) $\longrightarrow$ $NC-CH_2-CO_2Et$ (ECA)

By addition of CN in place of $-CO_2Et$ in DEM, it is further activated. As for example DEM has no reaction with keto carbonyl but ECA gives reaction with keto carbonyl.

Cyclopentanone $\xrightarrow{\overset{\ominus}{O}Et/\text{DEM}}$ No reaction

Cyclopentanone $\xrightarrow[\text{ECA}]{\overset{\ominus}{O}Et}$ 1-hydroxycyclopentyl–CH(CN)(CO$_2$Et) (OH, CO_2Et, CN)

15. Activation of benzene nucleus

(i) For electrophilic substitution

(a) Benzene nucleus cannot give electrophilic substitution without activation of electrophile as in the Friedel Craft reaction where $AlCl_3$, $FeBr_3$ or $GaBr_3$ acts as activator for electrophile. Benzene nucleus may be activated for electrophilic substitution by adding +M group in place hydrogen.

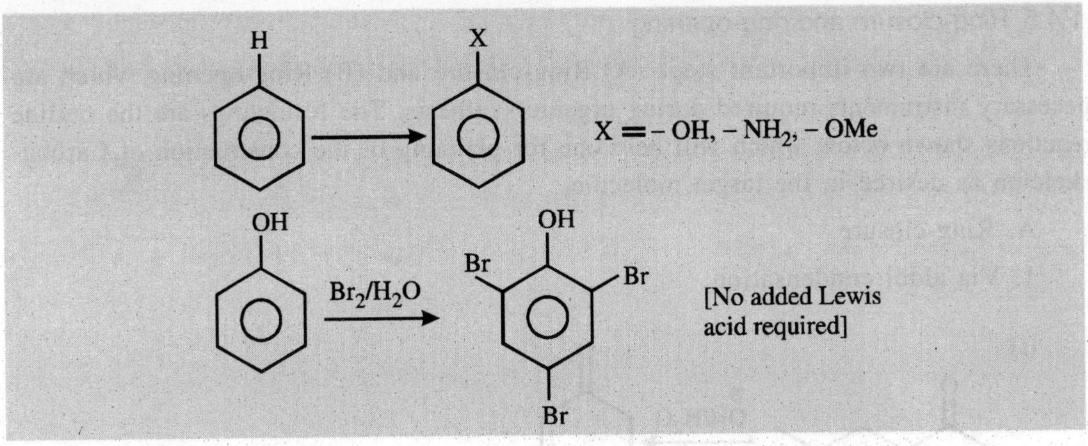
H
X
X = – OH, – NH2, – OMe
OH
Br2/H2O
OH
Br
Br
Br
[No added Lewis acid required]

In some cases, in addition to +M group in the benzene ring, Lewis acid is added for further activation and in that situation reaction rate will accelerate.

(b) – OH or – NH_2 group on the benzene ring may be deactivated with protection by adding acetyl or trimethyl silyl group in the process of electrophilic substitution.

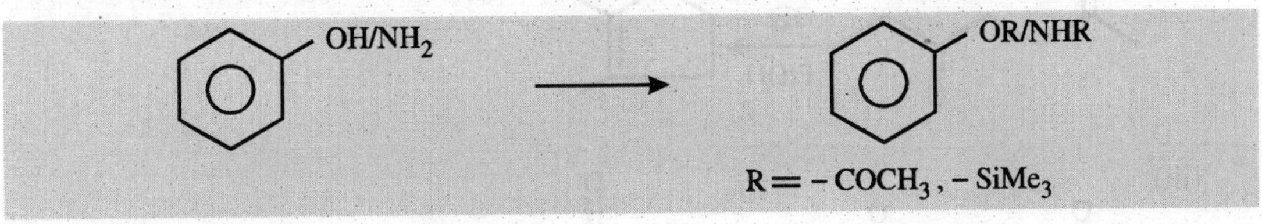
OH/NH2
OR/NHR
R = – COCH3, – SiMe3

(ii) For nucleophilic substitution

Benzene ring does not give nucleophilic substitution but the ring may be activated for nucleophilic substitution by adding –M group conjugated to ipso position i.e. ortho or para position with respect to ipso attacked carbon.

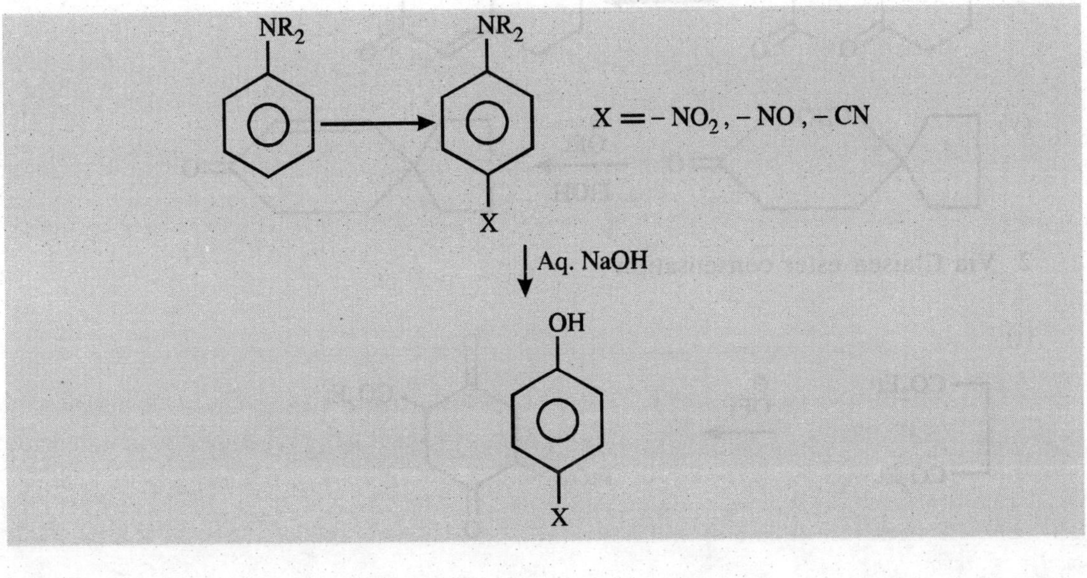
NR2
NR2
X = – NO2, – NO, – CN
X
Aq. NaOH
OH
X

1.4.5 Ring-closure and ring-opening

There are two important steps (A) Ring-closure and (B) Ring-opening which are necessary instruments required during organic synthesis. The followings are the outline reactions shown below which will help one for planning of the construction of Carbon-skeleton as desired in the target molecule.

A. Ring-closure

1. Via aldol condensation

(i) $\overset{\ominus}{O}H/H_2O$

(ii) $\overset{\ominus}{O}Et$, EtOH

(iii) $\overset{\ominus}{O}Et$, EtOH

(iv) $\overset{\ominus}{O}H/H_2O$

(v) CHO, O; $\overset{\ominus}{O}Et$, EtOH; O

2. Via Claisen ester consensation

(i) CO_2Et, CO_2Et; $\overset{\ominus}{O}Et$; O, CO_2Et, EtO_2C, O

(ii) [reaction scheme: diethyl oxalate + diethyl glutarate, $\overset{\ominus}{O}Et$ → cyclopentane-1,2-dione-3,5-dicarboxylate (CO_2Et)]

3. Via Dieckmann reaction

(i) [reaction scheme: diethyl pimelate, $\overset{\ominus}{O}Et$ → 2-carbethoxycyclohexanone (CO_2Et)]

(ii) [reaction scheme: diethyl adipate, $\overset{\ominus}{O}Et$ → 2-carbethoxycyclopentanone (CO_2Et)]

4. Via acyloin condensation (reductive coupling)

(i) $(CH_2)_4(CO_2Et)_2$ —(i) Na/Xylene; (ii) $H_3\overset{\oplus}{O}$→ 2-hydroxycyclohexanone (OH)

(ii) cyclohexane-1,2-di(CO_2Et) —(i) Na/Xylene, Me_3SiCl; (ii) $H_3\overset{\oplus}{O}$→ bicyclic hydroxy cyclobutanone (OH)

5. By Pinacolic reduction (reductive coupling)

$(CH_2)_4$ diketone —(i) Mg/C_6H_6; (ii) $H_3\overset{\oplus}{O}$→ cyclohexane-1,2-diol (OH, OH)

6. By SET reduction (reductive coupling)

stilbene diazonium salt ($\overset{\oplus}{N}_2$) —Cu, 110°→ phenanthrene

7. Via oxidative coupling*

(i) $HO(CH_2)_4OH \xrightarrow[C_6H_6]{Ag_2CO_3} [HO(CH_2)_3CO_2H] \xrightarrow{-H_2O}$ (5-membered lactone, C=O)

(ii) $HO(CH_2)_5OH \xrightarrow[C_6H_6]{Ag_2CO_3}$ (6-membered lactone, C=O)

(iii) $HC\equiv C-(CH_2)_6-C\equiv CH \xrightarrow[Py/O_2]{Cu(OAc)_2}$ cyclic $C\equiv C-C\equiv C$ with $(CH_2)_6$

8. Via Friedel-Crafts reaction

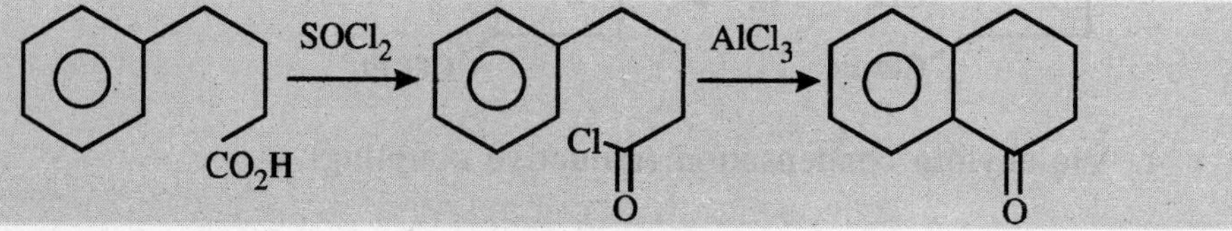

9. By heating of dicarboxylic acid with Ac_2O

(i) By heating of 1,4 and 1, 5-dicarboxylic acid with Ac_2O produces 5 and 6-membered cyclic anhydride respectively

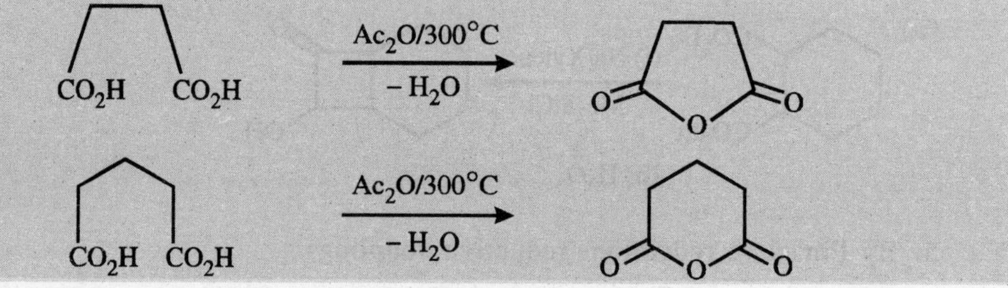

(ii) By heating of 1, 6 and 1, 7-dicarboxylic acid with Ac_2O produces 5 and 6-membered cyclic ketone respectively

$(CH_2)_4(CO_2H)_2 \xrightarrow[300°C]{Ac_2O}$ cyclopentanone

$(CH_2)_5(CO_2H)_2 \xrightarrow[300°C]{Ac_2O}$ cyclohexanone

*For detail see author's "Organic Synthesis" vol. 1

10. By heating hydroxy acid

(i) α-hydroxy acid gives lactide

(ii) γ-and δ-hydroxy acid produces γ and δ-lactone respectively

OH CO_2H $\xrightarrow[-H_2O]{\Delta}$ γ-lactone

OH CO_2H $\xrightarrow[-H_2O]{\Delta}$ δ-lactone

(iii) Lactone may be generated from bromo acid

Br OH $\xrightarrow[-HBr]{K_2CO_3}$

(iv) From Baeyer-villiger oxidation of ketone

CF_3CO_3H

(v) Cyclic ether may be generated from 1, 4 and 1, 5 dihydroxy

(a) OH OH $\xrightarrow[TsCl/Py]{1\ eq.}$ OTs OH

(b)

Py
– TsOH

OTs
OH

11. Via Diels-Alder reaction

(i)

Δ

(ii)

Δ

12. Via electrocyclic reaction

(i)

hν
Δ

(ii)

hν
Δ

13. Via carbene and nitrene

(i)

$:CH_2(S)$

(ii)

CH_2N_2

hν or
Cu/80°C

(iii) 2-nitrobiphenyl $\xrightarrow{(EtO)_3P}$ [nitrene, N:] → carbazole (N–H)

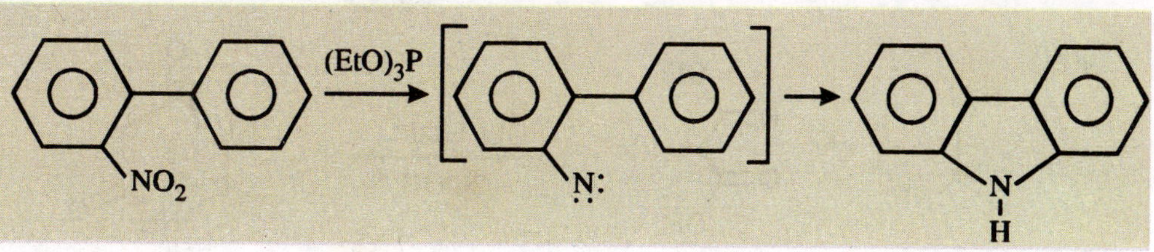

14. Via benzyne

(i) bromobenzene + furan $\xrightarrow{\overset{\ominus}{N}H_2}$ oxa-bridged adduct $\xrightarrow{H_3\overset{\oplus}{O}}$ 1-naphthol (OH)

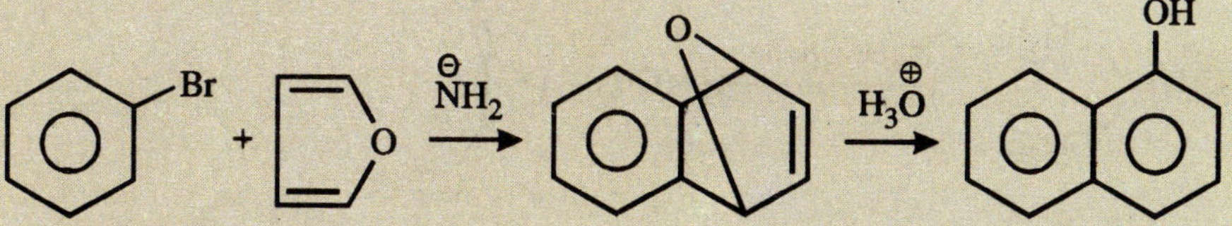

(ii) Br, NHMe $\xrightarrow{\overset{\ominus}{N}H_2}$ N-Me (indoline)

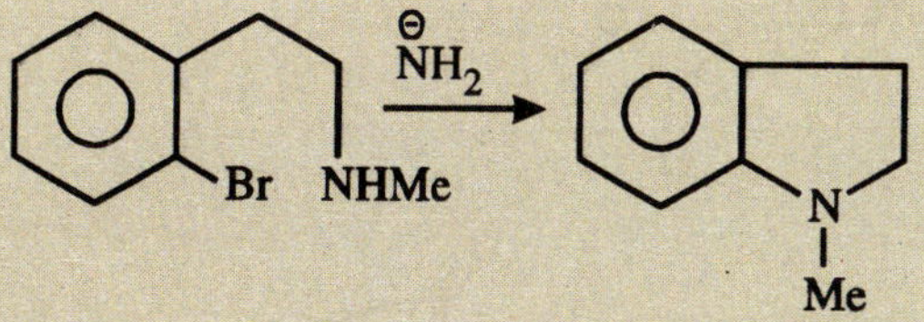

(iii) Br, CO_2Et $\xrightarrow{2\overset{\ominus}{N}H_2}$ CO_2Et (tetralin)

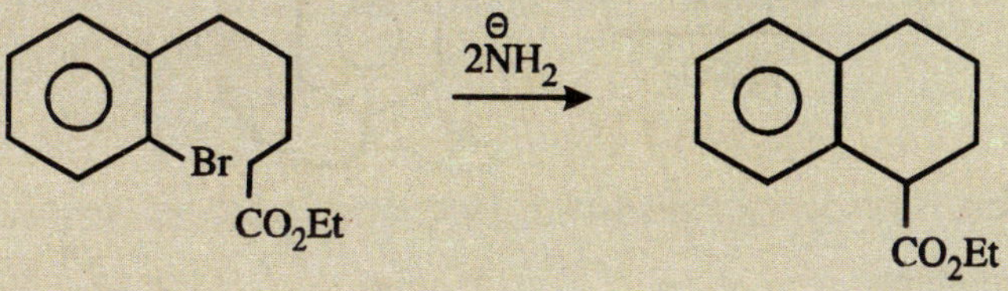

(iv) Br, N $\xrightarrow{\overset{\ominus}{N}H_2}$ N (phenanthridine)

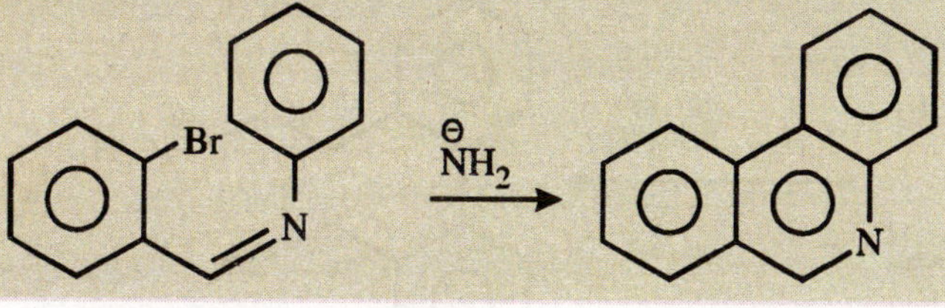

15. Via condensation reaction

(i) HO_2C $\xrightarrow[PPA]{HF\ or}$ O (tetralone)

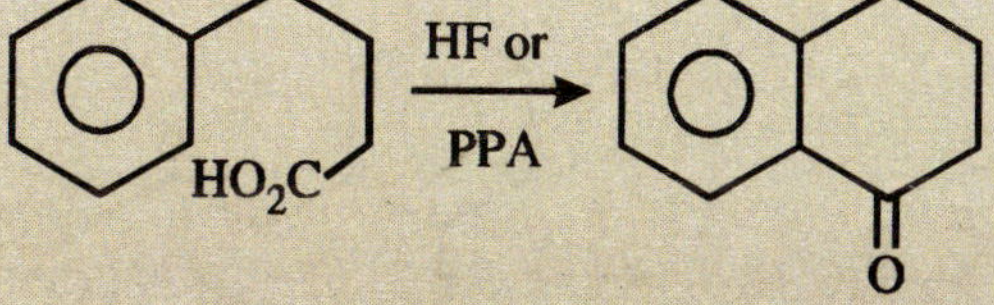

(ii) O, O ⇌ OH, HO $\xrightarrow{P_2O_5}$ O (2,5-dimethylfuran)

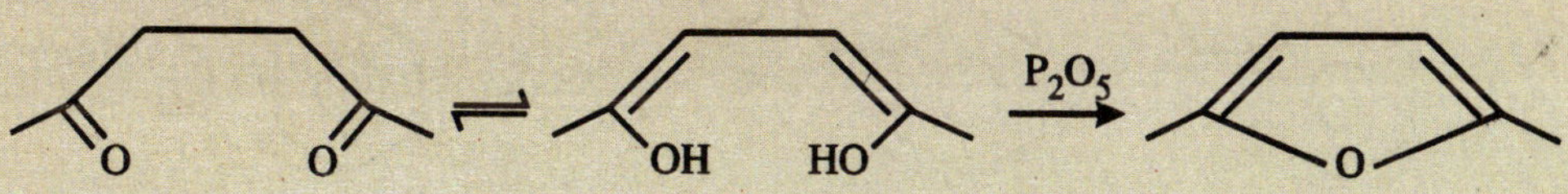

(iii) $H_2N-CO-NH_2$ + $CH_2(COOEt)_2$ $\xrightarrow[-2EtOH]{\overset{\ominus}{O}Et/EtOH}$ barbituric acid

16. With high dilution principle

(i) $(CH_2)_7(CO_2Et)_2$ $\xrightarrow{NaH/Xylene}$ $(CH_2)_6$, O, CO_2Et

(ii) $(CH_2)_9$, Br, CO_2H $\xrightarrow[CH_3COC_2H_5]{K_2CO_3}$ $(CH_2)_9$, O, O

(iii) $(CH_2)_{8-10}-COCl$ $\xrightarrow[PhNO_2]{AlCl_3}$ O, $(CH_2)_{8-10}$

17. By using wittig reagent

(i) Br, Br $\xrightarrow{(C_6H_5)_3\overset{\oplus}{P}-\overset{\ominus}{C}H_2}$ $\overset{\oplus}{P}(C_6H_5)_3$, Br $\xrightarrow{nBuLi}$ $\ominus$, $\overset{\oplus}{P}(C_6H_5)_3$, Br $\rightarrow$ H, $\overset{\oplus}{P}(C_6H_5)_3$ $\rightarrow$

(ii) o-phthalaldehyde ($C_6H_4(CHO)_2$) $+ (C_6H_5)_3P = CH - CH_2 - CH = P(C_6H_5)_3 \rightarrow$ benzocycloheptadiene

18. Via umpolung synthesis

H₂C=O + HS–(CH₂)₃–SH / BF_3 → 1,3-dithiane; (i) nBuLi (ii) Cl–(CH₂)₃–Br → 2-(3-chloropropyl)-1,3-dithiane; nBuLi → spiro cyclobutane dithiane; Hg^{+2}, $H_3O^{\oplus}$ → cyclobutanone

3 – 7 membered ring may be synthesized via umpolung

19. Benzo-fused hetero cyclic ring-closure

(i) o-phenylenediamine + benzil (C_6H_5–CO–CO–C_6H_5) → 2,3-diphenylquinoxaline

(ii) o-chloronitrobenzene $\xrightarrow{NH_2NH_2}$ o-nitrophenylhydrazine (NH–NH₂, NO₂) → 1-hydroxybenzotriazole (N–OH)

(iii) anthranilamide (CONH₂, NH₂) $\xrightarrow{HNO_2}$ 4-hydroxy-1,2,3-benzotriazine (OH)

(iv) o-nitrobenzoic acid (COOH, NO₂) $\xrightarrow{N_3H}$ o-aminobenzoyl azide ($C(=O)–\overset{\ominus}{N}–\overset{\oplus}{N}\equiv N$, NH₂) $\xrightarrow{\Delta}$ o-aminophenyl isocyanate (N=C=O, NH₂) → benzimidazol-2-one (NH, C=O, N–H)

20. Via Michael condensation followed by aldol

(i) $\overset{\ominus}{O}Et$ (ii)

$\overset{\ominus}{O}H/H_2O$ Δ

21. Via Michael condensation followed by Claisen ester condensation

CO_2Et CO_2Et (i) $\overset{\ominus}{O}Et$ (ii) EtO_2C OEt $\overset{\ominus}{O}H/H_2O$ Δ EtO_2C

22. Via aldol condensation followed by Michael

KOH/MeOH aldol

Michael

23. Via enamine

(i)

exo-enamine

$\overset{\ominus}{O}Et$

(ii)

endo-enamine

24. Via double Michael condensation

(i) Δ

(ii) $\overset{\ominus}{O}Et$ EtOH/Δ

H_5C_6 + C_6H_5

EtO_2C CO_2Et

H_5C_6 C_6H_5 EtO_2C CO_2Et

(iii) EtO_2C CO_2Et + EtO_2C CO_2Et

$\overset{\ominus}{O}Et$/EtOH/Δ reflux

EtO_2C CO_2Et EtO_2C CO_2Et

B. Ring-opening

Ring-opening is a very important step in the synthetic methodology part. There are various types of ring openings eg, carbocyclic, hetero cyclic, ketal or thioketal ring and epoxide ring. The followings are outline reactions.

1. Carbo cyclic ring

(i) Via electrocyclic (cycloreversion) reaction

(a) Δ or hν

(b) Δ or hν

(ii) Via retro Diels-Alder

(a) CHO; Δ, $- C_2H_4$; CHO

(b) O; CO_2Et; CO_2Et; Δ, $- C_2H_2$; EtO_2C; CO_2Et; O

(iii) Via Oxidation with fuming HNO_3

O; fuming HNO_3; $(CH_2)_4$ COOH COOH

(iv) Via Oxidation with Pb $(OAc)_4$

(a) H H OH OH; $Pb(OAc)_4$; $OHC - (CH_2)_4 - CHO$

(b) OH OH; $Pb(OAc)_4$; O O

(v) Via ozonolysis

(a) $\xrightarrow[H_2O_2]{O_3}$ O … COOH

(b) H O, H, O, O $\xrightarrow[\text{(ii) } H_3\overset{\oplus}{O}]{\text{(i) } O_3}$ H, H, H, H, HO_2C, COOH, HOOC, COOH

(vi) By haloform reaction

O, O $\xrightarrow[\text{(ii) } H_3\overset{\oplus}{O}]{\text{(i) } Cl_2/\overset{\ominus}{O}H}$ HOOC COOH

2. Ketal and thioketal

(a) O, O $\xrightarrow{H_3\overset{\oplus}{O}}$ =O

(b) S, S $\xrightarrow[\text{or } CH_3I-H_2O/CH_3CN/CaCO_3]{Hg^{+2}/H_3\overset{\oplus}{O}}$ =O

3. Heterocyclic ring

(i) Nitrogenous heterocyclic ring is opened via Hofmann exhaustive methylation

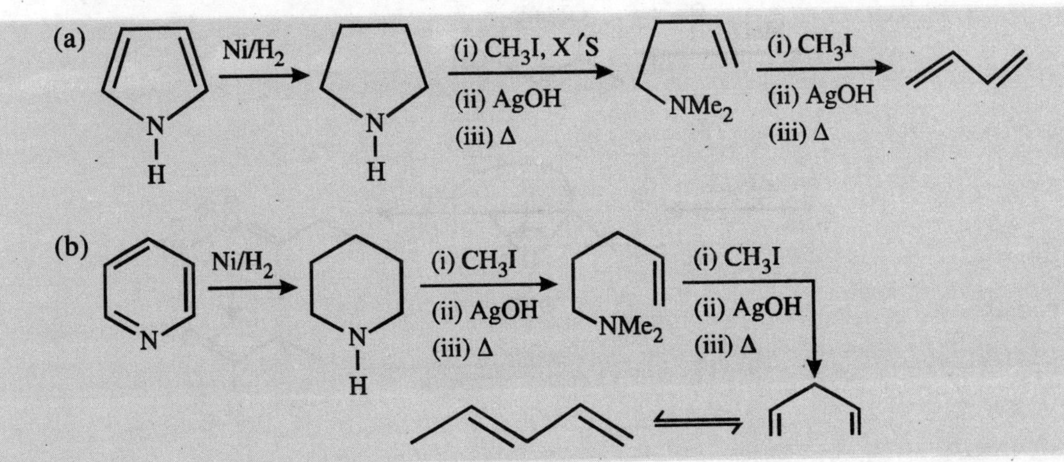

(ii) Furan ring is a bis-enol ether. It may be opened by aquous acid

$H_3\overset{\oplus}{O}$

CHO CHO

The epoxide

The epoxides are versatile intermediate in the synthesis of large number of organic derivatives. The epoxides are used extensively as the key intermediate for synthetic starting material. They are usually prepared by ring-opening of expoxides with a variety of nucleophiles.

(i)

O

(i) $LiAlH_4$ (ii) $H_3\overset{\oplus}{O}$ — OH

(i) CH_3MgBr (ii) $H_3\overset{\oplus}{O}$ — OH

NaN_3 / aq. EtOH — OH, N_3

Aq. HBr — OH, Br

$MeOH/\overset{\ominus}{O}Me$ — OH, OMe

$MeOH/\overset{\oplus}{H}$ — OMe, OH

BF_3/Δ — H, $O\oplus$, $\ominus BF_3$ → $\overset{\oplus}{O}$, $\overset{\ominus}{B}F_3$, H → O, H

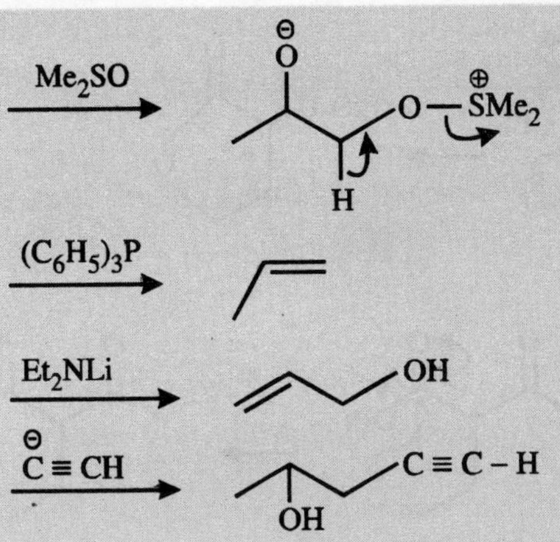

$\xrightarrow{Me_2SO}$ $\xrightarrow{(C_6H_5)_3P}$ $\xrightarrow{Et_2NLi}$ OH $\xrightarrow{\overset{\ominus}{C}\equiv CH}$ C ≡ C – H OH

(ii) Cl 1 eq. Me_2CuLi Me⊖ O (i)Me_2CuLi (ii) $H_3\overset{\oplus}{O}$ OH

(iii) $\overset{\ominus}{Me}$ O (i) Me_2CuLi (ii) $H_3\overset{\oplus}{O}$ OH

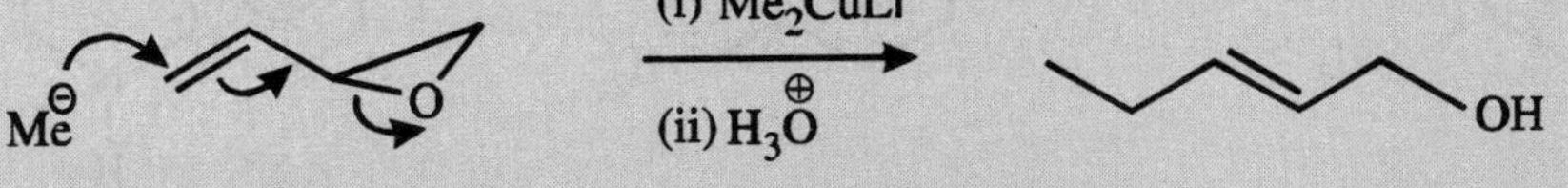

(iv) O H H EtOH/$\overset{\oplus}{H}$ OH OEt

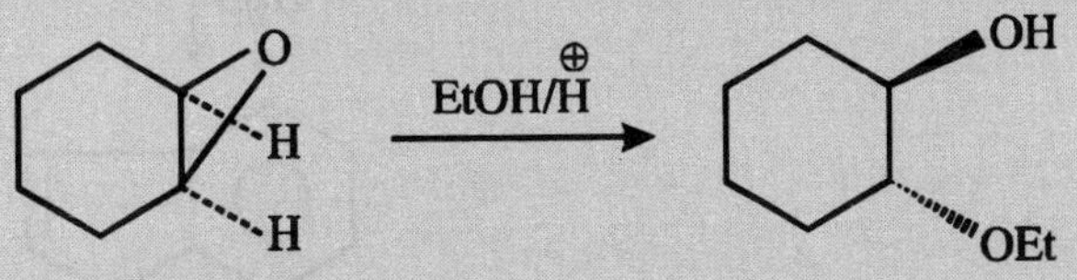

(v) O H $\overset{\ominus}{O}$ OH OH

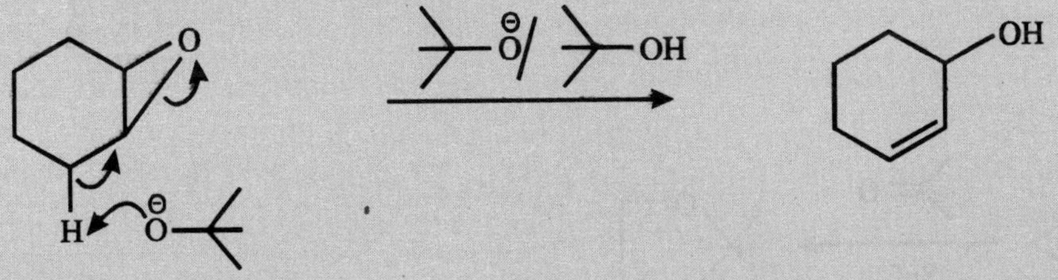

(vi) O $\overset{\oplus}{H}/H_2O$ H ⊕O :OH_2 $\overset{\ominus}{O}H/H_2O$ OH OH I OH OH II

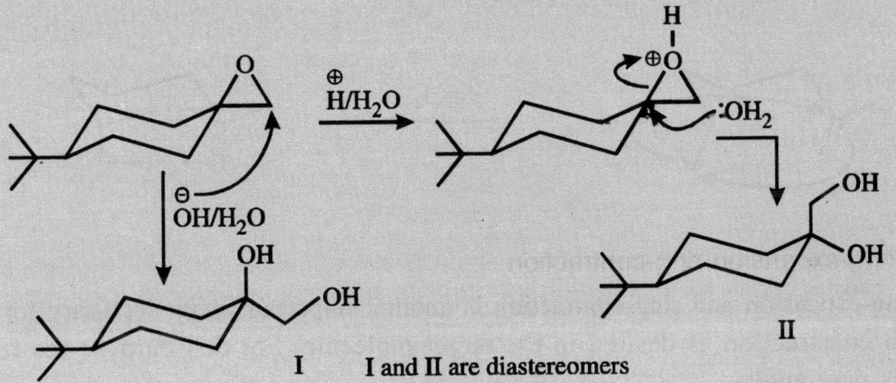

I and II are diastereomers

(vii)

CO_2Et CO_2Et (i) $\overset{\ominus}{O}Et$ (ii) epoxide → EtO_2C … OEt, $\overset{\ominus}{O}$ → EtO_2C … O

(viii)

CO_2Et (i) $\overset{\ominus}{O}Et$ (ii) epoxide → O, O, OEt, $\overset{\ominus}{O}$ → O, O, O

(ix)

O, NH_2 $Me_2\,S{=}CH_2$ → O, $\ddot{N}H_2$ → OH, N–H

$-\,H_2O$

N–H

(x)

O, =O, $SnCl_4$ → O, O

(xi)

O, O $SnCl_4$ → O, O

1.4.6 Ring-expansion ring-contraction

Ring-expansion and ring-contraction is another important steps necessary for carbon-skeleton construction as desired in the target molecule. For this purpose the following methods may apply.

Ring-expansion

(i) Via diazonium group or β-amino alcohol

CH_2N_2

$\overset{\ominus}{O}$ $\overset{\oplus}{N} \equiv N$

$CH_3NO_2/\overset{\ominus}{O}H$

$\overset{\ominus}{O}$ NO_2 Sn/HCl HO NH_2 HNO_2 HO $\overset{\oplus}{N} \equiv N$

(ii) By Pinacol-Pinacolone rearrangement

(a) OH OH H_2SO_4 O

(b) OH OH $HClO_4$ O

(iii) By Beckmann rearrangement

N – OH H_2SO_4 HN O

(iv) By Baeyer-Villiger Oxidation, ketone to lactone

O CF_3CO_3H O O

(v) Via Carbene

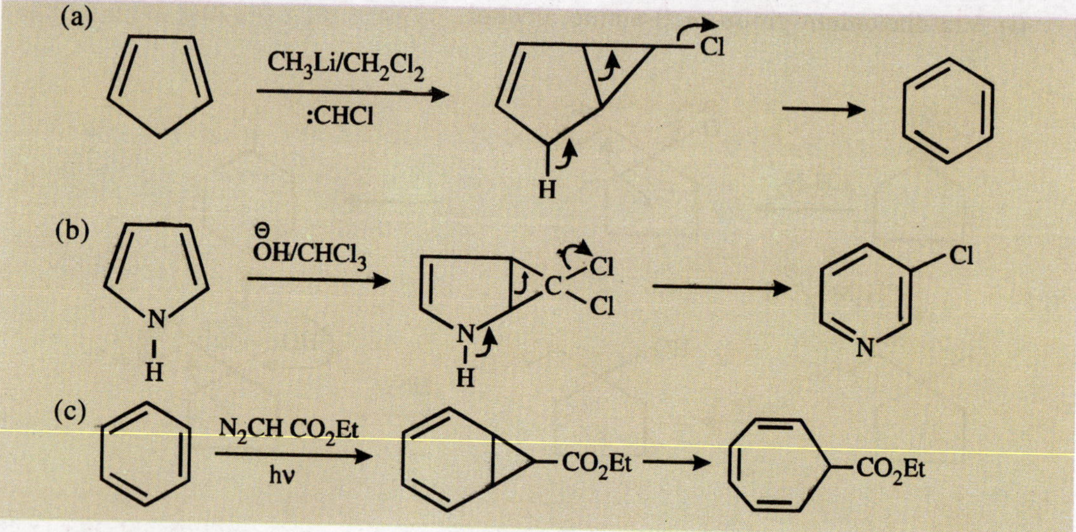

Ring-contraction

(i) By Benzillic acid rearrangement

(a)

O

O

(i) $\overset{\ominus}{O}H/H_2O$

(ii) $H_3\overset{\oplus}{O}$

CO_2H

OH

(b)

O

O

(i) $\overset{\ominus}{O}H/H_2O$

(ii) $H_3\overset{\oplus}{O}$

HO COOH

(ii) By Favorskii reaction

O

Cl

$\overset{\ominus}{O}H/H_2O$

O

$\overset{\ominus}{O}H$

$C\overset{\ominus}{O}_2$

$H_3\overset{\oplus}{O}$

CO_2H

(iii) Via diazo ketone

$$\text{2-aminocyclohexanone} \xrightarrow{HNO_2} \text{2-diazoniocyclohexanone} \xrightarrow{h\nu} [\ \ ] \rightarrow \text{cyclopentyl}{=}C{=}O \xrightarrow{H_2O} \text{cyclopentyl}-CO_2H$$

(iv) From dicarboxylic acid

$$\text{cyclohexanone} \xrightarrow{\text{fuming } HNO_3} (CH_2)_4(CO_2H)_2 \xrightarrow[300°C]{Ac_2O} \text{cyclopentanone}$$

1.4.7 Dehydrating agent

Dehydration is a very important part in the synthetic methodology. There are many dehydrating agents available e.g. conc H_2SO_4, Al_2O_3, $KHSO_4$, Na_2SO_4, P_2O_5, PCl_5, Ac_2O, $CaCl_2$, $ZnCl_2$, $C_6H_5-N=C=O$, DCC etc. The nature of the dehydrating agent employed depends on structure of the substrate and medium. The following dehydration reactions are some model examples.

(i) pinacol (OH OH) $\xrightarrow[400°,\ -2H_2O]{Al_2O_3}$ 2,3-dimethylbutadiene

(ii) (a) glycerol (OH OH OH) $\xrightarrow[-2H_2O]{KHSO_4} CH_2{=}CH-CHO$ dehydration with conc. H_2SO_4, self condensation of acraldehyde

(b) 1-vinylcyclohexanol (HO) $\xrightarrow[-H_2O]{KHSO_4}$ 1-vinylcyclohexene

(iii) $R-\underset{\underset{O}{\|}}{C}-NH_2 \xrightarrow[\text{or } PCl_5]{P_2O_5} R-C\equiv N$

(iv) Succinic acid $\xrightarrow{Ac_2O}$ succinic anhydride

(v) $R-CH=N-OH \xrightarrow{Ac_2O} R-C\equiv N$

(vi) $RCHO + 2ROH \xrightarrow{CaCl_2} R-CH(OR)_2$

(vii) $C_6H_5NH_2 + R-CHO \xrightarrow{ZnCl_2} C_6H_5N=CHR$

(viii) $CH_3-CH(OH)-CH_2-CHO \xrightarrow{ZnCl_2} CH_3-CH=CH-CHO$

[$\overset{\oplus}{H}$ dehydration may give further aldol]

(ix) $C_6H_6 + CH_2O + HCl \xrightarrow[-H_2O]{ZnCl_2} C_6H_5CH_2Cl$

(x) $CH_3-CH_2-NO_2 \rightleftharpoons CH_3-CH=\overset{\oplus}{N}(OH)\overset{\ominus}{O} \xrightarrow[-H_2O]{C_6H_5N=C=O} CH_3-C\equiv\overset{\oplus}{N}-\overset{\ominus}{O}$

(xi) 2,5-Hexanedione $\rightleftharpoons$ dienol (OH, HO) $\xrightarrow{P_2O_5}$ 2,5-dimethylfuran

(xii) $RCO_2NH_4 \xrightarrow[-H_2O]{Al_2O_3/\Delta} R-C(=O)-NH_2 \xrightarrow{Al_2O_3/\Delta} R-C\equiv N$

(xiii) $CH_3-N(N=O)-CH_2-COOH \xrightarrow[-H_2O]{Ac_2O} [CH_3-N(N=O)-CH=C=O] \xrightarrow{\text{ketene}}$ $CH_3-\overset{\oplus}{N}-\overset{\ominus}{C}H$ ring (sydnone: N=N–O–C(=O))

(xiv) DCC : Dicyclohexylcarbodiimide*

$C_6H_{11}-N=C=N-C_6H_{11}$

'DCC' is a very modern dehydrating agent of choice for direct conversion of acid into ester, amides and anhydride etc.

*For details see author's "Organic Synthesis" vol. 1

(a) $RCOOH \xrightarrow{DCC} (RCO)_2O + DCC - H_2O$

N=C=N NH − C=N O R H O R O O H R

NH − C − NH O + R O R O O NH − C=N O O H R O R O

(b) $RCOOH + ROH \xrightarrow{DCC} R-\underset{\underset{O}{||}}{C}-OR + DCC - H_2O$

(c) $C_6H_5OH + ROH \xrightarrow{DCC} C_6H_5-O-R + DCC - H_2O$

(d) $RCOOH + R-NH_2 \xrightarrow{DCC} R-\underset{\underset{O}{||}}{C}-NHR + DCC - H_2O$

The two C = N bonds make the central carbon atoms very reactive towards nucleophiles, particularly in the presence of acid. The resulting adduct, containing the grouping $-\underset{\underset{O}{||}}{C}-O-C{=}N-R$, is analogous to an anhydride and reacts readily with other nucleophiles.

1.4.8 Condensing agent

Condensation is another important step during organic synthesis. There are varities of condensing agents e.g. HF, PPA, $ZnCl_2$, I_2, $AlCl_3$, $POCl_3$ and DCC employed depending on the structure of condensing organic molecules.

The following reactions are some examples which illustrate these

(i) COOH $\xrightarrow[\text{or PPA}]{\text{HF}}$ O

(ii)

AlCl$_3$

(iii) EtO_2C ... $\overset{\ominus}{C}H$ + $\overset{\ominus}{C}H$ ... CO_2Et —I_2→ EtO_2C, EtO_2C, CO_2Et, CO_2Et

(iv)

Py

$POCl_3$

$-HOPOCl_2$

(v) Condensation with dehydration

$C_6H_5CH_2-O-CO-NH-CH_2-COOH + H_2N$

H_2N - protected glycine

– COOH group protected alanine

$-H_2O$ | DCC

$C_6H_5CH_2-O\ CONH-CH_2-C-NH$

(i) H_2/Pd (ii) anhydrous CF_3CO_2H (C – terminal deblocking)

(N - terminal deblocking)

glycinyl alanine

1.4.9 Functional group removal—Defunctionalisation*

In the process of organic synthesis functional group addition (FGA), functional group interconversion (FGI) are very important steps. Likewise, removal of functional group (FGR) i.e. defunctionalisation is another important step which is required for functional alteration (or adjustment of functionality) or to remove undesired functional group as necessary in the final target molecule.

The following outline reactions are some examples which illustrate this fact.

(i) Removal of carbonyl oxygen

(a) CH_3 CH_3 (C=O) CH_3 —Zn(Hg)/conc. HCl→ CH_3 CH_3

—NH_2-NH_2/ t-BuO$^{\ominus}$→ [CH_3 CH_3 C=N − NH_2] —− N_2→ CH_3 CH_3

HS, HS (ethanedithiol) /BF_3 → (dithiolane) —Raney Ni/H_2→ CH_3 CH_3

—$TsNHNH_2$→ CH_3 CH_3 C=N − NHTs —$NaBH_3CN$→ CH_3 CH_3

(b) HS, HS /BF_3 → S S —Raney Ni, H_2→

Unconjugated
C=O reacts faster

*For details see author's "Organic Synthesis" vol. 1

(c)

$\overset{\ominus}{OH}/H_2O$, $-HCO_2^{\ominus}$

(removal of -CHO function)

C_4H_9SH ; =CHSC$_4H_9$; Raney Ni, H_2 ; CH_3

(d)

$LiAlH_4$: R–CO–NMe$_2$ → R–CH$_2$–NMe$_2$

(e)

Zn/AcOH

Zn/AcOH

(f) CHO ; Ac_2O ; CHOAc ; O_3 ; Zn(Hg), Conc. HCl

(removal of -CHO function)

(ii) Removal of ester function of β-keto ester

CO_2Et ; dil HCl/Δ or NaCl/wet DMSO

CO_2Et ; dil HCl/Δ

(iii) Removal of carboxylic group

(a) gem. dicarboxyl: $(CH_3)_2C(CO_2H)_2$ $\xrightarrow{250°C}$ $(CH_3)_2CH{-}CO_2H$

(b) vic. dicarboxyl: cyclohexane-1,2-dicarboxylic acid (COOH, CO_2H) $\xrightarrow{Pb(OAc)_4}$ cyclohexene

(iv) Removal of oxygen function from epoxy, azoxy, nitro, nitrosyl, tertiary amine oxide and pyridine N-oxide

(a) $\xrightarrow{(C_6H_5)_3P}$

- cyclohexene oxide → cyclohexene
- $Ar-N{=}\overset{\oplus}{N}(O^{\ominus})-Ar$ → $Ar-N{=}N-Ar$
- $R_3\overset{\oplus}{N}-\overset{\ominus}{O}$ → R_3N
- pyridine $\overset{\oplus}{N}-\overset{\ominus}{O}$ → pyridine (N)

(b) $\xrightarrow{(EtO)_3P}$ $Ar-\ddot{N}:$

- $Ar-\overset{\oplus}{N}(=O)-\overset{\ominus}{O}$
- $Ar-N{=}O$

(v) Removal of nitro and nitrosyl

(a) $\xrightarrow[H_2O]{TiCl_3}$

- $(CH_3)_2CH-NO_2$ → $(CH_3)_2C{=}O$
- nitrocyclohexane (NO_2) → cyclohexanone (O)

(b) $R_2N-NO \xrightarrow{Ni/H_2} R_2NH$

(vi) Removal of halogen

(a) 3-bromobutan-2-one, 2-bromocyclohexanone $\xrightarrow[H_2O]{TiCl_3}$ butan-2-one, cyclohexanone

(b) pentane-2,4-dione $\xrightarrow{PCl_5}$ 4-chloropent-3-en-2-one; cyclohexane-1,3-dione $\xrightarrow{PCl_5}$ 3-chlorocyclohex-2-enone; $\xrightarrow[CH_3OH]{Zn(Hg)}$ pent-3-en-2-one, cyclohex-2-enone

(c) 7,7-dichlorobicyclo[4.1.0]heptane (Cl, Cl) $\xrightarrow{Li/liq\ NH_3}$ bicyclo[4.1.0]heptane

(d) epoxide with CH_3 and CH_2Br $\xrightarrow{NaBH_3CN}$ epoxide with CH_3 and CH_3

(e) 3-chlorobutan-2-one, 2-chlorocyclohexanone $\xrightarrow{Zn/AcOH}$ butan-2-one, cyclohexanone

(f) $R{-}CH_2{-}Cl \xrightarrow{LiAlH_4} R-CH_3$

(vii) Removal of hydroxyl, acetate, tosylate, halogen

(a) Zn/AcOH

X = OH, OAc, OTs, Cl

(b) P/HI

(c) $R-CH_2-OTs \xrightarrow{LiAlH_4} R-CH_3$

(viii) Removal of double bond

(a) Ni/H_2 Or $B_2H_6/AcOH$

(b) $H_2/Pd/CaCO_3/Pb(OAc)_2$ Lindlar catalyst

(c) H_2/Pd

(d) CO_2Et H_2/Pd CO_2Et

Unconjugated double bond reduced selectively

(e) $TiCl_3$ H_2O

(f) [iminium ion of 2-methylcyclohexanone and pyrrolidine] $\xrightarrow{NaBH_4}$ [N-(2-methylcyclohexyl)pyrrolidine]

(g) $CH_3-\overset{O}{\overset{\|}{C}}-NH_2 \xrightarrow{Me_3\overset{\oplus}{O}\overset{\ominus}{BF_4}} CH_3-\underset{OMe}{\underset{|}{C}}=\overset{\oplus}{N}H_2 \xrightarrow{NaBH_4} CH_3-\underset{OMe}{\underset{|}{CH}}-NH_2$

oxygen alkylation

(ix) Removal of functional groups from the benzenoid ring system

(a) Removal and shifting of carbonyl (oxygen) function adjacent to benzenoid ring system

[α-tetralone; acetophenone] $\xrightarrow[AlCl_3]{LiAlH_4}$ [tetralin; ethylbenzene]

($LiAlH_4$ + $AlCl_3$) is a selective reducing agent which reduces completely aryl alkyl ketone

[α-tetralone] $\xrightarrow{\overset{\ominus}{O}H/C_6H_5CHO}$ [2-benzylidene-1-tetralone, =CHC_6H_5] $\xrightarrow{LiAlH_4 + AlCl_3}$ [2-benzylidenetetralin, =CHC_6H_5] $\xrightarrow{O_3}$ [β-tetralone]

(Shifting of C = O group from one to two position)

(b) [$C_6H_5SO_3H$] $\xrightarrow[100°C]{dil\ H_2SO_4}$ [benzene]

(c) $C_6H_5CH_2\overset{\oplus}{N}Me_3 \xrightarrow[H_2O]{Na(Hg)} C_6H_5CH_3$

(d) $C_6H_5NO_2 \xrightarrow{H_2/Pd} C_6H_5NH_2 \xrightarrow[(i)\ H_3PO_2/\Delta]{(i)\ HNO_2} C_6H_6$

(e) $C_6H_5Br \xrightarrow[\text{alcoholic KOH}]{H_2/Pd/CaCO_3} C_6H_6$

$\xrightarrow{Mg/ether} C_6H_5MgBr \xrightarrow{H_2O} C_6H_6$

(f) $C_6H_5CH_2X \xrightarrow{H_2/Pd} C_6H_5CH_3$

$X = Cl, OR, OAc, NR_2$

(g) $m\text{-}CH_3COO\text{-}C_6H_4\text{-}CO_2Me \xrightarrow[\Delta]{MeOH/\overset{\oplus}{H}} m\text{-}HO\text{-}C_6H_4\text{-}CO_2Me$

In this reaction acyl ($COCH_3$) group is removed by ester exchange but — CO_2Me remain intact since MeOH is taken.

1.5 Principles and methodology of alkylation, acylation, formylation, benzoylation, carbethoxylation, arylation, vinylation and methylenylation O-alkylation versus C-alkylation

Alkylation or acylation are very important part, required during organic synthesis. Various techniques are available for alkylation or acylation depends on structure of the alkylating substrates or skeleton, alkylating agent and medium.

The following collections are a few usual alkylating substrates structure and usual alkylating or acylating agents required during organic synthesis.

1.5.1 Different alkylating substrates

1. Alkylation or acylation via enolate

(i) (a) Stabilised enolate

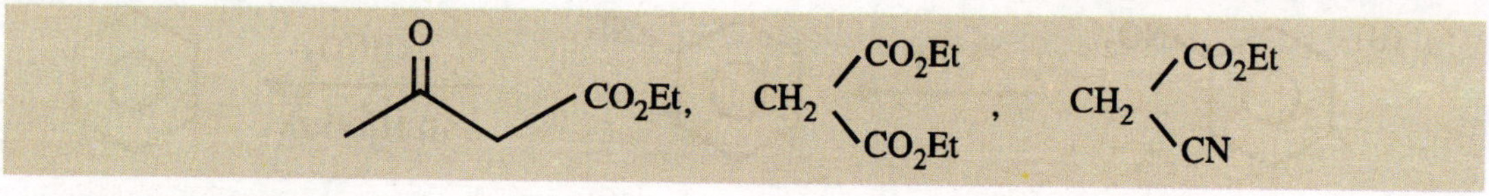

O; CO_2Et, $CH_2(CO_2Et)_2$, $CH_2(CO_2Et)(CN)$

(b) Ester enolate

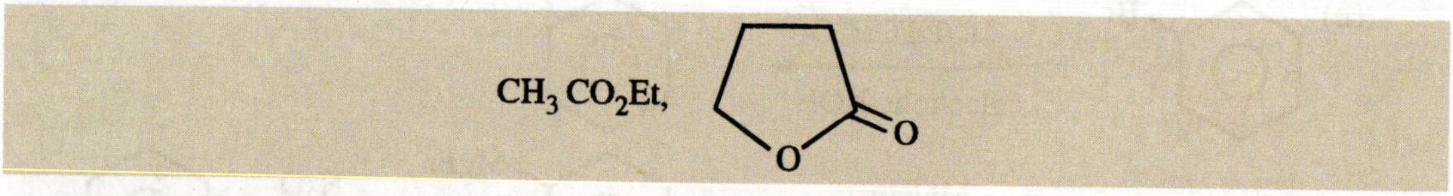

CH_3CO_2Et, O; O

(c) Carboxyl enolate

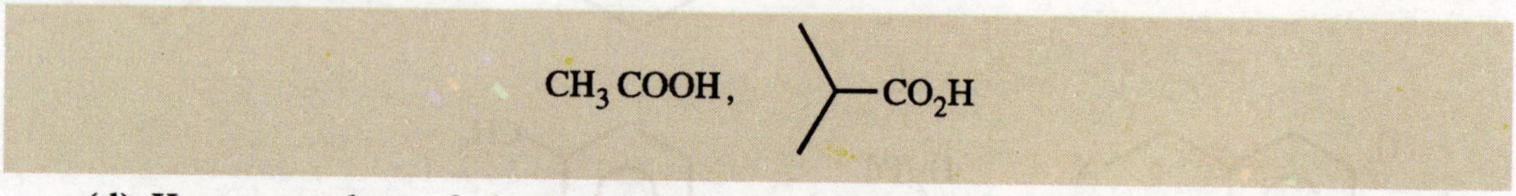

CH_3COOH, CO_2H

(d) Ketone enolate of single choice

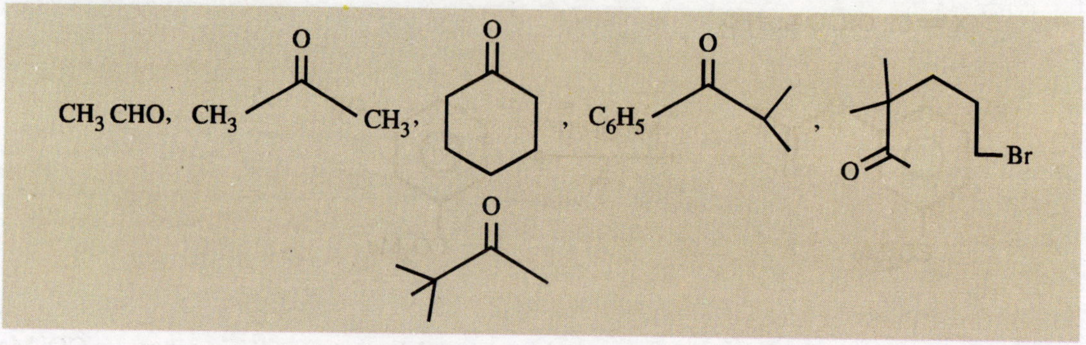

CH_3CHO, CH_3COCH_3, O; C_6H_5 O; O; Br; O

(ii) Ketone enolates of different choices

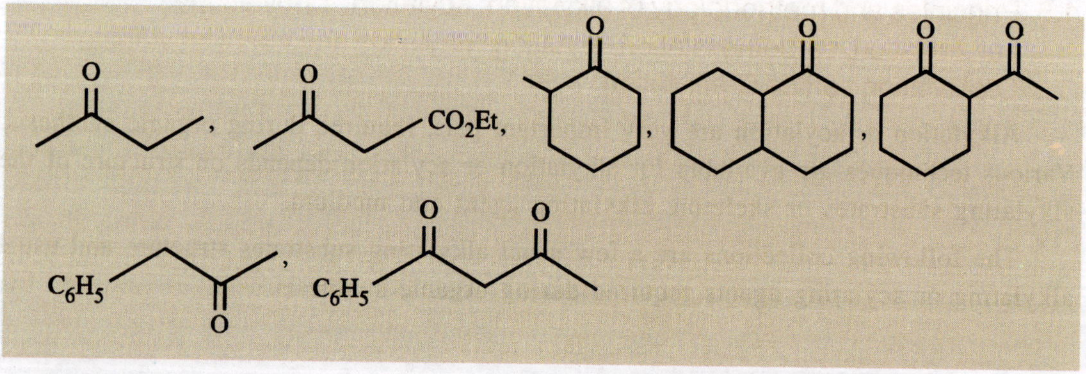

O; O CO_2Et, O; O; O O

C_6H_5 O; C_6H_5 O O

(iii) α β or β γ unsaturated carbonyl

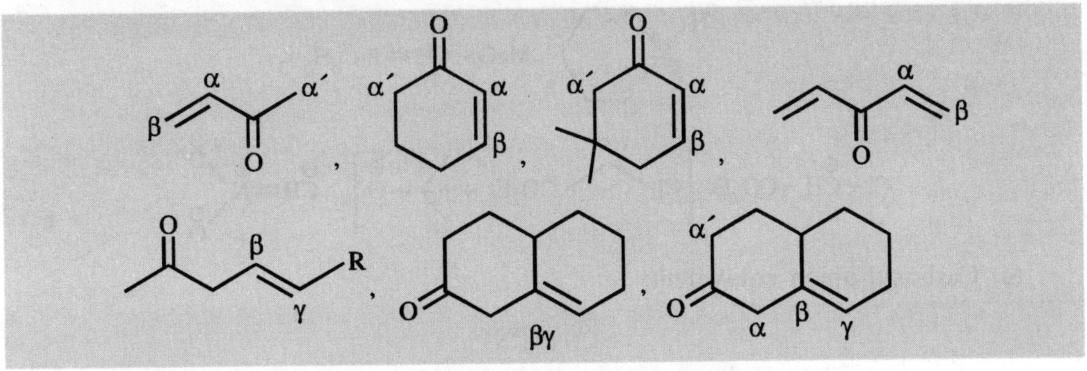

2. Alkylation via vinylogous substitution of 1, 3-dicarbonyl

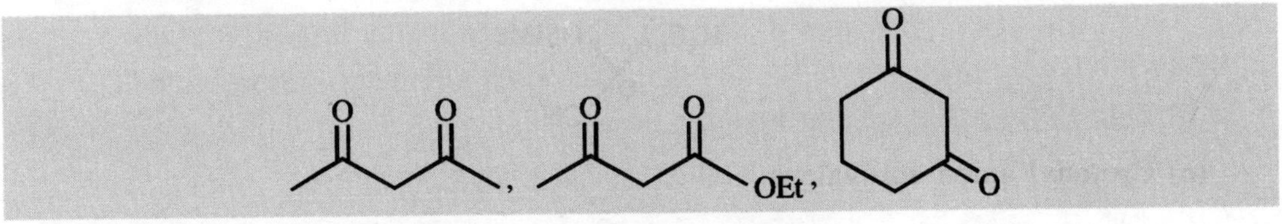

3. Alkylation of 1, 4-diketo at the middle carbons.

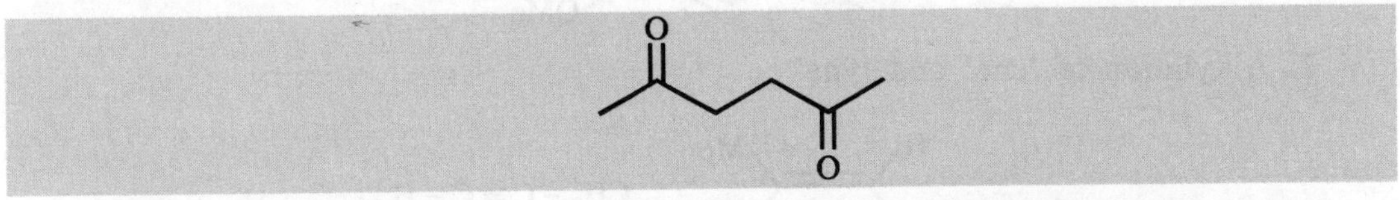

4. Alkylation via enamine, metalloenamine and N, N-dimethyl hydrazone

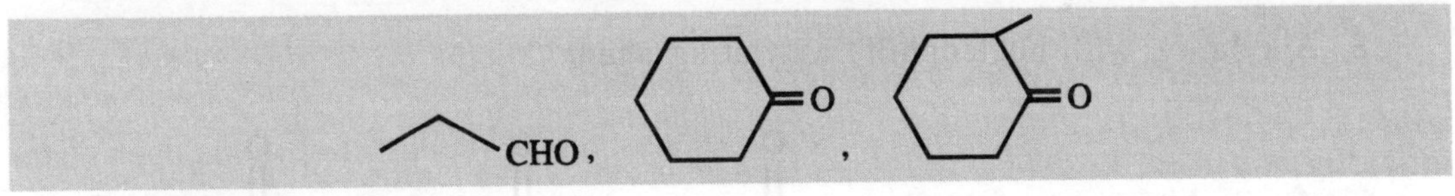

5. Alkylation of nitroalkane

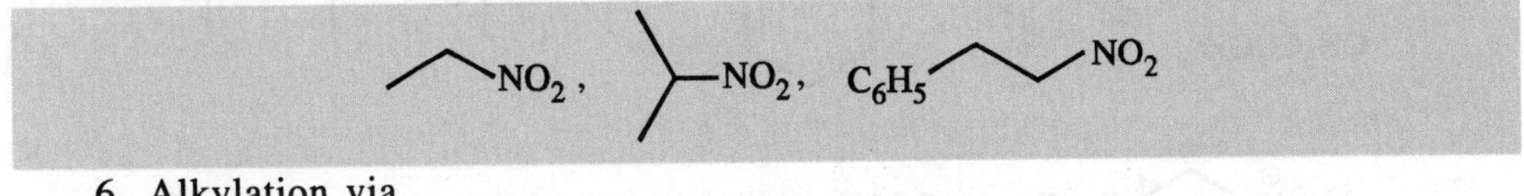

6. Alkylation via

(a) Acyl anion equivalent

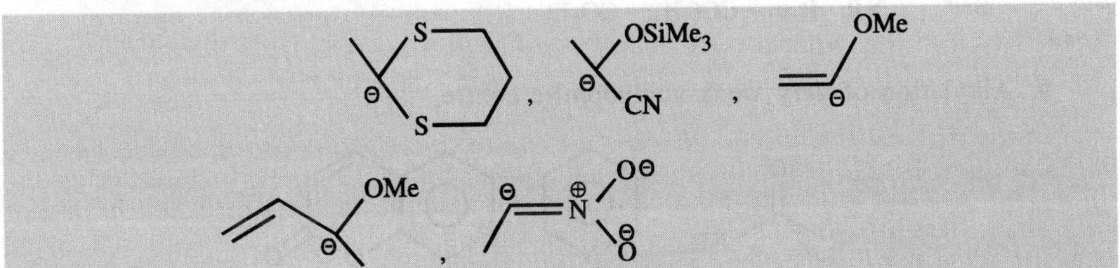

(b) Formyl anion equivalent

2-lithio-1,3-dithiane anion (H, ⊖ on C-2 of 1,3-dithiane), $MeO-CH{=}P(C_6H_5)_3$

$Cl-\overset{\ominus}{C}H-CO_2Et$ [Cl–CH$_2$CH$_2$–CO_2Et + t-BuO$^{\ominus}$], $\overset{\ominus}{C}H{=}\overset{\oplus}{N}(O^{\ominus})_2$

(c) Carboxyl anion equivalent

$\overset{\ominus}{C}N$ (KCN)

(d) Benzoyl anion equivalent

H_5C_6–$\overset{\ominus}{C}$(OSiMe$_3$)(CN)

(e) Crotonyl anion equivalent

$CH_2{=}CH-CH{=}\overset{\ominus}{C}-OMe$

7. Alkylation of 'ene' and 'yne'

H(Me)C=C(Me)Br, $CH_3-C\equiv C-H$

8. Alkylation with nucleophilic alkylating agent

CH_3COOH, methyl vinyl ketone, cyclohex-2-enone, di-*tert*-butyl ketone, di-isopropyl ketone

$N\equiv\overset{\oplus}{N}-\overset{\ominus}{C}H-R$ R= (–$COCH_3$, –CO_2Et, – CN)

Br–CH_2–R R= –$COCH_3$, –CO_2Et

9. Alkylation of very weak nucleophilic centre

$R-CO-NH_2$, pyridine, coumarin

10. Alkylation of benzene nucleus

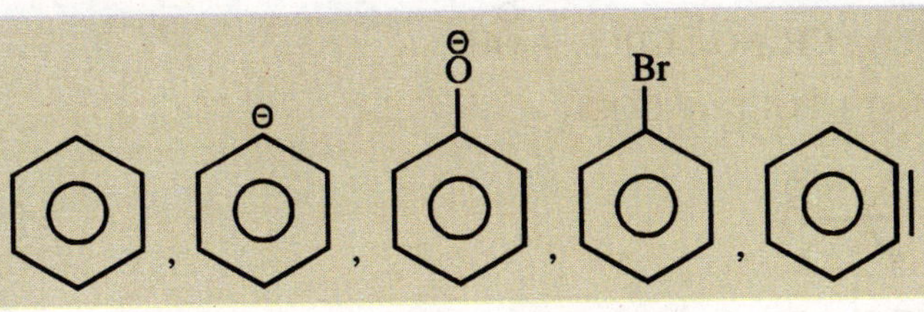

11. Alkylation of ambident nucleophile O-alkylation versus C-alkylation

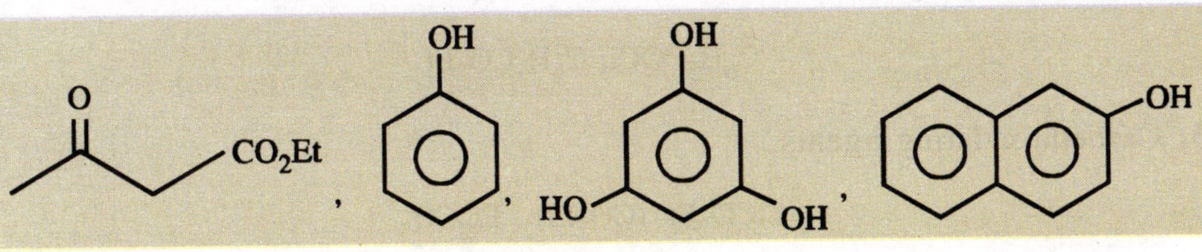

OH OH OH

CO_2Et , HO OH ,

1.5.2 Different alkylating and acylating agents (electrophilic and nucleophilic)

1. (a) Electrophilic alkylating agents

CH_3-Br, CH_3OTs

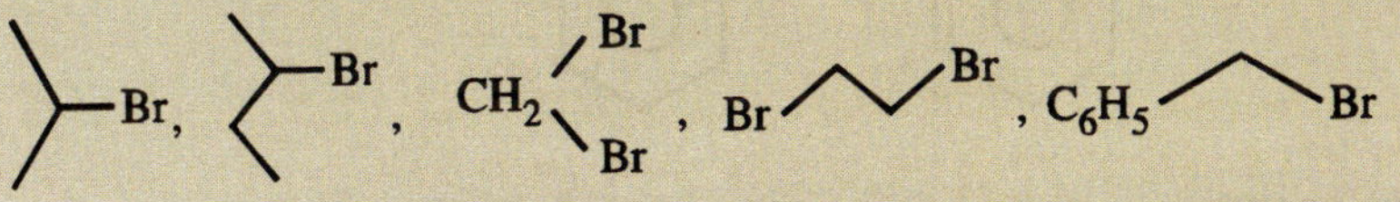

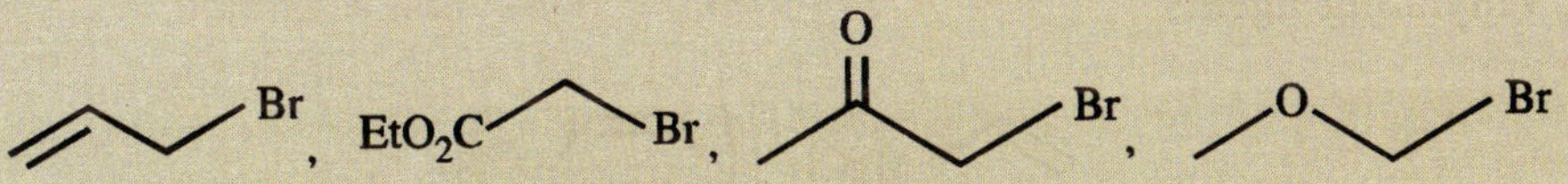

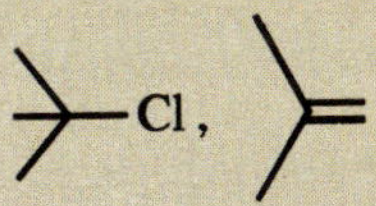

Br, Br, CH_2 Br Br, Br Br, C_6H_5 Br

Br, EtO_2C Br, O Br, O Br

Cl,

CH_2N_2, Me_2SO_4, Me – O – C(=O)–

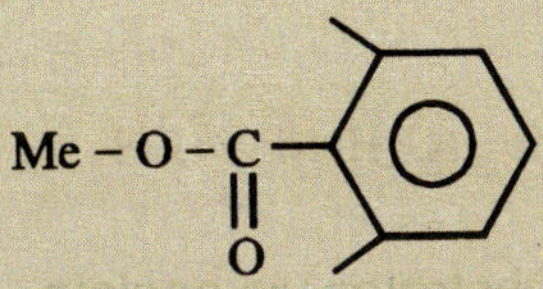

$CH_3OH + BF_3$ or $Al(OEt)_3$

O, $Me_3\overset{\oplus}{O}\ \overset{\ominus}{BF_4}$ ($Me_2O + MeF + BF_3$)

(b) Nucleophilic alkylating agents

MeMgBr, MeLi, Me_2Cd, Me_2CuLi, Me_3B, $\overset{\ominus}{C}H_2-\overset{\oplus}{S}(=O)Me_2$

2. Acetylating agents

Ac_2O, CH_3COCl, CH_3CO_2Et, $CH_3CN + ZnCl_2$, $ClCH_2COCl$

3. Formylating agents

HCO_2Et, $HC(OEt)_3$, CH_3NO_2, $CHCl_3$ + $\overset{\ominus}{O}H$

DMF ($HCONMe_2$) + $POCl_3$ or $COCl_2$

$ClCH_2CO_2Et$ + $(CH_3)_3C\text{–}\overset{\ominus}{O}$

$MeOCH_2Cl$ + $(C_6H_5)_3P$

4. Benzoylating agents

C_6H_5COCl, $C_6H_5CO_2Et$

5. Carbethoxylating agents

$ClCO_2Et$, $(CO_2Et)_2$, Et_2CO_3

6. Phenylating agents

Benzyne, $C_6H_5\overset{\oplus}{N}\equiv N\ \overset{\ominus}{Cl}$

7. Vinylating agents

$CH_2{=}CH\text{–}MgCl$, $(CH_2{=}CH)_2CuLi$

8. Methylenylating agents

$CH_2{=}\overset{\oplus}{N}Me_2$, $(C_6H_5)_3P{=}CH_2$

1.5.3 Alkylation of different alkylating substrates via enolate

During the alkylation via enolate there are two competitive processes, one is self condensation (e.g. aldol or Claisen ester condensation) and the other is O-alkylation.

Simple ketones e.g. CH_3COCH_3 or CH_3CHO and CH_3COOEt cannot be alkylated in weak basic medium in polar protic solvent because of self condensation nature.

The following examples are some outline reactions which illustrate the facts.

1. (a) Ethyl aceto acetate, Diethylmalonate and Ethyl cyano acetate are three synthetic starting materials which give stabilised enolate (resonance stabilised) and which are unable to give self condensation and alkylation or acylation are successful.

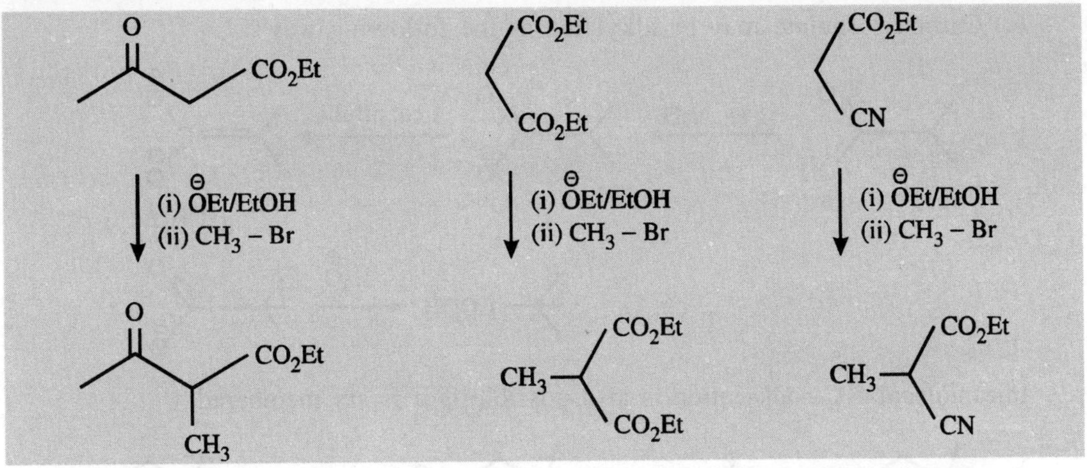

(b) Simple carbonyl compounds as for example acetone, acetaldehyde and ethyl acetate gives self condensation products with weak base. If strong and non-nucleophilic base is used in polar aprotic solvent as for example LDA in THF, convert acetone or ester to 100% enolate then self condensation may be avoided.

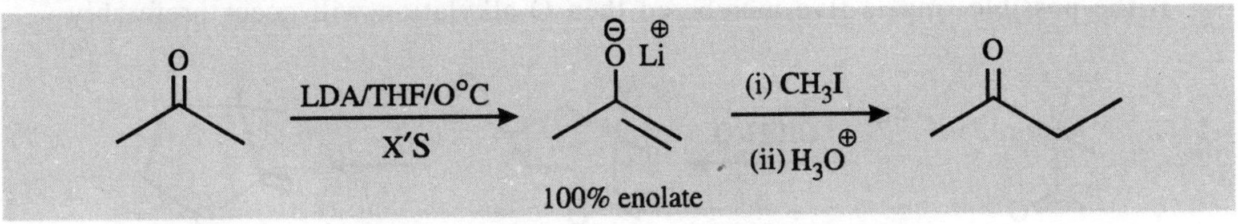

100% enolate

Slow addition of ketone to LDA and then addition of CH_3-I

γ-lactone is an ester where carbonyl group is blocked. This may be alkylated in a similar process

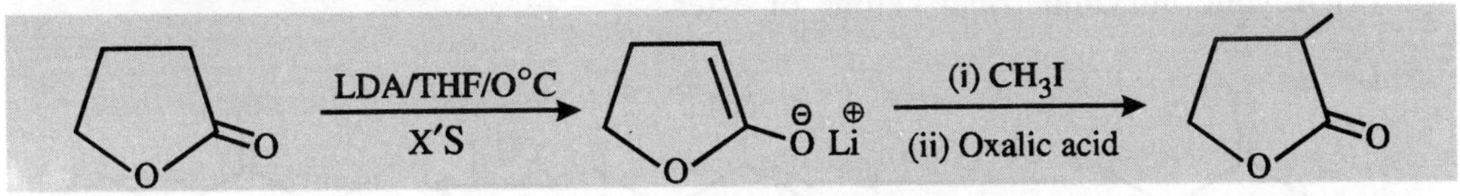

ketones which give symmetrical enolate or enolate of single choice may be alkylated as usual way

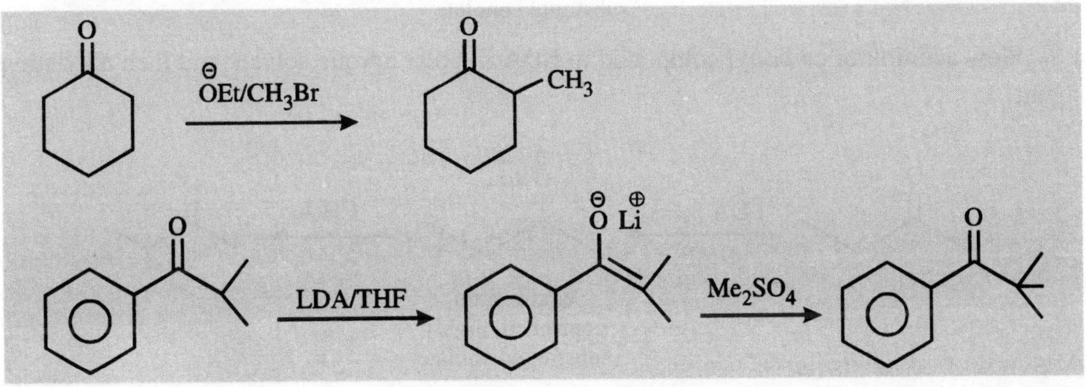

(c) Carboxyl enolate may be alkylated as the following way

1 eq. NaH
1 eq. nBuLi
CH_3Br
$H_3\overset{\oplus}{O}$
COOH

Intramolecular C—alkylation is also possible if it is six membered

Br
$\overset{\ominus}{O}H/H_2O$
Br

If the possible ring is five membered then O-alkylation will occur preferably

Br
$\overset{\ominus}{O}H/H_2O$
Br

The intramolecular ring size choice depends on better overlapping in the T.S.

2. Alkylation of unsymmetrical ketone via enolate

(a) Acyclic unsymmetrical ketone or ester

(i)

LDA (X'S)
DMF
Controlled
CH_3I
Kinetically controlled, less substituted enolate

Slow addition of carbonyl compound to LDA in polar aprotic solvent and then alkylating agent.

LDA (Less)
DMF
CH_3I
equilibrium controlled, highly substituted enolate

Slow addition of base LDA to carbonyl compound and then alkylating agent.

If some kinetically controlled enolate forms at all, at the equilibrium more substituted enolate favours

K. C enolate

T. C enolate

(ii)

$\overset{\ominus}{O}Et/EtOH$

CH_3Br

Here weak base is applied since the proton abstracted is benzylic and the enolate is conjugated with benzene ring.

LDA 2eq.

(i) CH_3Br (1eq)

(ii) $H_3\overset{\oplus}{O}$

LDA is a very strong base, first equivalent will abstract benzylic proton and the second equivalent will abstract keto methyl proton which is less acidic and alkylation occurs preferably at the end carbon because it is more nucleophilic.

$\overset{\ominus}{O}Et/EtOH$

CH_3Br

2LDA

CH_3Br 1eq.

more nucleophilic

diene-diolate dianion

$H_3\overset{\oplus}{O}$

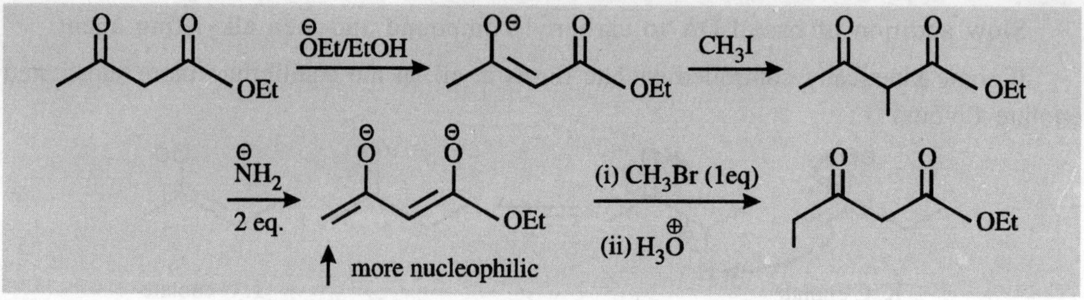

(b) Alkylation of unsymmetrical cyclohexanone, alkylation at the less substituted position

(i) Via kinetically controlled enolate formation

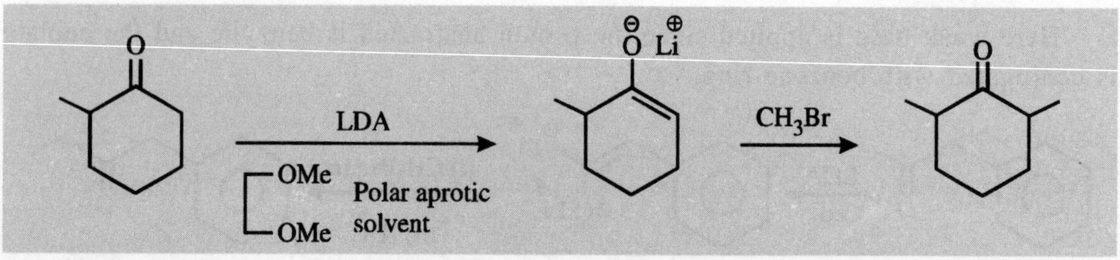

Slow addition of ketone to LDA base and then alkylating agent.

(ii) By activation of position

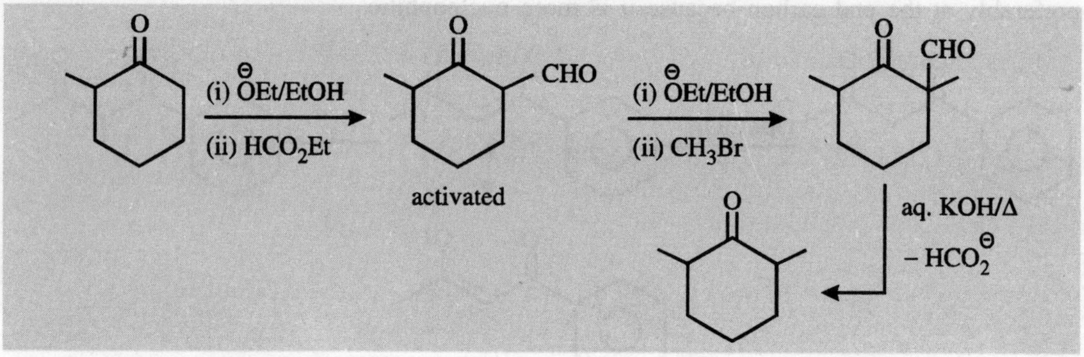

(iii) Via enamine

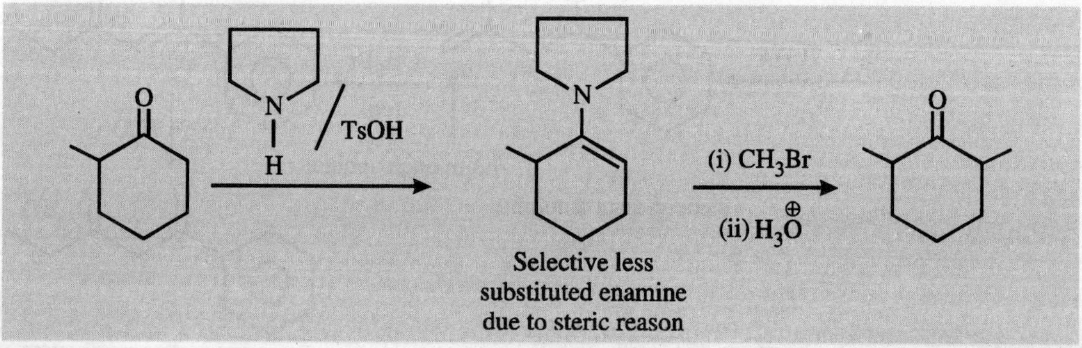

(iv) (i) $(CH_3)_3C\overset{\ominus}{O}$ / $(CH_3)_3C$–OH, (ii) CH_3Br

(c) Alkylation at the more substituted position

(i) Via specific enolate*

Me₃SiCl / Et₃N → CH₃Li, 1 eq. → CH₃I

Specific enolate since more substituted enol ether is formed specifically and preferably

OR CH_3I → $nBu_4\overset{\oplus\ominus}{NF}$

(ii) Via equilibrium controlled

LDA (Less), MeOCH₂CH₂OMe → Specific enolate → CH_3I

Slow addition of LDA base to the ketone in polar aprotic solvent.

*For details see author's "Organic Synthesis" vol. 1

If some kinetically controlled enolate forms at all, at the equilibrium more substituted enolate favours.

Via blocking

(a)

(i) $\overset{\ominus}{O}Et/EtOH$

(ii) HCO_2Et

CHO

=CH – OH

C_4H_9SH

=$CHSC_4H_9$

(i) $(CH_3)_3C\overset{\ominus}{O}$ / $(CH_3)_3COH$

(ii) CH_3Br

=$CHSC_4H_9$

aq. KOH Δ

(b)

(i) $\overset{\ominus}{O}Et/EtOH$

(ii) HCO_2Et

CHO

C_4H_9SH

=$CHSC_4H_9$

(i) $(CH_3)_3C\overset{\ominus}{O}$ / $(CH_3)_3COH$

(ii) CH_3I

=$CHSC_4H_9$

aq. KOH Δ

3. Alkylation of α, β and β, γ unsatuated carbonyl

(a) Alkylation of α, β unsaturated carbonyl

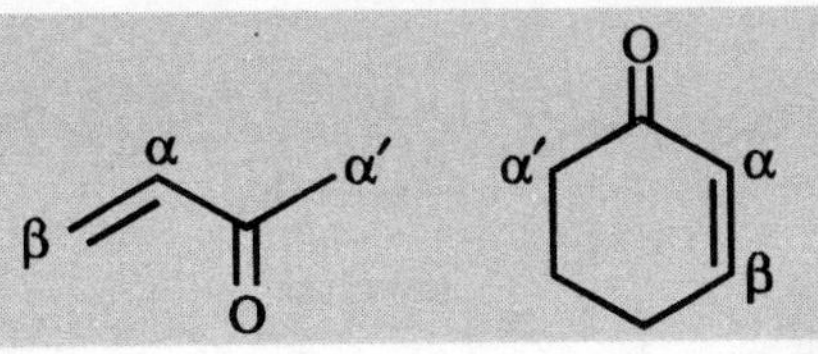

There are four non-equivalent position for alkylation of α, β unsaturated carbonyl α, β, α′ and carbonyl carbon

(i) Alkylation at α-position

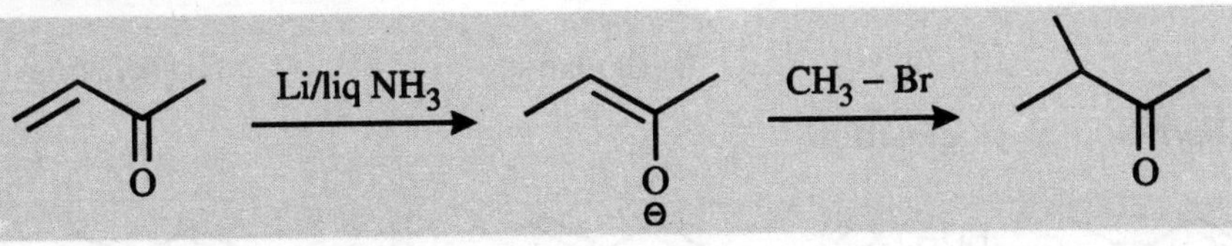

(ii) Alkylation at β-position

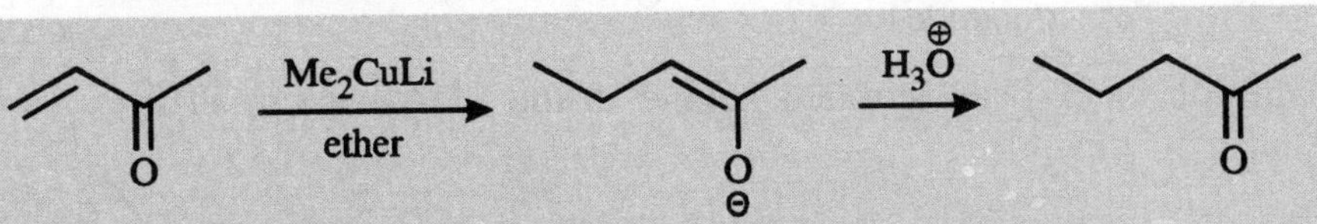

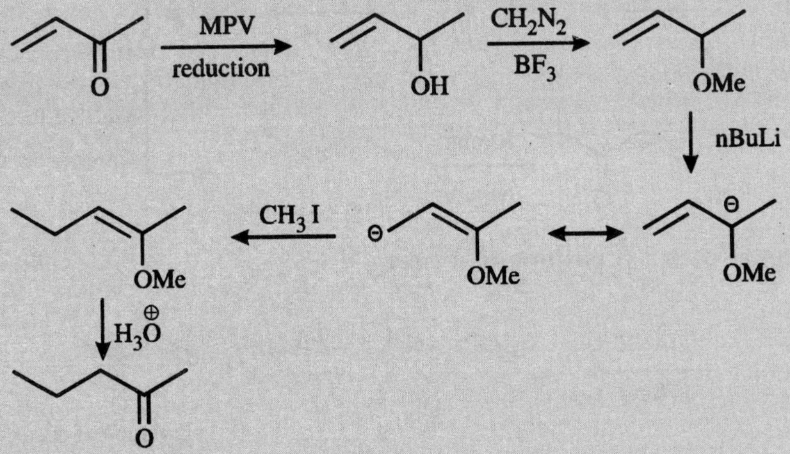

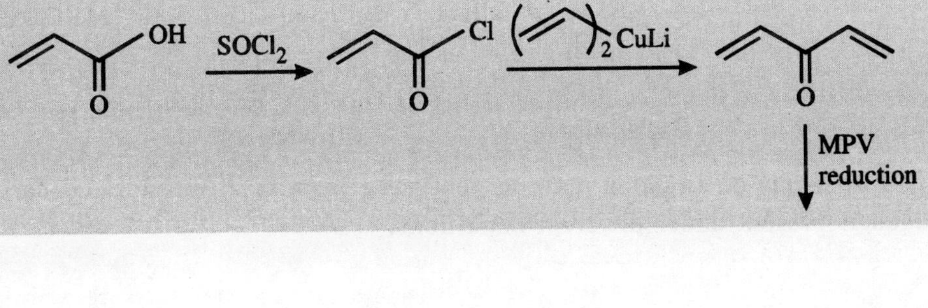

tBuLi; Me_3SiCl, Et_3N; O $SiMe_3$; OH; CH_3I; (i) $\overset{\ominus}{F}$ (ii) $H_3\overset{\oplus}{O}$

(iii) Alkylation at α-and β-position at a time

Me_2CuLi ether; Me I

β -alkylation α -alkylation

(iv) Alkylation at α′-position

(a) LDA (X′S) THF/O°C Controlled; CH_3I

Slow addition of α, β unsaturated ketone to the LDA base in THF

(b) $\overset{\ominus}{O}H/H_2O_2$; Me_3SiCl Et_3N; $\overset{\ominus}{Me}$; O $SiMe_3$; Me_2CuLi /ether, – 78°C; O⊖ O $SiMe_3$; Strong acid/Δ

(v) Alkylation at α, α′, β position at a time

LDA (X′S) THF/O°; CH_3I; alkylation at α′; Me_2CuLi; alkylation at β; CH_3I; alkylation at α

(vi) Two types of enolates may be generated from α, β unsaturated carbonyl if γ-hydrogen is present under different conditions.

(a) If α, β unsaturated carbonyl is treated with non-nucleophilic strong base in polar aprotic solvent at low temperature, α′-enolate is formed under controlled condition

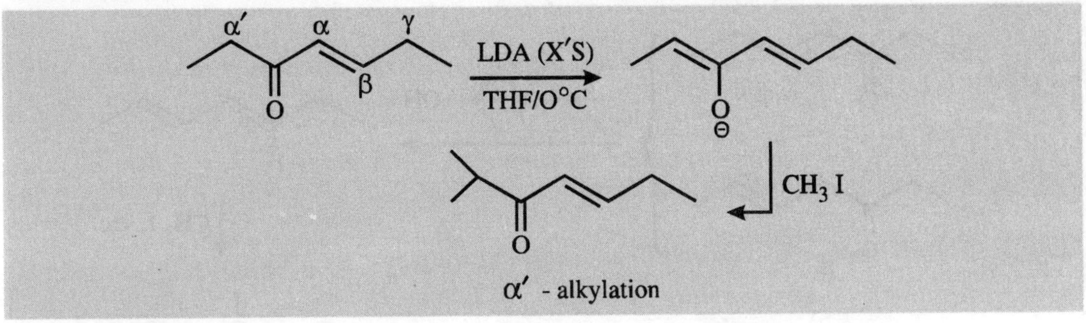

LDA (X′S)
THF/O°C
CH_3 I
α′ - alkylation

(b) If α, β-unsaturated carbonyl is treated with non-nucleophilic strong base in polar protic solvent dienolate is formed under equilibrium condition

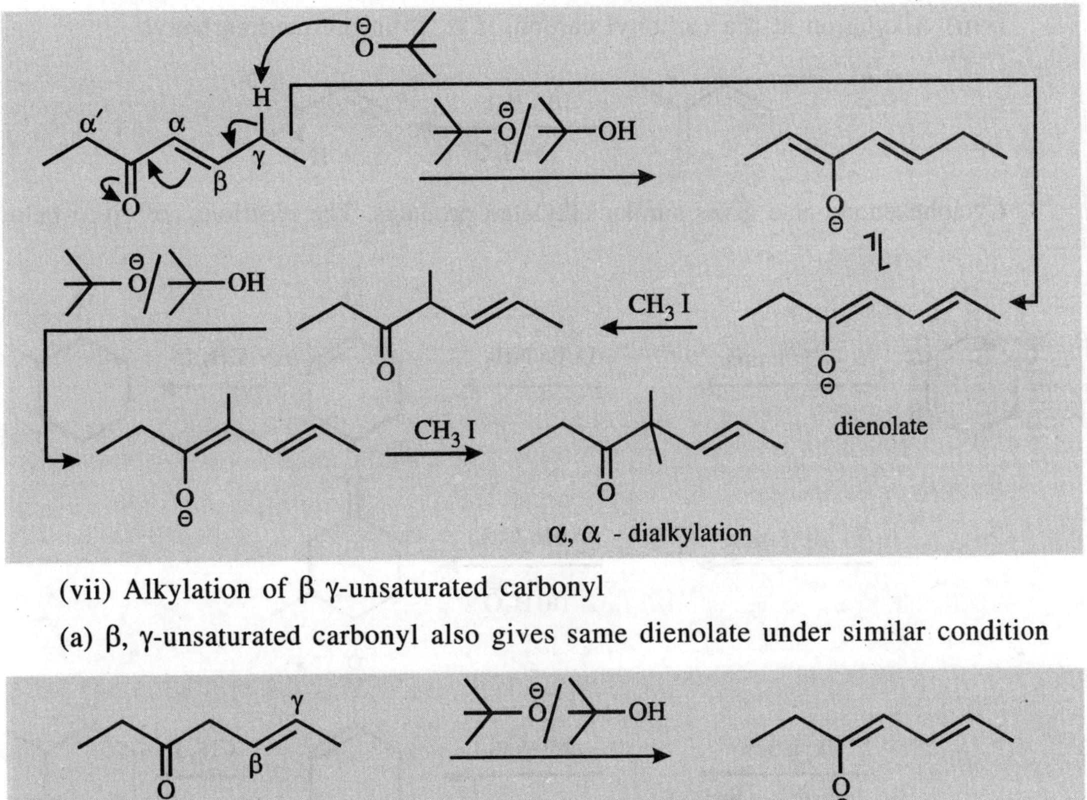

dienolate
CH_3 I
α, α - dialkylation

(vii) Alkylation of β γ-unsaturated carbonyl

(a) β, γ-unsaturated carbonyl also gives same dienolate under similar condition

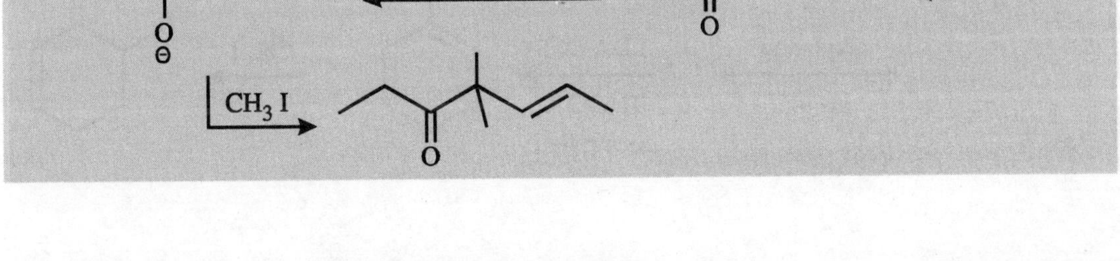

CH_3 I
CH_3 I

When α, β or β-γ-unsaturated carbonyl is treated with $tBuO^-/tBuOH$, same dienolate results and same dialkylated product obtained.

α, β -
+
β, γ -
OH
CH_3 I etc.

(viii) Alkylation at the carbonyl carbon of α, β-unsaturated carbonyl

(i) CH_3Li
(ii) H_2O
HO

Cyclohexenone also gives similar alkylated products. The reactions are given below.

α′ α β
α - alkylation
Li/liq NH_3
CH_3 I

β - alkylation
(i) Me_2CuLi
(ii) $H_3\overset{\oplus}{O}$

alkylation at α and β
Me_2CuLi
CH_3 I

alkylation at α′
LDA (X′S)
THF/O°C controlled
CH_3 I

alkylation at α′ → $\overset{\ominus}{O}H/H_2O_2$ → Me_3SiCl / Et_3N → Me_2CuLi → Strong HCl/Δ

O SiMe$_3$

alkylation at α, α′, β → LDA (X′S) THF/O°C → CH_3 I → Me_2CuLi → CH_3 I

alkylation at α, α → (i) t-BuO$^{\ominus}$/t-BuOH (ii) CH_3 I twice → CH_3 I → t-BuO$^{\ominus}$ / t-BuOH → CH_3 I

OR

dialkylation at α, α → (i) t-BuO$^{\ominus}$/t-BuOH (ii) CH_3 I twice

(ix) Alkylation via vinylogous substitution

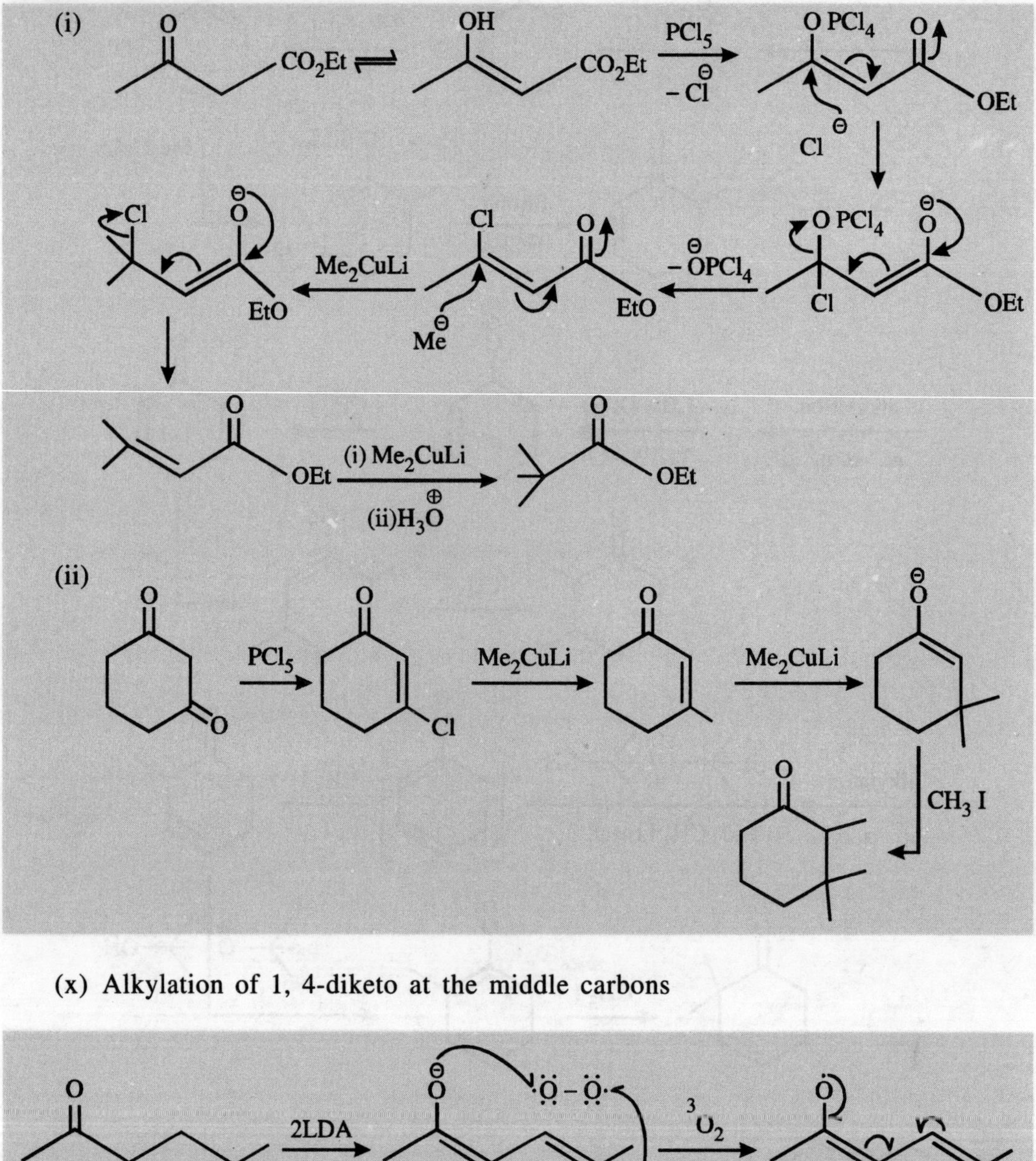

(i) O CO₂Et ⇌ OH CO₂Et $\xrightarrow[-Cl^{\ominus}]{PCl_5}$ O PCl₄ O OEt Cl⊖

O PCl₄ ⊖O Cl OEt $\xrightarrow{-\overset{\ominus}{O}PCl_4}$ Cl O EtO Me⊖ $\xrightarrow{Me_2CuLi}$ Cl ⊖O EtO

O OEt $\xrightarrow[(ii) H_3\overset{\oplus}{O}]{(i) Me_2CuLi}$ O OEt

(ii) O O $\xrightarrow{PCl_5}$ O Cl $\xrightarrow{Me_2CuLi}$ O $\xrightarrow{Me_2CuLi}$ ⊖O $\xrightarrow{CH_3I}$ O

(x) Alkylation of 1, 4-diketo at the middle carbons

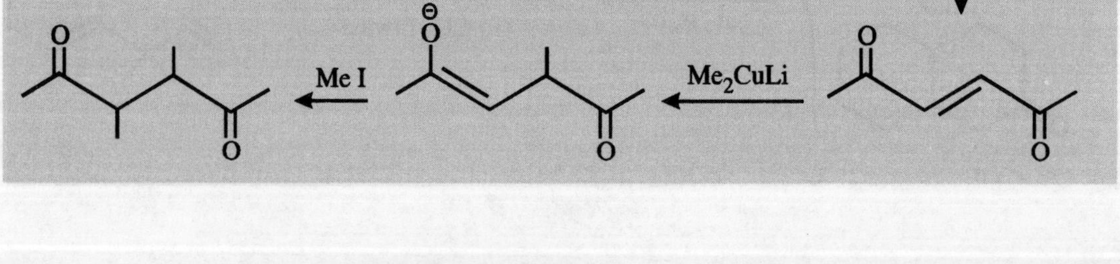

O O $\xrightarrow{2LDA}$ ⊖O O⊖ :Ö–Ö: $\xrightarrow[SET \text{ <o>},\ -O_2^{=}]{^3O_2}$ O· ·O

O O $\xrightarrow{Me_2CuLi}$ ⊖O O $\xrightarrow{Me\,I}$ O O

1.5.4 Alkylation via enamine, metalloenamine and NN-dimethyl hydrazone

(i) (a)

CHO $\xrightarrow[\text{TsOH}]{Me_2NH}$ NMe_2 (enamine) $\xrightarrow{CH_3I}$ $\overset{\oplus}{N}Me_2$ $\xrightarrow[(1eq.)]{H_3\overset{\oplus}{O}}$ CHO

x's acid may give aldol condensation of the product

(b)

O, N–H, TsOH, N, (i) CH_3I, (ii) $H_3\overset{\oplus}{O}$, O

exocyclic enamine/less substituted more favoured due to steric reason

(c)

O, N, H, endocyclic enamine, ⊕N, H, ⊖O, H, ⊕N, H, ⊖O, N, H, O

(ii)

CHO + H_2N– $\xrightarrow[\Delta]{TsOH}$ =N– $\xrightarrow{LDA/THF}$ $\ominus\overset{\oplus}{Li}$ =N– $\xrightarrow{CH_3I}$ =N– $\xrightarrow{H_3\overset{\oplus}{O}\text{/1eq.}}$ CHO

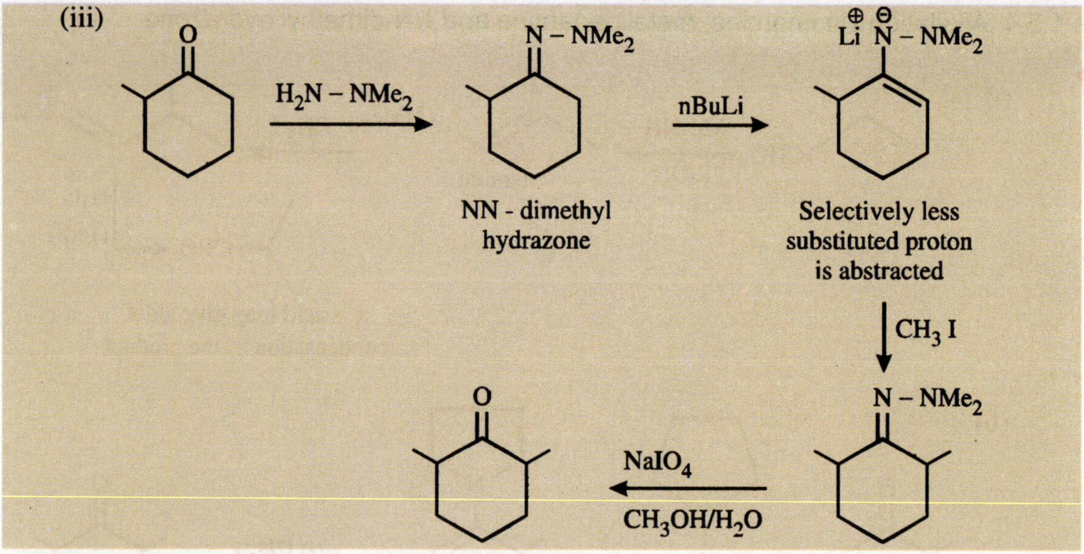

(iii)

H_2N-NMe_2 → N – NMe_2 (NN - dimethyl hydrazone) → nBuLi → $Li^{\oplus}$ $N^{\ominus}$ – NMe_2 (Selectively less substituted proton is abstracted) → CH_3I → N – NMe_2 → $NaIO_4$, CH_3OH/H_2O → O

1.5.5 Alkylation of nitroalkane

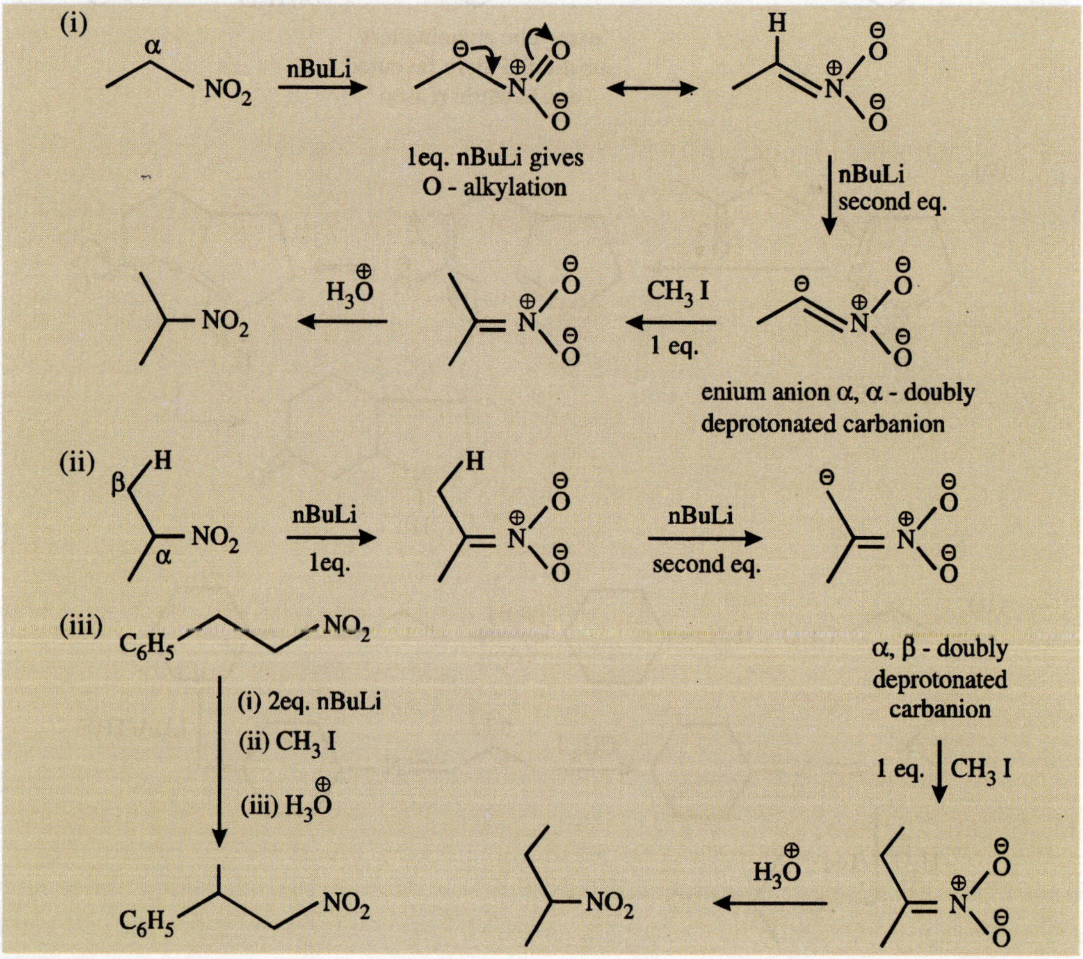

(i)

α, NO_2 → nBuLi →

1eq. nBuLi gives O - alkylation

↔ H

nBuLi second eq.

enium anion α, α - doubly deprotonated carbanion

CH_3I 1 eq. → $H_3\overset{\oplus}{O}$ → NO_2

(ii)

β, H, α, NO_2 → nBuLi 1eq. → H → nBuLi second eq. →

α, β - doubly deprotonated carbanion

1 eq. CH_3I → $H_3\overset{\oplus}{O}$ → NO_2

(iii)

C_6H_5, NO_2

(i) 2eq. nBuLi
(ii) CH_3I
(iii) $H_3\overset{\oplus}{O}$

C_6H_5, NO_2

1.5.6 Alkylation via acyl, formyl, carboxyl, benzoyl and crotonyl anion equivalent

(i) (a) Acyl anion equivalent

HS, HS / BF_3
H, O, S, S, H
nBuLi
acyl anion eq.
CH_3I
$Hg^{+2}/H_3\overset{\oplus}{O}$
O

$H-C\equiv C-H$
MeOK / MeOH
H, O - Me
tBuLi
O - Me
acyl anion eq.
CH_3I
$H_3\overset{\oplus}{O}$
O, O - Me

MPV / <H>
O, OH
CH_2N_2 / BF_3
OMe
nBuLi
OMe
acyl anion eq.
CH_3I
$H_3\overset{\oplus}{O}$
O, OMe, OMe

NO_2
2eq. / nBuLi
acyl anion eq.
CH_3I
$TiCl_3/H_2O$
O

(b) Formyl anion equivalent

$HS(CH_2)_3SH/BF_3$

nBuLi

CH_3I

formyl anion eq.

$Hg^{+2}/H_3\overset{\oplus}{O}$

CH_3-NO_2

2eq. nBuLi

$\overset{\ominus}{C}H=\overset{\oplus}{N}(O^{\ominus})_2$

CH_3I

formyl anion eq.

$TiCl_3/H_2O$

$MeOH + CH_2O + HCl \longrightarrow MeOCH_2Cl$

$(C_6H_5)_3P$

$MeOCH_2\overset{\oplus}{P}(C_6H_5)_3$

nBuLi

$MeO\overset{\ominus}{C}H\overset{\oplus}{P}(C_6H_5)_3$

formyl anion eq.

MeO

$H_3\overset{\oplus}{O}$

CHO

Cl CO_2Et

CO_2Et

formyl anion eq.

$H_3\overset{\oplus}{O}/\Delta$

CHO

(c) Carboxyl anion equivalent

$\overset{\ominus}{C}N$ —($C_6H_5CH_2Cl$, CH_3COCH_3)→ $C_6H_5CH_2CN$ —($H_3\overset{\oplus}{O}$)→ $C_6H_5CH_2COOH$

Carboxyl anion eq.

(d) Benzoyl anion equivalent

$C_6H_5CH{=}O$ —(Me_3SiCl, KCN, MeCN)→ $C_6H_5CH(OSiMe_3)CN$ —(nBuLi)→ $C_6H_5\overset{\ominus}{C}(OSiMe_3)CN$

benzoyl anion eq.

—(CH_3I)→ $C_6H_5C(CH_3)(OSiMe_3)CN$ —(H_2O, Δ)→ $CH_3COC_6H_5$

(e) Crotonyl anion equivalent

$CH_2{=}CHCH_2CHO$ —(CH_2N_2)→ $CH_2{=}CHCH{=}CH(OMe)$ —(tBuLi)→ $CH_2{=}CHCH{=}\overset{\ominus}{C}(OMe)$

Crotonyl anion eq.

—(CH_3I)→ $CH_2{=}CHCH{=}C(CH_3)OMe$ —($H_3\overset{\oplus}{O}$/Δ)→ $CH_2{=}CHCH_2COCH_3$ ⇌ $CH_3CH{=}CHCOCH_3$

1.5.7 Alkylation of 'ene' and 'yne'

(i) Alkylation of alkenyl halides via organo lithium or with organocuprates* give the substituted alkene with retention of configuration of the double bond, likewise, alkenyl cuprates reacts with retention of the geometry of the double bond.

The following examples will illustrate the facts.

(a) (E)-alkenyl bromide (H, Br) —(Li/ether)→ alkenyl carbanion $Li^{\oplus}$ —(CH_3I)→ methylated alkene (H, H; retention)

*For details see author's "Organic Synthesis" vol. 1

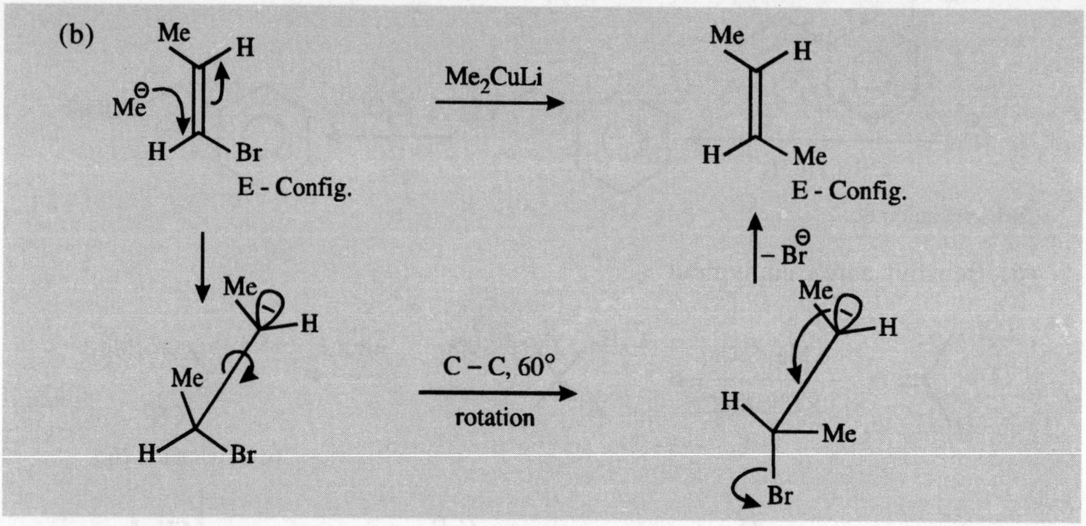

(b) Me, H, $\overset{\ominus}{Me}$, H, Br, E - Config. $\xrightarrow{Me_2CuLi}$ Me, H, H, Me, E - Config.

Me, H, Me, H, Br $\xrightarrow[\text{rotation}]{C-C,\ 60°}$ Me, H, H, Me, Br $\xrightarrow{-\overset{\ominus}{Br}}$

It is believed that this is because the nucleophile $\overset{\ominus}{Me}$ attack on the bromine containing carbon, undergoes rotation about the new C—C single bond in the direction shown as a result of the increase in the stabilising interaction between the occupied p-orbital and the C—Br bonding orbital.

(c) Acetylene itself reacts with organocuprate with almost complete syn stereoselectivity

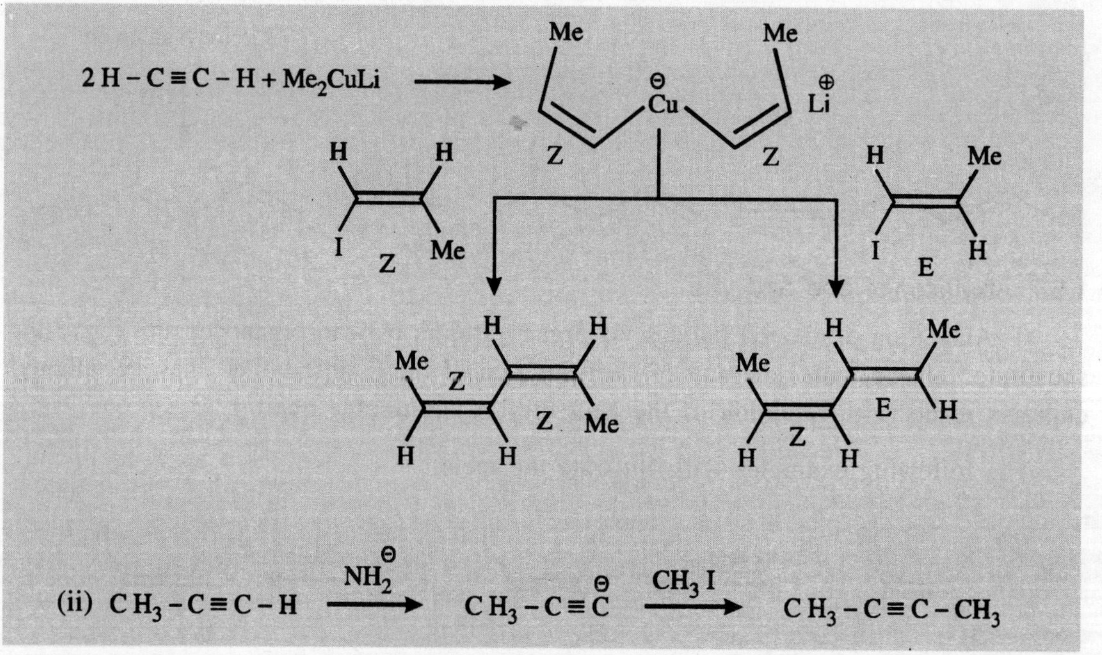

$2\,H-C\equiv C-H + Me_2CuLi \longrightarrow$ Me, Me, $\overset{\ominus}{Cu}$, $\overset{\oplus}{Li}$, Z, Z

H, H, I, Z, Me — H, Me, I, E, H

H, H, Me, Z, H, H, Z, Me — H, Me, Me, E, H, Z, H, H

(ii) $CH_3-C\equiv C-H \xrightarrow{\overset{\ominus}{NH_2}} CH_3-C\equiv \overset{\ominus}{C} \xrightarrow{CH_3I} CH_3-C\equiv C-CH_3$

1.5.8 Alkylation with nucleophilic alkylating agents

(i) $C_6H_5CO_2H \xrightarrow{CH_3Li} C_6H_5C(=O)O^{\ominus} \xrightarrow{CH_3Li} C_6H_5C(CH_3)(O^{\ominus})_2 \xrightarrow{H_3O^{\oplus}/\Delta} C_6H_5COCH_3$

(i) $(i)\ (CH_3)_2CH^{\ominus}Li^{\oplus}$; (ii) $H_3O^{\oplus}$

(i) $(i)\ (CH_3)_3C^{\ominus}Li^{\oplus}$; (ii) $H_3O^{\oplus}$

(ii) $CH_3COCl \xrightarrow{((CH_3)_3C)_2CuLi} CH_3COC(CH_3)_3$

(i) Me_2CuLi; (ii) $H_3O^{\oplus}$

(iii) $CH_3COCl \xrightarrow{Me_2Cd} CH_3COCH_3$

(iv) Ethylene oxide $\xrightarrow{(i)\ MeMgBr;\ (ii)\ H_3O^{\oplus}}$ Me–CH₂CH₂–OH

1.5.9 Alkylation of diazo and α-halo ketone or ester with Me_3B

(a)

$$Me_3B + N \equiv \overset{\oplus}{N} - \overset{\ominus}{C}H - CO_2Et \longrightarrow N \equiv \overset{\oplus}{N} - CH(\overset{\ominus}{B}Me_3) - CO_2Et \longrightarrow Me_2B - CH(Me) - CO_2Et \xrightarrow[\text{hydrolysis}]{\text{alkaline}} MeCH_2CO_2Et$$

(b)

BrCH$_2$COCH$_3$ (O, Br) → [tBuO$^{\ominus}$ / tBuOH] → enolate carbanion (O, Br, $\ominus$) → [Me_3B] → Br, Me–$\overset{\ominus}{B}Me_2$ intermediate → BMe_2 ketone → [tBuOH, Δ] → butanone

1.5.10 Alkylation of very weak nucleophilic centre

Alkylation of very weak nucleophilic centre may be done with a very strong electrophilic alkylating agent e.g. oxonium cation with a non-nucleophilic anion, $BF_4^{\ominus}$.

$Me_3\overset{\oplus}{O}\,\overset{\ominus}{BF_4}$ ($Me_2O + MeF + BF_3$)

Unstable with nucleophilic anion eg

$$Me_2\overset{\oplus}{O} - Me + \overset{\ominus}{Cl} \longrightarrow Me_2O + MeCl$$

$$Me_3\overset{\oplus}{O}\,\overset{\ominus}{BF_4} \xrightarrow{\text{pyridine}} Me - \overset{\oplus}{N}C_5H_5$$

$$\xrightarrow{CH_3CONMe_2} CH_3-\underset{\underset{OMe}{|}}{C}=\overset{\oplus}{N}Me_2 \xrightarrow{NaBH_4} CH_3-\underset{\underset{OMe}{|}}{CH}-NMe_2$$

Coumarin (benzopyran-2-one, O, =O) → 2-methoxybenzopyrylium ($\oplus$O, OMe)

1.5.11 Alkylation of benzene ring

(i) Alkylation via intermolecular reactions

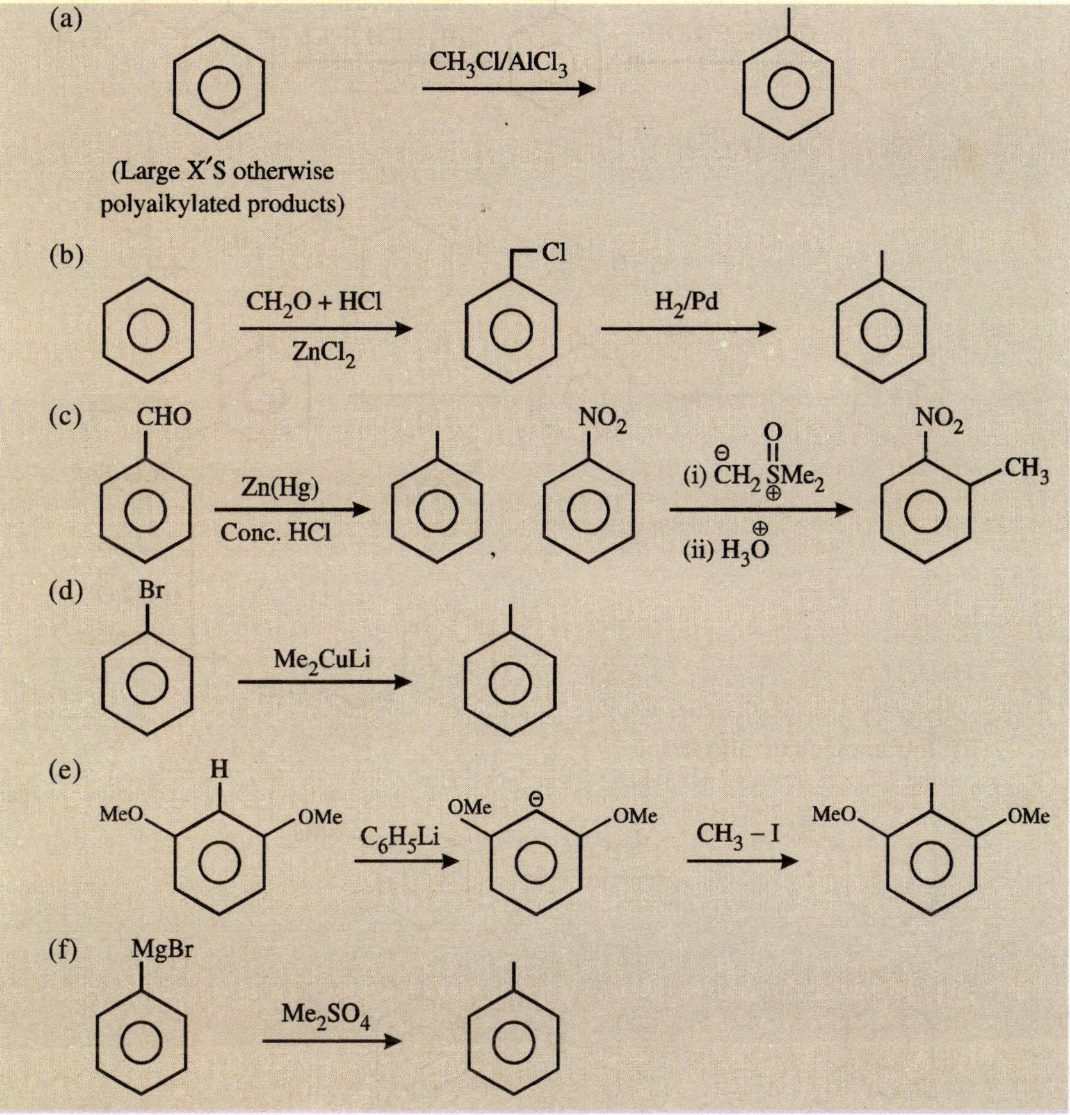

(g) OH — $CH_2O + Me_2NH$ → OH, NMe_2 — Me I → OH, $\overset{\oplus}{N}Me_3$ — Na/Hg, H_2O → OH

(h) OH — $\overset{\ominus}{OH}/CF_3CH_2OH$ → $\overset{\ominus}{O}$ — $C_6H_5 - CH_2 - Cl$ → $\overset{\ominus}{O}$, C_6H_5 — $H_3\overset{\oplus}{O}$ → OH, C_6H_5

(i) Br — $\overset{\ominus}{NH_2}$ → — $\overset{\ominus}{NH_2}$, CO_2Et, CO_2Et → CO_2Et, CO_2Et — (i) KOH, (ii) $H_3\overset{\oplus}{O}$, (iii) 250°C → CO_2H

(ii) Intramolecular alkylation

(a) $\overset{\oplus}{N}Me_3$ — $\overset{\ominus}{NH_2}/\Delta$ → NMe_2

(b) $\overset{\oplus}{N}Me_3$ — 300°C → NH_2

(c) [anisole] $\xrightarrow{AlCl_3/\Delta}$ [o-cresol] OH

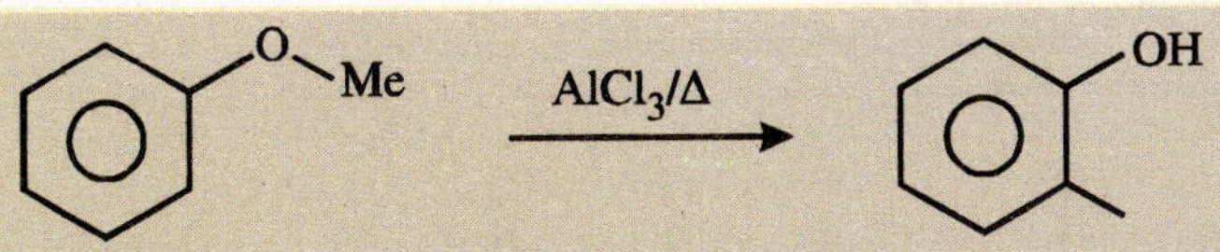

(d) [allyl phenyl ether] $\xrightarrow{\Delta}$ [2-allylphenol] OH

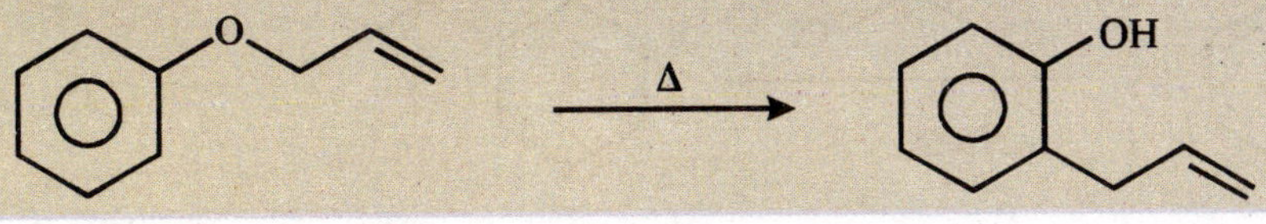

1.5.12 Acetylation, formylation, carbethoxylation, benzoylation, phenylation, vinylation and methylenylation

(i) Acetylation

(a) [acetone] $\xrightarrow[BF_3]{Ac_2O}$ [acetylacetone]

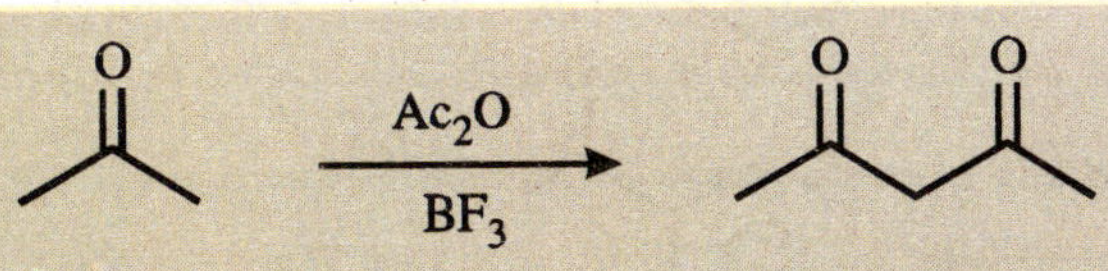

(b) [butanone] $\xrightarrow[BF_3]{Ac_2O}$ [3-methylpentane-2,4-dione]

(i) $\overset{\ominus}{O}Et$ (ii) CH_3COOEt ⟶ [heptane-2,4-dione]

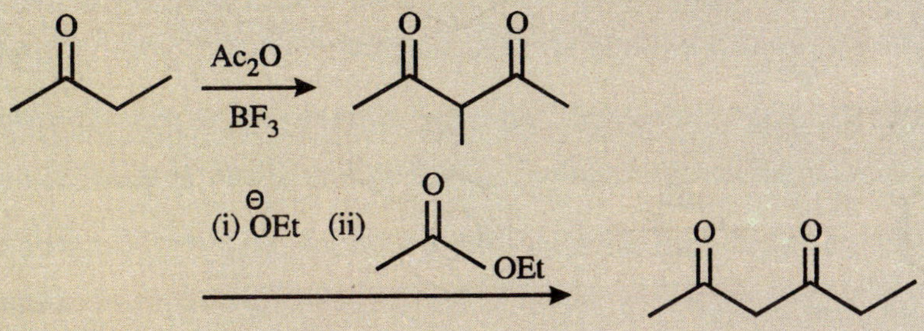

(c) CH_3COOEt $\xrightarrow{\overset{\ominus}{O}Et}$ CH_3COCH_2COOEt

CH_3COOEt

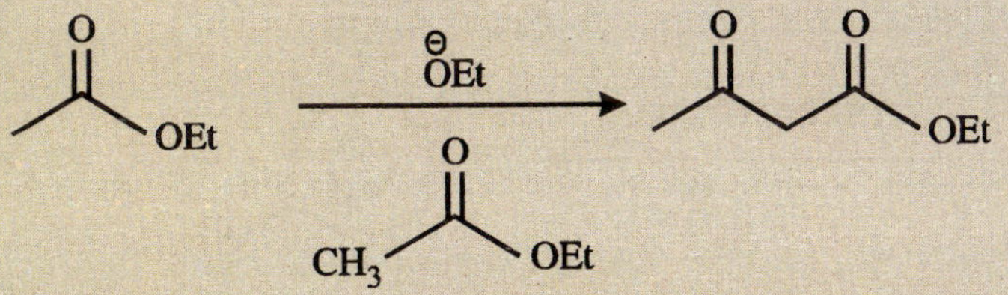

(d) $CH_3COCH_2CO_2Et$ $\xrightarrow[(ii)\ CH_3COCl]{(i)\ NaH/C_6H_6}$ $(CH_3CO)_2CHCO_2Et$

LDA 2eq. ⟶ $\overset{\ominus}{C}H_2COCH^{\ominus}CO_2Et$ $\xrightarrow{CH_3COCl}$ $CH_3COCH_2COCH^{\ominus}CO_2Et$ $\xrightarrow{aq.\ NH_4Cl}$ $CH_3COCH_2COCH_2CO_2Et$

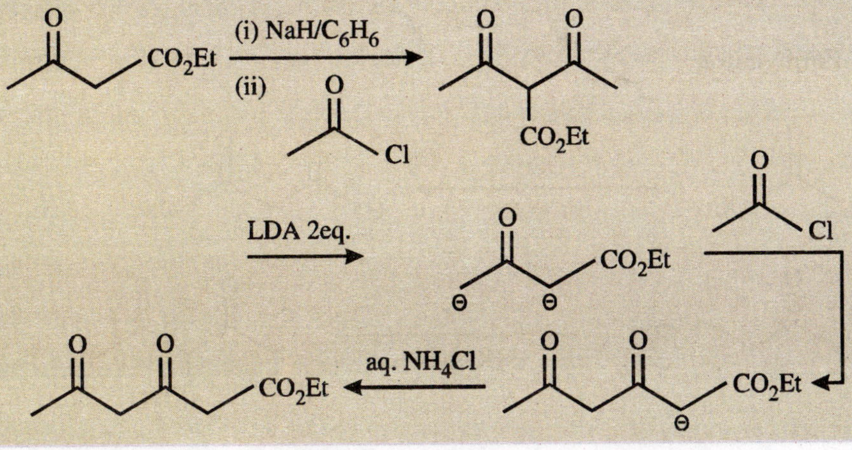

(e) (i) CH_3COCl (ii) $H_3O^{\oplus}$

(f) (i) $CH_3COCl/AlCl_3$

$ClCOCH_2Cl$, $AlCl_3$

(g) O $COCH_3$ — $AlCl_3$, Δ — OH, O

(h) OH, OH — (i) $CH_3CN/HCl/ZnCl_2$ (ii) $H_3O^{\oplus}$ — OH, OH, $COCH_3$

(ii) Formylation

(a) O, OEt — $^{\ominus}OEt$, HCO_2Et — H, O, O, OEt

(b) O, CO_2Et — $^{\ominus}OEt$, HCO_2Et — O, O, H, CO_2Et

(c) Cyclohexanone $\xrightarrow[\text{HCO}_2\text{Et}]{\overset{\ominus}{\text{O}}\text{Et}}$ 2-formylcyclohexanone

$\xrightarrow[\text{(ii) } H_3\overset{\oplus}{O}]{\text{(i) MeOCH}=\text{P}(\text{C}_6\text{H}_5)_3}$ cyclohexanecarbaldehyde (CHO)

$\xrightarrow[\text{(ii) } H_3\overset{\oplus}{O}/\Delta]{\text{(i) ClCH}_2\text{CO}_2\text{Et/ } t\text{-BuO}^{\ominus}}$ cyclohexanecarbaldehyde (CHO)

(d) 1-pyrrolidinylcyclohexene $\xrightarrow[\text{(ii) } H_3\overset{\oplus}{O}]{\text{(i) HCO}_2\text{Et}}$ 2-formylcyclohexanone

$\xrightarrow[\text{(ii) } H_3\overset{\oplus}{O}]{\text{(i) HCONMe}_2 + \text{POCl}_3}$ 2-formylcyclohexanone

(e) $CH_3-Br \xrightarrow[\text{(ii) Hg}^{+2}/H_3\overset{\oplus}{O}]{\text{(i) 2-lithio-1,3-dithiane anion (H, }\ominus\text{)}}$ CH_3-CHO

$\xrightarrow[\text{(ii) TiCl}_3/\text{H}_2\text{O}]{\text{(i) }\overset{\ominus}{C}H=\overset{\oplus}{N}(\overset{\ominus}{O})_2}$ CH_3-CHO

(f) Methyl vinyl ketone $\xrightarrow[\text{(ii) TiCl}_3/\text{H}_2\text{O}]{\text{(i) }\overset{\ominus}{C}H=\overset{\oplus}{N}(\overset{\ominus}{O})_2}$ H–CO–CH$_2$CH$_2$–CO–CH$_3$

(g) OH

(i) $\overset{\ominus}{O}H/CHCl_3$

(ii) $H_3\overset{\oplus}{O}$

OH CHO

(h) OH OH

(i) $HCONMe_2 + COCl_2$

(ii) $H_3\overset{\oplus}{O}$

OH OH CHO

(i) CH_3MgBr —(i) $HC(OEt)_3$ (ii) $H_3\overset{\oplus}{O}$→ CH_3-CHO

(iii) Carbethoxylation

(a) O —$\overset{\ominus}{O}Et$, $(CO_2Et)_2$→ O O CO_2Et —Δ→ O O OEt

(b) N —(i) $ClCO_2Et$ (ii) $H_3\overset{\oplus}{O}$→ O O OEt

(iv) Benzoylation

(a) O OEt —LDA→ $\overset{\ominus}{O}$ OEt —$C_6H_5CO_2Et$→ O C_6H_5 CO_2Et

(b) O CO_2Et —(i) NaH/C_6H_6 (ii) C_6H_5COCl→ O O C_6H_5 CO_2Et

—(i) LDA 2eq. (ii) C_6H_5COCl (iii) aq. NH_4Cl→ O O C_6H_5 CO_2Et

(c) Phenol $\xrightarrow[\text{Py}]{C_6H_5COCl}$ phenyl benzoate (C_6H_5–O–CO–C_6H_5)

(v) Phenylation

(a) Ethyl acetoacetate ($CH_3COCH_2CO_2Et$) + bromobenzene $\xrightarrow[\text{via benzyne}]{\overset{\ominus}{N}H_2\ 2eq.}$ $CH_3COCH(C_6H_5)CO_2Et$

(b) 1-Pyrrolidinocyclohexene $\xrightarrow[\text{(ii) } H_3\overset{\oplus}{O}]{\text{(i) benzyne}}$ 2-phenylcyclohexanone

(c) $C_6H_5CH{=}CH_2$ + $N\equiv\overset{\oplus}{N}$–$C_6H_5$ $\xrightarrow{Cu^+}$ $C_6H_5CH_2\dot{C}H$–CH(H)–C_6H_5 $\xrightarrow{Cu^{+2}}$ $C_6H_5CH{=}CHC_6H_5$

(d) $CH_2{=}CHCOCH_3$ + $N\equiv\overset{\oplus}{N}$–$C_6H_4$–Cl $\xrightarrow[\text{(ii) } H_2O]{\text{(i) } TiCl_3/\Delta}$ $CH_3COCH_2CH_2$–C_6H_4–Cl

(vi) Vinylation

(a) Cyclohexanone $\xrightarrow{CH_2{=}CHMgCl/THF}$ 1-vinylcyclohexanol (HO, vinyl)

(b) C_6H_5COCl $\xrightarrow{(CH_3CH{=}CH)_2CuLi}$ $C_6H_5COCH{=}CHCH_3$

(vii) Methylenylation

(a)

$CH_2=\overset{\oplus}{N}Me_2$ + cyclohexanone → 2-(CH_2NMe_2)cyclohexanone $\xrightarrow{CH_3I}$ 2-($CH_2\overset{\oplus}{N}Me_3$)cyclohexanone $\xrightarrow{\text{Base}\ \Delta}$ 2-methylenecyclohexanone

$C_6H_5COCH_3$ → $C_6H_5COCH=CH_2$

γ-butyrolactone → α-methylene-γ-butyrolactone

$CH_3COCH_2CO_2Et$ → $CH_3COC(=CH_2)CO_2Et$

(b)

$(C_6H_5)_3P=CH_2$ + acetone → $(CH_3)_2C=CH_2$

C_6H_5CHO → $C_6H_5CH=CH_2$

CHO written as "CIIO" on the page.

1.5.13 Alkylation of ambident nucleophile: O-alkylation versus C-alkylation

When ethylaceto acetate or any such active methylene compound is treated with a base, an anion is formed—when the negative charge may reside either on the C-atom or on the O-atom.

$$CH_3COCH_2CO_2Et \xrightarrow{Base} CH_3CO\overset{\ominus}{C}HCO_2Et \rightleftharpoons CH_3C(O^{\ominus})=CHCO_2Et$$

This type of anion is called ambident anion.

So if we alkylate an enolate anion, the alkyl group may enter either on C or on O-atom.

$$-\underset{\underset{O}{\|}}{C}-\overset{\ominus}{C}H- \xrightarrow{RX} -\underset{\underset{O}{\|}}{C}-\overset{R}{\overset{|}{C}H}- \quad \text{C - alkylation}$$

$$\rightleftharpoons$$

$$-\underset{\underset{O^{\ominus}}{|}}{C}=CH- \xrightarrow{RX} -\underset{\underset{O-R}{|}}{C}=CH- \quad \text{O - alkylation}$$

The nature of the alkylation process depends on the following factors:

1. Homogeneous or heterogeneous process
2. Solvent polarity
3. Choice of cations
4. Nature of the alkylating agent
5. Nature of the ambident substrate

O-alkylation

Various observations suggests that the reaction in solution of a particular enolate anion with a particular alkylating agent will give the greatest proportion of O-alkylation when the reaction conditions (solvent, cation, temperature) are chosen which allow the maximum amount of free enolate anion to be present. The enolate anion should be free and the negative charge must be located on oxygen exclusively rather than on carbon.

Solvent: For O-alkylation, the conditions may be satisfied by use of an aprotic solvent, because in protic solvent hydrogen bonding with enolate oxygen ion is possible (since free enolate anion is not possible). That is, maximum opportunity for O-alkylation exists when the oxygen atom is not shielded by association with a metal cation or a H-bonding solvent.

Cation: Large cations have a tendency to dissociate from the anions. So by using a large cation, the negative charge residing mainly on enolate oxygen. The size of the commonly used cation are in the order.

$$R_4\overset{\oplus}{N} > \overset{\oplus}{K} > \overset{\oplus}{Na} > \overset{\oplus}{Li}$$

Reaction condition: When the metal enolate is in solution i.e. a homogeneous reaction condition exists, there is no chance of forming a crystal lattice like $\overset{\oplus}{M}\ \overset{\ominus}{O}\ \overset{\oplus}{M}$ in which oxygen is shielded by associated metal atom in the crystal lattice. So in homogeneous reaction condition O-alkylation is favoured.

C-alkylation: In the same way we can say that C-alkylation is favoured.

(i) in a protic solvent which can form H-bond with the enolate oxygen anion.

(ii) with small cation which has a low tendancy to dissociate.

(iii) in heterogeneous reaction condition.

In cases when the metal enolate is not in solution, any heterogenous reaction of the alkylating agent with the solid enolate will necessary involve reaction with an anion in which the oxygen is shielded by the associated metal ion in the crystal lattice. Due to formation of the crystal lattice there is no free enolate oxygen anion but carbanion can be formed in free. So in heterogeneous reaction condition C-alkylation is favoured.

Effect of nature of the alkylating agent on the ambident ion in the alkylating process.

The following example illustrates typical proportion of O-alkylated versus C-alkylated products.

$$CH_3-\underset{\underset{O}{\|}}{C}-CH_2-CO_2Et + nC_4H_9X \xrightarrow[\text{Solvent}]{K_2CO_3/100^\circ} CH_3\,CO\underset{\underset{nC_4H_9}{|}}{CH}-CO_2Et + CH_3-\underset{\underset{O-nC_4H_9}{|}}{C}=CH-CO_2Et$$

Solvent	Dielectric constant	X	C-alkylation	O-alkylation
CH_3COCH_3	20	Cl	90%	10%
CH_3CN	38	Cl	81%	19%
Me_2SO	49	Cl	53%	47%
$HCONMe_2$	37	Cl	54%	46%
		Br	67%	33%
		I	99%	1%

Thus we see that for a given alkylating agent the amount of O-alkylation product increases with the increase of solvent polarity. Again, variations in the leaving group of alkylating agent are usually found to favour C-alkylation in the order: R—I> R—Br> R—Cl.

These observations can be explained on the basis of size, electronegativity and polarisability. For ambident enolate anion, oxygen is small, electronegative and less polarisable and carbon is larger, less electronegative and more polarisable.

A less polarisable O-atom coordinate to less polarisable alkyl carbon and more polarisable C-atom coordinate to more polarisable alkyl carbon.

Now we shall compare the O-and C-alkylated products for three primary alkylating agents.

(i) $C_6H_5—CH_2—Br$ (ii) $CH_2 = CH — CH_2 — Br$ (iii) n $C_4H_9 — Br$

The allyl and benzyl halides tend to give less O-alkylation than n-butyl halide. This is because benzyl and allyl halides are highly polarisable and n-butyl halide is less polarisable.

Solvent factor: Studies of alkylation in dipolar aprotic solvent such as DMF, DMSO and $MeOCH_2CH_2OMe$ have demonstrated that very substantial increase in the rates of reaction of enolate anion with alkylating groups results from their use in preference to alcohols or inert solvent. Their advantage over the protic solvent (eg ethanol) lies in the fact that they presumably do not solvate the enolate anion and consequently do not diminish its reactivity as a nucleophile. On the otherhand polar aprotic solvent do have the ability to solvate the cation, separating it from the cation enolate ion-pair and leaving a relatively free anion in the reaction mixture.

$\overset{\oplus}{Li}$ (chelated enolate, $O^{\ominus}$ … OEt) + Solvent $\rightleftharpoons$ $\overset{\oplus}{Li}$ ······ Solvent + (free enolate, $O^{\ominus}$, O, OEt)

The free anion is more reactive as nucleophile than are the associated ion-pair. This increase nucleophilic reactivity is accompanied by an increased tendency to form O-alkylated product.

Conditions imposed for the predominancy of O-alkylation over the C-alkylation or vice-versa are the control of regio selectivity because they are regiomers.

Combination of base and frequently used solvents.

Base	Solvent
$Me_3C - \overset{\ominus}{O}\overset{\oplus}{K}$	t-BuOH, DMSO, ether
$NaNH_2$	liq. NH_3, ether, benzene
NaH	ether, benzene
$(C_6H_5)_3\overset{\ominus}{C}\overset{\oplus}{Li}$	ether
$(Me_2CH)_2\overset{\ominus}{N}\overset{\oplus}{Li}$ (LDA)	ether

The dielectric constants of some commonly used solvents.

Solvents	Dielectric constant
Water (H_2O)	80
Dimethyl Sulphoxide (Me_2SO) DMSO	49
Methanol (MeOH)	33
Ethanol (EtOH)	24
Acetone (MeCOMe)	21
Acetic acid (MeCOOH)	6
Diethyl ether (Et – O – Et)	4
Benzene (C_6H_6)	2
Dimethyl formamide ($HCONMe_2$) DMF	37
Hexamethylphosphoric triamide, (HMPA) $(Me_2N)_3PO$	30

The following examples are some illustrations for O-alkylation versus C-alkylation.

(i) $\overset{\ominus}{O}\overset{\oplus}{K}$ (phenolate) $\xrightarrow[C_6H_6]{Br-CH_2CH=CH_2}$ $\overset{\ominus}{O}$ (ring) $CH_2CH=CH_2$

(a) In the phenolate anion carbon is more polarisable and allyl carbon is also highly polarisable and hence carbon alkylation is favoured.

(b) Benzene solvent is non-polar aprotic (inert) and hence heterogeneous phase and metal enolate oxygen will form crystal lattice i.e. enolate oxygen is highly solvated by lattice and therefore C—alkylation is favoured.

$\overset{\ominus}{O}\overset{\oplus}{K}$ (phenolate) $\xrightarrow[CH_3COCH_3]{Br-CH_2CH=CH_2}$ $O-CH_2CH=CH_2$ (ring)

Acetone solvent is a polar aprotic and hence solvate the cation $K^{\oplus}$ .... $O=C(Me)_2$,

separating it from the cation-enolate ion-pair and leaving free anion in the mixture and which is homogeneous mixture and O-alkylation is favoured.

(ii)

2-Naphthoxide ($O^{\ominus}Na^{\oplus}$) $\xrightarrow[DMSO]{C_6H_5-CH_2-Cl}$ 2-naphthyl–O–$CH_2-C_6H_5$

DMSO is a polar aprotic solvent which has a high solvating power for cation and poor solvating power for enolate-anion, thus leaving the negatively charged oxygen atom free for nucleophilic attack on benzyl chloride.

$Na^{\oplus}$ solvated by four DMSO molecules (S=O ······ $Na^{\oplus}$ ······ O=S)

2-Naphthoxide ($O^{\ominus}Na^{\oplus}$) $\xrightarrow[CF_3CH_2OH]{C_6H_5-CH_2-Cl}$ 1-($CH_2-C_6H_5$)-2-naphthol (OH)

Trifluoroethanol CF_3CH_2OH is an excellent solvent for enolate oxygen because of the electron withdrawing group CF_3 making it particularly efficient in H-bonding with the enolate oxygen and therefore enolate oxygen is highly shielded and consequently C-alkylation favours exclusively.

(iii)

C_6H_5–C($O^{\ominus}Li^{\oplus}$)=C(CH$_3$)$_2$ $\xrightarrow[ether]{Me_2SO_4}$ C_6H_5–CO–C(CH$_3$)$_3$

Ether is a slighly polar aprotic solvent and lithium ion is very smaller size so it will have a very low tendency to dissociate from enolate oxygen anion rather it is associated with enolate oxygen which is solvated and consequently makes heterogeneous phase and C-alkylation favours.

C_6H_5–C($O^{\ominus}Li^{\oplus}$)=C(CH$_3$)$_2$ $\xrightarrow[Me_2SO_4]{12\text{-}Crown-4}$ C_6H_5–C(O–Me)=C(CH$_3$)$_2$

12-Crown-4 is a cyclic ether and the cavity size of it is a good match with the Li^+ (ionic diameter 1.20Å) and consequently Li ion is exclusively solvated (binds) with the crown-ether and enolate anion is totally free and the reaction mixture is now homogeneous and O-alkylation favours.

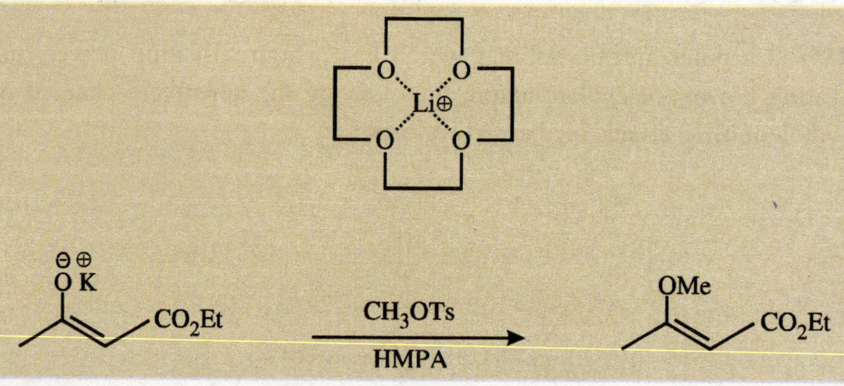

HMPA is a polar aprotic solvent and enolate ion is not solvated and is more reactive and the reaction mixture is homogeneous. Moreover HMPA solvate the cation.

$$(Me_2N)_3P=O\cdots\cdots\overset{\oplus}{K}$$

The alkylating agent CH_3OTs having very good leaving group alkylate enolate oxygen predominately. This can be rationalised by the Hammond postulate: the transition state occurs earlier in the reaction profile so that the new C—C and C—O bonds are less fully formed and the stabilities of the products are of less importance in determining reactivity than the distribution of charge in the enolate, which is greater at the oxygen atom.

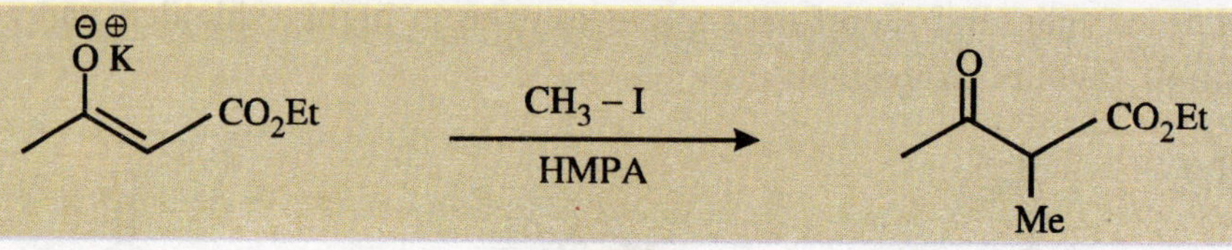

In the enolate anion carbon is more polarisable than oxygen and in CH_3—I, carbon is more polarisable than in CH_3OTs. On the other hand leaving ability of $\overset{\ominus}{I}$ is less than $\overset{\ominus}{O}Ts$ and consequently favours C-alkylation.

Enolate is a resonance hybrid where negative charge resides on carbon and oxygen and the resonance structure bearing negative charge on oxygen is more contributing and therefore, alkylation should at the more electronegative atom oxygen.

But in practice, C-alkylation is favoured in most cases. The reason for C-alkylation is thermodynamic since bond energy of (C = O + C—C) for C-alkylation is greater than bond energy of (C = C + C—O) for O-alkylation.

There are two exceptions to this generalisation in the polar protic solvent.

(i) Ring closure, when the ring formed by reaction at carbon is significantly strained whereas that formed by reaction at oxygen is not. In that situation alternative enolate oxygen alkylation is favoured.

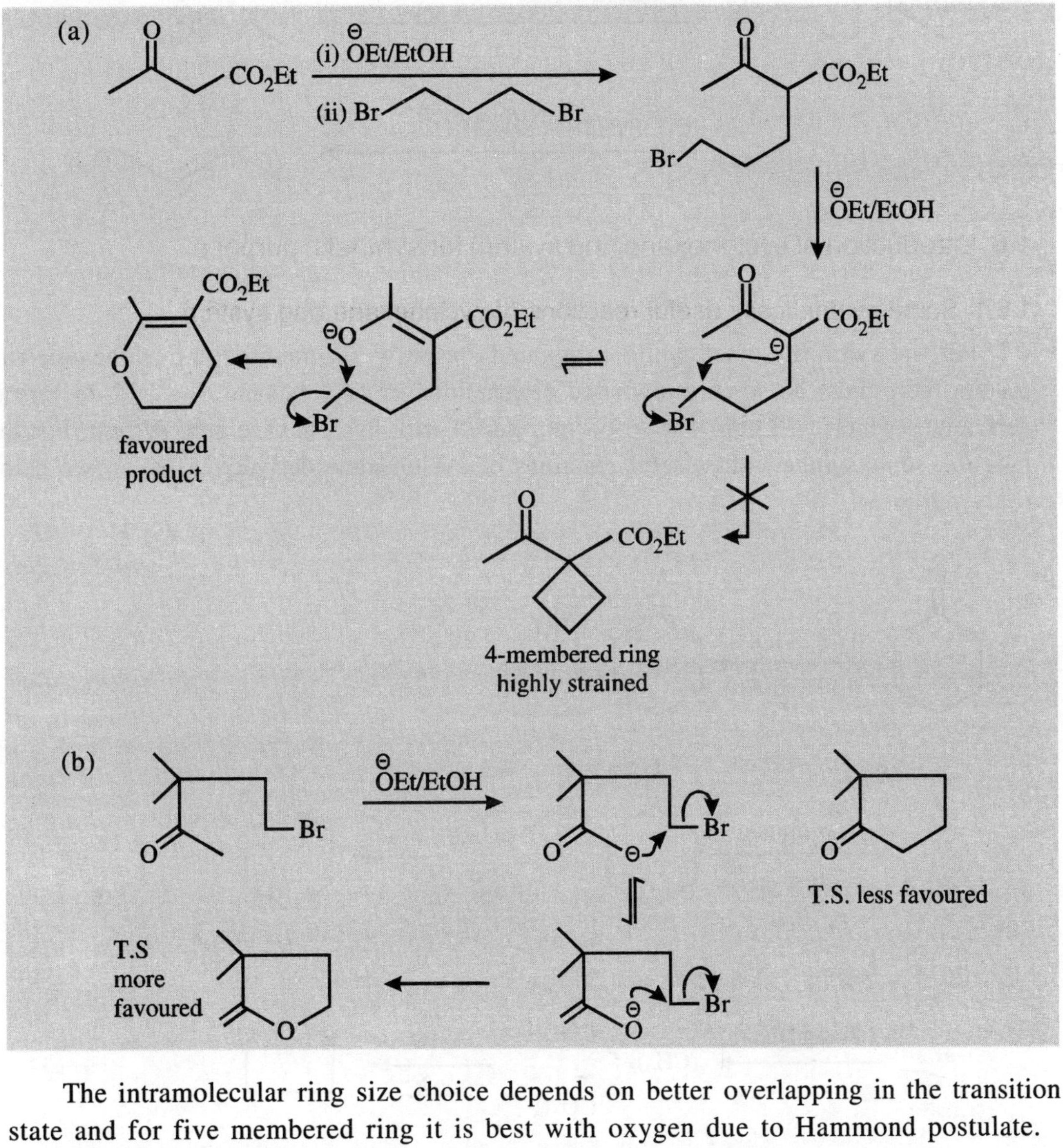

(a) (i) $\overset{\ominus}{O}Et/EtOH$ (ii) Br(CH$_2$)$_3$Br

$\overset{\ominus}{O}Et/EtOH$

favoured product

4-membered ring highly strained

(b) $\overset{\ominus}{O}Et/EtOH$

T.S. less favoured

T.S more favoured

The intramolecular ring size choice depends on better overlapping in the transition state and for five membered ring it is best with oxygen due to Hammond postulate.

With carbon, six membered ring size is the choice for best overlapping.

(ii) With very reactive alkylating agent, O-alkylation is favoured and it is again due to Hammond postulate.

(i) $\overset{\ominus}{O}Et/EtOH$ (ii) $CH_3OCH_2CH_2Cl$

(i) $\overset{\ominus}{O}Et/HMPA$ (ii) CH_3OTs

OMe

CO_2Et

1.6 Introduction of cyclohexane ring system for synthetic purpose

1.6.1 Some synthetically useful reactions of cyclohexane ring system

For the synthesis of alicyclic compound specially six-membered cyclohexane ring system there must be some knowledge of reactions of cyclohexane ring and its various derivatives which will help one to be introduced with cyclohexane ring system. For this purpose, some synthetically useful reactions of cyclohexane derivatives are shown below (only outlines).

(i) $\overset{\ominus}{O}Et/EtOH$ (ii) HCO_2Et → CHO

(i) $\overset{\ominus}{O}Et/EtOH$ (ii) $(CO_2Et)_2$ → CO_2Et —Δ→ CO_2Et

fuming HNO_3 → $(CH_2)_4(COOH)(CO_2H)$ —Ac_2O, 300°C→

HNO_2 → NO ⇌ N − OH —H_2/Pd OR Zn/CH_3COOH→ NH_2

$\xrightarrow{SeO_2}$ (cyclohexane-1,2-dione)

$\xrightarrow[CH_3CN]{KCN/Me_3SiCN}$ CN OSiMe$_3$ $\xrightarrow[H_3\overset{\oplus}{O}]{Strong}$ HO CO$_2$H $\xrightarrow{\Delta}$ COOH

LiAlH$_4$ → OH

(i) PCl$_5$ (ii) Mg/THF → MgCl

Me$_3$SiCl → SiMe$_3$

E⊕ → Me$_3$Si ⊕ E → CH$_2$ E

$\xrightarrow{(C_6H_5)_3P{=}CH_2}$ CH$_2$ exomethylene

N–H /TsOH → N → (i) $CH_2O + Me_2NH$ (ii) $H_3\overset{\oplus}{O}$ → O NMe$_2$

(i) Me I → O $\overset{\oplus}{N}Me_3$ → Base Δ → O CH$_2$

exomethylene ketone

(i) CH_3MgBr
(ii) $H_3\overset{\oplus}{O}$

endomethylene

O_3
H_2O_2

COOH

$NaBH_4$

OH

COOH

NH_2OH

N – OH

COOH

H₂/Pd

NH_2

COOH

NO

(i) $\overset{\ominus}{O}Et$
(ii) Et-O-NO

$\overset{\ominus}{O}Et$

$N-\overset{\ominus}{O}$

CO_2Et

$H_3\overset{\oplus}{O}$

NOH

CO_2Et

H_2/Pd

NH_2

CO_2Et

N

(i) Cl – CN
(ii) aq. tartaric acid

O

CN

O····H

O

CH_2N_2

OMe

$(C_6H_5)_3P=CH_2$

CH_2

O

aq. tartaric acid

CH_2

OMe

Na/ether → $H_3\overset{\oplus}{O}$ → $(C_6H_5)_3P{=}CH_2$, work up → CrO_3/Py

(i) $Pb(OAc)_4$ (ii) $H_3\overset{\oplus}{O}$ → $C_6H_5CH_2-Cl$, Py → $KMnO_4/\overset{\ominus}{O}H$ → $H_3\overset{\oplus}{O}$, Δ

1.6.2 Preparation of cyclohexenone

(i) OMe → Li/liqNH_3 → $H_3\overset{\oplus}{O}$ → Δ

(ii) (i) $Pb(OAc)_4$ (ii) $H_3\overset{\oplus}{O}$ → MnO_2

(iii) CrO_3/AcOH

(iv) Me_2NH / AcOH; CHO

(v) CO_2Et, CO_2Et — Na/Xylene → $H_3\overset{\oplus}{O}$ → OH — (i) TsCl (ii) Py/Δ

(vi) PCl_5 → Cl — Zn(Hg) / CH_3OH

(vii) H — Ra. / O_2 → H, O – O – H — $\overset{\ominus}{O}H$/E2 / – H_2O

1.6.3 Preparation of intermediates enol, enol ether, enol acetate, enolate, dienolate, bis-enolate, silyl enol ether, enamine and ene-dione

(i) $HClO_4$ → OH (enol) — CH_2N_2 → OMe (enol ether)

$Ac_2O/NaHCO_3$ → enol-acetate

(ii)

$\overset{\ominus}{O}Et$ → enolate ← Li/liqNH_3 ; t-BuO$^{\ominus}$ / t-BuOH → dienolate

(iii)

SeO_2 → 1, 2-dione; Li/liqNH_3 → bis-enolate

(iv)

Me_3SiCl / Et_3N → $OSiMe_3$

Silyl enol ether (specific enolate)

(v)

Pyrrolidine (N–H) /TsOH → enamine

(vi)

2 LDA → ; $^3O_2$ → ene-dione

Notes

2

Disconnection Design: The Retrosynthetic Analysis

2.1 A Few Definitions: Regarding Retrosynthesis

2.1.1 Disconnection (dix)

An analytical operation, which breaks a bond and converts a molecule into a possible readily available starting materials by dissection of a target molecule (TM). In short, it is an imaginary bond cleavage, corresponding to the reverse of a reaction.

Examples

(i) Ph–CH$_2$CH$_2$–OH $\xrightarrow{\text{dix}}$ $Ph^{\ominus}$ + $H_2C^{\oplus}$–CH_2–O–H

$Ph^{\ominus}$ ⇓ PhMgBr; $H_2C^{\oplus}CH_2OH$ ⇓ ethylene oxide

(ii) $CH_3CH(OH)CH_2CHO$ $\xrightarrow{\text{dix}}$ $CH_3CH^{\oplus}$–OH + $^{\ominus}CH_2CHO$

⇓ CH_3CHO; ⇓ CH_3CHO

2.1.2 Functional group interconversion (FGI)

Writing of one functional group (dummy) for another so that disconnection becomes possible to get the starting material available in the market.

Examples

(i) $-CH_2OH \xrightarrow[LiAlH_4]{FGI} -CO_2Et$

(ii) $-C(=O)-CH_2- \xrightarrow[Hg^{+2}/H_2SO_4]{FGI} -C\equiv C-$

(iii) $-COOH \xrightarrow[H_3O^{\oplus}]{FGI} -CN$

(iv) $-CH_2NMe_2 \xrightarrow[LiAlH_4]{FGI} -C(=O)-NMe_2$

(v) $CH_3COCH_2CO_2Et$ $\xrightarrow[H_3O^+]{\text{reversible FGI}}$ ethylene ketal (1,3-dioxolane) of $CH_3COCH_2CO_2Et$

(vi) cyclohexanone $\xrightarrow[H_3O^+]{\text{reversible FGI}}$ 1-pyrrolidinocyclohexene (enamine)

(vii) $HO(CH_2)_3CO_2H \xrightarrow{FGI}$ γ-butyrolactone

FGI is also necessary for functional group protection.

2.1.3 Synthon (SN)

A generalised fragment, usually an ion real or imaginary produced by disconnection. Synthon arises after imaginary bond cleavage may or may not be an intermediate in the real synthesis.

Examples

(i) $CH_3 \not| CH_2-OH \xrightarrow{dix} \underset{SN}{H_3\overset{\ominus}{C}} + \underset{SN}{\overset{\oplus}{C}H_2OH}$

(ii) $CH_3COCH_2 \not| CO_2Et$ … $\xrightarrow{dix} \underset{SN}{CH_3\overset{\oplus}{C}=O} + \underset{SN}{{}^{\ominus}CH_2CO_2Et}$

(iii) $R \not| CO_2H \xrightarrow{dix.} \overset{\oplus}{R} + \overset{\ominus}{C}O_2H$

(iv) $R \not| CH_2-NH_2 \xrightarrow{dix.} \overset{\oplus}{R} + \overset{\ominus}{C}H_2-NH_2$

Sometimes synthon may be a neutral species. Neutral synthons in the sense that these type of neutral synthons cannot be developed within a molecule directly but may be developed indirectly from other groupings or molecules (which are synthetic equivalents). Neutral synthons/synthetic equivalents are actually functional group interconvesion (FGI) and/or functional group addition.

Examples

(i) $H_3C-\underset{\underset{O}{\|}}{C}-CH_2-CH_3 \xrightarrow[Hg^{+2}/H_2SO_4]{FGI} H_3C-C\equiv C-CH_3$

SN SE

(ii) $CH_3-CH=CH-CH_3 \xrightarrow[Pb(OAc)_4]{FGI} CH_3-\underset{HO_2C}{CH}-\underset{CO_2H}{CH}-CH_3$

SN SE

(iii) (cyclohexenone) $\xrightarrow[(ii)\ H_3O^+/\Delta]{(i)\ Na/liq.NH_3}$ (anisole, OMe)

SN SE

Sometimes neutral intermediates which are not available in the market may be considered as neutral synthons.

(i) $:CCl_2 \Rightarrow H\overset{\ominus}{O}/CHCl_3$

SN SE

(ii) (benzyne) $\xrightarrow{HNO_2}$ (anthranilic acid, CO_2H, NH_2)

SN SE

2.1.4 Reconnection (rec)

During retroanalysis it is the join up a bond in the target molecule which will be broken during the forward synthesis and this operation is called reconnection.

Examples

(i) CHO (TM) $\xrightarrow[O_3/Me_2S]{rec.}$ (alkene)

(ii) (keto-aldehyde, O, CHO) TM $\xrightarrow[O_3/Me_2S]{rec.}$ (1-methylcyclohexene)

(iii) (cyclodecane-1,6-dione, O, O) TM $\xrightarrow[O_3]{rec.}$ (octalin)

(iv) (butadiene) TM $\xrightarrow[\Delta\ or\ hv]{rec.}$ (cyclobutene)

(v) (pentenal, H, O) TM $\xrightarrow[\Delta]{rec.}$ (allyl vinyl ether, O)

2.1.5 Synthetic equivalent (SE)

The actual compound (recognisable starting material which is available in the market or easy to derive) carrying out the function of a synthon (which is charged species usually) as it cannot itself be used because of its instability. Synthetic equivalent is a logical precursor of a target molecule.

Examples

(i) $CH_3 - CH_2 - OH \xRightarrow{dix} \overset{\ominus}{C}H_3 + \overset{\oplus}{C}H_2 - OH$

SN ⇓ CH_3MgBr (SE); SN ⇓ $H_2C=O$ (SE)

(ii) [aldol: 3-hydroxybutanal] $\xRightarrow{dix}$ [hydroxy-ethyl cation] + [enolate of acetaldehyde]

SN ⇓ CH_3-CHO (SE); SN ⇓ CH_3-CHO (SE)

(iii) [2,6-heptanedione] $\xRightarrow{dix}$ [acetone enolate anion] + [butanone cation]

SN ⇓ $CH_3COCH_2CO_2Et$ (SE); SN ⇓ $CH_2=CH-CO-CH_3$ (SE)

2.1.6 ⟹ Symbol (double lined hollow arrow)

Implies, the reverse of a chemical reaction. Usually FGI, FGA, **dix** written over it.

Examples

(i) $CH_3 - CH_2 - CH_2 - CH_3 \xRightarrow[H_2/Pd]{FGA} CH_3 - CH = CH - CH_3$

(ii) [benzene–H] $\xRightarrow[dil.H_2SO_4/\Delta]{FGA}$ [benzene–SO_3H]

(iii) $- CH_2OH \xRightarrow[LiAlH_4]{FGI} - CO_2Et$

(iv) $-\underset{\underset{O}{\parallel}}{C}-CH_2- \xRightarrow[Hg^{+2}/H_2SO_4]{FGI} -C \equiv C-$

2.1.7 Reagent

A compound which reacts to give an intermediate in the planned synthesis (or conversion) or to give the target molecule itself. It is sometimes synthetic equivalent of a synthon. In short reagent is a real chemical compound used as the equivalent of a synthon.

(i) $H^{\ominus} \Longrightarrow LiAlH_4$ (reagent)

(ii) $\overset{\ominus}{C}H_3 \Longrightarrow CH_3MgBr$ (reagent)

(iii) $Br^{\cdot} \Longrightarrow NBS$ (reagent)

(iv) $\overset{\ominus}{C}H_2 \Longrightarrow Ph_3\overset{+}{P}-\overset{-}{C}H_2$ (reagent)

The reagent is also a compound which may act in the planned synthesis or conversion as for (i) condensing agent, (ii) cyclisation, (iii) ring-opening, (iv) ring-contraction, (v) ring-expansion, (vi) redox agent, (vii) dehydrating agent, (viii) hydrating agent, etc.

2.1.8 Functional group addition (FGA)

(i) Acyclic or alicyclic saturated hydrocarbons are usually unreactive. Disconnection of such hydrocarbons or their desired derivative are difficult because hydrocarbon is a compound having no functional group which is a guiding signal for disconnection. Therefore, some functional groups should be added in place of hydrogen in a proper position so that after disconnection, recognisable starting material i.e. synthetic equivalents are available which may be subsequently converted to hydrocarbon or their desired derivative.

In short, the opposite of hydrogenolysis i.e. addition of a functional group in place of hydrogen is the FGA.

Examples

(a) *Retro-analysis*

(TM) H $\overset{FGA}{\Longrightarrow}$ OH $\Longrightarrow CH_3-\overset{\ominus}{C}H_2 + \overset{\oplus}{}$ OH

CH_3-CH_2MgBr (SE) $\quad CH_3CHO$ (SE)

$CH_3-CH_2MgBr \xrightarrow{FS} CH_3-CH_2-\underset{OH}{\underset{|}{C}H}-CH_3 \xrightarrow{(i)\ Al_2O_3/400°,\ (ii)\ Ni/H_2}$ (TM)

(b) *Retro-analysis*

(TM) $\overset{FGA}{\Longrightarrow}$ OH OH $\overset{FGI}{\Longrightarrow}$ H O O

H O SE $\Longleftarrow$ H O $\oplus$ + $\ominus$ O $\Longleftarrow$

EtO_2C O SE

Forward Synthesis

$\overset{\ominus}{O}Et$

CO_2Et

NaCl/DMSO

Krapcho decarboxylation

OHC

H

Zn(Hg)/
Conc. HCl

(c) FGA
Raney Ni/H_2

FGA
Li/liq NH_3

O OH OH OMe

FGI FGA FGI

(i) Na/liq. NH_3
(ii) H_3O^+/Δ

OMe

(ii) Some functional groups cannot be introduced directly into the benzene ring (e.g. $-NH_2$, $-CONH_2$, $-CO_2Et$) in that situation FGA is done in the benzene ring and subsequently this is converted to desired function.

Examples

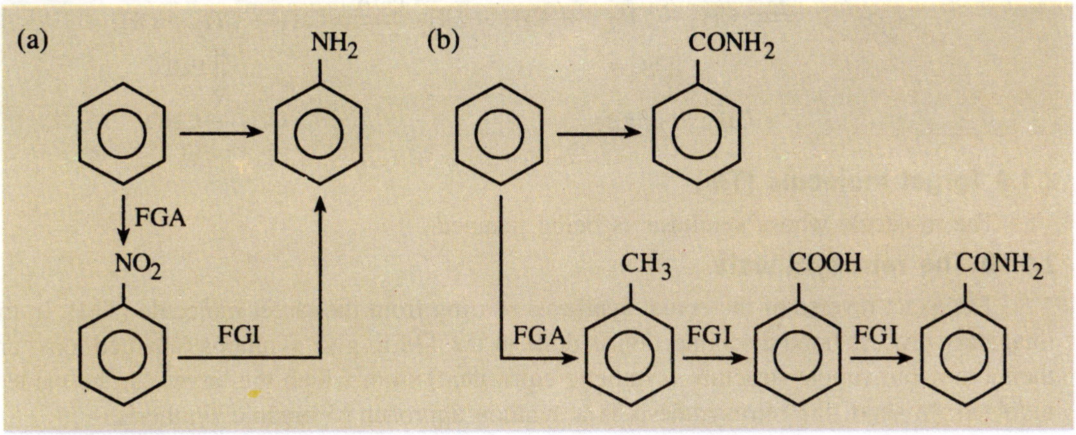

(a) Benzene → aniline (NH_2); via FGA to nitrobenzene (NO_2), then FGI to aniline.

(b) Benzene → benzamide ($CONH_2$); via FGA to toluene (CH_3), FGI to benzoic acid (COOH), FGI to benzamide ($CONH_2$).

(iii) FGA are of two types:

(a) FGA to the target molecule first and then disconnection is done which give recognisable synthetic equivalent.

(i) TM (3-methylbutan-2-one) $\xRightarrow{\text{FGA}}$ ethyl 2-methylacetoacetate (CO_2Et) $\xRightarrow{\text{dix}}$ enolate anion (CO_2Et) + $\overset{\oplus}{C}H_3 \Longrightarrow CH_3I$ (SE)

enolate anion $\Downarrow$ ethyl acetoacetate (CO_2Et), SE

(ii) 2-methylbutane (H) $\xRightarrow{\text{FGA}}$ butan-2-ol (OH) $\xRightarrow{\text{dix}}$ $CH_3-\overset{\ominus}{C}H_2$ + $\overset{\oplus}{C}H(OH)CH_3$

$CH_3-\overset{\ominus}{C}H_2 \Downarrow CH_3CH_2MgBr$ (SE)

$\overset{\oplus}{C}H(OH)CH_3 \Downarrow CH_3CHO$ (SE)

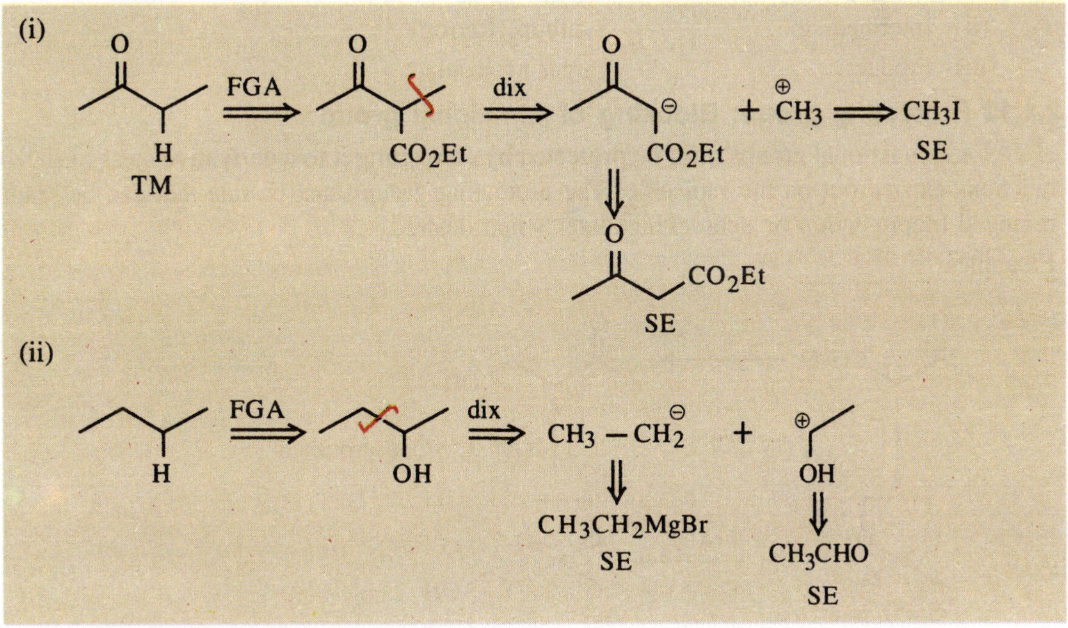

(b) Disconnection is done to the target molecule first and then FGA to the synthon so that recognisable synthetic equivalent is possible

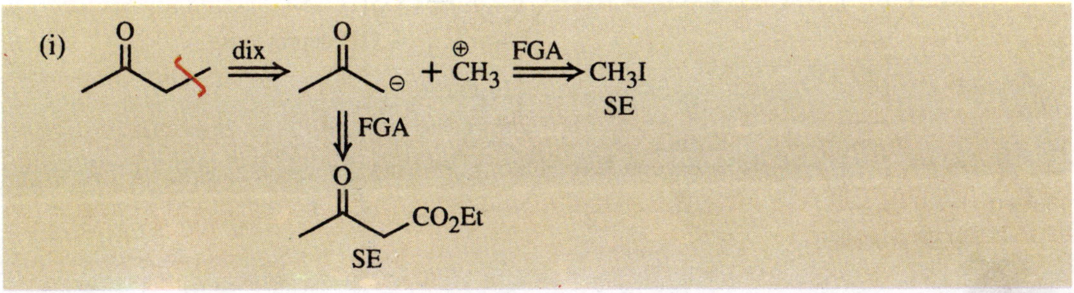

(i) butan-2-one $\xRightarrow{\text{dix}}$ acetonyl anion + $\overset{\oplus}{C}H_3 \xRightarrow{\text{FGA}} CH_3I$ (SE)

acetonyl anion $\Downarrow$ FGA: ethyl acetoacetate (CO_2Et), SE

(ii)

$$\text{(structure)} \overset{\text{dix}}{\Longrightarrow} CH_3-\overset{\ominus}{C}H_2 + CH_3-\overset{\oplus}{C}H_2 \overset{\text{FGA}}{\Longrightarrow} CH_3-CH_2-OH$$

$CH_3-\overset{\ominus}{C}H_2 \overset{\text{FGA}}{\Downarrow} CH_3CH_2MgBr$

$CH_3-CH_2-OH \overset{\text{FGI}}{\Downarrow} CH_3-CHO$

2.1.9 Target molecule (TM)

The molecule whose synthesis is being planned.

2.1.10 The retrosynthesis

The exact reverse of the actual synthesis starting from the target molecule (TM). In this imaginary process bond disconnection is done in the TM to give synthons (charged ions) and then a new but simple structure (synthetic equivalent) from which the target can be made is involved. In short, the retrosynthesis is a synthon approach to organic synthesis.

2.1.11 Direct synthesis versus retrosynthesis

	Direct Synthesis	Retrosynthesis
(i)	Starting material	Synthetic equivalent
(ii)	Intermediate	Synthon (Retron)
(iii)	Product	Target molecule

2.1.12 Protecting group: Blocking of functional group

Some functional groups must be protected by converting it to a derivative so as to survive reactions carried out on the molecule. The protecting group must be one that can be readily removed (deprotection or deblocking) later when desired.

Examples

(i) Keto ester (CO_2Et) ⟶ keto alcohol (OH)

Keto ester $\xrightarrow{\text{ethylene glycol (OH, OH) / dry HCl}}$ ketal ester (CO_2Et), C = O, protected $\xrightarrow{LiAlH_4}$ ketal alcohol (OH) $\xrightarrow{H_3\overset{\oplus}{O}}$ keto alcohol, C = O, deprotected

(ii) $HO-C_6H_4-CH_3 \longrightarrow HO-C_6H_4-CO_2H$

$HO-C_6H_4-CH_3 \xrightarrow{TsCl/Py} TsO-C_6H_4-CH_3$ (— OH protected) $\xrightarrow[\overset{\ominus}{O}H]{KMnO_4} TsO-C_6H_4-CO_2^{\ominus} \xrightarrow{H_3O^+} HO-C_6H_4-CO_2H$ (— OH deprotected)

2.1.13 Illogical Electrophile (IllE) and Illogical Nucleophile (IllNU)

Illogical means when normal polarity is reversed.

During disconnection of a covalent bond of a target molecule (TM), gives synthons which are usually charged ions may be either electrophilic or nucleophilic.

If the synthon either electrophilic or nucleophilic is in the right choice i.e. normal polarity is preserved than it is logical synthon either electrophilic or nucleophilic. The corresponding synthetic equivalent (SE) will be termed as logical synthetic equivalent.

If the synthon either electrophilic or nucleophilic is not in the right choice i.e. normal polarity is reversed then it is illogical synthon either electrophilic or nucleophilic. The corresponding synthetic equivalent will be termed as illogical synthetic equivalent.

Examples

Illogical electrophile

(i)

SN SE

O O Br CH_2

Br, H − C ≡ C , Br

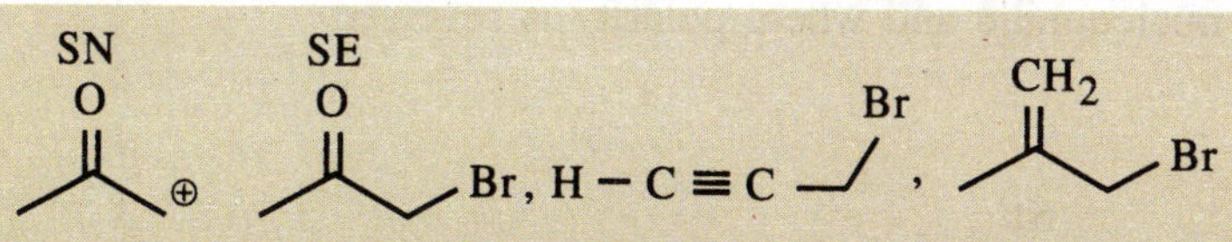

α-carbon adjacent to C = O is nucleophilic which is a normal polarity because of versatile enolate nature in base medium but polarity is reversed in monobromoacetone.

Propargyl bromide and 2-bromomethyl propene may be regarded as illogical electrophile in the sense, both the acetylenic and propene unit are FGI of acetyl group.

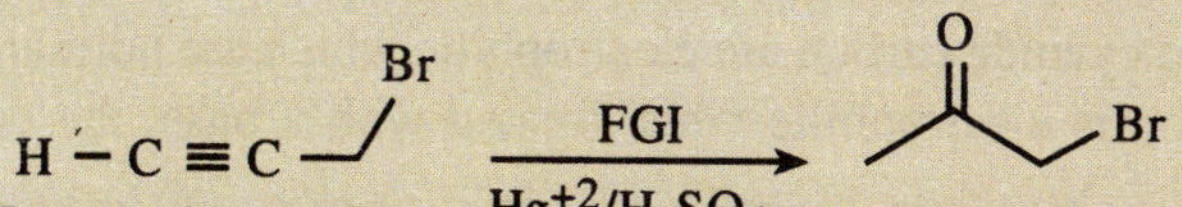

Br O

H − C ≡ C FGI → Br

Propargyl bromide Hg^{+2}/H_2SO_4

CH_2 O

Br FGI → Br

O_3

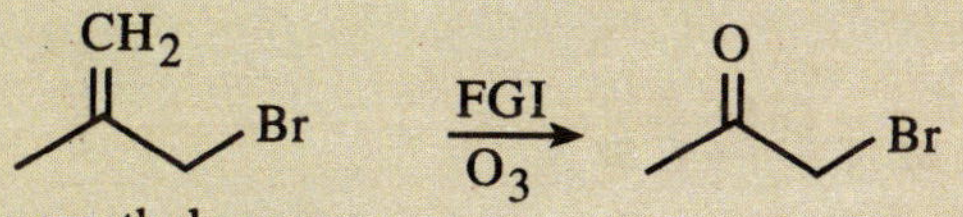

2-brmomethyl propene

(ii)

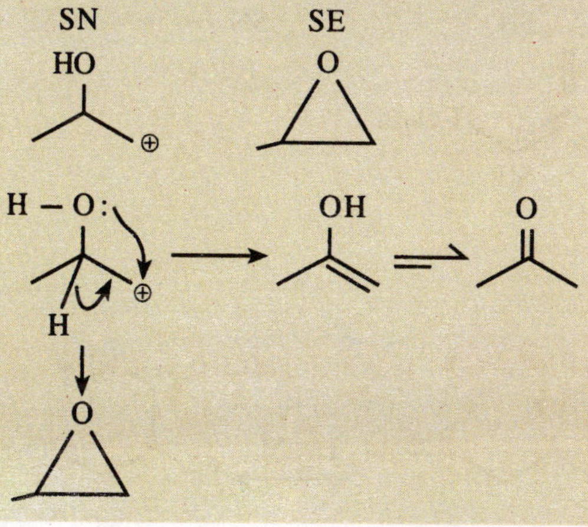

SN SE

HO O

H − O: OH O

H

O

'Ene' carbon or α-carbon of enol is nucleophilic which is a normal polarity but epoxidation of ene, the polarity is reversed since epoxy carbon is highly electrophilic.

Illogical Nucleophile

(i)

SN SE

C = O, carbon is electrophilic which is a normal polarity but when C = O is converted to cyclic thioketal and then strong base is treated, the C = O, carbon is converted to acylanion equivalent which is nucleophilic and where polarity is reversed.

(ii) SN SE

$\overset{\ominus}{C}O_2H$, $\overset{\ominus}{C}H_2-NH_2$ } $\overset{\ominus}{C}\equiv N$

Both the carboxyl carbon and cyanide carbon are electrophilic which are normal polarity but H − C ≡ N in which carbon is electrophilic when treated with a base, the conjugate base of hydrogen cyanide $\left(\overset{\ominus}{C}\equiv N\right)$ is nucleophilic.

$$H-\overset{\delta+}{C}\equiv N \xrightarrow{\overset{\ominus}{O}H} \overset{\ominus}{C}\equiv N$$

(TM) ⟹ Ph–⊕CH–CH$_2$–CO$_2$Et (SN) + $\overset{\ominus}{C}H_2-NH_2$ (SN) ⟹ $\overset{\ominus}{C}\equiv N$ (SE)

⇓

Ph–CH=CH–CO$_2$Et (SE)

Forward Synthesis

Ph–CH=CH–CO$_2$Et $\xrightarrow{\overset{\ominus}{C}N}$ Ph–CH(CN)–CH$_2$–CO$_2$Et $\xrightarrow{H_2/Pd}$ Ph–CH(CH$_2$NH$_2$)–CH$_2$–CO$_2$Et

(iii) SN SE

OMe

$Me_3C^{\ominus}\,Li^{\oplus}$

CH_3-Br

$H_3O^{\oplus}/\Delta$

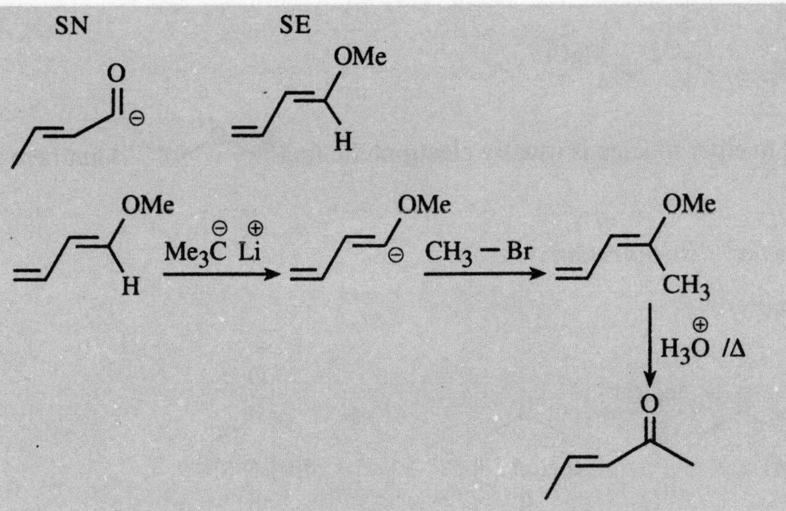

(iv) SN SE

OMe

OEt BrZn OEt

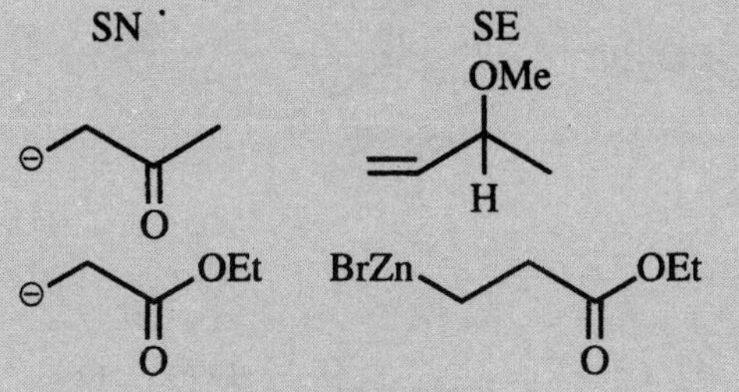

βC with respect to C = O is usually electrophilic either in Cl or in

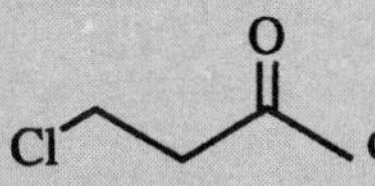

OMe

n BuLi

CH_3-I

H_3O^+/Δ

OEt HBr Br OEt Zn ether BrZn OEt

(i) CH_2O
(ii) H_3O^+

HO OEt

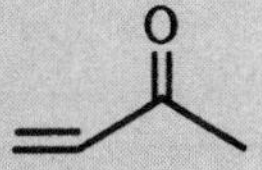

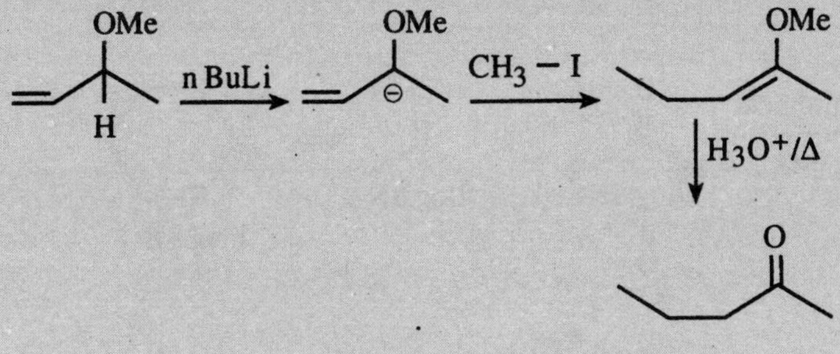

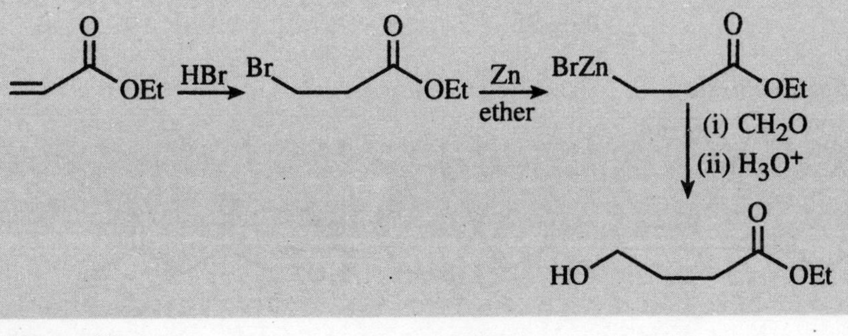

(v) 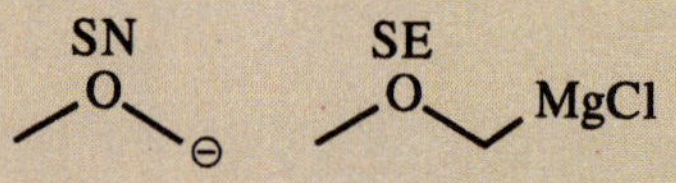 SN SE MgCl

αC adjacent to ether linkage is usually electrophilic (e.g. O Cl) but here polarity is reversed.

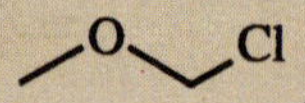

Logical and Illogical disconnection

(i) *Retro-analysis*

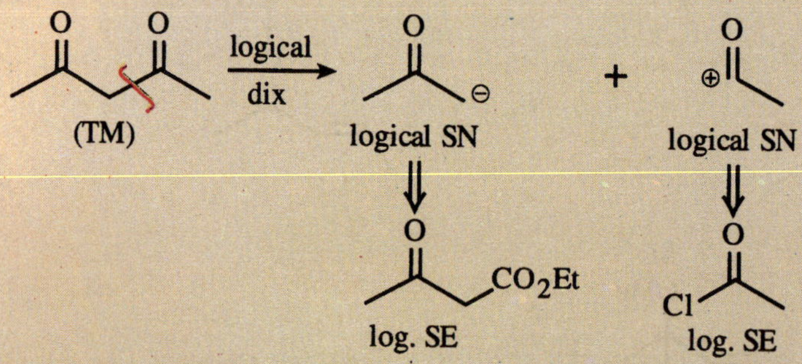

O O (TM) logical dix → O ⊖ logical SN + ⊕ O logical SN

⇓ O CO_2Et log. SE ⇓ O Cl log. SE

Forward Synthesis

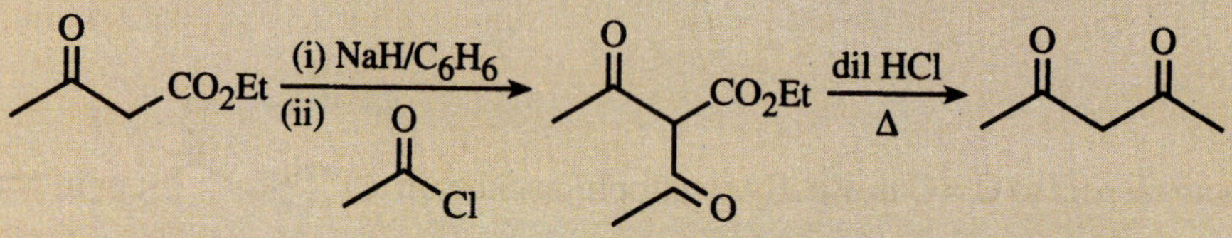

O CO_2Et (i) NaH/C_6H_6 (ii) O Cl → O CO_2Et O dil HCl Δ → O O

(ii) *Retro-analysis*

O O (TM) Illogical dix → O ⊕ Illog.SN + O ⊖ Illog.SN ⇒ S ⊖ S Illog.SE

⇓ O Cl Illog.SE

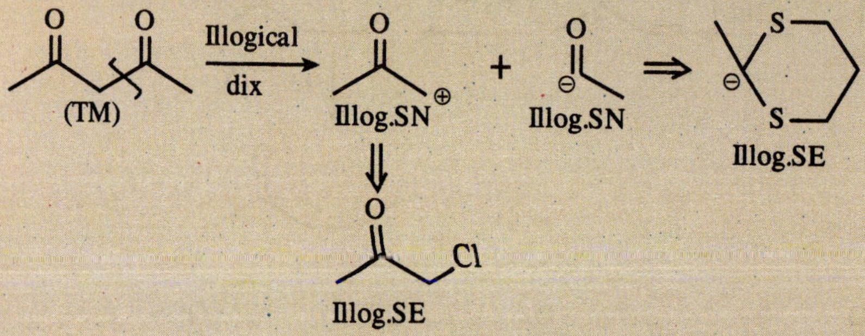

Forward Synthesis

O Cl + S ⊖ S → O S S Hg^{+2} H_3O^+ → O O

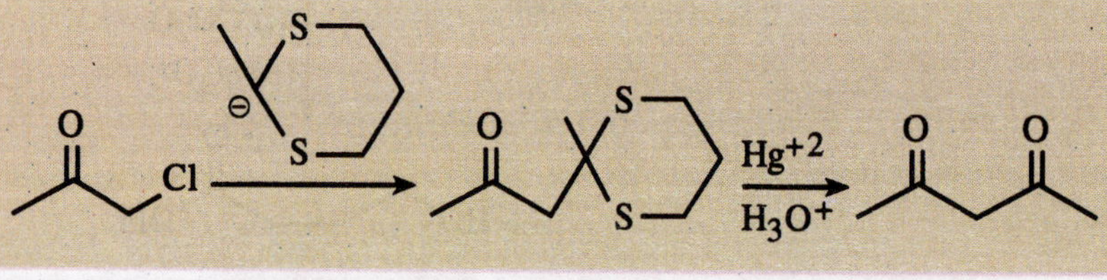

2. (i) *Retro-analysis*

(TM) —logical dix→ [$CH_3CO^{\oplus}$] log SN + [$^{\ominus}CH_2COOH$] log SN

log SN ⇓ CH_3COCl log SE; log SN ⇓ $CH_2(CO_2Et)_2$ log SE

Forward Synthesis

$CH_2(CO_2Et)_2$ —NaH/C_6H_6, CH_3COCl→ $CH_3COCH(CO_2Et)_2$ —(i) KOH (ii) H_3O^+/Δ→ CH_3COCH_2COOH

(ii) *Retro-analysis*

(TM) —Illogical dix→ [$CH_3COCH_2^{\oplus}$] Illog.SN + $^{\ominus}CO_2H$ Illog.SN

Illog.SN ⇓ CH_3COCH_2Cl Illog.SE; Illog.SN ⇓ KCN Illog.SE

Forward Synthesis

CH_3COCH_2Cl —KCN, CH_3COCH_3→ CH_3COCH_2CN —H_3O^+→ CH_3COCH_2COOH

3. (i) *Retro-analysis*

$CH_3COCH_2CH_2COOH$ —Mixed dix→ [$CH_3COCH_2^{\oplus}$] Illog.SN + $\overset{\ominus}{C}H_2-CO_2H$ log.SN ⇒ $CH_2(CO_2Et)_2$ log.SE

Illog.SN ⇓ CH_3COCH_2Cl Illog.SE

Forward Synthesis

$CH_2(CO_2Et)_2$ —(i) $\overset{\ominus}{O}Et$, (ii) CH_3COCH_2Cl→ $CH_3COCH_2CH(CO_2Et)_2$ —(i) KOH, (ii) H_3O^+/Δ→ $CH_3COCH_2CH_2COOH$

(ii) *Retro-analysis*

(TM) $CH_3COCH_2CH_2COOH$ ⇒ (Mixed dix) $CH_3COCH_2CH_2^{\oplus}$ (log.SN) + $\overset{\ominus}{C}O_2H$ (Illog.SN) ⇒ KCN (Illog.SE)

log.SN ⇓ $CH_3COCH{=}CH_2$ (log.SE)

Forward Synthesis

(TM) $CH_3COCH{=}CH_2$ —$\overset{\ominus}{C}N$ / $AlEt_3$→ $CH_3COCH_2CH_2CN$ —H_3O^+→ $CH_3COCH_2CH_2COOH$

(iii) *Retro-analysis*

(TM) $CH_3COCH_2CH_2COCH_3$ ⇒ (Mixed dix) $CH_3CO^{\oplus}$ (Illog.SN) + $^{\ominus}CH_2COCH_3$ (Log.SN)

Illog.SN ⇓ $HC{\equiv}C{-}CH_2Br$ Or $CH_2{=}C(CH_3)CH_2Br$ (Illog.SN)

Log.SN ⇓ $EtO_2CCH_2COCH_3$ (Log.SE)

Forward Synthesis

$CH_3COCH_2CO_2Et$ —(i) $\overset{\ominus}{O}Et$, (ii) $HC{\equiv}C{-}CH_2Br$→ $CH_3COCH(CO_2Et)CH_2C{\equiv}CH$ —Hg^{+2} / H_2SO_4→ $CH_3COCH_2CH_2COCH_3$

(i) $\overset{\ominus}{O}Et$ (ii) CH_2=C(CH$_3$)CH$_2$Br ; CO_2Et ; CO_2Et CH_2 ; (i) O_3 (ii) dil.HCl/Δ

4. (i) *Retro-analysis*

HO … OEt (TM) $\xrightarrow[\text{dix}]{\text{Mixed}}$ H–O… $\oplus$ + $\ominus$ … OEt ⇒ BrZn … OEt

Illog.SN log.SN log.SE

⇓

Illog.SE

Forward Synthesis

BrZn … OEt ; $\ominus$ O … OEt $\xrightarrow{H_3O^+}$ HO … OEt

(ii) *Retro-analysis*

HO … OEt $\xrightarrow[\text{dix}]{\text{Mixed}}$ HO – $\overset{\oplus}{C}H_2$ + $\ominus$ … OEt ⇒ BrZn … OEt

log.SN Illog.SN Illog.SE

⇓

CH_2O

log.SE

Forward Synthesis

BrZn … OEt $\xrightarrow{HCHO}$ $\overset{\ominus}{O}$ – CH_2 … OEt $\xrightarrow{H_3O^+}$ HO … OEt

2.1.14 Umpolung synthesis

The transformation of an electrophilic carbon (C = O) into a nucleophilic one (e.g. the deprotonated thiane) and the reverse process, alkylation of an aldehyde on the C = O, C using an electrophilic alkylating agent are said to involve umpolung synthesis.

$$RCHO \xrightarrow[BF_3]{HS(CH_2)_3SH} \text{2-R-1,3-dithiane} \xrightarrow{nBuLi} \text{2-R-1,3-dithian-2-yl anion} \xrightarrow{CH_3I} \text{2-R-2-}CH_3\text{-1,3-dithiane} \xrightarrow[H_3O^{\oplus}]{Hg^{+2}} R(H_3C)C{=}O$$

2.1.15 One-group disconnections

Disconnection of a target molecule aiming one particular functional group i.e. disconnection in which involve only one functional group is the one group disconnection.

Examples

(i)

$$CH_3CH_2COCH_3 \Rightarrow CH_3CH_2\overset{\oplus}{C}{=}O + \overset{\ominus}{C}H_3 \Rightarrow CH_3MgBr \text{ (SE)}$$

$$CH_3CH_2\overset{\oplus}{C}{=}O \Downarrow CH_3CH_2COOH \Rightarrow CH_3CH_2CN \text{ (SE)}$$

(ii)

$$CH_3CH_2COOH \Rightarrow CH_3-\overset{\oplus}{C}H_2 + \overset{\ominus}{C}O_2H \Rightarrow \overset{\ominus}{C}N \Rightarrow KCN \text{ (SE)}$$

$$CH_3-\overset{\oplus}{C}H_2 \Downarrow CH_3-CH_2-Br \text{ (SE)}$$

(iii)

$$CH_3CH_2C(CH_3)_2OH \Rightarrow CH_3-\overset{\ominus}{C}H_2 + \overset{\oplus}{C}(CH_3)_2OH \Rightarrow O{=}C(CH_3)_2$$

$$CH_3-\overset{\ominus}{C}H_2 \Downarrow CH_3-CH_2MgBr \text{ (SE)}$$

Forward Synthesis

$$(CH_3)_2C{=}O \xrightarrow[\text{(ii) } H_3O^+]{\text{(i) } CH_3CH_2MgBr} CH_3CH_2C(CH_3)_2OH$$

Why One Group?

New C – C bond is forming adjacent to functional group i.e. 1C away from functional group.

2.1.16 Two-group disconnections

When a target molecule contain two functional groups and if disconnection is done using the two together i.e. using of one functional group to help to disconnect another is the two-group disconnections.

Examples

(i)

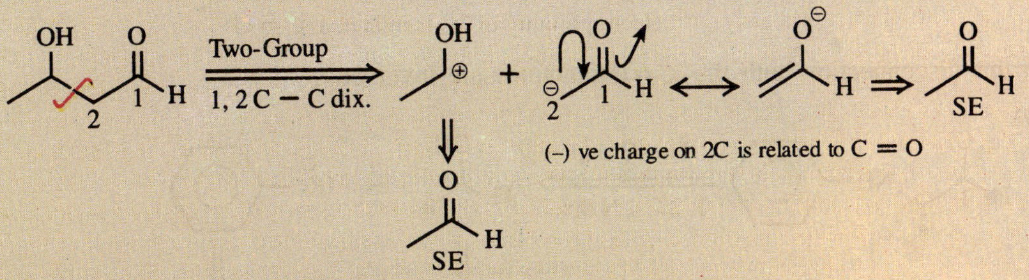

Two-Group 1, 2 C – C dix.

SE

(–) ve charge on 2C is related to C = O

In this disconnection, both the functional groups C = O and – OH are involved.

(ii)

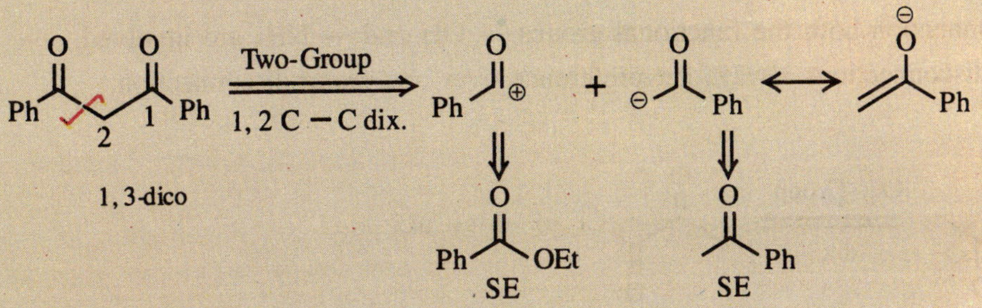

Two-Group 1, 2 C – C dix.

1, 3-dico

In this disconnection both the carbonyl functions are involved.

(iii)

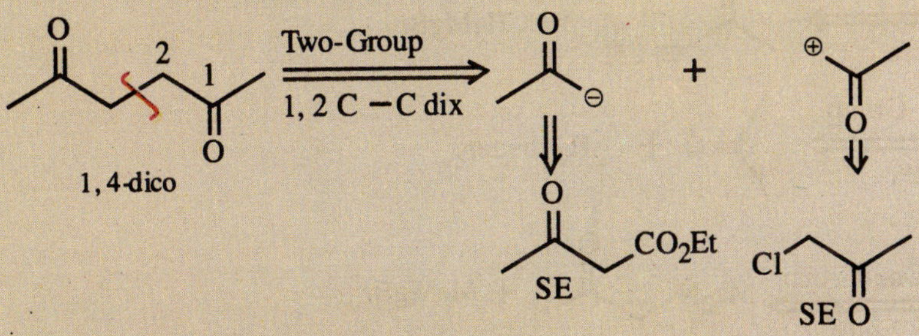

Two-Group 1, 2 C – C dix

1, 4-dico

In this disconnection both the carbonyl functions are involved.

electrophilicity of halo carbon is related to adjacent C=O

(iv) 1,5-dico —Two-Group, 1, 3 C−C dix.⟹ acetone enolate (⊖) ⟹ $CH_3COCH_2CO_2Et$ SE
+ ⊕$CH_2CH_2COCH_3$ ⟹ $CH_2{=}CHCOCH_3$ (3, 2, 1) SE

electrophilicity of 3C is related to C = O

In this disconnection both the C=O functions are involved

(v) $PhCH(OH)CH_2NH{-}C_6H_5$ —Two-Group, 1, 2 C – N dix.⟹ $PhCH(OH)CH_2^{\oplus}$ + $\overset{\ominus}{HN}{-}C_6H_5$

⇓ styrene oxide (Ph) SE; ⇓ $H_2N{-}C_6H_5$ SE

In this disconnection both the functional groups – OH and – NH_2 are involved.
Two-group disconnections always get preference over one-group disconnection.

Examples

(i) $CH_3COCOCH_3$ —One-Group dix⟹ $CH_3COCOCl$ + Me_2CuLi

$CH_3COCOCH_3$ —Two-Group dix⟹ 2 CH_3CO_2Et (better)

(ii) $(CH_3)_2C(OH)C{\equiv}CH$ —One-Group dix⟹ $CH_3COC{\equiv}CH$ + CH_3MgBr

$(CH_3)_2C(OH)C{\equiv}CH$ —Two-Group dix⟹ $(CH_3)_2C{=}O$ + C_2H_2 (better)

(iii) $Me_2NCH_2CH(OH)CH_3$ —One-Group dix⟹ Me_2NCH_2CHO (H) + MeMgBr

$Me_2NCH_2CH(OH)CH_3$ —Two-Group dix⟹ Me_2NH + propylene oxide (better)

2.1.17 Difference between conversion and synthesis

There are two main objectives of retrosynthetic analysis of a TM.

(i) to find out the most logical synthetic equivalent.

(ii) to find out the synthetic route or the path i.e. the sequence of reactions to be done during forward process which is indicated by synthons.

In the synthesis usually only TM is given but in the conversion both the starting material and TM are given. Of the various disconnection possibilities of a TM, only that disconnection is accepted which will give the synthetic equivalent correspond to the given starting material in the question in case of conversion.

Illustration

(TM) Ethyl methyl ketone $\xRightarrow{FGI}$ —C≡C— $\Rightarrow$ $2CH_3I$ (SE) + H—C≡C—H (SE)

$\Rightarrow$ $CH_3COCH_2CO_2Et$ (SE) + CH_3Br (SE)

$\Rightarrow$ CH_3COCH_2Br (SE) + Me_2CuLi (SE)

$\Rightarrow$ CH_3Br (SE) + CH_3CH_2Br (SE) + HCHO (SE) $\Rightarrow$ 1,3-dithiane (H, H)

$\Rightarrow$ CH_3CH_2Br (SE) + CH_3CHO (SE) $\Rightarrow$ 2-methyl-1,3-dithiane (H_3C, H) Or CH_3Br (SE) + CH_3CH_2CHO (SE) $\Downarrow$ 2-ethyl-1,3-dithiane (CH_3-H_2C, H)

$\Rightarrow$ CH_3CHO (SE) + CH_3CH_2MgBr (SE) Or CH_3CH_2CHO (SE) + CH_3MgBr (SE)

$\Rightarrow$ CH_3COCl (SE) + Et_2CuLi (SE) Or CH_3CH_2COCl (SE) + Me_2CuLi (SE)

$\Rightarrow$ CH_3CN (SE) + CH_3CH_2MgBr (SE) Or CH_3CH_2CN (SE) + CH_3MgBr (SE)

Retrosynthetic analysis shows that there are several methods (routes) with various synthetic equivalents for the synthesis of target molecule ethylmethyl ketone. But any one synthetic equivalent may be given as starting material for the conversion to ethylmethyl ketone and consequently one has to take the opportunity of that retroanalysis involving the synthetic equivalent (starting material) in the question for the forward conversion.

2.2 Activation and Controlling during disconnections

Purpose: Choice of right synthetic equivalent and proper technical condition for best possible overall yield in the synthetic route.

Activation and controlling are also necessary

(i) If there is a regioselective problems during disconnection.

(ii) If there is a chemoselective problems.

(iii) To avoid further (repeated) reaction.

(iv) To avoid self-condensation.

Examples

1. During disconnection, if alkyl group becomes negative charge synthon then it indicates activation requires (to abstract proton) and in that situation a protecting electron withdrawing group (e.g. – CO_2 Et) is added as FGA.

Suppose we want to synthesize the following two TM by Claisen ester condensation.

CO_2H (TM) A

(TM) B

For A

Retro-analysis

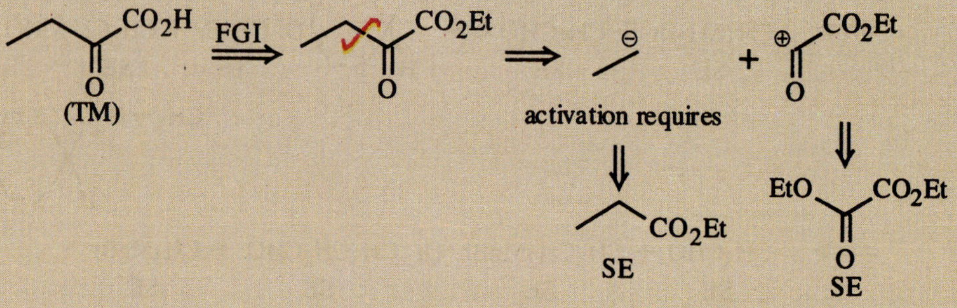

CO_2H (TM) $\xrightarrow{FGI}$ CO_2Et ⟹ ⊖ + ⊕ CO_2Et

activation requires ⇓ CO_2Et SE

⇓ EtO CO_2Et SE

Forward Synthesis

CO_2Et $\xrightarrow[CO_2Et-CO_2Et]{^{\ominus}OEt}$ CO_2Et, CO_2Et $\xrightarrow[\Delta]{(i)\ KOH,\ (ii)\ dil\ HCl}$ CO_2H

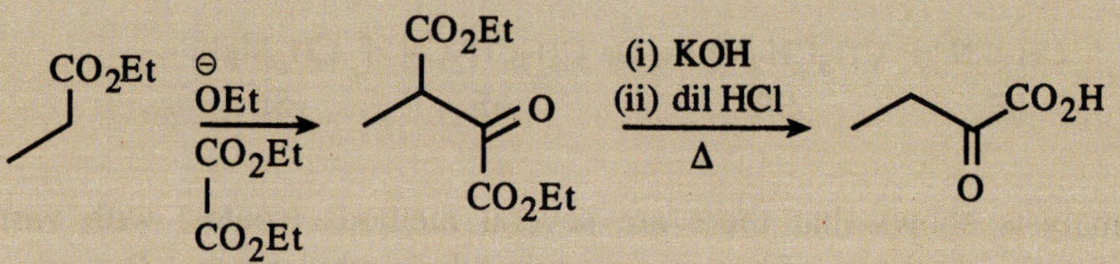

For B

Retro-analysis

(TM) ⟹ activation requires ⇓ FGA EtO_2C–CH₂CH₂CH₂–CO_2Et SE (Prepared from DEM) + ⊕ ⊕ O O ⇓ EtO OEt O O SE

Forward Synthesis

EtO_2C–CH₂CH₂CH₂–CO_2Et + EtO OEt O O $\xrightarrow{\overset{\ominus}{O}Et}$ EtO_2C CO_2Et O O $\xrightarrow[\Delta]{\text{dil HCl}}$ O O

2. Enolate and enamine are activation of 'ene'.

(i) During the disconnection of C=O compound at the αβ C – C bond, the negative charge synthon at the keto methylene group gives enolate which is activated ene. But this enolate cannot be alkylated with alkyl halides because of self condensation nature of enolate (aldol condensation). Therefore, here control is required by converting it to stabilised enolate (to arrest the self-condensation) by adding protecting Ew group (– CO_2Et) which is FGA.

Actually, stabilised enolate is further activation of normal enolate and controlling.

O α β (TM) ⟹ O ⊖ further activation required + $\overset{\oplus}{C}H_3$ ⟹ CH_3 – Br SE

⇓ FGA

O CO_2Et doubly activated right SE

O wrong SE

Forward Synthesis

(TM) (i) $\overset{\ominus}{O}Et$ (ii) CH_3-Br → CO_2Et → dil HCl, Δ

This also explain why highly activated stabilised enolate or carbanion is needed for Michael condensation where milder condition is necessary because of, in addition to avoid self condensation, it will also stop the subsequent aldol or Claisen ester condensation.

(ii) CO_2Et (TM) ⇒ ⊖ + ⊕ CO_2Et ⇓ Cl CO_2Et SE

N right SE   O⊖ wrong SE

Now if we want to alkylate cyclohexanone enolate with the α-haloester in base medium, Darzen reaction will occur preferentially rather than alkylation.

This is because of more acidic nature of α-haloester than cyclohexanone.

H Cl OEt O $\overset{\ominus}{O}Et$ → Cl OEt O⊖ → CO_2Et O⊖ Cl → O CO_2Et

Darzen product

O⊖ H + Cl OEt O ⇌ O + Cl OEt O⊖

Therefore, highly activated and protected-one, here enamine is the best suitable SE.

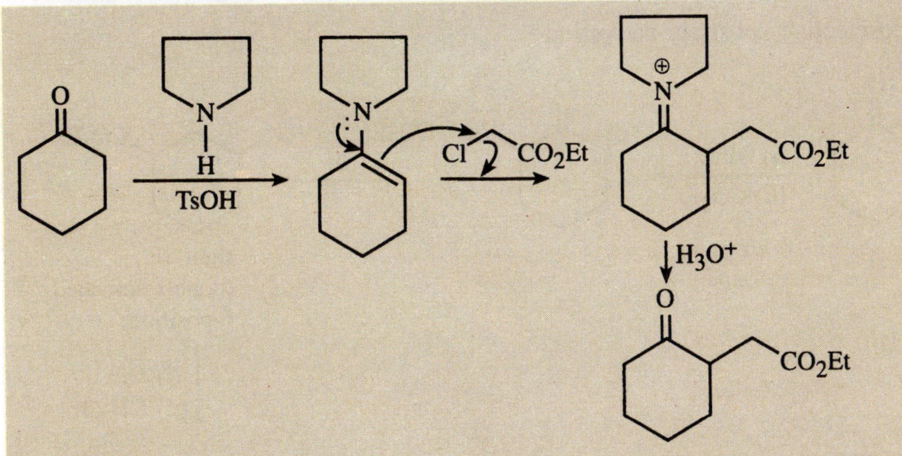

TsOH; H_3O^+; $Cl\frown CO_2Et$

3. TM $\xrightarrow{FGI}$ ... $\Rightarrow$ ... + $^{\ominus}$—CO_2H $\Rightarrow$ SE

right SE (ECA) Or wrong SE (DEM)

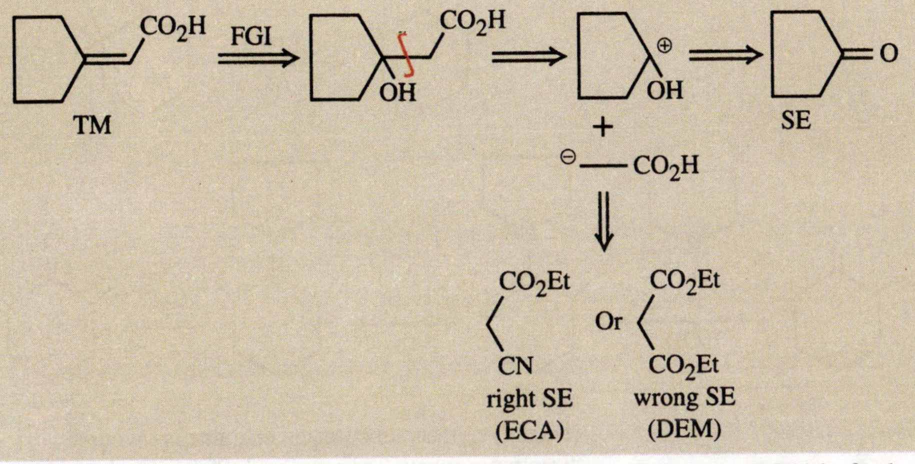

'DEM' is activated but it has no reaction with keto carbonyl. 'ECA' is further activated than 'DEM', 'ECA' has a reaction with keto carbonyl.

$\overset{\ominus}{OEt}$; (i) KOH; (ii) $H_3O^{\oplus}/\Delta$

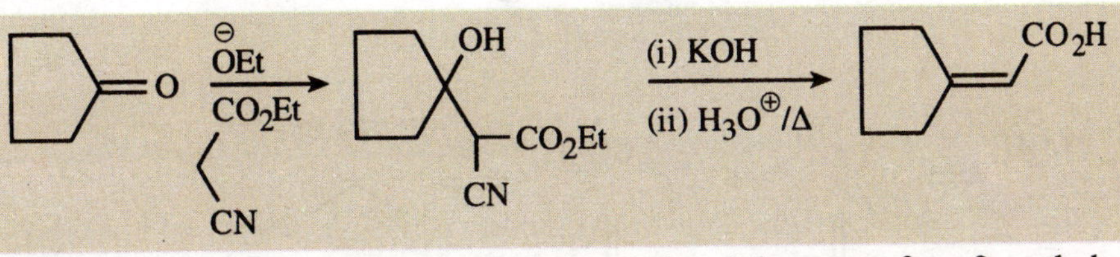

4. Suppose, we want to synthesize 2, 6-dimethyl cyclohexanone from 2-methyl cyclohexanone via enolate. During the alkylation in addition to the desired TM gem dialkylated product is also possible. Therefore, regioselective control is necessary.

(i) Technical controlled: LDA (X'S), THF /–78°C; $CH_3 - Br$

right SE (Kinetically controlled enolate)

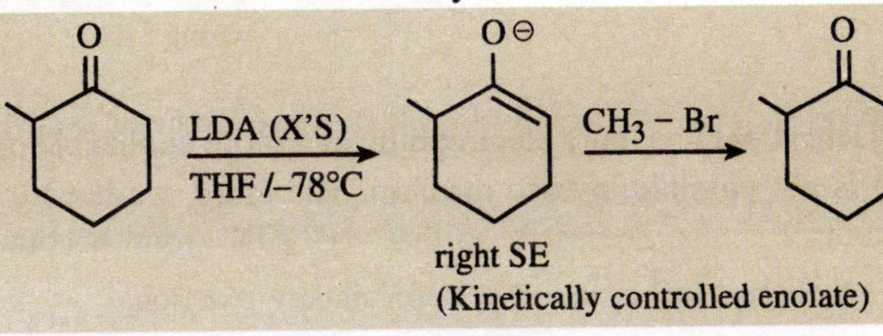

Slow addition of ketone to bulky LDA base with a definite time interval and then CH_3Br.

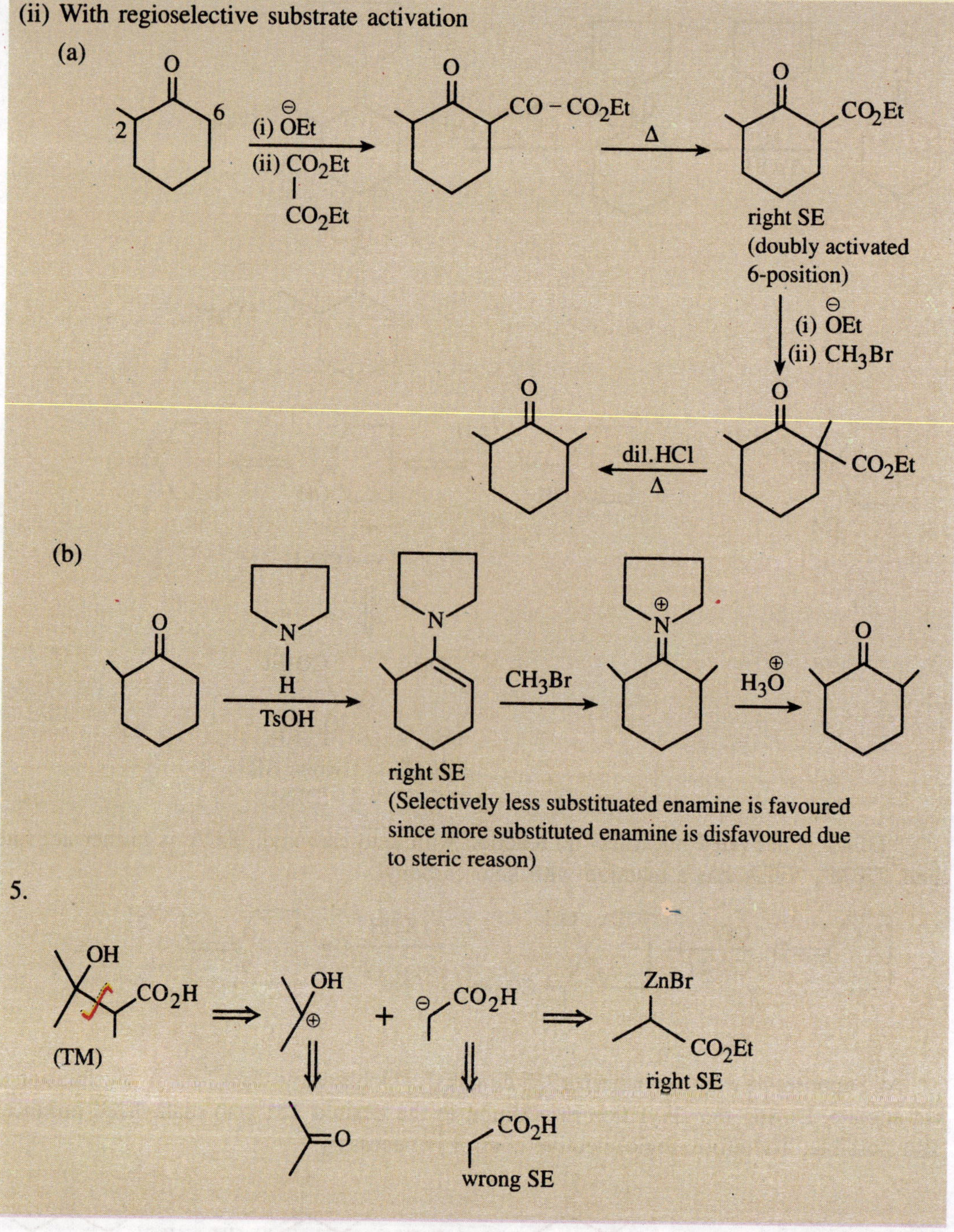

Here C=O is more electrophilic and more enolizable power than the acid. So, carboxylation is not possible in base medium. Therefore, we have to activate the α-position of acid and control is also necessary so that – COOH group becomes inert.

All these can be done by Reformatsky reaction.

$CH_3CH_2CO_2H \xrightarrow{P/Br_2} CH_3CH(Br)CO_2H \xrightarrow{EtOH/H^{\oplus}} CH_3CH(Br)CO_2Et \xrightarrow{Zn/ether} CH_3CH(ZnBr)CO_2Et$

$\xrightarrow{(CH_3)_2C=O} (CH_3)_2C(\overset{\ominus}{O}\overset{\oplus}{Zn}Br)CH(CH_3)CO_2Et \xrightarrow{(i)\ KOH,\ (ii)\ H_3O^{\oplus}} (CH_3)_2C(OH)CH(CH_3)CO_2H$

2.3 Synthons And Synthetic Equivalents

2.3.1 Nucleophilic synthons and synthetic equivalents

Synthon **Synthetic Equivalent**

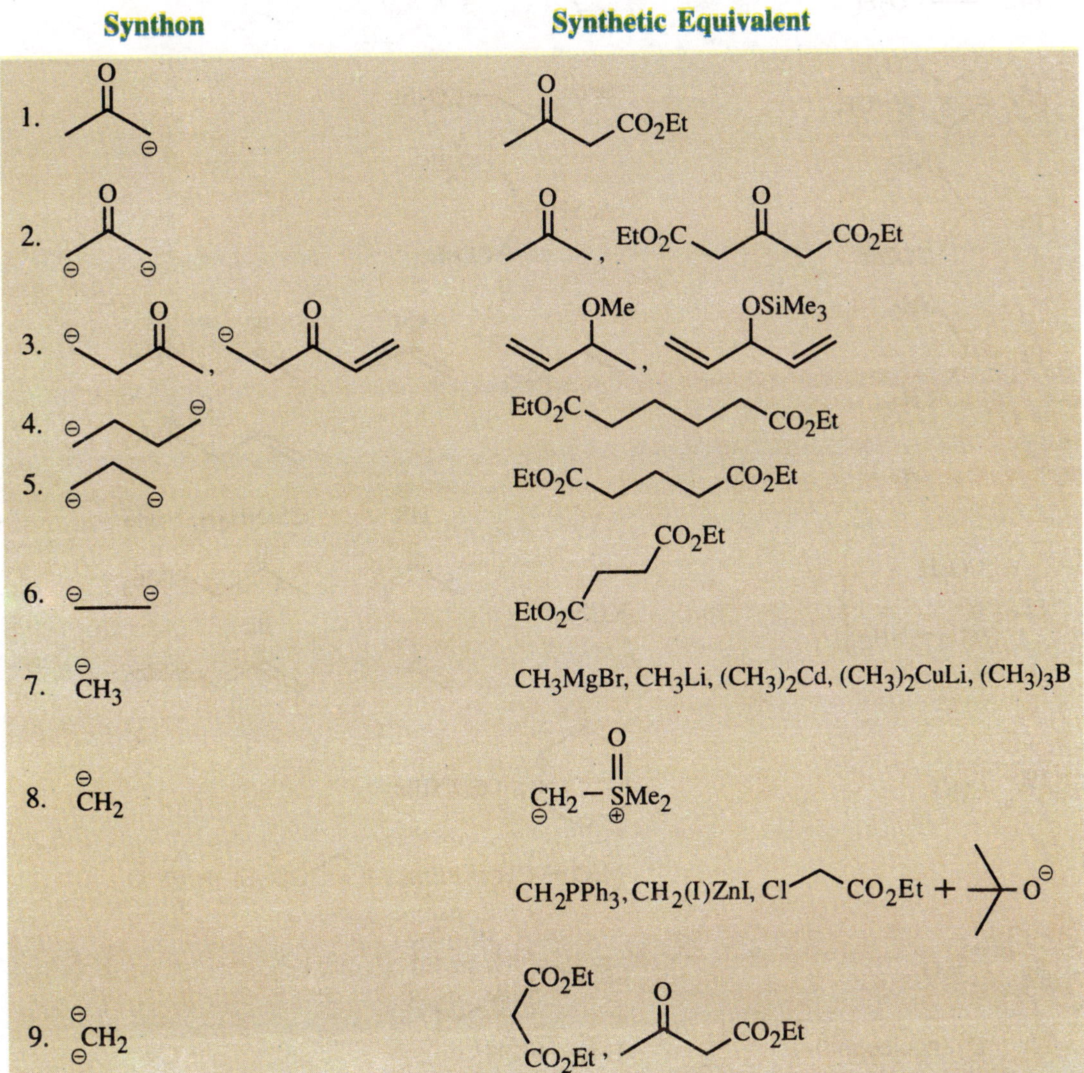

10. $CH_2=\overset{\ominus}{C}H$	$CH_2=CHMgCl$, $H-C\equiv\overset{\ominus}{C}$, $CH_2=CHSiMe_3$
11. $CH_3-\overset{\ominus}{C}=O$	$H-C\equiv\overset{\ominus}{C}$, 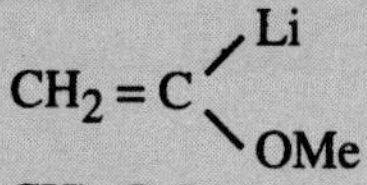 $CH_2=C(Li)(OMe)$
12. $CH_3-CH=\overset{\ominus}{C}H$	$(CH_3-CH=CH)_2CuLi$
13. $^{\ominus}CH_2-CO_2H$	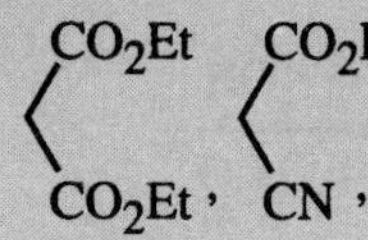 $CH_2(CO_2Et)_2$, $NC-CH_2-CO_2Et$, $(CH_3CO)_2O$, 2,4,4,6-tetramethyl-5,6-dihydro-4H-1,3-oxazine* *(commercially available)
$^{\ominus}CH_2-CHO$	$\overset{\oplus}{Li}\overset{\ominus}{C}\equiv COEt$
14. $^{\ominus}CH_2-CO_2Et$	$BrZn-CH_2-CO_2Et$ 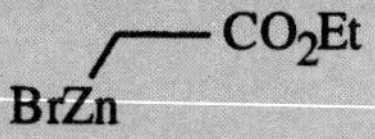
15. $^{\ominus}CH_2CH_2-CO_2Et$ or $^{\ominus}CH_2CH_2CH_2-OH$ 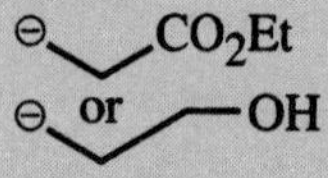 	$BrZn-CH_2CH_2-CO_2Et$
16. $H_2N-\overset{\ominus}{C}H-COOH$ 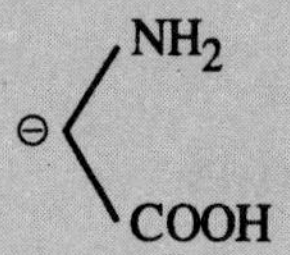 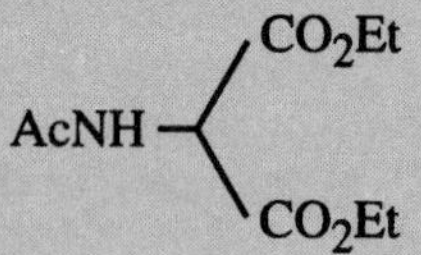	$AcNH-CH(CO_2Et)_2$
17. $H_2N-\overset{\ominus}{C}H-CH_3$	$CH_3CH_2NO_2$; SN: $CH_2=\overset{\ominus}{C}H$; SE: $CH_2=CH-CH_2MgCl$, $CH_2=CH-CH_2SiMe_3$
	$\overset{\ominus}{HS}$; $CS(NH_2)_2$, NaHS
18. $\overset{\ominus}{C}O_2H$ $\overset{\ominus}{C}H_2-NH_2$	KCN; 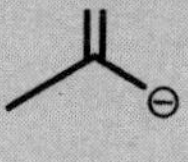   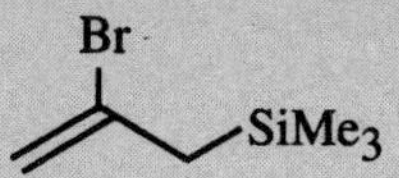 $CH_3-C(=CH_2)-\overset{\ominus}{C}H_2$, $CH_3-C(=CH_2)-CH_2SiMe_3$; $CH_2=\overset{-}{C}-\overset{-}{C}H_2$, $CH_2=C(Br)-CH_2SiMe_3$
19. $\overset{\ominus}{C}HO$	1,3-dithiane, $\overset{\ominus}{O}H/CHBr_3$ $MeO-CH=PPh_3$, $Cl-CH_2-CO_2Et + (CH_3)_3C-\overset{\ominus}{O}$
20. $\overset{\ominus}{\underset{\ominus}{C}}=O$	CH_3NO_2, 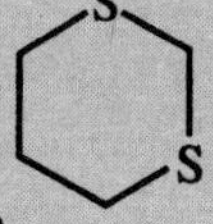 1,3-dithiane
21. 'O' (nucleophilic)	$H_2O_2 + \overset{\ominus}{O}H$

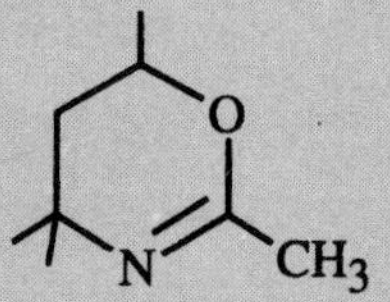

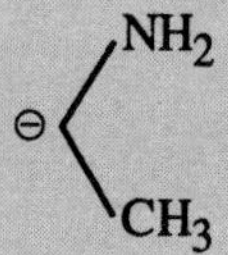

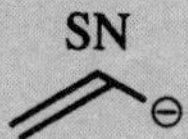

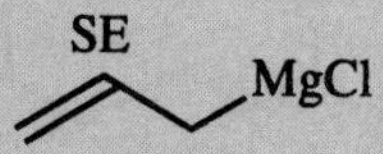

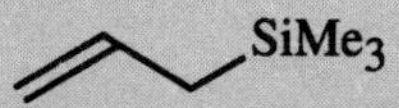

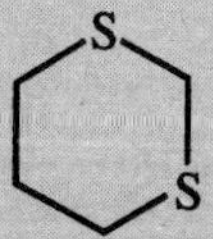

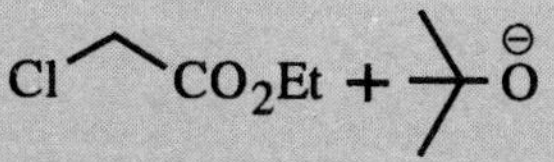

	Synthon	Synthetic equivalent
22.	$H^{\ominus}$	$LiAlH_4$, $LiAlH_4 + AlCl_3$, $NaBH_4$ B_2H_6, Me_2BH $Al(i-Pro)_3$, $LiAlH(OEt)_3$, $LiAlH(OCMe_3)_3$ DIBAL $(CH_2O + \overset{\ominus}{O}H)$, HCOOH
23.	$D^{\ominus}$	$LiAlD_4$, $NaBD_4$, B_2D_6, $Al(-OCDMe_2)_3$
24.	$\overset{\ominus}{C}Cl_3$	$CHCl_3 + \overset{\ominus}{O}H + LiCl$, $CCl_4 + Ph_3P$
25.	$^{\ominus}=\overset{\oplus}{N}(O^{\ominus})_2$	CH_3NO_2
26.	$H-O-\overset{\ominus}{C}H_2$	$MeO-CH_2MgCl$, $\overset{\ominus}{O}-\overset{\ominus}{C}H_2 2\overset{\oplus}{Li}$
27.	$\overset{\ominus}{N}H_2$	(phthalimide), CH_3CN (Ritter reaction)
28.	$Br^{\ominus}$	PBr_5, KBr + 18-Crown-6

Application of Nucleophilic Synthon and Synthetic Equivalent in Retrosynthetic Analysis

1. *Retro-analysis*

(TM) $\Rightarrow$ SN + SE (Cl–CH₂–CO–CH₃)

SN $\Downarrow$ SE ($CH_3COCH_2CO_2Et$)

Forward Synthesis

$CH_3COCH_2CO_2Et \xrightarrow[\text{(ii) } ClCH_2COCH_3]{\text{(i) } \overset{\ominus}{O}Et} CH_3COCH(CO_2Et)CH_2COCH_3 \xrightarrow[\Delta]{\text{dilHCl}} CH_3COCH_2CH_2COCH_3$

2. *Retro-analysis*

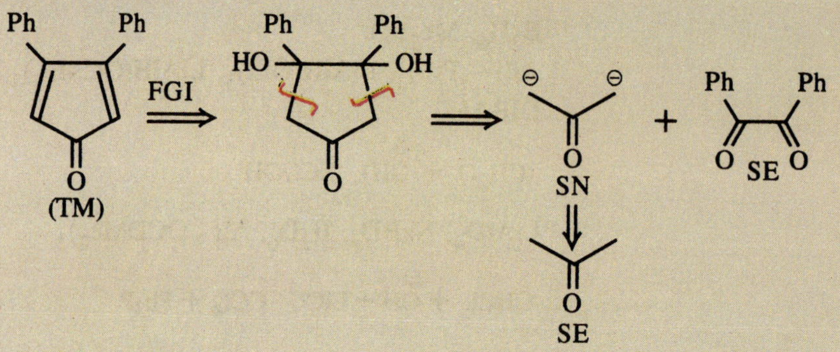

Ph Ph (TM) FGI ⟹ Ph Ph HO OH O ⟹ ⊖ ⊖ O SN ⇓ O SE + Ph Ph O SE O

Forward Synthesis

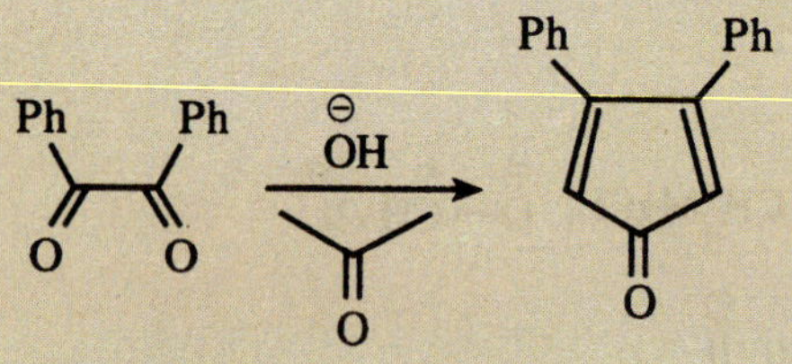

Ph Ph O O —$\overset{\ominus}{OH}$, acetone→ Ph Ph O

3. *Retro-analysis*

O O O (TM) ⟹ O ⊖ ⊖ SN ⇓ O EtO_2C CO_2Et SE + O O EtO OEt SE

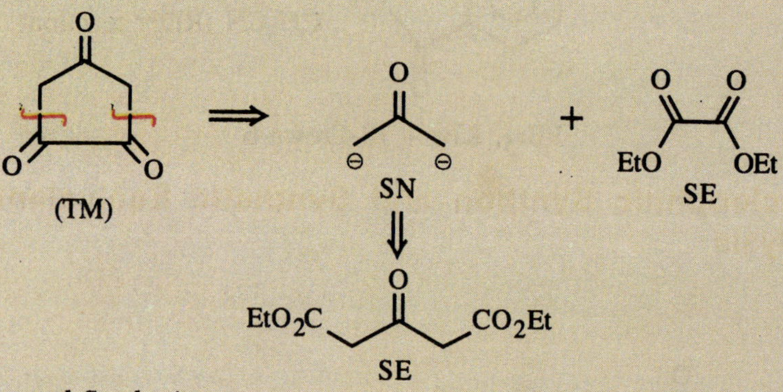

Forward Synthesis

O EtO_2C CO_2Et —$\overset{\ominus}{O}$Et(2eq), O O EtO OEt→ O EtO_2C CO_2Et O O —dil HCl, Δ→ O O O

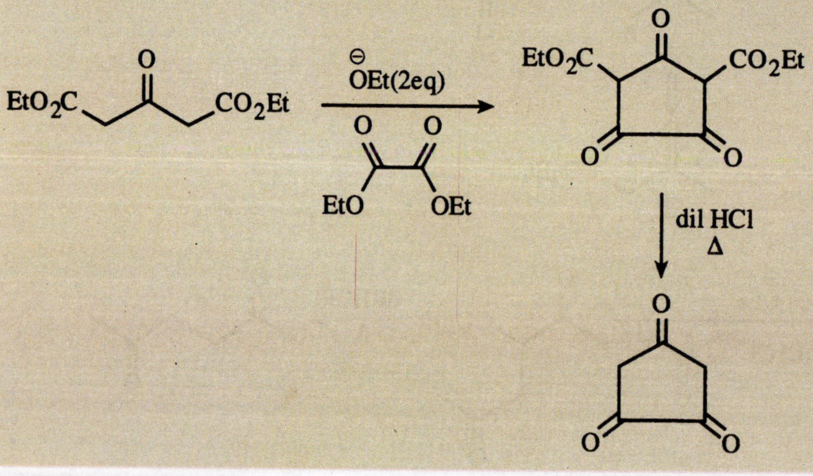

4. *Retro-analysis*

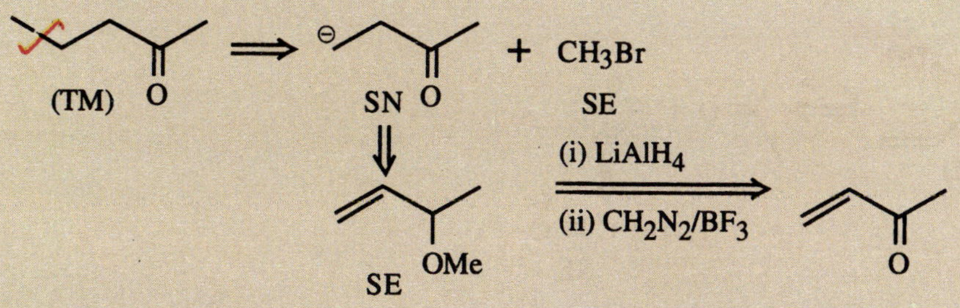

(TM) ⇒ $^{\ominus}CH_2CH_2COCH_3$ (SN) + CH_3Br (SE)

SN ⇓ $CH_2=CH-CH(OMe)CH_3$ (SE) $\xrightarrow[\text{(ii) } CH_2N_2/BF_3]{\text{(i) } LiAlH_4}$ $CH_2=CH-COCH_3$

Forward Synthesis

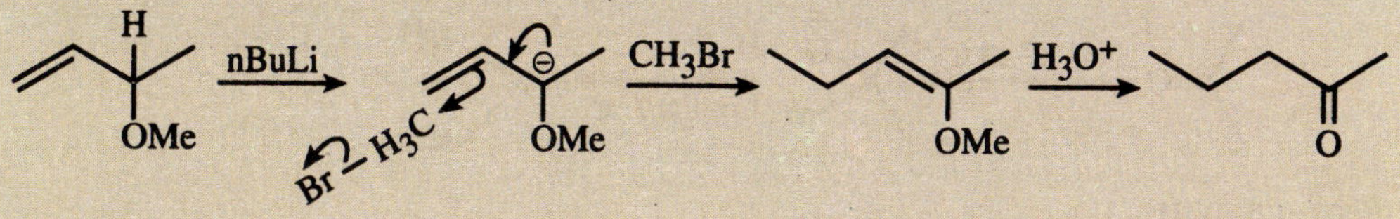

$CH_2=CH-CH(H)(OMe)CH_3$ $\xrightarrow{nBuLi}$ carbanion + H_3C-Br $\xrightarrow{CH_3Br}$ enol ether (OMe) $\xrightarrow{H_3O^+}$ ketone

5. *Retro-analysis*

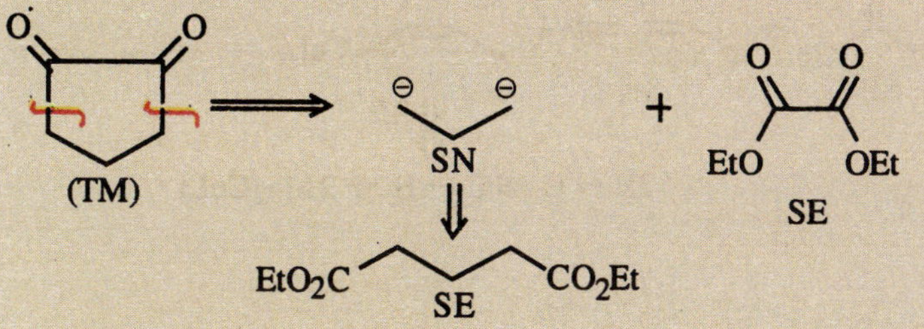

(TM) ⟹ dianion (SN) + EtO_2C-CO_2Et (EtO, OEt; SE)

SN ⇓ $EtO_2C-CH_2CH_2CH_2-CO_2Et$ (SE)

Forward Synthesis

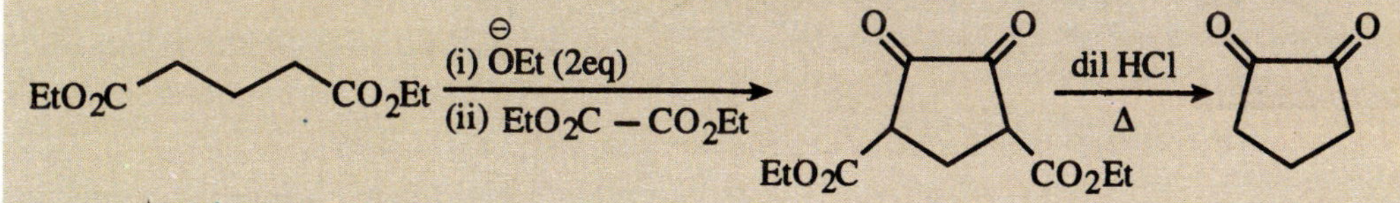

$EtO_2C(CH_2)_3CO_2Et$ $\xrightarrow[\text{(ii) } EtO_2C-CO_2Et]{\text{(i) } \overset{\ominus}{O}Et \text{ (2eq)}}$ diester cyclopentane-1,2-dione (EtO_2C, CO_2Et) $\xrightarrow[\Delta]{\text{dil HCl}}$ cyclopentane-1,2-dione

6. (i) *Retro-analysis*

OH (TM) ⇒ $(CH_3)_2C=O$ (SE) + $CH_3-\overset{\ominus}{C}=O$ (SN)

⇓

$H-C\equiv\overset{\ominus}{C}$ (SE) $\overset{\overset{\ominus}{N}H_2}{\Longrightarrow}$ $H-C\equiv C-H$

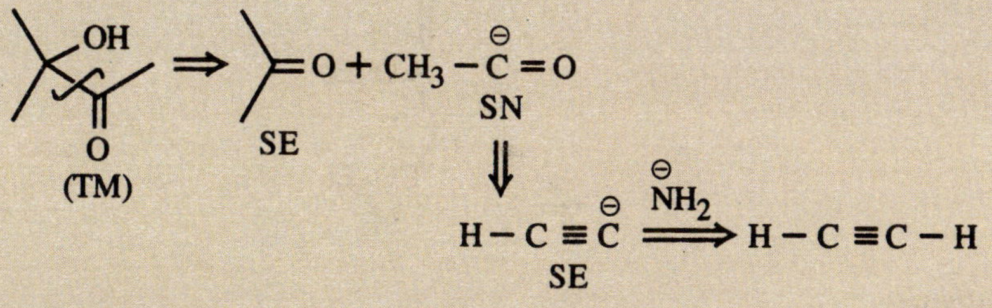

Forward Synthesis

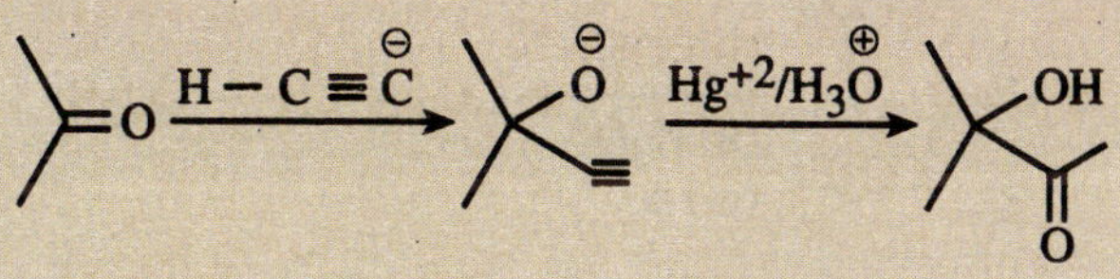

$(CH_3)_2C=O$ $\xrightarrow{H-C\equiv\overset{\ominus}{C}}$ $(CH_3)_2C(\overset{\ominus}{O})C\equiv CH$ $\xrightarrow{Hg^{+2}/H_3\overset{\oplus}{O}}$ $(CH_3)_2C(OH)COCH_3$

(ii) *Retro-analysis*

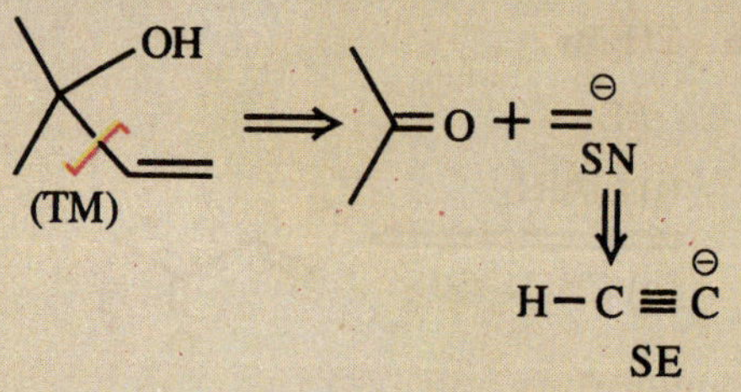

(TM) ⇒ acetone + $=^{\ominus}$ (SN) ⇓ $H-C\equiv\overset{\ominus}{C}$ SE

Forward Synthesis

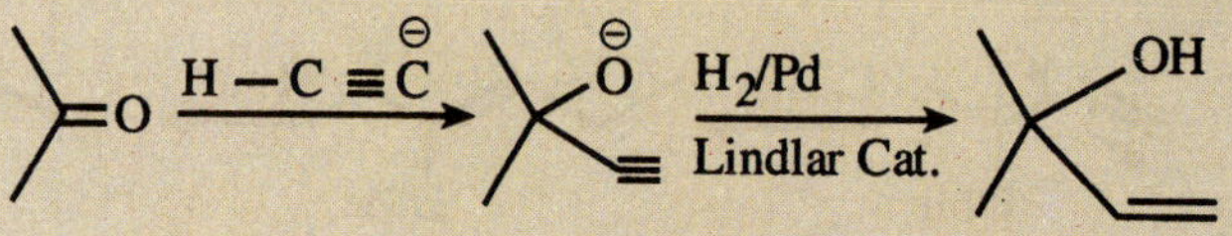

acetone $\xrightarrow{H-C\equiv\overset{\ominus}{C}}$ alkoxide $\xrightarrow[\text{Lindlar Cat.}]{H_2/Pd}$ OH product

7. *Retro-analysis*

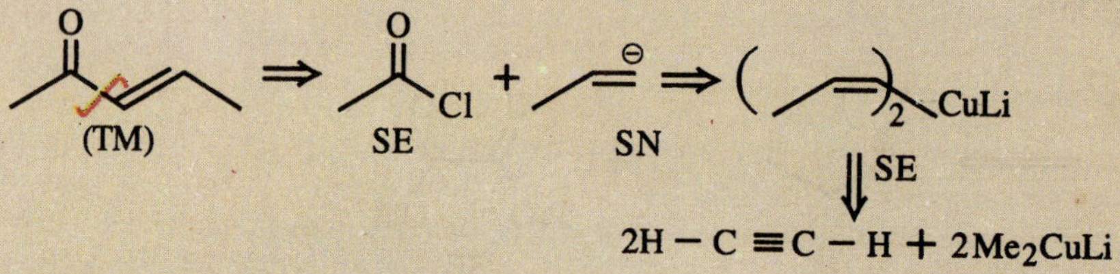

(TM) ⇒ acetyl chloride (SE) + vinyl anion (SN) ⇒ $(\,\text{propenyl})_2CuLi$ ⇓ SE $2H-C\equiv C-H + 2Me_2CuLi$

Forward Synthesis

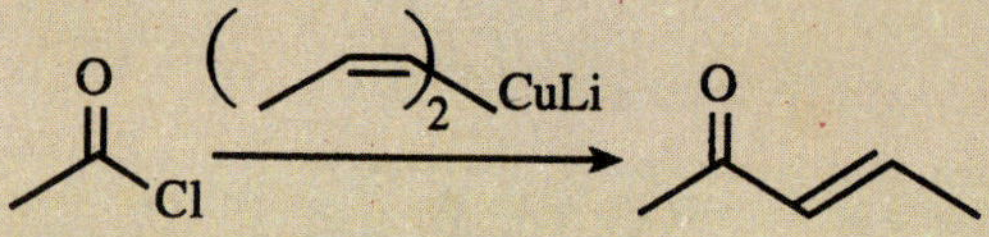

acetyl chloride $\xrightarrow{(\text{propenyl})_2CuLi}$ pent-3-en-2-one

8. *Retro-analysis*

$CH_3 - CH_2 - CO_2H \Rightarrow CH_3Br + \overset{\ominus}{C}H_2 - CO_2H$

(TM) SE SN

⇓

$CH_2(CO_2Et)_2$

SE

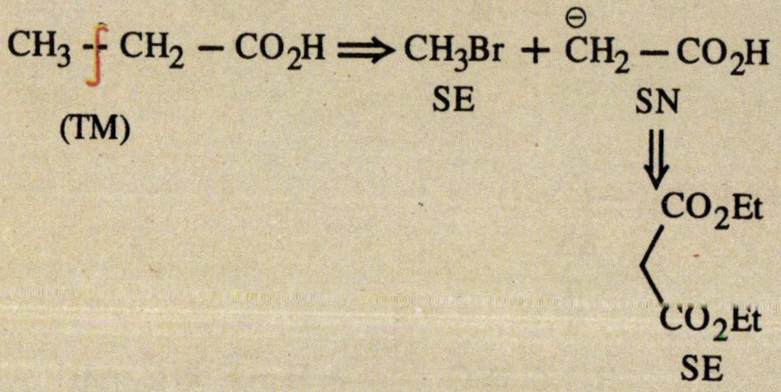

Forward Synthesis

$CH_2(CO_2Et)_2 \xrightarrow[\text{(ii) } CH_3Br]{\text{(i) } \overset{\ominus}{O}Et} H_3C-CH(CO_2Et)_2 \xrightarrow[\text{(ii) } H_3O^+ \text{ (iii) } \Delta]{\text{(i) KOH}} CH_3 - CH_2 - CO_2H$

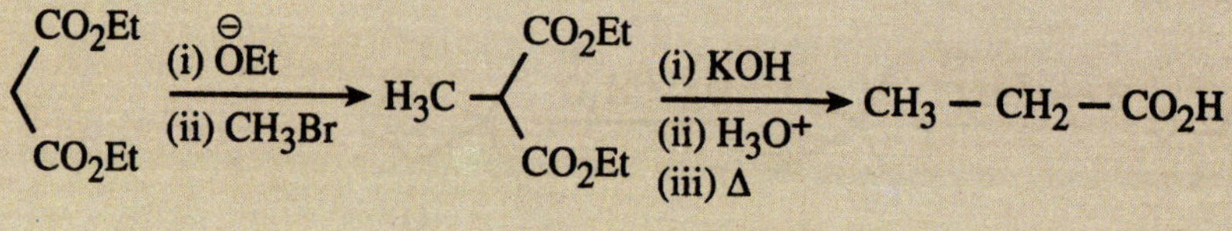

9. *Retro-analysis*

(TM) $\Rightarrow$ CH_3Br (SE) + ⊖ (SN) $\Rightarrow$ AcHN–CH(CO_2Et)$_2$ (SE) $\xrightarrow[\text{(ii) Zn/AcOH/}Ac_2O]{\text{(i) }HNO_2}$ $CH_2(CO_2Et)_2$

NH$_2$, CO$_2$H

Forward Synthesis

AcHN–CH(CO_2Et)$_2$ $\xrightarrow[\text{(ii) }CH_3Br]{\text{(i) }\ominus OEt}$ AcHN–C(CH_3)(CO_2Et)$_2$ $\xrightarrow[\text{(ii) }H_3O^+ \text{ (iii) }\Delta]{\text{(i) KOH}}$ $CH_3CH(NH_2)CO_2H$

10. *Retro-analysis*

Ph–CH_2–CH(CH_3)–NH_2 (TM) $\Rightarrow$ Ph–CH_2–Br (Br) + ⊖$CH(CH_3)$–NH_2 (SN)

$\Downarrow$

$CH_3CH_2NO_2$ (SE)

Forward Synthesis

$CH_3CH_2NO_2$ $\xrightarrow[\text{2 eq.}]{\text{nBuLi}}$ $CH_3-\overset{\ominus}{C}=\overset{\oplus}{N}(O^{\ominus})_2$ $\xrightarrow{\text{Ph–}CH_2\text{–Br}}$ $CH_3-C(CH_2Ph)=\overset{\oplus}{N}(O^{\ominus})_2$

$\downarrow \overset{\oplus}{H_3O}$

Ph–CH_2–CH(CH_3)–NH_2 $\xleftarrow[\text{Conc. HCl}]{\text{Zn(Hg)}}$ $CH_3-CH(CH_2Ph)-NO_2$

11. (i) *Retro-analysis*

$CH_3CH_2CH_2$–COOH (TM) $\Rightarrow$ $CH_3CH_2CH_2$–Br + $\overset{\ominus}{C}O_2H$ (SN)

$\Downarrow$

KCN

Forward Synthesis

$CH_3CH_2CH_2Br$ $\xrightarrow[CH_3COCH_3]{\text{KCN}}$ $CH_3CH_2CH_2CN$ $\xrightarrow{H_3O^+}$ $CH_3CH_2CH_2COOH$

(ii) *Retro-analysis*

$$\underset{\text{(TM)}}{Ph\text{-}CH(CH_2NH_2)\text{-}CH_2\text{-}CO_2Et} \Rightarrow \underset{\text{SE}}{Ph\text{-}CH{=}CH\text{-}CO_2Et} + \underset{\text{SN}}{\overset{\ominus}{C}H_2-NH_2}$$

SN ↓ $\overset{\ominus}{C}N$ (SE)

Forward Synthesis

$$Ph\text{-}CH{=}CH\text{-}CO_2Et \xrightarrow[AlEt_3]{\overset{\ominus}{C}N} Ph\text{-}CH(CN)\text{-}CH_2\text{-}CO_2Et \xrightarrow{H_2/Pd} Ph\text{-}CH(CH_2NH_2)\text{-}CH_2\text{-}CO_2Et$$

12. (i) *Retro-analysis*

$$\underset{\text{(TM)}}{CH_3CH_2CH_2\text{-}CHO} \Rightarrow \underset{\text{SE}}{CH_3CH_2\text{-}Br} + \underset{\text{SN}}{\overset{\ominus}{C}HO}$$

SN ⇓ 1,3-dithiane (H_2C with two S, SE)

Forward Synthesis

$$\text{1,3-dithiane } (H_2C(S)_2) \xrightarrow[\text{(ii) } CH_3-CH_2-Br]{\text{(i) nBuLi}} CH_3-H_2C(H)C(S)_2 \xrightarrow[H_3\overset{\oplus}{O}]{Hg^{+2}} CH_3CH_2CH_2\text{-}CHO$$

(ii) *Retro-analysis*

$$\underset{\text{(TM)}}{\text{cyclohexyl-CHO}} \Rightarrow \underset{\text{SE}}{\text{cyclohexanone }(=O)} + \underset{\text{SN}}{\overset{\ominus}{C}HO}$$

SN ⇓ $Me-O-CH=PPh_3$ (SE)

Forward Synthesis

$$\text{cyclohexanone} \xrightarrow[\text{wittig reaction}]{\text{(i) } Me-O-CH=PPh_3} \text{cyclohexylidene}=CH-OMe \xrightarrow{H_3\overset{\oplus}{O}} \text{cyclohexyl-CHO}$$

(iii) *Retro-analysis*

(TM) CHO ⇒ =O (SE) + $\overset{\ominus}{C}HO$ (SN) ⇓ Cl CO$_2$Et (SE)

Forward Synthesis

=O + Cl CO$_2$Et, $\overset{\ominus}{O}$ (Darzen) → O CO$_2$Et → $H_3\overset{\oplus}{O}$, Δ → CHO

Darzen

13. *Retro-analysis*

O (TM) ⇒ Cl Br (SE) + $\overset{\ominus}{C} = O$ (SN, $\ominus$)

⇓

S S (SE) ⇒ CH_2O + HS SH + BF_3

Forward Synthesis

S S → (i) nBuLi (ii) Cl Br → S S Cl → nBuLi → S S → $Hg^{+2}/H_3\overset{\oplus}{O}$ → O

14. *Retro-analysis*

O O (TM) ⇒ O (SE) + 'O' SN (nucleophilic)

⇓

H_2O_2 + $\overset{\ominus}{O}H$

SE

Forward Synthesis

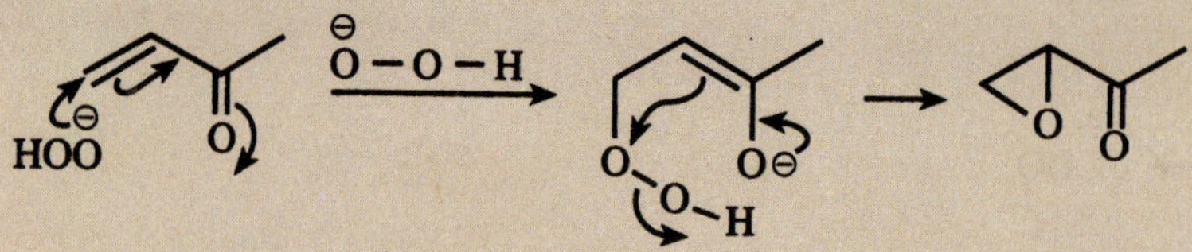

HOO$^{\ominus}$ ... $\overset{\ominus}{O}-O-H$ → ... O–O–H ... O$^{\ominus}$ →

15. *Retro-analysis*

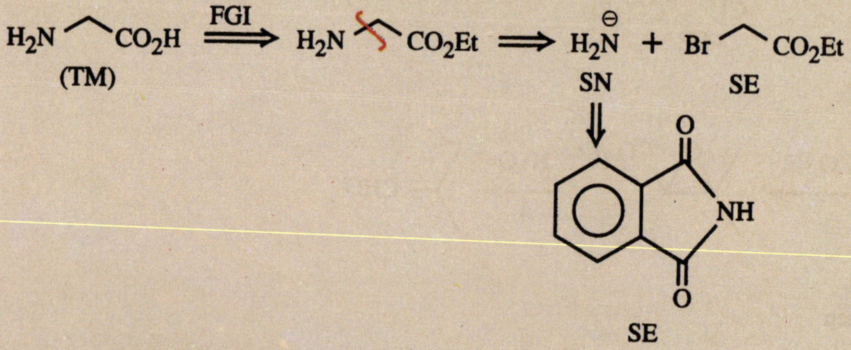

$H_2N\frown CO_2H$ (TM) $\overset{FGI}{\Longrightarrow}$ $H_2N\frown CO_2Et$ ⟹ $H_2\overset{\ominus}{N}$ (SN) + $Br\frown CO_2Et$ (SE)

⇓

NH (SE)

Forward Synthesis

NH $\xrightarrow[\text{(ii) } Br\frown CO_2Et]{\text{(i) } \overset{\ominus}{O}Et}$ N$\frown CO_2Et$ $\xrightarrow[\text{(ii) } H_3\overset{\oplus}{O}]{\text{(i) NaOH}}$ $H_2N\frown CO_2H$

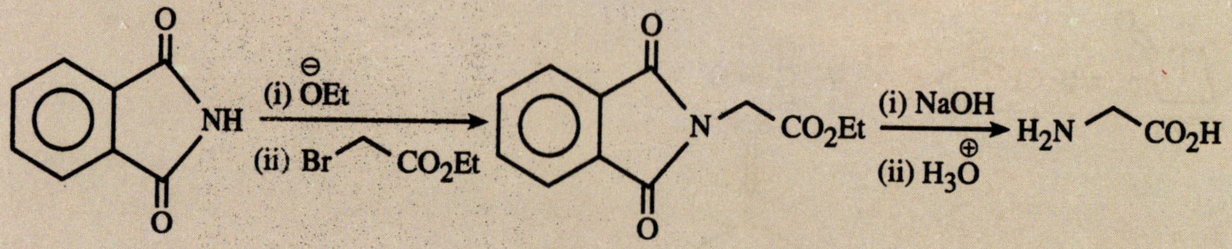

2.3.2 Electrophilic synthons and synthetic equivalents

Synthon — **Synthetic Equivalent**

1. 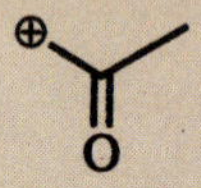 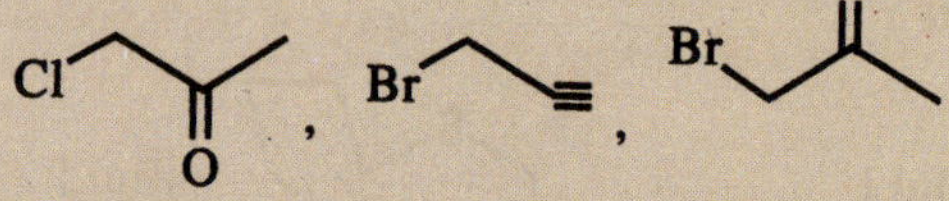 ⊕ (acetonyl cation) — Cl (chloroacetone), Br (propargyl bromide), Br (methallyl bromide)

2. 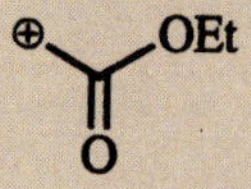 ⊕ ... OEt — Cl ... OEt

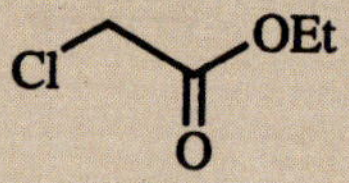

3. 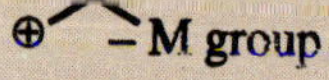 ⊕⌒ –M group — ⟋⟍–M group

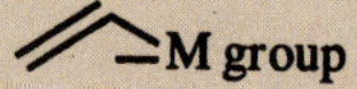

(–M = –CN, –C = O, –CO_2Et, –NO_2 etc.)

(a) 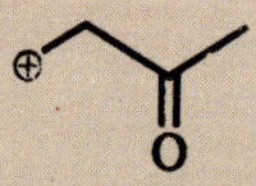 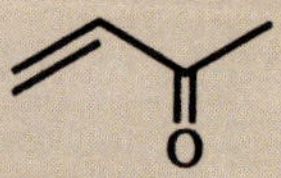 ⊕ ... O — methyl vinyl ketone (O)

(b) 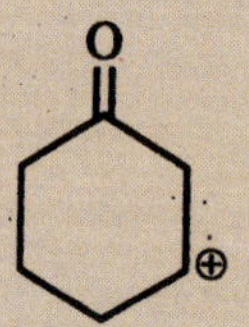 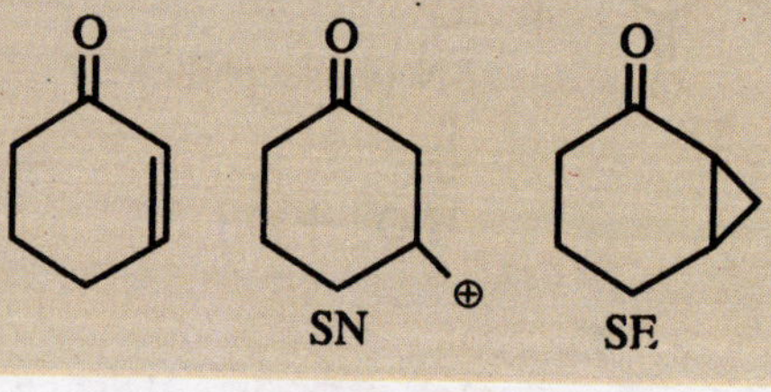 O ... ⊕ — O (cyclohexenone), O ... ⊕ (SN), O (SE)

	Synthon	Synthetic Equivalent
4.	$\overset{\oplus}{C}H_2 - \overset{\oplus}{C} = O$	$ClCH_2COCH_3$, $ClCH_2CHO$, $ClCH_2COCl$
5.	$\overset{\oplus}{C}H_2$	$CH_2 = \overset{\oplus}{N}Me_2$, $CH_3COC(=CH_2)CO_2Et$, 2-methylenecyclohexanone, $CH_2=CHCOCH_3$, $CH_3COC(=CH_2)CH_3$, $CH_2=CH\overset{\oplus}{P}Ph_3$
6.	$\overset{\oplus}{\underset{\oplus}{C}}H_2$	CH_2Br_2, CH_2O
7.	$\overset{\oplus}{\underset{\oplus}{C}}H$	$Me-O-CH=\overset{\oplus}{N}Me_2$, $Me-O-CH=CH_2$
8.	$\overset{\oplus\oplus\oplus}{C}H$	$H-C(OEt)_3$
9.	$\overset{\oplus}{C}H_3$	CH_3-Br, CH_3OTs, CH_3OSO_2Me, $(CH_3)_2SO_4$, 2,6-dimethylphenyl$-CO_2Me$, $(CH_3)_3\overset{\oplus}{O}\overset{\ominus}{B}F_4$, $CH_3NH_2(CH_3\overset{\oplus}{N}_2)$
10.	$\overset{\oplus}{C}H_2-\overset{\oplus}{C}H_2$	$Br-CH_2-CH_2-Br$
11.	$\overset{\oplus}{C}H_2CH_2OH$	ethylene oxide
12.	$\overset{\oplus}{C}H_2CH_2CH_2OH$	$CH_2=CHCHO$, $CH_2=CHCO_2Et$
13.	$\overset{\oplus}{C}H_2OH$	CH_2O
14.	$\overset{\oplus}{C}H(OH)Me$	MeCHO
15.	$\overset{\oplus}{C}Me_2OH$	Me_2CO
16.	$CH_3-CH=\overset{\oplus}{C}H$	$CH_3-CH=CH-Br$
17.	$CH_2=\overset{\oplus}{C}H$	$CH_3CH(OEt)_2$
18.	$CH_2=\overset{\oplus}{C}(OEt)$	$CH_3-C(OEt)_3$

	Synthon	Synthetic Equivalent
19.	$CH_3-\overset{\oplus}{C}=O$	CH_3COCl, CH_3CN, CH_3CO_2Et, $(CH_3CO)_2O$
20.	$H-\overset{\oplus}{C}=O$	HCO_2Et, HC $(OEt)_3$
		$CHCl_3 + \overset{\ominus}{O}H$ (via Carbene, Reimer-Tiemann reaction)
		$Me_2N-CHO + POCl_3$ or $COCl_2$
		(for activated arene or hetero arene or enamine)
		$CO + HCl + AlCl_3$ / $HCN + HCl + AlCl_3$ } for benzene
21.	$Et-O-\overset{\oplus}{C}=O$	$Cl-CO_2Et$
22.	$HO-\overset{\oplus}{C}=O$ ($\overset{\oplus}{C}O_2H$)	CO_2, $Cl-CO_2Et$
23.	$H_2N-\overset{\oplus}{C}=O$ (HNCO)	$KNCO + H_2O$, $NH_4NCO + \Delta$
24.	$\overset{\oplus}{\underset{\oplus}{C}}=O$	$O=C(OEt)_2$, $O=CCl_2$
25.	$O=\overset{\oplus}{C}-\overset{\oplus}{C}=O$	EtO_2C-CO_2Et
26.	$\overset{\oplus}{C}(=O)-CH_2-\overset{\oplus}{C}(=O)$	$CH_3COCH_2COCH_3$, $CH_3COCH_2CO_2Et$, $EtO_2CCH_2CO_2Et$
27.	$\overset{\oplus}{C}(=O)-CH_2CH_2-\overset{\oplus}{C}(=O)$	succinic anhydride, $CH_3COCH_2CH_2COCH_3$
28.	$\overset{\oplus}{C}HCl_2$	$CHCl_3$ (Via. : CCl_2)
29.	'O' (electrophilic)	$F_3C\,CO_3H$, m-chloroperbenzoic acid ($ClC_6H_4CO_3H$) { (i) H_2O_2 (ii) $\overset{\ominus}{O}H$
30.	$\overset{\oplus}{C}N$	$Cl-CN$

	Synthon	Synthetic Equivalent
31.	$\overset{\oplus}{N}H_2$	NH_2-Cl, NH_2OCH_3, $NH_2-OSO_3^{\ominus}$
32.	$\overset{\oplus}{C}H_2-NMe_2$	$Cl-C(=O)-NMe_2$
33.	$Me-O-\overset{\oplus}{C}H_2$	$Me-O-CH_2-Cl$
34.	$CH_3-C\equiv\overset{\oplus}{C}$	$CH_3-C\equiv C-Br$
35.	$\overset{\oplus}{Cl}$	Cl_2, Cl_2+FeCl_3, $ClCH_2CH_2Cl$, N-chlorosuccinimide ($N-Cl$) $+ H^+$
36.	$\overset{\oplus}{S}O_3H$	Conc. H_2SO_4
37.	$\overset{\oplus}{S}O_2Cl$	$ClSO_3H$
38.	$CH_3COCH_2\overset{\oplus}{C}=O$	O (Ketene dimer), $CH_3COCH_2CO_2Et$

Application of Electrophilic Synthon and Synthetic Equivalent in Retro-synthetic Analysis

1. (a) *Retro-analysis*

(TM) $CH_3COCH_2CH_2COCH_3 \Rightarrow CH_3COCH_2CO_2Et + \overset{\oplus}{C}H_2COCH_3$

SN $\Downarrow$

$ClCH_2COCH_3 \xRightarrow[AcOH]{Cl_2} CH_3COCH_3$

Forward Synthesis

$CH_3COCH_2CO_2Et \xrightarrow[\text{(ii) } ClCH_2COCH_3]{\text{(i) } \overset{\ominus}{O}Et} CH_3COCH(CO_2Et)CH_2COCH_3 \xrightarrow[\Delta]{\text{dil HCl}} CH_3COCH_2CH_2COCH_3$

(b) *Retro-analysis*

(TM) $CH_3COCH_2CH_2CO_2H$ ⇒ $CH_3COCH_2^{\oplus}$ (SN) + $CH_2(CO_2Et)_2$

SN ⇓ $H-C\equiv C-CH_2-Br$ (SE)

Forward Synthesis

$CH_2(CO_2Et)_2$ —(i) $\overset{\ominus}{O}Et$; (ii) $H-C\equiv C-CH_2-Br$ → $H-C\equiv C-CH_2-CH(CO_2Et)_2$

—(i) KOH; (ii) H_3O^+/Δ → $H-C\equiv C-CH_2-CH_2-COOH$

—Hg^{+2}/H_3O^+ → $CH_3COCH_2CH_2COOH$

2. *Retro-analysis*

(TM) $CH_3CH_2CH_2COCH_3$ ⇒ $(CH_3)_2CuLi$ + $^{\oplus}CH_2CH_2COCH_3$ (SN)

SN ⇓ $CH_2=CHCOCH_3$ (SE)

Forward Synthesis

$CH_2=CHCOCH_3$ —(i) $(CH_3)_2CuLi$; (ii) H_3O^+ → $CH_3CH_2CH_2COCH_3$

3. *Retro-analysis*

(TM) ⇒ HO OH $\overset{FGI}{\Rightarrow}$

$$\Downarrow$$

SE $CO_2Et + \overset{\oplus}{C}H_2 - \underset{\underset{O}{\|}}{C} - CH_3 \left(\equiv \overset{\oplus}{C}H_2 - \underset{\underset{O}{\|}}{\overset{\oplus}{C}}\right)$

SN

$$\Downarrow$$

Cl

SE

Forward Synthesis

(i) $\overset{\ominus}{O}Et$ (ii) Cl

$-H_2O$

4. (a) *Retro-analysis*

NMe_2 (TM) $\Rightarrow$ $+ \overset{\oplus}{C}H_2 - NMe_2 \longleftrightarrow CH_2 = \overset{\oplus}{N}Me_2 \left(\equiv \overset{\oplus}{C}H_2\right)$

SE

$$\Downarrow$$

$$CH_2O + Me_2NH + HCOOH$$

Forward Synthesis

$\xrightarrow[\text{Via enol}]{CH_2 = \overset{\oplus}{N}Me_2}$ NMe_2

(b) *Retro-analysis*

CO_2Et CO_2Et (TM) FGI ⟹ HO, H, CO_2Et CO_2Et ⟹ O, PPh_3, CO_2Et CO_2Et

⟹ O, CO_2Et, CO_2Et + $Ph_3P = CH - \overset{\oplus}{C}H_2$ ≡ $Ph_3\overset{\oplus}{P} - CH = CH_2$ SE

⟸ O, Cl + CO_2Et, CO_2Et

Forward Synthesis

CO_2Et, CO_2Et —(i) $\overset{\ominus}{O}Et$ (ii) Cl, O→ O, CO_2Et, CO_2Et —(i) $\overset{\ominus}{O}Et$, $Ph_3\overset{\oplus}{P} - CH = CH_2$→ O, CO_2Et, CO_2Et, ⊖, $\oplus PPh_3$ → O, Ph_3P, CO_2Et, CO_2Et —$- Ph_3PO$→ CO_2Et, CO_2Et

5. *Retro-analysis*

EtO_2C, EtO_2C, CO_2Et, CO_2Et (TM) ⟹ 2 CO_2Et, CO_2Et + $\overset{\oplus}{C}H_2\overset{\oplus}{}$ SN ⟹ CH_2 Br Br

Forward Synthesis

2 CO_2Et, CO_2Et —(i) $OEt^{\ominus}$ (ii) Br⌃Br→ EtO_2C, EtO_2C, CO_2Et, CO_2Et

6. *Retro-analysis*

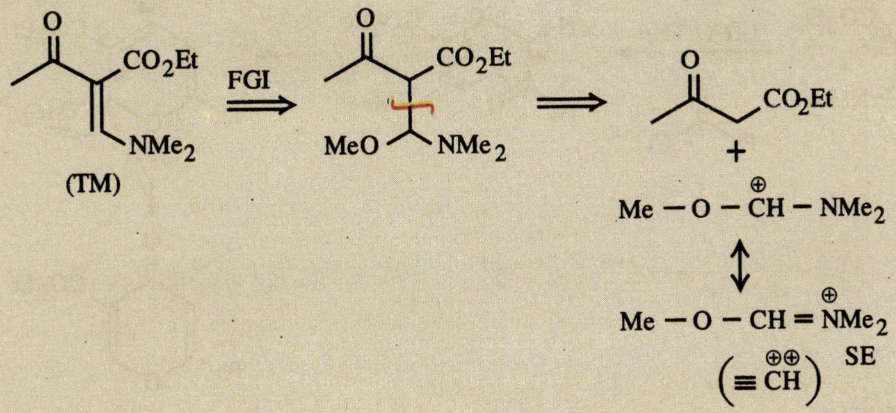

(TM) FGI ⟹ MeO NMe$_2$ ⟹ CO_2Et + $Me-O-\overset{\oplus}{C}H-NMe_2$ ↕ $Me-O-CH=\overset{\oplus}{N}Me_2$ $(\equiv \overset{\oplus\oplus}{CH})$ SE

Forward Synthesis

CO_2Et $\xrightarrow[Me-O-CH=\overset{\oplus}{N}Me_2]{Et_3N}$ H CO_2Et MeO NMe$_2$ ↓ − MeOH CO_2Et NMe$_2$

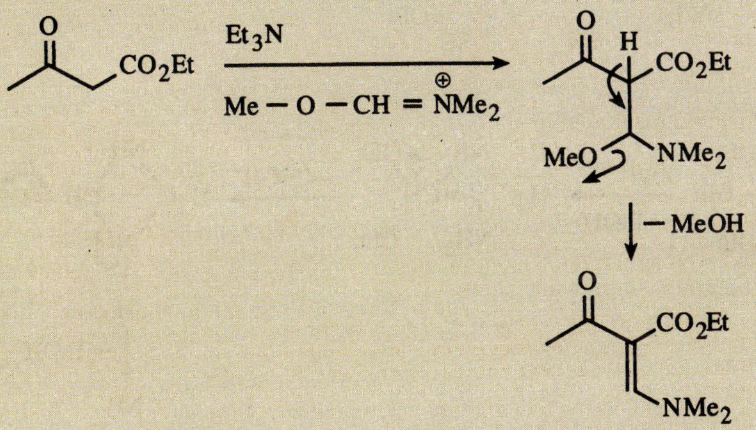

Application

CO_2Et Ph O (TM) FGI ⟹ CO_2Et Ph O NMe$_2$ (TM) ⟹ CO_2Et Ph O NMe$_2$ ⇓ Ph Cl + CO_2Et NMe$_2$

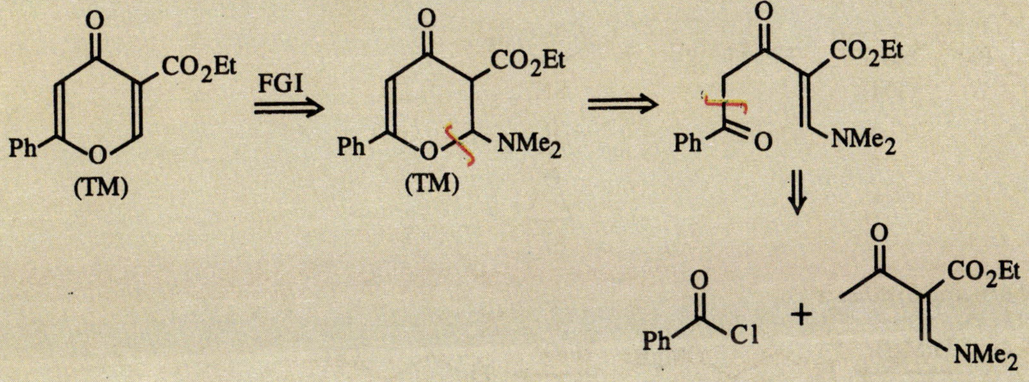

Forward Synthesis

LDA / THF, PhCOCl → (TM) →

7. ***Retro-analysis***

(TM) ⇒ $H_2C(NH_2)_2$ + $\overset{\oplus\oplus\oplus}{CH}$ ⇒ $HC(OEt)_3$

SN SE

Forward Synthesis

$H_2C(NH_2)_2$ + $H-C(OEt)_3$ —($^{\ominus}$OEt, − EtOH)→ — (− EtOH) → — (− EtOH) →

8. ***Retro-analysis***

Ph⌇CH₂CH₂OH (TM) ⇒ PhMgBr + $^{\oplus}$CH₂CH₂OH (SN) ⇓ ethylene oxide (SE)

Forward Synthesis

ethylene oxide —PhMgBr→ $PhCH_2CH_2OMgBr$ —$H_3O^{\oplus}$→ $PhCH_2CH_2OH$.

9. *Retro-analysis*

(TM) OH ⇒ $(CH_3)_2CuLi$ + $\oplus$ OH (SN) ⇒ CHO (SE)

Forward Synthesis

CHO $\xrightarrow{Me_2CuLi}$ CHO $\xrightarrow{LiAlH_4}$ OH

10. *Retro-analysis*

(TM) ⇒ OH + $\oplus$ (SN) ⇒ $CH_3CH(OEt)_2$ (SE)

Forward Synthesis

H–O + $CH_3-CH(OEt)_2$ $\xrightarrow[\Delta]{-EtOH}$ O–CH(OEt)–CH_2-H $\xrightarrow{-EtOH}$ O–CH=CH_2

11. *Retro-analysis*

(TM) O–C(OEt)=CH_2 ⇒ OH + $\oplus$ OEt (SN) ⇒ $CH_3C(OEt)_3$ (SE)

Forward Synthesis

OH + $CH_3-C(OEt)_3$ $\xrightarrow[\Delta]{-EtOH}$ O–C(OEt)(OEt)–CH_2–H $\xrightarrow{-EtOH}$ O–C(OEt)=CH_2

12. *Retro-analysis*

COOH (TM) ⇒ CH_3-CH_2MgBr + $\overset{\oplus}{C}O_2H$ (SN) ⇒ CO_2 or $ClCO_2Et$

Forward Synthesis

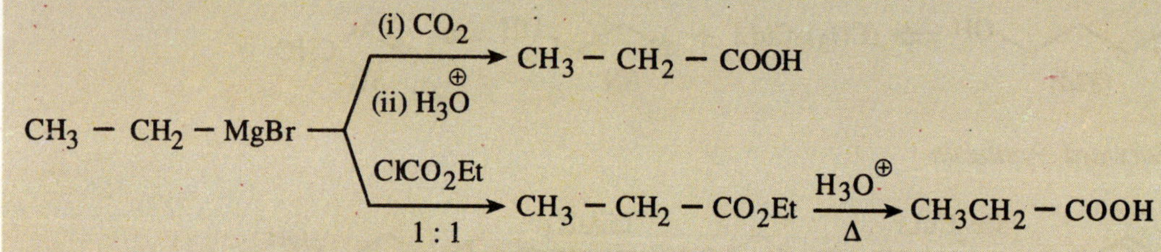
(i) CO2
(ii) H3O⊕
CH3 − CH2 − COOH
CH3 − CH2 − MgBr
ClCO2Et
1 : 1
CH3 − CH2 − CO2Et
H3O⊕
Δ
CH3CH2 − COOH

13. *Retro-analysis*

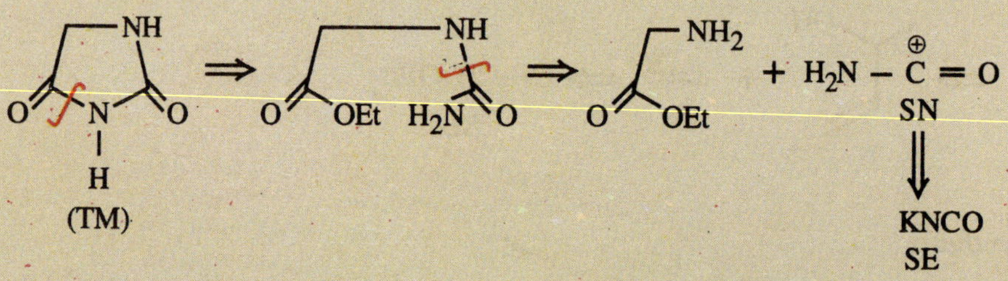
NH
O
N
O
H
(TM)
NH
O
OEt
H2N
O
NH2
O
OEt
+ H2N − C⊕ = O
SN
KNCO
SE

Forward Synthesis

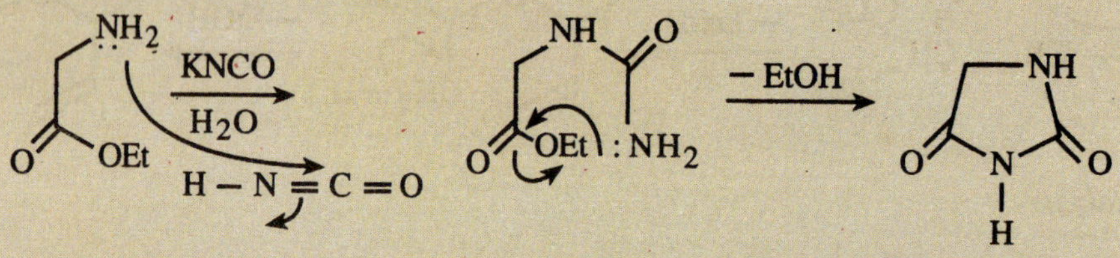
NH2
O
OEt
KNCO
H2O
H − N = C = O
NH
O
O
OEt
: NH2
− EtOH
NH
O
N
O
H

14. (i) *Retro-analysis*

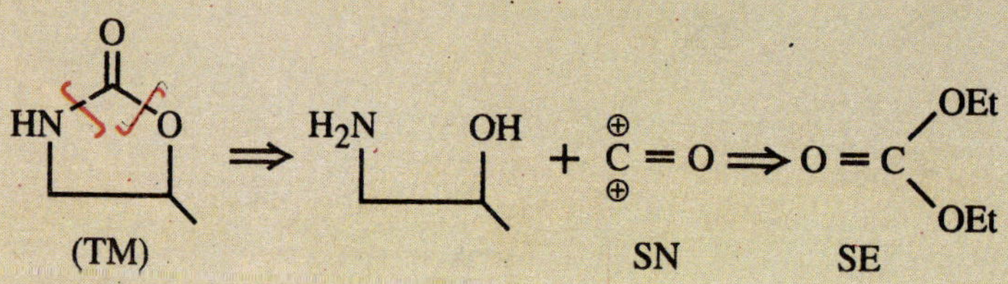
O
HN
O
(TM)
H2N
OH
+ ⊕C⊕ = O
O = C
OEt
OEt
SN
SE

Forward Synthesis

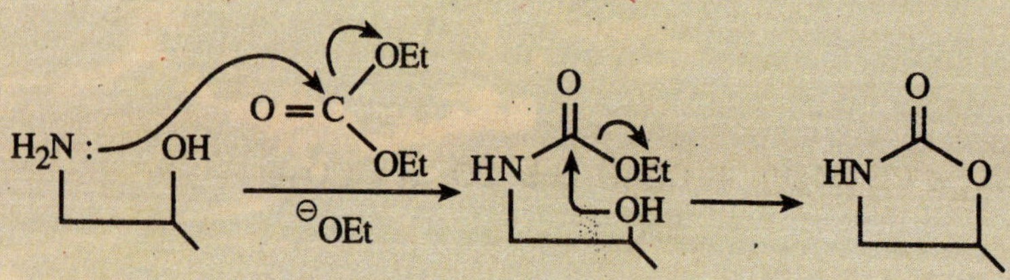
H2N :
OH
O = C
OEt
OEt
⊖OEt
O
HN
OEt
OH
O
HN
O

(ii) *Retro-analysis*

$EtO_2C-CH_2-CO-CH_2-CO_2Et$ (TM) $\Rightarrow$ 2 $CH_2(CO_2Et)_2$ + $\overset{\oplus}{C}=O$ (with two ⊕ charges; SN) $\Rightarrow$ $O=C(OEt)_2$ (SE)

Forward Synthesis

$CH_2(CO_2Et)_2$ $\xrightarrow[O=C(OEt)_2]{\overset{\ominus}{O}Et}$ $(EtO_2C)_2CH-CO-CH(CO_2Et)_2$ $\xrightarrow{\text{(i) KOH; (ii) } H_3O^+/\Delta}$ $HO_2C-CH_2-CO-CH_2-CO_2H$ $\xrightarrow{EtOH/H^+}$ $EtO_2C-CH_2-CO-CH_2-CO_2Et$

15. *Retro-analysis*

1,3-Dioxolane (TM) $\Rightarrow$ $HOCH_2CH_2OH$ + $\overset{\oplus}{C}H_2$ (with two ⊕ charges; SN) $\Rightarrow$ CH_2O (SE)

Forward Synthesis

$HOCH_2CH_2OH$ + HCHO $\xrightarrow{\text{HCl gas}}$ $HOCH_2CH_2-O-CH_2-OH$ $\xrightarrow{-H_2O}$ 1,3-dioxolane

16. *Retro-analysis*

3,5-Diphenylcyclopentane-1,2,4-trione (TM) $\Rightarrow$ $PhCH_2COCH_2Ph$ + $\overset{\oplus}{C}(=O)-\overset{\oplus}{C}(=O)$ (SN) $\Rightarrow$ $EtO-CO-CO-OEt$ (SE)

Forward Synthesis

Ph, Ph, O; (i) $\overset{\ominus}{O}Et$; (ii) CO_2Et | CO_2Et; O, O, Ph, Ph, O

17. *Retro-analysis*

O, HN, N, H, O, O, TM ⇒ H_2N, O, NH_2 + O, ⊕, ⊕, O, SN ⇒ EtO, O, EtO, O, SE

Forward Synthesis

H_2N, O, NH_2, O, EtO, EtO, O; $\overset{\ominus}{O}Et$, EtOH; O, HN, O, NH, O

18. *Retro-analysis*

O, O, (TM) ⇒ + O, ⊕, ⊕, O, SN ⇒ O, O, O, SE

Forward Synthesis

O, O, O; $AlCl_3$; O, HO, O; (i) $SOCl_2$; (ii) $AlCl_3$; O, O

19. *Retro-analysis*

(TM) 4-methyl-4-(dichloromethyl)cyclohexa-2,5-dienone ⇒ 4-methylcyclohexadienone anion + $\overset{\oplus}{C}HCl_2$ (SN) ⇒ $CHCl_3$ + $\overset{\ominus}{O}H$ (:CCl_2) (SE)

⇓ $O^{\ominus}$ (p-methylphenoxide) ⇓ OH (p-cresol) SE

Forward Synthesis

OH (p-cresol) $\xrightarrow{\overset{\ominus}{O}H/CHCl_3}$ (dienone with $\overset{\ominus}{C}Cl_2$) $\xrightarrow{H_2O}$ (dienone with $CHCl_2$)

20. (i) *Retro-analysis*

(TM) epoxide ⇒ alkene + 'O' (electrophilic) SN

⇓

$F_3C-C(=O)-O-OH$

SE

Forward Synthesis

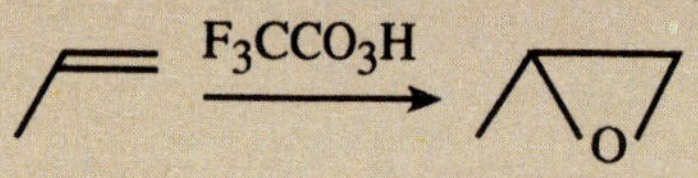

F_3CCO_3H

(ii) *Retro-analysis*

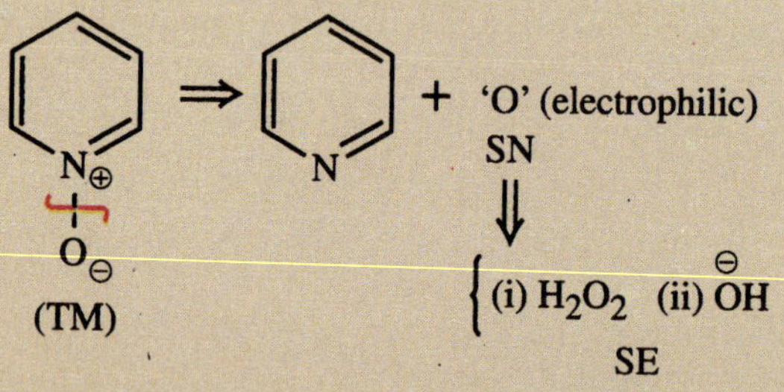

⇒ + 'O' (electrophilic)
SN
⇓
(i) H_2O_2 (ii) $\overset{\ominus}{O}H$
SE

(TM)

Forward Synthesis

+ H – O – O – H ⟶ OH

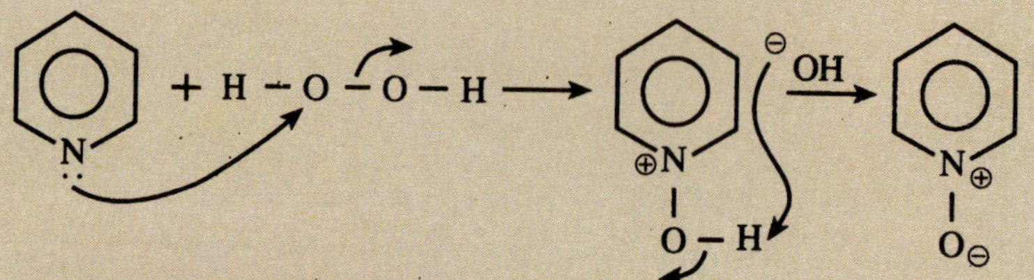

21. *Retro-analysis*

CN ⇒ + $\overset{\oplus}{C}N$ ⇒ Cl – CN
SN SE

(TM)

SE

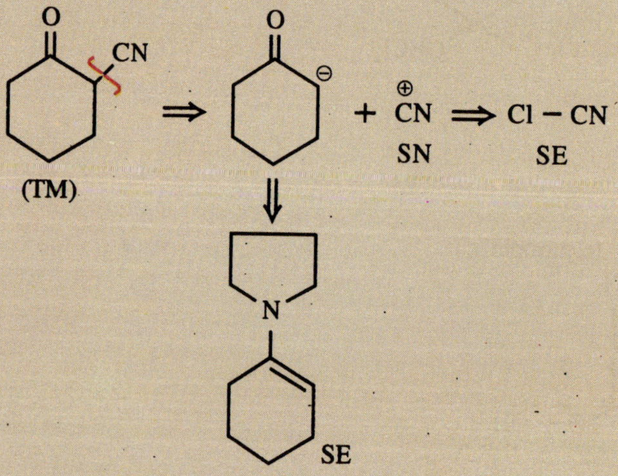

Forward Synthesis

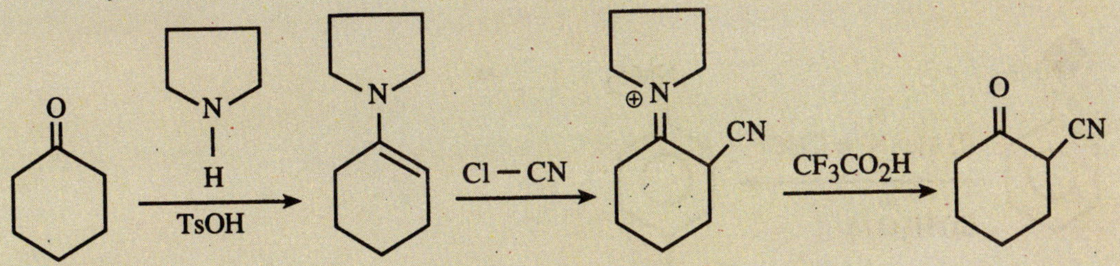

Cyclohexanone + pyrrolidine (N–H), TsOH → enamine; $Cl-CN$ → iminium ($\oplus$N) with CN; CF_3CO_2H → 2-cyanocyclohexanone (O, CN)

22. *Retro-analysis*

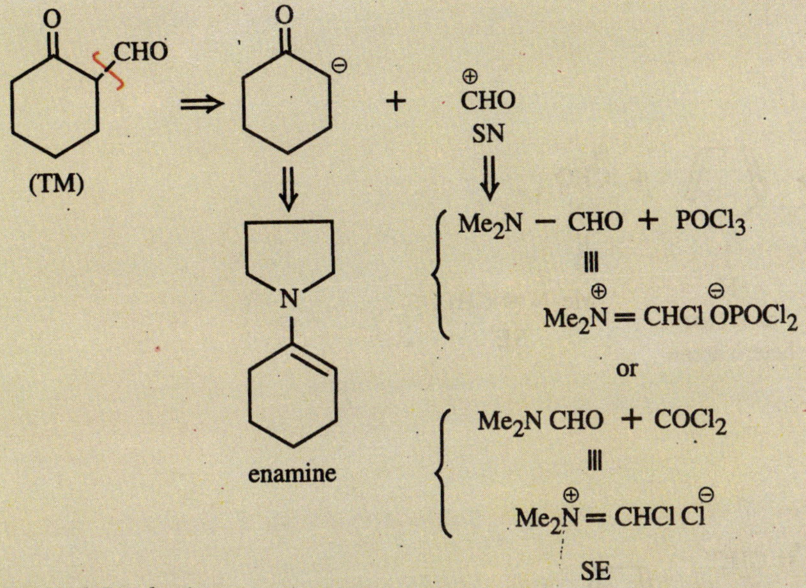

2-formylcyclohexanone (O, CHO) (TM) ⇒ cyclohexanone enolate ($\ominus$) + $\overset{\oplus}{CHO}$ SN

⇓ enamine

⇓

$Me_2N - CHO + POCl_3$

⫼

$Me_2\overset{\oplus}{N} = CHCl\ \overset{\ominus}{O}POCl_2$

or

$Me_2N\ CHO + COCl_2$

⫼

$Me_2\overset{\oplus}{N} = CHCl\ \overset{\ominus}{Cl}$

SE

Forward Synthesis

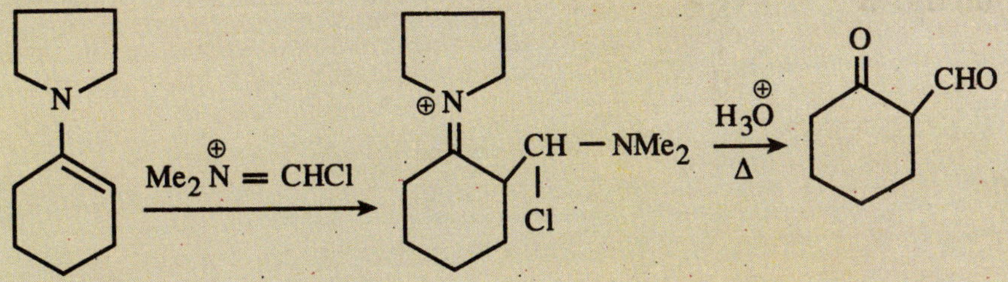

enamine, $Me_2\overset{\oplus}{N} = CHCl$ → iminium ($\oplus$N) with $CH(Cl) - NMe_2$; $H_3\overset{\oplus}{O}$, Δ → 2-formylcyclohexanone (O, CHO)

23. *Retro-analysis*

NMe_2 / CHO (TM) ⇒ NMe_2 activated arene + $\overset{\oplus}{CHO}$ SN

⇓

$Me_2\overset{\oplus}{N} = CHCl\ \overset{\ominus}{Cl}$

SE

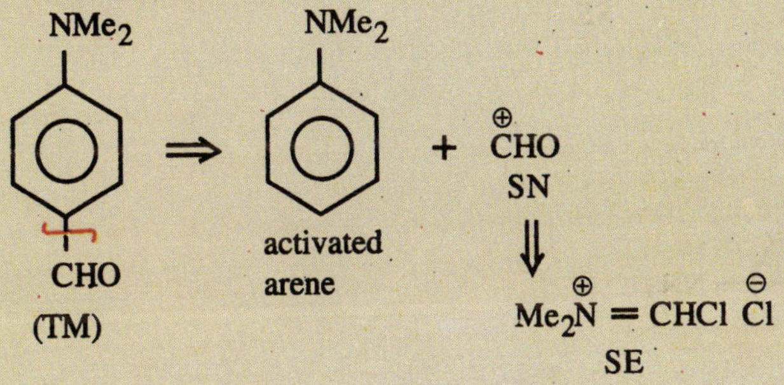

Forward Synthesis

$$\text{PhNMe}_2 \xrightarrow[\text{(ii) } H_3\overset{\oplus}{O}/\Delta]{\text{(i) } Me_2\overset{\oplus}{N}=CHCl} p\text{-}Me_2N\text{-}C_6H_4\text{-}CHO$$

24. *Retro-analysis*

(TM) pyrrole-2-CHO ⇒ pyrrole (N–H; activated hetero arene) + $\overset{\oplus}{C}HO$ (SN) ⇓ $Me_2\overset{\oplus}{N}=CHCl$ (SE)

Forward Synthesis

$$\text{pyrrole} \xrightarrow[\text{(ii) } H_3\overset{\oplus}{O}/\Delta]{\text{(i) } Me_2\overset{\oplus}{N}=CHCl} \text{pyrrole-2-CHO}$$

25. *Retro-analysis*

(TM) $(CH_3)_3C\text{–}NH_2 \Rightarrow (CH_3)_3C\text{–}MgCl + \overset{\oplus}{N}H_2$ (SN) $\Rightarrow H_2N-OCH_3$ (SE)

Forward Synthesis

$$(CH_3)_3C\text{–}MgCl + H_2N-OCH_3 \longrightarrow (CH_3)_3C\text{–}NH_2$$

26. *Retro-analysis*

(TM) ⇒ NH_2, CO_2Et + $\overset{\oplus}{}$ (SN)

FGI ⇓ NO, CO_2Et — SN ⇓ SE

⇓

CO_2Et + HNO_2

SE

CO_2Et $\xrightarrow[\text{(ii) Zn/AcOH}]{\text{(i) } HNO_2}$ CO_2Et, $\ddot{N}H_2$ ⟶ $H^{\oplus}$

⇓

NH, CO_2Et

2.3.3 Mixed electrophilic and nucleophilic synthons and synthetic equivalents

Mixed synthetic equivalent in the sense in any synthetic equivalent either same carbon will behave as electrophilic as well as nucleophilic centre or different carbons will behave as electrophilic and nucleophilic centres in the same molecule.

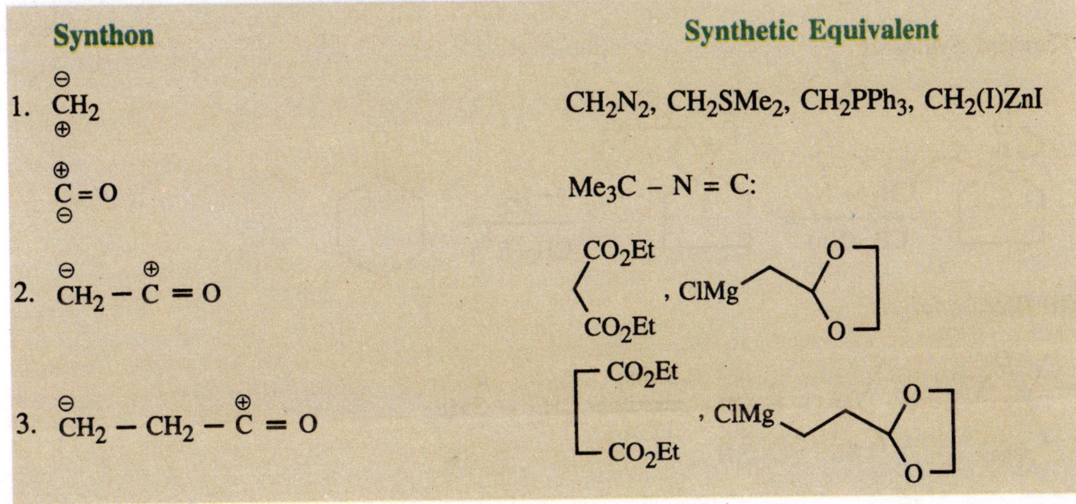

Synthon	Synthetic Equivalent
1. $\overset{\ominus}{\underset{\oplus}{CH_2}}$	CH_2N_2, CH_2SMe_2, CH_2PPh_3, $CH_2(I)ZnI$
$\overset{\oplus}{\underset{\ominus}{C}}=O$	$Me_3C-N=C:$
2. $\overset{\ominus}{CH_2}-\overset{\oplus}{C}=O$	CO_2Et, CO_2Et, ClMg, O, O
3. $\overset{\ominus}{CH_2}-CH_2-\overset{\oplus}{C}=O$	CO_2Et, CO_2Et, ClMg, O, O

Synthon	Synthetic Equivalent
4. $\overset{\ominus}{C}H_2 - CH_2 - CH_2 - \overset{\oplus}{C}H_2$	$(EtO_2C)_2CH-CH_2CH_2CH_2Br$
5. $\overset{\ominus}{C}H_2CH_2CH_2CH_2\overset{\oplus}{C}{=}O$	$EtO_2C-CH_2CH_2CH_2CH_2-C(=O)OEt$
6. $\overset{\ominus}{C}H_2C(=O)CH_2CH_2CH_2\overset{\oplus}{C}{=}O$	$CH_3C(=O)CH_2CH_2CH_2C(=O)OEt$

Application of mixed synthon and synthetic equivalent in retrosynthetic analysis

1. (i) *Retro-analysis*

(TM) $\Rightarrow$ $\Rightarrow$ + $\overset{\ominus}{C}H_2$ $\overset{\oplus}{}$ SN

FGA $\Downarrow$ $N \equiv N$

O SE

$\overset{\ominus}{C}H_2 - \overset{\oplus}{N} \equiv N$

SE

Forward Synthesis

$\overset{\ominus}{C}H_2 - \overset{\oplus}{N}_2$ / $CH_2(Nu)$ $\longrightarrow$ $\overset{\ominus}{O}$, $\overset{\oplus}{N}_2$ $\xrightarrow[CH_2(E^{\oplus})]{-N_2}$ O

(ii) *Retro-analysis*

(TM) $\Rightarrow$ $=O$ (SE) + $\overset{\ominus}{C}H_2$ $\overset{\oplus}{}$ (SN) $\xrightarrow[FGA]{SMe_2}$ $\overset{\ominus}{C}H_2 - \overset{\oplus}{S}Me_2$ (SE)

Forward Synthesis

$$(CH_3)_2C{=}O + {}^{\ominus}CH_2-\overset{\oplus}{S}Me_2 \xrightarrow{CH_2(Nu)} (CH_3)_2C(O^{\ominus})-CH_2-\overset{\oplus}{S}Me_2 \xrightarrow{CH_2(E^+)} \text{epoxide}$$

(iii) *Retro-analysis*

(TM) $\Rightarrow$ SE + $\overset{\ominus}{C}H_2$ / $\overset{\oplus}{}$ SN $\xrightarrow[FGA]{PPh_3}$ $\overset{\ominus}{C}H_2-\overset{\oplus}{P}Ph_3$ SE

$\Downarrow$

$CH_2(I)ZnI$
SE

Forward Synthesis

(TM) + $\overset{\ominus}{C}H_2-\overset{\oplus}{P}Ph_3$ $\xrightarrow{CH_2(Nu)}$ $\xrightarrow{CH_2(E^+)}$

Or

(TM) + $CH_2(I)ZnI$ $\xrightarrow{CH_2(Nu)}$ $\xrightarrow{CH_2(E^+)}$

(iv) *Retro-analysis*

(TM) CO_2Et $\Rightarrow$ $(CH_3)_2C{=}O$ + $\overset{\ominus}{C}H-CO_2Et$ ($\oplus$) SN

FGA $\Downarrow$ HCl

$Cl-CH_2-CO_2Et$
SE

Forward Synthesis

Cl⁀CO₂Et/⊖OEt → CO₂Et, Cl, CH₂(Nu), CH₂(E⁺) → CO₂Et

2. *Retro-analysis*

EtO₂C (TM) ⇒ EtO₂C, SN + SE ⇓ EtO₂C, O, OEt, SE

Forward Synthesis

EtO₂C, O, OEt; ⊖OEt / O; CH₂(Nu) → EtO₂C, O, OEt, O; ⊖OEt ↓ EtO₂C, O, OEt⊖, O; C=O (E⁺) ← EtO₂C, O, O

3. *Retro-analysis*

EtO_2C O O CO_2Et (TM) ⇒ 2 EtO_2C ⊖ ⊕ O SN ⇓ O OEt OEt O SE

Forward Synthesis

EtO_2C OEt O $\xrightarrow{\ominus OEt}$ EtO_2C ⊖ EtO O ⊖ CO_2Et OEt O

CH_2(Nu)
C = O (E^+)

O EtO_2C CO_2Et O

4. *Retro-analysis*

CO_2Et CO_2Et (TM) ⇒ CO_2Et CO_2Et ⊕ SN ⇓ CO_2Et CO_2Et Br ⇒ Br Br SE + CO_2Et CO_2Et SE

Forward Synthesis

CO_2Et CO_2Et ; $\overset{\ominus}{O}Et$; Br Br ; CO_2Et CO_2Et Br ; $\overset{\ominus}{O}Et$ CH_2 (Nu) CH_2 (E^+) ; CO_2Et CO_2Et

5. ***Retro-analysis***

CO_2Et ⇒ O ⊖ ⊕ CO_2Et

(TM)

SN

⇓

CO_2Et O OEt SE

Forward Synthesis

CO_2Et O OEt ; $\overset{\ominus}{O}Et$ CH_2(Nu) C = O (E^+) ; CO_2Et O

6. ***Retro-analysis***

O O ⇒ O ⊖ ⊕ O ⇒ O OEt O

(TM) SN SE

Forward Synthesis

SE (keto ester, OEt) —[$\overset{\ominus}{O}Et$; Claisen cond.; $CH_2(Nu)$; $C=O\ (E^+)$]→ cyclohexane-1,3-dione

2.3.4 Ambident electrophile

Synthon	Synthetic Equivalent
$\overset{\oplus}{C}H_2-CH_2-C(=O)-CH_3$	$CH_2=CH-C(=O)-CH_3$
$\overset{\oplus}{C}H=CH-C(=O)-CH_3$	$CH\equiv C-C(=O)-CH_3$

2.3.5 Ambident nucleophile

Synthon	Synthetic Equivalent
$\overset{\ominus}{N}-C-\overset{\ominus}{C}$	$CH_3C(NH_2)=CHCO_2Et$
$\overset{\ominus}{O}-C-\overset{\ominus}{C}$	$CH_3COCH_2CO_2Et$, phenol (OH)
$\overset{\ominus}{O}-\overset{\ominus}{N}$	$\overset{\ominus}{O}-N=O$
$\overset{\ominus}{N}-\overset{\ominus}{C}$	$\overset{\ominus}{C}\equiv N$

2.3.6 Electrophilic nitrogen

Synthon	Synthetic Equivalent
$\overset{\oplus}{N}O_2$	$(HNO_3 + H_2SO_4)$, $(HNO_3 + Ac_2O)$, $(AgNO_3 + BF_3 + CH_3CN)$, $(EtO-NO_2 + \overset{\ominus}{O}Et)$
$\overset{\oplus}{N}O$	HNO_2, $(Et-O-NO + \overset{\ominus}{O}Et)$
$-\overset{\oplus}{N}\equiv N$	$C_6H_5-\overset{\oplus}{N}\equiv N$, $\overset{\ominus}{C}H_2-\overset{\oplus}{N}\equiv N$

Synthon	Synthetic Equivalent
$\overset{\oplus}{N}H_2$	$NH_2 - OCH_3$, $NH_2 - Cl$
$H - \ddot{N}$, $R - \ddot{N}$	$N_3H + h\nu$, $R - N = C = O + h\nu$, $C_6H_5NO_2 + (EtO)_3P$, $R - N_3 + \Delta$
$R - C(=O) - \ddot{N}$	$(RCONH_2 + Br_2 + \overset{\ominus}{O}H)$, $RCON_3 + h\nu$ or Cu/Δ
$>=N^{\oplus}$	$>=N - OH + PCl_5$

2.3.7 Nucleophilic nitrogen and synthetic equivalents

H_3N, CH_3NH_2, Me_2NH, Me_3N

NaN_3, KCN, KNO_2, NH_2OH, CH_3CN

$NH_2 - NH_2$, $PhNH_2$, $PhNHNH_2$, $NH_2 - C(=O) - NH_2$, $NH_2 - C(=O) - NH - NH_2$

$NH_2 - C(=NH) - NH_2$, $NC - CH_2 - CO_2Et$, phthalimide (N–H), $TsNHNH_2$

o-phenylenediamine ($C_6H_4(NH_2)_2$), 2-aminobenzamide ($H_2N-C_6H_4-CONH_2$), $C_6H_5NHNHC_6H_5$

2.3.8 Neutral synthons and synthetic equivalents

Neutral synthons and synthetic equivalents are usually FGI. The knowledge of this type of synthon and synthetic equivalent is very essential in the retrosynthetic analysis. This is because of the fact some functional groups or groupings cannot be introduced directly into a molecule or to an intermediate which are present in the target molecule. In that case a dummy group (SE) is created and then FGI is done as in the TM. These are synthetic equivalents for neutral synthons. This type of synthetic equivalent is the latent functionality.

Synthon		Synthetic Equivalent
1. (i) $CH_3 - C(=O) - CH_2 - CH_3$	Hg^{+2}/H_3O^+	$Me - C \equiv C - Me$
(ii) cis-2-butene (H, H)	$H_2/Pd/BaSO_4$, Lindlar Cat	
(iii) trans-2-butene (H, H)	Li/liqNH_3	
2. (i) cis-$HOCH_2CH=CHCH_2OH$ (H, H; HO, OH)	(i) $\overset{\ominus}{N}H_2$ (ii) $2CH_2O$ (iii) H_3O^+ (iv) Lindlar cat	$H - C \equiv C - H$
(ii) $H - \equiv - CH_2Br$	(i) $\overset{\ominus}{N}H_2$ (ii) CH_2O (iii) H_3O^+ (iv) PBr_3	
(iii) 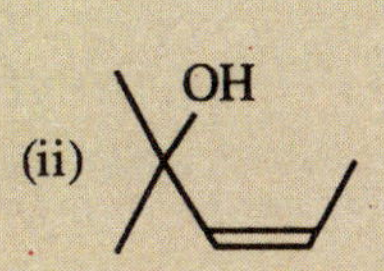	(i) $\overset{\ominus}{N}H_2$ (ii) $(CH_3)_2C{=}O$ (iii) Hg^{+2}/H_3O^+	
3. (i) $- \equiv - Br$	(i) Br_2(1eq.) (ii) KOH	$Me - C \equiv C - H$
(ii) $(CH_3)_2C(OH)CH{=}CHCH_3$ (cis)	(i) $\overset{\ominus}{N}H_2$ (ii) $(CH_3)_2C{=}O$ (iii) Lindlar cat.	
(iii) $- \equiv - CH_2CH(OH)CH_3$	(i) $\overset{\ominus}{N}H_2$ (ii) propylene oxide (iii) $H_3\overset{\oplus}{O}$	
(iv) $CH_3COCH_2COCH_3$	(i) $\overset{\ominus}{N}H_2$ (ii) CH_3COCl (iii) $Hg^{+2}/H_3\overset{\oplus}{O}$	

	Synthon		Synthetic Equivalent
4.			
(i)	CH_3COCH_3	(i) CH_3MgBr(leq), (ii) $H_3\overset{\oplus}{O}$ ⟹	$CH_3-C\equiv N$
(ii)	$-COOH$	$H_3\overset{\oplus}{O}$ ⟹	$CH_3-C\equiv N$
(iii)	$CH_3CH_2NH_2$	H_2/Pd ⟹	$CH_3-C\equiv N$
(iv)	$(CH_3)_2CHNH_2$	(i) CH_3MgBr(leq), (ii) Li/liq NH_3 ⟹	$CH_3-C\equiv N$
(v)	CH_3CONH_2	$\overset{\ominus}{O}H/H_2O_2$ ⟹	$CH_3-C\equiv N$
(vi)	$(CH_3)_3C-NH_2$	$(CH_3)_3C-OH/H_2SO_4$ ⟹	$CH_3-C\equiv N$
5.			
(i)	CH_3CHO	DIBAL ⟹	CH_3CONH_2
		$LiAlH(-O\,Bu^t)_3$ ⟹	CH_3COCl
		$LiAlH(OEt)_3$ ⟹	$-\equiv N$
		DIBAL ⟹	CH_3COOEt
(ii)	$CH_3CH_2NMe_2$	$LiAlH_4$ ⟹	CH_3CONMe_2
(iii)	$CH_3CH_2NH_2$	$LiAlH_4$ or Zn/AcOH ⟹	$CH_3CH{=}NOH$

	Synthon	Reagent	Synthetic Equivalent
6.	$CH_3CH_2NH_2$	$Br_2/\overset{\ominus}{O}H$ ⟹	$CH_3CH_2CONH_2$
		H_2/Pd ⟹	$CH_3CH_2N_3$
		(i) NH_3 (ii) H_2/Pd ⟹	CH_3CHO
7.	CH_3CH_2COOH	(i) $Cl_2/\overset{\ominus}{O}H$ (ii) $H_3\overset{\oplus}{O}$ ⟹	$CH_3CH_2COCH_3$
		(i) O_3 (ii) H_2O_2 ⟹	$CH_3CH_2CH=CH_2$
8.	$(CH_3)_2C=O$	$TiCl_3$, H_2O ⟹	$(CH_3)_2CH-NO_2$
		O_3, H_2O_2 ⟹	$(CH_3)_2C=CH_2$
9.	CH_3CH_2CHO	Conc H_2SO_4 ⟹	$CH_3CH_2CH(OH)COOH$
10.	$R-C(CH_3)_2-OH$	(i) $2CH_3MgBr$ (ii) $H_3\overset{\oplus}{O}$ ⟹	RCO_2Et
11.	$(CH_3)_3C-CO-CH_3$	(i) Mg/C_6H_6 (ii) $H_3\overset{\oplus}{O}$ (iii) PCl_5 ⟹	$(CH_3)_2C=O$, $\xrightarrow{((CH_3)_3C)_2CuLi}$ CH_3COCl

Synthon	Synthetic Equivalent

12. CH_2N_2 $\xrightarrow{KOH}$ $CH_3-N(NO)-OTs$

13. (alkene with two H) / (dimethyl succinic anhydride) $\xrightarrow{Pb(OAc)_4}$ / $\xrightarrow[\Delta]{Ac_2O}$ (2,3-dimethylsuccinic acid: OH, OH)

14. $H-\equiv-OEt$ $\xrightarrow[\text{(ii) } \overset{\ominus}{N}H_2]{\text{(i) EtOH/HCl}}$ $Cl\text{—}CHO$

15. $O=C(CO_2Et)_2$ $\xrightarrow[\text{(ii) } H_3O^{\oplus}]{\text{(i) } HNO_2}$ $CO_2Et\text{—}CO_2Et$

EtO_2C-lactone $\xrightarrow[\text{(ii) epoxide}]{\text{(i) } \overset{\ominus}{O}Et}$

16. lactone $\xrightarrow[\text{(ii) } HCO_2H/\Delta]{\text{(i) } \overset{\ominus}{O}H/H_2O}$ EtO_2C-lactone

17. $Cl\text{—}COOH$ $\xrightarrow{HCl}$ lactone $\xleftarrow{EtOH}$ $HO\text{—}CO_2Et$

18.

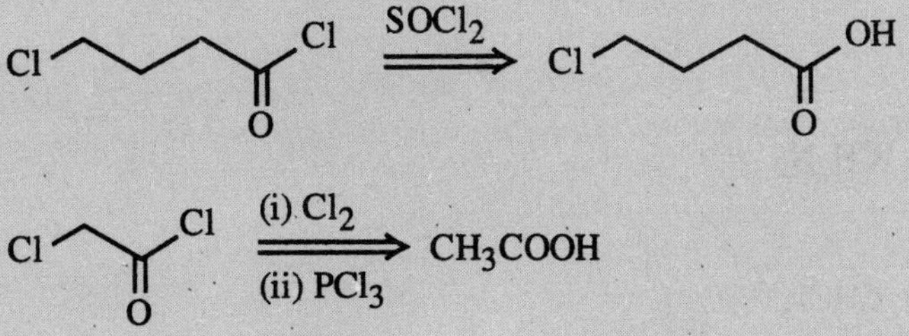

Cl SOCl$_2$ Cl OH

O O

Cl (i) Cl_2 CH_3COOH

Cl (ii) PCl_3

O

O O

(i) $\overset{\ominus}{O}Et$

19. O O CO$_2$Et

(ii) O

O O

O HBr

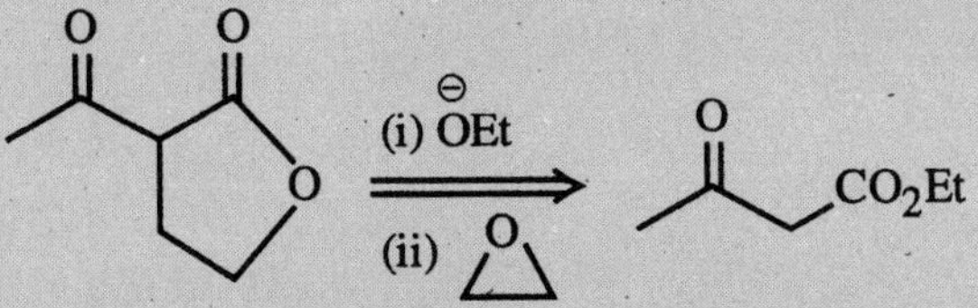

20. Br O

Δ

O

O (i) MeOH

O

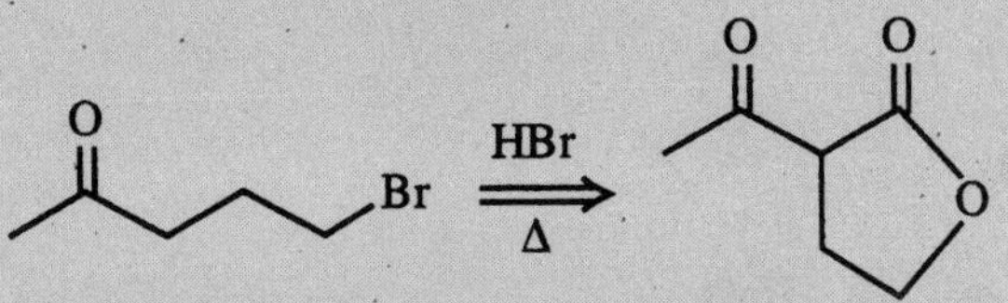

21. MeO$_2$C Cl (ii) SOCl$_2$

O

COOH HO COOH

fuming

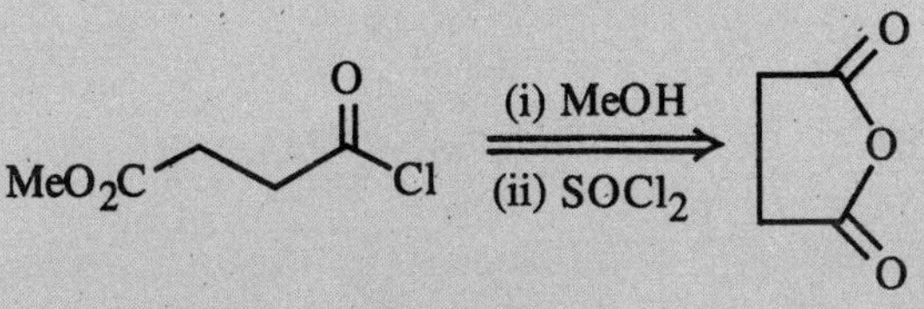

22. O

COOH H$_2$SO$_4$ HO$_2$C COOH

(Citric acid)

Cl (i) HOCl Cl

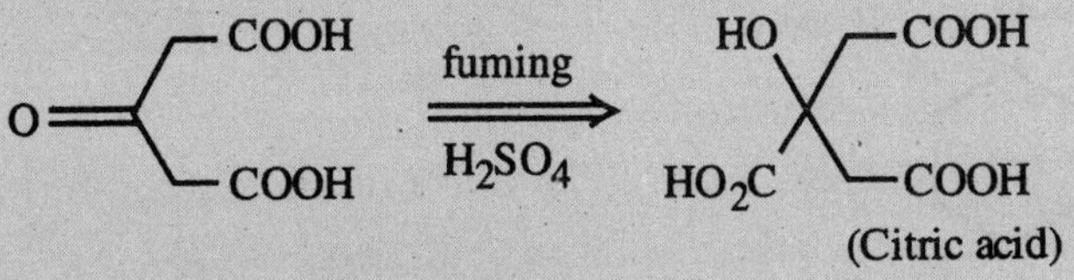

23. O

(epichlorohydrin) (ii) NaH

O

(EtO)$_3$P

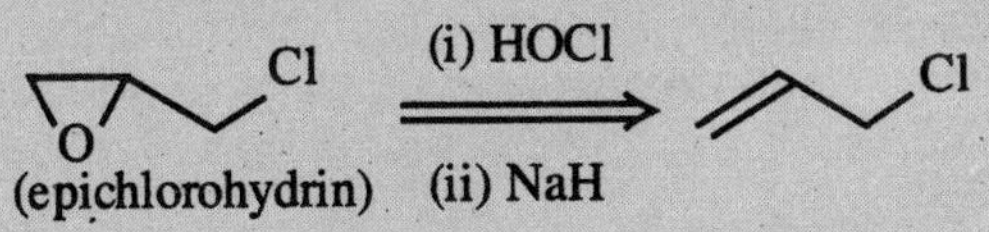

24. $H_5C_6 - C \equiv C - C_6H_5$ H_5C_6 C_6H_5

OR O

(i) N_2H_4

(ii) HgO or $Pb(OAc)_4$

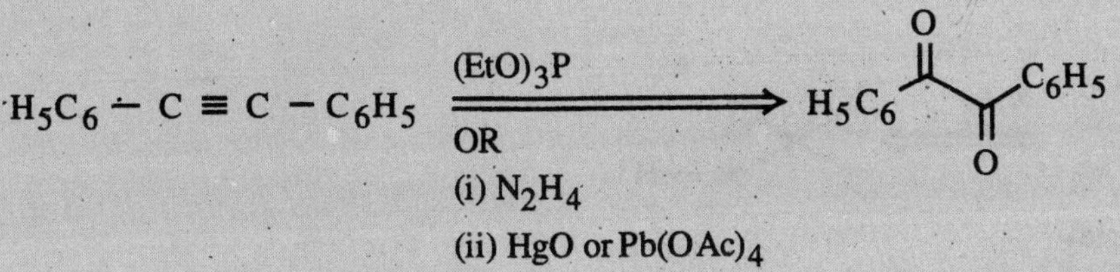

25.

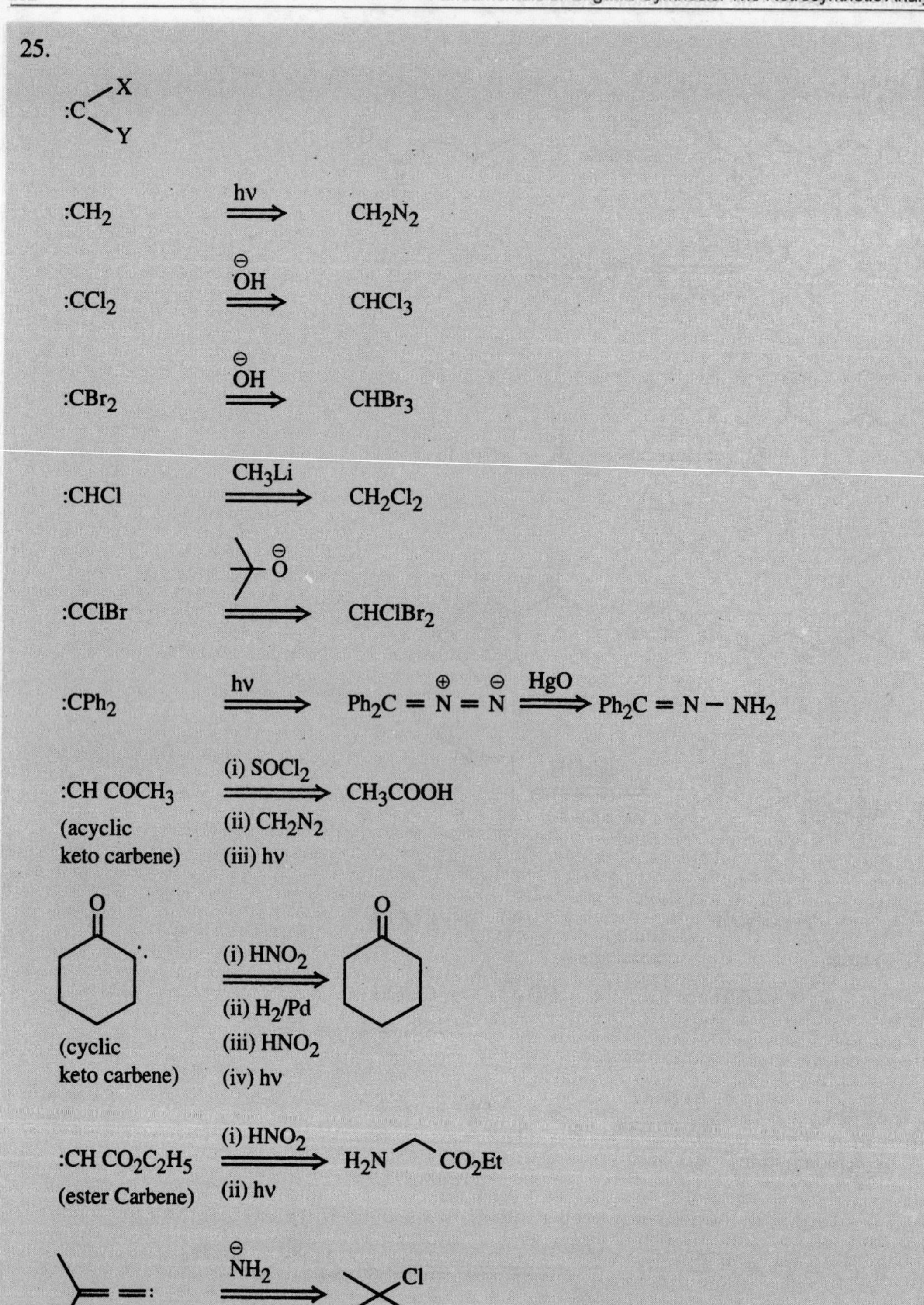
:C X Y
:CH2 hν CH2N2
:CCl2 OH CHCl3
:CBr2 OH CHBr3
:CHCl CH3Li CH2Cl2
:CClBr O CHClBr2
:CPh2 hν Ph2C = N = N HgO Ph2C = N − NH2
:CH COCH3
(acyclic
keto carbene)
(i) SOCl2
(ii) CH2N2
(iii) hν
CH3COOH
O
(i) HNO2
(ii) H2/Pd
(iii) HNO2
(iv) hν
(cyclic
keto carbene)
:CH CO2C2H5
(ester Carbene)
(i) HNO2
(ii) hν
H2N CO2Et
NH2
Cl
− H
(allenic carbene)

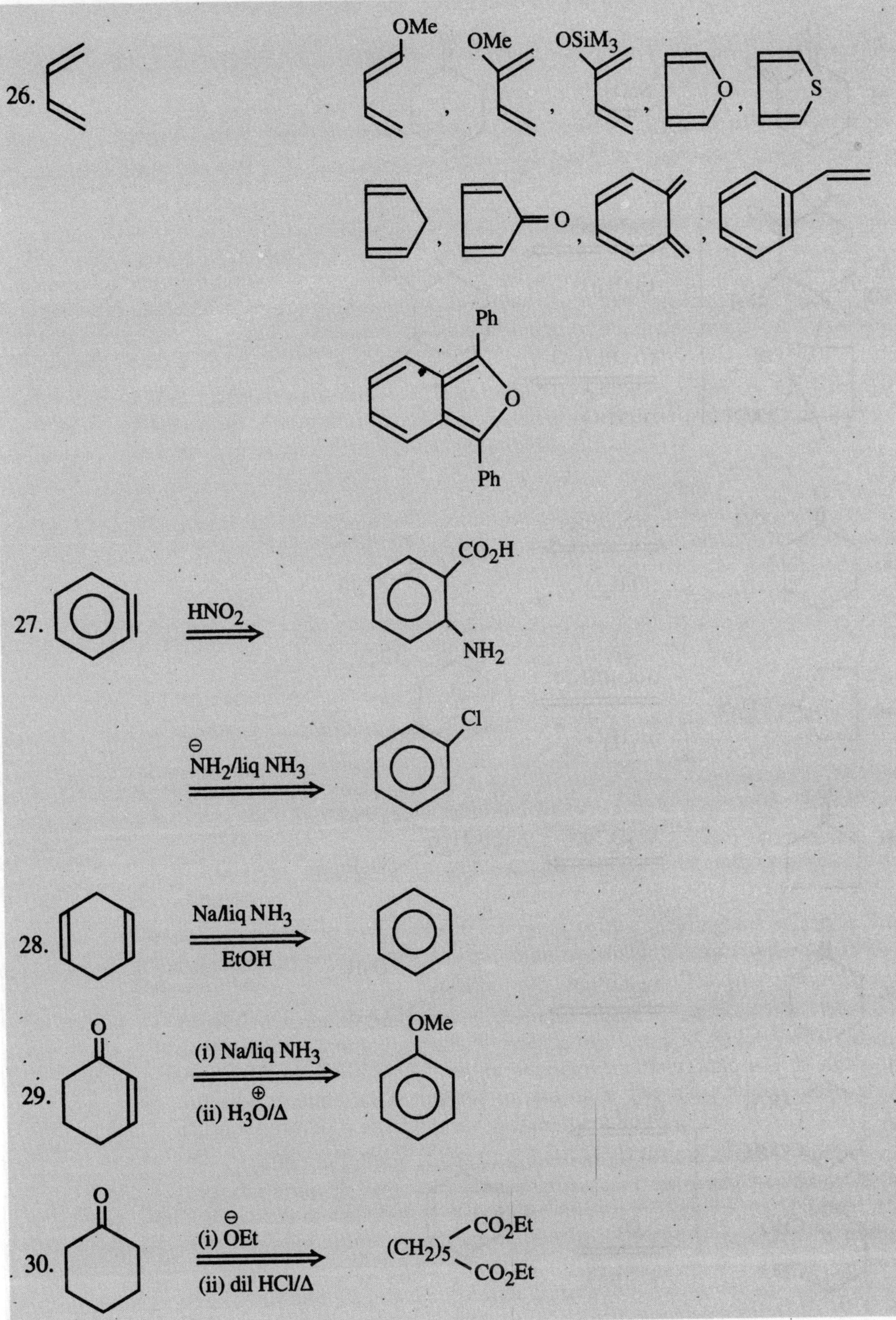

26.
OMe
OMe
OSiM3
O
S
O
Ph
O
Ph
27.
HNO2
CO2H
NH2
NH2/liq NH3
Cl
28.
Na/liq NH3
EtOH
29.
O
(i) Na/liq NH3
(ii) H3O/Δ
OMe
30.
O
(i) OEt
(ii) dil HCl/Δ
(CH2)5
CO2Et
CO2Et

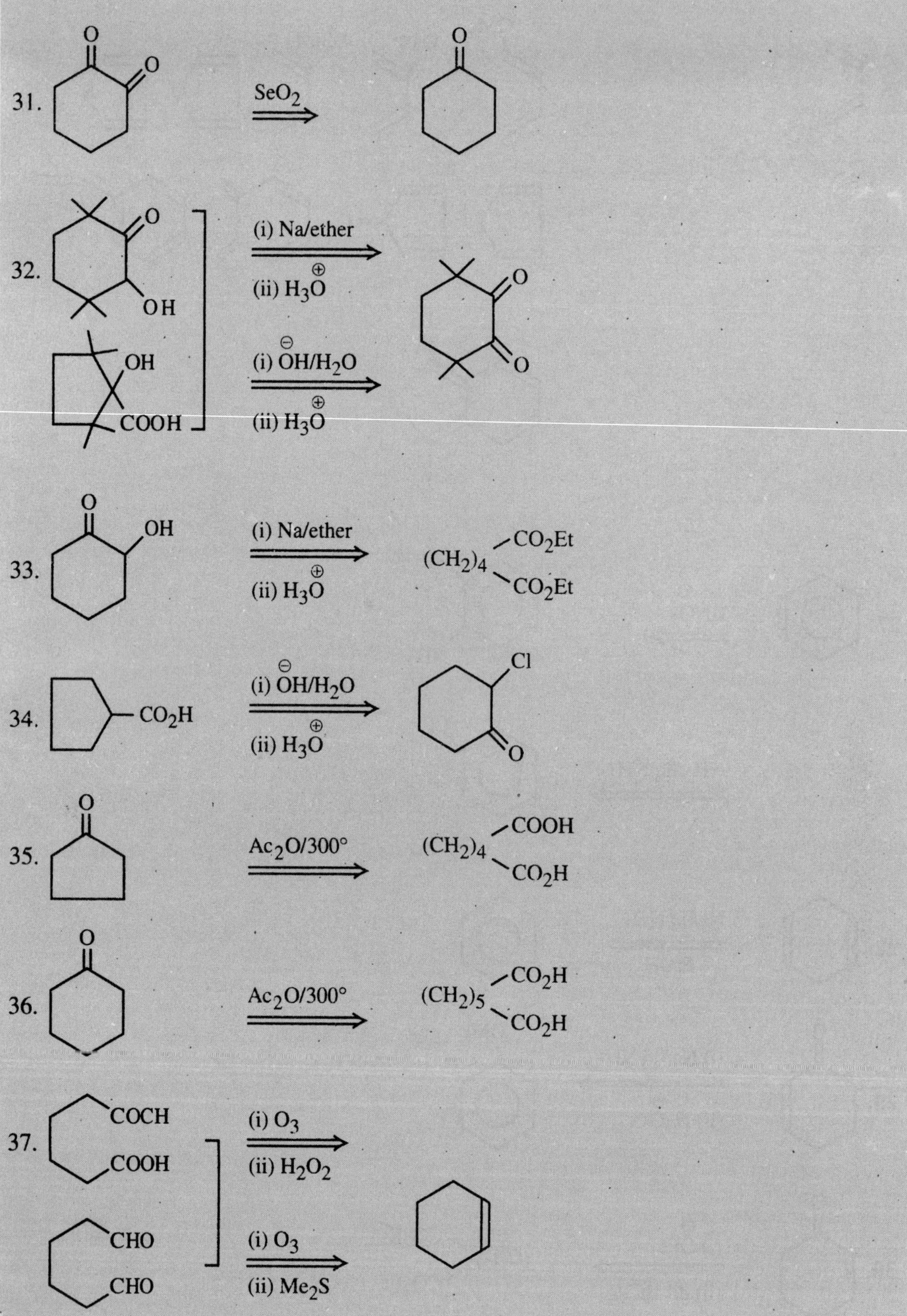

31.
SeO_2
32.
OH
(i) Na/ether
(ii) $H_3\overset{\oplus}{O}$
OH
COOH
(i) $\overset{\ominus}{O}H/H_2O$
(ii) $H_3\overset{\oplus}{O}$
33.
OH
(i) Na/ether
(ii) $H_3\overset{\oplus}{O}$
$(CH_2)_4$
CO_2Et
CO_2Et
34.
CO_2H
(i) $\overset{\ominus}{O}H/H_2O$
(ii) $H_3\overset{\oplus}{O}$
Cl
35.
$Ac_2O/300°$
$(CH_2)_4$
COOH
CO_2H
36.
$Ac_2O/300°$
$(CH_2)_5$
CO_2H
CO_2H
37.
COCH
COOH
(i) O_3
(ii) H_2O_2
CHO
CHO
(i) O_3
(ii) Me_2S

38. (hexanedial: CHO, CHO) $\xRightarrow{Pb(OAc)_4}$ (cyclohexane-1,2-diol: OH, OH)

39. (pentanedial: CHO, CHO) $\xRightarrow{Pb(OAc)_4}$ (cyclopentane-1,2-diol: OH, OH)
$\xRightarrow{H_3\overset{\oplus}{O}}$ (2-methoxy-dihydropyran: O, OMe)

40. (butanedial: CHO, CHO) $\xRightarrow[\text{(ii) } H_2/Ni \quad \text{(iii) } H_3\overset{\oplus}{O}]{\text{(i) } Br_2/MeOH}$ (furan: O) via cyclic bis-acetal (MeO, O, OMe)

41. (4-tert-butylcyclohexanone: O) $\xRightarrow[\text{(ii) } CrO_3]{\text{(i) Raney Ni}/H_2}$ (4-tert-butylphenol: OH)

42. (cyclohexane-1,3-dione: O, O) $\xRightarrow[\text{(ii) } CrO_3/AcOH]{\text{(i) Raney Ni}/H_2}$ (resorcinol: OH, OH)

43. (cyclohexane-1,4-dione: O, O) $\xRightarrow[\text{(ii) } CrO_3/AcOH]{\text{(i) Raney Ni}/H_2}$ (p-benzoquinone: O, O)

44. (benzotrifluoride: CF_3) $\xRightarrow[\text{(ii) } SbF_5]{\text{(i) } Cl_2 + PCl_5}$ (toluene: CH_3)

45. (prenyl chloride: Cl) $\xRightarrow{HCl}$ (isoprene)

46. 3-Chlorocyclopentene $\overset{HCl}{\Longrightarrow}$ cyclopentadiene

47. $Br-CH_2CH_2CH_2CH_2-Br$ $\xrightarrow{\text{(i) } Br_2 \text{ (ii) } H_2/Pd}$ 1,3-butadiene

48. $Me-O-CH_2-Cl$ $\Longrightarrow$ MeOH + CH_2O + HCl

49. Maleic anhydride $\Longrightarrow$ $EtO_2C-C\equiv C-CO_2Et$

$CH_2=CH-CH=C=O \Longrightarrow$ 3-hydroxy-2-pyranone (OH, =O, O)
(Vinyl Ketene)

50. $CHCl_2-CO_2H$ $\xrightarrow[\text{(ii) } H_3\overset{\oplus}{O}]{\text{(i) } \overset{\ominus}{O}H/Cl_2}$ $CH_3COCH_2CO_2Et$

51. $Cl_2C=C=O$ $\xrightarrow[\text{(ii) } Et_3N]{\text{(i) } SOCl_2}$ $CHCl_2-CO_2H$

52. $(CH_3)_2C=C=O$ $\xrightarrow[\text{(ii) } Et_3N]{\text{(i) } SOCl_2}$ $(CH_3)_2CH-CO_2H$

53. Some neutral synthons cannot be used directly during the synthesis because of their self-condensation nature under the given reaction condition.

(i) $CH_2=CH-CHO$ $\xrightarrow[\text{or Conc. } H_2SO_4]{KHSO_4}$ $CH_2(OH)-CH(OH)-CH_2(OH)$
SN SE
(acid sensitive self-condensation)

Aniline (NH_2) + glycerol (OH, OH, OH) + $PhNO_2$ + Conc.H_2SO_4 $\longrightarrow$ quinoline (N)

Conc. H_2SO_4

more reactive than self-condensation

$\overset{\oplus}{H}$

$-H_2O$

$PhNO_2$ <O>

(ii) Me$_2$NH SN SE + CH_2O + Me_2NHCl

Self-cond. in base medium

$Me_2\overset{\oplus}{N}H$ Δ

(iii) H_2N CHO H_2N SN Self-cond. SE (i) Br_2 / (ii) NH_3 CH_3CHO

(a) H_2N H NH_2 $\overset{\oplus}{H_3O}$ H_2N H

H NH_2 (more reactive than self-cond.)

N N H

(b)

$+$ $\xrightarrow{\Delta}$ $\xrightarrow[\Delta]{H_3\overset{\oplus}{O}}$

O O H₂N CH₃ O O N CH₃ N CH₃

(iv)

CHO $\overset{\oplus}{H}$ $\Longrightarrow$ CH_3CHO

SN SE

Self-cond.

$+ \ CH_3CHO \ + \ H_2SO_4 \xrightarrow{\Delta}$

NH_2 N CH_3

$CH_3CHO \xrightarrow{\overset{\oplus}{H}}$ H O CH_3 $\xrightarrow{NH_2}$ OH NH CH_3

(more reactive than self-cond.)

$\downarrow H^{\oplus}$

N CH_3 $\xleftarrow{\langle O \rangle}$ H N H

(v)

$OHC \diagdown\!\!\diagup CHO \Longrightarrow (MeO)_2CH \diagdown\!\!\diagup CH(OMe)_2 \xleftarrow[HCl]{MeOH} OHC \diagdown\!\!\diagup CHO$

SN SE

(Self-cond.)

$(MeO)_2CH \diagdown\!\!\diagup CH(OMe)_2 \xrightarrow[NH_2 - NH_2]{H_3O^{\oplus}}$ N N H

$- 2H_2O$

(vi)

OHC⌒CHO ⟹ MeO–(tetrahydrofuran)–OMe; (i) Br_2/MeOH, (ii) Ni/H_2 → furan

SN (Self-cond.) — SE (Cyclic bis-acetal)

MeO–(tetrahydrofuran)–OMe + HS(CH$_2$)$_3$SH, HCl → bis(1,3-dithiane) (S, S, H, H, S, S)

(i) n BuLi (ii) CH_3I

S, S, CH_3, H, S, S

(i) n BuLi (ii) CH_3I

S, S, CH_3, CH_3, S, S

$Hg^{+2}/H_3\overset{\oplus}{O}$

O= ... =O

(vii)

HO_2C⌒CHO, Conc. H_2SO_4 ⟹ HO_2C–CH$_2$–CH(OH)–COOH

SN (Self-cond.) — (Malic acid) SE

Phenol (OH) + HO_2C–CH$_2$–CH(OH)–COOH + Conc. H_2SO_4 $\xrightarrow{\Delta}$ coumarin

$\overset{\oplus}{O}$–H, H, COOH $\xrightarrow{\Delta}$ OH, OH, OH, O $\xrightarrow{-2H_2O}$ coumarin (O, O)

2.3.9 Various carbanion sources and synthetic equivalents

CH_3MgBr	Me_2CuLi	$\overset{\ominus}{C} \equiv N$
CH_3-O-CH_2MgBr	Ph_2CuLi	Me_3B
$CH_2=CHMgBr$	$(Me-CH=CH)_2CuLi$	$CH_2(ZnI)I$
$CH_3-C\equiv CMgBr$	CH_3Li	$\overset{\ominus}{C}H_2-\overset{\oplus}{S}Me_2$
$H_5C_2O-C\equiv CMgBr$	$Me_2C=\overset{\ominus}{C}Me\overset{\oplus}{Li}$	$\overset{\ominus}{C}H_2-\overset{\oplus}{P}Ph_3$
	Me_2Cd	$\overset{\ominus}{\underset{\ominus}{C}}H-\overset{\oplus}{P}Ph_3$
$Cl-\overset{\ominus}{C}H-CO_2Et$		$^{\ominus}CH(OMe)(\overset{\oplus}{P}Ph_3)$
$\overset{\oplus}{N}_2-\overset{\ominus}{C}H-CO_2Et$	N (1-cyclohexenyl-pyrrolidine)	$\overset{\ominus}{C}H_2-\overset{\oplus}{S}(=O)Me_2$
		$\overset{\ominus}{C}H_2-S(=O)Me$
$BrZnCH_2-CO_2Et$	$Me-\overset{\ominus}{C}H-CH=N-C_6H_{11}$ $\overset{\oplus}{Li}$	$\overset{\ominus}{C}H_2-\overset{\oplus}{N}Me_3$
$BrZnCH_2-CH_2-CO_2Et$	O $SiMe_3$	$CH_2=C(Li)(OMe)$
		OMe, Li (o-lithioanisole)
$CH_2=C(O^{\ominus})H$	$O^{\ominus}$ (arene carbanion)	
$O^{\ominus}$ (cyclohexenolate)	N–H (heteroarene carbanion)	

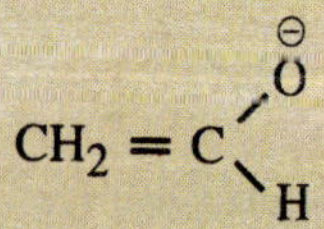

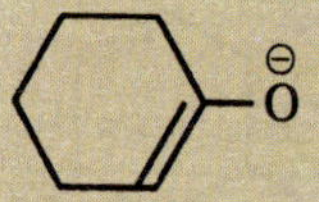

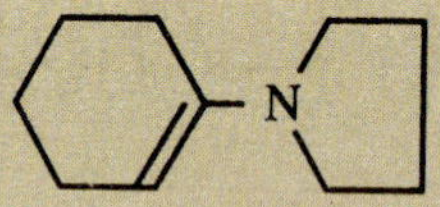

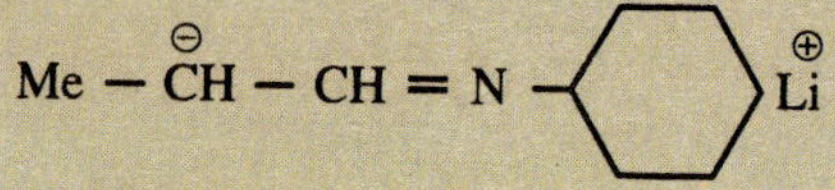

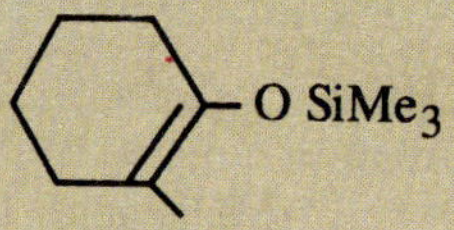

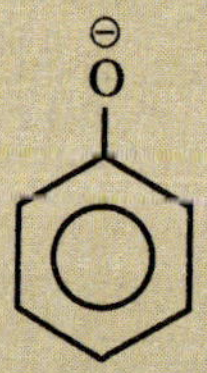

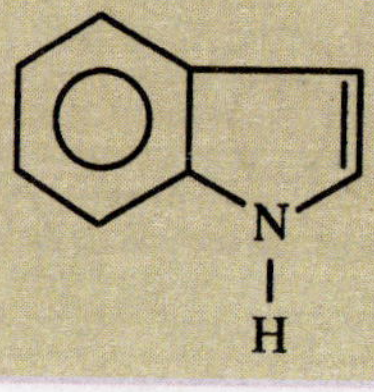

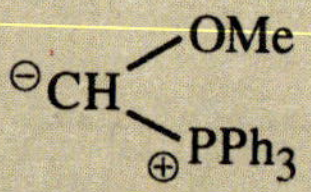

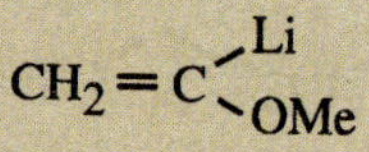

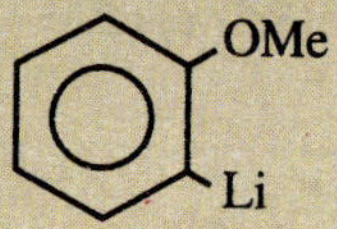

$$\left.\begin{array}{l} \text{(1,3-dithiane anion: } \overset{\ominus}{C}H\text{ between two S, ring)} \\ \overset{\ominus}{C}H = \overset{\oplus}{N}(\overset{\ominus}{O})\text{—}\overset{\ominus}{O} \end{array}\right\} \equiv H-\overset{\ominus}{C}=O$$

$$\left.\begin{array}{l} CH_3\overset{\ominus}{C}\text{(1,3-dithiane)} \text{ or } CH_3\overset{\ominus}{C}(OSiMe_3)CN \\ CH_3-\overset{\ominus}{C}=\overset{\oplus}{N}(\overset{\ominus}{O})\text{—}\overset{\ominus}{O} \\ CH_2=\overset{\ominus}{C}\text{—}OMe \\ H-C\equiv\overset{\ominus}{C} \end{array}\right\} \equiv CH_3-\overset{\ominus}{C}=O$$

$$\left.\begin{array}{l} Ph\overset{\ominus}{C}(CN)OH \\ Ph\overset{\ominus}{C}(OSiMe_3)CN \end{array}\right\} \equiv Ph-\overset{\ominus}{C}=O \qquad \left\{\begin{array}{l} CH_2=CH-CH=\overset{\ominus}{C}\text{—}OMe \\ \equiv CH_3-CH=CH-\overset{\ominus}{C}=O \end{array}\right.$$

$$CH_2=CH-\overset{\ominus}{C}H(OSiMe_3)-CH=CH_2 \equiv \overset{\ominus}{C}H_2-CH_2-C(=O)-CH=CH_2 \quad \left\{\begin{array}{l} CH_2=CH-\overset{\ominus}{C}(OMe)Me \\ \equiv \overset{\ominus}{C}H_2-CH_2-C(=O)CH_3 \end{array}\right.$$

$\overset{\ominus}{C}H(X)Y$

$X = Y = -CO_2Et$ (DEM)

$X = -CN \quad Y = -CO_2Et$ (ECA)

$X = -COCH_3 \quad Y = -CO_2Et$ (EAA)

2.4 Guidelines for disconnections

Regarding disconnection of a target molecule, the following points are to be noted.

1. (i) Synthetic equivalent should be readily market available and cheapest.

(ii) Synthetic equivalent should be easy to handle.

2. (i) Disconnection and synthetic equivalent should be chosen in such a way in which least number of steps involved in the real synthesis for highest possible overall yield and purest grade of products.

(ii) Disconnection and synthetic equivalent should be chosen in such a manner in which in the real synthetic steps it should have a reasonable mechanism either correspond to known reliable reaction or name reaction.

3. Use of two-group disconnections always preferred wherever possible.

4. There are two important jobs for disconnection.

(i) It is the analytical and logical searching to find out the best synthetic equivalent. More the different kinds of disconnections of a TM more the different kinds of synthons and hence more corresponding synthetic equivalents. Now, choice will have to be done (points 1 and 2).

That's why the retrosynthesis is termed as the synthon approach to organic synthesis. There are usually a very large number of possible synthetic routes for the synthesis of a given target molecule. It is impractical actually to try each method in the laboratory to see which is the best as this would be costly, time consuming and frustrating.

(ii) Disconnection also gives information for more reasonable synthetic route which is indicated by synthons and which are intermediates in the real synthetic route. During disconnection, synthon is chosen not only for C-skeleton but also for desired functional group(s) at the required site.

Therefore, choice of best synthetic equivalents so that the real synthetic route is free from chemo-, regio- or undesired stereo problems, but with desired carbon-skeleton and also functional groups at the required site is the ultimate target in the retrosynthetic analysis.

The above points are the "LOGIC" of disconnections.

It is to be noted, there is no 'right' or 'best' synthesis for any given target molecule. One may be able to devise shorter, more stereochemically controlled, higher yielding, pure, more versatile and short-cut method to synthesize any given target molecule—but it does not mean 'right' or 'best' synthesis because synthesis depends on one's laboratory conditions, availability of starting materials and reagents or sub-synthesis of starting materials and lastly costing which is economic interest.

Retrosynthetic analysis of a target molecule may give various synthons and their corresponding synthetic equivalents and consequently vast database potential transformations are possible for a given target molecule. Therefore, the subject remains largely a matter of personal opinion and interpretation. In this respect organic synthesis is like the game of 'chess', in which the rules are simple to learn but the effective application may take many years to perfect.

It is the aim of this chapter to teach the rules and to provide some hints and guidelines for their application.

Guideline 1

Before disconnection of a TM, FGA or FGI may be done (specially oxygen based functional groups) to facilitate C—C disconnection so that recognisable starting materials i.e. synthetic equivalents are available in the market.

Examples

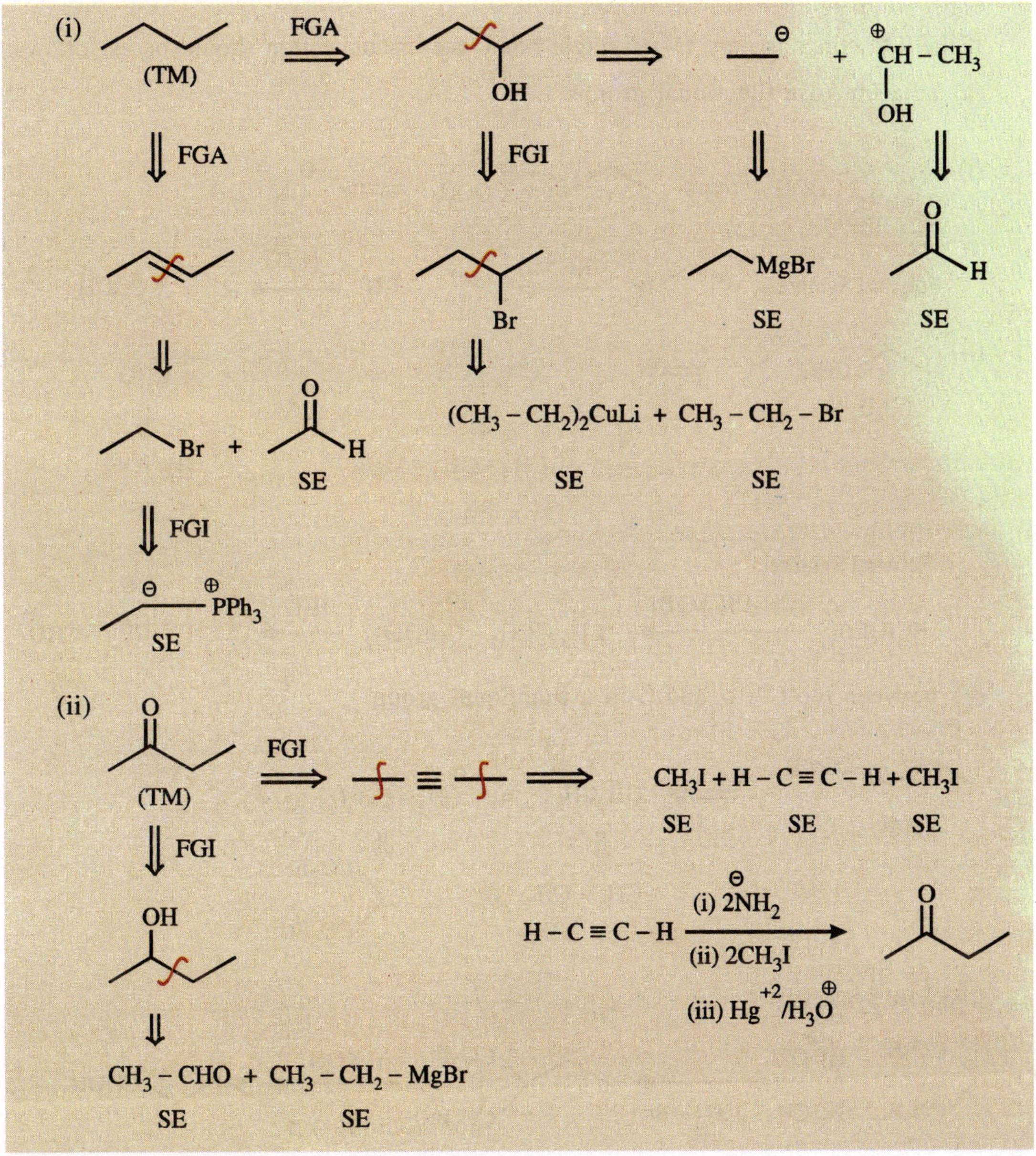

(i) (TM) $\xrightarrow{FGA}$ (OH) $\Rightarrow$ $^{\ominus}$ + $\overset{\oplus}{CH}-CH_3$ (OH)

FGA ⇓ (Br + H, SE) FGI ⇓ (Br) ⇓ MgBr SE ⇓ (O, H) SE

$(CH_3-CH_2)_2CuLi + CH_3-CH_2-Br$

SE SE

⇓ FGI

$\overset{\ominus}{\ } \overset{\oplus}{PPh_3}$

SE

(ii) (TM) $\xrightarrow{FGI}$ $\equiv$ $\Rightarrow$ $CH_3I + H-C\equiv C-H + CH_3I$

SE SE SE

⇓ FGI

OH

⇓

$CH_3-CHO + CH_3-CH_2-MgBr$

SE SE

$$H-C\equiv C-H \xrightarrow[\text{(ii) } 2CH_3I \quad \text{(iii) } Hg^{+2}/H_3\overset{\oplus}{O}]{\text{(i) } 2\overset{\ominus}{N}H_2}$$

Forward Synthesis

$CH_3-CHO + CH_3-CH_2-MgBr \longrightarrow$ (OMgBr) $\xrightarrow{H_3\overset{\oplus}{O}}$ (OH) $\xrightarrow{CrO_3 \mid Py}$ (O)

Guideline 2

If the TM contains only C—C single bond, the disconnection should be as follows.

(a) adjacent to a functional groups

(i) COOH (TM) $\Longrightarrow$ Br + $\overset{\ominus}{C}OOH$ $\Longrightarrow$ $\overset{\ominus}{C}N$

Forward Synthesis, Br $\xrightarrow{KCN}$ CN $\xrightarrow{H_3\overset{\oplus}{O}}$ COOH

(ii) CHO (TM) $\Longrightarrow$ $CH_3-\overset{\ominus}{C}H_2$ + $\overset{\oplus}{C}HO$

$\Downarrow$ $\Downarrow$

CH_3-CH_2-MgBr $HC(OEt)_3$

SE SE

Forward Synthesis

$HC(OEt)_3 \xrightarrow{CH_3CH_2MgBr} CH_3-CH_2-CH(OEt)_2 \xrightarrow{H_3\overset{\oplus}{O}} CH_3-CH_2-CHO$

(b) between the C's α and β to a functional group

COOH (TM) $\Longrightarrow$ $CH_3\overset{\oplus}{C}H_2$ + $\overset{\ominus}{C}H_2-CO_2H$

$\Downarrow$ $\Downarrow$

CH_3-CH_2-Br $CH_2(CO_2Et)_2$

SE SE

Forward Synthesis

$CH_2(CO_2Et)_2 \xrightarrow[\text{(ii) } CH_3-CH_2-Br]{\text{(i) } \overset{\ominus}{O}Et}$ (CO₂Et)₂ $\xrightarrow[\text{(ii) } H_3\overset{\oplus}{O}/\Delta]{\text{(i) KOH}}$ COOH

(c) between the C's β and γ to a functional groups

Ph–CO–CH₂–CH₂ ⌇ CN (TM) ⟹ PhCOCH₂CH₂⊕ + ⊖CN

⇓ PhCOCH=CH₂ (SE) ⇓ KCN (SE)

Forward Synthesis

$$PhCOCH=CH_2 \xrightarrow[AlEt_3]{KCN} PhCOCH_2CH_2CN$$

(d) adjacent to a branching carbon in a carbon chain

$(CH_3)_2C(OH)CH_2CH_3 \Longrightarrow 2\overset{\ominus}{C}H_3 + EtO_2C\text{–}CH_2CH_3$ (SE)

⇓ $2CH_3MgBr$ (SE)

Forward Synthesis

$$CH_3CH_2CO_2Et \xrightarrow[\text{(ii) } H_3\overset{\oplus}{O}]{\text{(i) } 2CH_3MgBr} (CH_3)_2C(OH)CH_2CH_3$$

Guideline 3

(i) If the TM contains only C—C single bonds and two functionalised carbon atoms close together (separated by not more than three other carbons) one should try to disconnect in between the functional groups.

Example

(a) CH_3COCH_2 ⌇ $COCH_3$ (TM) ⟹ $CH_3COCH_2^{\ominus}$ + $CH_3\overset{\oplus}{C}O$

⇓ $CH_3COCH_2CO_2Et$ (SE) ⇓ CH_3COCl (SE)

Forward Synthesis

$$CH_3COCH_2CO_2Et \xrightarrow[\text{(ii) } CH_3COCl]{\text{(i) } NaH/C_6H_6} CH_3COCH(COCH_3)CO_2Et \xrightarrow[\Delta]{\text{dil HCl}} CH_3COCH_2COCH_3$$

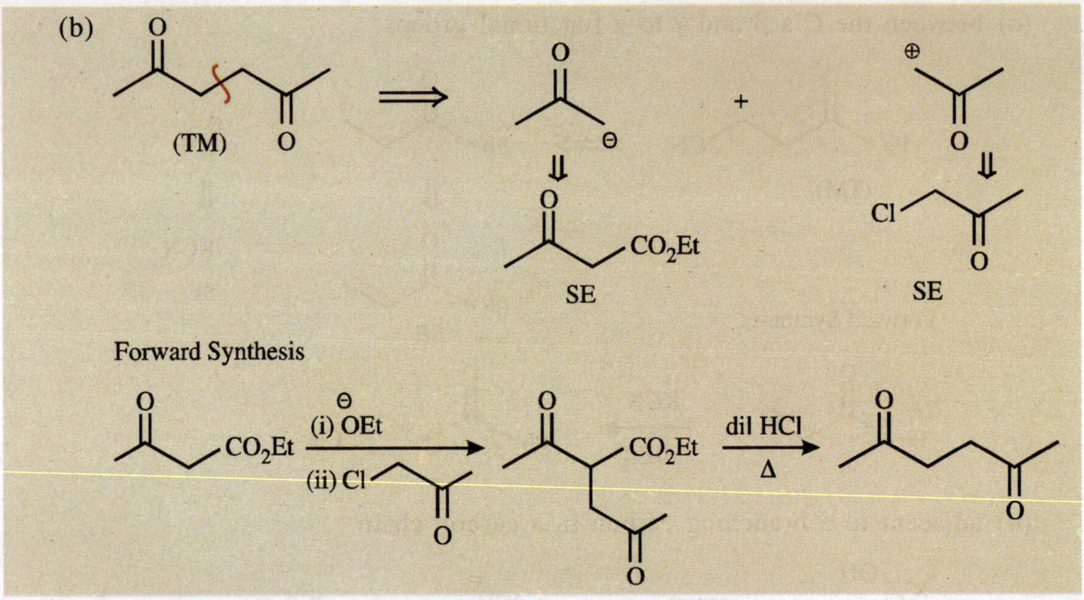

(b)

(TM)

SE SE

Forward Synthesis

(i) OEt
(ii) Cl

dil HCl
Δ

(ii) If the functionalised carbons are further apart, separate disconnections should be done as in Guide line 1.

Guideline 4

For compounds (TM) consisting of two parts joined by a heteroatom (O, N, S) disconnect next to the hetero atom.

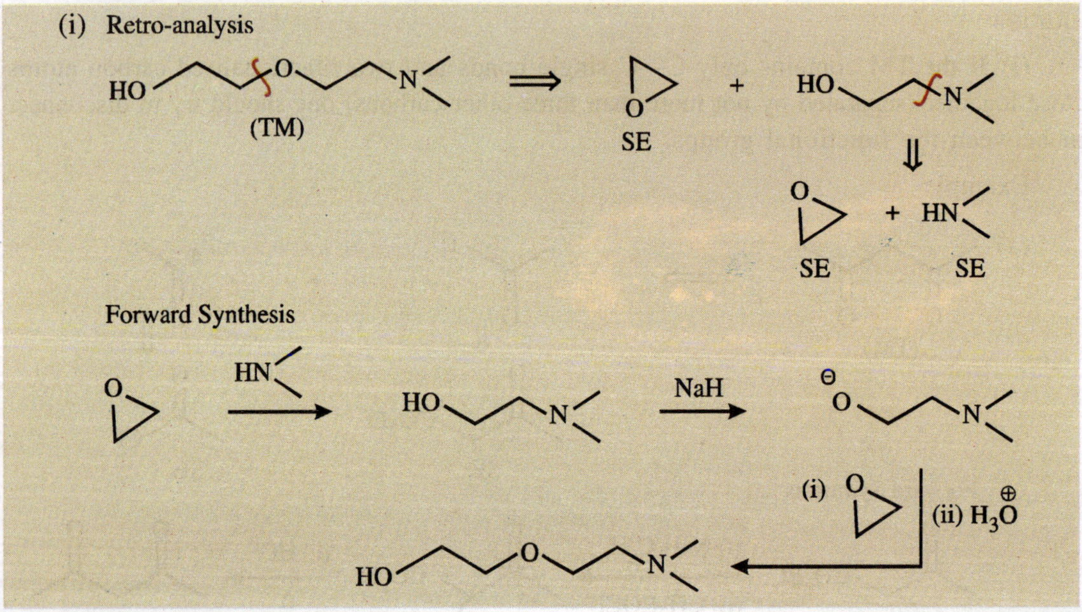

(i) Retro-analysis

HO (TM)

SE

HO

HN

SE SE

Forward Synthesis

HN

HO

NaH

(i)

(ii) $H_3\overset{\oplus}{O}$

HO

(ii) Retro-analysis

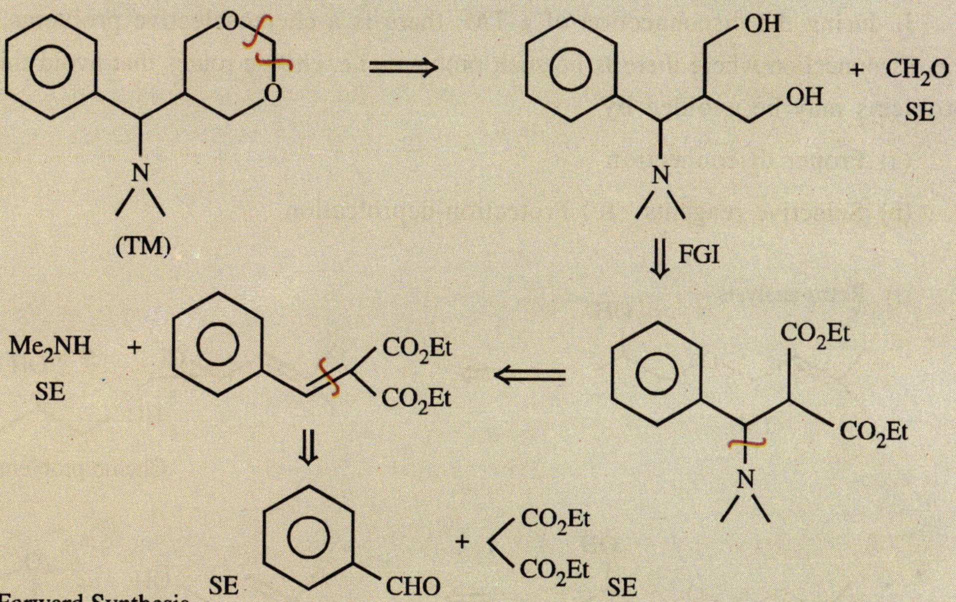

(TM) ⟹ + CH_2O SE

⇓ FGI

Me_2NH SE + ... CO_2Et CO_2Et ⟸ ... CO_2Et CO_2Et

⇓

SE CHO + CO_2Et CO_2Et SE

Forward Synthesis

CHO → ($CH_2(CO_2Et)_2$, $\ominus$OEt) → CO_2Et CO_2Et → (Me_2NH) → CO_2Et CO_2Et → ($LiAlH_4$) → OH OH → (CH_2O, BF_3) → TM

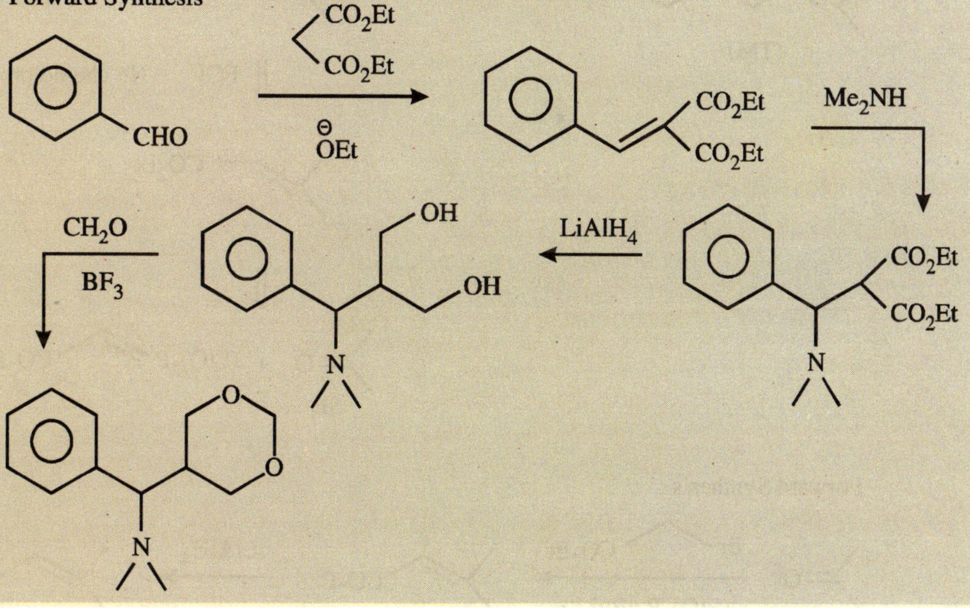

Review synthetic problems

With retrosynthetic analysis how can you synthesis the following TM.

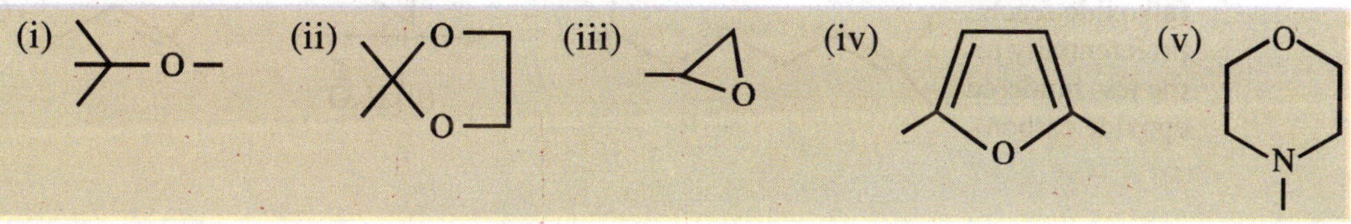

(i) (ii) (iii) (iv) (v)

Guideline 5

If during the disconnection of a TM, there is a chemoselective problems, consider the disconnection where there is no such problems i.e. choose routes that avoid this. Chemo problems may be avoided by

(a) Proper disconnection

(b) Selective reagents (c) Protection-deprotection

(i) Retro-analysis

OH

(TM)

Cl

:OH

HÖ

Chemo problems

OH

(TM)

OH + O

SE

FGI No chemo problems

CO_2Et

O + $(EtO)_3P$ CO_2Et

SE

Forward Synthesis

O — Br CO_2Et; $(EtO)_3P$,nBuLi wittig → CO_2Et — $LiAlH_4$ → OH

NaH

(alkoxide reacts preferentially to the less hindered epoxide carbon) OH O ← (i) epoxide (ii) $H_3O^{\oplus}$ — $O^{\ominus}$

(ii) Retro-analysis

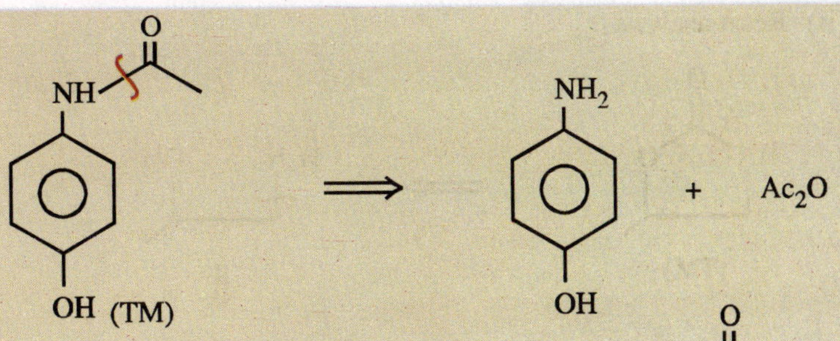

p-HO–C_6H_4–NHCOCH$_3$ (TM) $\Longrightarrow$ p-HO–C_6H_4–NH_2 + Ac_2O

Forward Synthesis

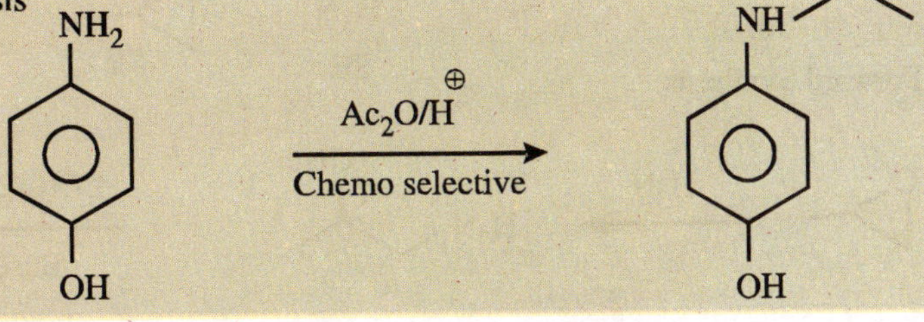

p-HO–C_6H_4–NH_2 $\xrightarrow[\text{Chemo selective}]{Ac_2O/H^{\oplus}}$ p-HO–C_6H_4–NHCOCH$_3$

Review synthetic problems

Give the forward synthesis with retro-analysis of the following TM.

(i) CO_2Me, CO_2Et

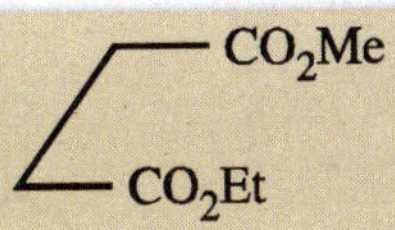

(ii) CHO

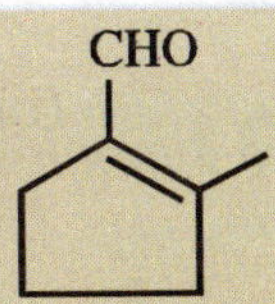

(iii) OH

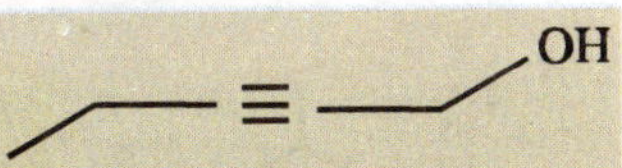

(iv) OH, O, Ph, OH

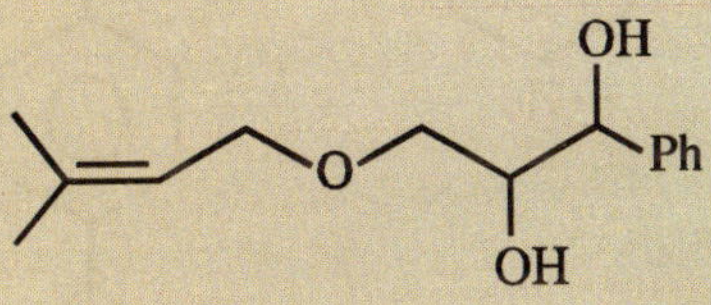

(v) Ph, O, O, Ph, O

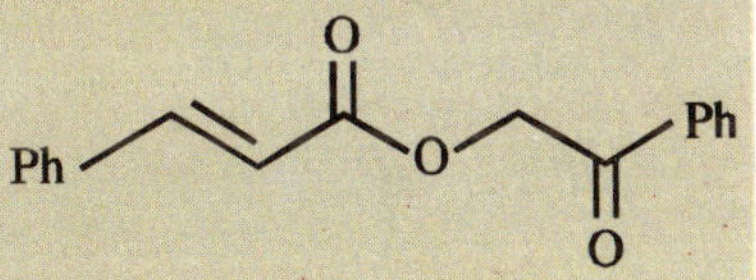

(vi) O, O

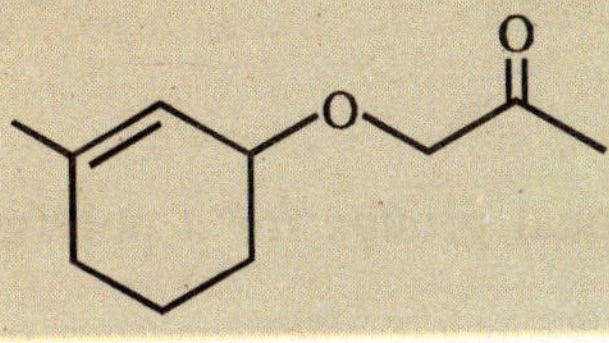

(vii) H, O, O, O

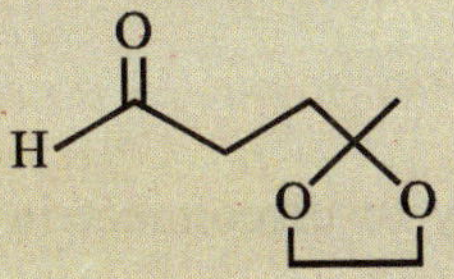

Guideline 6

Sometimes double disconnection gives short-cut and simple synthetic equivalent.

(i) Retro-analysis

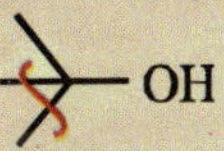

$(CH_3)_3C$–OH (TM) $\Longrightarrow$ $2CH_3MgBr + CH_3CO_2Et$

3° alcohol with two identical groups

SE always 2eq. same grignard reagent and ester

Forward Synthesis, $CH_3CO_2Et \xrightarrow[\text{(ii) } H_3\overset{\oplus}{O}]{\text{(i) } 2CH_3MgBr}$ $(CH_3)_3C$–OH

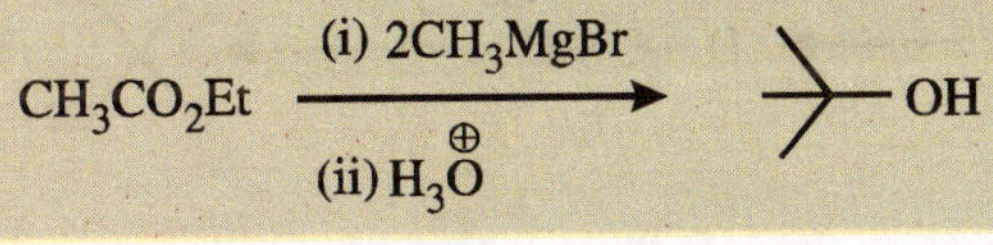

(ii) Retro-analysis

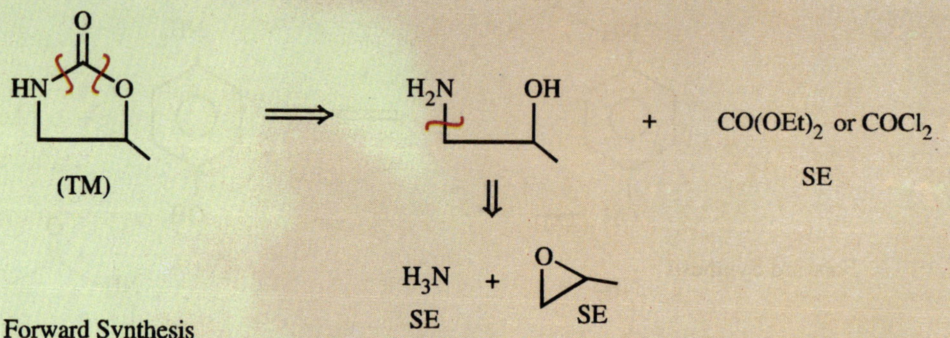

(TM) ⟹ H_2N–CH$_2$–CH(OH)–CH$_3$ + $CO(OEt)_2$ or $COCl_2$ (SE)

⇓

H_3N (SE) + propylene oxide (SE)

Forward Synthesis

propylene oxide $\xrightarrow{H_3N}$ H_2N–CH$_2$–CH(OH)–CH$_3$ $\xrightarrow[\text{Base}]{CO(OEt)_2}$ H–N oxazolidinone

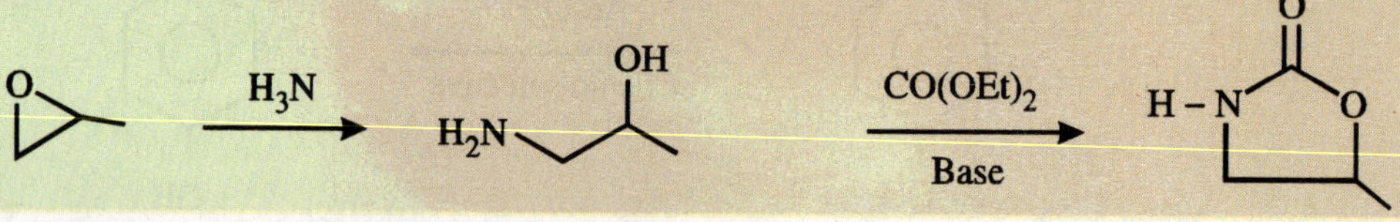

Review synthetic problems

Give the forward synthesis with retro-analysis.

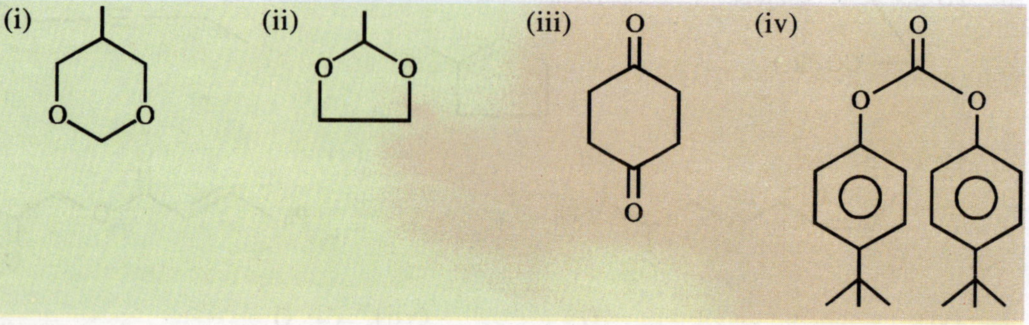

(i) (ii) (iii) (iv)

Guideline 7

Always try to disconnect in between functional groups for a polyfunctional TM.

(i) Retro-analysis

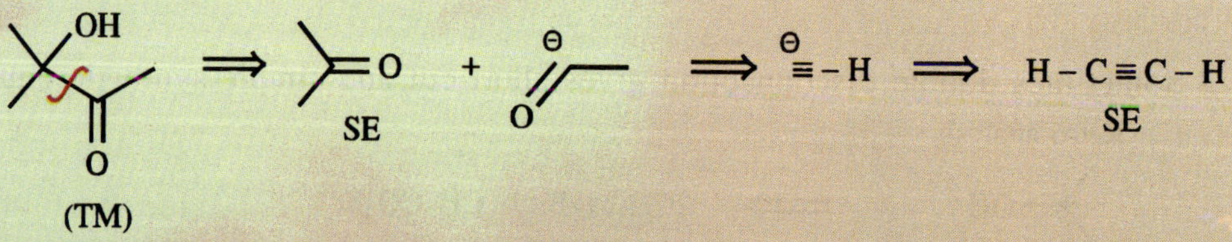

(TM) ⟹ acetone (SE) + acyl anion ⟹ $^{\ominus}$≡–H ⟹ H–C≡C–H (SE)

Forward Synthesis

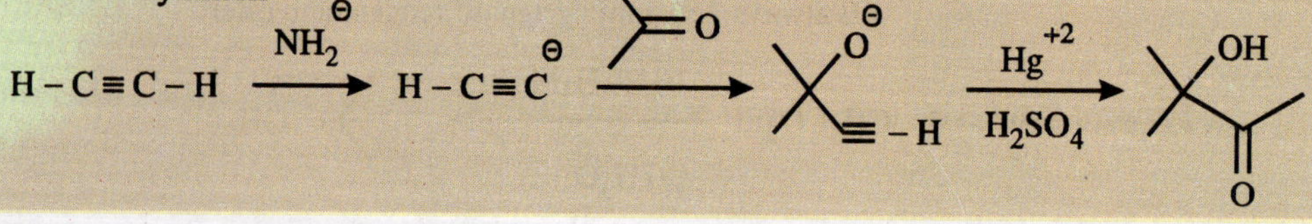

$$H-C\equiv C-H \xrightarrow{NH_2^{\ominus}} H-C\equiv C^{\ominus} \xrightarrow{\text{acetone}} (CH_3)_2C(O^{\ominus})-C\equiv C-H \xrightarrow[H_2SO_4]{Hg^{+2}} (CH_3)_2C(OH)-CO-CH_3$$

(ii) Retro-analysis

$$\text{(TM)}\ CH_3CH(OH)COOH \Rightarrow CH_3-CHO + \overset{\ominus}{C}OOH \Rightarrow \overset{\ominus}{C}N$$

SE SE

Forward Synthesis

$$CH_3CHO \xrightarrow{KCN} CH_3-CH(OH)CN \xrightarrow{H_3\overset{\oplus}{O}} CH_3CH(OH)COOH$$

Review synthetic problems

Give the forward synthesis with retro-analysis.

(i) $CH_3COCOOH$ (ii) $CH_3COCH_2CO_2Et$

Guideline 8

Symmetrical disconnection i.e. disconnection should be done in such a way so that same synthetic equivalents are obtained (to avoid side products).

(i) Retro-analysis

$$\text{(TM)}\ (CH_3)_2CH-CH(OH)-C(CH_3)_2-CHO \Rightarrow (CH_3)_2CH-CHO + (CH_3)_2CH-CHO \Rightarrow (CH_3)_2C{=}O + MeO-CH{=}PPh_3$$

SE

$$\Downarrow$$

$$(CH_3)_2C{=}O + MeO-CH{=}PPh_3$$

SE

Forward Synthesis

$$(CH_3)_2C{=}O \xrightarrow[\text{(ii) } H_3\overset{\oplus}{O}]{\text{(i) } MeO-CH=PPh_3} (CH_3)_2CH-CHO \xrightarrow[TsOH]{\text{pyrrolidine (N–H)}} (CH_3)_2C{=}CH-N\text{(pyrrolidine)}$$

$$(CH_3)_2C{=}CH-N\text{(pyrrolidine)} \xrightarrow[\text{(ii) } H_3\overset{\oplus}{O}]{\text{(i) } (CH_3)_2CH-CHO} (CH_3)_2CH-CH(OH)-C(CH_3)_2-CHO$$

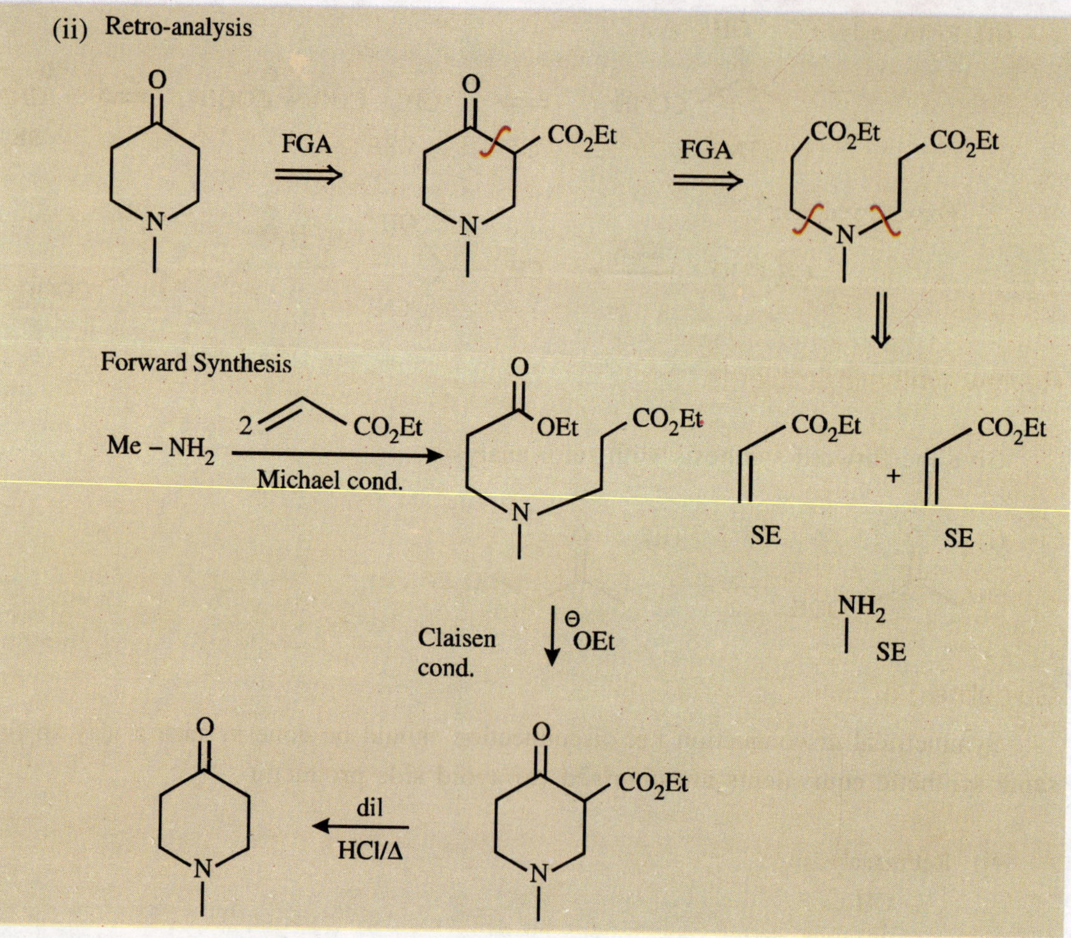

Review synthetic problems

Give the forward synthesis with retro-analysis.

(i) OH, CHO

(ii) O, Ph, Ph, OH

(iii) CO_2Et, Ph, Ph, O

(iv) O, O

(v) O, O, CO_2Et

(vi) HO, CO_2H, COOH, CO_2H

Guideline 9

If a TM having C—C double bonds, the retrosynthetic analysis are as shown below.

(i) If the TM contains isolated double bonds disconnection should be done in such a way that wittig reaction can be tried for its forward synthesis.

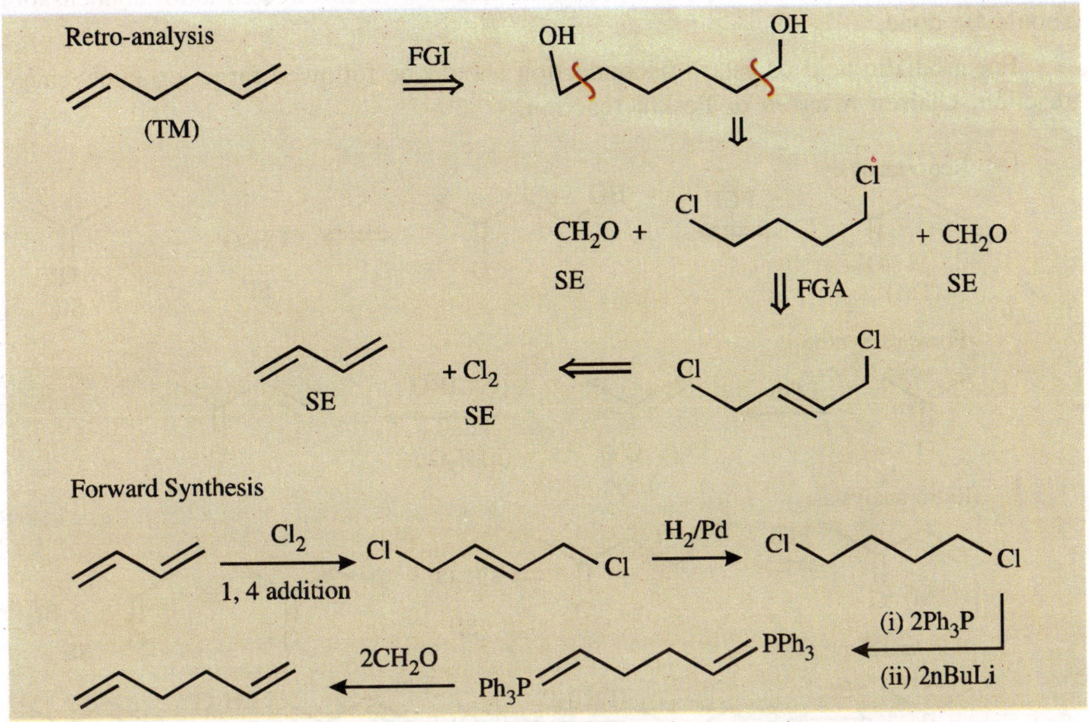

Retro-analysis

(TM) $\xrightarrow{FGI}$ OH ... OH ⇓ CH_2O + Cl ... Cl + CH_2O
SE SE
⇓ FGA
SE + Cl_2 ⟸ Cl ... Cl
SE

Forward Synthesis

$\xrightarrow[\text{1, 4 addition}]{Cl_2}$ Cl ... Cl $\xrightarrow{H_2/Pd}$ Cl ... Cl

(i) $2Ph_3P$
(ii) 2nBuLi

Ph_3P ... PPh_3 $\xrightarrow{2CH_2O}$

(ii) If double bonded carbons are attached to hydrogen and cis or trans alkene required then one should try via corresponding alkyne.

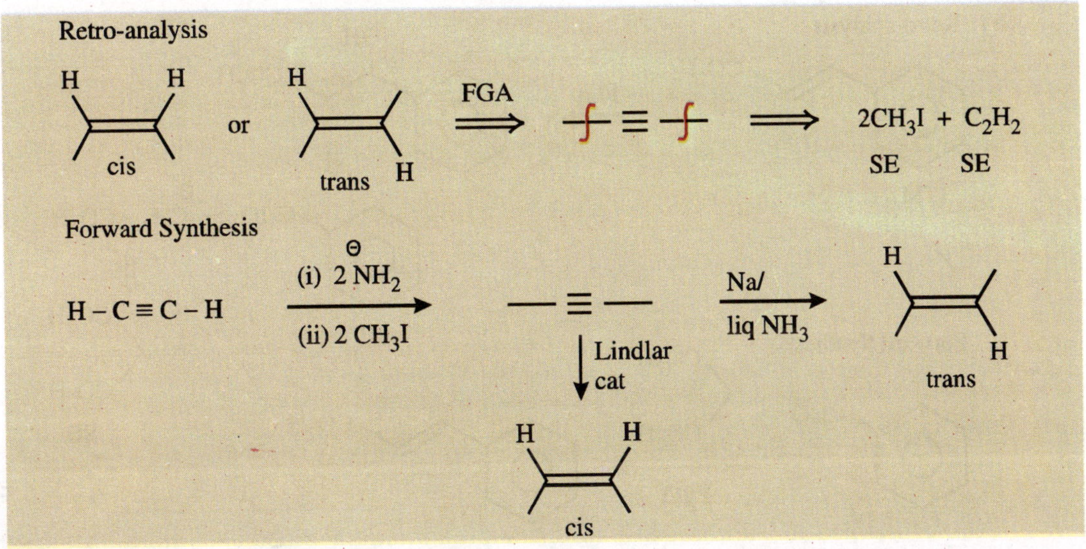

Retro-analysis

H H cis or H H trans $\xrightarrow{FGA}$ —≡— ⟹ $2CH_3I + C_2H_2$
SE SE

Forward Synthesis

$H-C\equiv C-H$ $\xrightarrow[\text{(ii) } 2\, CH_3I]{\text{(i) } 2\, NH_2^{\ominus}}$ —≡— $\xrightarrow[\text{liq } NH_3]{Na/}$ H H trans

Lindlar cat

H H cis

(iii) If the TM having a C—C double bond conjugated to a – M groups (carbonyl, ester, cyanides, nitro), disconnection should be followed targeting aldol condensation, Mannich reaction, knoevenagel reaction, Reformatsky reaction or wittig reaction with stabilised ylide (more nucleophilic than basic).

For aldol type disconnection, if same SE (i.e. same two carbonyl) then normal aldol condensation and if different SE (i.e. different carbonyl) then directed aldol condensation should be done.

For aromatic acid or ester, disconnection should be followed targeting knoevenagel reaction, Claisen reaction or Perkin reaction.

(a) Retro-analysis

(TM) FGI ⟹ HO ... ⟹ CH_2O (SE) + (SE)

Forward Synthesis

LDA → O⊖ 100% → (i) CH_2O (ii) $H_3\overset{\oplus}{O}/\Delta$ →

Retro-analysis

(TM) FGI ⟹ HO ... ⟹ CH_2O (SE) + Br ... ⟹ ... + Br_2 (SE)

Forward Synthesis

Br_2 / HOAc → Br ... → $(EtO)_3P$ / nBuLi → $(EtO)_3P$ ... → CH_2O / $H_3\overset{\oplus}{O}$ →

(b) Retro-analysis

(TM) CO_2H FGI ⟹ OH ... CO_2H ⟹ CHO (SE) + $\overset{\ominus}{C}H_2 - CO_2H$ ⟹ CO_2H / CO_2H (SE)

Forward Synthesis

CHO → CO_2H / CO_2H, Py/Δ → CO_2H

(iv) If the TM having a C—C double bond at the βγ position with a – M group, disconnection should be followed targeting wittig reaction.

If – M group is – CO_2Et then Reformatsky reaction may also be tried.

Retro-analysis

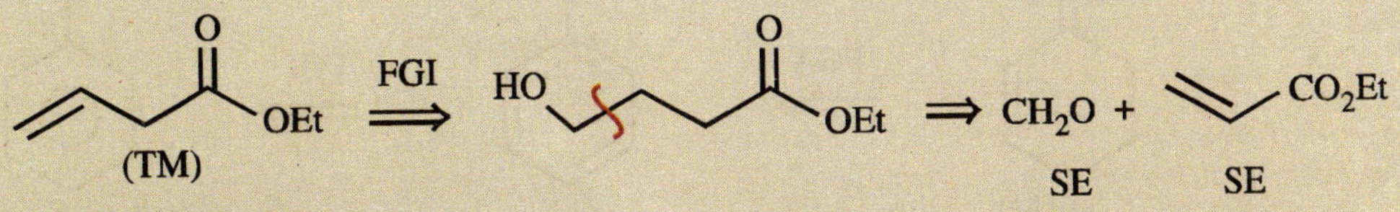

(TM) $\xrightarrow{\text{FGI}}$ HO...OEt $\Rightarrow$ CH_2O + ...CO_2Et

SE SE

Forward Synthesis

...CO_2Et $\xrightarrow{\text{HBr}}$ Br...CO_2Et $\xrightarrow{\text{Zn/ether}}$ BrZn...CO_2Et

(i) CH_2O (ii) $H_3\overset{\oplus}{O}/\Delta$

...CO_2Et

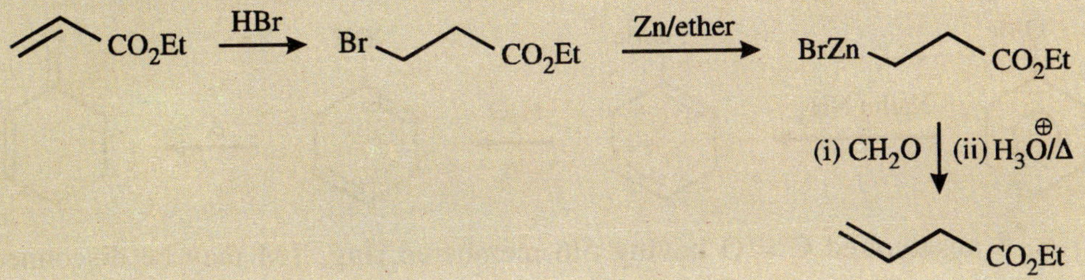

(v) If the TM having C—C double bond at the γδ position with a – M group, disconnection should be done targeting condensation of EAA with allyl halide.

Retro-analysis (TM) $\Rightarrow$ ...Cl + EtO_2C...

SE SE

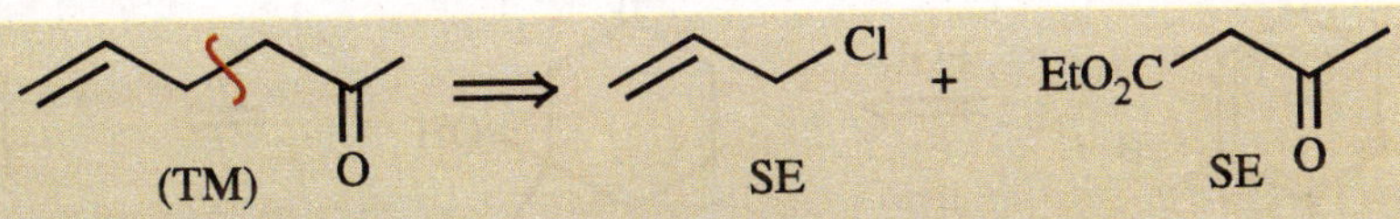

Forward Synthesis

EtO_2C... SE $\xrightarrow{\overset{\ominus}{O}Et}$ (...Cl) ...CO_2Et $\xrightarrow[\Delta]{\text{dil HCl}}$ ...

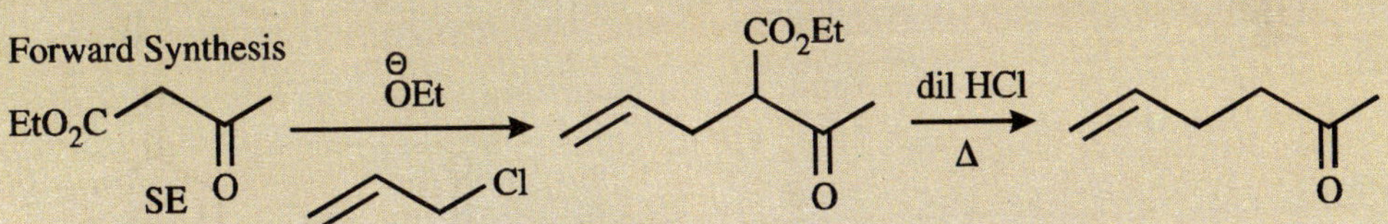

Review synthetic problems

Show the forward synthesis with retrosynthetic analysis of the following TM.

1.

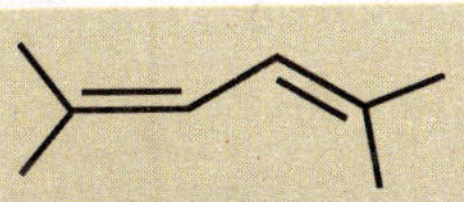

2. 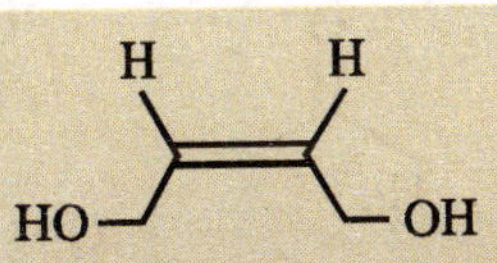 H H HO OH

3.

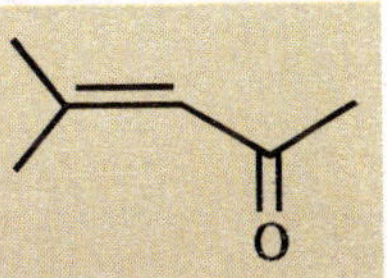

4.

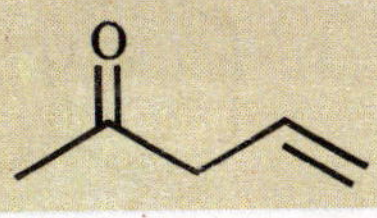

5.

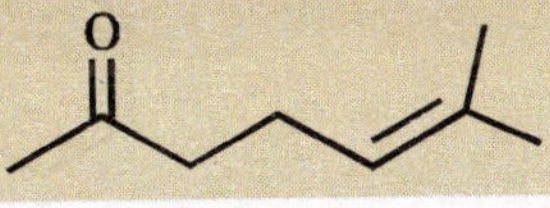

Guideline 10

(a) If the TM is 6-membered ring having α, β unsaturated ketone, first of all it should be converted to (by FGA and FGI) suitable benzenoid derivative which should be the synthetic equivalent and then Birch reduction is tried.

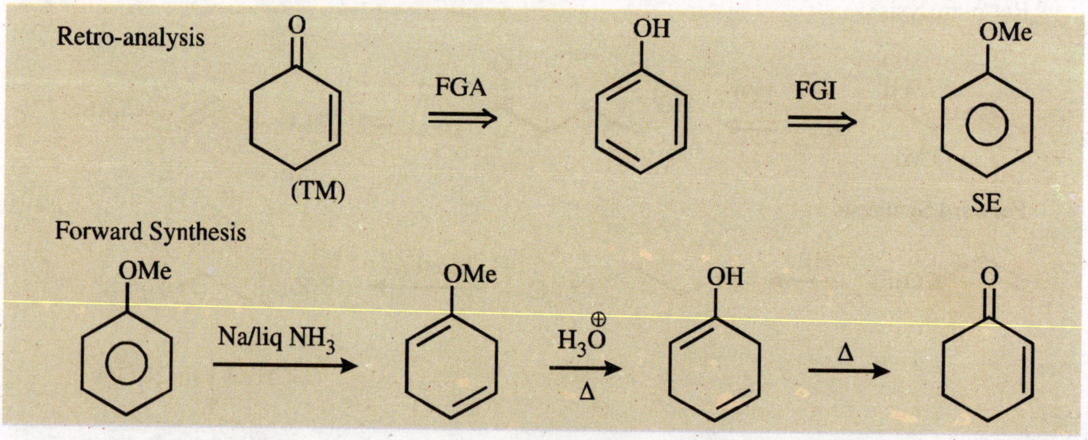

Retro-analysis

(TM) FGA ⟹ OH FGI ⟹ OMe SE

Forward Synthesis

OMe —Na/liq NH_3→ OMe —$H_3\overset{\oplus}{O}$, Δ→ OH —Δ→ O

(b) α, β unsaturated C = O having 5/6 membered ring, TM may be disconnected targeting aldol condensation.

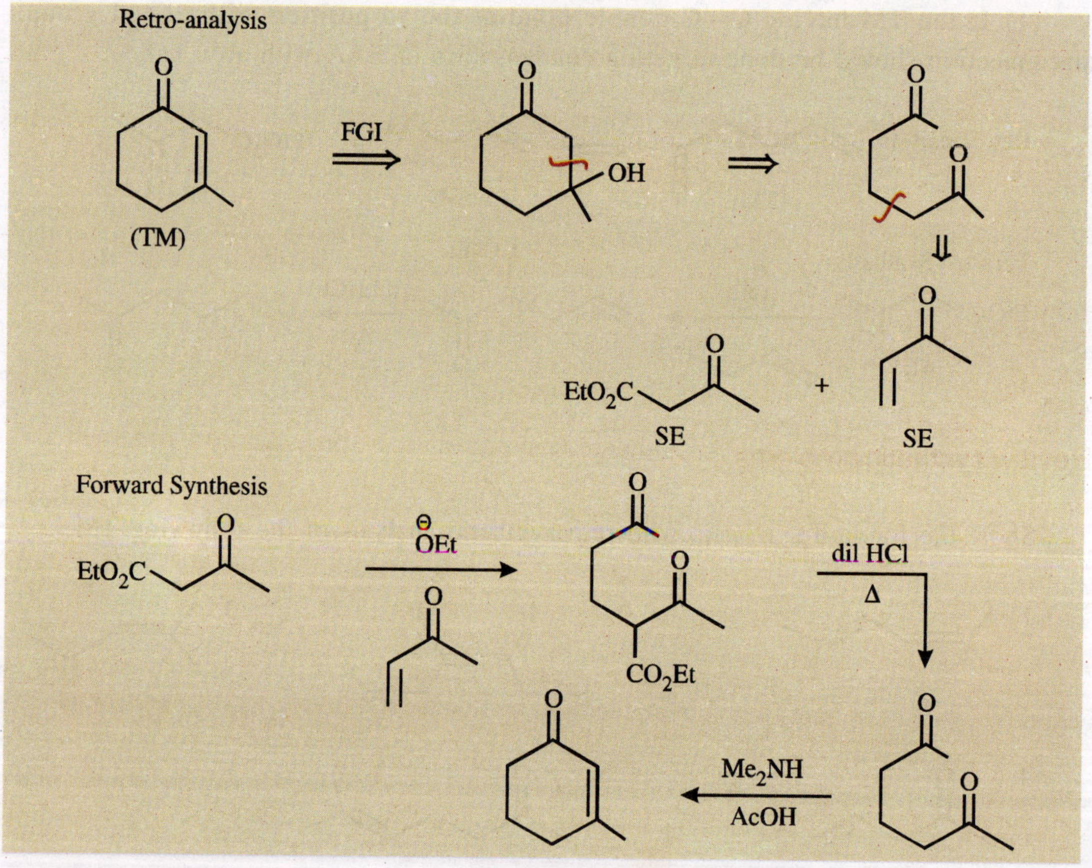

Retro-analysis

(TM) FGI ⟹ OH ⟹ ⟹ EtO_2C (SE) + (SE)

Forward Synthesis

EtO_2C —$\overset{\ominus}{O}Et$→ CO_2Et —dil HCl, Δ→ —Me_2NH, AcOH→

(c) If the TM is bicyclic ring having α, β unsaturated ketone then Robinson annelation should be tried (the sequence of Michael condensation followed by cyclisation, usually by aldol is known as Robinson annelation).

Retro-analysis

(TM) $\overset{\text{FGI}}{\Longrightarrow}$ OH $\Longrightarrow$ $\Downarrow$

SE + SE

Forward Synthesis

$\xrightarrow{\overset{\ominus}{\text{O}}\text{Et}}$ $\xrightarrow[\Delta]{\text{Me}_2\text{NH/AcOH}}$

Review synthetic problems

Show the forward synthesis with retro-analysis of the following TM.

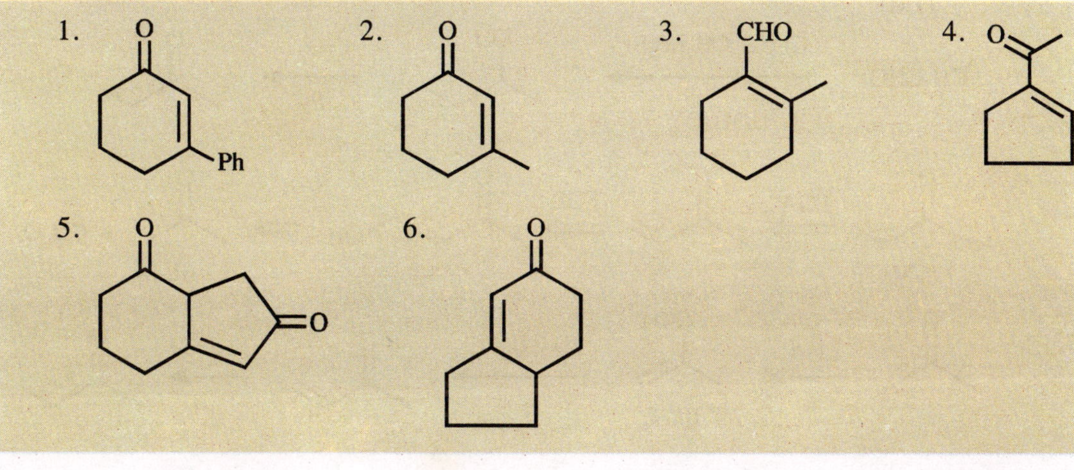

Guideline 11

A. Disconnection of acyclic monoketones.

The different retrosynthetic analysis and forward synthesis are shown below.

(i) (TM) $\xRightarrow{FGI}$ $-\!\!\equiv\!\!-$ $\Longrightarrow$ $2CH_3I + H-C\equiv C-H$ (SE, SE)

$H-C\equiv C-H \xrightarrow[\text{(ii) } CH_3I]{\text{(i) } \overset{\ominus}{N}H_2}$ $-\equiv-$ $\xrightarrow{Hg^{+2}/H_2SO_4}$ butanone

(ii) (TM) $\Longrightarrow$ $CH_3COCH_2CO_2Et$ (SE) + CH_3I (SE)

$CH_3COCH_2CO_2Et \xrightarrow[\text{(ii) } CH_3I]{\text{(i) } \overset{\ominus}{O}Et}$ $CH_3COCH(CH_3)CO_2Et$ $\xrightarrow[\Delta]{\text{dil HCl}}$ butanone

(iii) (TM) $\Longrightarrow$ CH_3COOH + CH_3CH_2MgBr (SE)

$\Downarrow$

SE CH_3CN

$CH_3CN \xrightarrow[\text{(ii) } H_3\overset{\oplus}{O}]{CH_3CH_2MgBr}$ butanone

(iv) (TM) $\xRightarrow{FGI}$ butan-2-ol $\Longrightarrow$ CH_3CHO (SE) + CH_3CH_2MgBr (SE)

$CH_3CHO \xrightarrow[\text{(ii) } H_3\overset{\oplus}{O}]{\text{(i) } CH_3CH_2MgBr}$ butan-2-ol (OH) $\xrightarrow[Py]{CrO_3}$ butanone

(v) (TM) $\xRightarrow{FGA}$ $CH_3COCH=CH_2$ $\xRightarrow{FGI}$ $CH_3COCH_2CH_2OH$ $\Longrightarrow$ CH_3COCH_3 (SE) + CH_2O (SE)

$CH_3COCH_3 \xrightarrow{LDA}$ enolate ($O^{\ominus}$) 100% $\xrightarrow[\text{(ii) } H_3\overset{\oplus}{O}/\Delta]{\text{(i) } CH_2O}$ $CH_3COCH=CH_2$ $\xrightarrow{H_2/Pd}$ butanone

(vi) Butanone ⟹ acetyl anion + $CH_3CH_2^{\oplus}$ ⟹ CH_3CH_2Br SE; acetyl anion ⇓ 2-lithio-2-methyl-1,3-dithiane anion

CH_3CHO —(i) HS–CH₂CH₂CH₂–SH /HCl, (ii) nBuLi→ 2-methyl-1,3-dithiane anion —(i) CH_3CH_2Br, (ii) $Hg^{+2}/H_3O^{\oplus}$→ butanone

B. Disconnection of cyclic monoketones

(i) (a) For 5/6-membered ring, acyl bond is disconnected followed by creation of – CO_2Et group at both the ends are recommended and Dieckmann condensation should be tried.

Cyclopentanone (TM) ⟹ EtO_2C–CH₂CH₂CH₂CH₂–CO_2Et ⟹ 2DEM + Br–CH₂CH₂CH₂–Br

EtO_2C–(CH₂)₄–CO_2Et —$^{\ominus}$OEt→ 2-ethoxycarbonylcyclopentanone —dil HCl, Δ→ cyclopentanone

(b) 5/6-membered ring may be disconnected targeting acyloin condensation.

Cyclopentanone (TM) —FGA⟹ cyclopentane-1,2-dione ⟹ EtO_2C–CH₂CH₂CH₂–CO_2Et ⟹ 2DEM + Br – CH_2 – Br

Forward Synthesis

EtO_2C–(CH₂)₃–CO_2Et —(i) Na/xylene, (ii) $H_3O^{\oplus}$→ 2-hydroxycyclopentanone —Zn, CH_3COOH→ cyclopentanone

(c) first convert 5/6-membered ring to α, β unsaturated CO by FGA and then disconnection is done targeting aldol and then 1, 5 dico targeting Michael condensation.

FGA
aldol dix
Michael dix
+
CO_2Et
(TM)

Forward Synthesis

CO_2Et
(i) $\overset{\ominus}{O}Et$
(ii)
CO_2Et
dil HCl
Δ
Me_2NH | AcOH
H_2/Pd

(d) 6-membered cyclic monoketone may be obtained from benzenoid SE.

OH
OH
FGI
FGA
(TM)
SE

Forward Synthesis

OH
OH
Raney Ni/H_2
CrO_3

(ii) 4-membered cyclic monoketone may be disconnected either targeting umpolung or acyloin condensation.

(a)
Cl
Br
$+ CH_2O$
SE
SE

(i) HS–CH$_2$CH$_2$CH$_2$–SH /HCl, (ii) $Ph_3C^{\ominus}$; Cl~~~Br; (i) $Ph_3C^{\ominus}$; Hg^{+2}, $H_3O^{\oplus}$

(b)

FGA ⟹ SE ⟹ CO$_2$Et / CO$_2$Et

Forward Synthesis

CO$_2$Et / CO$_2$Et → Na/xylene, TMSCl → OTMS / OTMS → $H_3O^{\oplus}$ → O / OH → Zn | AcOH → O

(iii) 3-membered cyclic monoketone is disconnected targeting umpolung.

⟹ Cl~~Br + CH_2O

Forward Synthesis

CH_2O → (i) HS–CH$_2$CH$_2$CH$_2$–SH /HCl, (ii) nBuLi → Cl~~Br → nBuLi → Hg^{+2}, $H_3O^{\oplus}$

Review synthetic problems

With retrosynthetic analysis show the forward synthesis for the following TM.

1. 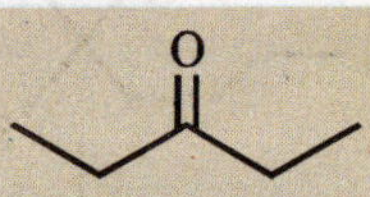 2. 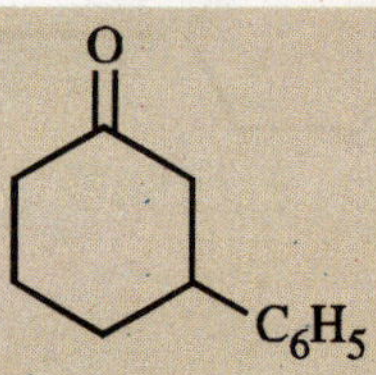 C_6H_5 3. C_6H_5

4. O 5. O

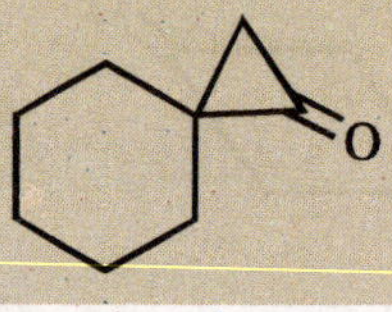

Guideline 12

Disconnection of 1, 2-dioxygen system.

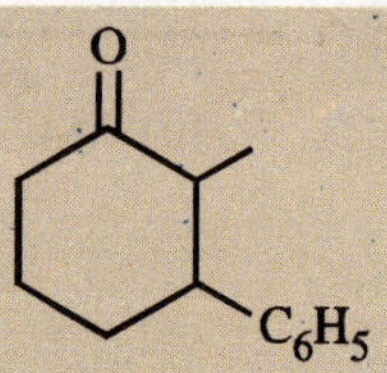

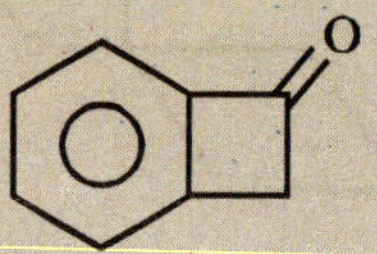

Dioxygen system may be dihydroxy, dicarbonyl, diester or diacid, keto ester or acid, hydroxy carbonyl or ester or acid.

A. Acyclic system

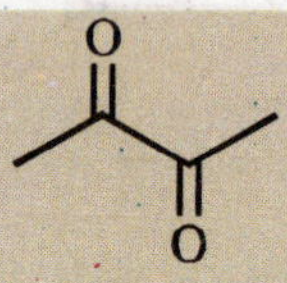

O O (TM)

The different retro-analysis and forward synthesis are shown below.

(i) (TM) $\xRightarrow{FGI}$ $\xRightarrow{FGI}$ $\Rightarrow 2CH_3I + H-C\equiv C-H$

Forward Synthesis

$H-C\equiv C-H \xrightarrow[(ii)\ CH_3I]{(i)\ \overset{\ominus}{N}H_2}$ $\xrightarrow{Hg^{+2}/H_2SO_4}$ $\xrightarrow{HNO_2}$ N – OH $\xrightarrow{H_3\overset{\oplus}{O}}$ diketo

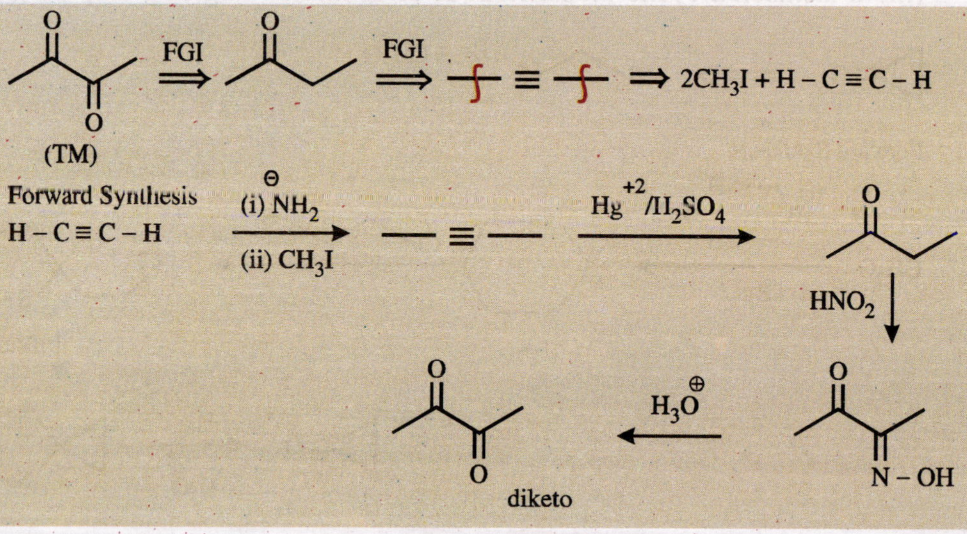

(ii) (TM) ⇒ $CH_3\overset{\oplus}{C}O$ + $CH_3\overset{\ominus}{C}O$ ⇒ $\overset{\ominus}{C}\equiv C-H$ ⇒ C_2H_2

$CH_3\overset{\oplus}{C}O$ ⇓ CH_3CHO

Forward Synthesis

$H-C\equiv C-H$ —(i) $\overset{\ominus}{N}H_2$; (ii) CH_3CHO→ OH (alkynyl alcohol) —Hg^{+2}/H_2SO_4→ OH, O hydroxy carbonyl —CrO_3/Py→ diketone (O, O)

(iii) (TM) ⇒ $2CH_3CO_2Et$

Forward Synthesis

CH_3COOEt (O, OEt) —(i) Na/xylene; (ii) $H_3\overset{\oplus}{O}$→ O, OH —Cu^{+2}/Py→ diketone (O, O)

(iv) (TM) ⇒ acyl cation ($\oplus$) + acyl anion ($\ominus$) ⇒ dithiane anion (S, S, $\ominus$)

acyl cation ⇓ CH_3CHO (O, H)

Forward Synthesis

CH_3CHO —(i) $HS(CH_2)_3SH/BF_3$; (ii) nBuLi→ dithiane anion (S, S, $\ominus$) —(i) CH_3CHO; (ii) $Hg^{+2}/H_3\overset{\oplus}{O}$→ OH, O —$CrO_3$ | Py→ diketone (O, O)

B. Cyclic system

Disconnect in between two carbonyl groups and creating – CO_2Et in both the ends and then acyloin condensation should be tried.

Forward Synthesis

(i) $\overset{\ominus}{O}Et$

(ii) Br–$(CH_2)_2$–Br

$(EtO_2C)_2$ ... $(CO_2Et)_2$

(i) KOH (ii) HCl (iii) 250° (iv) $EtOH/H^{\oplus}$

$EtO_2C-(CH_2)_4-CO_2Et$

(i) Na/xylene (ii) $H_3O^{\oplus}$

Cu^{+2} Py

Cyclic monoketone may selectively be oxidised to cyclic 1, 2-dione with SeO_2.

Conclusion

1, 2-dioxygen system may be synthesized via.

(1) 'yne', (2) umpolung, (3) acyloin, (4) selective oxidation of monoketone by SeO_2.

Review synthetic problems

Show the forward synthesis of the following TM with retrosynthetic analysis.

1. Ph–CO–CO–Ph

2. (hydroxy ketone, OH)

3. (keto acid, OH)

4. Ar–CH₂–CO–CO–CH₂–Ar

5. (cyclopentane-1,2-dione)

Guideline 13

Disconnection of 1, 3-dioxygen system.

A. Acyclic system

(a) Diketo

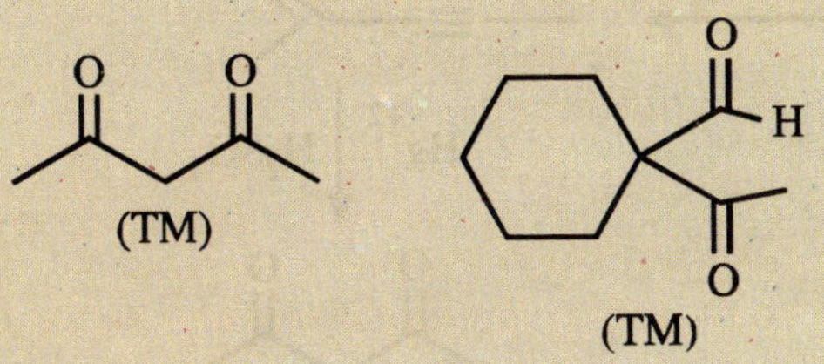

(TM) (TM)

Route - 1

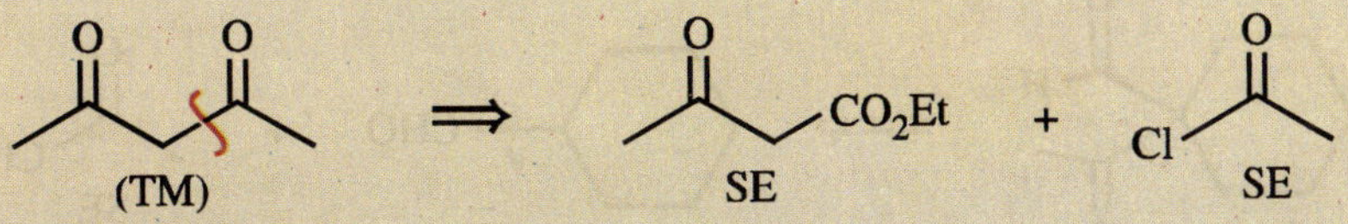

(TM) ⟹ SE CO_2Et + Cl SE

Forward Synthesis

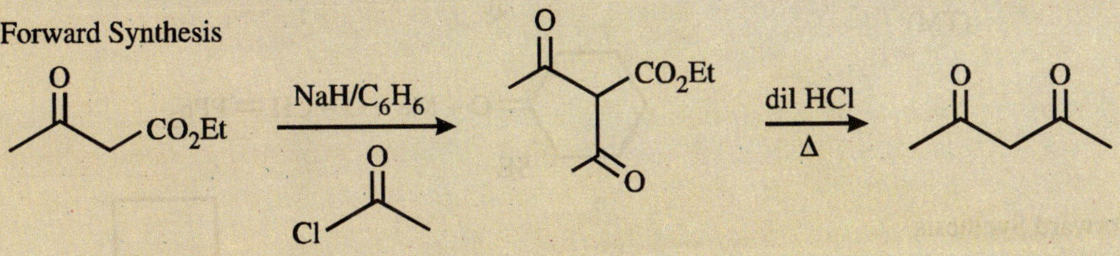

CO_2Et $\xrightarrow{NaH/C_6H_6}$ CO_2Et $\xrightarrow[\Delta]{\text{dil HCl}}$

Route - 2

(TM) ⟹ SE Cl + ⊖ ⟹ ⊖ S S SE

⇓

NO_2 SE

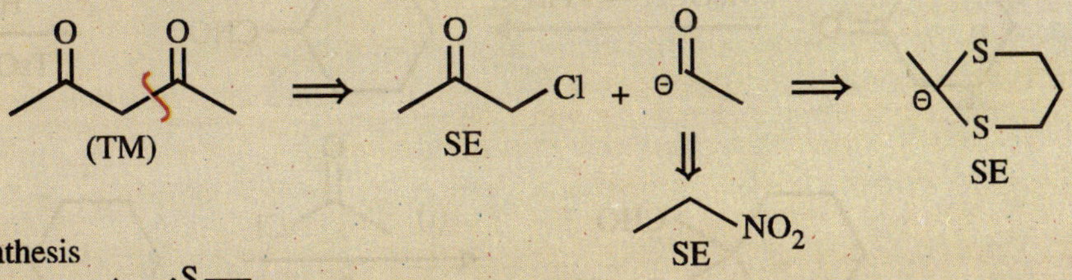

Forward Synthesis

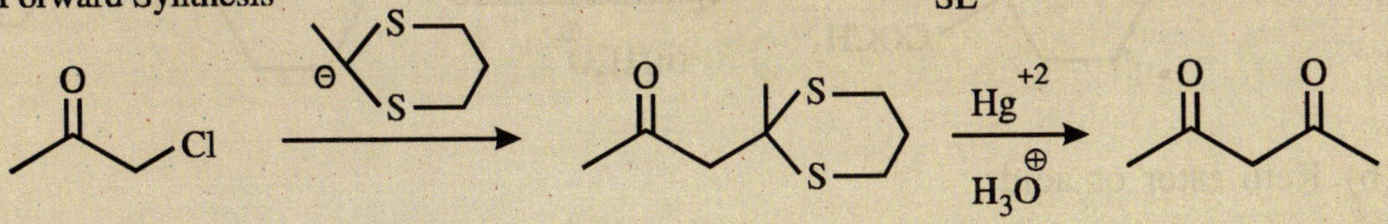

Cl ⊖ S S → S S $\xrightarrow[H_3O^{\oplus}]{Hg^{+2}}$

NO_2 $\xrightarrow{nBuLi}$ $=N^{\oplus}(O^{\ominus})_2$ → NO_2

$TiCl_3$ | H_2O

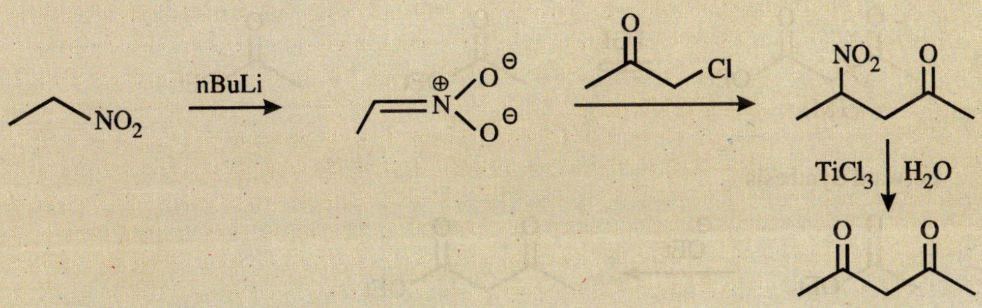

Route - 3

(TM)

FGI

$—≡—H + ClCOCH_3$

Forward Synthesis

$—≡—H \xrightarrow{\ominus NH_2} —≡^{\ominus} \xrightarrow{CH_3COCl} —≡—COCH_3 \xrightarrow{Hg^{+2},\ H_2SO_4}$

(TM)

CHO + SE

$=O + Me-O-CH=PPh_3$

SE

Forward Synthesis

$\xrightarrow{MeOCH=PPh_3}$ CHO $\xrightarrow[TsOH]{\text{pyrrolidine (N–H)}}$

(i) CH_3COCl (ii) $H_3O^{\oplus}$

CHO

$COCH_3$

(b) Keto ester or acid

(TM) OEt

FGI

OEt + OEt

Forward Synthesis

OEt $\xrightarrow{\ominus OEt}$ OEt

(c) Diester

(TM)

Forward Synthesis

(d) hydroxy ester or acid or aldehyde

(TM)

Forward Synthesis

(i) BrZn

(ii) $H_3O^{\oplus}$

(TM)

FGI

SE

Forward Synthesis

CF_3CO_3H

EtOH

HCO_2H, Δ

DIBAL

DIBAL is Diisobutyl aluminium hydride which is a selective reducing agent.

Conclusion: Acyclic 1, 3-dioxygen system may be synthesised via.

(i) Claisen ester condensation, (ii) Aldol condensation, (iii) Reformatsky reaction, (iv) enamine, (v) Umpolung.

B. Cyclic system

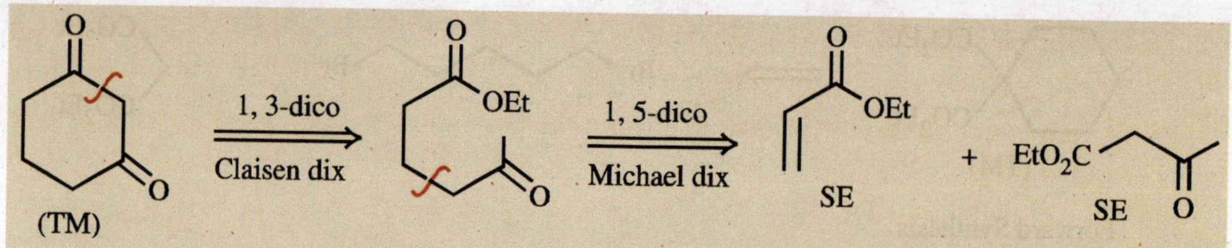

(TM) $\xRightarrow[\text{Claisen dix}]{\text{1, 3-dico}}$ $\xRightarrow[\text{Michael dix}]{\text{1, 5-dico}}$ SE + EtO_2C SE

Forward Synthesis

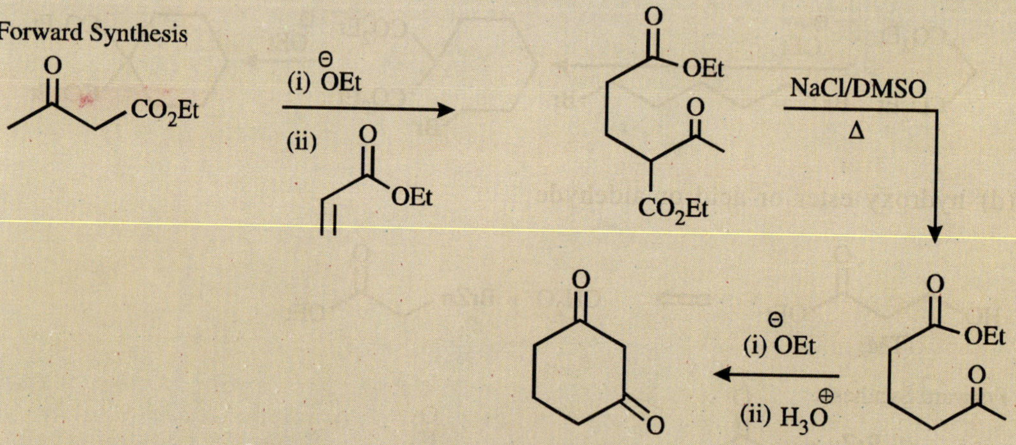

CO_2Et (i) $\overset{\ominus}{O}Et$ (ii) OEt → OEt CO_2Et NaCl/DMSO Δ → OEt

(i) $\overset{\ominus}{O}Et$ (ii) $H_3\overset{\oplus}{O}$

(TM) ⟹ OH OH SE

Forward Synthesis

OH OH (i) Raney Ni/H_2 (ii) CrO_3/Py

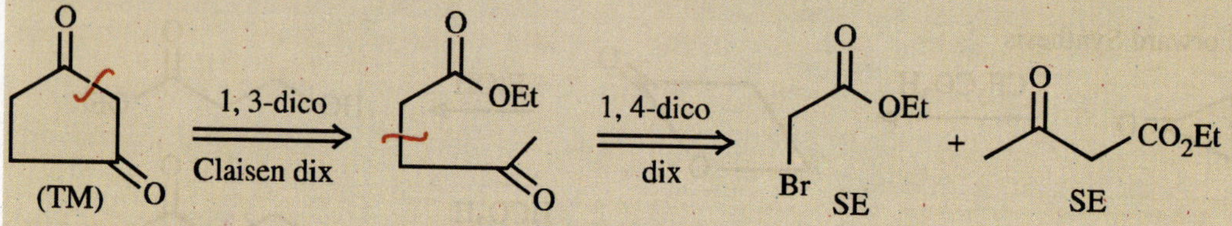

(TM) $\xRightarrow[\text{Claisen dix}]{\text{1, 3-dico}}$ OEt $\xRightarrow[\text{dix}]{\text{1, 4-dico}}$ OEt Br SE + CO_2Et SE

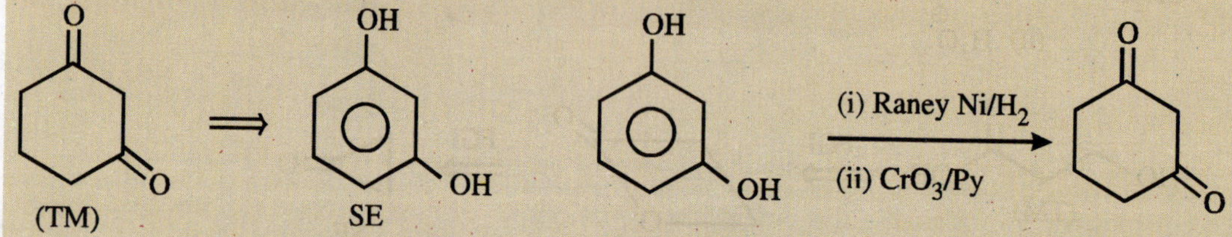

Forward Synthesis

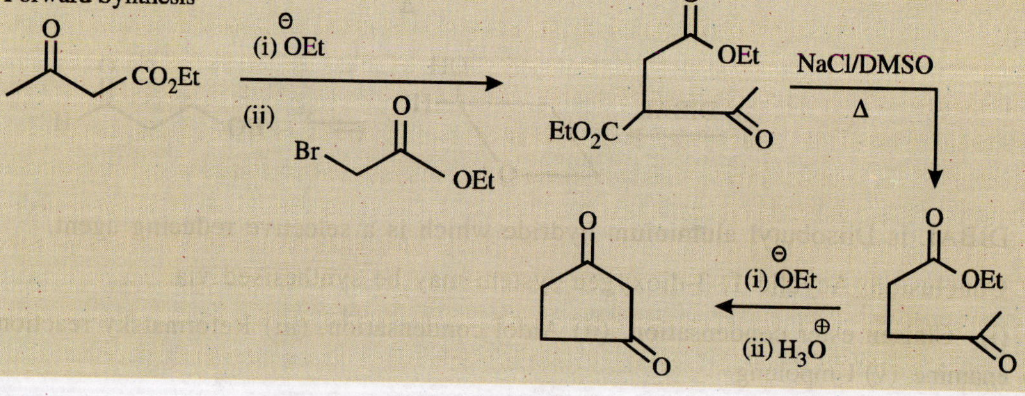

CO_2Et (i) $\overset{\ominus}{O}Et$ (ii) Br OEt → OEt EtO_2C NaCl/DMSO Δ → OEt

(i) $\overset{\ominus}{O}Et$ (ii) $H_3\overset{\oplus}{O}$

Conclusion:

(i) cyclic six membered 1, 3 dioxygen system may be synthesized from Michael condensation followed by Claisen condensation or from suitable benzenoid derivative.

(ii) cyclic 5-membered 1, 3-dioxygen system is synthesized by EAA + α-haloester + Claisen ester condensation.

Review synthetic problems

Show the forward synthesis for the following TM with retro-analysis.

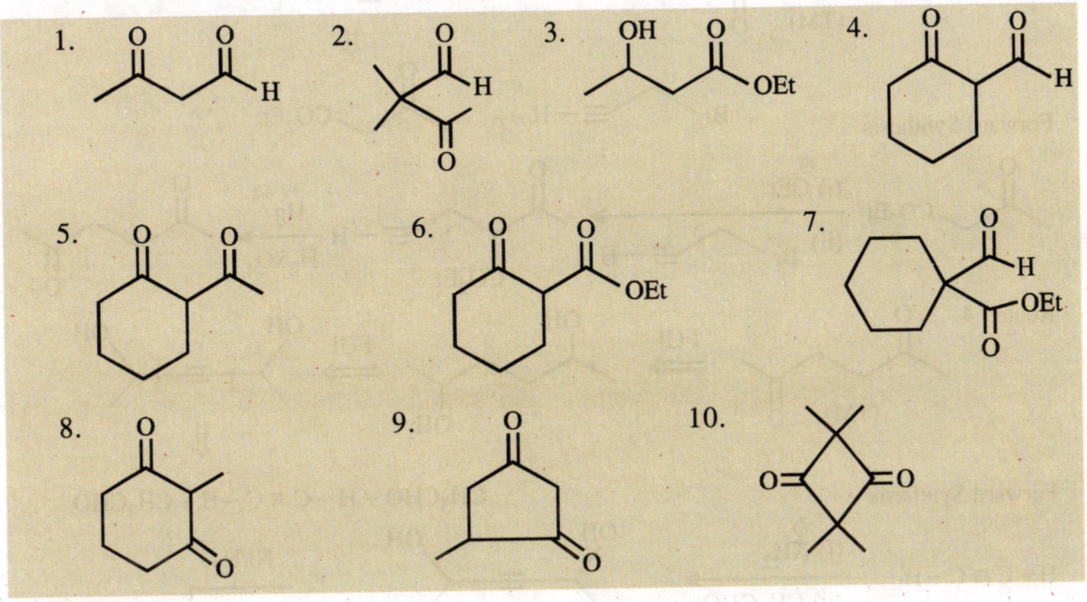

1. 2. 3. OH O OEt 4. H 5. 6. OEt 7. H OEt 8. 9. 10.

Guideline 14

Disconnection of 1, 4-dioxygen system.

A. Acyclic system

(a) Diketo

Route - 1

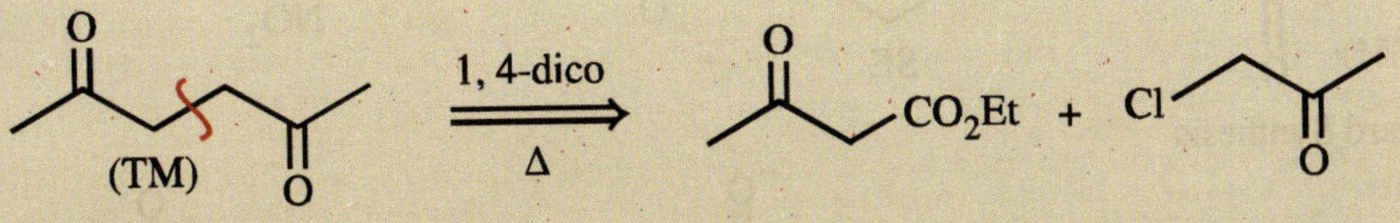

(TM) $\xrightarrow[\Delta]{\text{1, 4-dico}}$ CO_2Et + Cl

Forward Synthesis

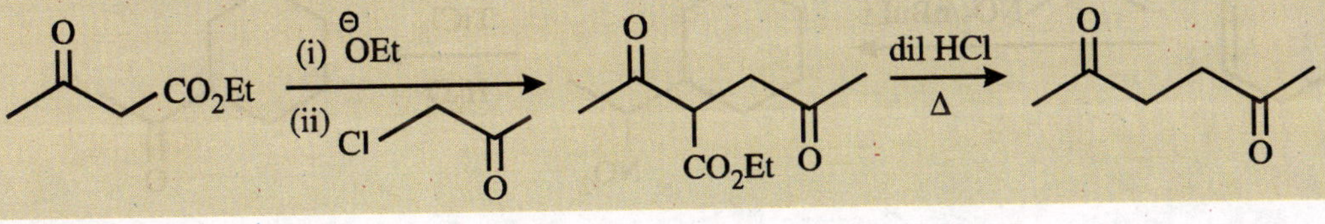

CO_2Et $\xrightarrow[\text{(ii) Cl}]{\text{(i) } \overset{\ominus}{\text{O}}\text{Et}}$ CO_2Et $\xrightarrow[\Delta]{\text{dil HCl}}$

Route - 2

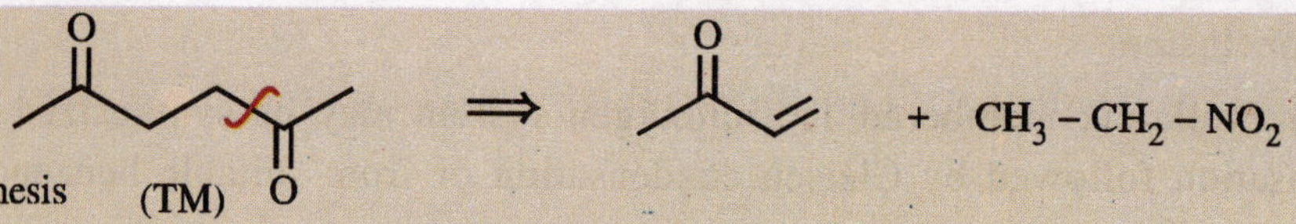

(TM) ⟹ + $CH_3-CH_2-NO_2$

Forward Synthesis

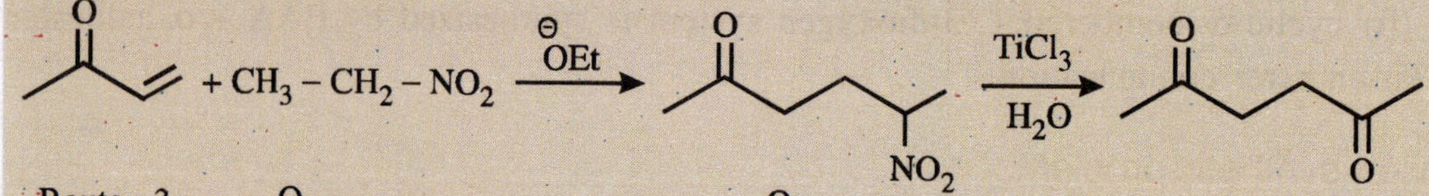

+ $CH_3-CH_2-NO_2$ $\xrightarrow{\ominus OEt}$ NO_2 $\xrightarrow[H_2O]{TiCl_3}$

Route - 3

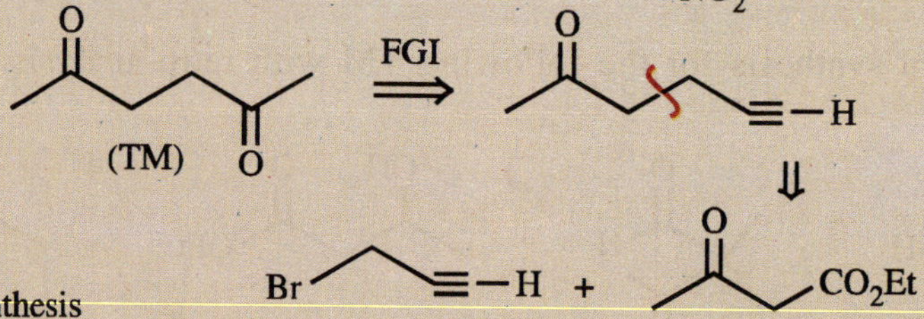

(TM) $\overset{FGI}{\Longrightarrow}$ ≡–H

⇓

Br ≡–H + CO_2Et

Forward Synthesis

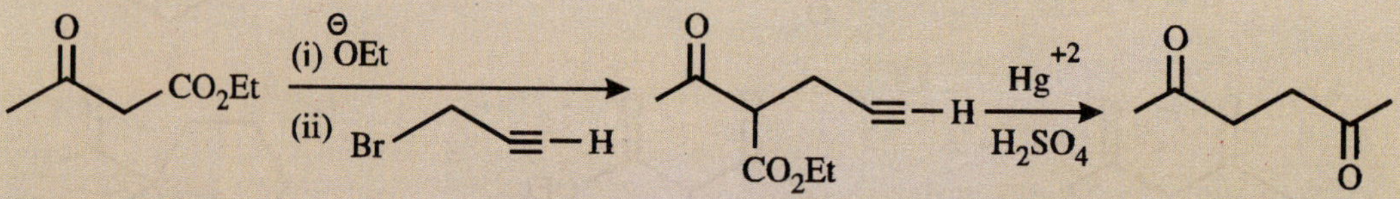

CO_2Et $\xrightarrow[(ii)\ Br \equiv -H]{(i)\ \ominus OEt}$ ≡–H, CO_2Et $\xrightarrow[H_2SO_4]{Hg^{+2}}$

Route - 4

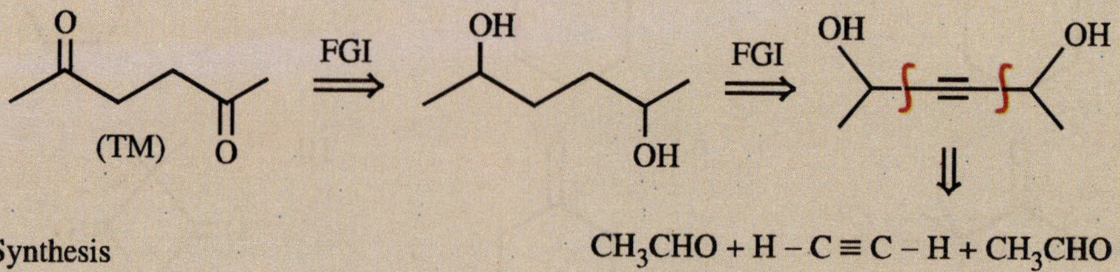

(TM) $\overset{FGI}{\Longrightarrow}$ OH, OH $\overset{FGI}{\Longrightarrow}$ OH, OH

⇓

$CH_3CHO + H-C\equiv C-H + CH_3CHO$

Forward Synthesis

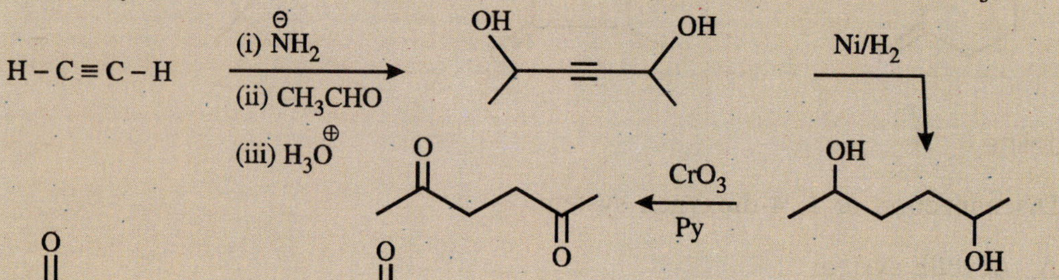

$H-C\equiv C-H$ $\xrightarrow[(ii)\ CH_3CHO\ (iii)\ H_3O^{\oplus}]{(i)\ \overset{\ominus}{N}H_2}$ OH, OH $\xrightarrow{Ni/H_2}$ OH, OH $\xrightarrow[Py]{CrO_3}$

(TM) ⟹ SE + $^{\ominus}$ (O) ⟹ $^{\ominus}$ NO_2 ⟹ NO_2

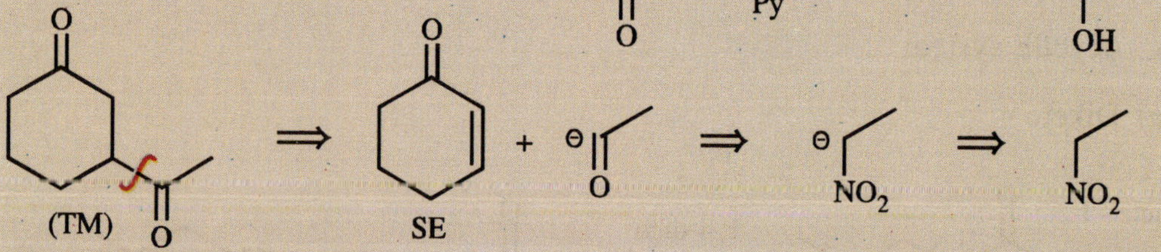

Forward Synthesis

$\xrightarrow{\ \ NO_2/nBuLi}$ NO_2 $\xrightarrow[H_2O]{TiCl_3}$

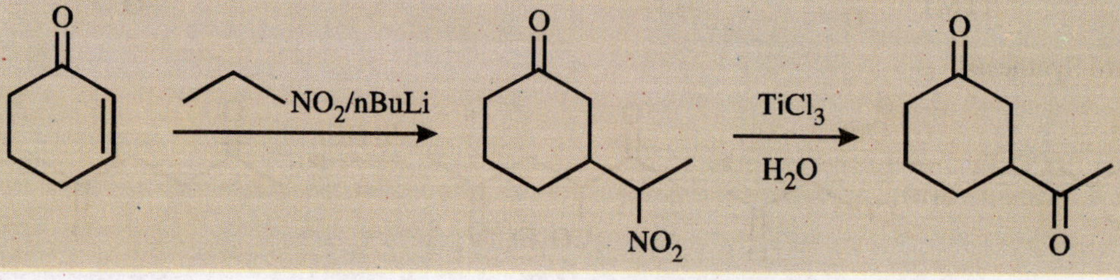

(b) Keto ester or acid

Route - 1

(TM) ⟹ CO_2Et + Br OEt

Forward Synthesis

(i) $\overset{\ominus}{O}Et$

(ii) Br OEt

(iii) KOH

(iv) HCl/Δ (v) EtOH/$H^{\oplus}$

Route - 2

⟹ Cl + CO_2Et CO_2Et

Forward Synthesis

(i) $\overset{\ominus}{O}Et$ (ii) Cl

(i) KOH (ii) HCl/Δ (iii) EtOH/$H^{\oplus}$

(c) Diester or Diacid

EtO OEt (TM) ⟹ 2 CO_2Et CO_2Et

Forward Synthesis

(i) $\overset{\ominus}{O}Et$ (ii) Br_2

Br CO_2Et CO_2Et

EtO_2C CO_2Et EtO_2C CO_2Et

(i) KOH

(ii) $H_3\overset{\oplus}{O}$/250°C

(iii) EtOH/$H^{\oplus}$

CO_2H OH O ⟹ CO_2H + $\overset{\ominus}{C}O_2H$ ⟹ $\overset{\ominus}{C}N$

⇓

=O + CN CO_2Et (Not DEM)

Forward Synthesis

CN CO_2Et (i) $\overset{\ominus}{O}Et$ (ii) =O → CN CO_2Et KCN → CN CN CO_2Et (i) KOH (ii) $H_3\overset{\oplus}{O}$, 250°C → CO_2H CO_2H

(d) Hydroxy ester, acid or aldehyde

HO OEt O $\overset{FGI}{⟹}$ HO OH O $\overset{FGI}{⟹}$ O O

⇓ FGI

O + $CH_2(CO_2Et)_2$ SE SE

O SE

Forward Synthesis

O CF_3CO_3H → O O $\overset{\ominus}{O}Et$ EtOH → HO OEt O

CO_2Et CO_2Et + O $\overset{\ominus}{O}Et$ → CO_2Et O O (i) KOH (ii) HCO_2H ↑

HCO_2H → HO OH O

DIBAL → HO H O

Conclusion:

Acyclic 1, 4-dioxygen system may be synthesized via.

(i) Michael condensation, (ii) Umpolung, (iii) Starting from EAA and DEM.

B. Cyclic system

Route - 1

O, O (TM) —FGA⇒ O, CO_2Et, EtO_2C, O ⇒ [CO_2Et CO_2Et + EtO_2C EtO_2C]

Forward Synthesis

2[CO_2Et CO_2Et] —$\overset{\ominus}{O}Et$→ O, CO_2Et, EtO_2C, O —dil HCL, Δ→ O, O

Route - 2

O, O (TM) —FGI⇒ OH, OH —FGI⇒ OH, OH SE

Forward Synthesis

OH, OH —Raney Ni/H_2→ OH, OH —CrO_3/Py→ O, O

Route - 3

O, O (TM) —FGI⇒ O, NO_2 —FGI⇒ OH, NO_2 ⇒

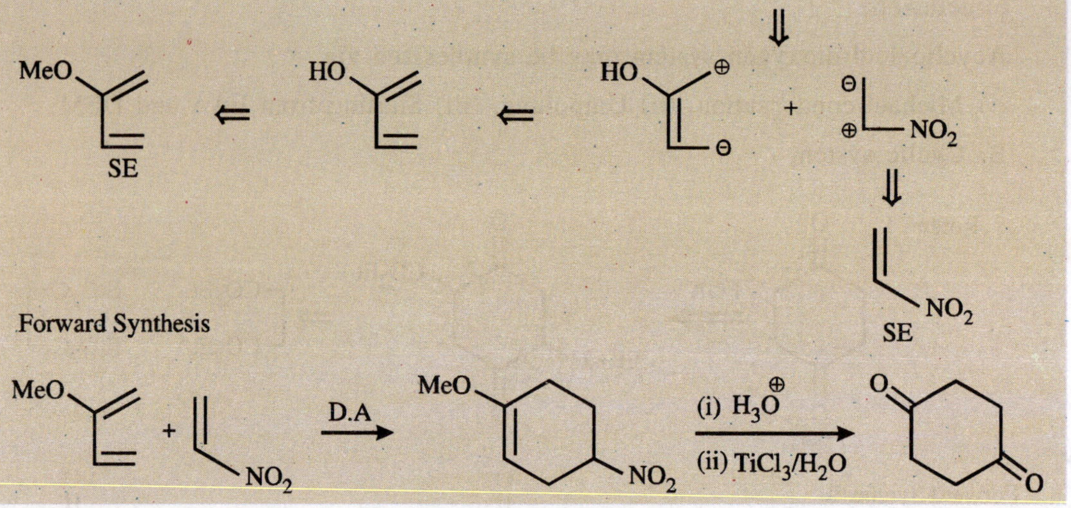

MeO SE ⇐ HO ⇐ HO ⊕ ⊖ + ⊖ ⊕ NO_2 ⇓ SE NO_2

Forward Synthesis

MeO + NO_2 —D.A→ MeO NO_2 —(i) $H_3\overset{\oplus}{O}$ (ii) $TiCl_3/H_2O$→ O O

Conclusion:

Cyclic 1, 4-dioxygen system may be synthesized via (i) Dieckmann condensation, (ii) DA reaction, (iii) from suitable benzenoid derivative.

Review synthetic problems

Show the forward synthesis for the following TM with retrosynthetic analysis.

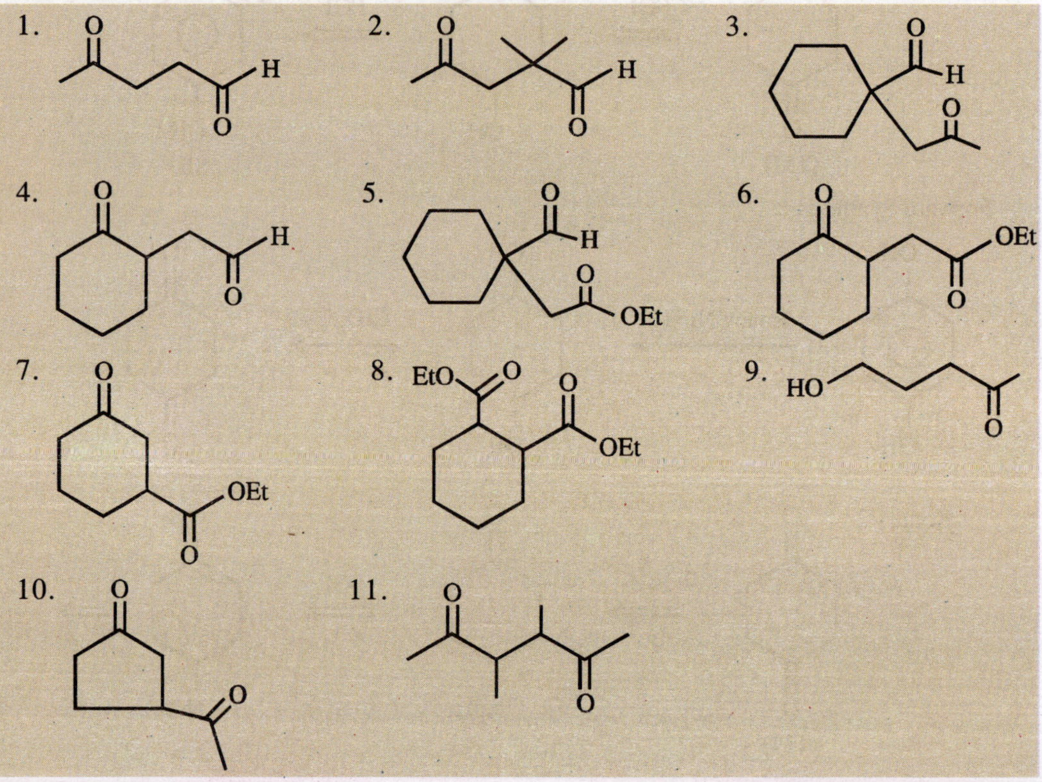

1. O H O
2. O H O
3. O H O
4. O H O
5. O H O OEt
6. O OEt O
7. O OEt O
8. EtO O O OEt
9. HO O
10. O O
11. O O

Guideline 15

Disconnection of 1, 5-dioxygen system.

A. Acyclic system

(a) Diketo

Route - 1

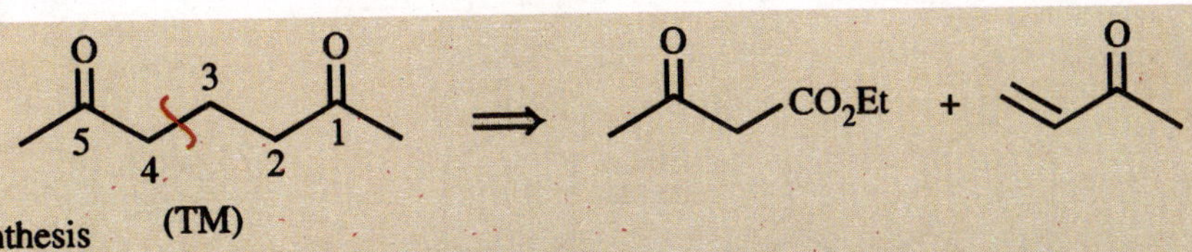

(TM)

$\Rightarrow$ CO_2Et +

Forward Synthesis

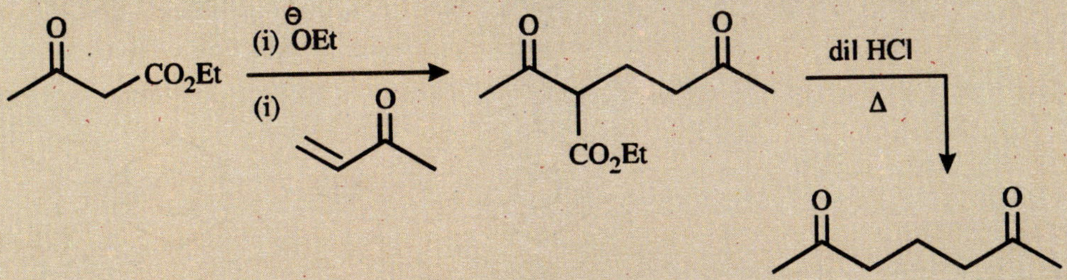

CO_2Et (i) $\overset{\ominus}{O}Et$ (i) $\rightarrow$ CO_2Et dil HCl Δ

Route - 2

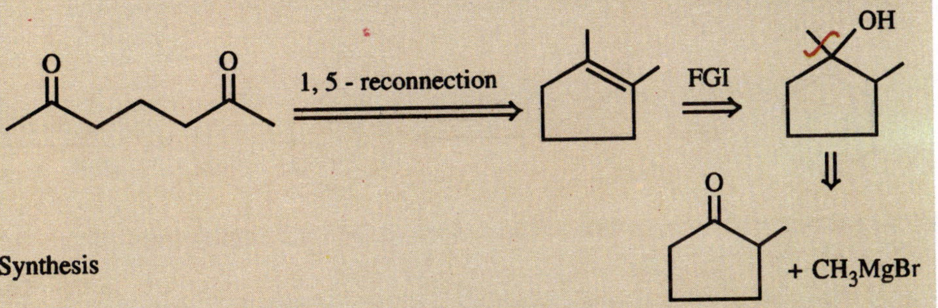

1, 5 - reconnection $\Rightarrow$ FGI $\Rightarrow$ OH

$\Downarrow$

+ CH_3MgBr

Forward Synthesis

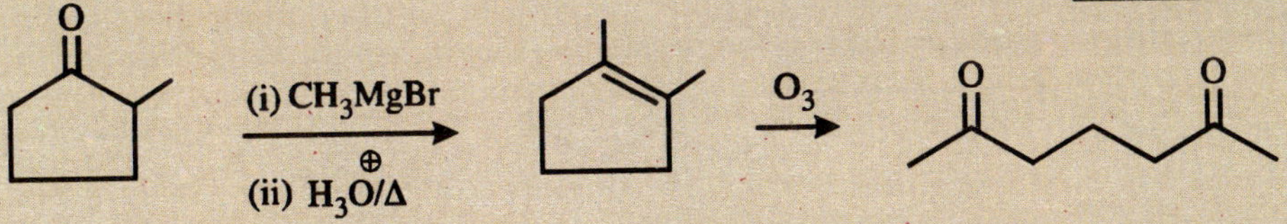

(i) CH_3MgBr (ii) $H_3\overset{\oplus}{O}/\Delta$ $\rightarrow$ O_3 $\rightarrow$

Route - 3

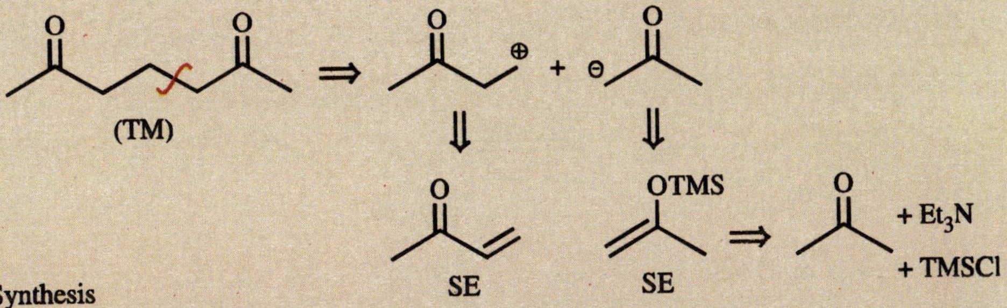

(TM) $\Rightarrow$ $\oplus$ + $\ominus$

$\Downarrow$ $\Downarrow$

SE OTMS SE $\Rightarrow$ + Et_3N + TMSCl

Forward Synthesis

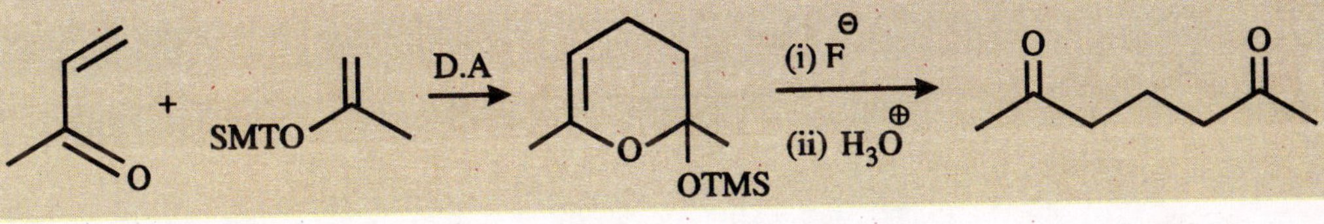

+ SMTO D.A $\rightarrow$ OTMS (i) $\overset{\ominus}{F}$ (ii) $H_3\overset{\oplus}{O}$ $\rightarrow$

Route - 1

(TM) 1 2 3 4 5

3, 4 dix
Michael dix.

SE

Route - 2

+ CO_2Et

FGI

OH

FGI

(MT)

(i) $\overset{\ominus}{O}Et$

dil HCl
Δ

OMe

CO_2Et

(i)

Br

SE

OMe

+ CO_2Et

SE

Forward Synthesis

HO OMe

Br

+

CO_2Et

$\overset{\ominus}{N}H_2$(2eq)

OMe

CO_2Et

dil HCl/Δ

CH_3MgBr +

OMe

(i) Raney Ni/H_2
(ii) CrO_3/Py
or H_2/Pd

(i) Li/liqNH_3
(ii) $H_3\overset{\oplus}{O}$/Δ

(b) Ketoester or acid

Michael
dix

OEt

CO_2Et (MT) + OEt

(TM)

OTMS

Et_3N + TMSCl

SE SE

(c) Diester or Diacid

EtO

(TM)

(i) $\overset{\ominus}{F}$
(ii) H_3O

OEt

Michael
dix.

OTMS

D.A
DEM

OTMS

OEt

(d) hydroxy ester or acid or aldehyde

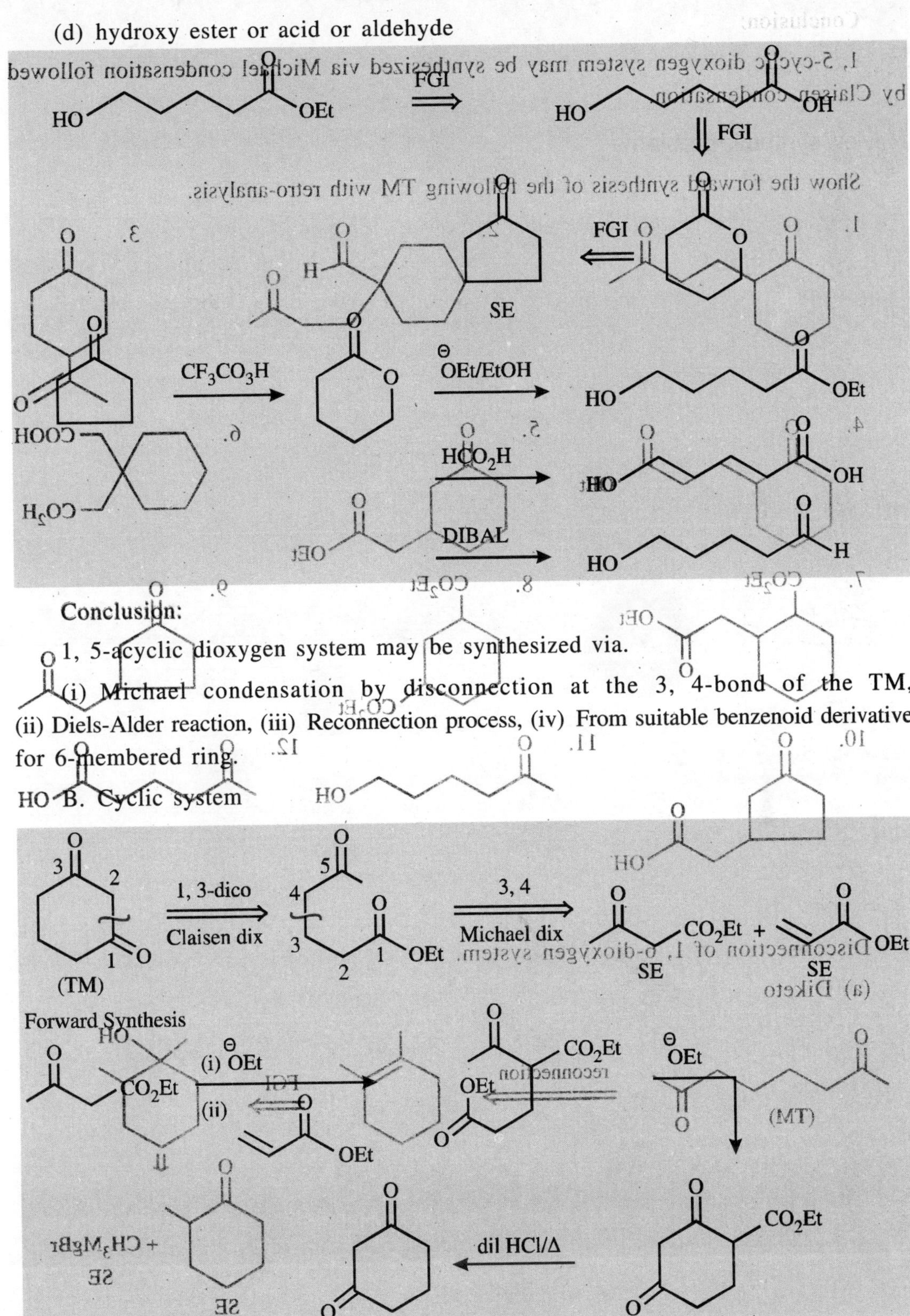

Conclusion:

1, 5-acyclic dioxygen system may be synthesized via.

(i) Michael condensation by disconnection at the 3, 4-bond of the TM, (ii) Diels-Alder reaction, (iii) Reconnection process, (iv) From suitable benzenoid derivative for 6-membered ring.

B. Cyclic system

Conclusion:

1, 5-cyclic dioxygen system may be synthesized via Michael condensation followed by Claisen condensation.

Review synthetic problems

Show the forward synthesis of the following TM with retro-analysis.

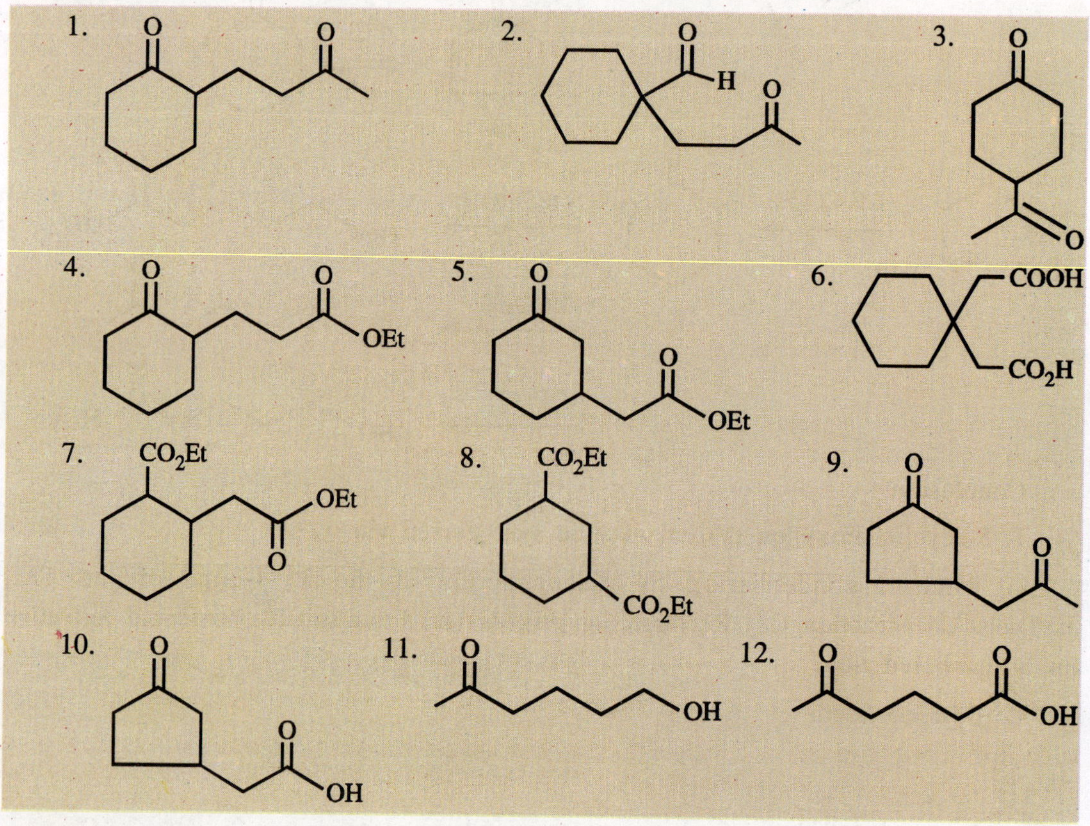

1. O O

2. O H O

3. O O

4. O O OEt

5. O O OEt

6. COOH CO_2H

7. CO_2Et OEt O

8. CO_2Et CO_2Et

9. O O

10. O O OH

11. O OH

12. O O OH

Guideline 16

Disconnection of 1, 6-dioxygen system.

(a) Diketo

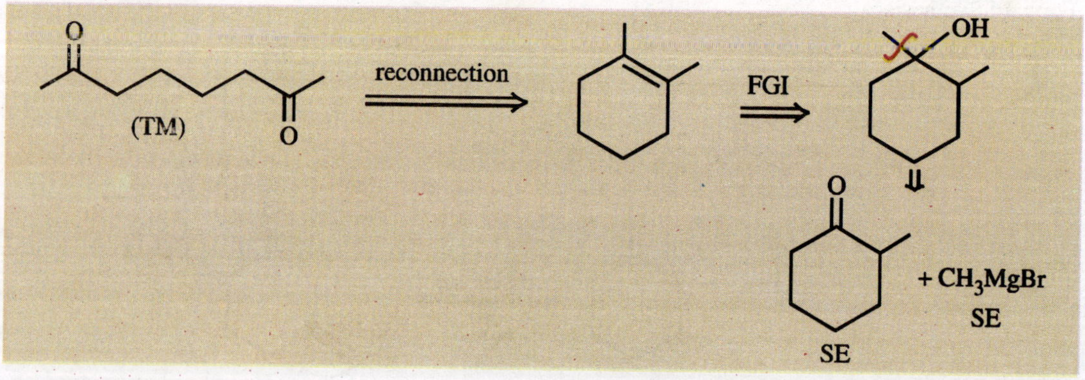

O (TM) O reconnection FGI OH O + CH_3MgBr SE SE

Forward Synthesis

O

(i) CH_3MgBr

(ii) $H_3\overset{\oplus}{O}/\Delta$

O_3

O

O

O

(TM)

O

OH

OH

≡

OH

OH

reverse ⇓ oxy-Cope

O

OH

reverse

pinacol

OH

O

SE

Forward Synthesis

O

(i) $Mg/Hg/C_6H_6$

(ii) oxalic acid

OH

OH

Δ

oxy-Cope

OH

OH

O

O

O

H

reconnection

O

(TM)

DA

dix

+ ||

SE

FGI

HO

$H-C\equiv\overset{\ominus}{C}$ +

SE

O

SE

Forward Synthesis

(i) $H-C\equiv C^{\ominus}$
(ii) $H_3O^{\oplus}$

$H_2/Pd/CaCO_3/Pb(OAc)_2$
lindlar cat

$KHSO_4$

DA

O_3/Me_2S

(b) Keto ester or acid

(TM)

reconnection

FGI

SE

(i) Mg/Hg/C_6H_6
(ii) oxalic acid

+ CH_3MgBr

Forward Synthesis

(i) CH_3MgBr
(ii) $H_3O^{\oplus}/\Delta$

O_3
H_2O_2

(TM)

FGI

FGA

FGA

SE

+

SE

Forward Synthesis

Ni/H_2

CrO_3 / AcOH

(c) Diester or Diacid

reconnect ⟹ (TM) ⟹ SE

FGI

Forward Synthesis

DA

O_3 / H_2O_2

fuming HNO_3

(d) hydroxy ester or acid or aldehyde

reconnect

FGI

SE

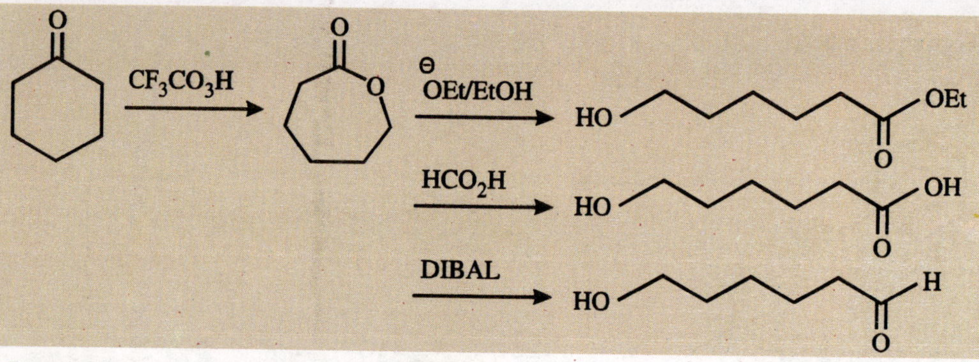

CF_3CO_3H → ; $\ominus$OEt/EtOH → HO...OEt ; HCO_2H → HO...OH ; DIBAL → HO...H

Conclusion:

1. When the TM is 1, 6-dico system, then first reconnect through dicarbonyl creating a C—C double bond between carbonyl carbons which will give cyclohexene derivative which may be available in the market or may be synthesized easily and then on ozonolysis (or any suitable oxidation) gives the desired 1, 6-dicazbonyl system.
2. Starting from suitable benzenoid derivative.
3. Starting from DEM.

Review synthetic problems

Show the forward synthesis of the following TM with retro-analysis.

1. 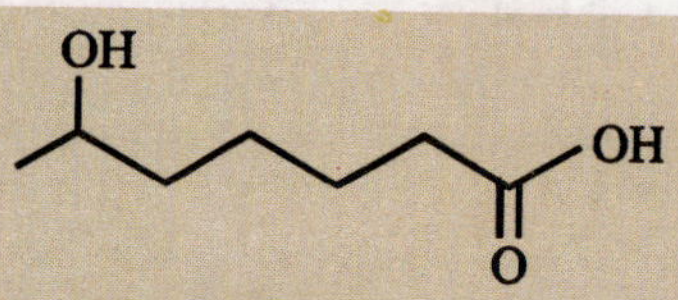 OH, OH, O
2. 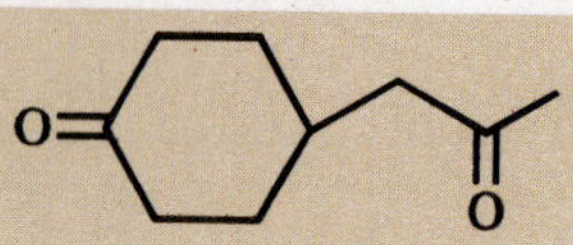 O=, O
3. 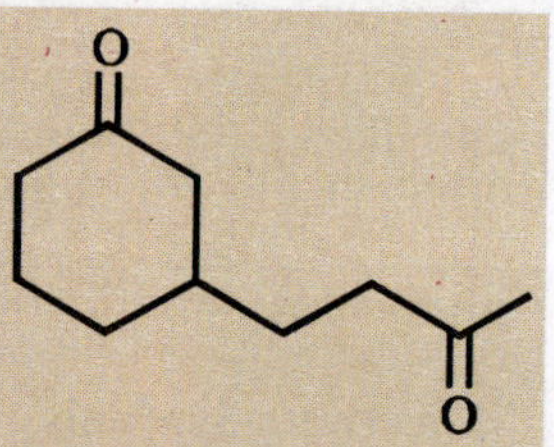 O, O
4. 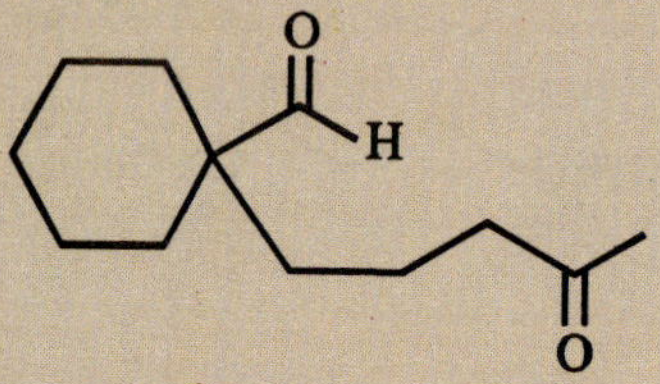 O, H, O
5. 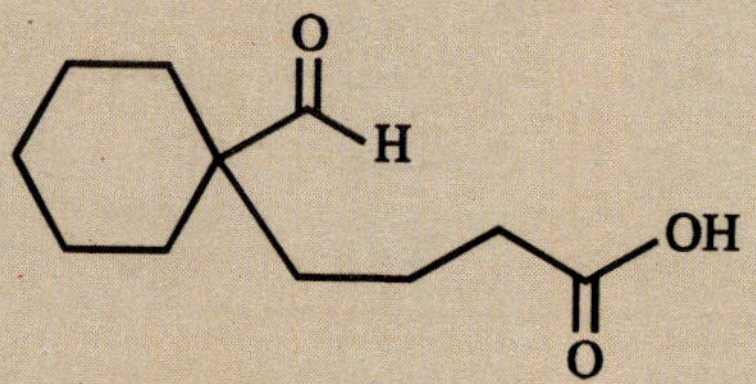 O, H, OH, O
6. 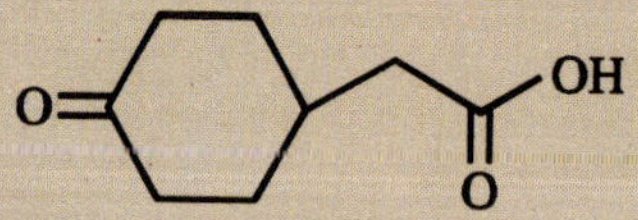 O=, OH, O
7. 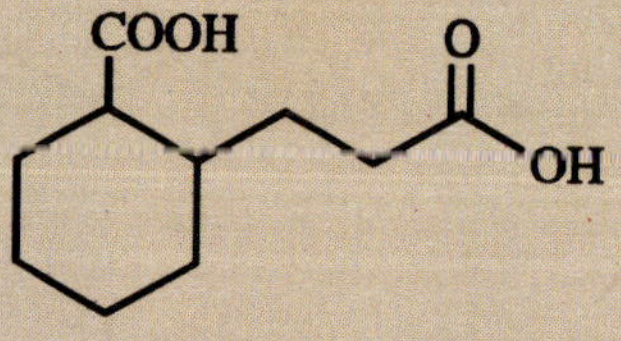 COOH, O, OH
8. 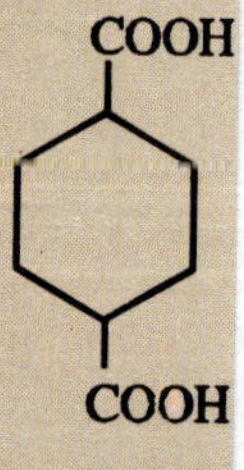 COOH, COOH
9. 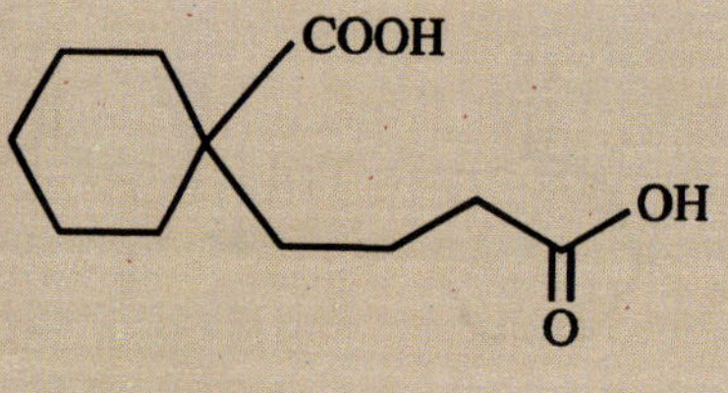 COOH, OH, O
10. 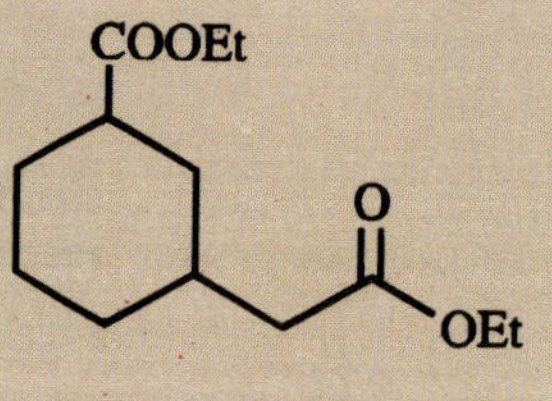 COOEt, O, OEt
11. 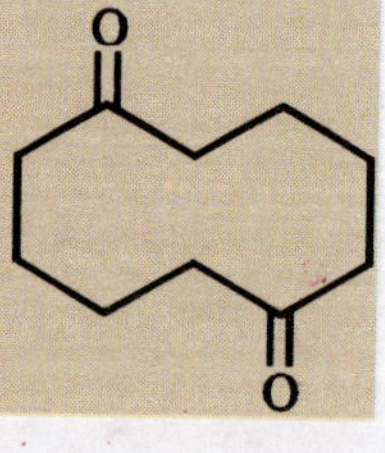 O, O

Guideline 17

If the side chain is attached to a ring then, at first side chain is disconnected.

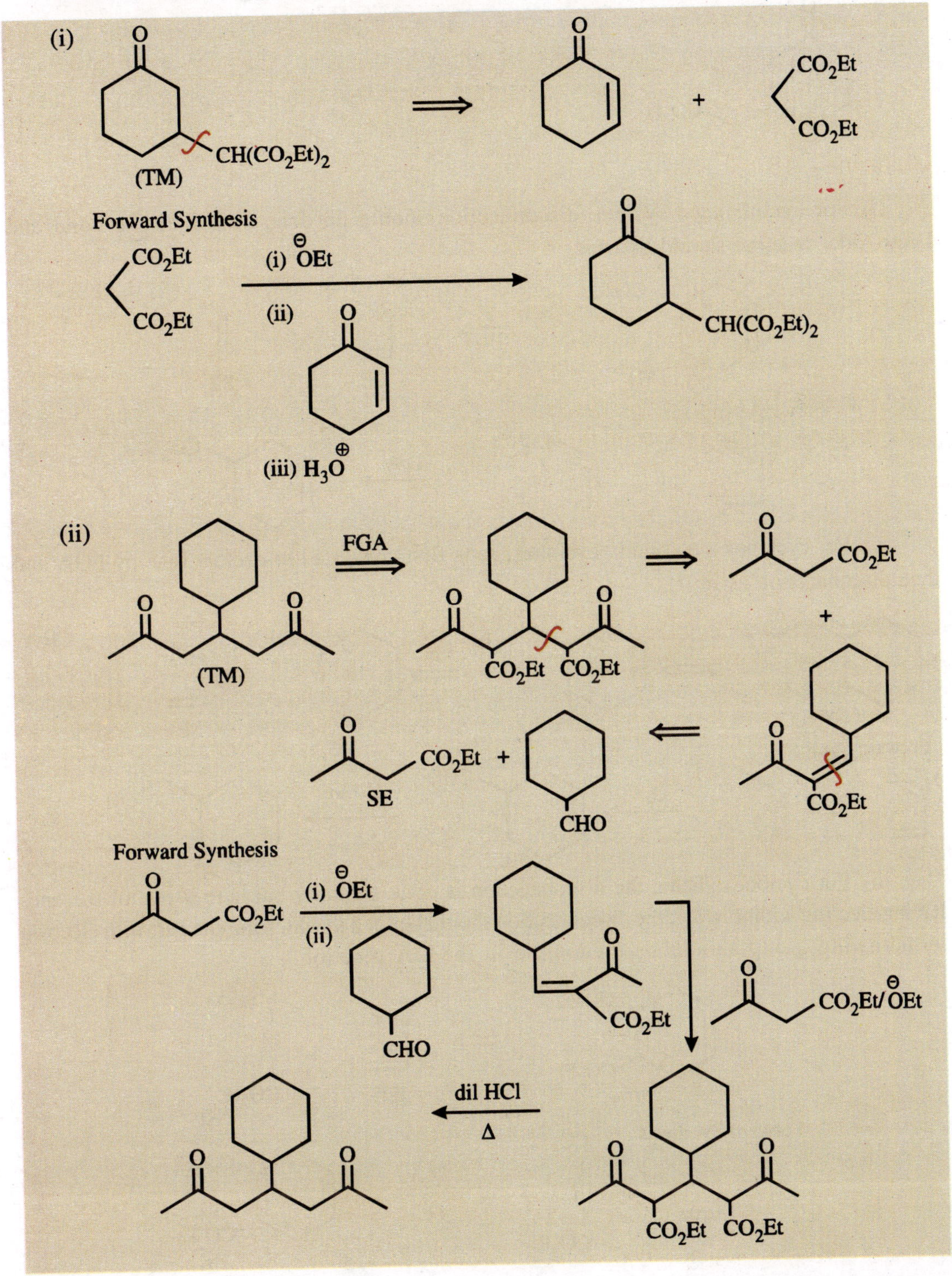

(i) O (TM) $CH(CO_2Et)_2$ ⟹ O + CO_2Et CO_2Et

Forward Synthesis

CO_2Et CO_2Et (i) $\overset{\ominus}{O}Et$ (ii) O (iii) $H_3\overset{\oplus}{O}$ → O $CH(CO_2Et)_2$

(ii) O O (TM) ⟹(FGA) O O CO_2Et CO_2Et ⟹ O CO_2Et + O CO_2Et

O CO_2Et SE + CHO ⟸

Forward Synthesis

O CO_2Et (i) $\overset{\ominus}{O}Et$ (ii) CHO → O CO_2Et → O $CO_2Et/\overset{\ominus}{O}Et$ → O O CO_2Et CO_2Et → dil HCl, Δ → O O

Review synthetic problems

Show the forward synthesis with retro-analysis of the following TM.

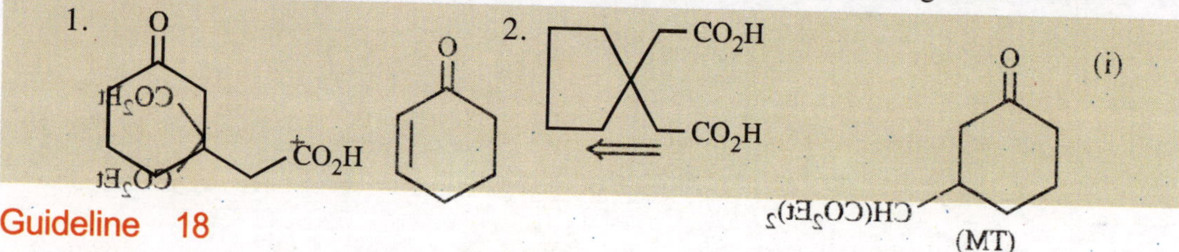

Guideline 18

(i) For cyclohexene system, disconnection should be done on the allylic bond and Diels-Alder reaction should be tried.

Forward Synthesis

(ii) For cyclohexane system, cyclohexane is first converted to cyclohexene by FGA and then disconnection as in (i).

Forward Synthesis

(iii) For cyclohexadiene, the disconnection is done similarly but here if one of the enes having electron withdrawing, the disconnection should be such in which the ene part with electron withdrawing group should be dienophile in the DA reaction.

Forward Synthesis

Review synthetic problems

Show the forward synthesis of the following TM with retro-analysis.

1. 2. 3. 4. 5.

Guideline 19

Bridgehead ring structure is disconnected similarly and in such a design so that it can be synthesised by DA reaction.

CO_2H $\xrightarrow{FGI}$ CO_2Et $\xrightarrow{FGA}$ CO_2Et $\Rightarrow$ SE + SE CO_2Et

Forward Synthesis

+ CO_2Et $\xrightarrow{D.A}$ CO_2Et $\xrightarrow{H_2/Pd}$ CO_2Et $\xrightarrow{(i)\ \overset{\ominus}{O}H \quad (ii)\ H_3\overset{\oplus}{O}}$ CO_2H

Review synthetic problems

Show the forward synthesis of the following TM with retro-analysis.

1. 2. 3.

Guideline 20

Benzene nucleus or its derivative may be developed by retrosynthetic analysis using DA reaction.

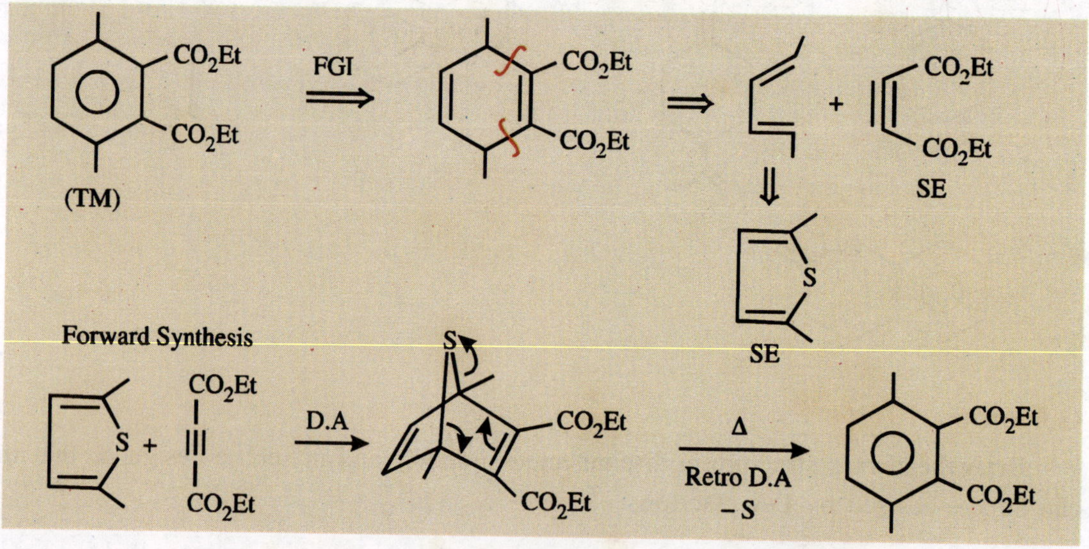

(TM) CO_2Et CO_2Et FGI ⟹ CO_2Et CO_2Et ⟹ + CO_2Et CO_2Et SE ⇓ S SE

Forward Synthesis

S + CO_2Et ≡ CO_2Et —D.A→ S CO_2Et CO_2Et —Δ, Retro D.A, – S→ CO_2Et CO_2Et

Review synthetic problems

With retrosynthetic analysis show the forward synthesis for the following TM.

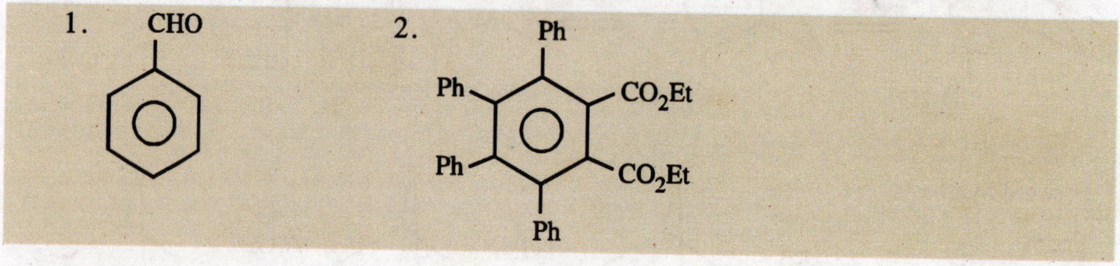

1. CHO

2. Ph, Ph, Ph, Ph, CO_2Et, CO_2Et

Guideline 21

Cyclohexene or cyclohexadiene derivative may be developed by retrosynthetic analysis of which the SE is benzenoid derivative by utilising Birch-reduction.

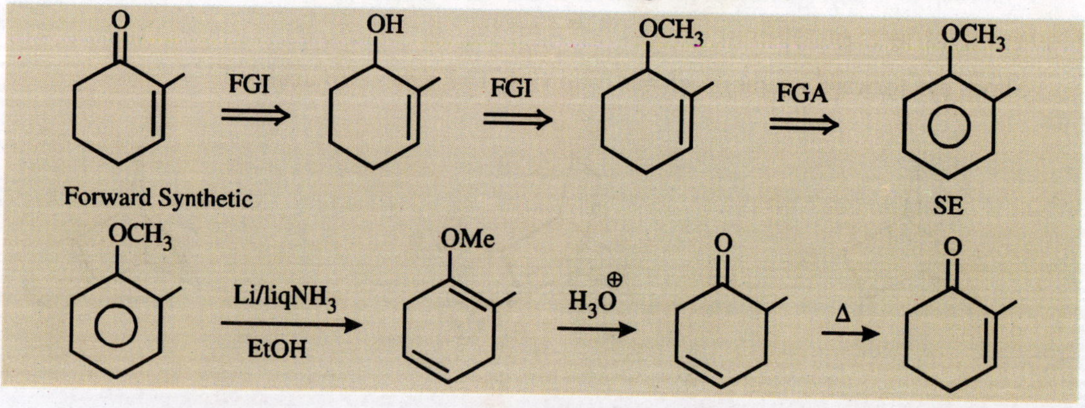

O FGI ⟹ OH FGI ⟹ OCH_3 FGA ⟹ OCH_3 SE

Forward Synthetic

OCH_3 —Li/liqNH_3, EtOH→ OMe —$H_3O^{\oplus}$→ O —Δ→ O

Review synthetic problems

Show the forward synthesis of the following TM with retro-analysis.

1. 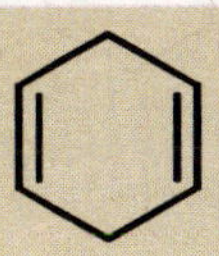 2. 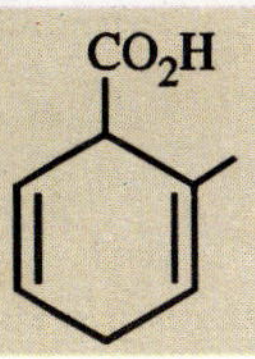 CO_2H

Guideline 22

A ketone α or β to quaternary carbon may be synthesised from the following retro-analysis.

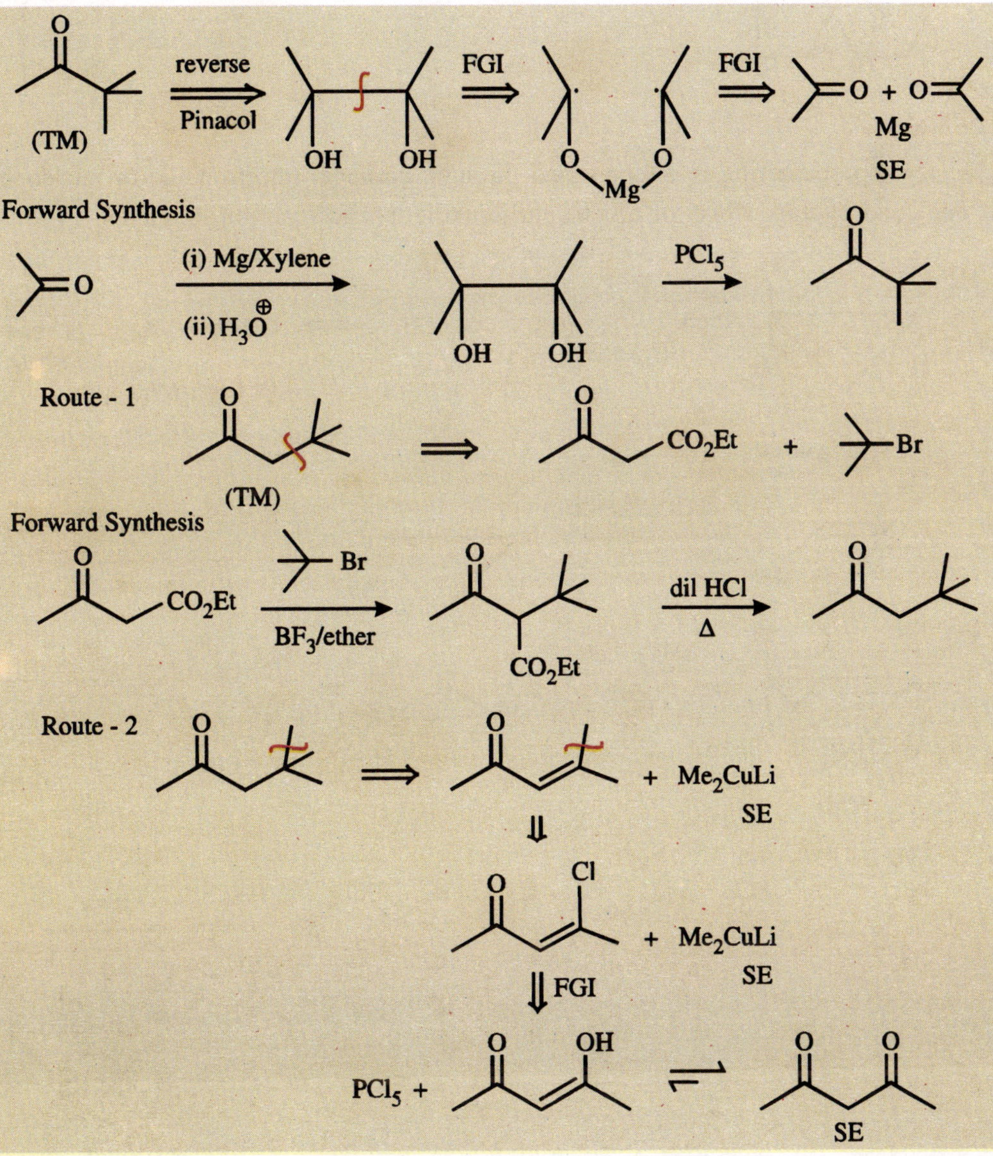

(TM) reverse Pinacol ⟹ OH OH FGI ⟹ O Mg O FGI ⟹ O + O Mg SE

Forward Synthesis

O (i) Mg/Xylene (ii) $H_3O^{\oplus}$ → OH OH PCl_5 → O

Route - 1

O (TM) ⟹ O CO_2Et + Br

Forward Synthesis

O CO_2Et Br BF_3/ether → O CO_2Et dil HCl Δ → O

Route - 2

O ⟹ O + Me_2CuLi SE

⇓

O Cl + Me_2CuLi SE

⇓ FGI

PCl_5 + O OH ⇌ O O SE

Forward Stnthesis

$\xrightarrow{PCl_5}$ $\xrightarrow{Me_2CuLi}$

Review synthetic problems

Show the forward synthesis of the following TM with retro-analysis.

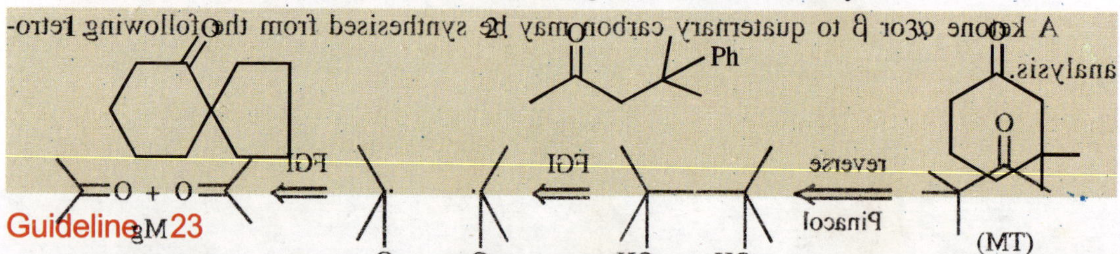

Guideline 23

A. Cyclopropane ring is disconnected through 'carbene' (electrophilic or nucleophilic) and 'ene' and cycloaddition or electrophilic/nucleophilic reaction is tried.

(i) Me H H Me (TM) ⇒ Me Me + :CH_2 ⇒ CH_2N_2/UV/O_2 (singlet carbene) Or CH_2I_2/Zn–Cu

SE

Forward Synthesis

Me Me $\xrightarrow{CH_2I_2/Zn–Cu}$ Me H H Me (retention of the configuration due to cyclic one step mech.)

(ii) Me Me H H Ph Cl (TM) ⇒ Me Me SE + Ph–C̈–Cl ⇒ Ph Cl C H Cl ⇒ Ph C=O H SE

Forward Synthesis

Ph C=O H $\xrightarrow{PCl_5}$ Ph C Cl H Cl $\xrightarrow{CH_3Li}$ Ph Cl C Cl Li (carbenoid) ≡ Ph C: Cl $\xrightarrow{\text{Me Me (cis)}}$ Me Me H H Ph Cl

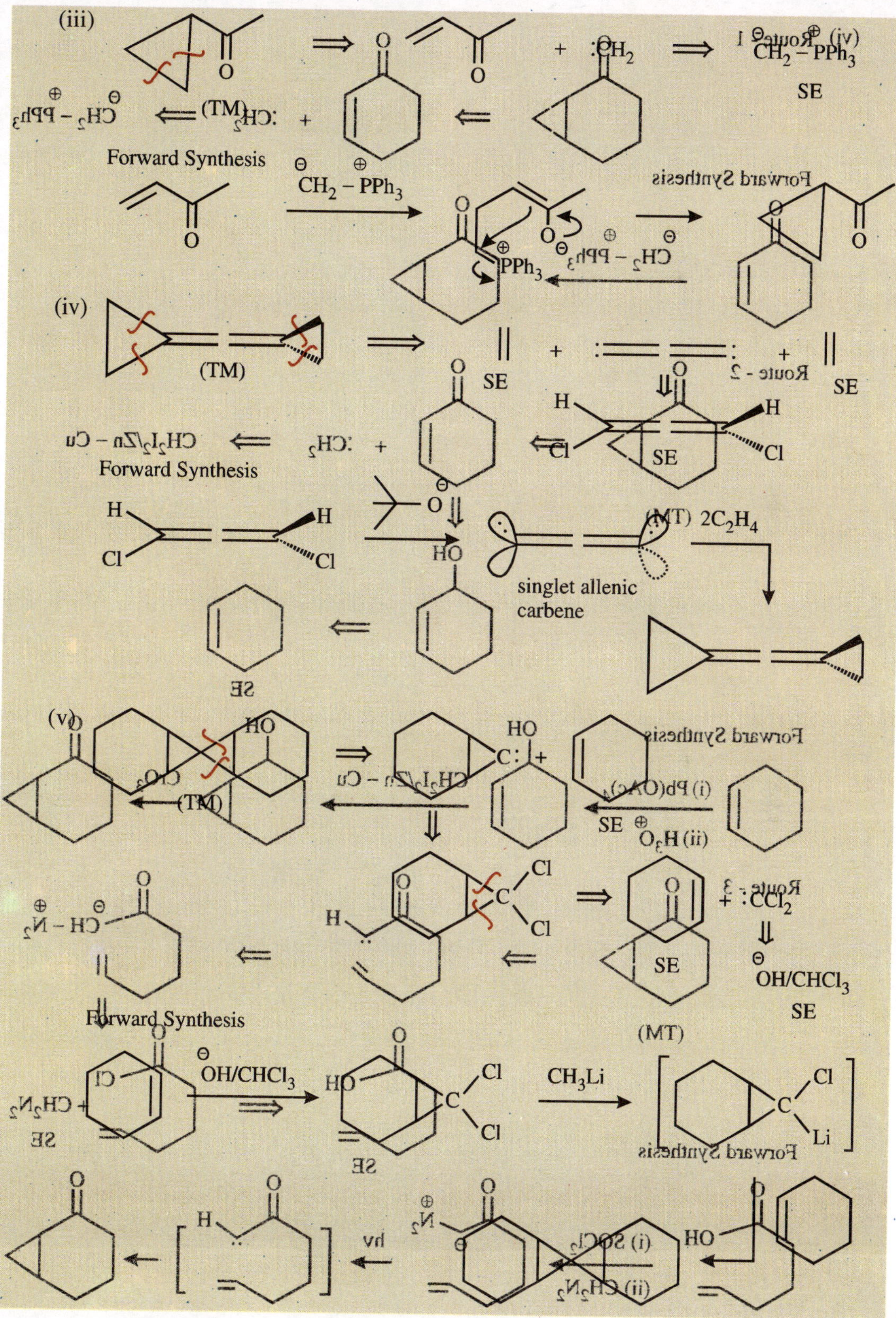

(iii)
(TM)
Route 1 (iv)
CH_2-PPh_3
SE
Forward Synthesis
$CH_2 - PPh_3$
Forward Synthesis
(iv)
(TM)
SE
Route - 2
SE
$CH_2I_2/Zn - Cu$
Forward Synthesis
SE
(MT) $2C_2H_4$
singlet allenic carbene
SE
(v)
(TM)
CrO_3
$CH_2I_2/Zn - Cu$
Forward Synthesis
(i) $Pb(OAc)_4$
(ii) H_3O^+
SE
Route - 3
$:CCl_2$
$OH/CHCl_3$
SE
SE
Forward Synthesis
$OH/CHCl_3$
CH_3Li
(MT)
CH_2N_2
SE
SE
Forward Synthesis
hν
(i) $SOCl_2$
(ii) CH_2N_2

(vi) Route - 1

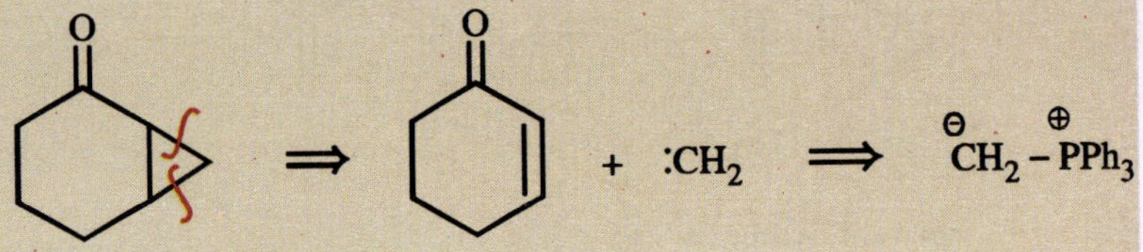

$\Rightarrow$ + :CH_2 $\Rightarrow$ $\overset{\ominus}{C}H_2 - \overset{\oplus}{P}Ph_3$

Forward Synthesis

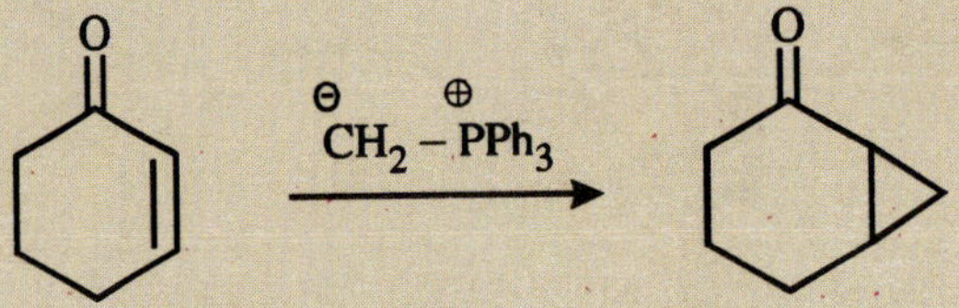

$\xrightarrow{\overset{\ominus}{C}H_2 - \overset{\oplus}{P}Ph_3}$

Route - 2

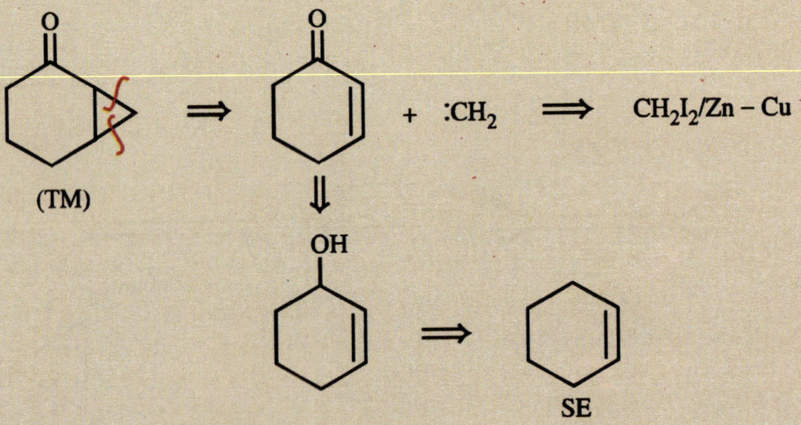

$\Rightarrow$ + :CH_2 $\Rightarrow$ $CH_2I_2/Zn - Cu$

(TM)

OH

$\Rightarrow$

SE

Forward Synthesis

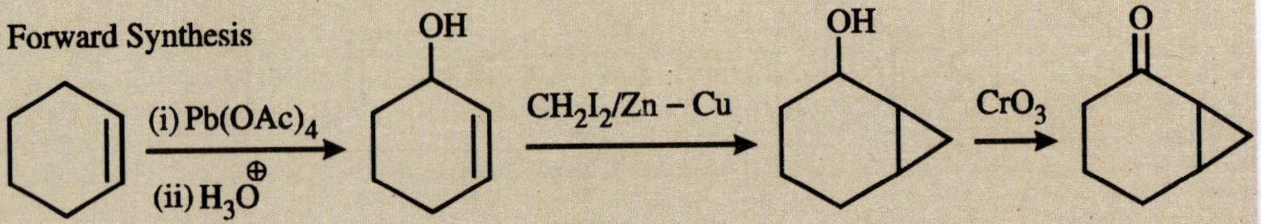

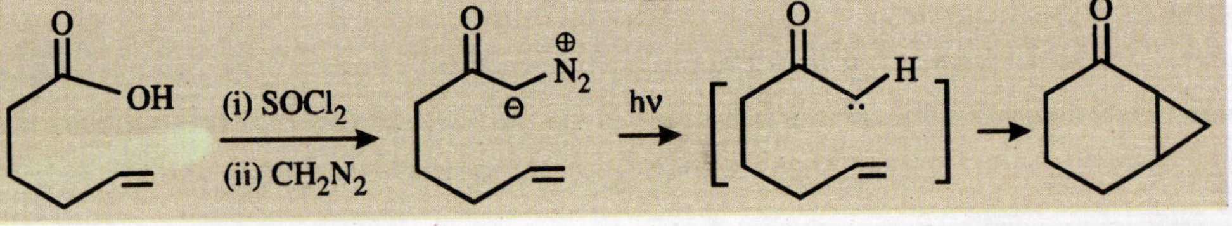

(i) $Pb(OAc)_4$ (ii) $H_3\overset{\oplus}{O}$ → OH → $CH_2I_2/Zn - Cu$ → OH → CrO_3 → O

Route - 3

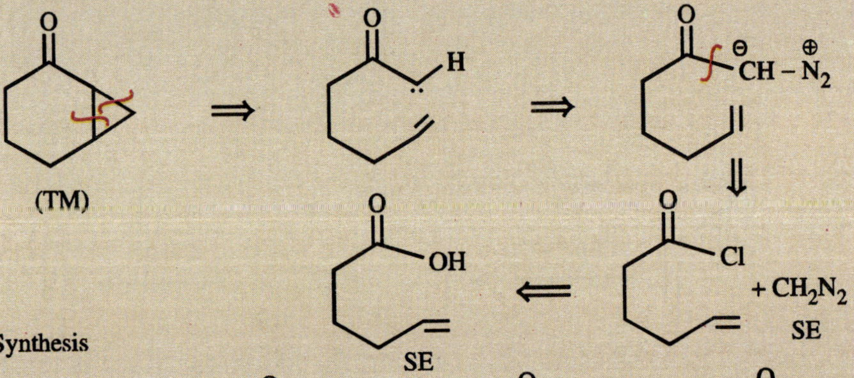

$\Rightarrow$ H $\Rightarrow$ $\overset{\ominus}{C}H - \overset{\oplus}{N_2}$

(TM)

OH $\Leftarrow$ Cl + CH_2N_2

SE SE

Forward Synthesis

OH (i) $SOCl_2$ (ii) CH_2N_2 → $\overset{\oplus}{N_2}$ $\ominus$ → hν → [H] →

B. Cyclobutane ring is disconnected through double 'ene' and photochemical cycloaddition is tried.

Ph Ph
H H H H
COOH COOH
(TM)

$\Longrightarrow$ Ph–CH=CH–Ph + CO_2H–CH=CH–CO_2H

Forward Synthesis

Ph–CH=CH–Ph + CO_2H–CH=CH–CO_2H $\xrightarrow[\pi^2 s + \pi^2 s]{h\nu}$ cyclobutane (Ph, Ph, H, H, H, H, CO_2H, CO_2H)

Review synthetic problems

Show the forward synthesis of the following TM with retro-analysis.

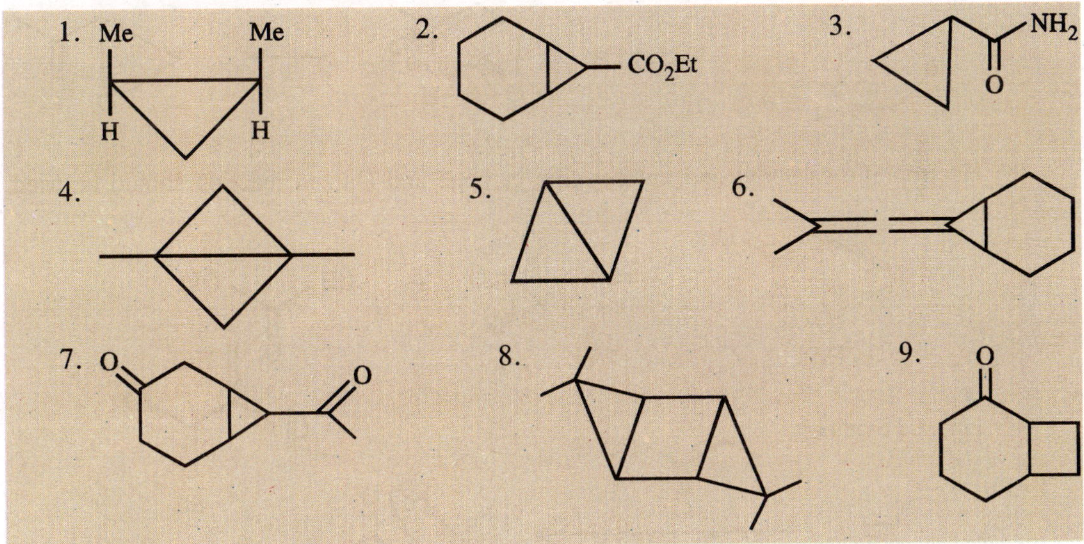

Guideline 24

Epoxides may be disconnected as the following rules.

(i) For isolated epoxides the disconnection is either to neutral 'O' (which is electrophilic) and 'ene' or to carbonyl and nucleophilic carbene.

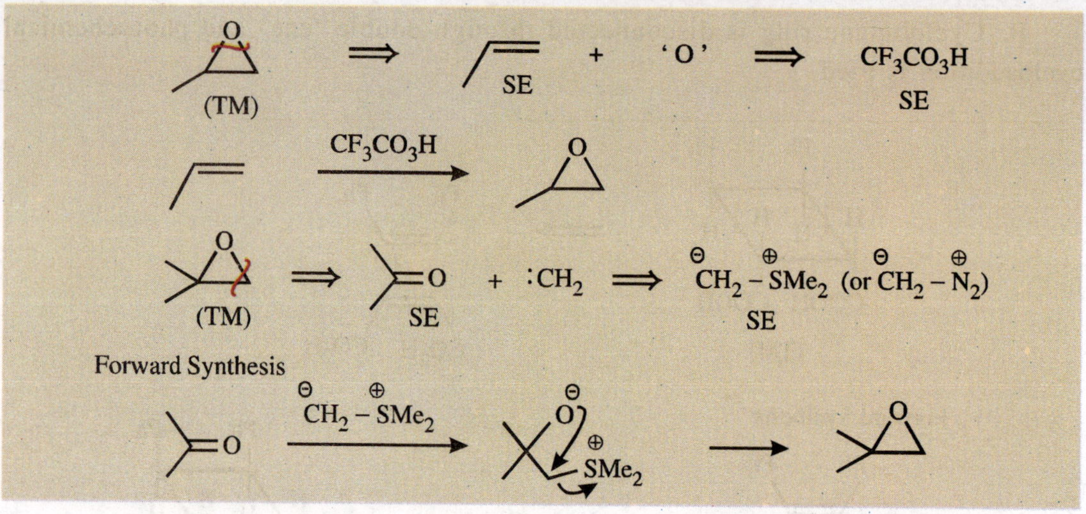

(TM) ⟹ SE + 'O' ⟹ CF_3CO_3H SE

CF_3CO_3H

(TM) ⟹ SE + :CH_2 ⟹ $\overset{\ominus}{C}H_2 - \overset{\oplus}{S}Me_2$ (or $\overset{\ominus}{C}H_2 - \overset{\oplus}{N}_2$) SE

Forward Synthesis

$\overset{\ominus}{C}H_2 - \overset{\oplus}{S}Me_2$ → $\overset{\oplus}{S}Me_2$ →

(ii) For epoxide adjacent to a – M group may be disconnected as either to.

(a) electrophilic α, β unsaturated double bond with a – M group and nucleophilic oxygen.

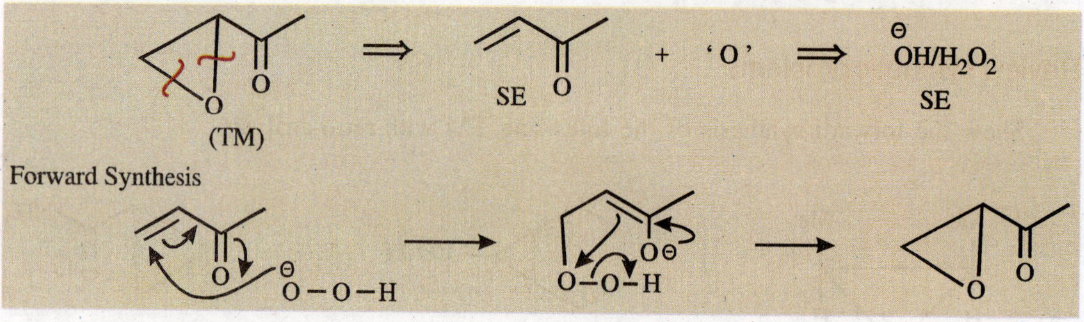

(TM) ⟹ SE + 'O' ⟹ $\overset{\ominus}{O}H/H_2O_2$ SE

Forward Synthesis

$\overset{\ominus}{O}-O-H$ → O–O–H →

(b) electrophilic carbonyl and nucleophilic carbene and Darzen reaction should be tried.

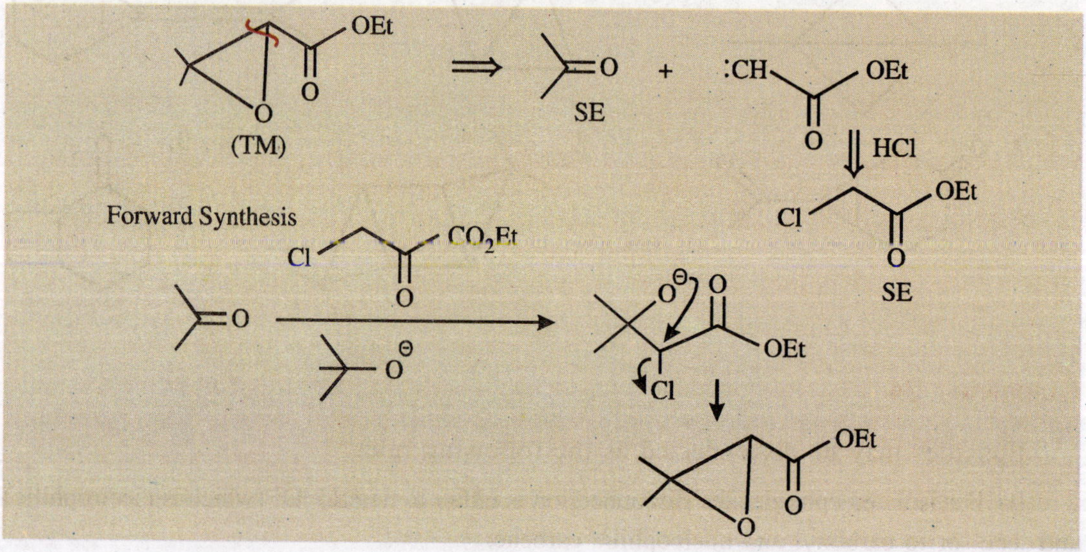

(TM) ⟹ SE + :CH–CO–OEt ⟸ HCl Cl–CH₂–CO–OEt SE

Forward Synthesis

Cl–CH₂–CO–CO_2Et, $(CH_3)_3C\overset{\ominus}{O}$ → OEt, Cl → OEt

Review synthetic problems

Show the forward synthesis of the following TM with retro-analysis.

1. 2. CO_2Et 3. OEt O
4. 5. O O 6. Cl O

Guideline 25

Disconnection by reconnection and reverse (Name) reaction.

One of the most important features is to find out the best synthetic equivalent for a given TM by retrosynthetic analysis. This disconnection (retro-analysis) may be done through neutral synthon without creating conventional charged synthons. This type of disconnection or retro-analysis may be termed as disconnection by reconnection and reverse (Name) reaction.

The following examples will clarify this concept.

A. (i) O OH (TM) O — 1, 6-dico reconnection ⟹ Neutral SN — FGA ⟹ OH — reverse dix of grignard ⟹ O SE + CH_3MgBr SE

Forward Synthesis

O — (i) CH_3MgBr (ii) $H_3\overset{\oplus}{O}/\Delta$ → — O_3/H_2O_2 → O OH O

(ii)

(TM) $\xRightarrow[\text{reconnection}]{\text{1, 6-dico}}$ SN $\xRightarrow{\text{FGA}}$ (diacid) $\Downarrow$ FGI

CO_2H, CO_2H

CO_2Et + CO_2Et $\xLeftarrow[\text{dix of DA}]{\text{reverse}}$ CO_2Et, CO_2Et $\xLeftarrow{\text{FGA}}$ CO_2Et, CO_2Et

Forward Synthesis

CO_2Et + CO_2Et $\xrightarrow{\text{D.A}}$ CO_2Et, CO_2Et $\xrightarrow{H_2/Pd}$ CO_2Et, CO_2Et

(i) KOH

(ii) $H_3\overset{\oplus}{O}$

$\xleftarrow{O_3}$ $\xleftarrow{Pb(OAc)_4}$ CO_2H, CO_2H

(iii)

MeO, H (TM) $\xRightarrow[\text{reconnection}]{\text{1, 6 dico}}$ OMe SN

$\Downarrow$ reverse Birch reduction

OMe SE

Forward Synthesis

OMe $\xrightarrow[\text{EtOH}]{\text{Na/liq } NH_3}$ OMe

$\downarrow O_3/Me_2S$

MeO, H

(Ozonolysis is selective with more electron rich ene)

(iv)

HO (TM) O O

reverse lactonisation

HO 6 4 3 2 5 1 O OH OH

1, 6-dico recon.

reverse dix of D.A

OH SN

FGI

SE CO_2Et + OH SE

Forward Synthesis

CO_2Et + D.A → EtO_2C — $LiAlH_4$ → OH

O_3/H_2O_2

COOH CO_2H OH

HCO_2H/Δ ←

HO O O

(γ-lactone is more favoured than δ-lactone)

(v)

O O (TM)

reverse ketalisation dix

OH HO O SN

FGI

O OEt EtO O 6 O 5 1 2 3 4

reconnection by reverse Claisen ester condensation

O 5 4 3 2 1 O OEt H O

1, 5 dico reverse dix of Michael cond.

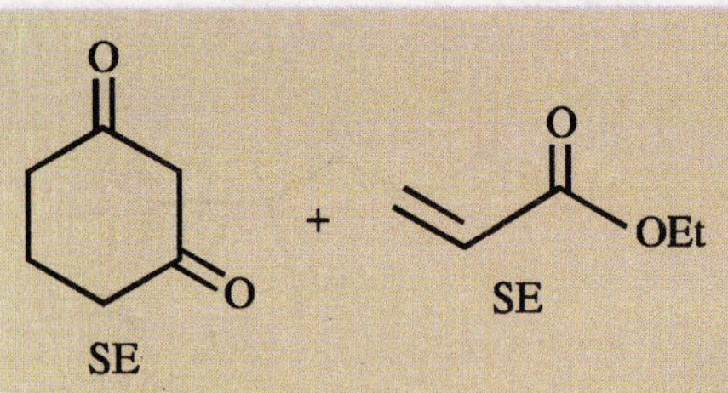

SE + SE

Forward Synthesis

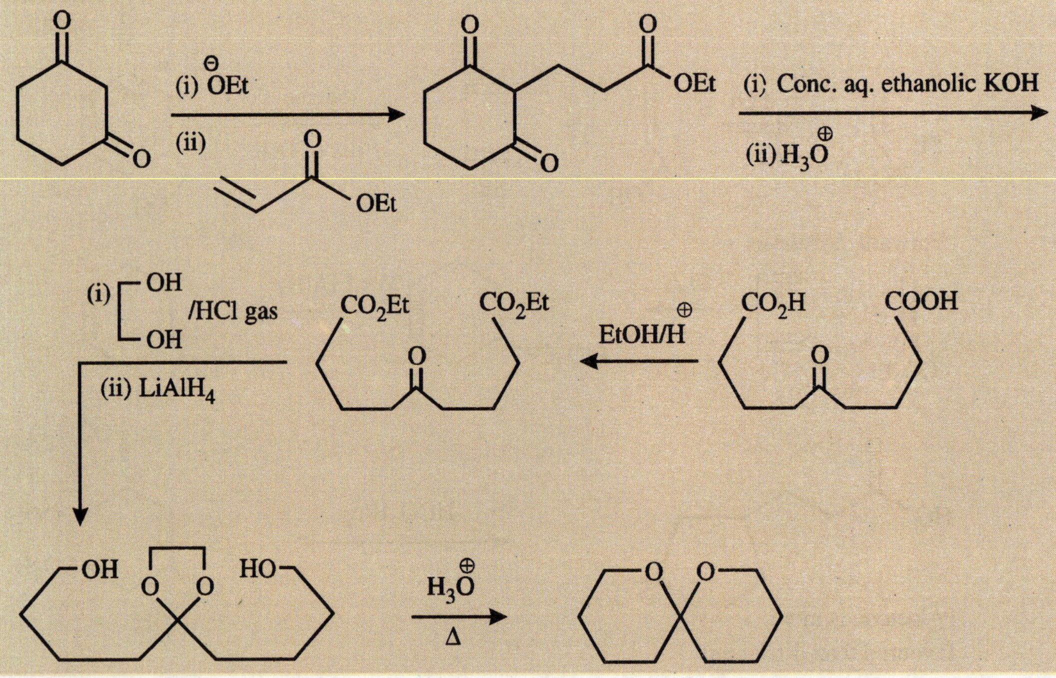

(i) $\overset{\ominus}{O}Et$ (ii) ethyl acrylate → (i) Conc. aq. ethanolic KOH (ii) $H_3\overset{\oplus}{O}$

$EtOH/\overset{\oplus}{H}$

(i) ethylene glycol /HCl gas (ii) $LiAlH_4$

$H_3\overset{\oplus}{O}$, Δ

Review synthetic problems

Find out he synthetic equivalent by reconnection and reverse disconnection of the following TM and then show the forward synthesis.

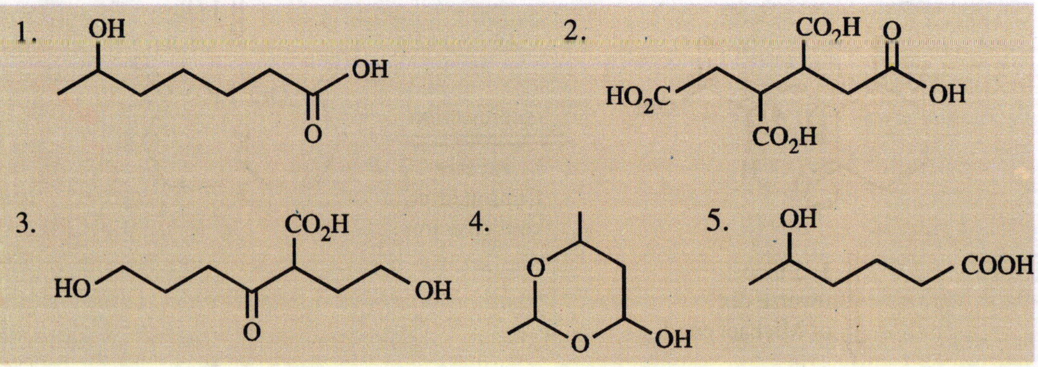

1\.

2\.

3\.

4\.

5\.

B. Synthetic equivalent may be find out by disconnection with a reverse name reaction.

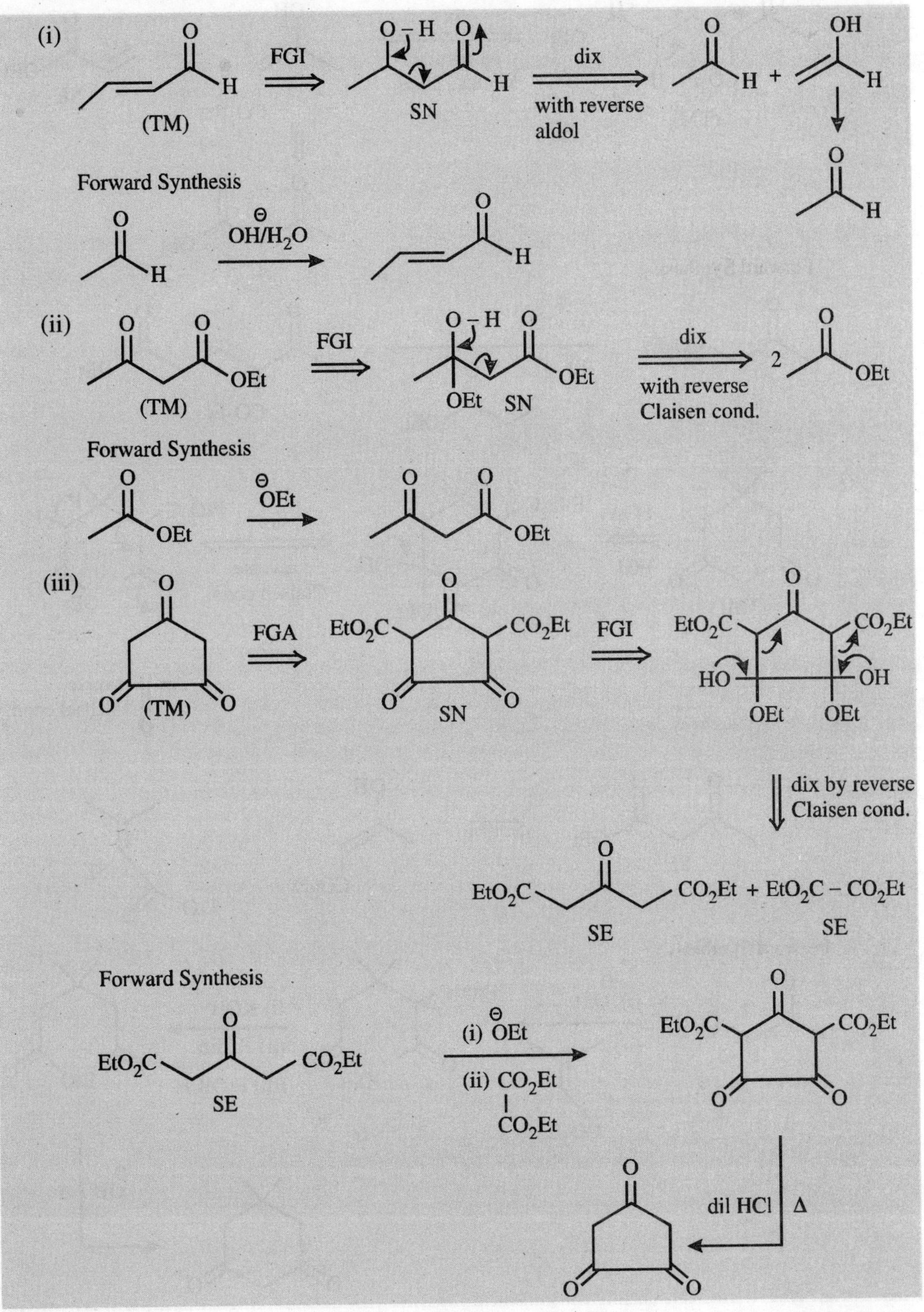

(i)

(TM) FGI ⟹ SN; dix with reverse aldol ⟹ + ; ↓

Forward Synthesis

$\overset{\ominus}{O}H/H_2O$ →

(ii)

(TM) FGI ⟹ OEt SN; dix with reverse Claisen cond. ⟹ 2 OEt

Forward Synthesis

$\overset{\ominus}{O}Et$ →

(iii)

(TM) FGA ⟹ EtO_2C, CO_2Et, SN; FGI ⟹ EtO_2C, CO_2Et, HO, OH, OEt, OEt

⇓ dix by reverse Claisen cond.

EtO_2C–CH₂–CO–CH₂–CO_2Et (SE) + EtO_2C-CO_2Et (SE)

Forward Synthesis

EtO_2C–CH₂–CO–CH₂–CO_2Et (SE) —(i) $\overset{\ominus}{O}Et$; (ii) CO_2Et–CO_2Et → EtO_2C, CO_2Et (cyclic trione)

dil HCl | Δ →

(iv)

(TM)

dix by reverse Michael cond.

OH, CO_2Et + OEt SE

SE

Forward Synthesis

CO_2Et (i) $\overset{\ominus}{O}Et$ (ii) OEt → OEt CO_2Et

(v)

(TM) FGA FGI EtO_2C OH OEt

dix reverse Claisen cond. EtO_2C H OEt

dix reverse Michael cond.

SE OEt ⟸ OH CO_2Et + SE EtO

Forward Synthesis

CO_2Et (i) $\overset{\ominus}{O}Et$ (ii) CO_2Et → EtO_2C OEt

(i) KOH (ii) HCl/Δ (iii) EtOH/$H^{\oplus}$ → EtO

$\overset{\ominus}{O}H$ Δ

(vi)

(TM) $\xrightarrow[\text{reverse Claisen cond.}]{\text{dix}}$ $\xRightarrow{\text{FGI}}$

$\xleftarrow[\text{reverse. Fries rearrangement}]{\text{dix}}$ + Ph–CO–Cl

SE

⇓ FGI

OH + CH₃COCl

SE SE

Forward Synthesis

$\xrightarrow[\text{Py}]{\text{MeCOCl}}$ $\xrightarrow[\Delta]{\text{AlCl}_3}$

PhCOCl Py

$\overset{\ominus}{\text{O}}\text{Et}$

(vii)

(TM) $\xRightarrow[\text{reverse Pinacol}]{\text{dix}}$ OH OH $\xRightarrow[\text{reverse Pinacolic <H>}]{\text{dix}}$ 2 O

Forward Synthesis

O $\xrightarrow[\text{(ii) } H_3\overset{\oplus}{O}]{\text{(i) } Mg/C_6H_6}$ OH OH $\xrightarrow[\text{(ii) } H_3\overset{\oplus}{O}]{\text{(i) } PCl_5/\Delta}$ O

(viii)

OH (TM) $\rightleftharpoons$ O $\xRightarrow[\text{reverse Claisen rearrangement}]{\text{dix}}$ O SE

Forward Synthesis

O $\xrightarrow{\Delta}$ OH

(ix)

Me H H Me H Me Me H (TM) $\xRightarrow[\text{reverse (2 + 2) photo cycloaddition}]{\text{dix by}}$ Me Me SE + Me Me SE

Forward Synthesis

Me Me SE + Me Me SE $\xrightarrow[\pi^2 s + \pi^2 s]{h\nu}$ Me H H Me H Me Me H

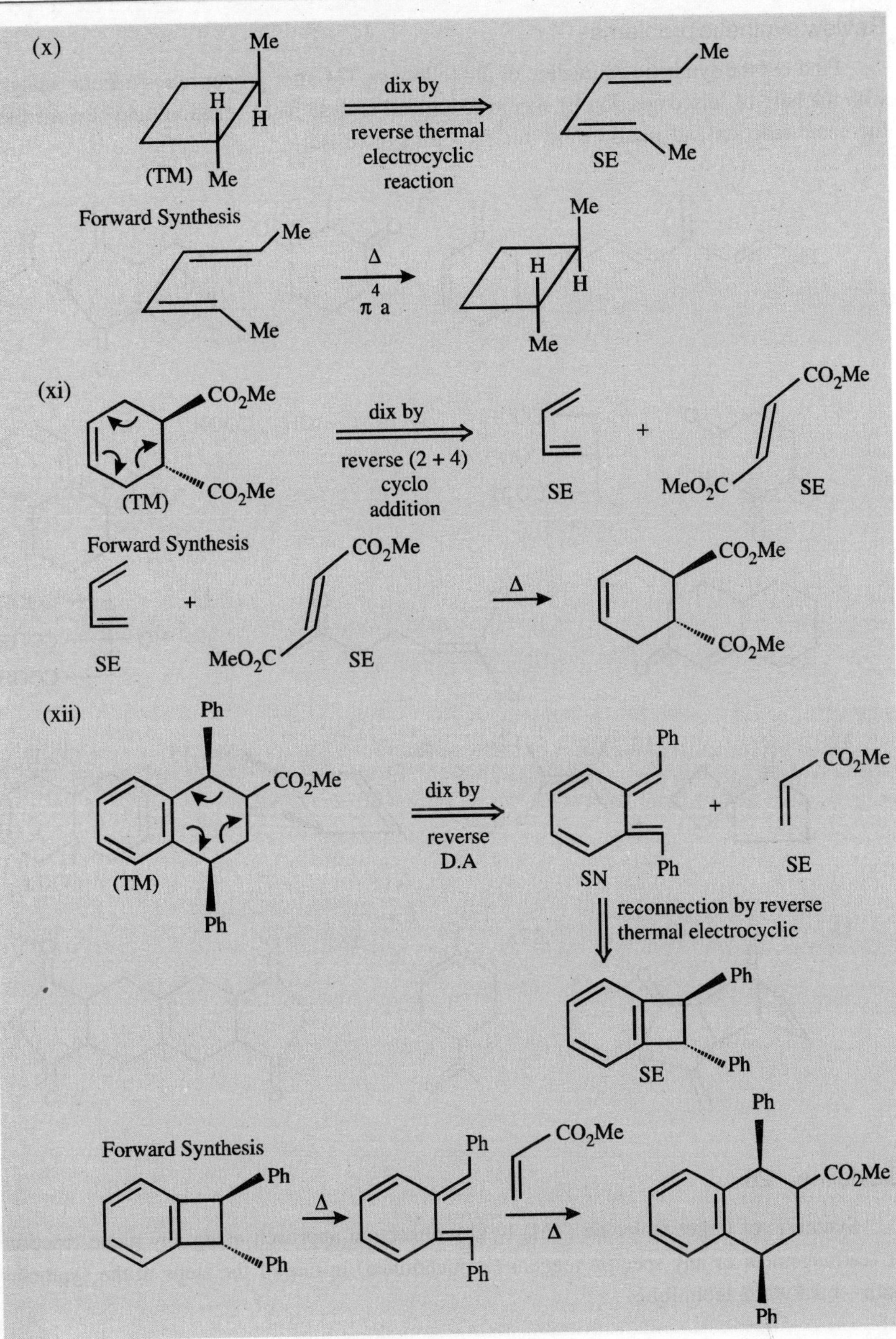
(x)
Me
H
H
(TM) Me
dix by
reverse thermal
electrocyclic
reaction
Me
SE
Me
Forward Synthesis
Me
Δ
4
π a
Me
Me
H
H
Me
(xi)
CO2Me
(TM)
CO2Me
dix by
reverse (2 + 4)
cyclo
addition
SE
+
CO2Me
MeO2C
SE
Forward Synthesis
SE
+
CO2Me
MeO2C
SE
Δ
CO2Me
CO2Me
(xii)
Ph
CO2Me
(TM)
Ph
dix by
reverse
D.A
Ph
SN
Ph
+
CO2Me
SE
reconnection by reverse
thermal electrocyclic
Ph
SE
Ph
Forward Synthesis
Ph
Ph
Δ
Ph
Ph
CO2Me
Δ
Ph
CO2Me
Ph

Review synthetic problems

Find out the synthetic equivalent of the following TM after proper retrosynthetic analysis with the help of 'disconnection by reconnection and reverse name reaction' and also mention the name reaction and finally show the forward synthesis.

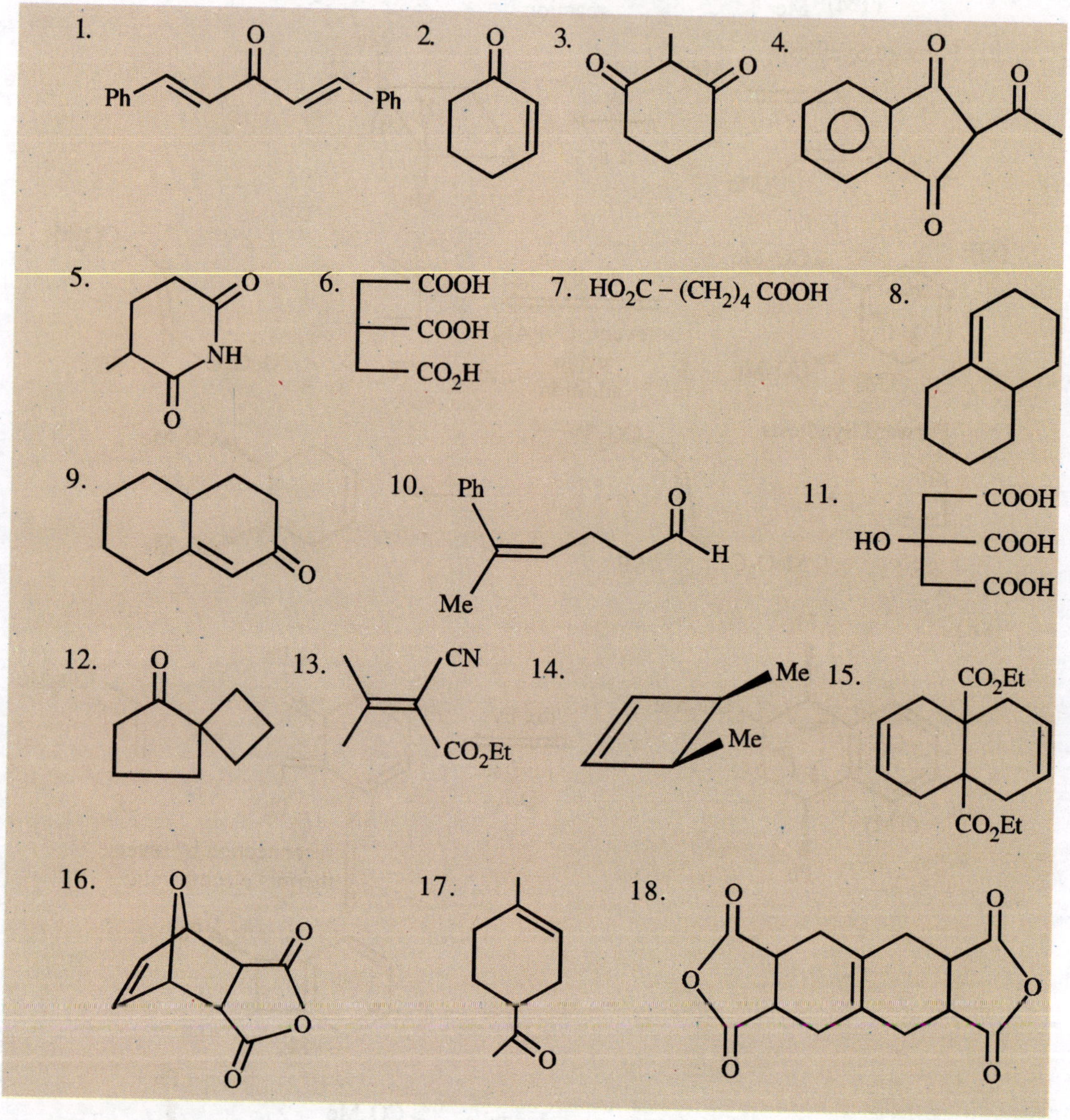

1. Ph, O, Ph
2. O
3. O, O
4. O, O, O
5. O, NH, O
6. COOH, COOH, CO_2H
7. $HO_2C-(CH_2)_4COOH$
8.
9. O
10. Ph, O, H, Me
11. HO, COOH, COOH, COOH
12. O
13. CN, CO_2Et
14. Me, Me
15. CO_2Et, CO_2Et
16. O, O, O, O
17. O
18. O, O, O, O, O, O

Guideline 26

Synthesis of target molecule (TM) by disconnection approach using any name reaction or rearrangement or any specific reagent (as mentioned) in one of the steps in the synthetic path—backward technique.

(i) Using Beckmann rearrangement

MeO–C$_6$H$_4$–NH–C(=O)–CH$_3$ (TM) $\xrightarrow[\text{Beckmann}]{\text{reverse}}$ MeO–C$_6$H$_4$–C(=NOH)–CH$_3$

⇓ FGI

MeO–C$_6$H$_4$ + CH$_3$COCl ⇐ MeO–C$_6$H$_4$–C(=O)–CH$_3$ + NH_2OH

Forward Synthesis

MeO–C$_6$H$_5$ $\xrightarrow[AlCl_3]{CH_3COCl}$ MeO–C$_6$H$_4$–C(=O)–CH$_3$ $\xrightarrow{NH_2OH}$ Me–O–C$_6$H$_4$–C(=N–OH)–CH$_3$ $\xrightarrow[\text{(ii) } H_3O^{\oplus}]{\text{(i) } PCl_5/\Delta}$ Me–O–C$_6$H$_4$–NH–C(=O)–CH$_3$

(ii) Using Perkin reaction

C$_6$H$_5$–CH=CH–C(=O)OH (TM) $\overset{\text{FGA}}{\Longrightarrow}$ C$_6$H$_5$–CH=CH–C(=O)–O–C(=O)–CH$_3$

⇓ FGI

C$_6$H$_5$CHO (SE) + $^{\ominus}$CH$_2$–C(=O)–O–C(=O)–CH$_3$ + $^{\oplus}$C(=O)–CH$_3$ ⇐ C$_6$H$_5$–CH(OC(=O)CH$_3$)–CH$_2$–C(=O)–O–C(=O)–CH$_3$

$^{\ominus}$CH$_2$–C(=O)–O–C(=O)–CH$_3$ ⇓ Ac_2O SE

$^{\oplus}$C(=O)–CH$_3$ ⇓ Ac_2O SE

Forward Synthesis

C$_6$H$_5$CHO $\xrightarrow[\text{(ii) } H_3O^{\oplus}]{\text{(i) } Ac_2O,\ CH_3CO_2Na}$ C$_6$H$_5$–CH=CH–CO_2H

(iii) Using Benzidine rearrangement

Biphenyl (TM) $\xRightarrow{FGA}$ $H_2N-C_6H_4-C_6H_4-NH_2$

reverse ⇓ Benzidine rearr.

$C_6H_5-NO_2$ (SE) $\xLeftarrow{FGI}$ $C_6H_5-NH-NH-C_6H_5$

Forward Synthesis

$C_6H_5-NO_2 \xrightarrow{Zn/NaOH} C_6H_5-NH-NH-C_6H_5$

↓ HCl/Δ

$H_2N-C_6H_4-C_6H_4-NH_2 \xrightarrow[(ii)\ H_3PO_2]{(i)\ HNO_2}$ $C_6H_5-C_6H_5$

(iv) Using wittig reaction

2-methylpent-2-ene $\xRightarrow{FGI}$ 2-methylpentan-2-ol (OH) $\xRightarrow{FGI}$ ⟶ $(CH_3)_2C=O + Ph_3P=CHCH_2CH_3$

C=O and halide (which will be ylide) should be chosen in such a way better S_N2 and nucleophile arises.

Forward Synthesis

$CH_3CH_2CH_2Br \xrightarrow[(ii)\ nBuLi]{(i)\ Ph_3P} CH_3CH_2CH=PPh_3 \xrightarrow{(CH_3)_2C=O}$ 2-methylpent-2-ene

(v) (a) Using Hofmann's rearrangement

CH_3NH_2 (TM) $\xRightarrow[\text{Hofmann}]{\text{reverse}}$ $CH_3-\underset{\underset{O}{\|}}{C}-NH_2$ $\Rightarrow$ $CH_3COOH + NH_3$

Hof. Starting material (SE)

$CH_3CONH_2 \xrightarrow{Br_2/KOH} CH_3NH_2$

(b) Using Lossen rearrangement

$$CH_3NH_2 \text{ (TM)} \xRightarrow{FGI} CH_3-\underset{\underset{O}{||}}{C}-NH_2 \xRightarrow{FGA} CH_3-\underset{\underset{O}{||}}{C} \wr NH-OH$$

Lossen starting material

$$\Downarrow$$

$$CH_3CO_2Et + NH_2-OH$$

Forward Synthesis

$$CH_3CO_2Et \xrightarrow{NH_2OH} CH_3-\underset{\underset{O}{||}}{C}-NHOH \xrightarrow[\Delta]{\overset{\oplus}{H}} CH_3NH_2$$

(c) Using Curtius rearrangement

$$CH_3NH_2 \text{ (TM)} \xRightarrow{FGI} CH_3-\underset{\underset{O}{||}}{C}-NH_2 \xRightarrow{FGA} CH_3-\underset{\underset{O}{||}}{C} \wr NH-NH_2$$

Curtius starting material

$$\Downarrow$$

$$CH_3CO_2Et + NH_2-NH_2$$

Forward Synthesis

$$CH_3CO_2Et \xrightarrow{NH_2-NH_2} CH_3-\underset{\underset{O}{||}}{C}-NH-NH_2 \xrightarrow[\Delta]{HNO_2} CH_3NH_2$$

(d) Using Schmidt rearrangement

$$CH_3NH_2 \text{ (TM)} \xRightarrow{FGI} CH_3-\underset{\underset{O}{||}}{C}-NH_2 \xRightarrow{FGA} CH_3-\underset{\underset{O}{||}}{C}-NH-NH_2$$

$$\Downarrow \text{FGA}$$

$$CH_3COOH + N_3H \Leftarrow CH_3-\underset{\underset{O}{||}}{C} \wr NH-\overset{\oplus}{N}\equiv N$$

Schmidt starting material

Forward Synthesis

$$CH_3COOH \xrightarrow{N_3H} CH_3-\underset{\underset{O}{||}}{C}-\overset{\overset{H}{|}}{N}-\overset{\oplus}{N}\equiv N \xrightarrow{\Delta} CH_3NH_2$$

(vi) Using Benzilic acid rearrangement

Ph—C(OH)(Ph)—COOH (TM) $\xrightarrow[\text{Benz. acid rearr.}]{\text{reverse}}$ Ph—CO—C(Ph)(OH)—$O^{\ominus}$ ⟹ Ph—CO—CO—Ph

Ph—CO—CO—Ph $\xrightarrow[\text{(ii) } H_3O^{\oplus}]{\text{(i) KOH}}$ $Ph_2C(OH)COOH$

(vii) Using Michael condensation

(TM) $\xRightarrow{\text{FGA}}$ (diol, OH … OH) $\xRightarrow{\text{FGI}}$ H—C(=O)— … —C(=O)—

1, 5 dico Michael dix ⟹ H—C(=O)—CH=CH—CH_3 (SE) + CO_2Et—CH_2—CO—CH_3 (SE)

Forward Synthesis

$CH_3COCH_2CO_2Et$ $\xrightarrow[\text{(ii) } CH_3CH{=}CHCHO]{\text{(i) } \overset{\ominus}{O}Et}$ H—C(=O)— … —CH(CO_2Et)—C(=O)— $\xrightarrow{\text{NaCl/DMSO}}$ H—C(=O)— … —C(=O)— $\xrightarrow[\text{Conc. HCl}]{\text{Zn(Hg)}}$ (TM)

(viii) Using Claisen ester condensation

(TM) $\xRightarrow{\text{FGA}}$ EtO_2C … CO_2Et (cyclic dione diester) ⟹ EtO_2C—$(CH_2)_3$—CO_2Et + EtO—C(=O)—C(=O)—OEt

Forward Synthesis

EtO_2C—$(CH_2)_3$—CO_2Et $\xrightarrow[\text{(ii) EtO—CO—CO—OEt}]{\text{(i) } \overset{\ominus}{O}Et}$ (CO_2Et, CO_2Et cyclopentane-1,2-dione) $\xrightarrow[\Delta]{\text{dil HCl}}$ cyclopentane-1,2-dione

(ix) Using Favorskii rearrangement

(x) Using Claisen rearrangement

(xi) Using Cope rearrangement

(xii) Using Hofmann exhaustive methylation

Number of possible tautomers = Number of possible starting materials(SE) but less substituted tautomer is always preferred since Hofmann elimination goes via less crowded TS.

Forward synthesis

SE (pyridine) —Ni/H_2→ piperidine (N–H) —(i) X´S MeI (ii) AgOH (iii) Δ→ N—Me, Me —(i) CH_3I (ii) AgOH (iii) Δ→ diene

(b)

+ NH_3 ⇒ FGA ... :NH_2 → N–H → N–H SE

equivalent tautomers

Forward synthesis

N–H —Ni/H_2→ N–H —(i) X´S CH_3I (ii) AgOH (iii) Δ→ N, Me Me —(i) CH_3I (ii) AgOH (iii) Δ→ ≡

Review synthetic problems

Show the forward synthesis after proper retrosynthetic analysis or via disconnection or backward technique or reconnection and reverse reaction technique of the following TM as directed either with any name reaction or specific reagent indicates within the bracket in one of the synthetic steps for each case.

1. Ph (wittig reaction)

2. O O (Knoevenagel)

3. N–H O (Beckmann)

4. $CONHCH_3$ (Hofmann)

5. CO_2H H_5C_6 (Perkin)

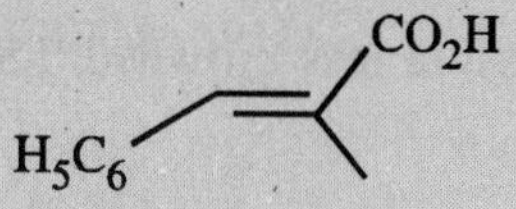

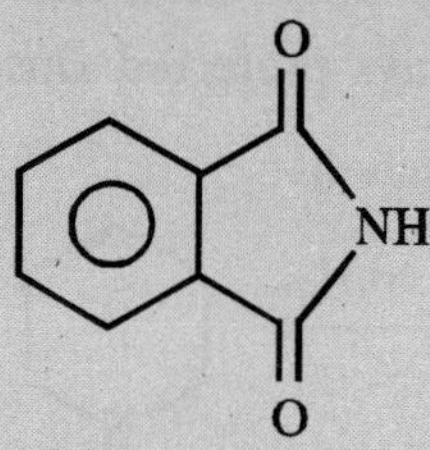

6. O NH O (Schmidt)

7. O (Pinacol-Pinacolone rearrangement)

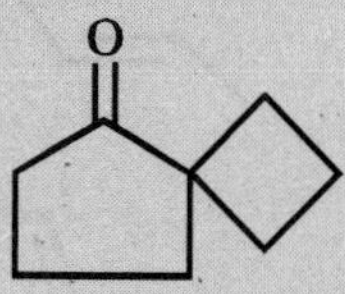

8. HO O OH (Baeyer-Villiger)

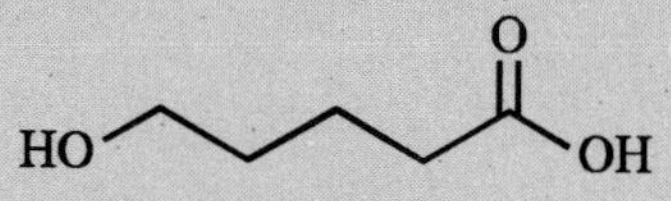

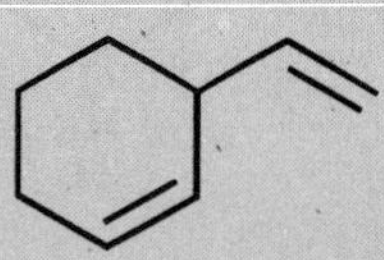

9. , (Hofmann exhaustive)

10. Cl Cl (Benzidine rearr.)

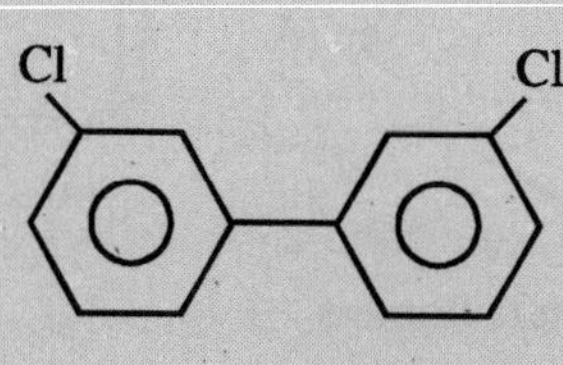

11. Ph Ph Ph Ph N Ph (Diels Alder)

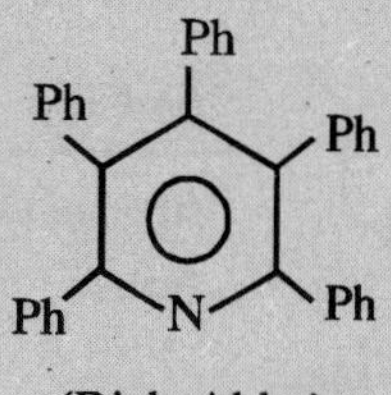

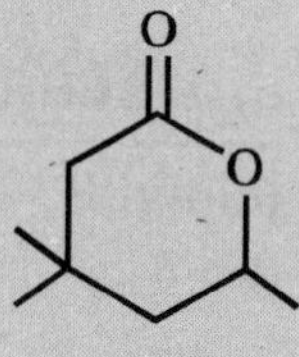

12. O O (Michael cond.)

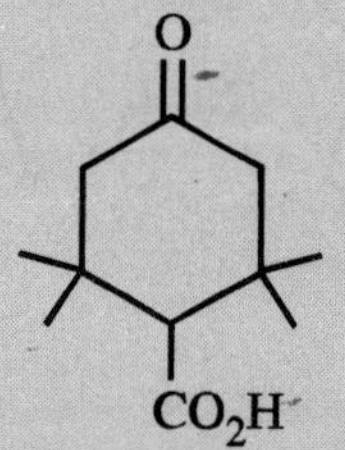

13. O CO_2H (double Michael cond.)

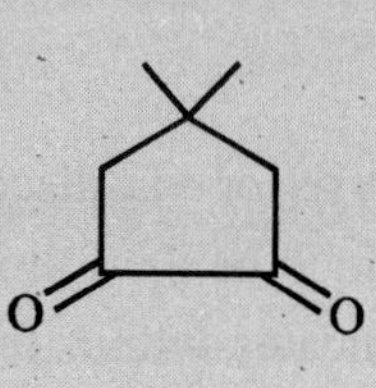

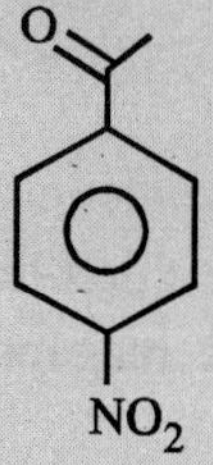

14. O O , O NO_2 (Claisen ester cond.)

15. O CO_2Et O (Michael followed by Claisen ester cond.)

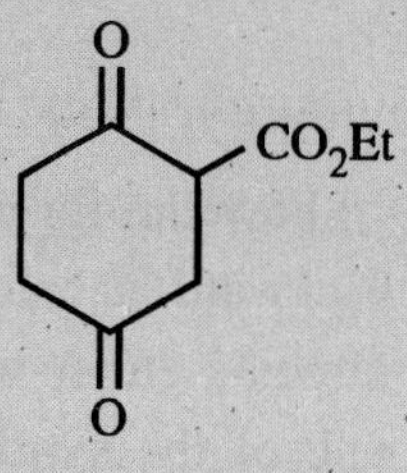

16. O CO_2Et (Michael followed by aldol)

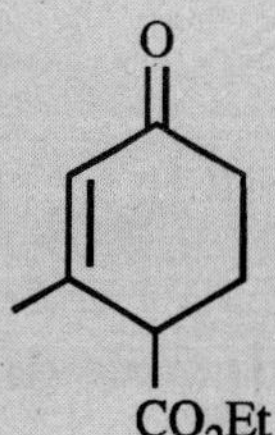

17. CO_2H (Wittig followed by Michael cond.)

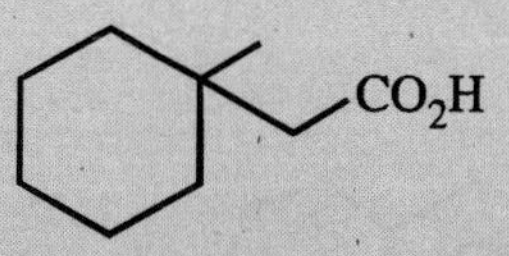

18. 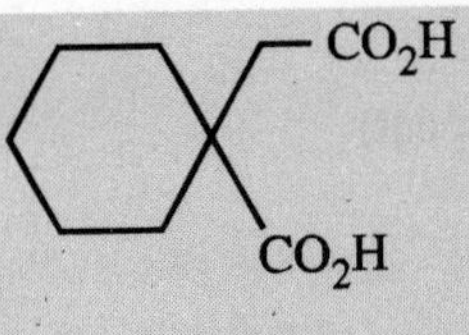 CO_2H CO_2H (Knoevenagel followed by Michael cond.)

19. 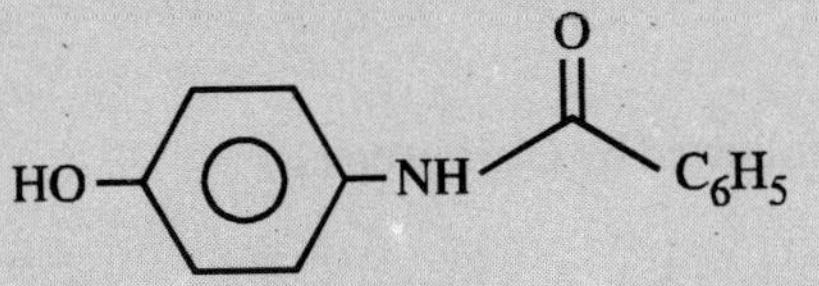 HO–NH–C(=O)–C_6H_5 (Fries followed by Beckmann)

20. 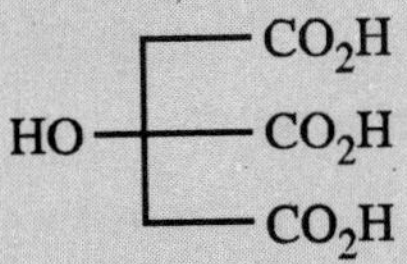 HO, CO_2H, CO_2H, CO_2H (Reformatsky or Benzilic acid type rearrangement)

21. 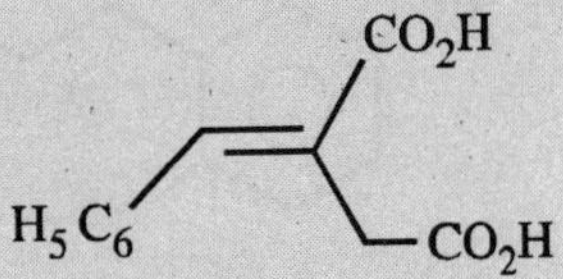 H_5C_6, CO_2H, CO_2H (Wittig or Aldol)

22. 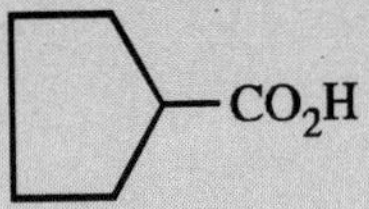 CO_2H (Wolff rearrangement or Favorskii)

23. 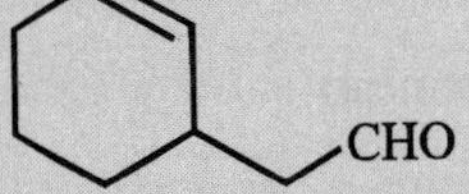 CHO (Claisen rearrangement)

24. 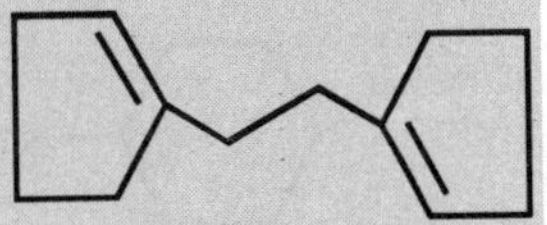 (Cope rearrangement)

25. 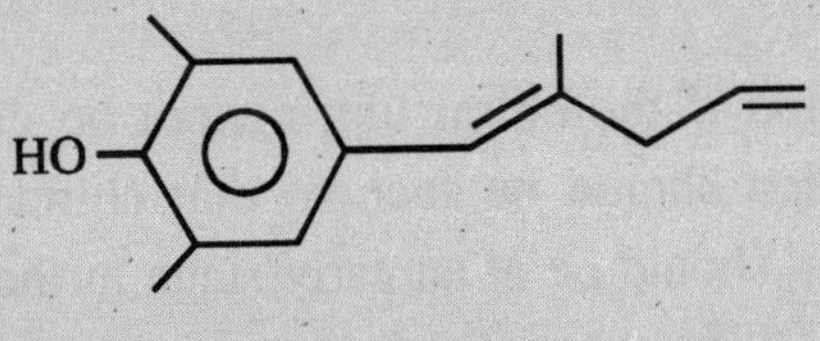 HO (Claisen rearrangement followed by double Cope)

26. 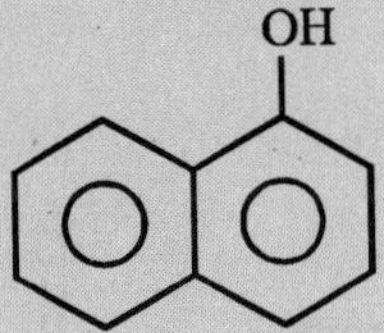 OH (Using NH_2 CO_2H)

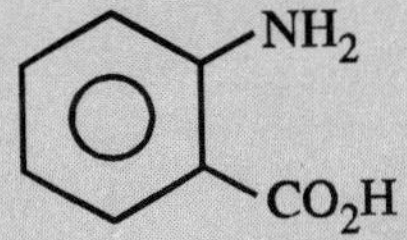

27. 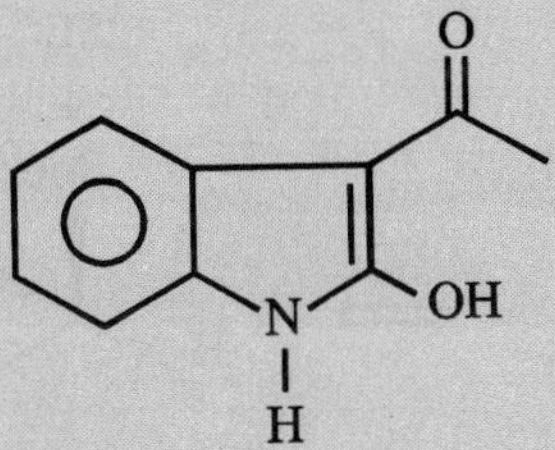 O, OH, N–H ; O, O (Using O, O)

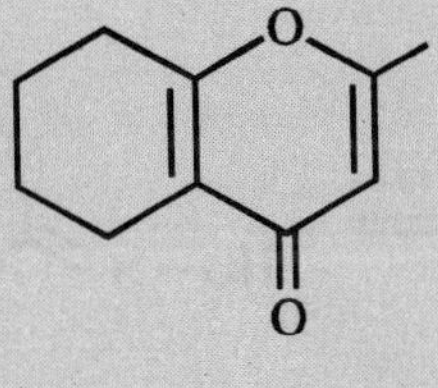

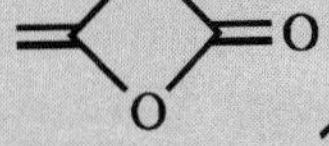

28. 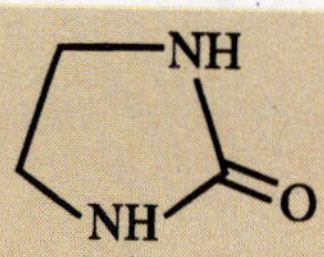 (Hofmann followed by Schmidt rearrangement)

29. 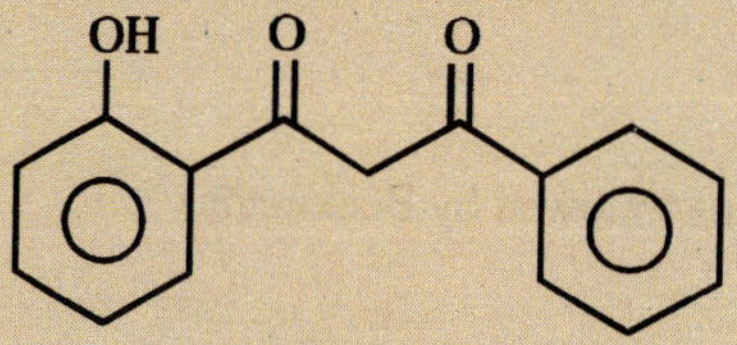 (Fries followed by Claisen cond.)

30. 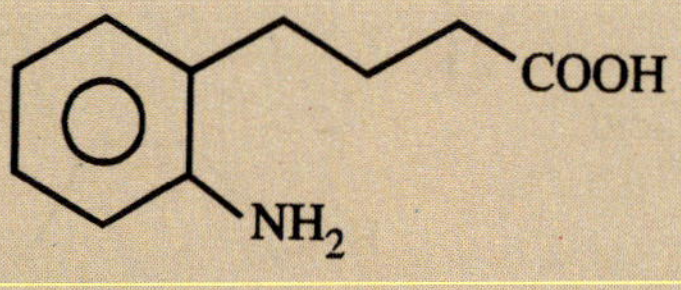 COOH, NH_2 (Using Beckmann rearrangement)

31. 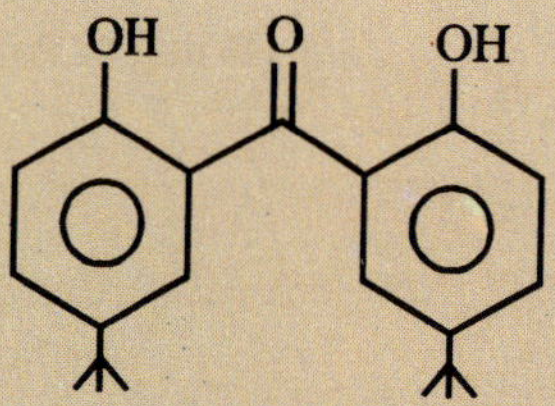 (Using Photo Fries rearrangement)

32. 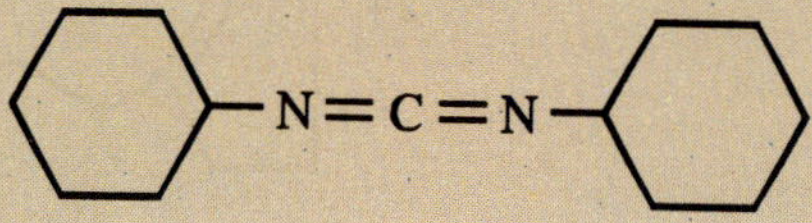 N=C=N (Hofmann rearrangement)

Guideline 27

If a target molecule having a chiral unit and if the chiral unit cannot be developed with asymmetric induction then the disconnection should be such so that chiral unit can be added in pure enantiomeric form i.e. resolution should be at an early stage in the forward synthesis as this will avoid unwanted material—the wrong enantiomer.

Retro-analysis

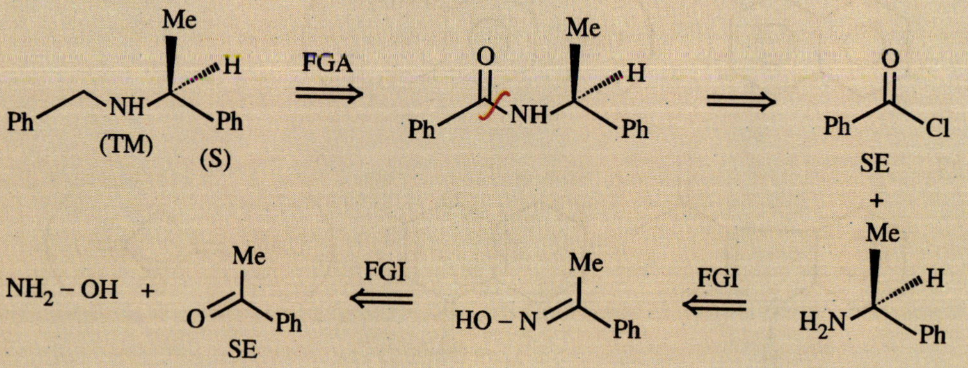

(TM) (S) $\xrightarrow{\text{FGA}}$ ⟹ Ph–COCl SE + H_2N–CH(Me)Ph

NH_2-OH + PhCOMe (SE) $\xLeftarrow{\text{FGI}}$ HO – N=C(Me)Ph $\xLeftarrow{\text{FGI}}$ H_2N–CH(Me)Ph

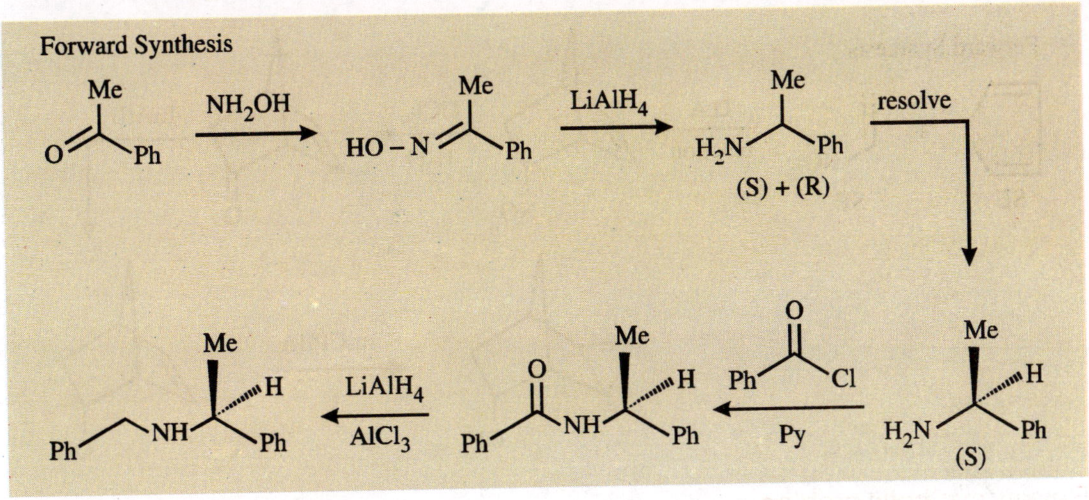

Forward Synthesis

Me, O, Ph —NH_2OH→ HO – N, Me, Ph —$LiAlH_4$→ H_2N, Me, Ph (S) + (R) — resolve → H_2N, Me, H, Ph (S) —Ph(C=O)Cl, Py→ Ph(C=O)NH, Me, H, Ph —$LiAlH_4$, $AlCl_3$→ Ph, NH, Me, H, Ph

Review synthetic problems

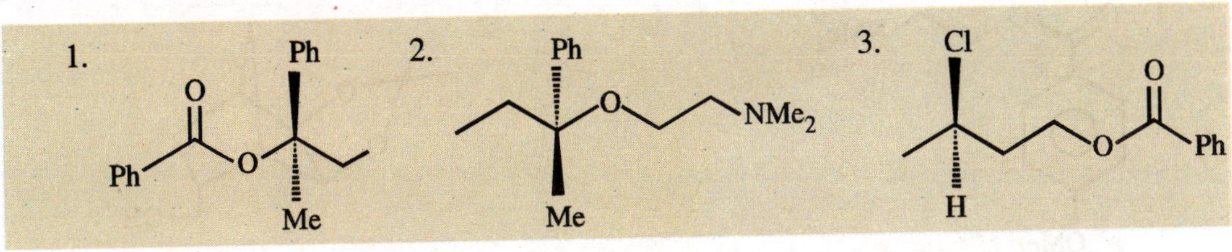

1. Ph, O, Ph, O, Me

2. Ph, O, NMe_2, Me

3. Cl, O, O, Ph, H

Guideline 28

If the target molecule contain more than one reactive functional groups then disconnection should be such in which the more reactive group should be introduced in the late stage (and hence reverse disconnection) in the forward synthesis.

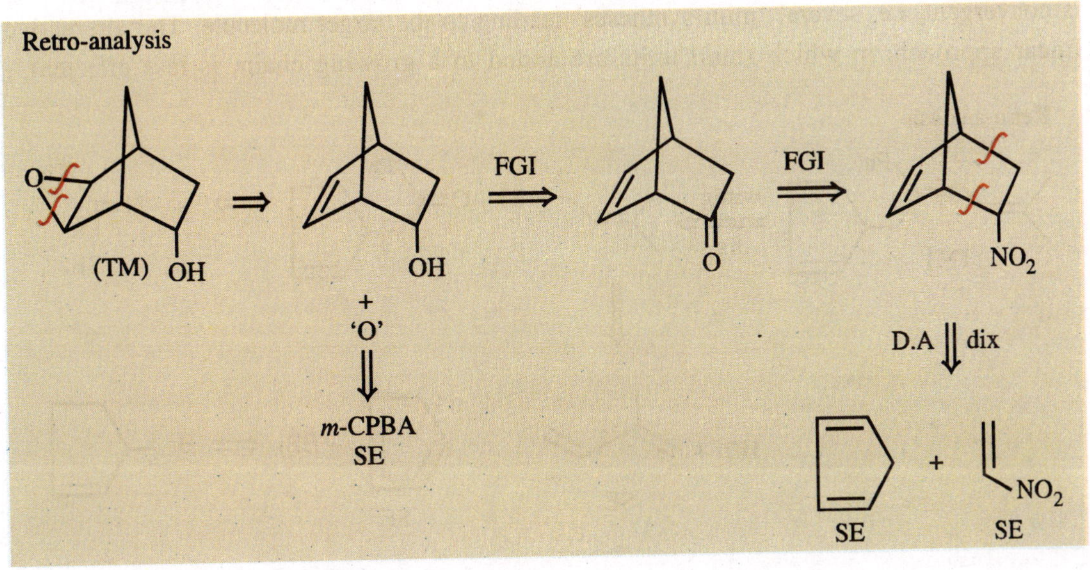

Retro-analysis

O, (TM) OH ⇒ OH + 'O' ⇓ *m*-CPBA SE —FGI⟹ O —FGI⟹ NO_2 — D.A ⇓ dix → SE + NO_2 SE

Forward Synthesis

SE + SE —D.A reaction→ (NO$_2$) —$TiCl_3$, H_2O→ (O) —$LiAlH_4$→ (OH) —m-CPBA→ (O, OH)

Review synthetic problems

1. (O, O, NMe_2, OMe)

2. (O, O, OH)

Guideline 29

Convergent synthesis is always better than linear approach. For this purpose in the retrosynthetic analysis try to divide the TM into approximately equal halves, and thereafter aim to do the same with each subsequent disconnection. This will lead to synthesis which is convergent, i.e. several 'mini-syntheses' leading to the target molecule. The alternative linear approach, in which small units are added to a growing chain, is less efficient.

Retro-analysis

(Ph) (TM) ⇒ wittig dix ⇒ Br + O= (Ph) ⇒ + O= (Ph) SE + Br– ⇐ + HBr SE

⇓ HBr + SE

Forward Synthesis

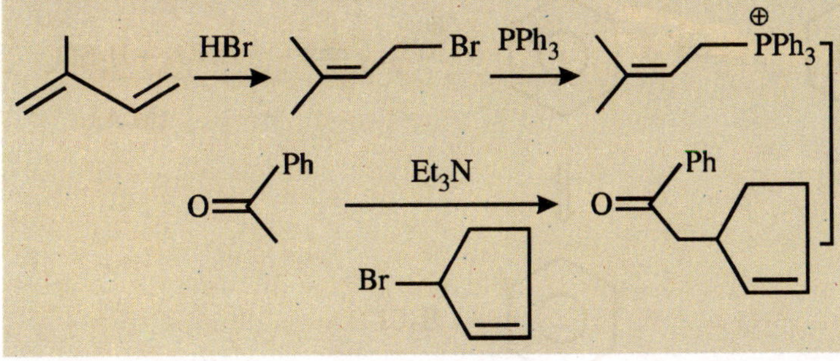

HBr; Br; PPh_3; $\overset{\oplus}{P}Ph_3$; nBuLi; Ph; Ph; Et_3N; O; O; Ph; Br

Review synthetic problems

1.

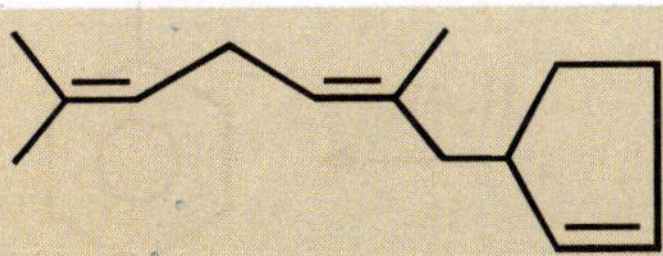

2.

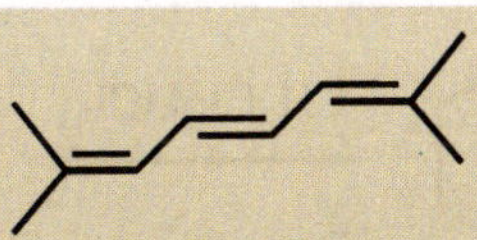

3. O; Ph; O; Ph

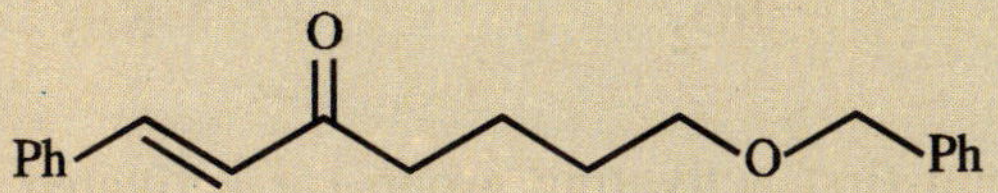

Guideline 30

Disconnection or retrosynthetic analysis of benzenoid target molecules.

During the disconnection of benzenoid target molecules, one should consider the following points.

1. Substitution pattern

Substitution pattern in benzene nucleus are of two types.

(a) Right substitution pattern

(i) O/p-directing group with O/p-substitution.

(ii) m-directing group with m-substitution

(b) Wrong substitution pattern

(i) O/p-directing group with m-substitution.

(ii) m-directing group with O/p-substitution.

(i) If there are two substitution in the TM, the group which is easier to introduce should be done first and consequently disconnection should be reverse.

NH_2 (TM) $\xrightarrow{FGI}$ ... NO_2 ⇒ + $\overset{\oplus}{N}O_2$ ⇒ $HNO_3 + H_2SO_4$ (M.A)

⇓

+ CH_3Cl

SE

Forward Synthesis

$\xrightarrow{CH_3Cl/AlCl_3}$ $\xrightarrow{M.A}$ NO_2 $\xrightarrow{H_2/Pd}$ NH_2

(ii) For electrophilic substitution add first activating group (or activated FGI of deactivating group) and then deactiving group and disconnection should be reverse.

COOH, NO_2 (TM) ⇒ COOH + M.A

⇓ FGI

(methyl group is activated FGI of deactivating carboxylic group)

Me ⇒ + $CH_3Cl/AlCl_3$

Forward Synthesis

$\xrightarrow[AlCl_3]{CH_3Cl}$ Me $\xrightarrow{KMnO_4}$ COOH $\xrightarrow{M.A}$ COOH, NO_2

(TM) O_2N NO_2 ⟹ M.A + ⟹ +

Forward Synthesis

/$AlCl_3$ → M.A → O_2N NO_2

(iii) (a) For nucleophilic substitution add first deactivating group and then nucleophile and disconnection should be reverse.

NO_2 CN NO_2 (TM) ⟹ NO_2 NO_2 + KCN ⇓ + M.A

Forward Synthesis

M.A → NO_2 NO_2 KCN → NO_2 CN NO_2

(b) Nucleophile may be added to the benzene ring via diazo salt or sulphonation.

CN (TM) ⟹ KCN + $\overset{\oplus}{N}\equiv N$ ⟹ NH_2

Forward Synthesis

NH_2 HNO_2 → $\overset{\oplus}{N}\equiv N$ KCN / CuCN → CN

OH (TM) FGI SO_3H $+ H_2SO_4$

conc. H_2SO_4 SO_3H (i) $\overset{\ominus}{O}H$ (ii) $H_3\overset{\oplus}{O}$ OH

(c) Nucleophile may also be added via benzyne.

OMe O (TM) OMe + HO

OMe Br

Forward Synthesis

OMe Br $\overset{\ominus}{N}H_2$ OMe HO OMe O

(iv) Substitution pattern may be changed in the TM.

(a) Methyl group which is O/p directing may be converted to m-directing by conversion it to – COOH, – CCl_3 or – CF_3.

(b) m-directing group to O/p-directing group.

$-CO_2H \Longrightarrow -CH_3$

CO_2H, NO_2 FGI $\Longrightarrow$ NO_2 $\Longrightarrow$ + M.A

(wrong subst.)

(TM)

Forward Synthesis

M.A $\longrightarrow$ NO_2 $KMnO_4$ $\longrightarrow$ COOH, NO_2

$-NO_2 \Longrightarrow -NH_2$

NO_2, NH_2 FGI $\Longrightarrow$ NO_2, NO_2 $\Longrightarrow$ + M.A

(TM) (wrong subst.)

Forward Synthesis

M.A $\longrightarrow$ NO_2, NO_2 NaHS / CH_3OH $\longrightarrow$ NO_2, NH_2

(v) Wrong substitution may be done by using dummy functional group.

Cl FGA $\Longrightarrow$ Cl, NH_2 $\Longrightarrow$ Cl_2 + NH_2 FGR $\Longrightarrow$

(TM) (wrong subst.) (dummy)

Forward Synthesis

M.A $\longrightarrow$ NO_2 H_2 / Pd $\longrightarrow$ NH_2 (i) AcOH (ii) Cl_2 $\longrightarrow$ Cl, NHAc (i) H_3O^+ (ii) HNO_2 (iii) H_3PO_2 $\longrightarrow$ Cl

(vi) During introduction of any group into benzene ring, always take care so that other group should not be effected by either redox agent or by any other reagent.

OMe, Cl, NO_2 (TM) $\xrightarrow{\text{M.A}}$ OMe, Cl $\xrightarrow[\text{ZnCl}_2]{\text{CH}_2\text{O/HCl}}$ OMe

not right disconnection path since
M.A oxidises – CH_2Cl ⟶ – COOH

$\Downarrow$ $CH_2O/HCl/ZnCl_2$

OMe, NO_2 $\xrightarrow{\text{M.A}}$ OMe

right disconnection path

Therefore, oxidisable group should be disconnected first and should be introduced last.

(vii) When there is a O/p substitution involved, one single isomer may be achieved by blocking/deblocking without separation of isomers.

NH_2, Br, Br (TM) $\xrightarrow{\text{FGA}}$ NH_2, Br, Br ⇒ Br_2 + NH_2 ⇒ NH_2 (SE) +

Forward Synthesis

NH_2 $\xrightarrow{\text{/AlCl}_3}$ NH_2 $\xrightarrow[\text{H}_2\text{O}]{\text{Br}_2}$ NH_2, Br, Br $\xrightarrow[\Delta]{\text{HF}}$ NH_2, Br, Br

Review synthetic problems

After proper retrosynthetic analysis find out the synthetic equivalent and then show the forward synthesis of the following benzenoid TM.

1. 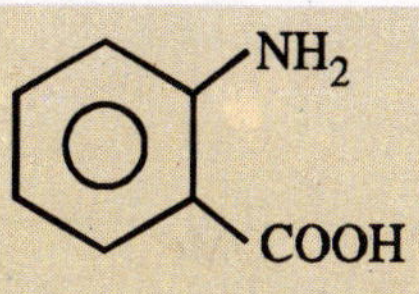 NH_2, COOH
2. 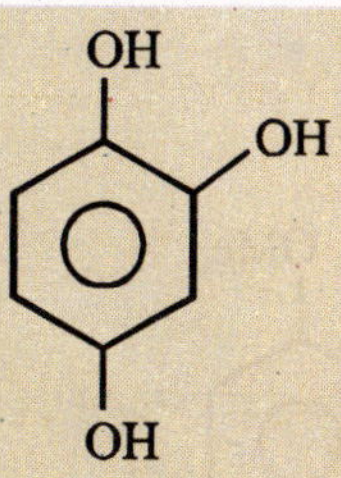 OH, OH, OH
3. 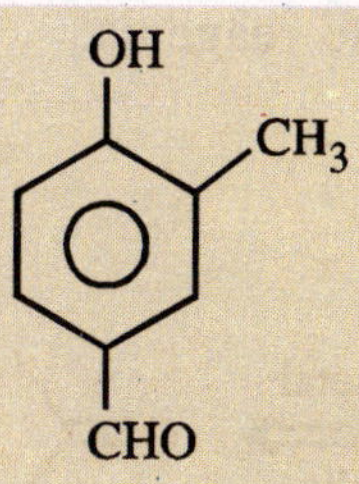 OH, CH_3, CHO
4. 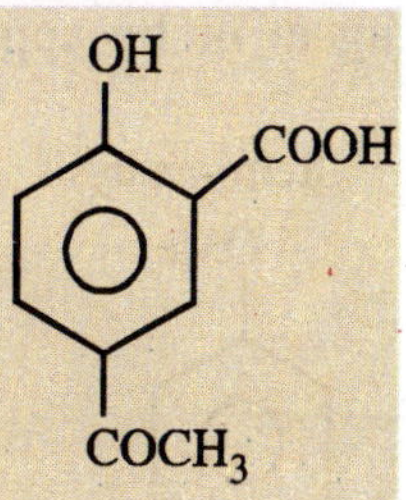 OH, COOH, $COCH_3$
5. 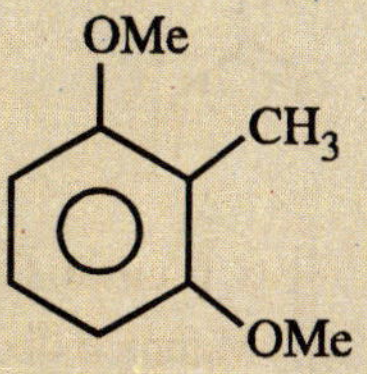 OMe, CH_3, OMe
6. 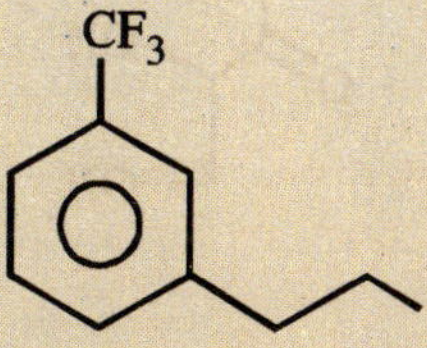 CF_3
7. 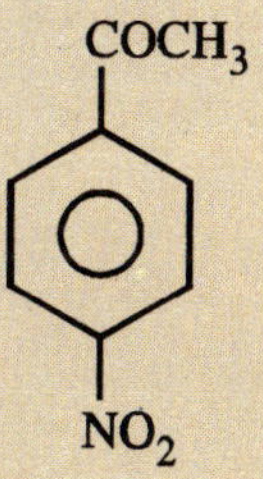 $COCH_3$, NO_2
8.  OH, NO
9. 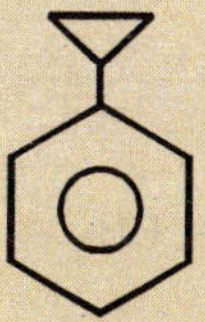
10. 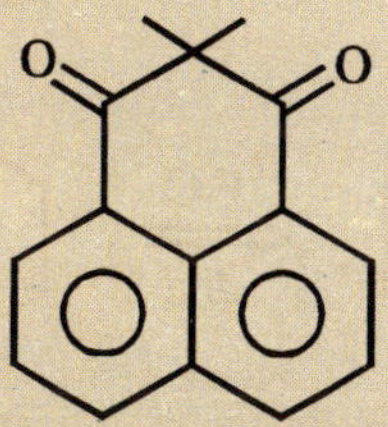 O, O
11. 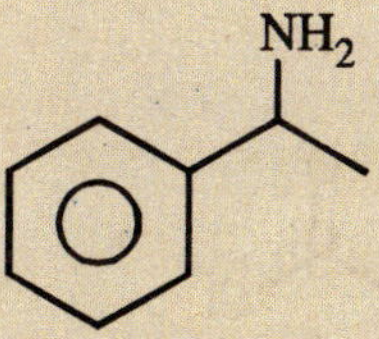 NH_2
12. 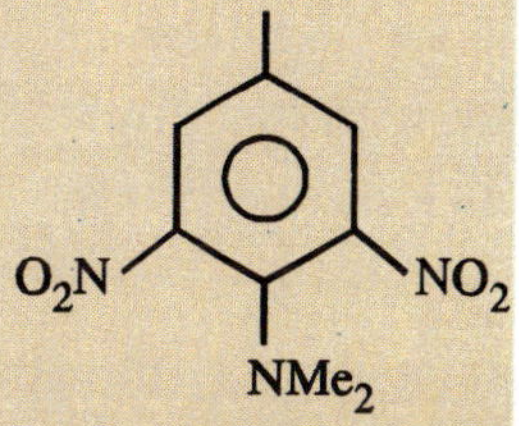 O_2N, NO_2, NMe_2
13. 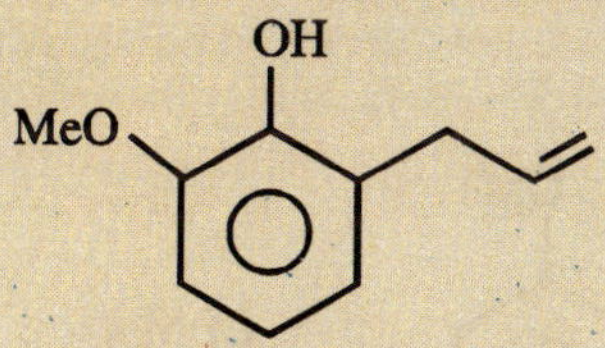 OH, MeO
14. 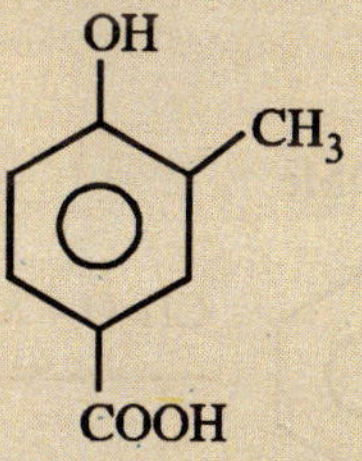 OH, CH_3, COOH
15. 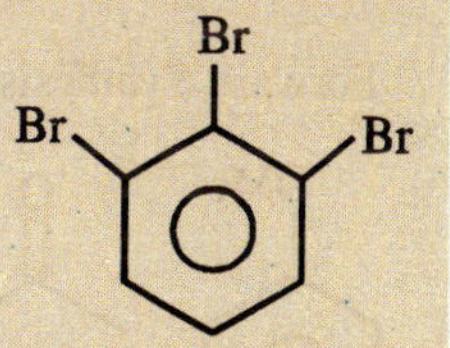 Br, Br, Br
16. 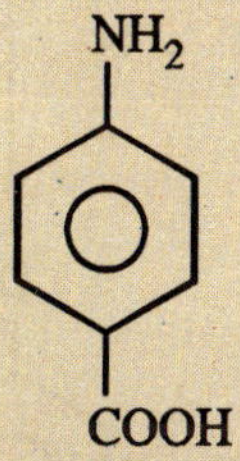 NH_2, COOH
17. 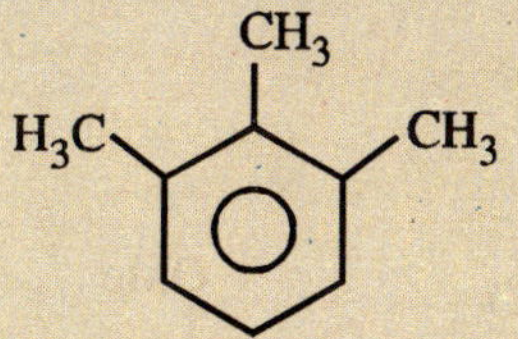 CH_3, H_3C, CH_3
18. 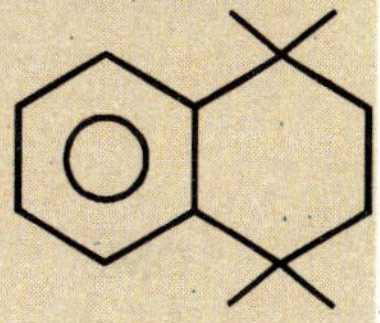
19. 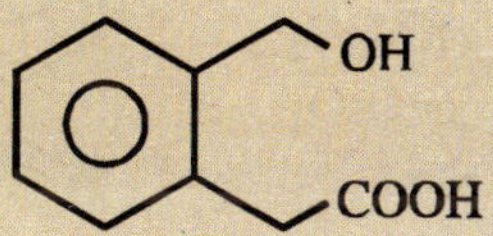 OH, COOH
20. 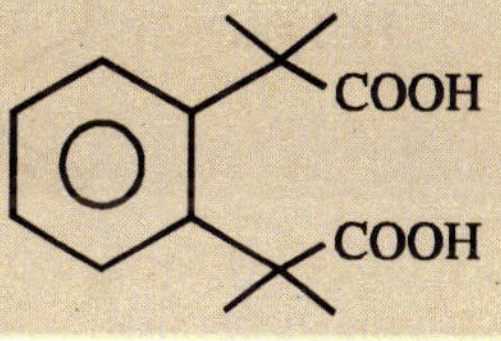 COOH, COOH

2. If the side chain is not functional group or any group which cannot be added directly into the benzene ring then first add functional group or any simple side chain (which is actually synthetic equivalent group for the desired side chain in the TM) to the benzene ring and then functional group or side chain modification is done either by redox agent or by appropriate reagents.

Retro-analysis

OMe

(TM)

ketal dix

O= + OMe OH OH

FGI

OMe

α, β unsat. C=O Mannich dix.

OMe + $CH_2 = \overset{\oplus}{N}Me_2$

F.C. dix

O Cl SE + OMe SE

Forward Synthesis

OMe O Cl $AlCl_3$ → OMe

$CH_2O + Me_2NH$ → OMe NMe_2

(i) MeI (ii) $\overset{\ominus}{O}Et/\Delta$

OMe

1% alkaline $KMnO_4$ or OsO_4

OMe OH OH

O, HCl gas

OMe

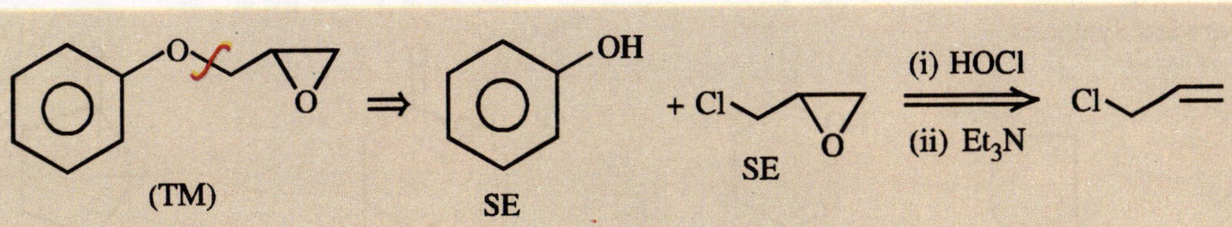

(TM) ⇒ OH SE + Cl SE (i) HOCl (ii) Et_3N Cl

Forward Synthesis

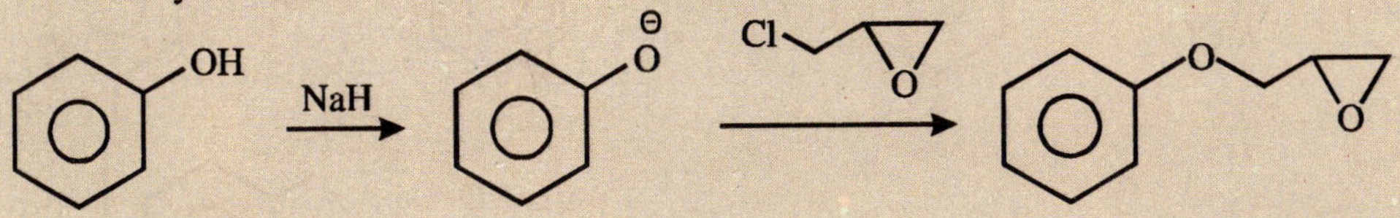

OH NaH → $O^{\ominus}$ Cl O → O O

Retro-analysis

O (TM) ⇒ HO O ⇒ H SE + HO O HO

⇓ FGI

O O SE ⇐ HO O HO O + SE ⇐ HO O O

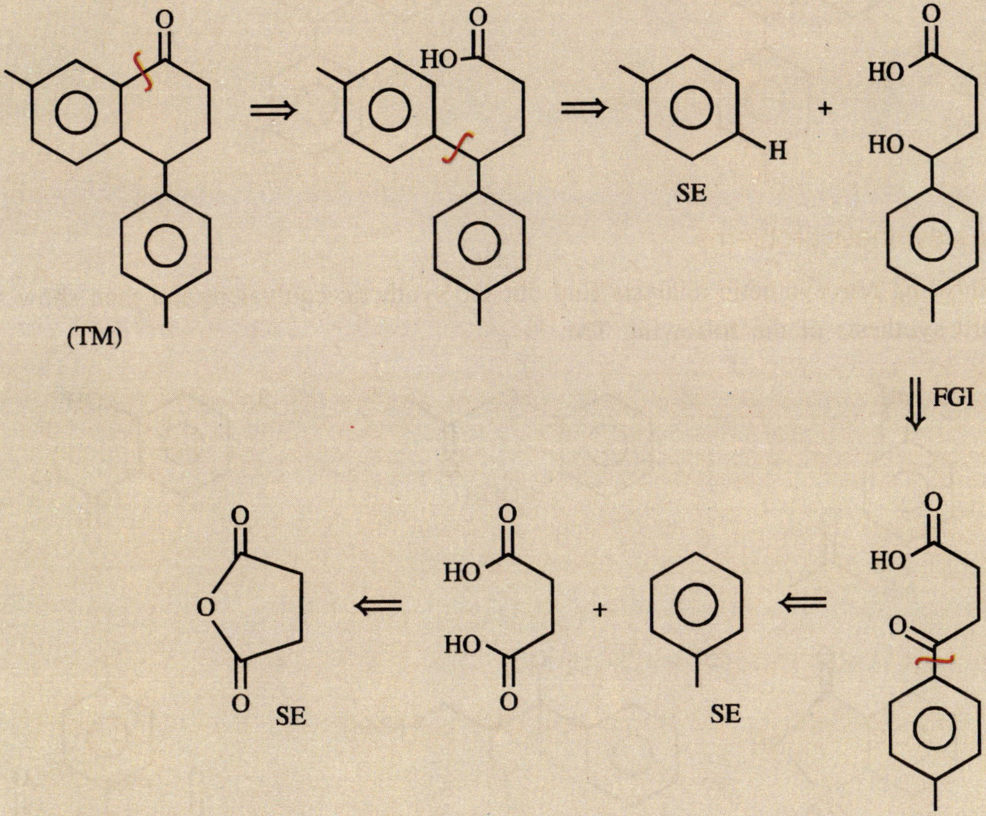

Forward Synthesis

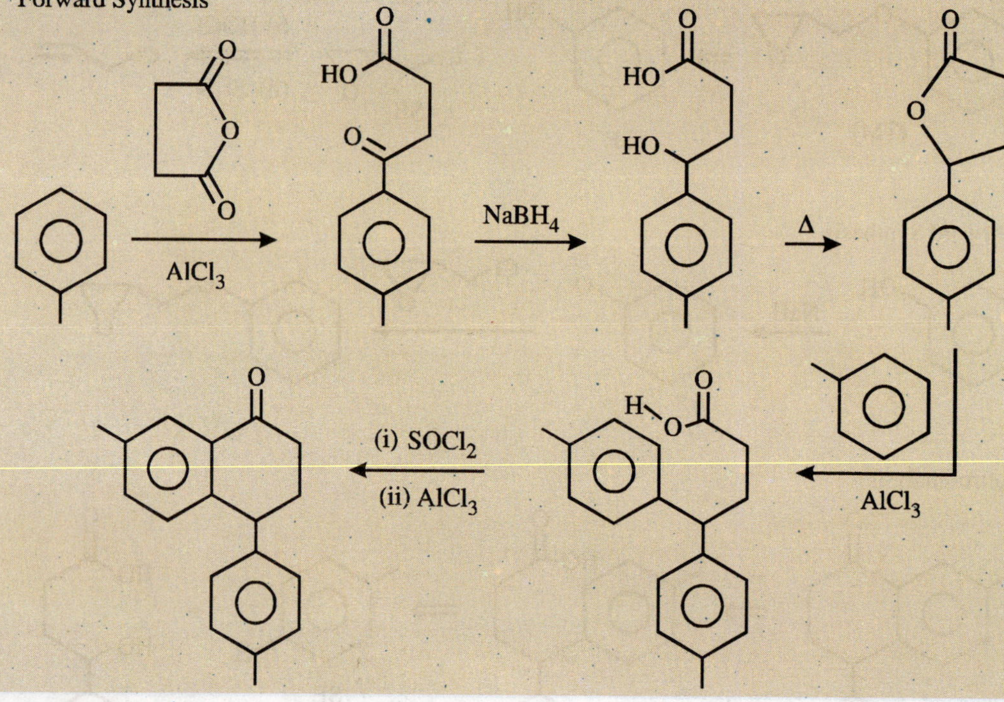

NaBH$_4$ AlCl$_3$ Δ (i) SOCl$_2$ (ii) AlCl$_3$ AlCl$_3$

Review synthetic problems

Showing retrosynthetic analysis find out the synthetic equivalent and then show the forward synthesis of the following TM.

1. 2. 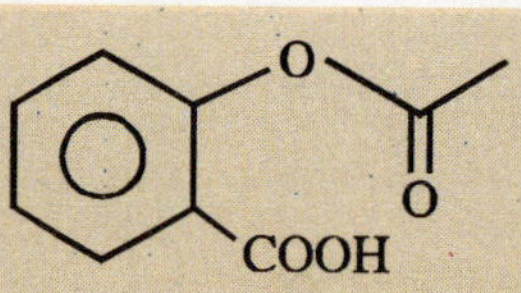 3. 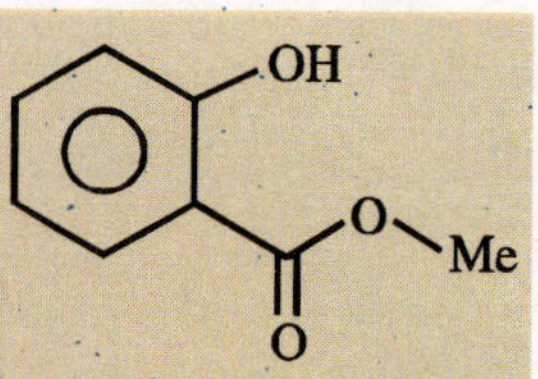 4. 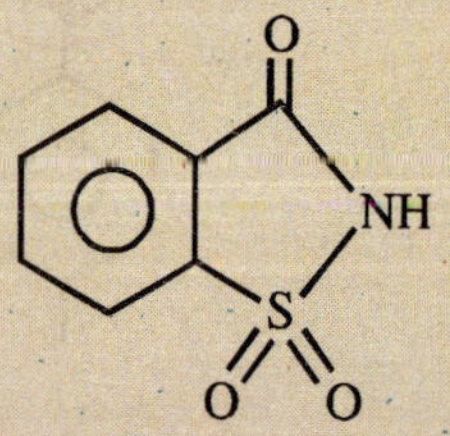 5. 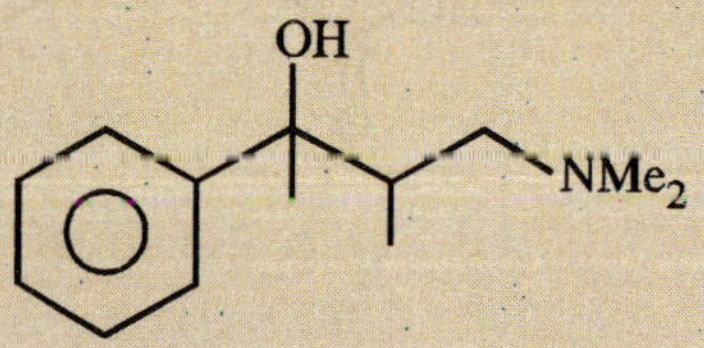 6. 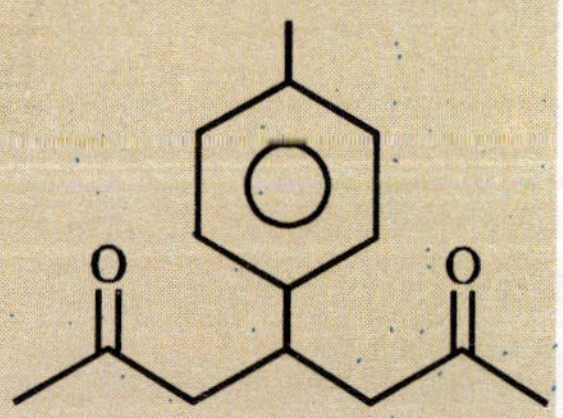 7. 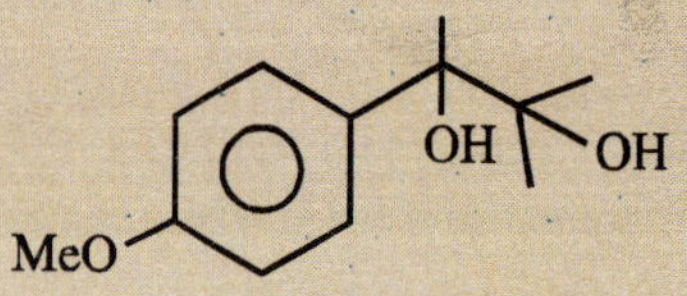 8.

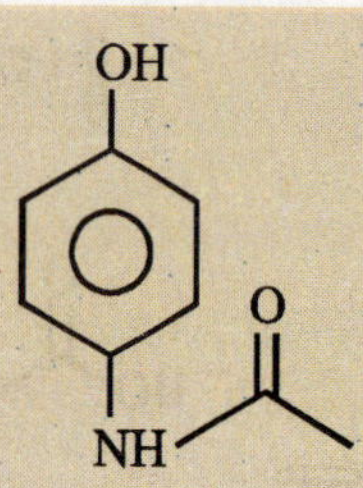

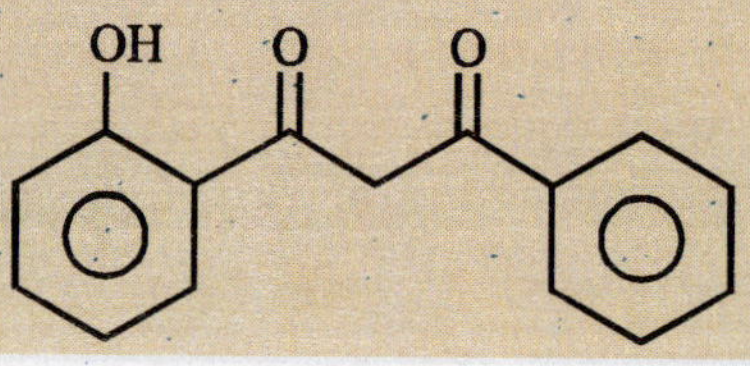

9. Me_2N ... O

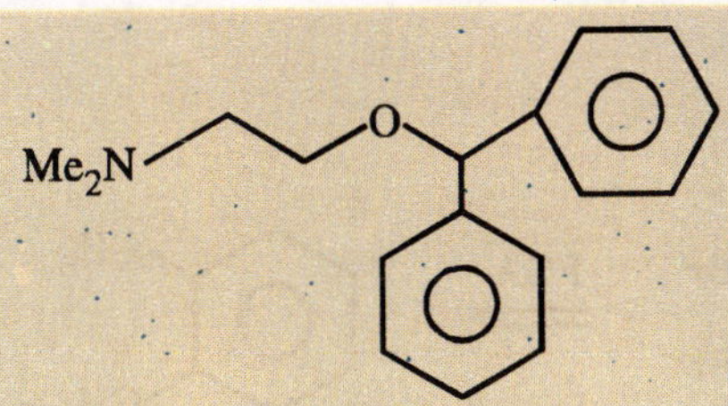

10. OH ... CO_2H

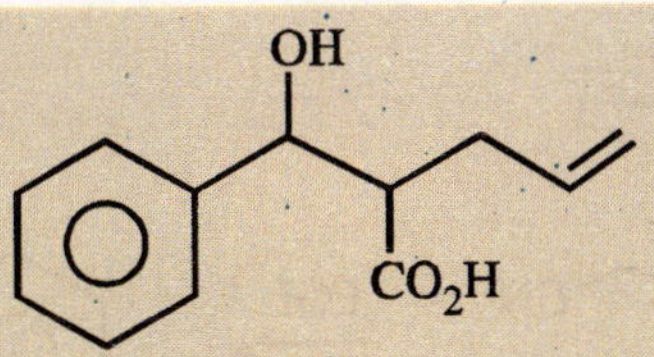

11. OH ... O ... NH

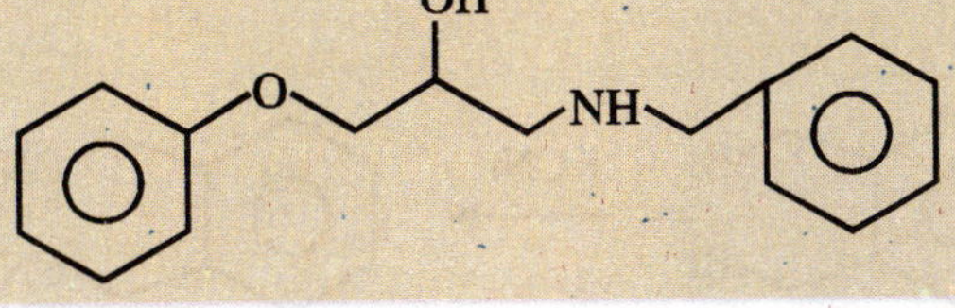

12. N ... N

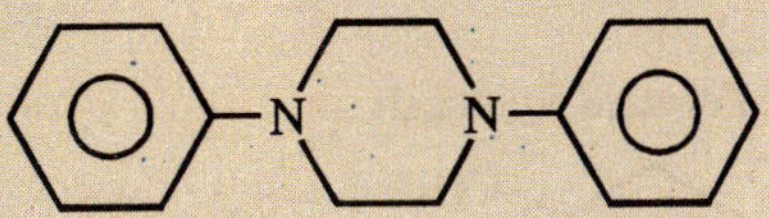

3. Benzofused carbocyclic or heterocyclic compounds are usually synthesized from a mono or diortho substitution of benzene.

Benzofused carbocyclic compounds are disconnected either next to a ring junction or one atom removed from a ring junction.

For benzofused heterocyclic compounds the correct disconnection is usually that of a carbon-hetero atom bond. If in the benzo-fused heterocyclic compound contain hetero atom—hetero atom bond in that case nucleophilic nitrogen is $-NH_2$ and electriphilic nitrogens are $-NO_2$, $-N{=}O$, $-\overset{\oplus}{N}\equiv N$ and $-\ddot{N}$: Disconnection of benzofused TM should be such in which synthetic equivalent should be preformed benzene derivative.

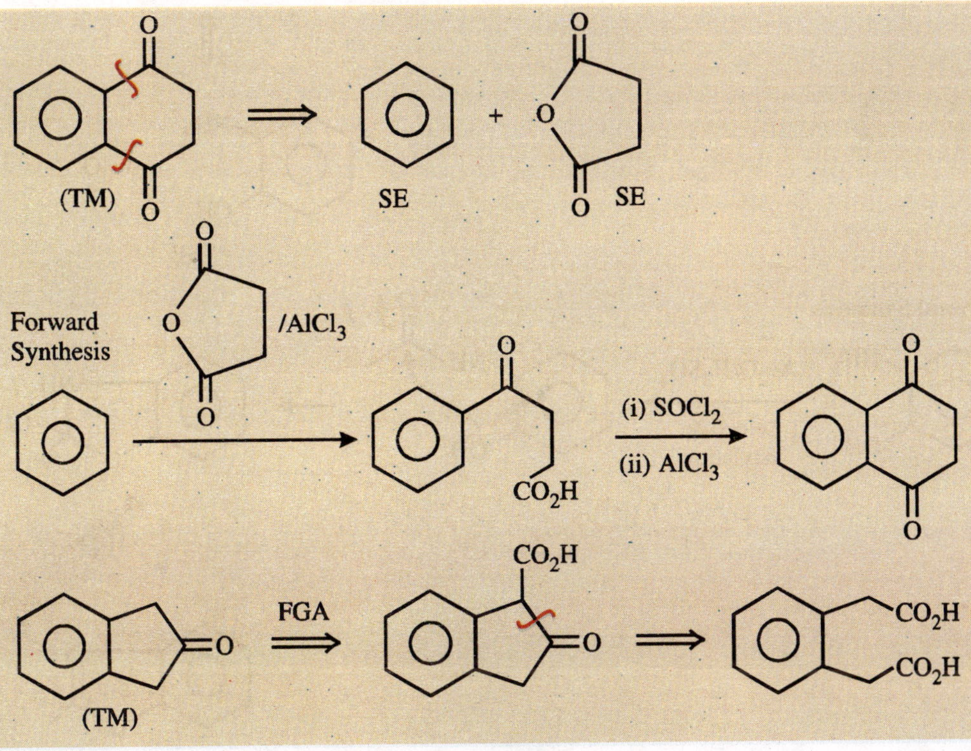

(TM) ⟹ SE + SE

Forward Synthesis

/$AlCl_3$

CO_2H

(i) $SOCl_2$

(ii) $AlCl_3$

(TM) FGA ⟹ CO_2H ⟹ CO_2H, CO_2H

⇓ FGI

FGI ⟹ FGA ⟹

SE

Forward Synthesis

2eq. NBS → KCN →

$H_3O^{\oplus}$

Ac_2O 300°C

(TM)

SE

+ Ac_2O

Forward Synthesis

Ac_2O/H_2SO_4 selective acylation

$- H_2O$

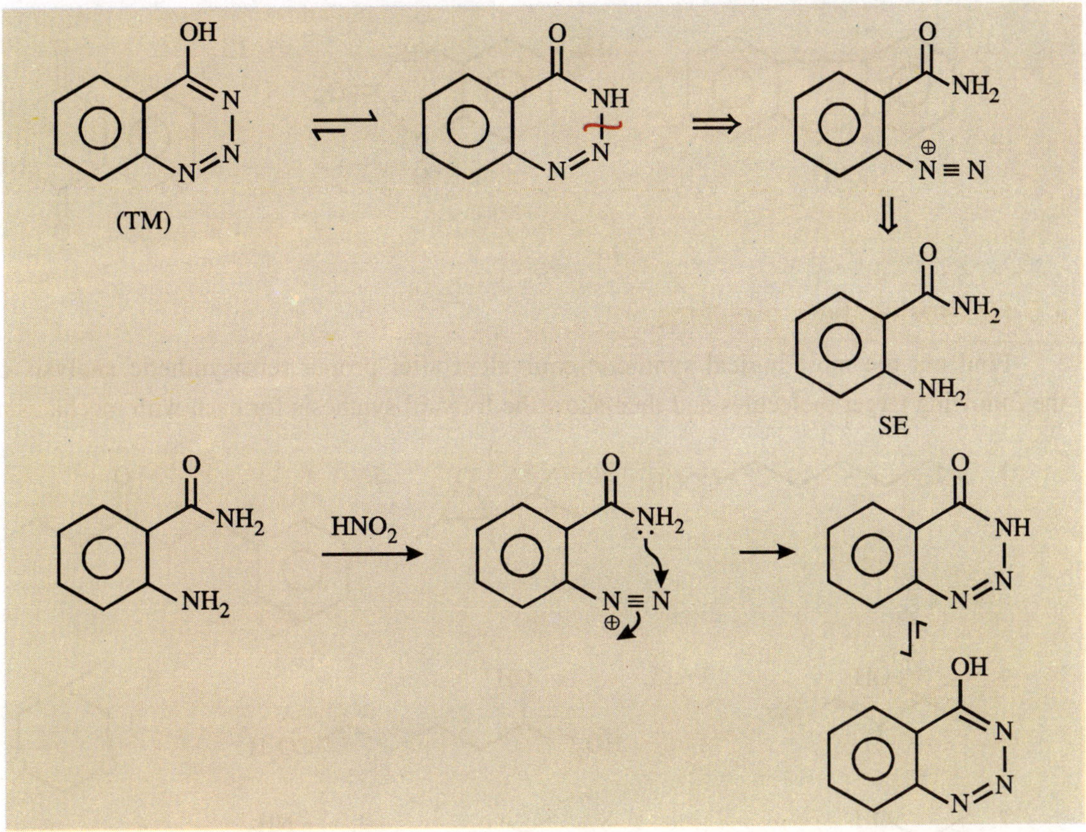

OH N N N (TM) ⇌ O NH N N ⇒ O NH_2 ⊕ N≡N ⇓ O NH_2 NH_2 SE

O NH_2 NH_2 $\xrightarrow{HNO_2}$ O NH_2 N≡N ⊕ → O NH N N ⇌ OH N N N

Review synthetic problems

Write down the synthetic route for the following TM with retrosynthetic analysis.

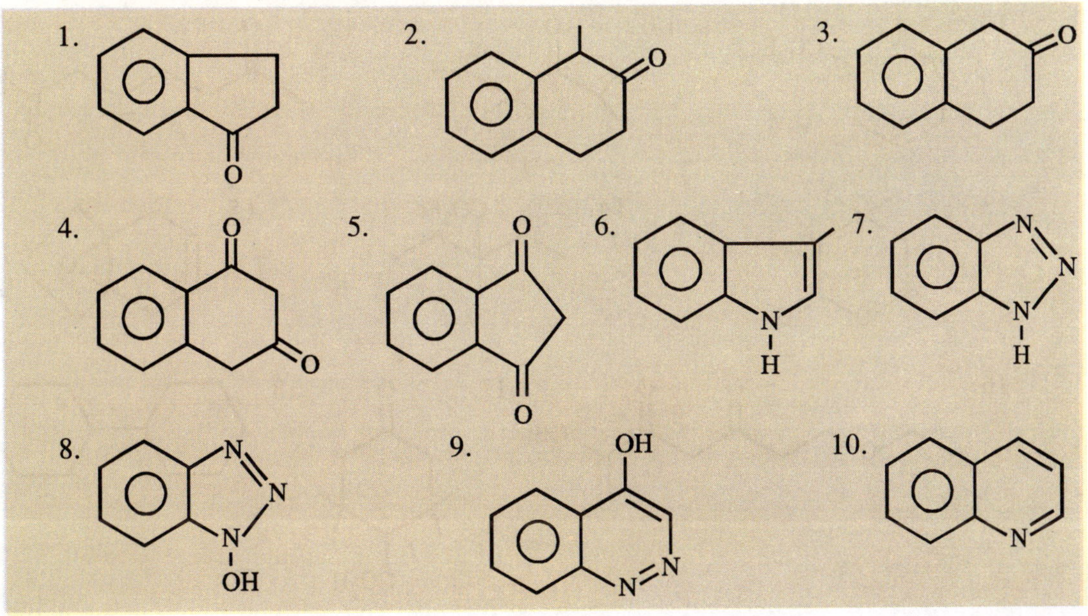

1. O
2. O
3. O
4. O O
5. O O
6. N H
7. N N N H
8. N N N OH
9. OH N N
10. N

11. 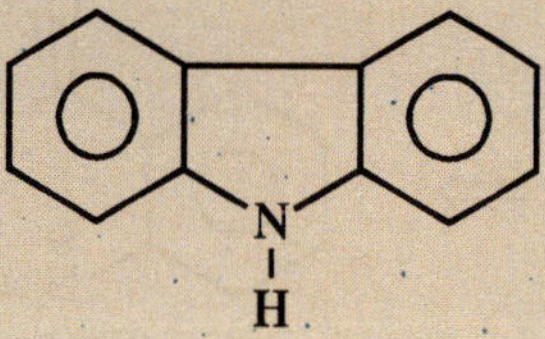 N-H 12. 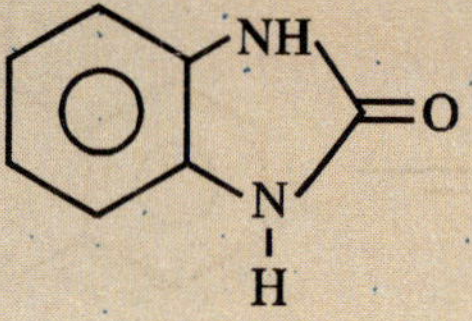 NH, =O, N-H 13. 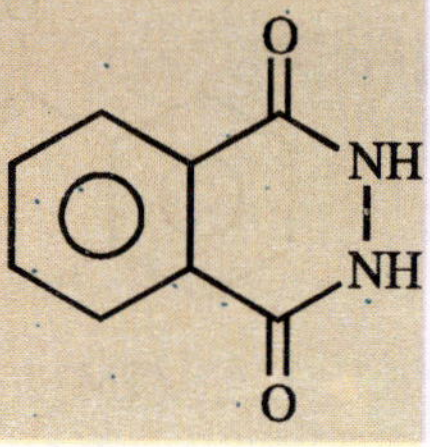 O, NH, NH, O

2.5 Review synthetic problems

Find out the most logical synthetic equivalent after proper retrosynthetic analysis of the following target molecules and then show the forward synthesis for each with mechanism.

1. 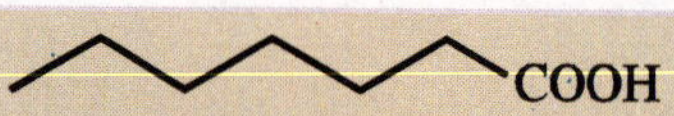 COOH 2. 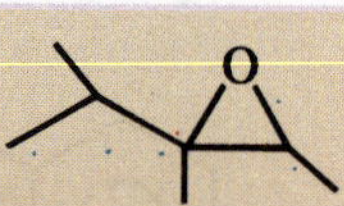 O 3. 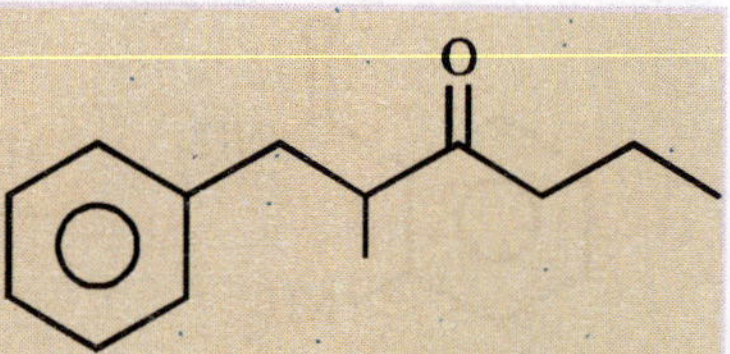 O

4. 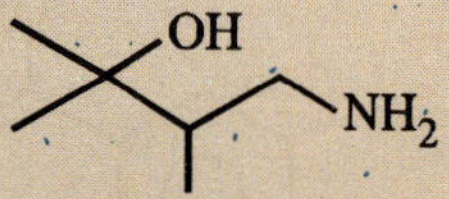 OH, NH_2 5. 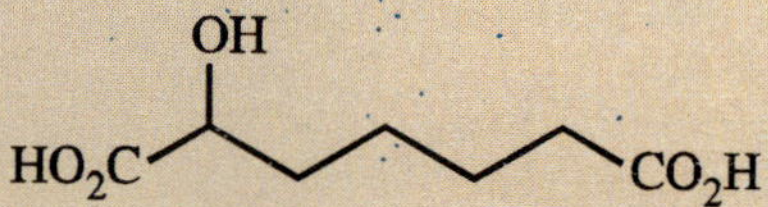 OH, HO_2C, CO_2H 6. 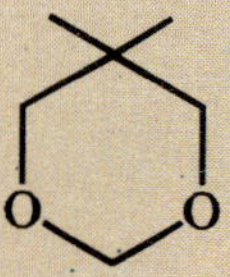 O, O

7. 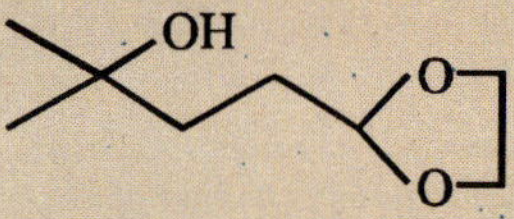 OH, O, O 8. 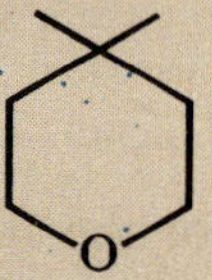 O 9. 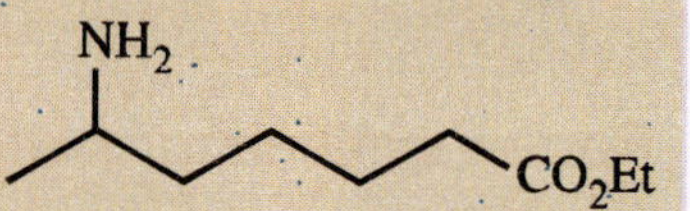 NH_2, CO_2Et

10. 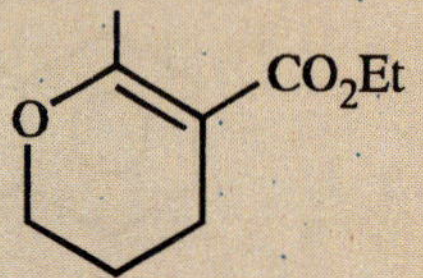 O, CO_2Et 11. 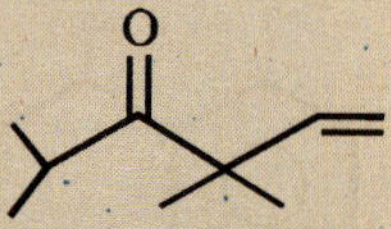 O 12. 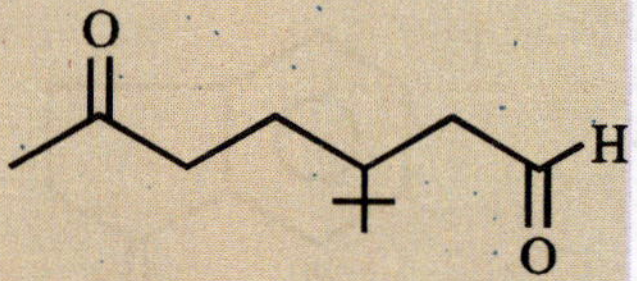 O, H, O

13. 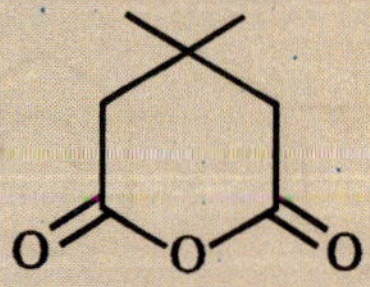 O, O, O 14. 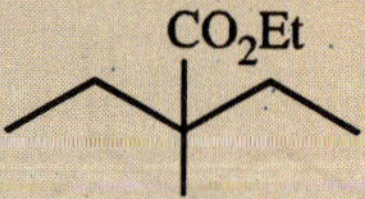 CO_2Et 15. 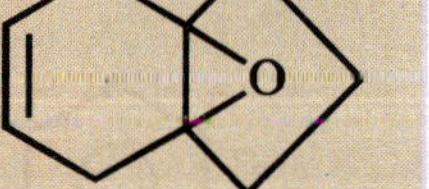 O

16. 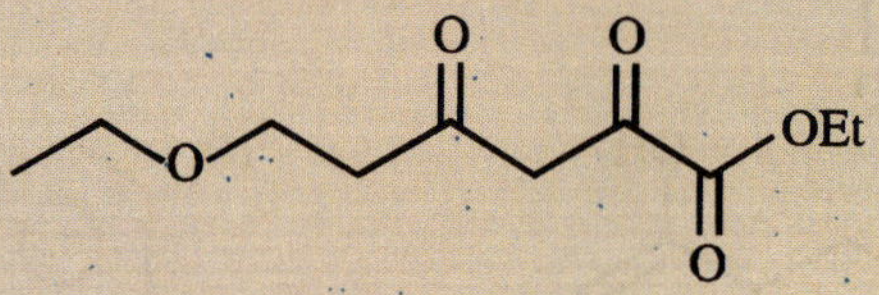 O, O, O, OEt, O 17. 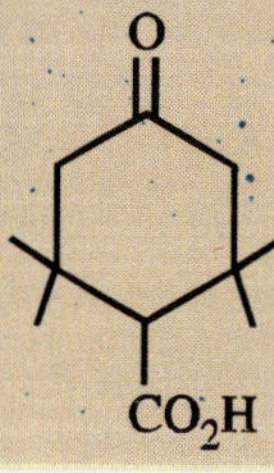 O, CO_2H 18.

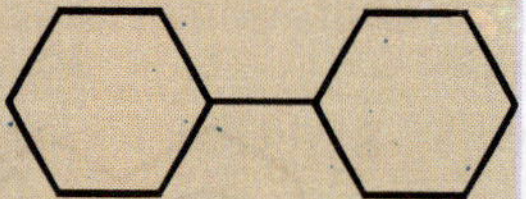

19. 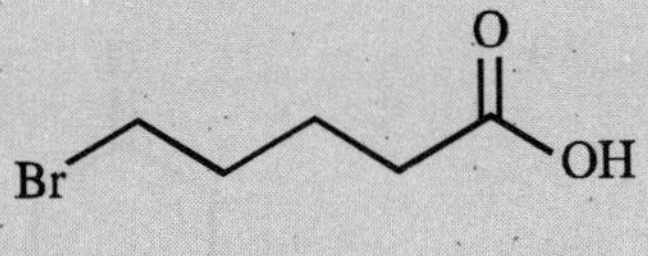 Br OH O

20. 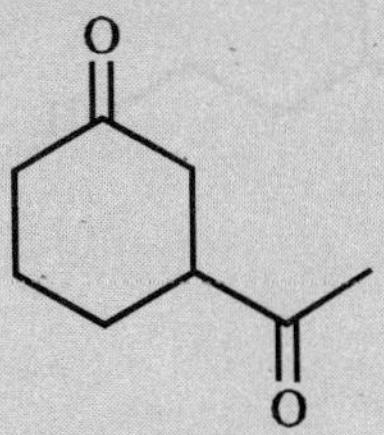 O O

21. 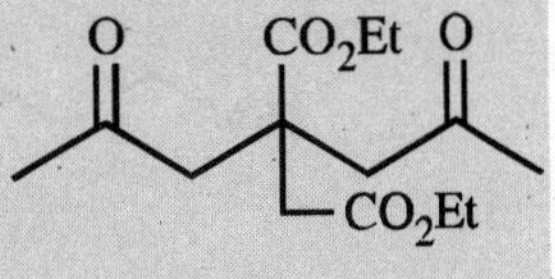 O CO_2Et O CO_2Et

22. 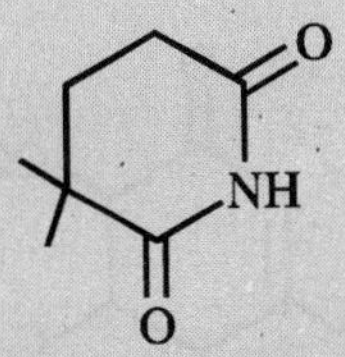 O NH O

23. 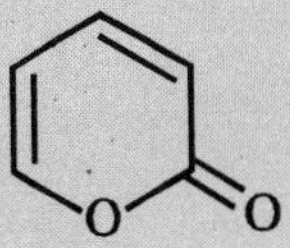 O O

24.

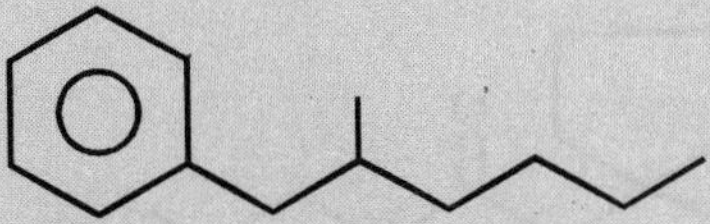

25. 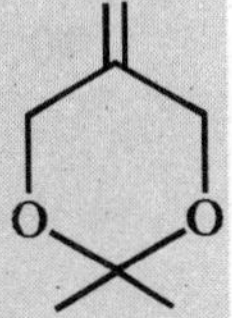 O O

26. 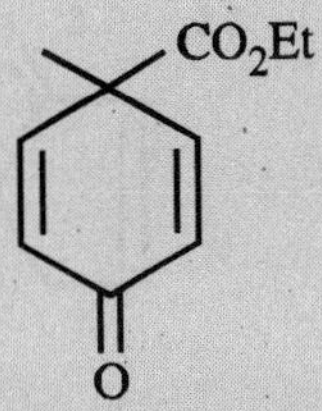 CO_2Et O

27. 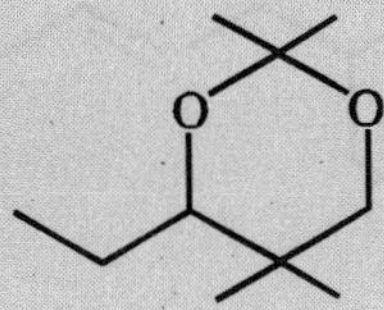 O O

28. 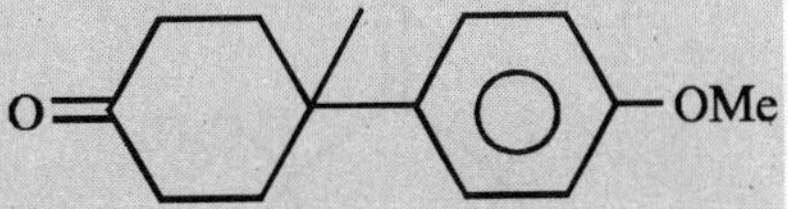 O OMe

29. 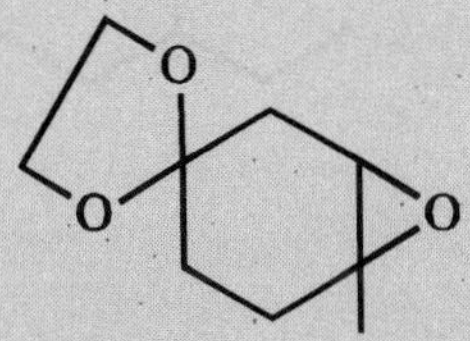 O O O

30. 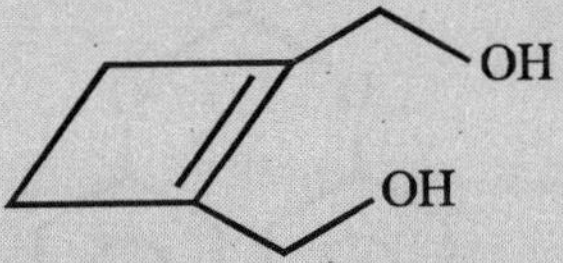 OH OH

31. 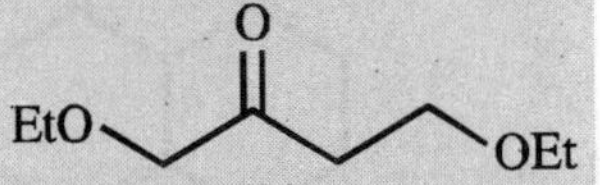 O EtO OEt

32. 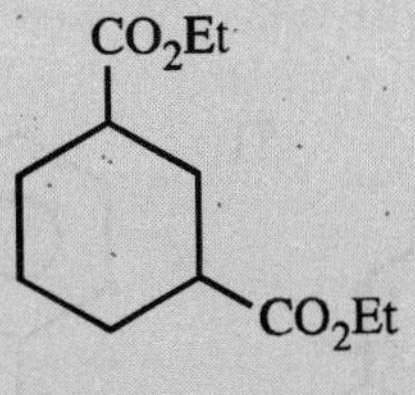 CO_2Et CO_2Et

33. 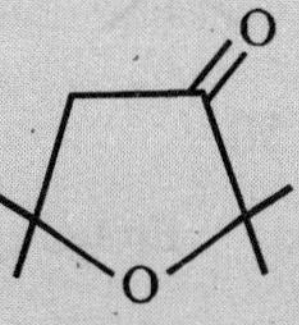 O O

34. 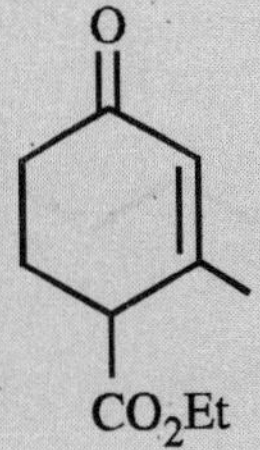 O CO_2Et

35. 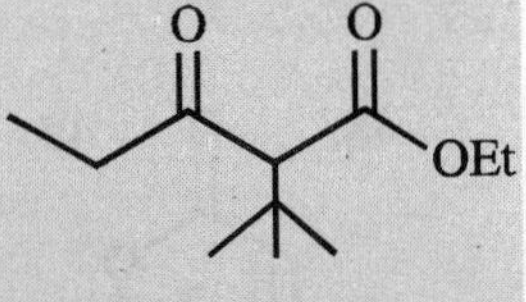 O O OEt

36. 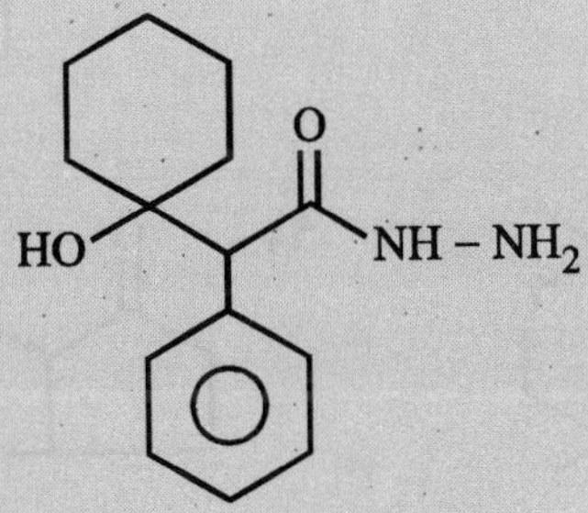 O HO NH – NH_2

37. 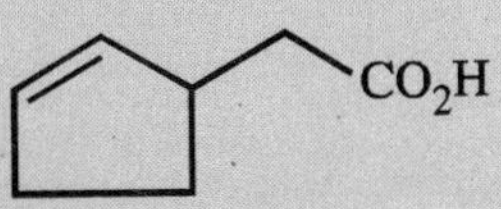 CO_2H

38. CO_2H HO OH O

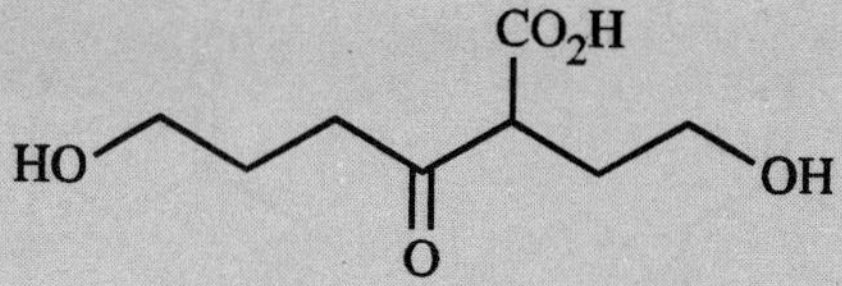

39. O CO_2H

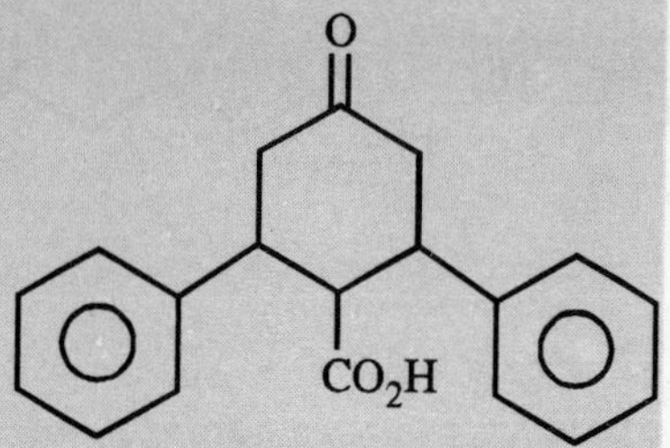

40. HO O O NEt_2

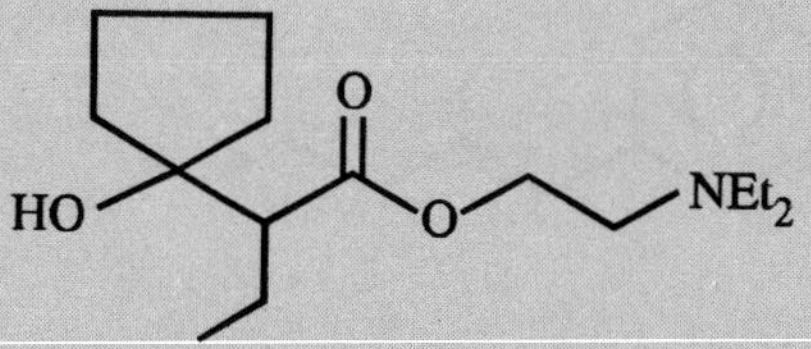

41. O CO_2Et H_5C_6 C_6H_5

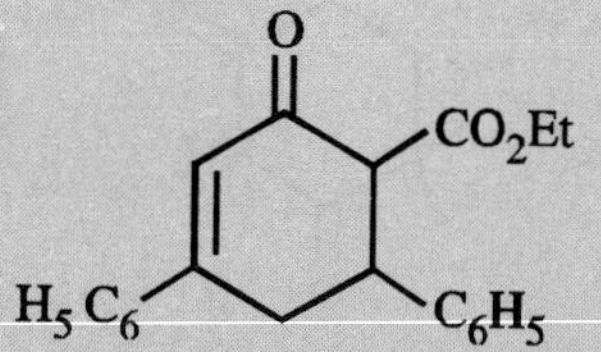

42. O C_6H_5 C_6H_5 OH OH

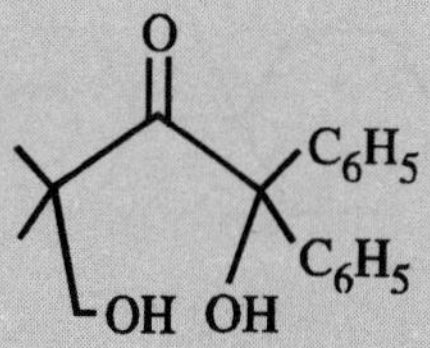

43.

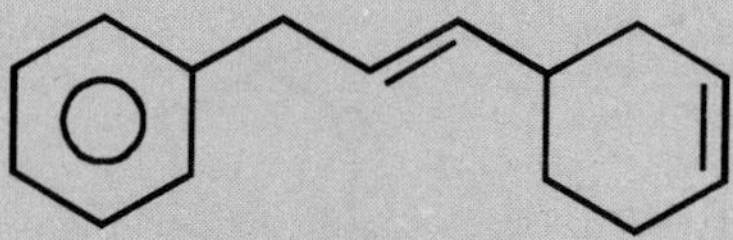

44. O CO_2Et

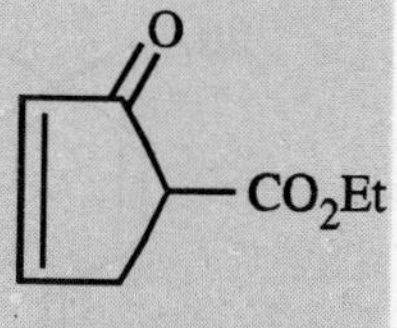

45. O O

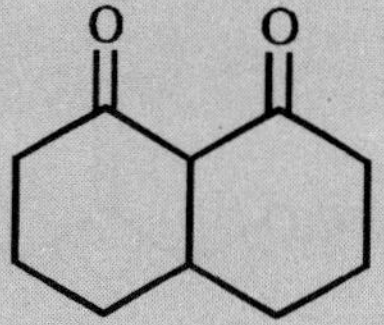

46. O O O

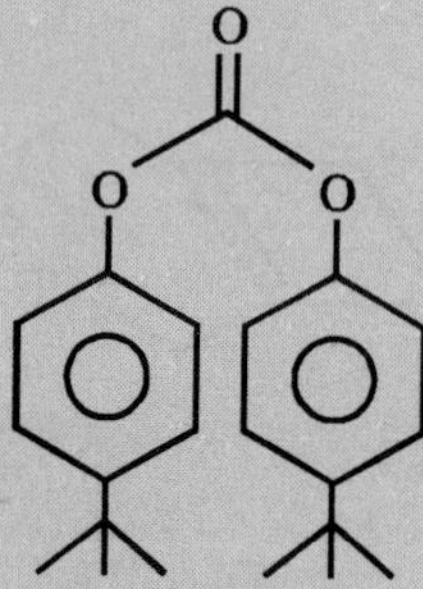

47. MeO OMe OH

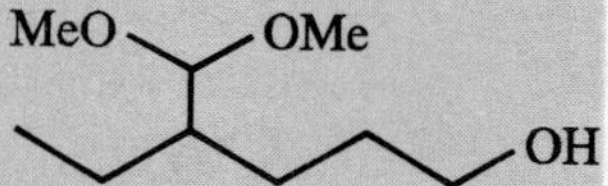

48. O O

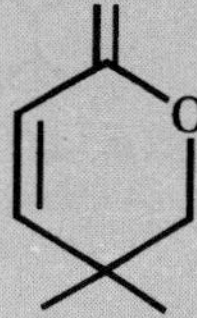

49. OH

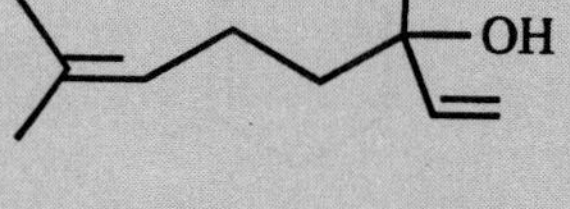

50. O EtO_2C CO_2Et

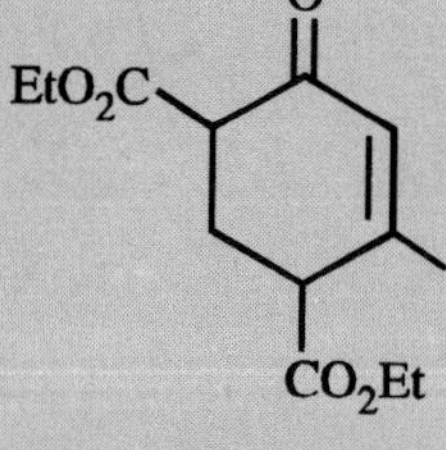

51.

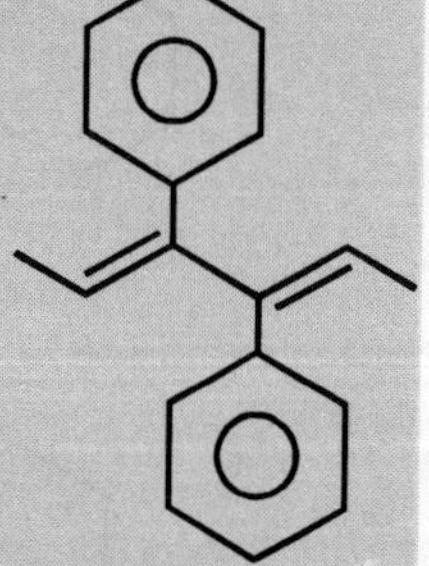

52. O

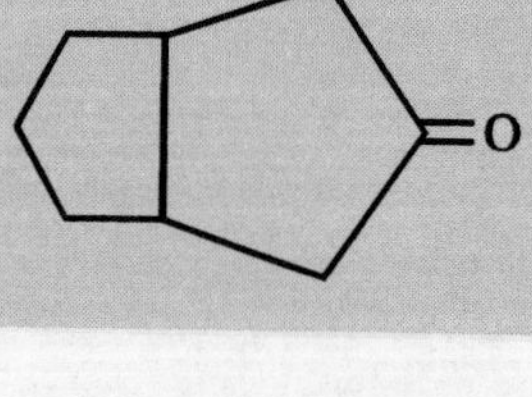

53. O O

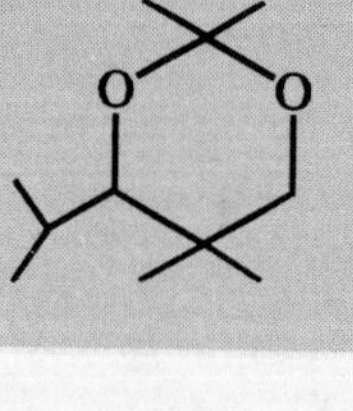

54. O O

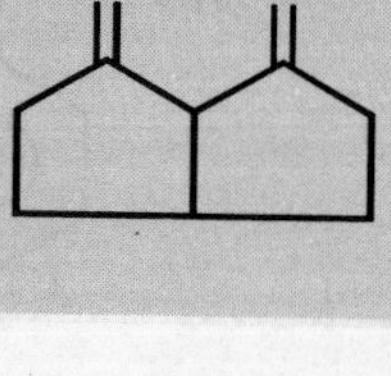

55. CO_2Et CO_2Et

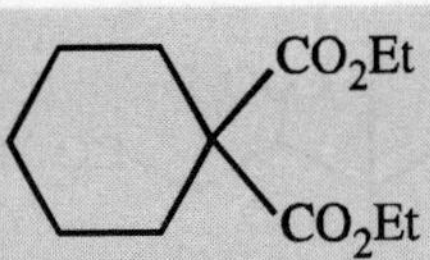

56. O O

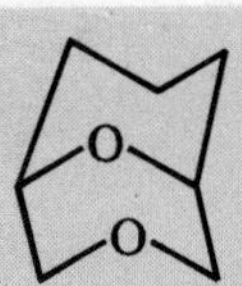

57. Me_2N OMe OH

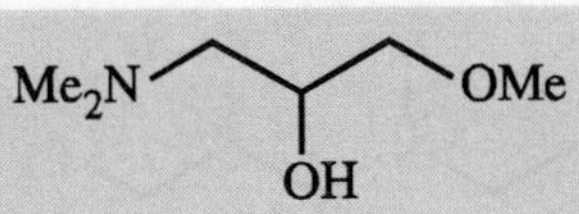

58. O

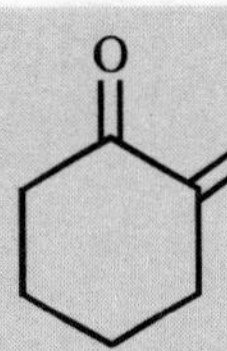

59. H_2N CO_2H

60. O

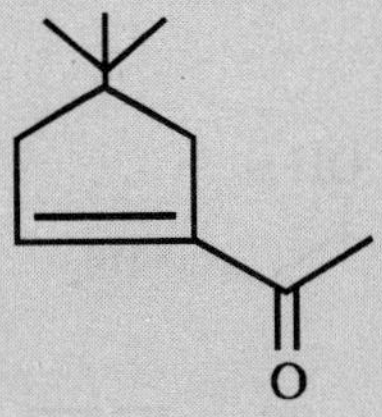

61. H_5C_6 O C_6H_5

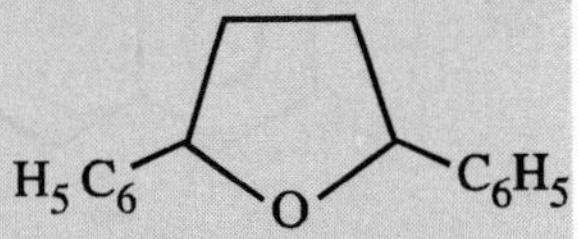

62. CO_2H

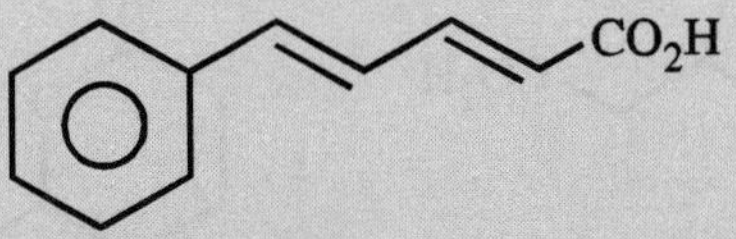

63. O O

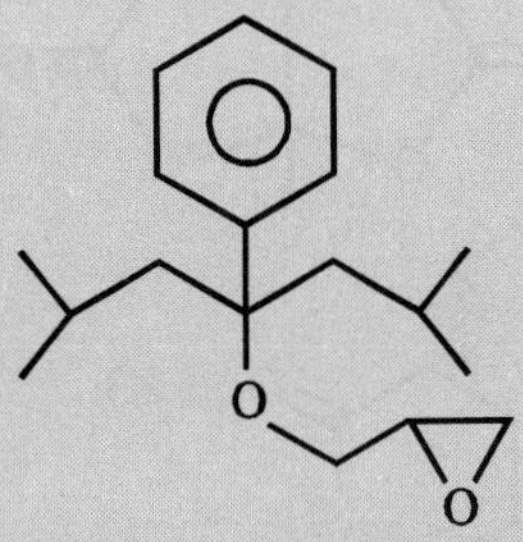

64. H_5C_6 O C_6H_5

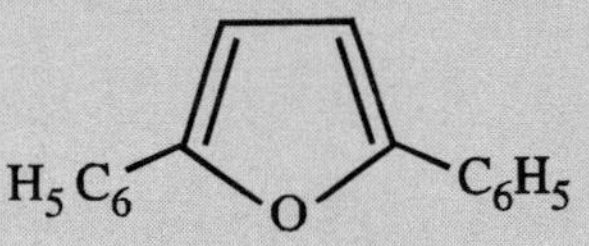

65. EtO_2C O OEt CO_2Et

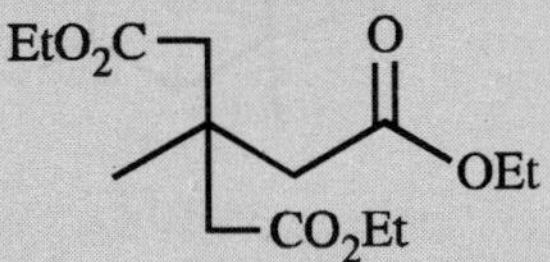

66. O

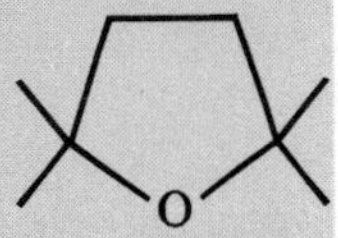

67. C≡C

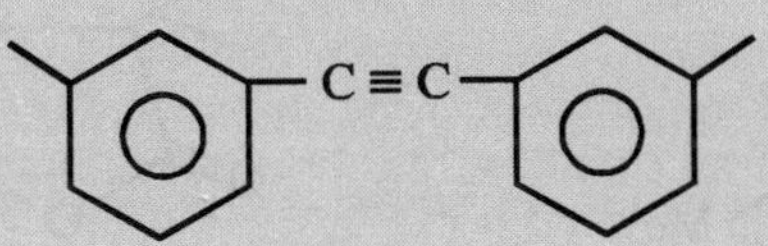

68. O

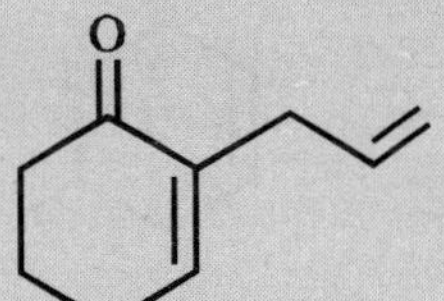

69. O CHO

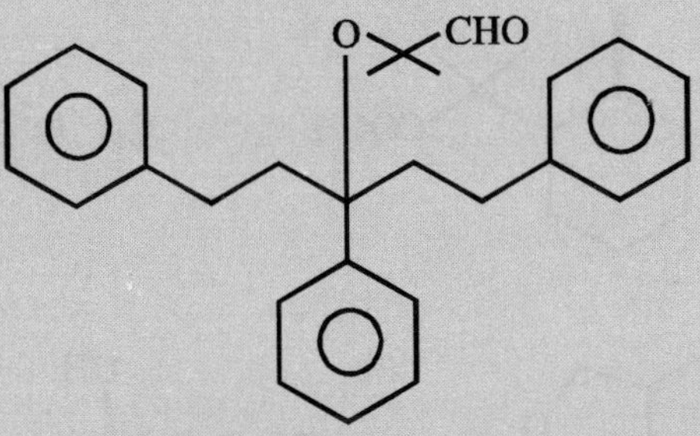

70. O NH N O H

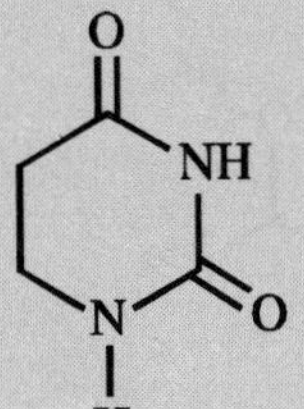

71. O

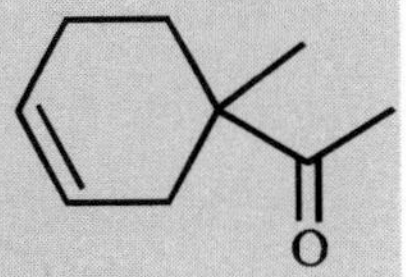

72. O O

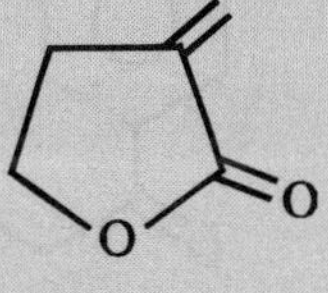

73. OH

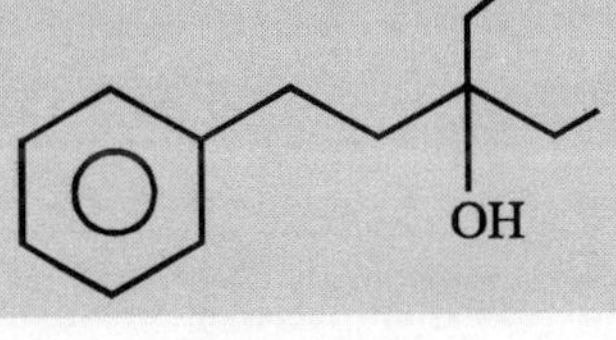

74. O O O

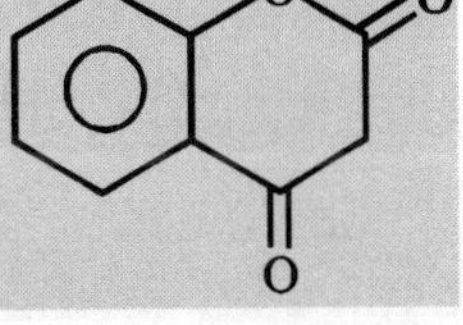

75. Me, O–H, HO, N, Cl

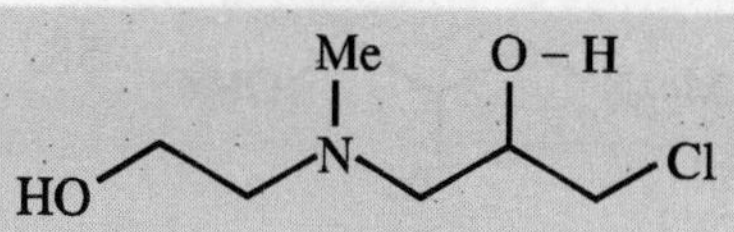

76. O, COOH, CO_2H

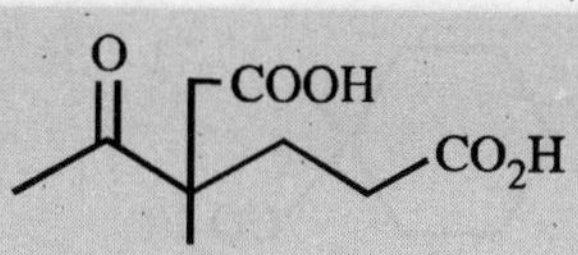

77. OH, O, NMe_2

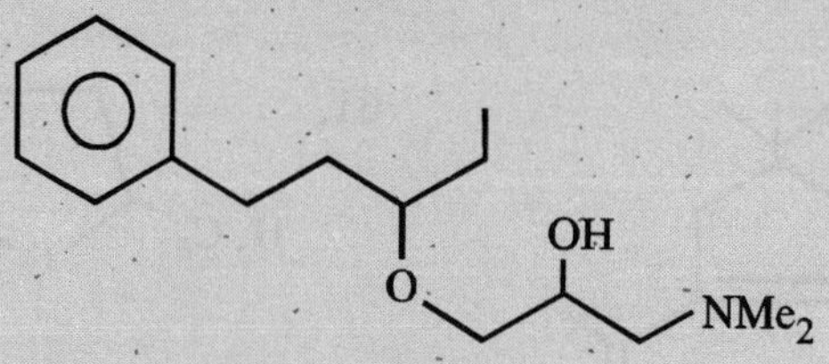

78. O, O, O

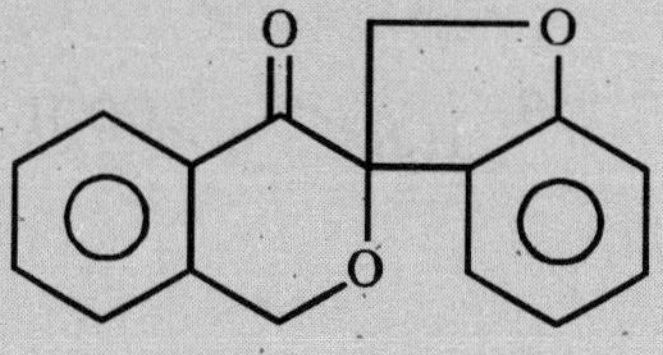

79. O, O, O, O

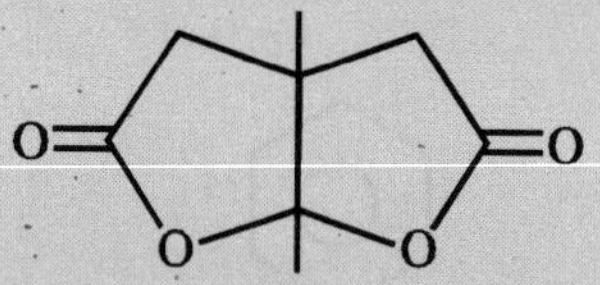

80. O, CO_2H

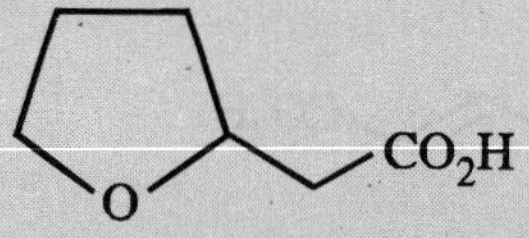

81. O, O, CO_2Et

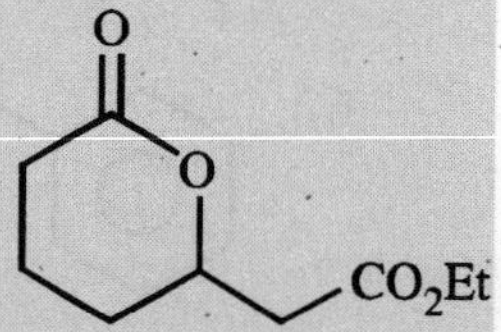

82. O, NH, O

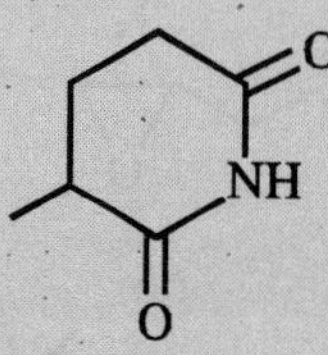

83. O, O

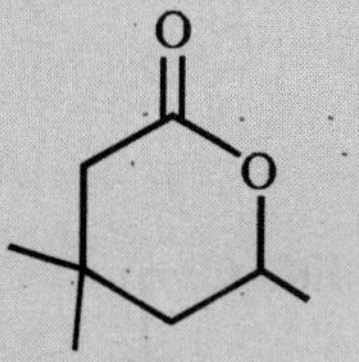

84. O, O, O

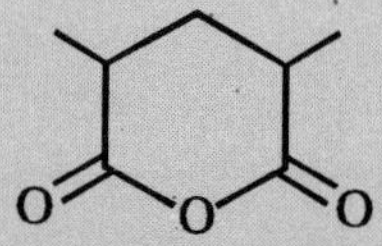

85.

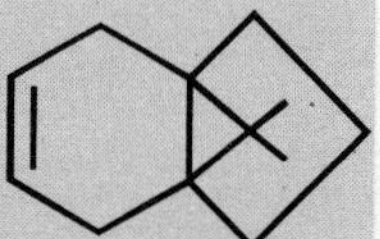

86. O, O

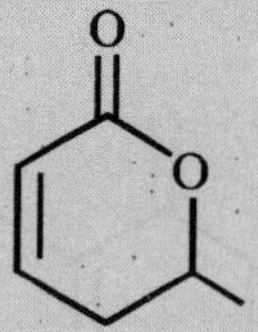

87. O, O, Me, Me

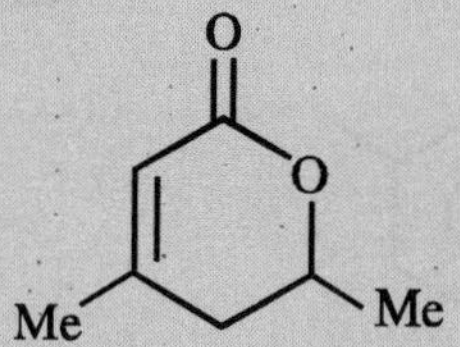

88. O, O, C_6H_5

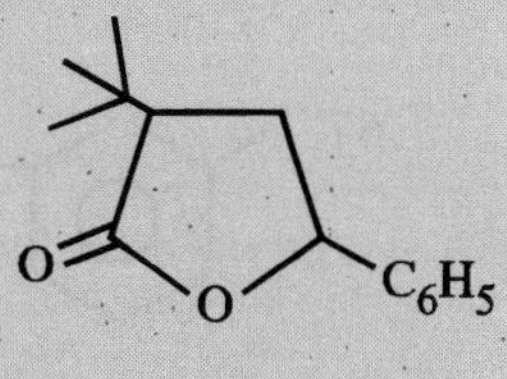

89. O, O, MeO, Me

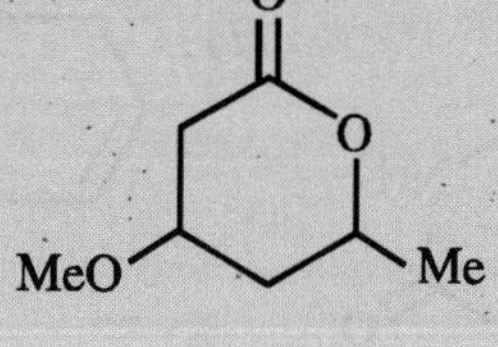

90. OH, COOH

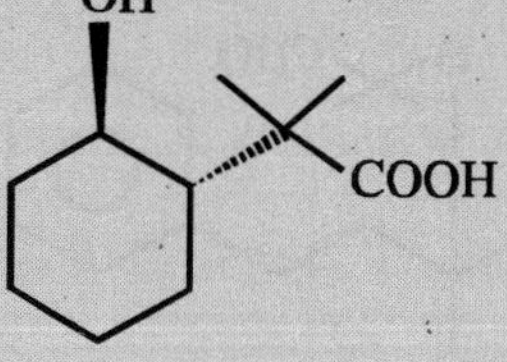

91. O, O

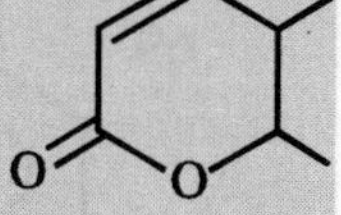

92. O, O

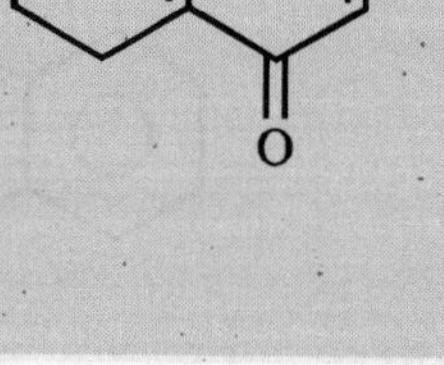

93. O

94. OH, HO_2C, O, NH

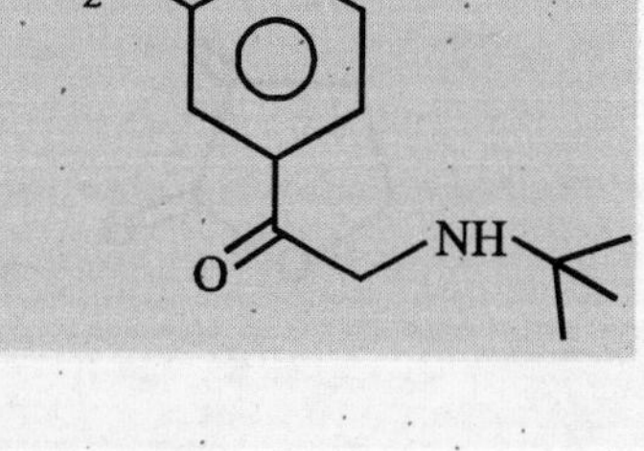

95. COOH

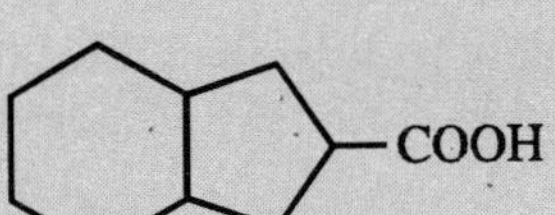

96. O

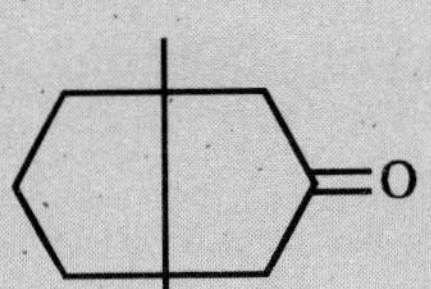

97. Me_2N O

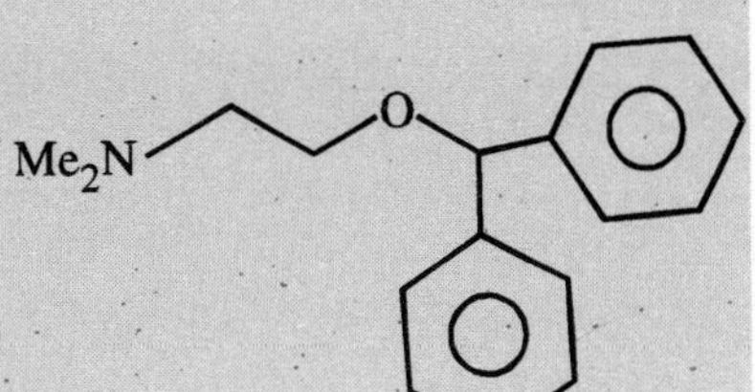

98. O O COOH

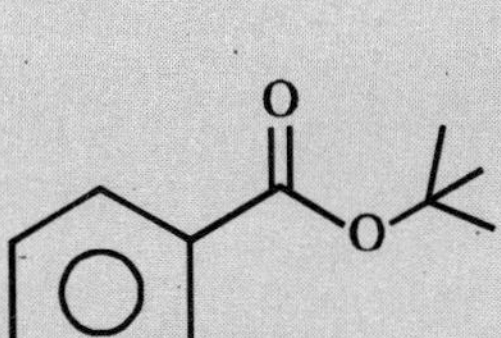

99. Ph Ph O O O

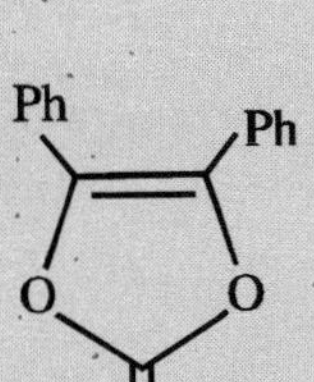

100.

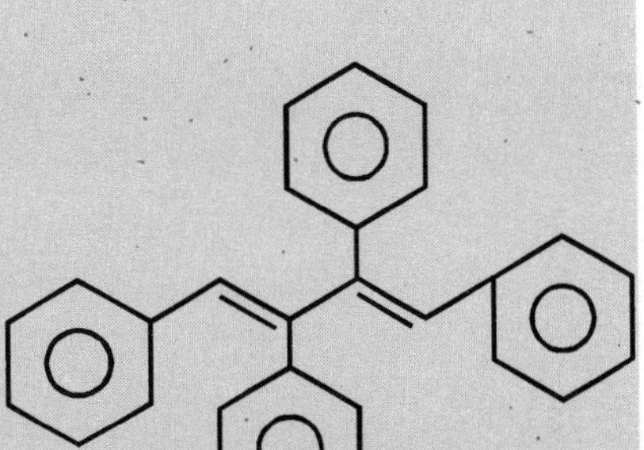

101. COOH N

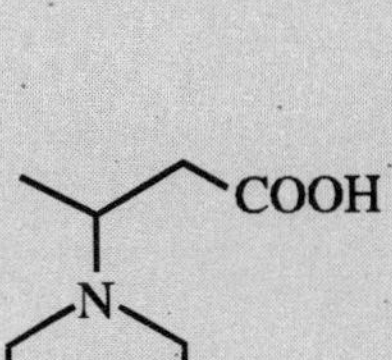

102. $(CH_2)_6$

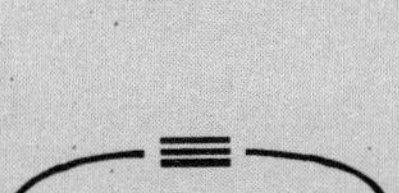

103. EtO_2C CO_2Et EtO_2C CO_2Et O

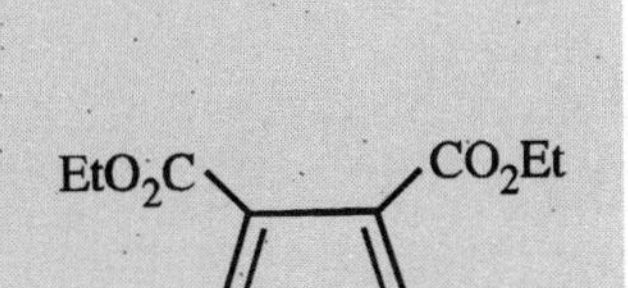

104. O

105. O OEt O

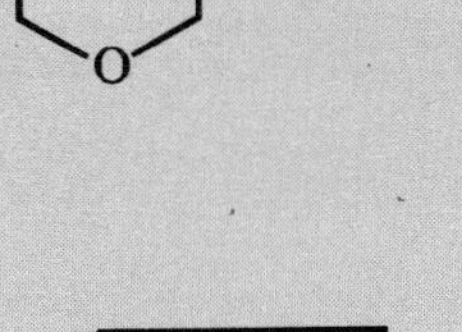

106. O CO_2Et

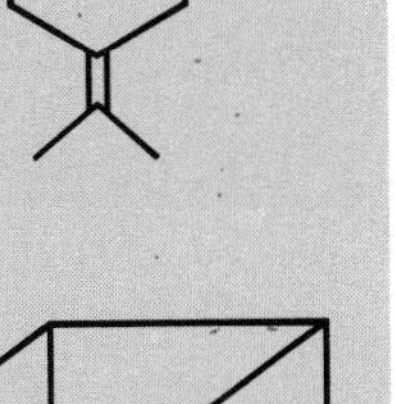

107. Me Me N Me

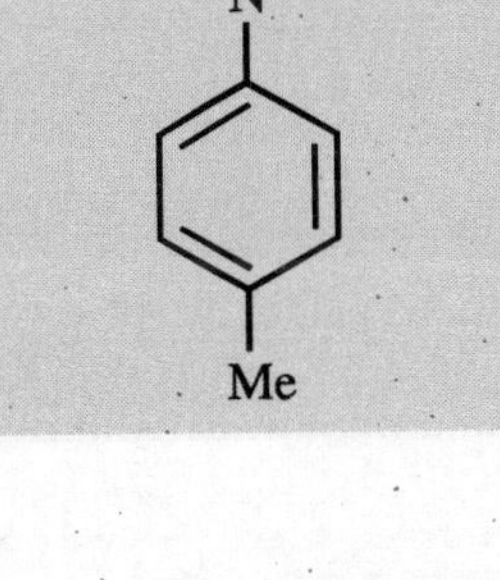

108.

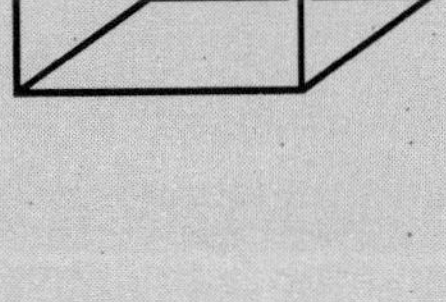

109.

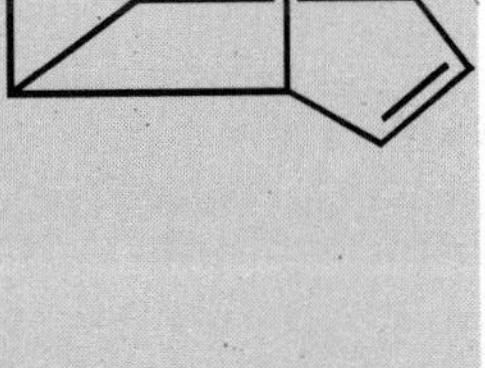

Note

3

Synthetic Application of Michael Condensation, Claisen Ester Condensation and Diels-Alder Reaction

3.1 Michael condensation

3.1.1 What is Michael condensation?

The conjugate addition of stabilised carbanion (which is termed as Michael donor) to an α, β unsaturated compound with a —M group (—CN, —NO_2, —CHO, —$COCH_3$, —COPh, —CO_2Et, etc.) which is termed as Michael acceptor, is known as Michael condensation. The Michael addition results in the formation of a new C—C bond.

Example:

$\overset{\ominus}{O}Et/EtOH$ — CO_2Et — Michael donor (resonance stabilised carbanion) — Michael acceptor — EtO_2C — Et – O – H

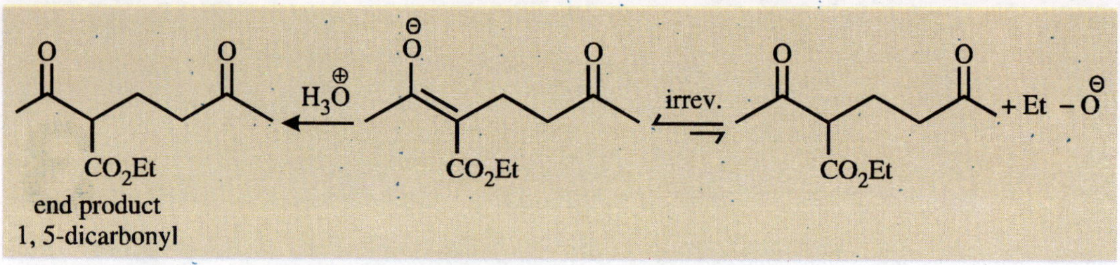

end product
1, 5-dicarbonyl

3.1.2 Mechanistic explanation of Michael condensation

(i) In the Michael condensation only catalytic amount of base is used i.e. per reaction one mole of base is required i.e. 1 : 1 starting material and base are used. In this condensation last step is irreversible because of resonance stabilised anion formation.

(ii) Weak basic catalyst is used in Michael condensation e.g. pyridine, piperidine, Et_3N etc. If strong base e.g. NaOEt, LDA, Bu^tOK $NaNH_2$, NaH etc. then only 0.1 to 0.3 equivalent of base is used.

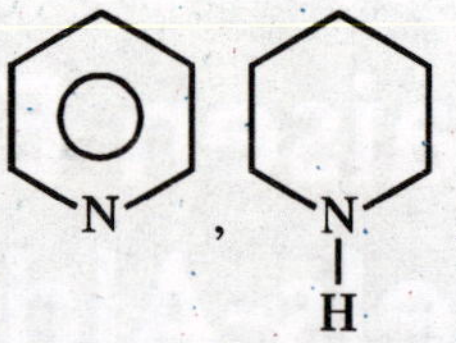

(iii) Michael condensation is reversible and hence excess base gives, back reaction in exchange of condensation product.

(iv) If Michael acceptor contain ketomethyl group then strong base will give self-condensation of Michael acceptor (aldol condensation).

(v) Polar protic solvent is used in Michael condensation.

(vi) For best Michael condensation (a) mild condition, (b) excess Michael donor, (c) weak and catalytic amount of base and (d) low temperature are required.

(vii) The use of excess base or full equivalent of strong base, high temperature, long reaction time promotes serveral parallel reactions, as for examples.

(a) Retro Michael condensation

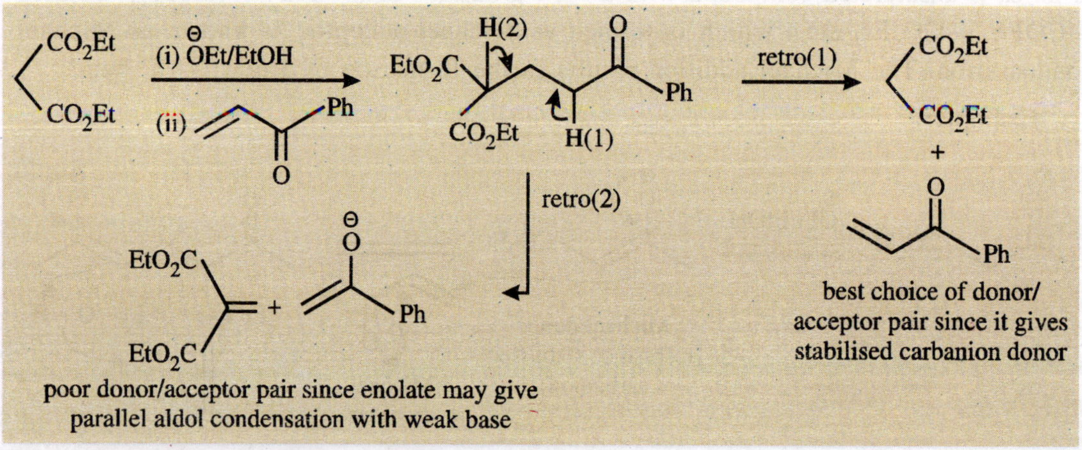

best choice of donor/ acceptor pair since it gives stabilised carbanion donor

poor donor/acceptor pair since enolate may give parallel aldol condensation with weak base

(b) If in the retro Michael condensation there is a possibility of two parallel stabilised carbanion, it may give different pair of Michael donor/acceptor with equal facility.

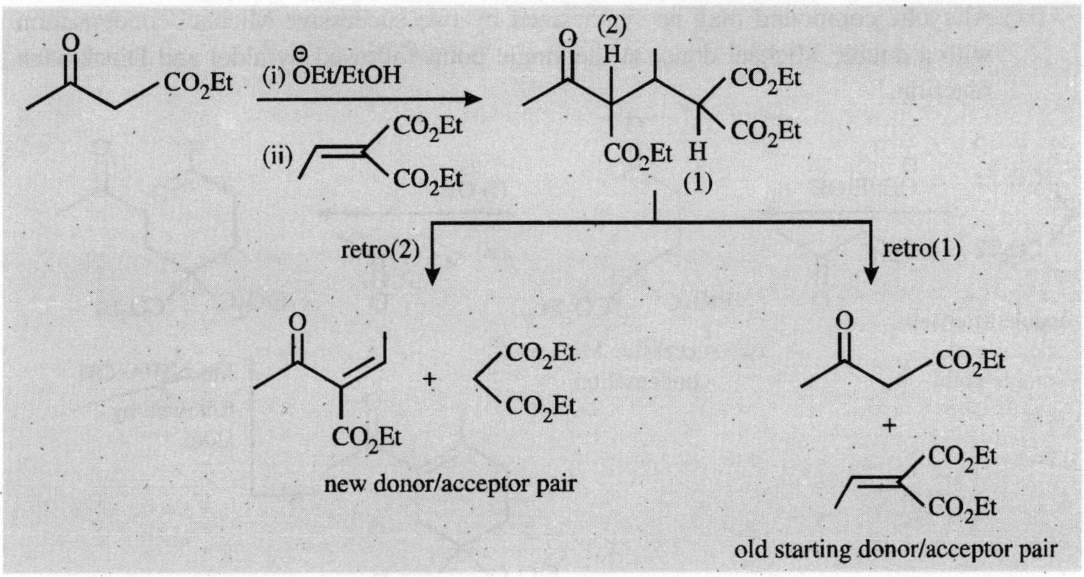

CO2Et
(i) OEt/EtOH
(ii)
CO2Et
CO2Et
(2)
H
CO2Et
CO2Et
CO2Et H
(1)
retro(2)
retro(1)
CO2Et
+
CO2Et
CO2Et
CO2Et
new donor/acceptor pair
CO2Et
+
CO2Et
CO2Et
old starting donor/acceptor pair

(c) May give further transformation such as aldol condensation or Claisen ester condensation of the initial adduct.

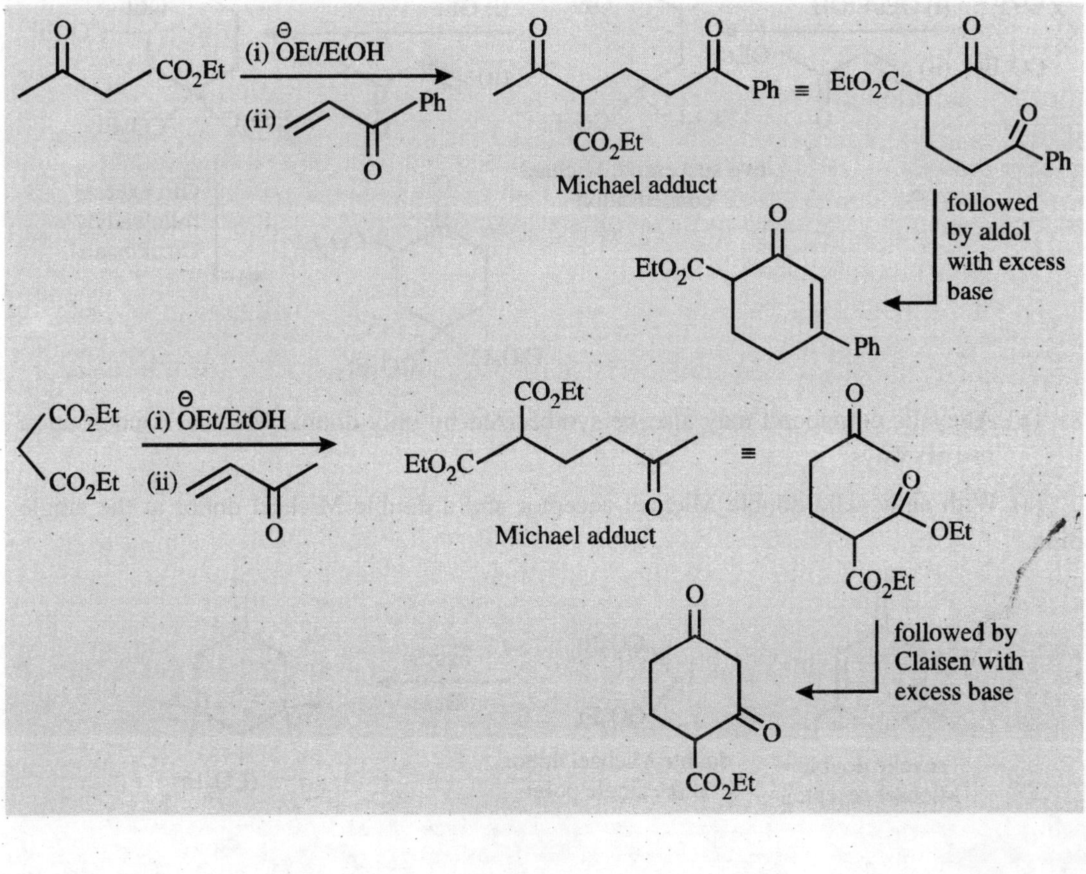

CO2Et
(i) OEt/EtOH
(ii)
Ph
CO2Et
Michael adduct
Ph
EtO2C
Ph
followed by aldol with excess base
EtO2C
Ph
CO2Et
CO2Et
(i) OEt/EtOH
(ii)
CO2Et
EtO2C
Michael adduct
OEt
CO2Et
followed by Claisen with excess base
CO2Et

(viii) The formation of six-membered alicyclic ring by Michael condensation followed by aldol or Claisen or Dieckmann reactions are very important synthetically.

(ix) Alicyclic compound may be synthesized by two successive Michael condensation with a double Michael donor at the single point followed by aldol and Dieckmann reaction.

double Michael donor at the single point; (i) $\overset{\ominus}{O}Et/EtOH$ (ii) methyl vinyl ketone → two successive Michael condensation; (i) $\overset{\ominus}{O}Et$ (ii) methyl vinyl ketone → $Me_2NH/AcOH$ followed by aldol

(i) $\overset{\ominus}{O}Et/EtOH$ (ii) ethyl acrylate (OEt) → two successive Michael condensation; (i) $\overset{\ominus}{O}Et$ (ii) ethyl acrylate (OEt) → $\overset{\ominus}{O}Et$ excess followed by Dieckmann

(x) Alicyclic compound may also be synthesized by only double Michael condensation in two ways.

(a) With an acyclic double Michael acceptor and a double Michael donor at the single point.

acyclic double Michael acceptor + double Michael donor at the single point (CO$_2$Et / CO$_2$Et) → excess Base → $(CO_2Et)_2$

(b) With a cyclic double Michael acceptor and a double Michael donor at the two different points of the same molecule.

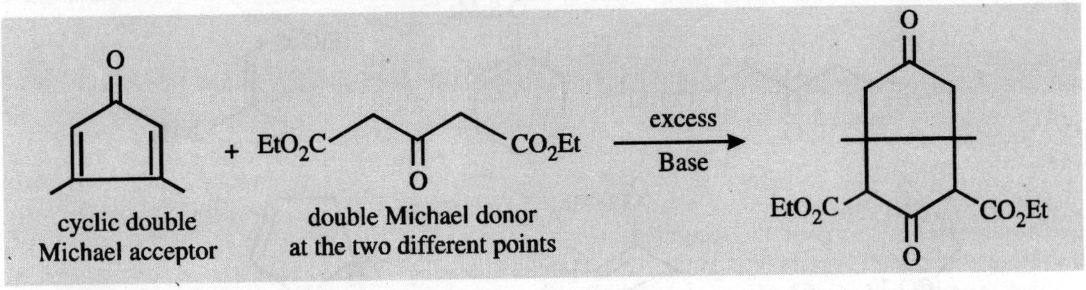

(xi) Alicyclic unit may also be generated by intramolecular Michael condensation.

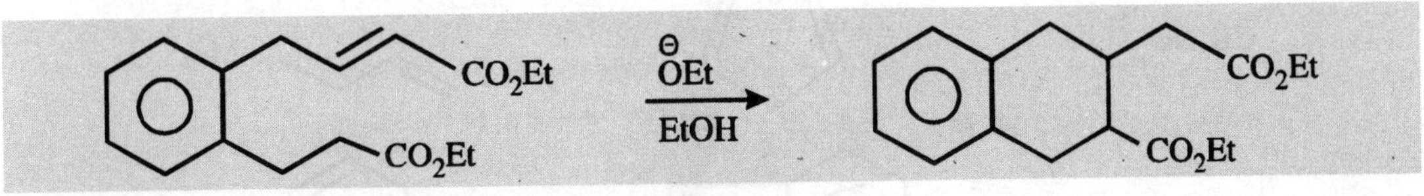

(xii) The main utility of Michael condensation is the synthesis of 1, 5-dicarbonyl compound.

(xiii) Stabilised carbanion is termed as "Michael donor" and α, β unsaturated compound having —M group is termed as "Michael acceptor".

3.1.3 Illustrations of some commonly used 'Michael donor'

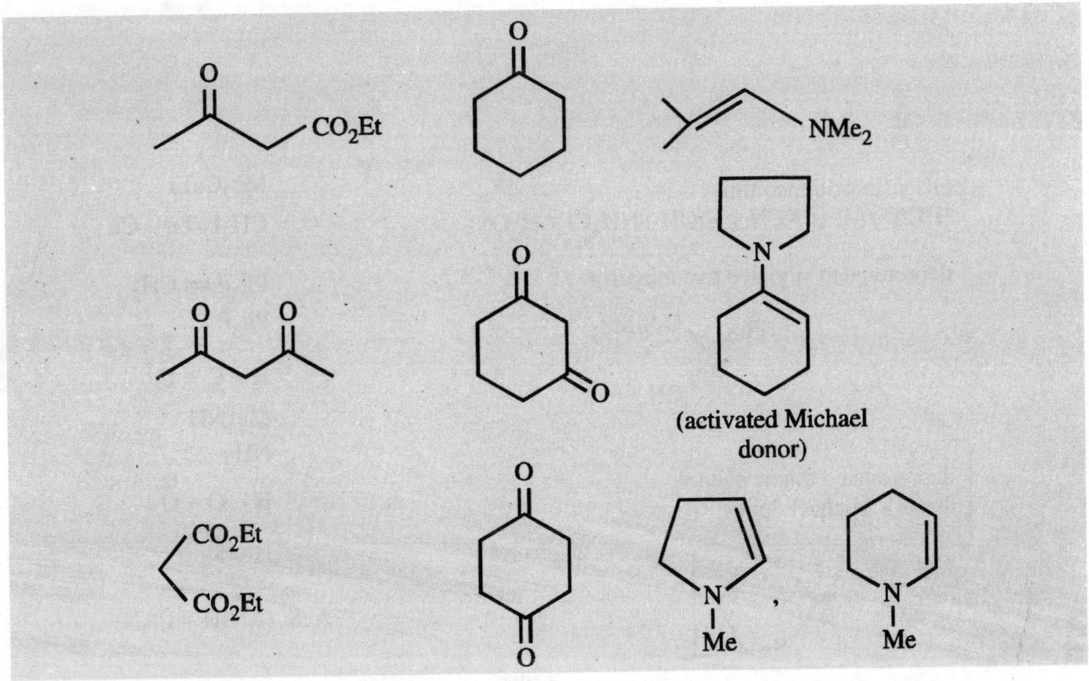

EtO_2C CO_2Et O

O $SiMe_3$

EtO_2C NH_2

CN CO_2Et

O $SiMe_3$

N Me

CH_3NO_2

$SiMe_3$

CH_3-CN

N Li

(potentially enamine)

H_5C_6 CN

H_5C_6 NO_2

HCN (in acid medium)
(HCN +$AlEt_3$/KCN + EtOH +NH_4Cl + H_2O)

deprotonated silylated cyanohydrin

Ph O $SiMe_3$ ⊖ CN

[deprotonated thiane cannot act as a Michael donor

S S ⊖]

Me_2CuLi

$CH_2I_2/Zn-Cu$

$Ph_3P=CH_2$

Ph_3P

$H-C\equiv\overset{\ominus}{C}$

Me_2NH

NH_3

$H-O-\overset{\ominus}{O}$

$H-\overset{\ominus}{S}$

$Ac\overset{\ominus}{S}$ (AcSH + Et_3N)

3.1.4 Some illustrative examples for the preparation of Michael donor

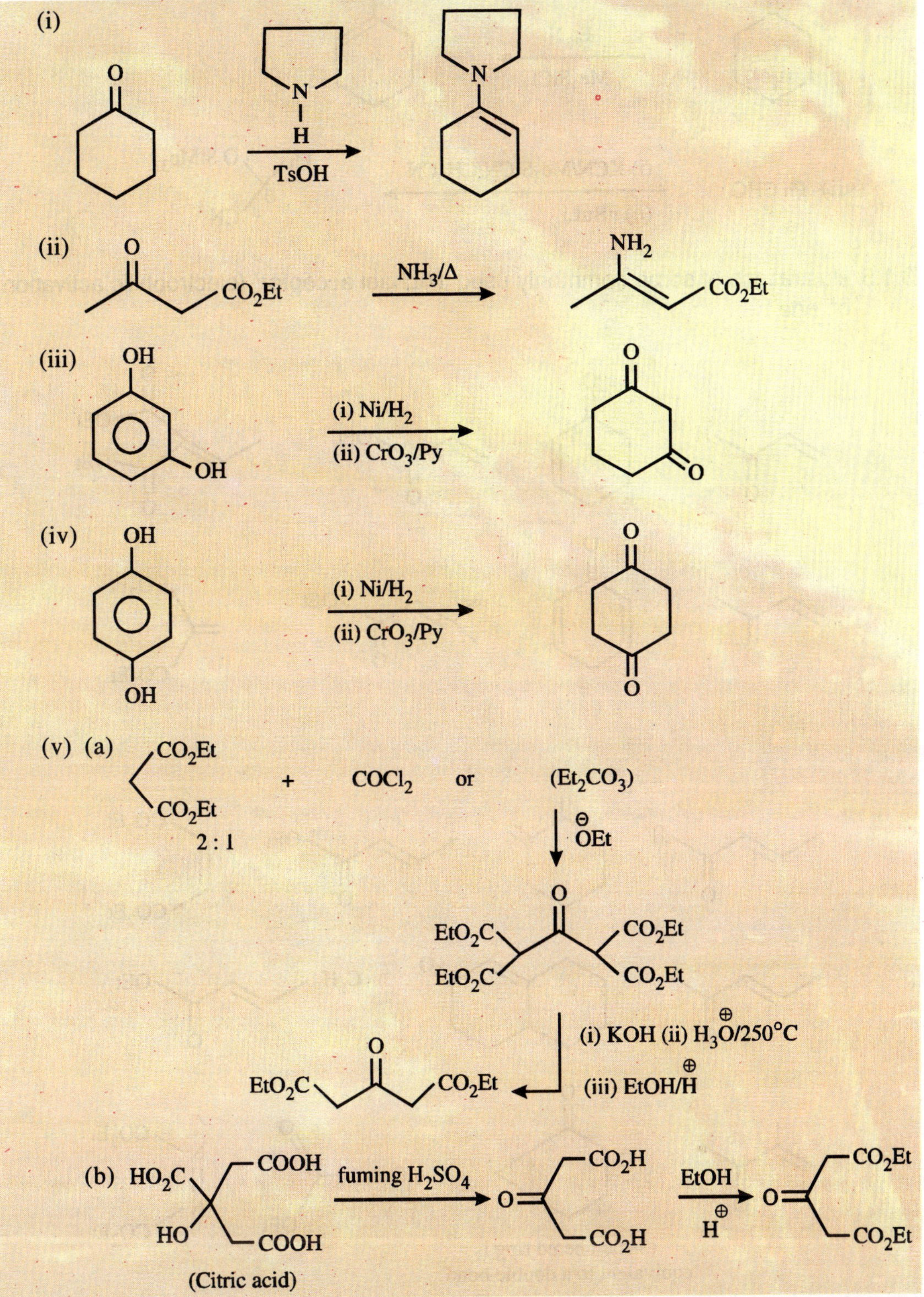

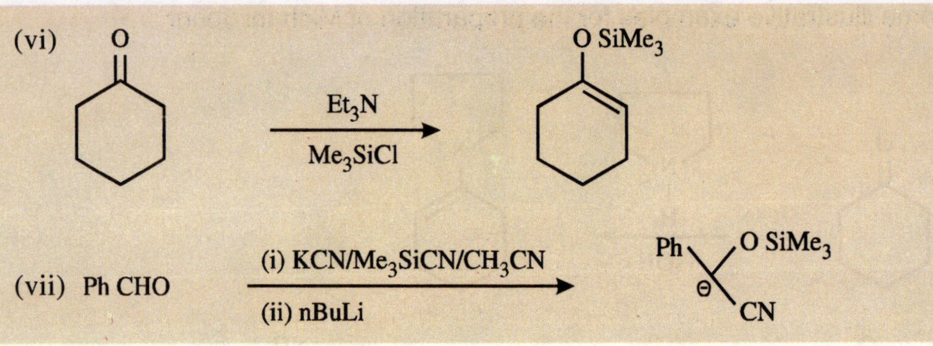

(vi) Cyclohexanone $\xrightarrow[Me_3SiCl]{Et_3N}$ 1-(trimethylsilyloxy)cyclohexene (O $SiMe_3$)

(vii) Ph CHO $\xrightarrow[\text{(ii) nBuLi}]{\text{(i) KCN/}Me_3SiCN/CH_3CN}$ Ph–C$^{\ominus}$(O $SiMe_3$)(CN)

3.1.5 Illustrations of some commonly used 'Michael acceptor' (electrophilic activation of 'ene')

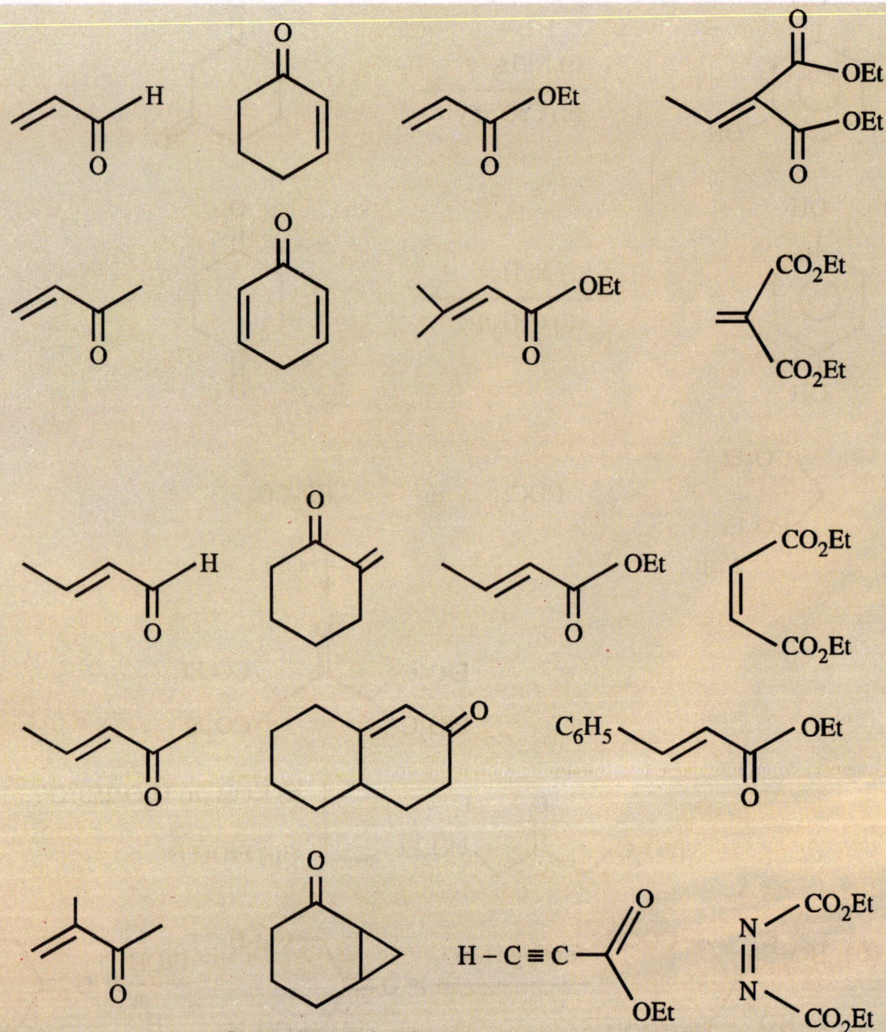

H, O, OEt, OEt, OEt, O, O, O

O, OEt, CO_2Et, CO_2Et, O, O

H, O, OEt, CO_2Et, CO_2Et, O, O

O, C_6H_5, OEt, O, O

O, O, H–C≡C–, OEt, N=N, CO_2Et, CO_2Et, O

(3-membered ring is equivalent to a double bond)

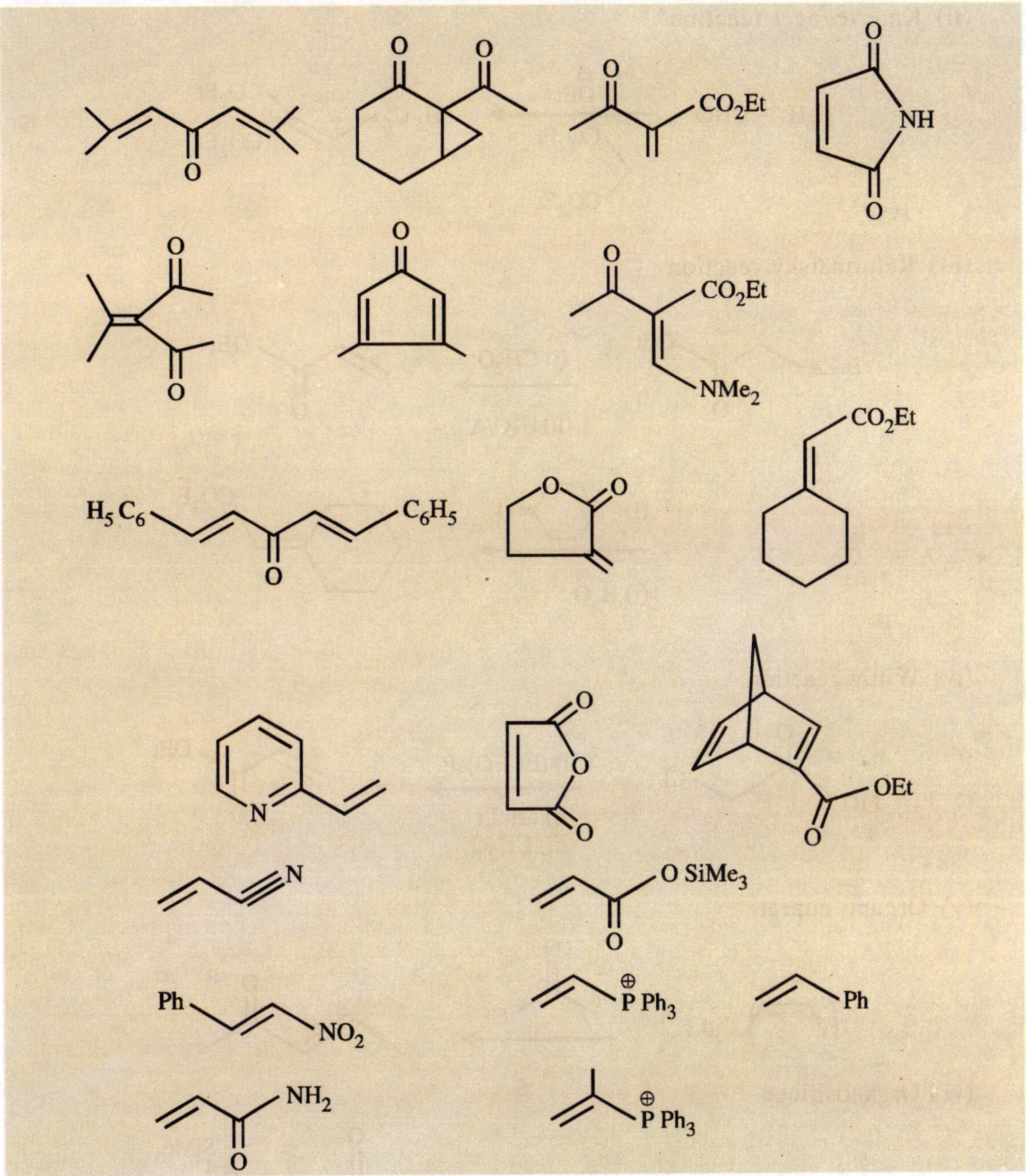

3.1.6 Some illustrative examples for the preparation of Michael acceptors

(i)

$$\xrightarrow[\text{(ii) } H_3\overset{\oplus}{O}/\Delta]{\text{(i) } \overset{\ominus}{O}H/CH_2O}$$

$$\xrightarrow[\Delta]{\text{dil HCl}}$$

(ii) Knoevenagel reaction

$C_6H_5 - CHO$ → ($^{\ominus}OEt$; $CH_2(CO_2Et)_2$) → H_5C_6–CH=C(CO_2Et)$_2$

(iii) Reformatsky reaction

BrZn–CH_2–C(=O)OEt → (i) CH_2O; (ii) $H_3O^{\oplus}/\Delta$ → CH_2=CH–C(=O)OEt

→ (i) cyclohexanone; (i) $H_3O^{\oplus}$ → cyclohexylidene–CH–CO_2Et

(iv) Wittig reaction

EtO–C(=O)–CH_2–Cl → (i) $(Et-O)_3P$; (ii) nBuLi; (iii) CH_2O → CH_2=CH–C(=O)OEt

(v) Organo cuprate

(CH₃CH=CH)$_2$CuLi → (CH_3C(=O)Cl) → CH_3CH=CH–C(=O)–CH_3

(vi) Organosilicon

CH_2=CH–CH_2–MgCl → (Me_3SiCl, THF) → CH_2=CH–CH_2–$SiMe_3$ → (CH_3C(=O)Cl, THF) → $^{\oplus}CH_2$–CH($SiMe_3$)–C(=O)–CH_3 → CH_2=CH–C(=O)–CH_3

(vii) Mannich reaction

($Me_2NH + CH_2O$) $Me_2\overset{\oplus}{N}{=}CH_2$ → (acetone) → (enol, OH) + $Me_2\overset{\oplus}{N}{=}CH_2$ → $CH_3COCH_2CH_2NMe_2$ →(Δ) $CH_3COCH{=}CH_2$

(butan-2-one) → $CH_3COC(=CH_2)CH_3$

$CH_3COCH_2CO_2Et$ → $CH_3COC(=CH_2)CO_2Et$

(cyclohexanone) → (2-methylenecyclohexanone)

(γ-butyrolactone) →(Et_3N) (α-methylene-γ-butyrolactone)

(viii) $CH_3COCH_2CO_2Et$ →($MeO-CH{=}\overset{\oplus}{N}Me_2$) $CH_3COCH(CO_2Et)CH(OMe)NMe_2$ →(Δ, − MeOH) $CH_3COC(CO_2Et){=}CHNMe_2$

(ix) $CH_3CHO + CH_2O$ →($\overset{\ominus}{HO}/H_2O$) $CH_2{=}CHCHO$ →(Ag_2O) $CH_2{=}CHCOOH$ →((i) $SOCl_2$ (ii) NH_3) $CH_2{=}CHCONH_2$

(x) OH → Et_3N / Me_3SiCl → $OSiMe_3$
O O

(xi) Cl–OMe → $\overset{\oplus}{Ph_3P}-\overset{\ominus}{CH_2}$ → $\overset{\oplus}{Ph_3P}$ H OMe → Δ / – MeOH → $\overset{\oplus}{Ph_3P}$

(xii) O → CH_2I_2/Zu-Cu → O

(xiii) H + CO_2Et → Δ → CO_2Et

(xiv) O O → (i) $\overset{\ominus}{OH}$ (ii) =O → O O

(xv) CO_2Et CO_2Et → (i) $\overset{\ominus}{OEt}$ (ii) CH_2O → H_2C= O OEt OEt O

(xvi) 2 Ph CHO + O → $\overset{\ominus}{OH}/H_2O$ Δ → Ph O Ph

(xvii) R O O R + O → $\overset{\ominus}{OH}/H_2O$ Δ → O R R [R = CH_3, C_6H_5–]

(xviii) O + O → HCl → O

3.1.7 Synthesis of target molecules using Michael condensation with retrosynthetic analysis

1. TM $\xrightarrow[\text{Michael dix}]{\text{1, 4 dico}}$ SE $+ \overset{\ominus}{C}O_2H \Longrightarrow \overset{\ominus}{C}N$ SE

SE $\Downarrow$ α, β-unsaturated C=O Mannich dix.

$+ Me_2NH + CH_2O$

Forward Synthesis

$\xrightarrow{HCN/AlEt_3}$ CN $\xrightarrow{H_3\overset{\oplus}{O}}$ COOH

When $\overset{\ominus}{C}N$ ion is used as a Michael donor it should be used in acid medium either with $HCN/AlEt_3$ or with ($KCN + EtOH + NH_4Cl + H_2O$). In basic medium carbonyl reactivity will increase to give cyanohydrin or parallel aldol condensation.

2. Me_2N–CO_2H (TM) $\overset{\text{FGI}}{\Longrightarrow}$ Me_2N–$OSiMe_3$

$\Downarrow$

Me_2NH + $OSiMe_3$

SE SE

Forward Synthesis

$OSiMe_3$ $\xrightarrow{Me_2NH}$ Me_2N–$OSiMe_3$ $\xrightarrow{\text{(i) AgF, (ii) } H_3\overset{\oplus}{O}}$ Me_2N–OH

Free acrylic acid cannot act as a Michael acceptor since Michael donor Me_2NH which is also a base interact with the free acid to give salt.

3.

O

O

O ⟹ + 'O' (nucleophilic)

⇓

TM SE $\overset{\ominus}{O}H/H_2O_2$

SE

Forward Synthesis

$\overset{\ominus}{O}-O-H$

$\overset{\ominus}{O}-O-H$ is a stabilised anion due to –I effect of oxygen

4.

⟹ + :CH_2(nucleophilic carbene)

⇓

TM SE $Ph_3\overset{\oplus}{P}-\overset{\ominus}{C}H_2$

SE

Forward Synthesis

$\overset{\ominus}{C}H_2-\overset{\oplus}{P}Ph_3$

5.

⟹ + $\overset{\ominus}{Me}$ ⟹ Me_2CuLi

TM SE SN SE

Forward Synthesis

+ $Cu^{\ominus}$(Me)(Me) $\overset{\oplus}{Li}$ → $\overset{\ominus}{O}\overset{\oplus}{Li}$ → $\overset{\ominus}{O}\overset{\oplus}{Li}$ + MeCu

$H_3\overset{\oplus}{O}$

Me_2CuLi is a stabilised carbanion supplier since negative charge is not on methyl but on the Cu atom which has got a larger size and the negative charge is non-classical.

6.

TM ⟹ CO_2Et CO_2Et ⟹ CO_2Et (anion) + CO_2Et (cation)

activation required

CH_3Br + $CH_2(CO_2Et)_2$ ⟸ $CH_3CH(CO_2Et)_2$

SE SE

$CH_2=CHCO_2Et$ SE

CH_2O + BrZn–CH_2CO_2Et

Forward Synthesis

$CH_2(CO_2Et)_2$ —(i) $\overset{\ominus}{O}Et$, (ii) CH_3Br→ $CH_3CH(CO_2Et)_2$ —(i) $\overset{\ominus}{O}Et$, (ii) $CH_2=CHCO_2Et$→ CO_2Et CO_2Et CO_2Et

(i) KOH

(ii) $H_3\overset{\oplus}{O}$

(iii) 250°C

OH OH (Ac_2O, Δ) →

7.

TM ⟹ CO_2Et CO_2Et ⟹ $CH_3CH=CHCO_2Et$ + $\overset{\ominus}{C}H_2CO_2Et$

SE

CH_3CHO + BrZn–CH_2CO_2Et

$CH_2(CO_2Et)_2$

SE

Forward Synthesis

CO_2Et / CO_2Et (i) $\overset{\ominus}{O}Et$ (ii) CO_2Et → CO_2Et CO_2Et CO_2Et (i) KOH (ii) $H_3\overset{\oplus}{O}$ /250°C → O OH OH O → Ac_2O/Δ

8. TM ⟹ α, β unsat. C =O aldol dix ⟹ 1, 5-dico Michael dix at 3, 4 ⟹ SE + SE

CH_2O + ⟸ α, β unsat. C= O, aldol dix.

Forward Synthesis

LDA/THF − 78°C → K. C. enolate → Me₂NH/AcOH

The last step is intramolecular aldol cond. Since the reactant is unsymmetrical dione it may give other intramol. aldol product in the usual base medium ($\overset{\ominus}{O}H/H_2O$), (since methyl hydrogen is more acidic) in addition to linear cond. (intermol. aldol). To avoid all these things $Me_2NH(AcOH)$ is used in presence of which the reaction will go via enol which is preferred with more subst. enol and also linear cond. is arrested.

9.

TM $\Rightarrow$ $\overset{\ominus}{C}H_3$ + (cyclohexylidene)acetic acid $\xrightarrow{FGI}$ ethyl (cyclohexylidene)acetate

$\overset{\ominus}{C}H_3 \Downarrow Me_2CuLi$ (SE)

wittig dix. $\Downarrow$ Cyclohexanone (SE) + $(EtO)_3P = CHCO_2Et$ (SE)

$\Downarrow$ $(EtO)_3P$ + $BrCH_2CO_2Et$

Forward Synthesis

Cyclohexanone $\xrightarrow[(ii)\ H_3\overset{\oplus}{O}]{(i)\ (EtO)_3P\ =\ CHCO_2Et}$ ethyl (cyclohexylidene)acetate $\xrightarrow{Me_2CuLi}$ ethyl (1-methylcyclohexyl)acetate $\xrightarrow[(ii)\ H_3\overset{\oplus}{O}]{(i)\ KOH}$ (1-methylcyclohexyl)acetic acid (CO_2H)

10.

TM (hexane-2,5-dione) $\Rightarrow$ methyl vinyl ketone (SE) + $CH_3\overset{\ominus}{C}=O$ (SN) $\Rightarrow$ $CH_3CH_2NO_2$ (SE)

Forward Synthesis

$CH_3CH_2NO_2$ $\xrightarrow[(ii)\ CH_3COCH=CH_2]{(i)\ \overset{\ominus}{O}Et}$ $CH_3COCH_2CH_2CH(NO_2)CH_3$ $\xrightarrow[H_2O]{TiCl_3}$ $CH_3COCH_2CH_2COCH_3$

11.

TM: $CH(CO_2H)$, $CH(CO_2H)$, $CH_2(CO_2H)$ $\xrightarrow{FGI}$ triester (1 CO_2Et at 2, CO_2Et at 3, 4 CO_2Et 5) $\xrightarrow[at\ 3,\ 4]{Michael\ dix}$ $EtO_2CCH=CHCO_2Et$ (SE) + $CH_2(CO_2Et)_2$ (SE)

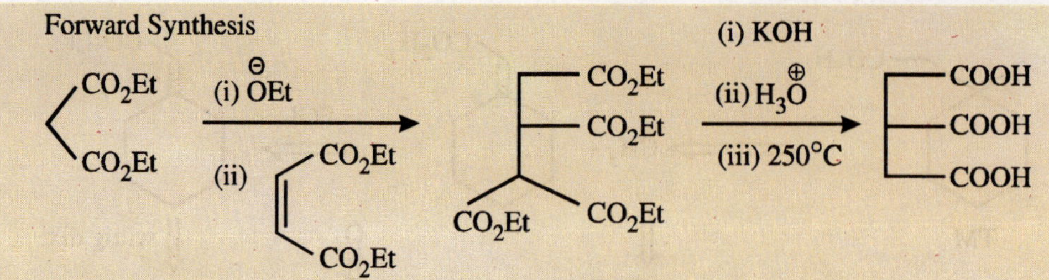

Forward Synthesis

(i) $\overset{\ominus}{O}Et$ (ii) EtO₂C–CH=CH–CO₂Et; (i) KOH (ii) $H_3\overset{\oplus}{O}$ (iii) 250°C

12.

α, β unsat. C=O aldol dix; 1, 5 -dico dix. at 3, 4

SE SE

1, 5 -dico dix. at 3, 4

SE SE

$+ Me_2NH + CH_2O$

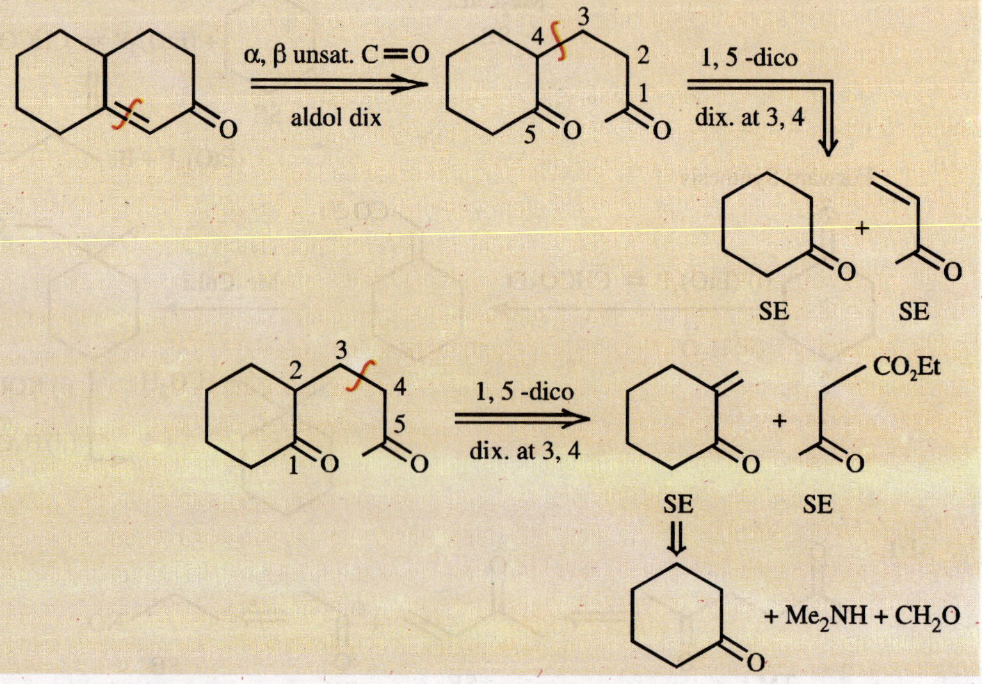

Since 1, 5-dicarbonyl is unsymmetrical and hence two types of disconnections are possible of which methyl vinyl ketone is the better enone (Michael acceptor) than exomethylene ketone since it may tautomerizes to another enone (in base medium) which is more stable and therefore possibility of parallel products.

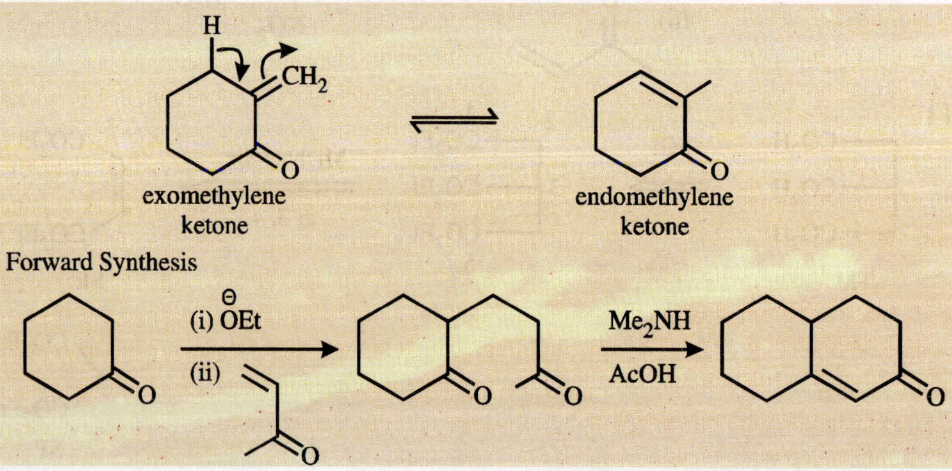

exomethylene ketone ⇌ endomethylene ketone

Forward Synthesis

(i) $\overset{\ominus}{O}Et$ (ii) methyl vinyl ketone; Me_2NH AcOH

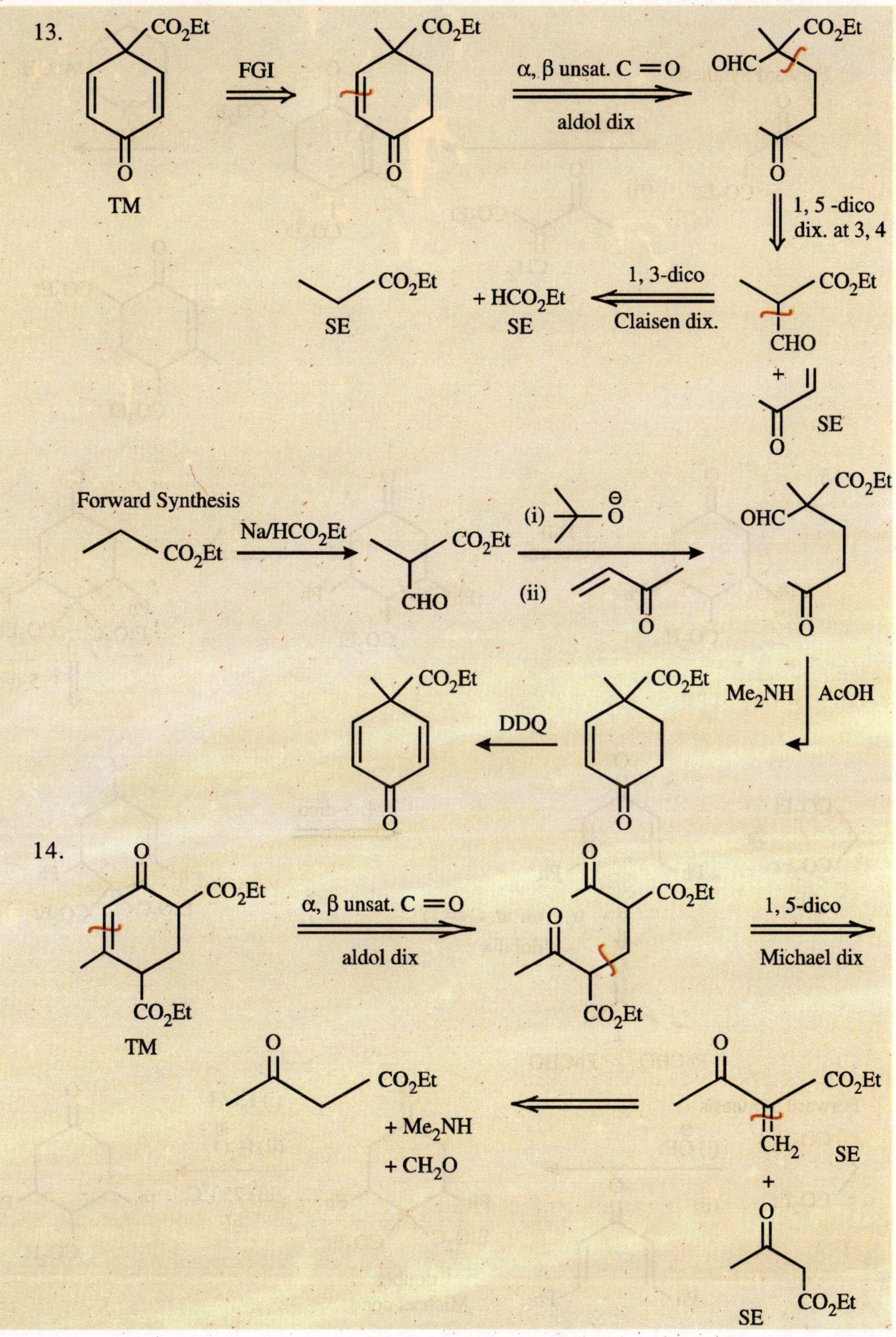
13.
CO_2Et
TM
FGI
α, β unsat. C =O
aldol dix
OHC
1, 5 -dico
dix. at 3, 4
CO_2Et
SE
+ HCO_2Et
SE
1, 3-dico
Claisen dix.
CHO
SE
Forward Synthesis
Na/HCO_2Et
(i)
(ii)
Me_2NH
AcOH
DDQ
14.
α, β unsat. C =O
aldol dix
1, 5-dico
Michael dix
+ Me_2NH
+ CH_2O
CH_2
SE

Forward Synthesis

O, CO_2Et — (i) $\overset{\ominus}{O}Et$ (ii) O, CO_2Et, CH_2 → O, CO_2Et, O, CO_2Et — piperidine (N–H) /AcOH → O, CO_2Et, CO_2Et

15. O, Ph, Ph, CO_2H $\xrightarrow{\text{FGI}}$ O, Ph, Ph, CO_2Et $\xrightarrow{\text{FGA}}$ O, Ph, Ph, EtO_2C, CO_2Et

⇓ 1, 5-dico

O, Ph, Ph, EtO_2C, CO_2Et $\xLeftarrow{\text{1, 5-dico}}$ CO_2Et, CO_2Et SE + O, Ph, Ph SE

⇓ α, β unsat. C=O aldol dix

O + PhCHO PhCHO

Forward Synthesis

CO_2Et, CO_2Et — (i) $\overset{\ominus}{O}Et$ (ii) O, Ph, Ph → O, Ph, Ph, EtO_2C, CO_2Et double Michael cond. — (i) KOH (ii) $H_3\overset{\oplus}{O}$ (iii) 250°C → O, Ph, Ph, CO_2H

16. TM $\xRightarrow{\text{FGA}}$ $\xRightarrow[\text{aldol dix}]{\alpha, \beta \text{ unsat. } C{=}O}$

$\xRightarrow[\text{Michael dix.}]{\text{1, 5-dico}}$ SE + CHO

OMe, OMe, CHO, OMe

$\Downarrow$ wittig dix

$Ph_3\overset{\oplus}{P}-\overset{\ominus}{C}H-OMe$ + (SE) $\xRightarrow[\text{craft dix}]{\text{Friedel}}$ CH_3COCl (SE) + OMe (SE)

Forward Synthesis

OMe $\xrightarrow[AlCl_3]{CH_3COCl}$ OMe $\xrightarrow[\text{(ii) } H_3\overset{\oplus}{O}]{\text{(i) } Ph_3P=CH-OMe}$ CHO, OMe

$\xrightarrow[\text{(ii) methyl vinyl ketone}]{\text{(i) } \overset{\ominus}{O}Et}$ CHO, OMe

piperidine /AcOH → OMe

$\downarrow H_2/Pd$

OMe

17.

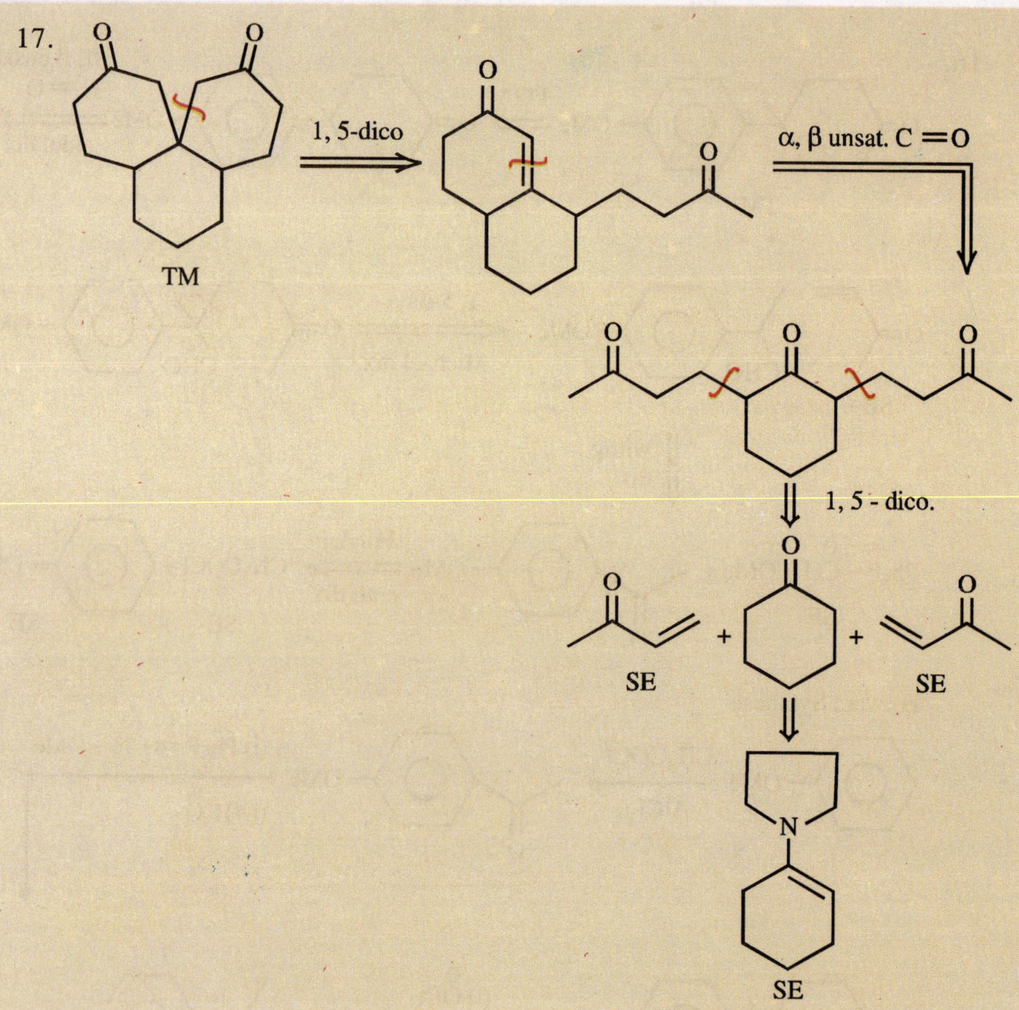

TM

1, 5-dico

α, β unsat. C =O

1, 5 - dico.

SE + + SE

SE

Forward Synthesis

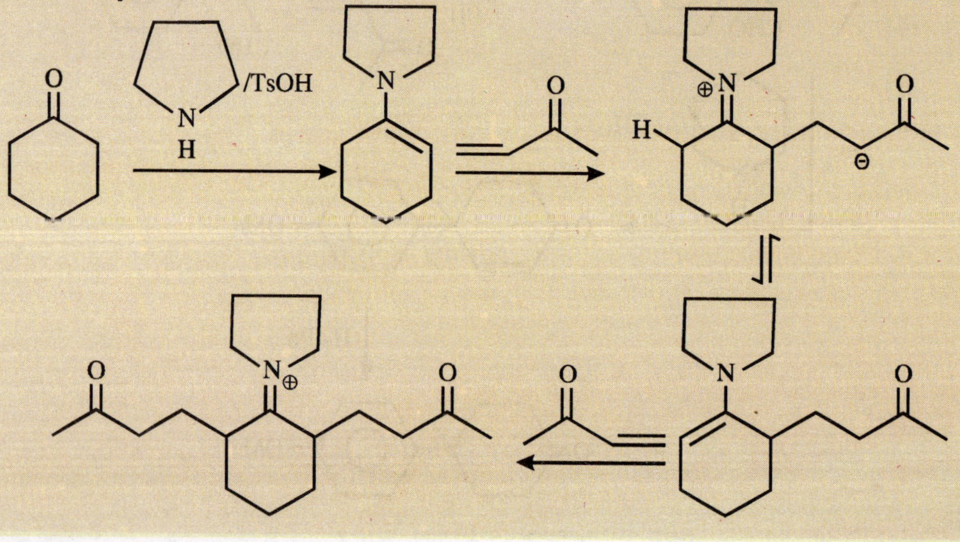

/TsOH

N

H

$MeCO_2H/AcONa$

KOH / MeOH / aldol cond.

Michael cond.

This is an example of aldol followed by Michael condensation

18.

COOH

CO_2H

TM

FGI

O

wittig

dix

$CH_3 - CHO$ +

SE

H

O

$Ph_3P^{\oplus}$

Ph_3P +

SE

SE

Forward Synthesis

Ph_3P

Michael cond.

H

$Ph_3P^{\oplus}$

CH_3CHO

COOH

$H_3\overset{\oplus}{O}$

CO_2H

19. CO_2Et CO_2Et ⟹ wittig dix CO_2Et CO_2Et O $Ph_3\overset{\oplus}{P}$ ⊖

⇓

O Cl + CO_2Et CO_2Et ⟸ CO_2Et CO_2Et O + $Ph_3\overset{\oplus}{P}$ SE

SE SE

Forward Synthesis

CO_2Et CO_2Et (i) $\overset{\ominus}{O}Et$ (ii) Cl O → CO_2Et CO_2Et O (i) NaH (ii) $Ph_3\overset{\oplus}{P}$ Micheal cond.

CO_2Et CO_2Et ← wittig CO_2Et CO_2Et O PPh_3

20. O N – H O TM ⟹ 1, 5-dico O N – H O ⟹ C – N imide O NH_2 SE + OEt O SE

Forward Synthesis

OEt O (i) $\overset{\ominus}{O}Et$ (ii) NH_2 O Michael cond. → O NH_2 OEt O NaOEt – EtOH → O NH O

21.

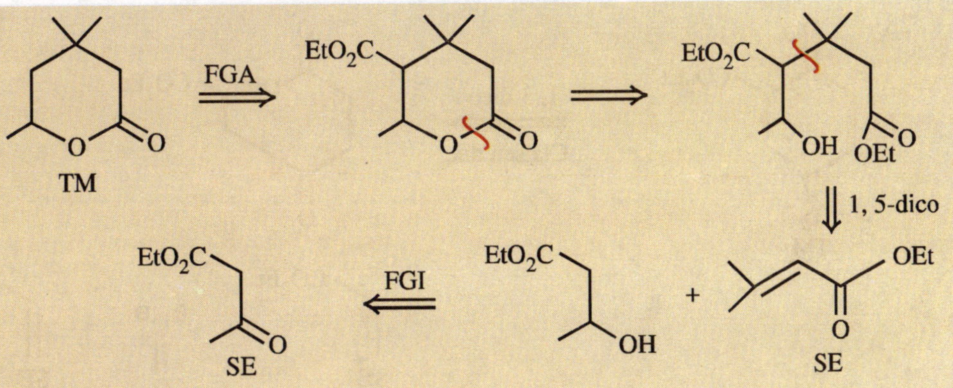

TM FGA $\Rightarrow$ EtO_2C $\Rightarrow$ EtO_2C OH OEt O

1, 5-dico

EtO_2C O SE $\Leftarrow$ FGI EtO_2C OH + OEt O SE

Forward Synthesis

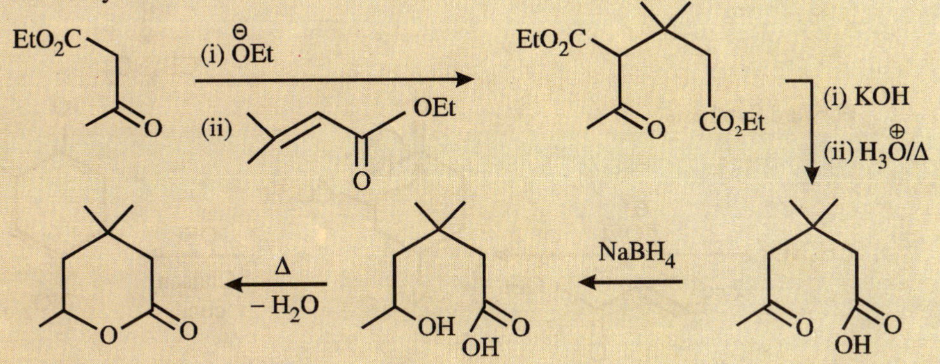

EtO_2C O (i) $\overset{\ominus}{O}Et$ (ii) OEt O $\rightarrow$ EtO_2C O CO_2Et

(i) KOH (ii) $H_3\overset{\oplus}{O}/\Delta$

O O $\leftarrow$ Δ $-H_2O$ OH OH O $\leftarrow$ $NaBH_4$ O OH O

22.

N Me O TM $\Rightarrow$ $\oplus$N Me $O^{\ominus}$ $\Rightarrow$ $\oplus$N Me O

N Me SE $\Leftarrow$ $\ominus$ $\oplus$N Me + O SE

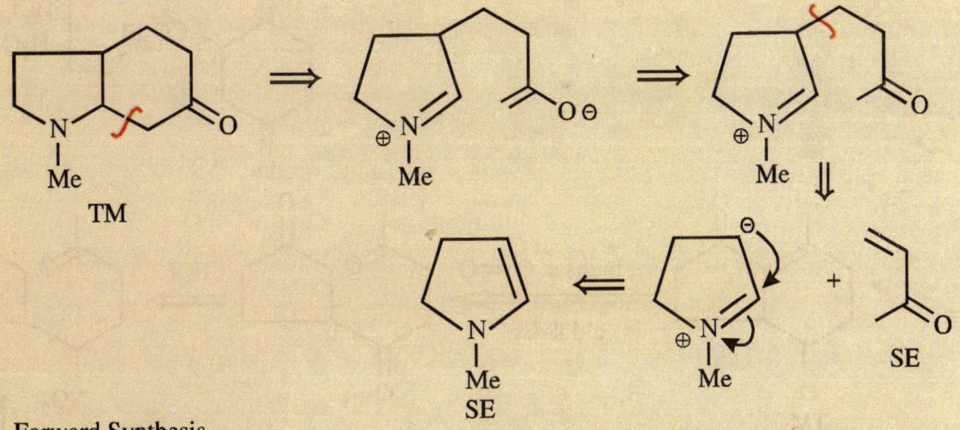

Forward Synthesis

N Me + O $\rightarrow$ $\oplus$N Me $\overset{\ominus}{O}$ $\rightleftharpoons$ $\oplus$N Me $O^{\ominus}$

N Me O

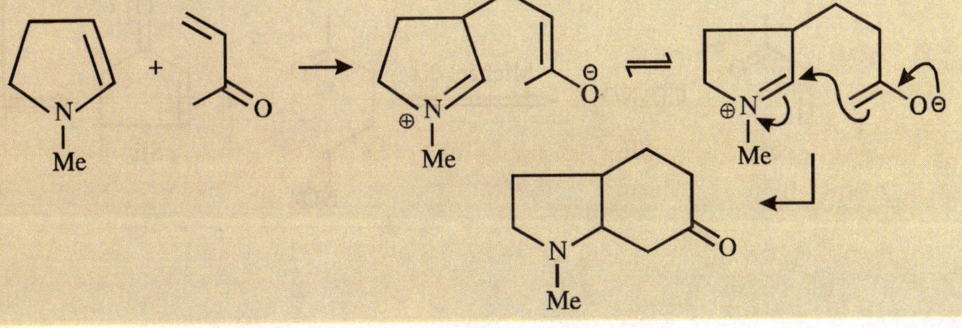

23.

TM

1, 3-dico

Claisen dix.

OEt CO_2Et

CO_2Et + C (SN) + CO_2Et

SE SE

CH_3NO_2

Forward Synthesis

CH_3NO_2 — $\overset{\ominus}{2\ OEt}$, 2 eq. (ethyl acrylate), stepwise → (NO_2, CO_2Et, OEt) — $\overset{\ominus}{OEt}$ Claisen cond. → (CO_2Et, NO_2) — $TiCl_3$/H_2O → (CO_2Et)

24.

TM

α, β unsat. C ═O

aldol dix

FGI

NO_2

SE + CH_3NO_2 SE ⇐ Michael dix (NO_2) + SE

Forward Synthesis

CH_3NO_2 —(i) $\overset{\ominus}{O}Et$, (ii) methyl vinyl ketone→ 5-nitropentan-2-one (NO_2) —(i) $\overset{\ominus}{O}Et$, (ii) cyclopent-2-enone→ product with NO_2

α-H adjacent to $-NO_2$ group is more acidic and hence preferred nitro enolate

—$\overset{\ominus}{O}Et$, aldol cond.→ bicyclic enone with NO_2 —$TiCl_3$, H_2O→ bicyclic dione

25. TM (3-oxocyclohexyl)acetic acid (CO_2H) ⟹ 3-oxocyclohexyl cation (⊕) + $\overset{\ominus}{C}O_2H$ ⟹ $\overset{\ominus}{C}N$ SE

⇓

3-oxocyclohexylmethyl $\overset{\oplus}{P}Ph_3$ ⟹ cyclohex-2-enone (SE) + $\overset{\ominus}{C}H_2 - \overset{\oplus}{P}Ph_3$ SE

Forward Synthesis

cyclohex-2-enone —$\overset{\ominus}{C}H_2 - \overset{\oplus}{P}Ph_3$, Michael cond.→ bicyclic cyclopropane ketone —$\overset{\ominus}{C}N$, Michael cond.→ enolate ($O^{\ominus}$) with CH_2CN —$H_3\overset{\oplus}{O}$→ (3-oxocyclohexyl)acetic acid (COOH)

≡ 3-oxocyclohexyl cation (⊕)

3.1.8 Review synthetic problems

Find out the synthetic equivalent after proper retrosynthetic analysis of the following target molecules and then show the forward synthesis using Michael condensation.

1. O O
2. O
3. O
4. O N H
5. O O O
6. O O
7. 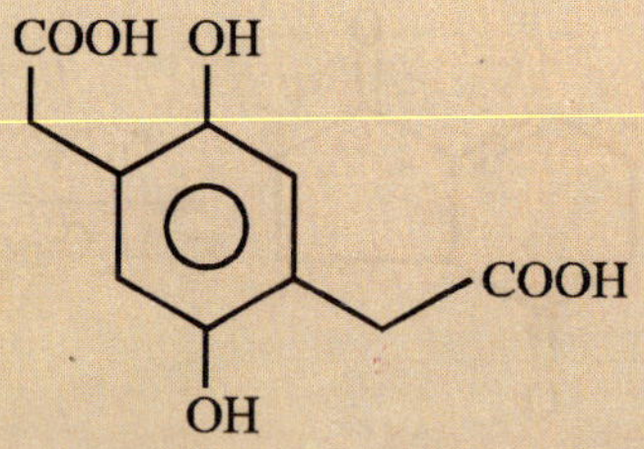 COOH OH COOH OH
8. O COOEt
9. O O
10. O
11. N H COOH COOH
12. O
13. O O O
14. O O EtO_2C CO_2Et
15. O O
16. O O
17. HO CO_2Et N H
18. O Cl CN Cl CN O
19. O NH O
20. O O

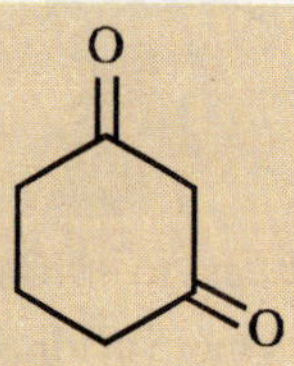

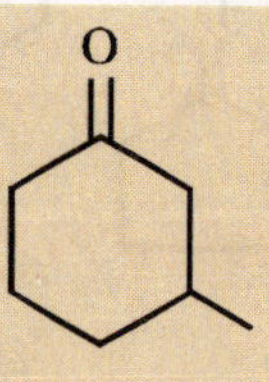

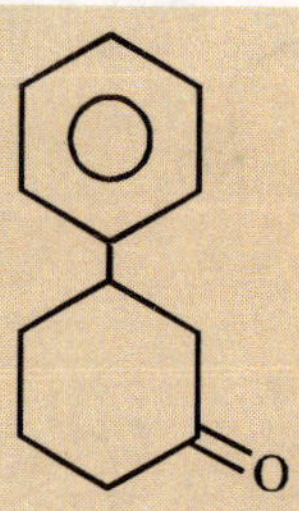

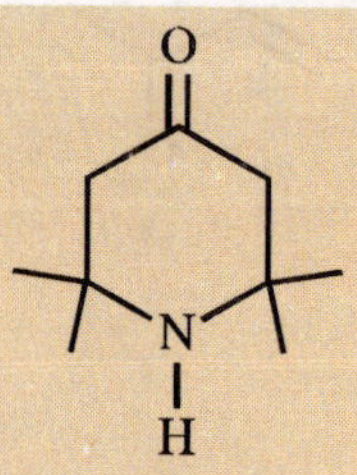

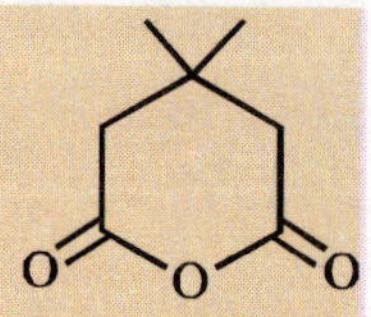

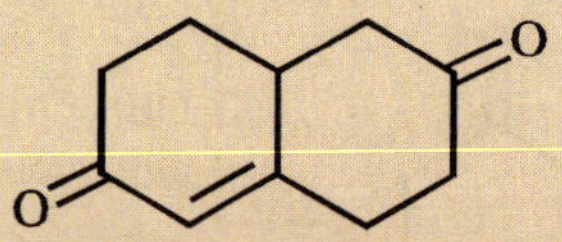

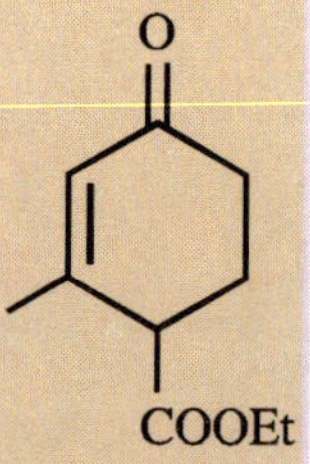

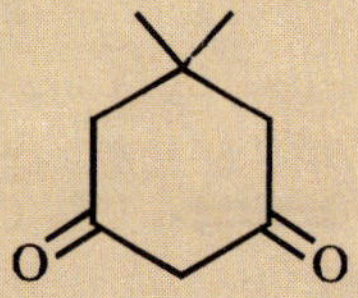

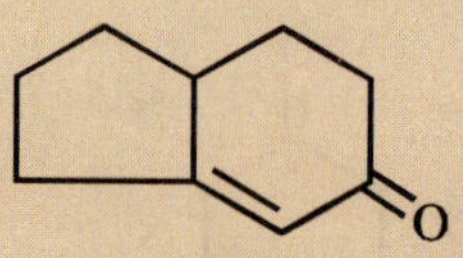

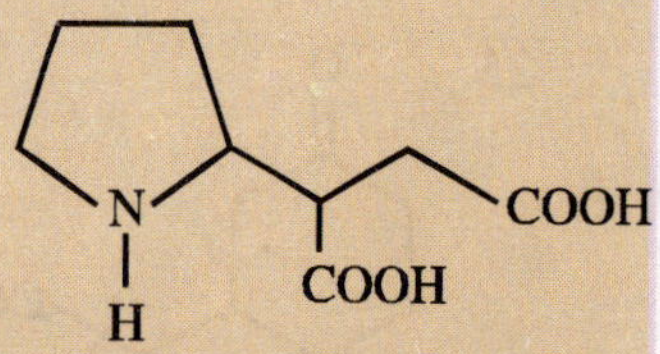

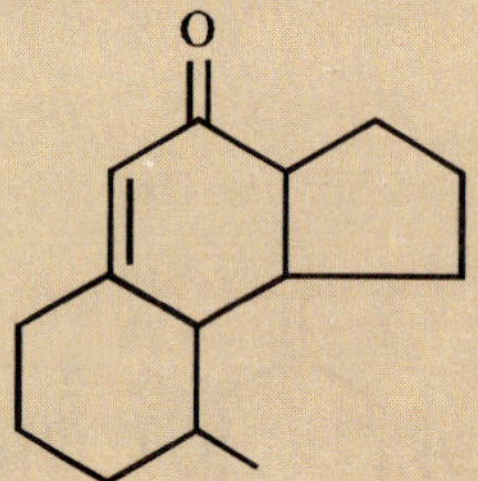

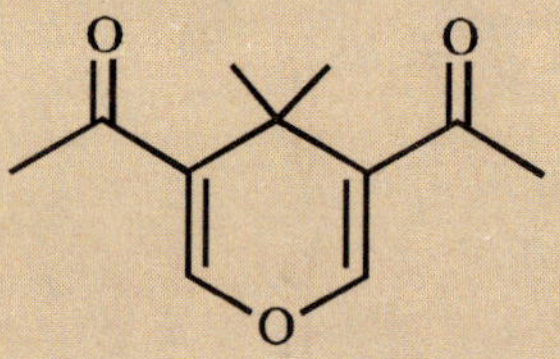

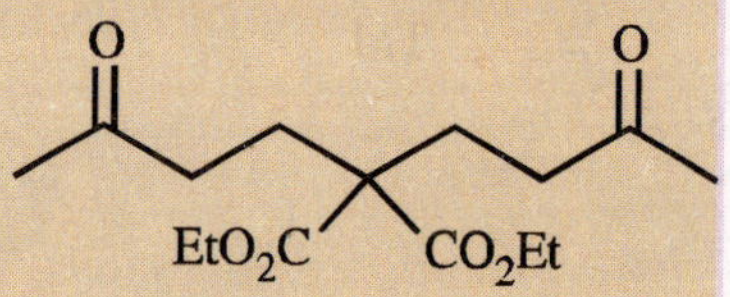

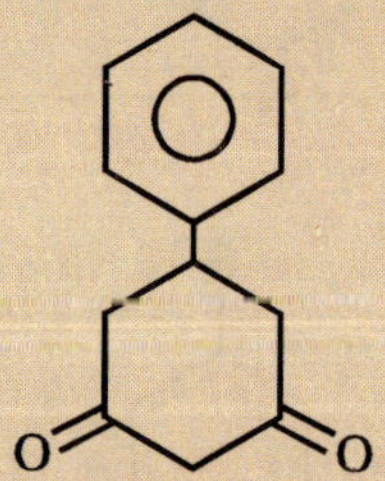

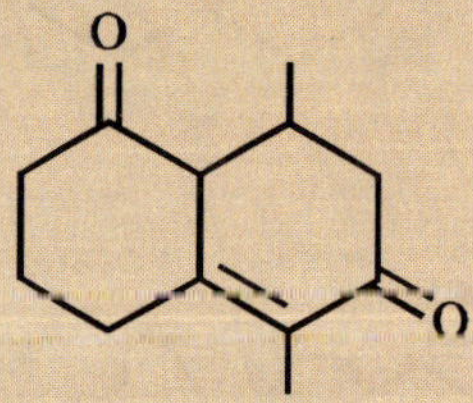

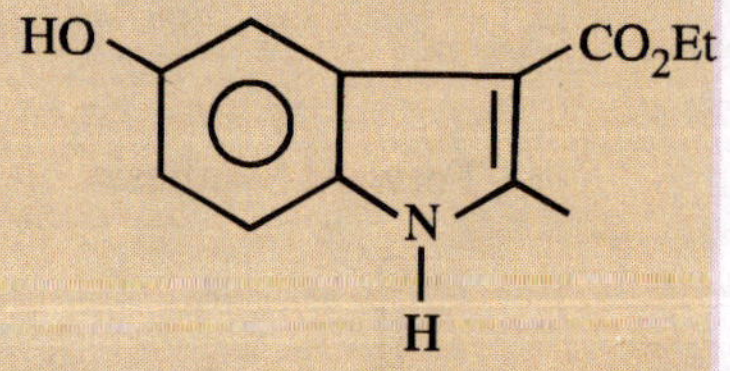

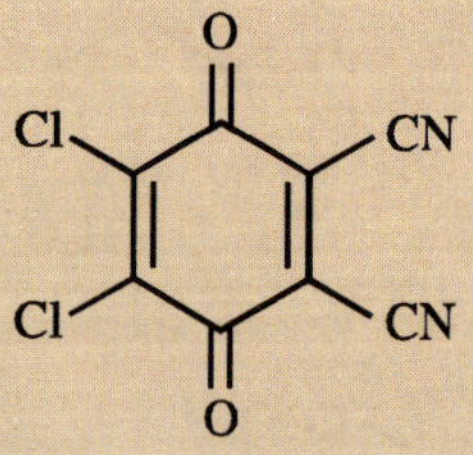

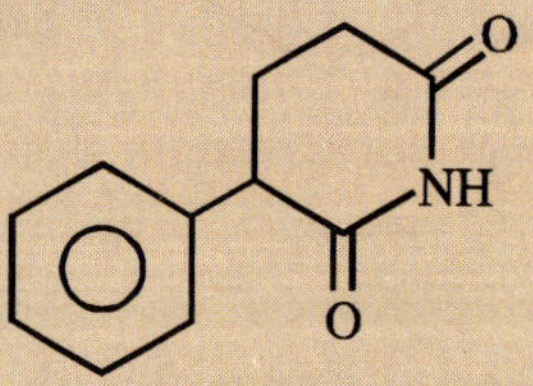

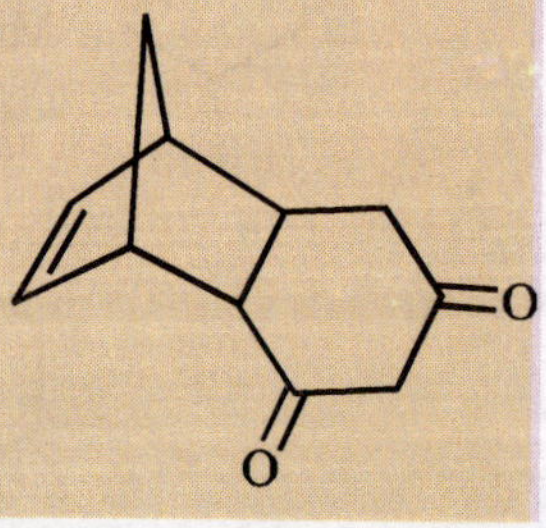

21. 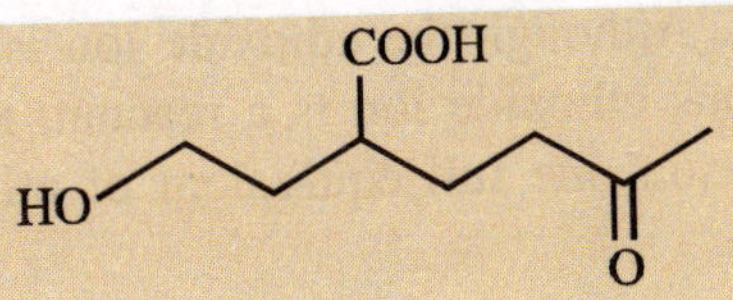 (structure: HO, COOH, O)

22. 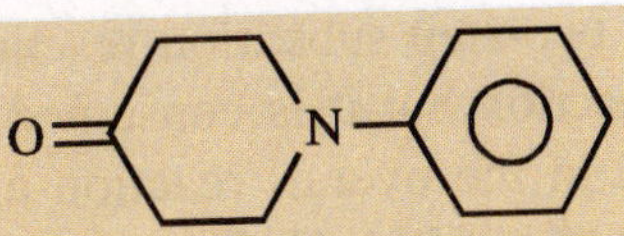 (structure: O=, N)

23. 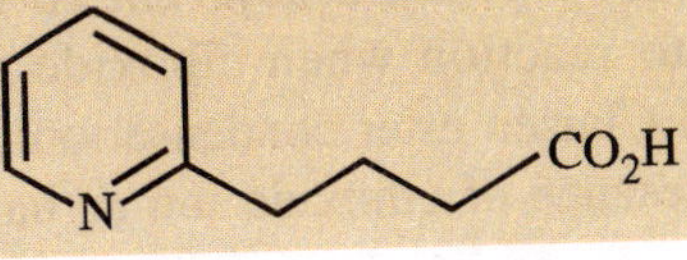 (structure: N, CO_2H)

24. 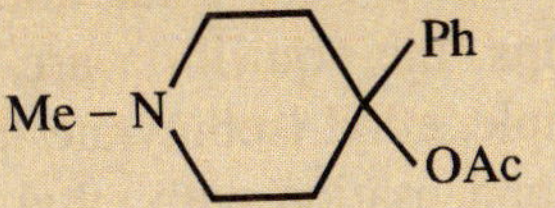 (structure: Me – N, Ph, OAc)

3.2 Claisen ester condensation

3.2.1 What is Claisen ester condensation?

The base-promoted condensation of two ester molecules which contain an α-H atom is known as Claisen ester condensation. The simple example is the formation of ethyl aceto acetate from ethyl acetate catalysed by ethoxide ion.

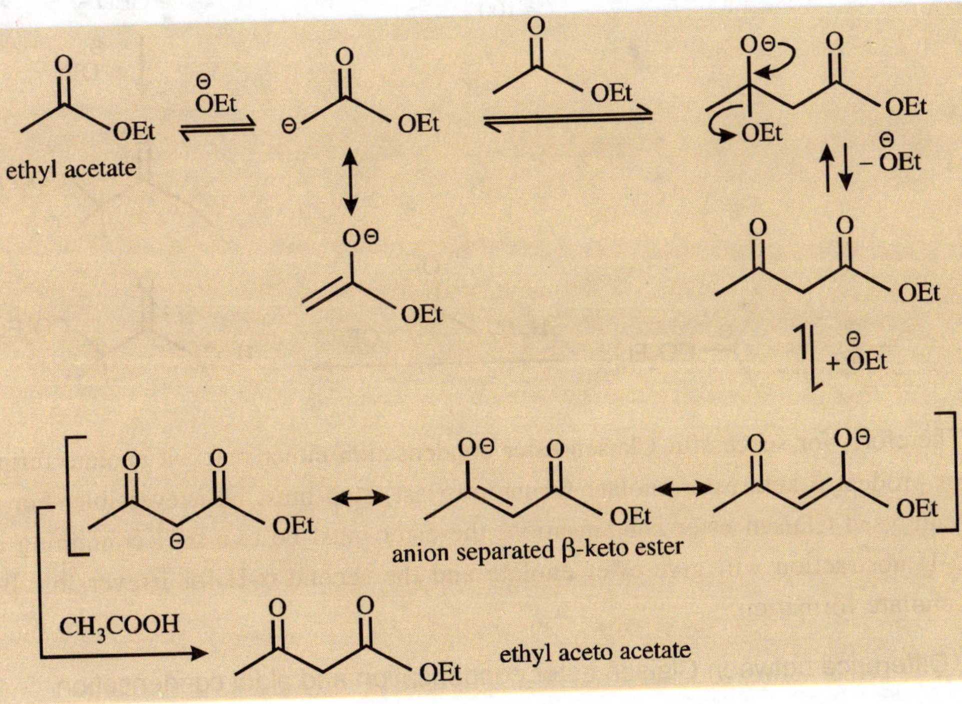

ethyl acetate — $\overset{\ominus}{O}Et$ — OEt — anion separated β-keto ester — CH_3COOH — ethyl aceto acetate — $-\overset{\ominus}{O}Et$ — $+\overset{\ominus}{O}Et$

3.2.2 Mechanistic explanation of Claisen ester condensation

Each step in the Claisen ester condensation is reversible but ethyl aceto acetate can be isolated in significant yield because it is completely removed from the equilibrium by conversion into its anion with ethoxide ion (anion separated β-keto ester, irreversibly).

Although the ester enolate ion is formed in very low concentration but it undergoes a nucleophilic acyl substitution reaction with a second molecule of ester and consequently the usual two-step substition reaction is observed. Although the ethoxide ion is a catalyst for the reaction but it is consumed ultimately. Thus ethoxide ion is a reactant rather than a catalyst in the overall reaction and for this reason one full equivalent of ethoxide ion must be used in the Claisen ester condensation.

If the product β keto ester does not contain a C—H bond adjacent to both keto and ester groups, the equilibria are unfavourable to reaction when ethoxide ion is the base. For example, ethyl isobutyrate fails to give self Claisen ester condensation or crossed Claisen ester condensation with ethyl benzoate in presence of ethoxide ion as base (which is a weak base) but condensation may be done with very strong base $(C_6H_5)_3\overset{\ominus}{C}\ \overset{\oplus}{Na}$ in which first step is almost irreversible.

$(CH_3)_2CH{-}CO_2Et$ → [$(C_6H_5)_3\overset{\ominus}{C}$, $-(C_6H_5)_3C-H$] → $(CH_3)_2\overset{\ominus}{C}{-}CO_2Et$ (~ 100%) → [$(CH_3)_2CH{-}CO_2Et$] → tetrahedral intermediate (O⊖, OEt, CO_2Et) → [$-\overset{\ominus}{O}Et$] → $(CH_3)_2CH{-}CO{-}C(CH_3)_2{-}CO_2Et$

$(CH_3)_2\overset{\ominus}{C}{-}CO_2Et$ + $H_5C_6{-}CO{-}OEt$ → $H_5C_6{-}CO{-}C(CH_3)_2{-}CO_2Et$

Therefore, for successful Claisen ester condensation either reactant enolate formation step or product β-keto ester enolate formation (last) step must be irreversible. For weak base catalysed Claisen ester condensation, the ester must be two α-H containing ester, one α-H abstraction will give ester enolate and the second α-H for irreversible β-keto ester enolate formation.

3.2.3 Difference between Claisen ester condensation and aldol condensation

In the synthetic procedure Claisen ester condensation sometimes parallel the aldol condensation during inter or intramolecular Claisen ester condensation and therefore, difference of the two condensation should not be confused.

(i) The Claisen condensation differs from the aldol reaction only after the oxyanion has been formed by addition of the enolate to the carbonyl group, in the Claisen condensation this anion eliminates ethoxide ion to give β-keto ester; whereas in the aldol condensation the anion gains proton from water to give a β-hydroxy aldehyde or ketone.

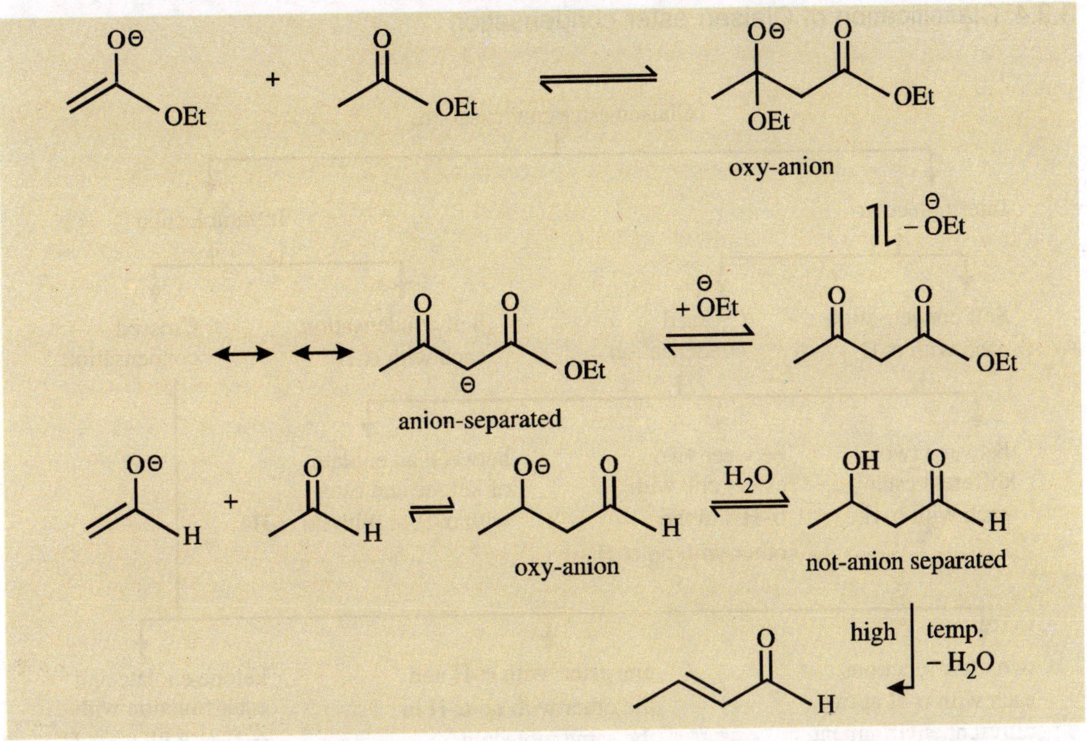

oxy-anion

$\overset{\ominus}{OEt}$ − OEt

$+ \overset{\ominus}{OEt}$

anion-separated

H_2O

oxy-anion

not-anion separated

high temp. $- H_2O$

(ii) The aldol condensation is an addition reaction of enolate ion or an enol with an aldehyde or ketone followed by dehydration. The Claisen condensation is a nucleophilic acyl substitution reaction of an enolate ion with an ester group.

(iii) The aldol condensation is catalysed by both base and acid. The Claisen condensation requires a full equivalent of base and is not catalysed by acid.

(iv) The aldol addition requires only one α-H. A second α-H is required for the dehydration step of the aldol condensation. In the Claisen condensation the ester must have at least two α-H, one for ester enolate formation and the other for β-keto ester enolate.

3.2.4 Classification of Claisen ester condensation

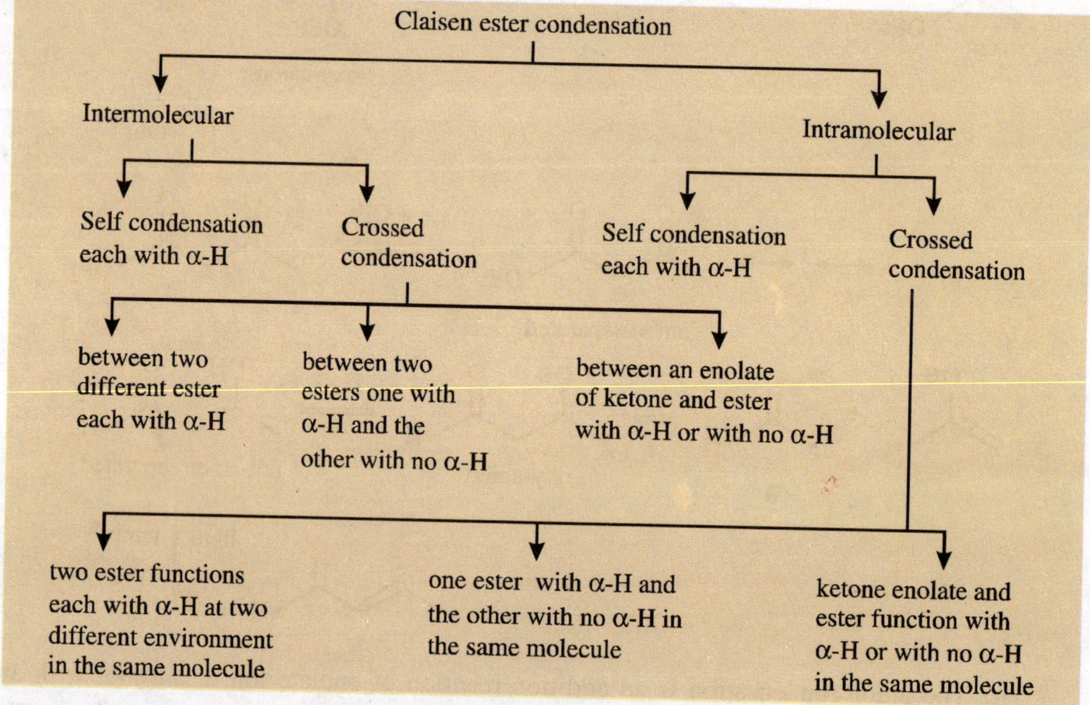

3.2.5 Intermolecular Claisen ester condensation

1. Self condensation: When condensation occurs between two same α-H containing ester, it is self Claisen ester condensation. It has great synthetic value.

Some examples are given below:

(a) O, OEt → ($^{\ominus}$OEt, EtOH) → O, O, OEt

(b) OEt, O → ($^{\ominus}$OEt, EtOH) → OEt, O, O

(c) OEt, EtO_2C, O, EtO, O, CO_2Et → ($^{\ominus}$OEt, EtOH) → O, EtO_2C, CO_2Et, O

2. Crossed condensation

(i) When condensation occurs between two different α-H containing ester—since this type of condensation gives four different products and therefore it has no synthetic value.

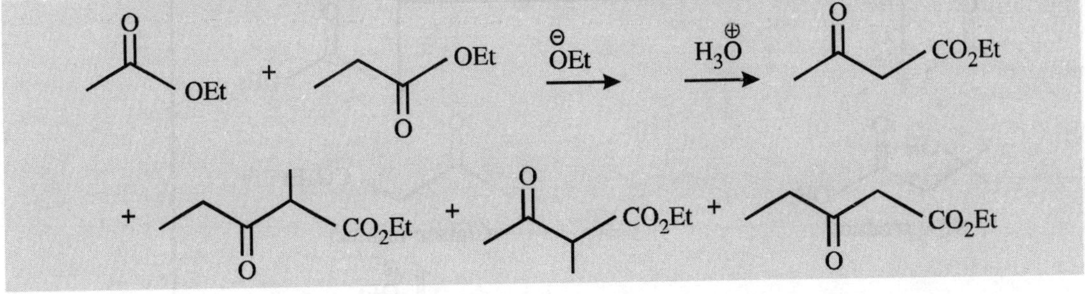

(ii) When condensation occurs between two esters one with α-H and the other ester with no α-H—it has a great synthetic value.

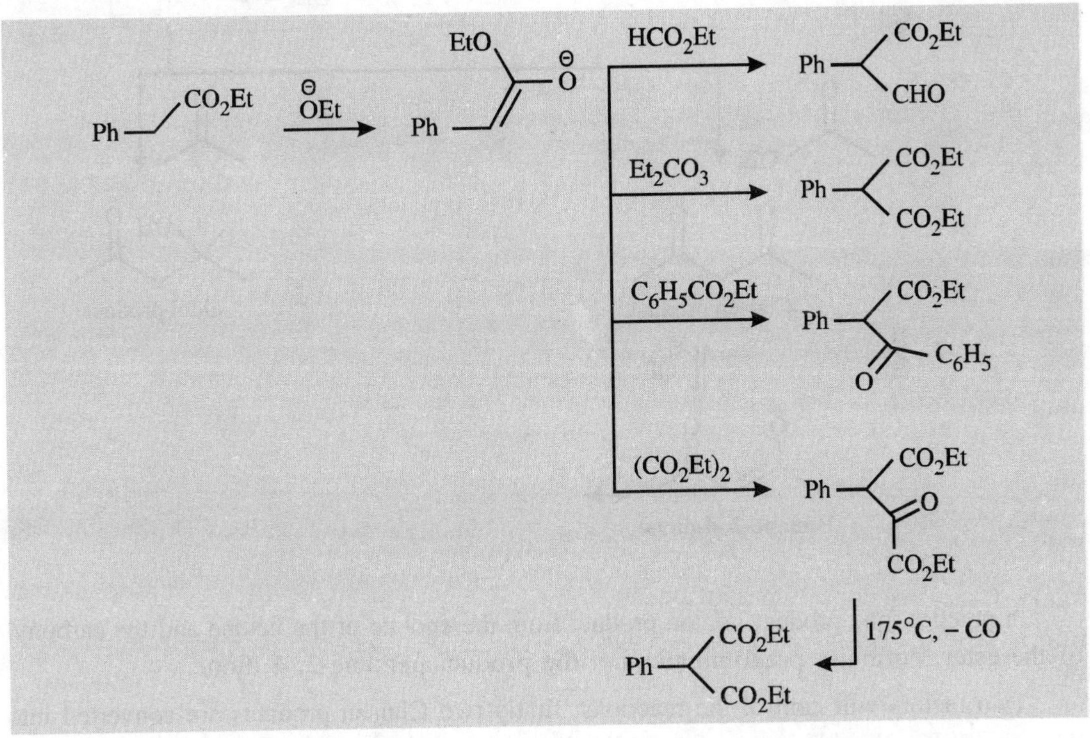

(iii) When condensation occurs between an enolate of ketone (not aldehyde) and an ester with α-H or with no α-H—it has a great synthetic value.

(a) When a ketone having α-H (CH_3COCH_3) are subjected to base promoted reaction then four products are expected (two aldol and two Claisen products) since each of the two enolates (ketone and ester enolate) reacts with each of the two carbonyl groups.

aldol product

Claisen product

Claisen product

aldol product

Pentane-2, 4-dione

In practice, one product i.e. the product from the enolate of the ketone and the carbonyl of the ester, normally predominates i.e. the product pentane-2, 4 dione.

Two factors will control the reactions: firstly two Claisen products are converted into their anions by ethoxide ion so that equilibria are favourable to the Claisen products but not to the aldol products. Secondly, acetone and pentane-2, 4-dione are more acidic than CH_3CO_2Et and ethyl aceto acetate respectively so that the formation of pentane-2, 4-dione as its anion is the predominant product.

(b) Condensation of ketone enolate and an ester with α-H or with no α-H may give directed aldol type, Claisen ester condensation.

(i) $(C_6H_5)_3\overset{\ominus}{C}$ ~ 100% OEt $(CO_2Et)_2$ CO_2Et

(ii) (i) $\overset{\ominus}{O}Et$ (ii) $(CO_2Et)_2$ CO_2Et 175°C CO_2Et

3.2.6 Intramolecular Claisen ester condensation

Intramolecular Claisen ester condensation is unique and versatile application for the synthesis of alicyclic and heterocyclic ring (when nitrogen is nucleophilic centre and ester carbonyl is electrophilic centre).

1. Self-condensation: Condensation on the diester of C_6 and C_7 occurs intramolecularly (Dieckmann condensation) to give 5 or 6-membered cyclic β-keto ester respectively.

(a) OEt CO_2Et $\overset{\ominus}{O}Et$ EtOH CO_2Et

(b) OEt CO_2Et $\overset{\ominus}{O}Et$ EtOH CO_2Et

2. Crossed condensation

(i) When two ester function each with α-H at two different environment.

OEt EtO_2C CO_2Et $\overset{\ominus}{O}Et$/EtOH EtO_2C CO_2Et

In the protic solvent EtOH, the equilibrium controlled and the thermodynamically more stable cyclopentanone derivative results.

(ii) When an ester with α-H and an ester with no α-H in the same molecule.

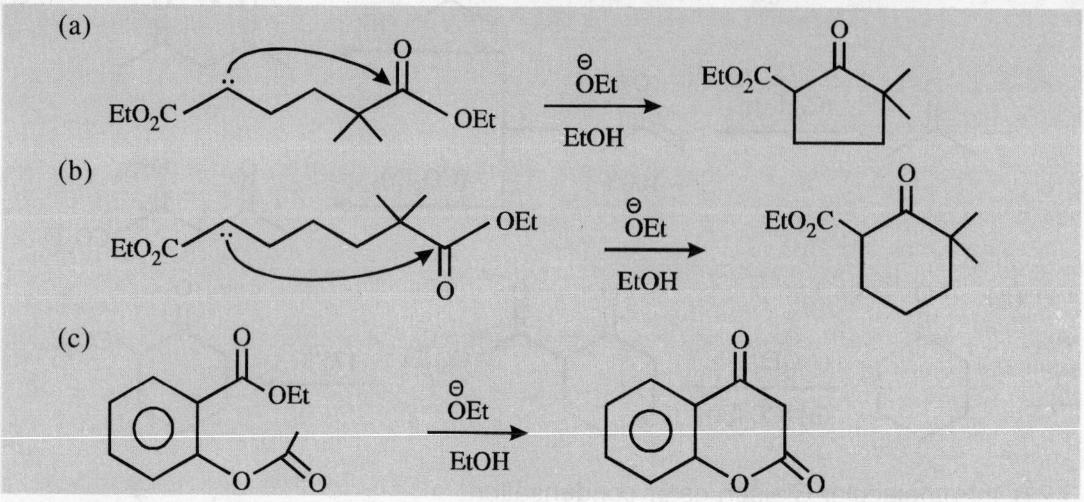
(a)
EtO2C
OEt
OEt
EtOH
EtO2C
(b)
EtO2C
OEt
OEt
EtOH
EtO2C
(c)
OEt
OEt
EtOH

(iii) Ketone enolate and ester function with α-H or with no α-H in the same molecule.

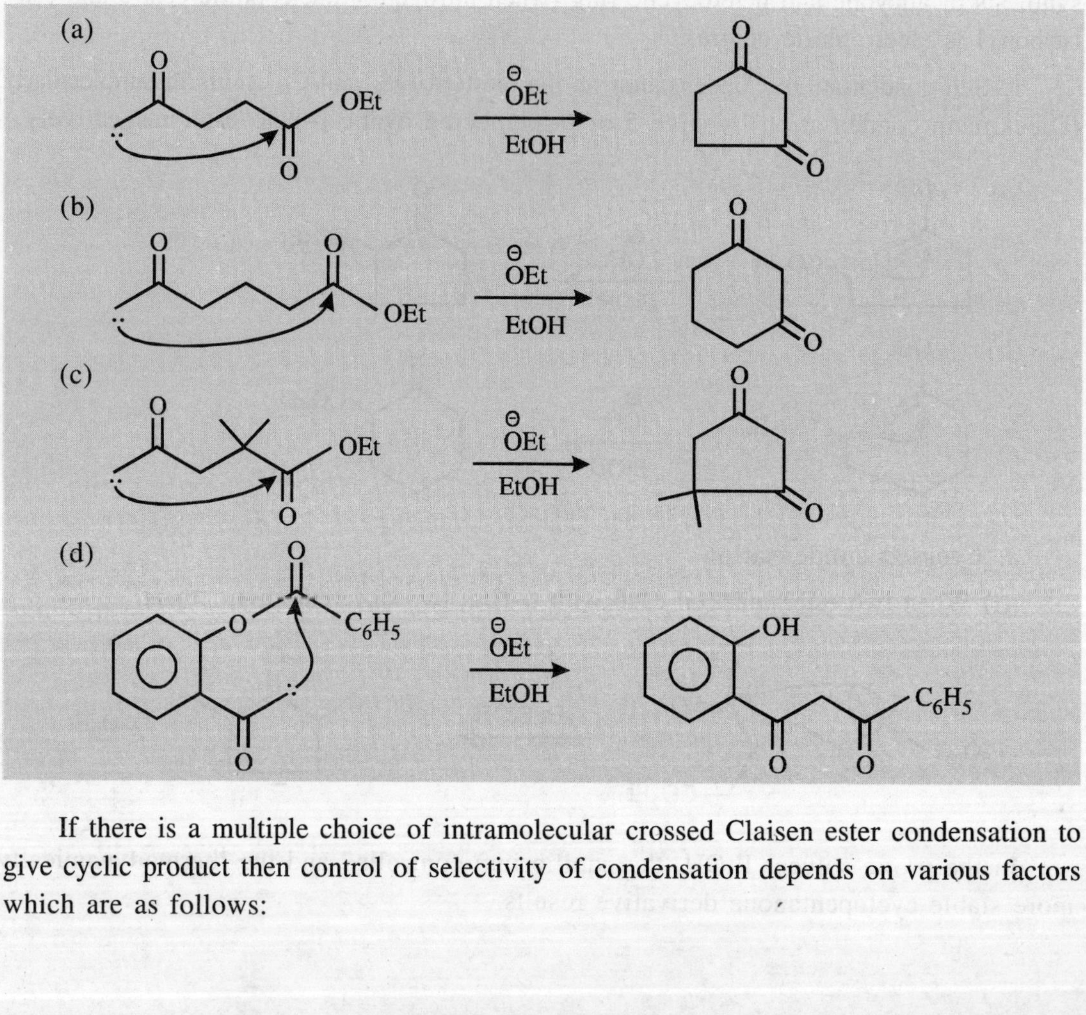
(a)
OEt
OEt
EtOH
(b)
OEt
OEt
EtOH
(c)
OEt
OEt
EtOH
(d)
C6H5
OEt
EtOH
OH
C6H5

If there is a multiple choice of intramolecular crossed Claisen ester condensation to give cyclic product then control of selectivity of condensation depends on various factors which are as follows:

(a) ring size, (b) acidity of comparable α-H of the reactant, (c) preferential stability of reactant enolate ion, (d) preferential stability of enolate ion of the cyclic product (β-keto ester or β-diketo), (e) preference of Claisen ester condensation over aldol, (f) Nature of the solvent.

Some examples are depicted as follows:

1. (a) If there is a choice between four and six membered ring then only six memberedring is the observe product since four membered ring is highly strained although both the rings are enolate ion separated.

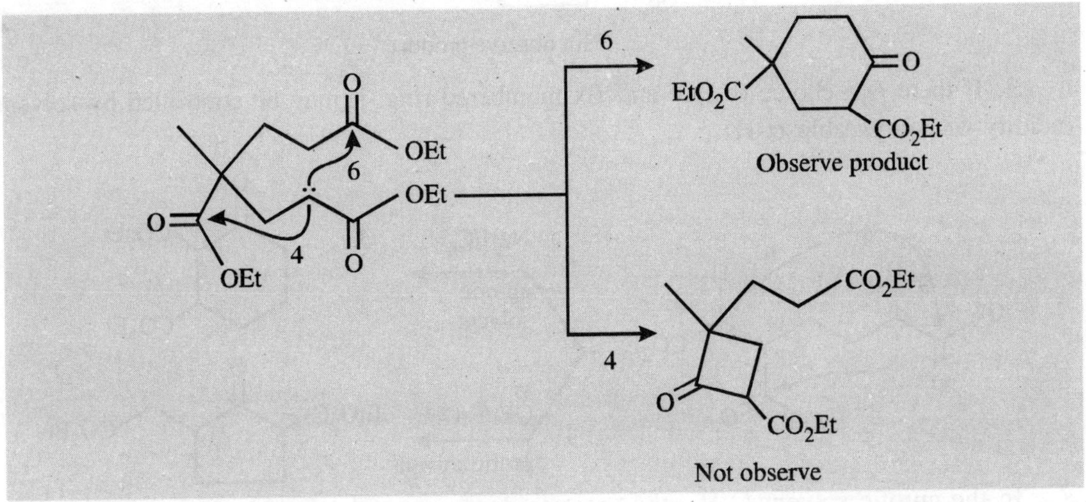

Observe product

Not observe

(b) If there is a choice between five and seven membered ring then only five membered ring is the observe product since seven membered ring is more strained although both the rings are ion separated.

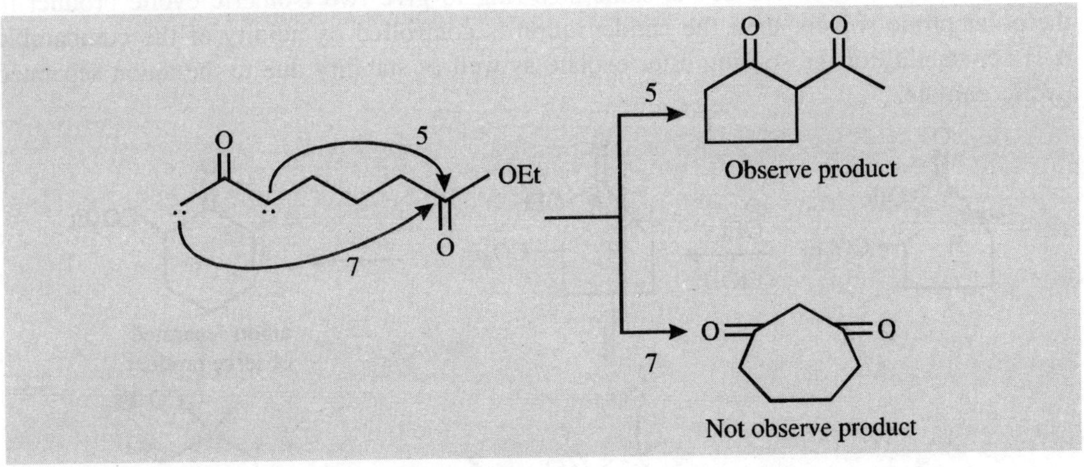

Observe product

Not observe product

(c) If there is a choice in between two five membered isomeric ring then observe-product is the one which is enolate ion separated.

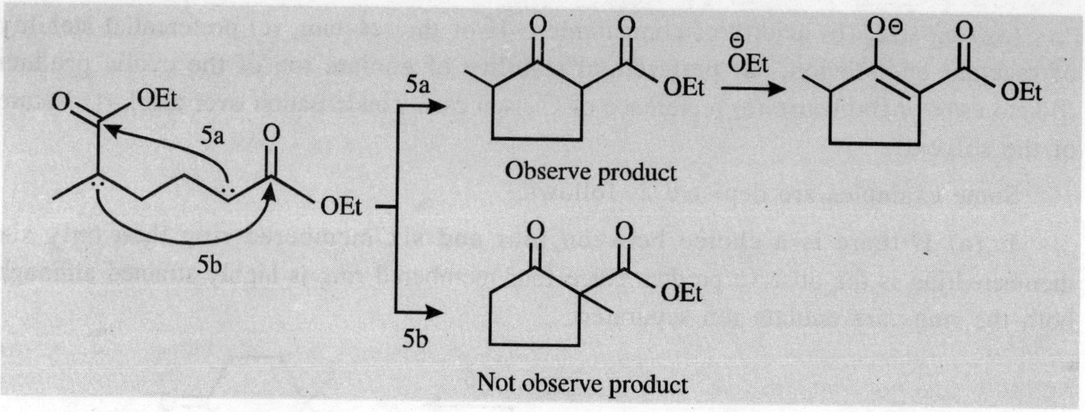

Observe product

Not observe product

2. If there is a choice of five and six membered ring, it may be controlled by solvent (acidity of comparable α-H).

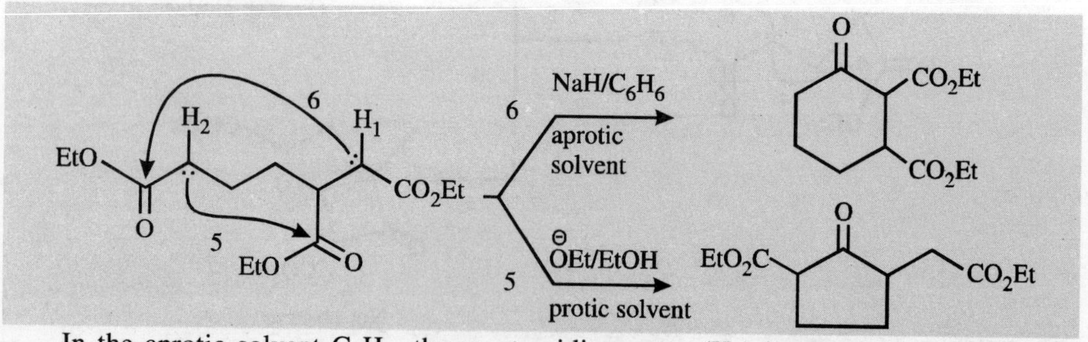

NaH/C_6H_6 aprotic solvent

$\overset{\ominus}{O}Et/EtOH$ protic solvent

In the aprotic solvent C_6H_6, the most acidic proton (H_1) is extracted preferentially, leading predominantly to the formation of a cyclohexanone derivative—a kinetically controlled product. In protic solvent EtOH, the establishment of equilibrium results in the thermodynamically stable cyclopentanone derivative.

3. If there is a choice of the same size ring to give two isomeric cyclic product in the polar protic solvent then the condensation is controlled by acidity of the comparable α-H i.e. stability of the starting ester enolate as well as stability due to the anion separated cyclic enolate.

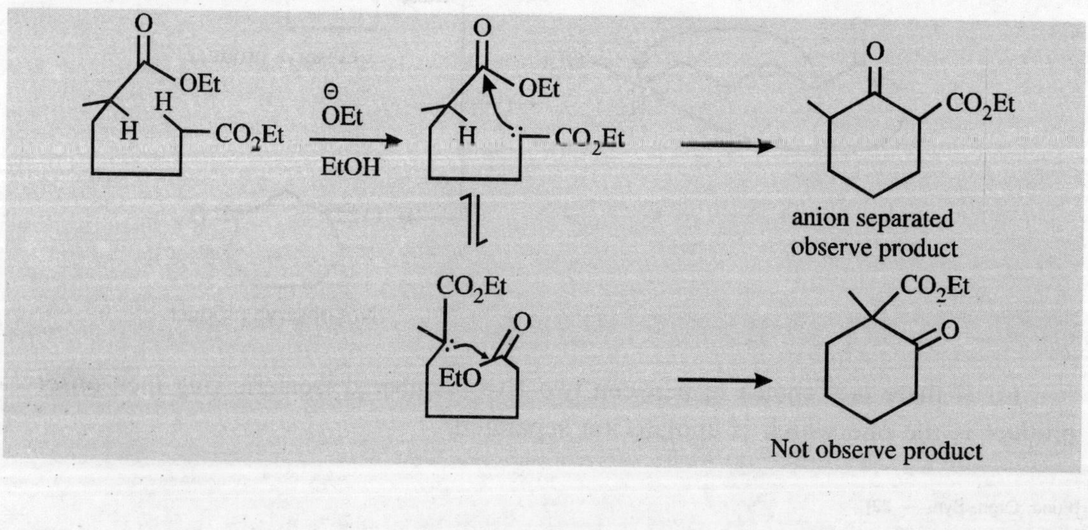

anion separated observe product

Not observe product

4. If there is a choice between intramolecular Claisen ester condensation and intra aldol condensation of the same size ring then the condensation is controlled by preferential stability of the ketone enolate (rather than ester enolate) and anion separated cyclic product is observed.

(a) $\overset{\ominus}{O}Et/EtOH$

(b) $\overset{\ominus}{O}Et/EtOH$

Preferential ketone enolate

anion separated observe cyclic product (Claisen product)

$\overset{\ominus}{O}Et/EtOH$

(aldol product)
Not observe product due to reversibility of aldol to the reactant

(c) $\overset{\ominus}{O}Et/EtOH$

observe Claisen product

$\overset{\ominus}{O}Et/EtOH$

Not observe, aldol product

However, aldol condensation may be done selectively by controlling the condensing reagent as shown below.

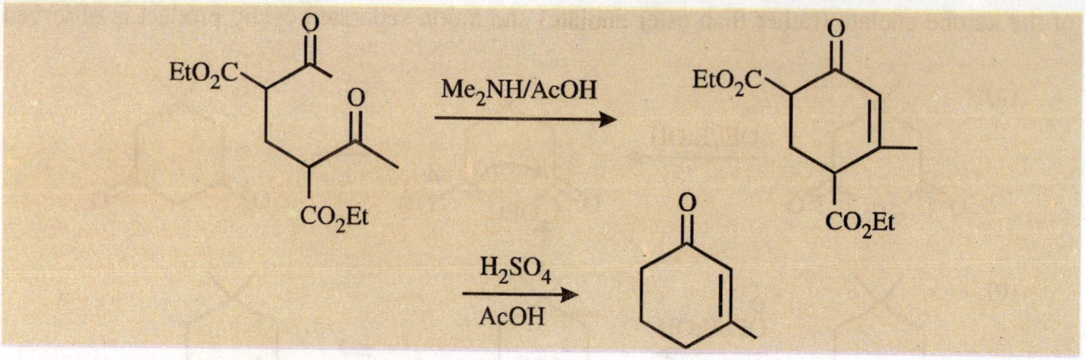

$Me_2NH/AcOH$

H_2SO_4 / AcOH

3.2.7 Definition of Claisen ester condensation

Claisen ester condensation may be defined as the condensation between enolate of either keto carbonyl or ester carbonyl (which are carbanion donor) and ester carbonyl (which is the only carbanion acceptor) inter or intramolecularly to give either acyclic or cyclic β-keto ester or β-diketo by acylation.

3.2.8 α-H containing esters as synthetic equivalent

Illustration of some α-H containing esters which are used as synthetic equivalent in the Claisen ester condensation.

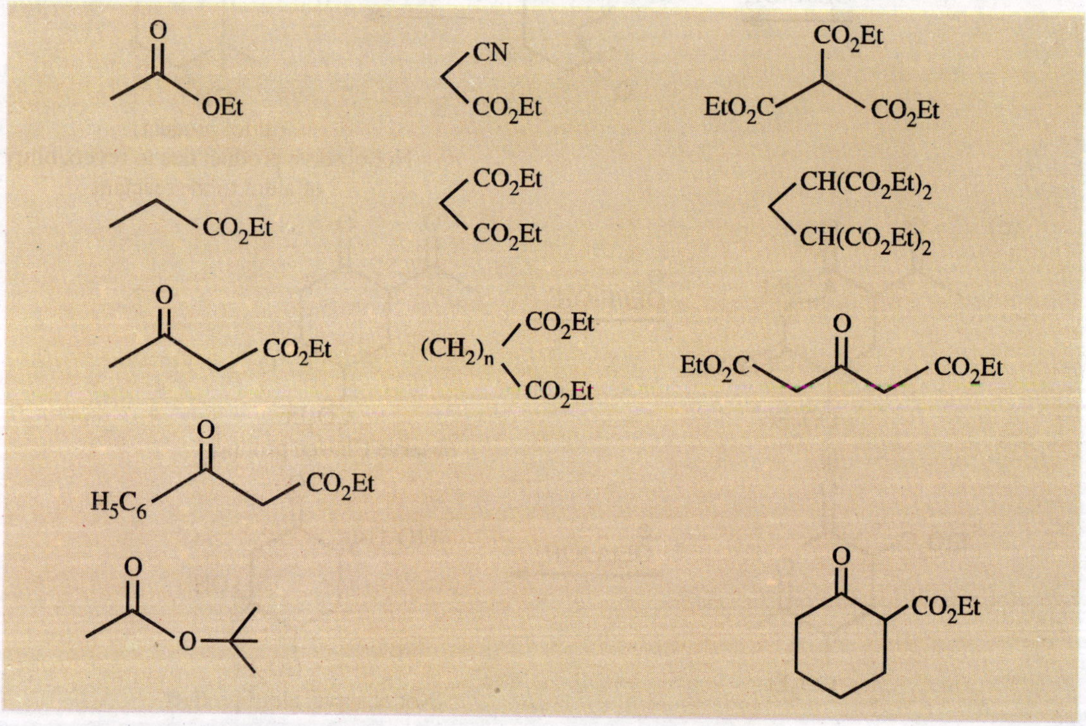

$HCOCH_2CO_2Et$

$CH_3COCH(CO_2Et)CH_2CO_2Et$

$EtO_2CCH_2COCO_2Et$

2-(ethoxalyl)cyclohexanone (CO_2Et)

$(EtO_2C)_2CH$–$CH(CH_2CO_2Et)(CO_2Et)$ [EtO$_2$C, CO$_2$Et, EtO$_2$C, CO$_2$Et]

3.2.9 Preparation of α-H containing esters

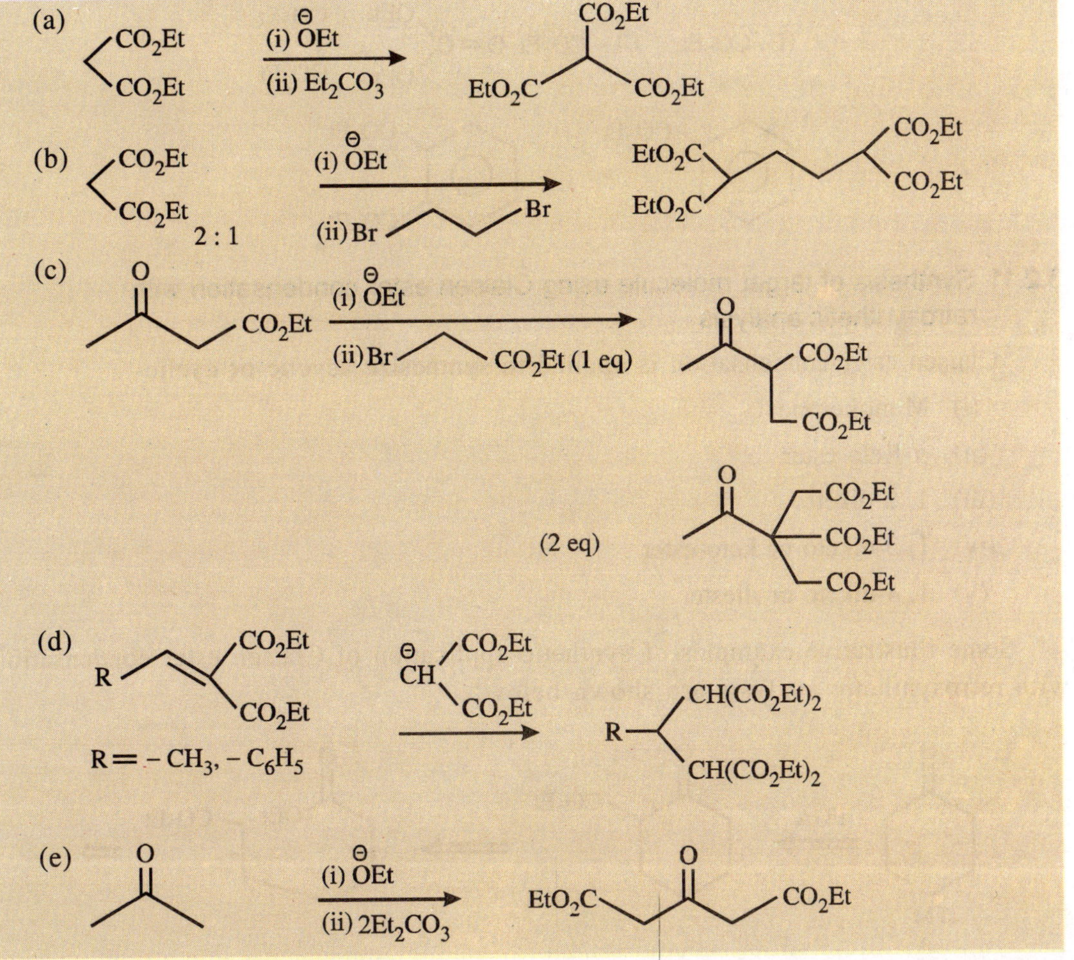

(f) Ethyl acetate $+ (CO_2Et)_2 \xrightarrow{\overset{\ominus}{O}Et}$

1 :1 X'S ⟶ $EtO_2C-CO-CH_2-CO_2Et$

2 :1 X'S ⟶ $EtO_2C-CH_2-CO-CO-CH_2-CO_2Et$

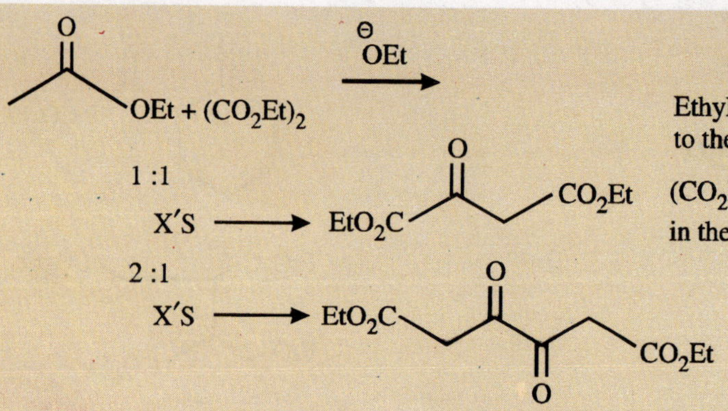

Ethylacetate is added to the mixture of $(CO_2Et)_2 + \overset{\ominus}{O}Et$ in the required ratio

(g) $EtO_2C-CH=CH-CO_2Et + CH_2(CO_2Et)_2 \xrightarrow{\overset{\ominus}{O}Et} (EtO_2C)_2CH-CH(CO_2Et)-CH_2-CO_2Et$

3.2.10 Synthetic equivalent having no α-H containing esters

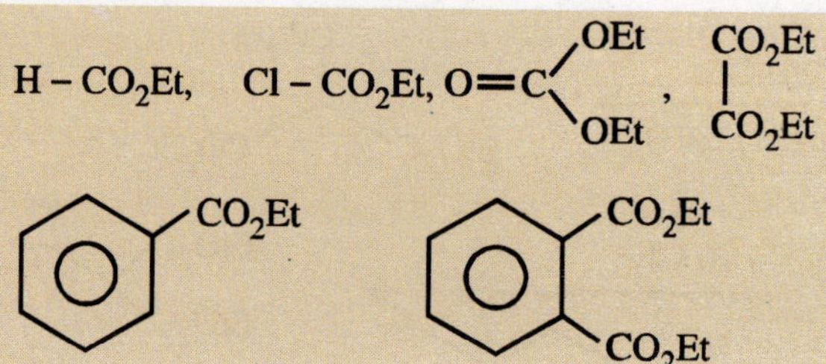

$H-CO_2Et$, $Cl-CO_2Et$, $O=C(OEt)_2$, $(CO_2Et)_2$ (diethyl oxalate), $C_6H_5CO_2Et$ (ethyl benzoate), $C_6H_4(CO_2Et)_2$ (diethyl phthalate)

3.2.11 Synthesis of target molecule using Claisen ester condensation with retrosynthetic analysis

Claisen ester condensation is applied to synthesize acyclic or cyclic

(i) Monoketone

(ii) α-Keto ester

(iii) 1, 2 diketo

(iv) 1, 3-diketo or keto-ester

(v) 1, 4-diketo or diester

Some illustrative examples of synthetic application of Claisen ester condensation with retrosynthetic analysis are shown below.

1. Cyclohexanone (TM) $\xRightarrow{FGA}$ 2-ethoxycarbonylcyclohexanone $\Rightarrow$ $EtO_2C-(CH_2)_5-CO_2Et$ (diethyl heptanedioate) $\Rightarrow$

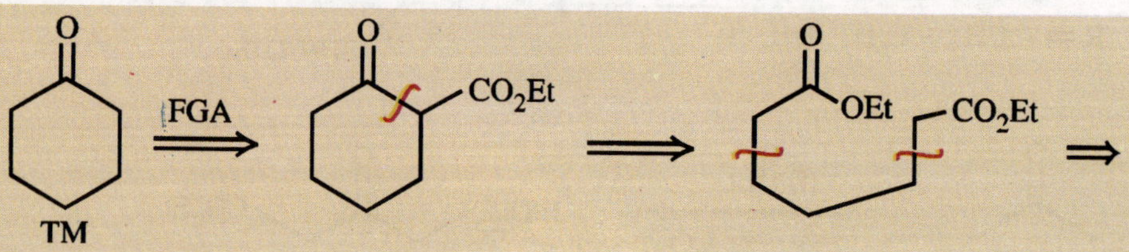

$\Rightarrow 2\overset{\ominus}{C}H_2 - CO_2Et$ + $^{\oplus}$CH$_2$CH$_2$CH$_2$$^{\oplus}$ (SN) $\Rightarrow$ Br–CH$_2$CH$_2$CH$_2$CH$_2$–Br (SE)

$\Downarrow$

2 $CH_2(CO_2Et)_2$

SE

Forward Synthesis

2 $CH_2(CO_2Et)_2$ —(i) $\overset{\ominus}{O}Et$; (ii) Br–(CH$_2$)$_4$–Br→ $(EtO_2C)_2HC$–(CH$_2$)$_4$–$CH(CO_2Et)_2$

(i) KOH
(ii) $H_3\overset{\oplus}{O}$/250°C
(iii) EtOH/$\overset{\oplus}{H}$

→ EtO_2C–(CH$_2$)$_6$–CO_2Et —$\overset{\ominus}{O}Et$, EtOH, reflux→ 2-carbethoxycyclohexanone (CO_2Et)

dil HCl/Δ → cyclohexanone

2. TM (cyclopentane-1,2-dione... 1,3-dione) $\Rightarrow$ $^{\ominus}$CH$_2$CH$_2$CH$_2$$^{\ominus}$ activation requires $\Rightarrow$ EtO_2C–(CH$_2$)$_3$–CO_2Et SE

\+ EtO–C(=O)–C(=O)–OEt SE

Forward Synthesis

EtO_2C–(CH$_2$)$_3$–CO_2Et (diethyl Glutarate) —$\overset{\ominus}{O}Et$, EtO–C(=O)–C(=O)–OEt (diethyl oxalate)→ diketone ring with EtO_2C and CO_2Et

dil. HCl / Δ ↓

cyclopentane-1,2-dione product (O, O)

Glutarate and oxalate are mixed in 1:1 ratio but oxalate should be in XS technique: in a mixture of oxalate and base, Glutarate is added dropwise

3.

O O O TM ⟹ EtO_2C O CO_2Et SE + EtO OEt O O SE

Forward Synthesis

EtO_2C O CO_2Et EtO OEt O O $\xrightarrow{\ominus OEt}$ EtO_2C O CO_2Et O O $\xrightarrow[\Delta]{\text{dilHCl}}$ O O O

4.

O O O TM $\xRightarrow{\text{FGA}}$ OH H O O O ⟹ O O + O O SE

Forward Synthesis

O O + O O $\xrightarrow[\text{EtOH}]{\ominus OEt}$ O O HO O $\xrightarrow[\text{E1cB}]{\ominus OEt}$ O O O

5.

O O TM ⟹ O OH OH $\xRightarrow{\text{FGI}}$ O H O O H $\xRightarrow{\text{FGI}}$ O CO_2H CO_2H $\Downarrow$ FGI

O CO_2Et CO_2Et $\xLeftarrow{\text{FGI}}$ O EtO_2C O O CO_2Et ⟸ O + SE $2EtO_2C-\underset{\underset{O}{\|}}{C}-OEt$ SE

Forward Synthesis

$\overset{\ominus}{O}Et$ / EtOH

EtO_2C OEt O · OEt O CO_2Et → EtO_2C O O CO_2Et ⇌ EtO_2C OH OH CO_2Et

H_2SO_4

O

6. TM ⟹ OEt OEt + EtO_2C SE CO_2Et ⟹ CH_3I + OEt OEt SE

Forward Synthesis

EtO_2C CO_2Et (i) $\overset{\ominus}{O}Et$ (ii) CH_3 I → OEt OEt EtO_2C CO_2Et $\overset{\ominus}{O}Et$ → EtO_2C CO_2Et

dilHCl/Δ

7. CO_2H O TM FGI ⟹ CO_2Et O FGA ⟹ CO_2Et CO_2Et O

⇓

CO_2Et + OEt CO_2Et O

SE SE

Forward Synthesis

CO_2Et (i) $\overset{\ominus}{O}Et$ (ii) CO_2Et–CO_2Et → CO_2Et, CO_2Et (i) KOH (ii) $H_3\overset{\oplus}{O}$ → CO_2H, COOH, O

HCl/Δ → COOH, O

Selectively β-keto acid decarboxylated

8. TM (O, O) $\xRightarrow{\text{FGA}}$ EtO_2C, CO_2Et, O, O ⇒ CO_2Et CO_2Et + EtO, OEt, O, O SE

⇓

CO_2Et, CO_2Et SE ⇐ $\overset{\ominus}{C}H_2-CO_2Et$ + OEt, O SE

Forward Synthesis

CO_2Et, CO_2Et (i) $\overset{\ominus}{O}Et$ (ii) OEt, O (iii) $H_3\overset{\oplus}{O}$ → EtO_2C, CO_2Et CO_2Et

(i) KOH (ii) HCl/Δ → HO_2C, CO_2H

$EtOH/\overset{\oplus}{H}$ → EtO_2C, CO_2Et + EtO, OEt, O, O

(not linear combination of the α-H containing ester because of entropy advantage for the cyclic product)

(i) $\overset{\ominus}{O}Et$ (ii) $(CO_2Et)_2$ → EtO_2C, CO_2Et, O, O

dilHCl/Δ → O, O

9.

FGA ⟹ EtO_2C ... CO_2Et ⟹ EtO_2C ... CO_2Et, EtO_2C, CO_2Et

TM

EtO_2C ... Br + EtO_2C ... CO_2Et (SE) + Br ... CO_2Et

SE SE

Forward Synthesis

SE —HBr / EtOH→ Br ... CO_2Et

CO_2Et / CO_2Et —(i) $\overset{\ominus}{O}Et$ (ii) 2eq. Br ... CO_2Et→ EtO_2C ... CO_2Et, EtO_2C, CO_2Et

—$\overset{\ominus}{O}Et$→ EtO_2C ... CO_2Et —dilHCl/Δ→

10.

CO_2Et ⟹ CO_2Et, CO_2Et (SE) + $CH_3\ CO_2Et$ (SE)

TM

Forward Synthesis

OEt, OEt + $\overset{\ominus}{O}Et$ + $CH_3\ CO_2Et$ ⟶ CO_2Et

1 :1 ethyl Phthalate X'S
i,e ethyl acetate (1 mole) is added slowly over the mixture of ethyl Phthalate (1 mole) and base

11. TM (1,2-diphenylpyrazolidine-3,5-dione) ⟹ diethyl malonate (SE) + H_5C_6–NH–NH–C_6H_5 (SE)

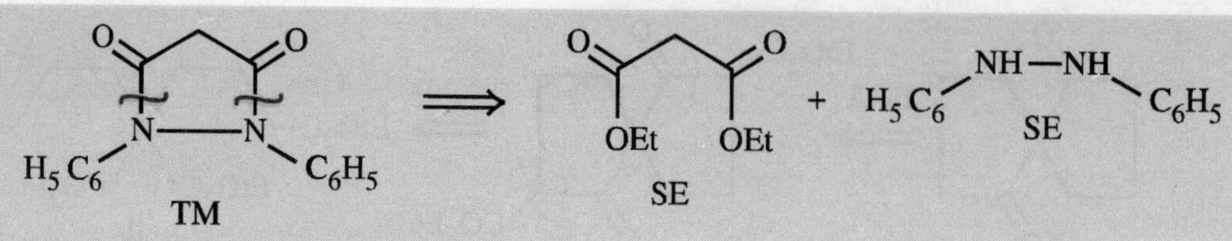

Forward Synthesis

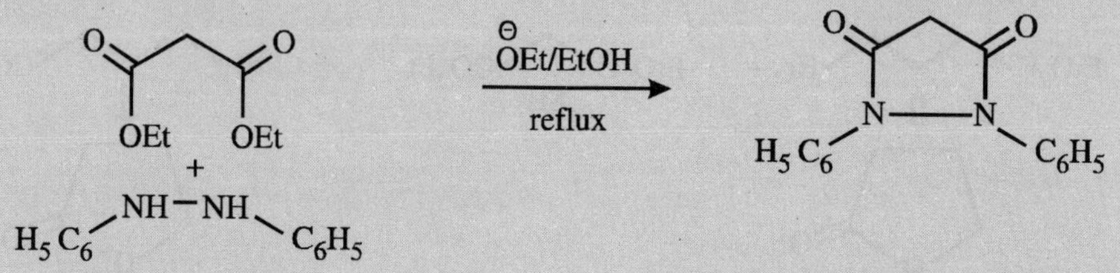

Diethyl malonate + H_5C_6–NH–NH–C_6H_5 $\xrightarrow[\text{reflux}]{\overset{\ominus}{O}Et/EtOH}$ 1,2-diphenylpyrazolidine-3,5-dione

12. (TM) $\overset{FGI}{\Longrightarrow}$ ketone $\overset{FGA}{\Longrightarrow}$ β-keto ester (CO_2Et) ⟹ N-(CH_2CO_2Et) piperidine-4-CO_2Et ⟹ Br–CH_2–CO_2Et (SE) + piperidine-4-CO_2Et (SE)

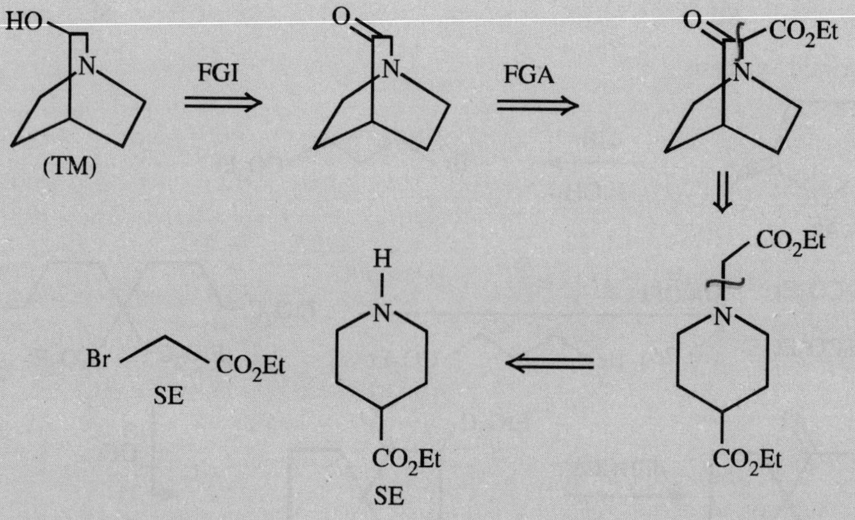

Forward Synthesis

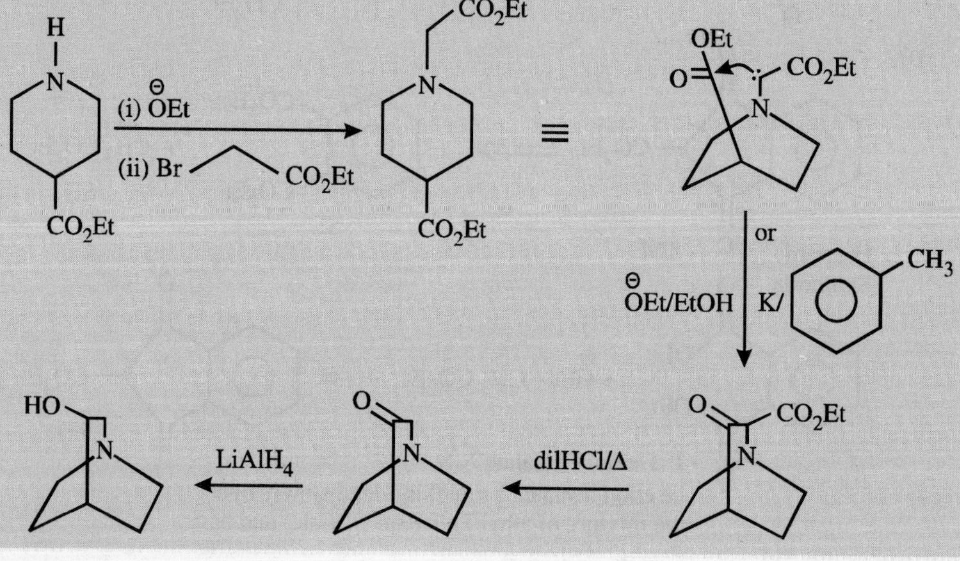

Piperidine-4-CO_2Et $\xrightarrow[\text{(ii) Br–}CH_2\text{–}CO_2Et]{\text{(i) }\overset{\ominus}{O}Et}$ N-(CH_2CO_2Et) piperidine-4-CO_2Et ≡ (OEt, O=, CO_2Et cyclization form)

$\overset{\ominus}{O}Et/EtOH$ or K/toluene (CH_3) ⟶ β-keto ester (CO_2Et) $\xrightarrow{\text{dilHCl}/\Delta}$ ketone $\xrightarrow{LiAlH_4}$ HO– (TM)

13.

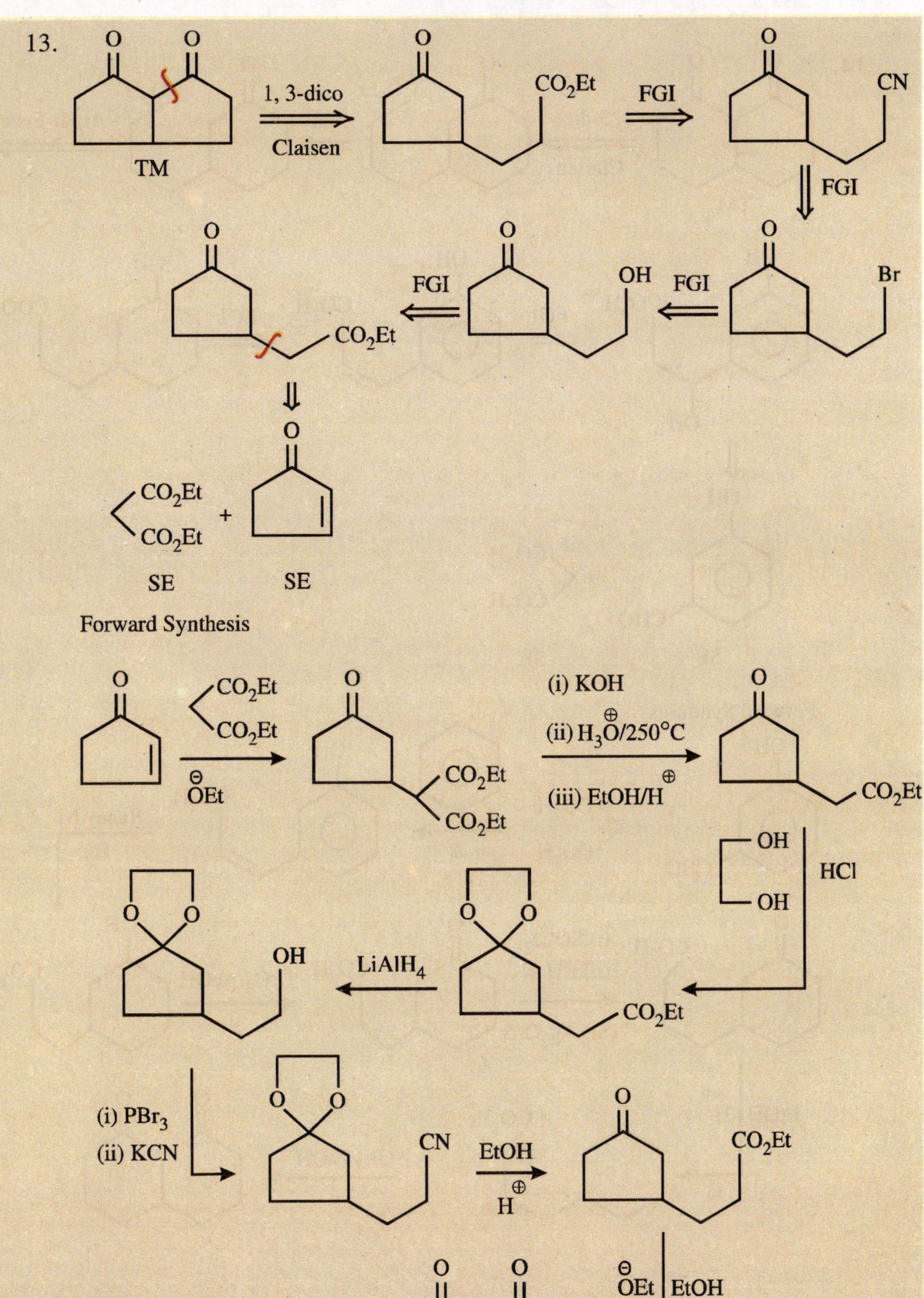
O
O
TM
1, 3-dico
Claisen
CO_2Et
FGI
CN
FGI
Br
FGI
OH
FGI
CO_2Et
CO_2Et
CO_2Et
+
SE
SE
Forward Synthesis
CO_2Et
CO_2Et
$\overset{\ominus}{O}Et$
CO_2Et
CO_2Et
(i) KOH
(ii) $H_3\overset{\oplus}{O}$/250°C
(iii) EtOH/$\overset{\oplus}{H}$
CO_2Et
OH
OH
HCl
OH
$LiAlH_4$
CO_2Et
(i) PBr_3
(ii) KCN
CN
EtOH
$\overset{\oplus}{H}$
CO_2Et
$\overset{\ominus}{O}Et$
EtOH

14.

TM

1, 3-dico Claisen ⟹ FGI ⟹ Arndt-Eistert ⟹

CO_2Et CO_2H

OH CO_2H FGI ⟸ OH CO_2H FGI ⟸ OH COOH

OH

⇓

OH CHO + CO_2H CO_2H

SE SE

Forward Synthesis

OH CHO + CO_2H CO_2H —(pyridine, Δ)→ OH CO_2H —(Raney Ni, H_2)→ OH CO_2H

—(CrO_3/AcOH)→ CO_2H —((i) $SOCl_2$ (ii) CH_2N_2 (iii) Ag_2O/Δ)→ CO_2H

—(EtOH, $H^{\oplus}$)→ CO_2Et —($\overset{\ominus}{O}Et$/EtOH)→

3.1.12 Review synthetic problems

After proper retrosynthetic analysis show the forward synthesis of the following target molecules using Claisen ester condensation.

1. 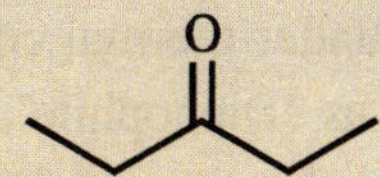 2. 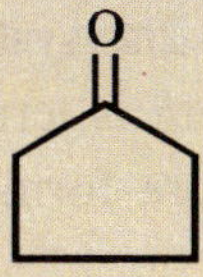 3. 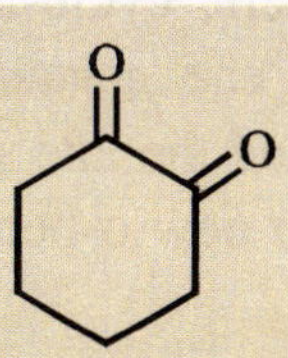 4. 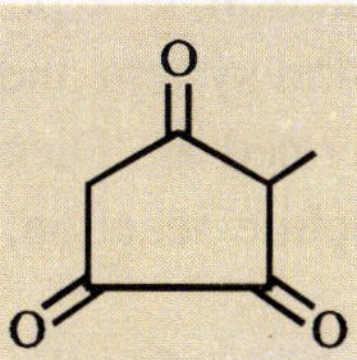 5. 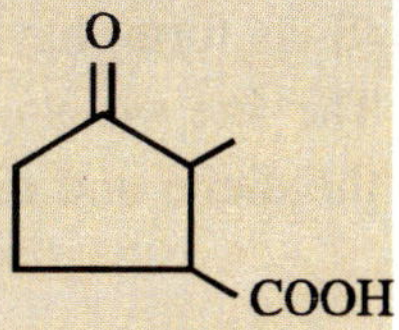 COOH

6. 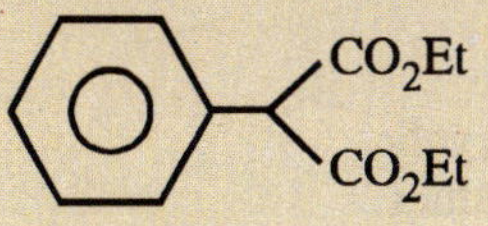 CO_2Et CO_2Et 7. 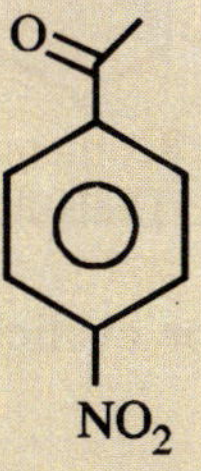 NO_2 8. 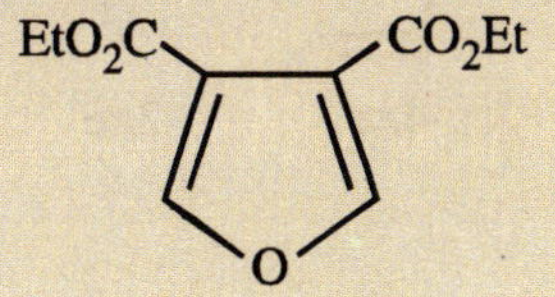 EtO_2C CO_2Et O 9.

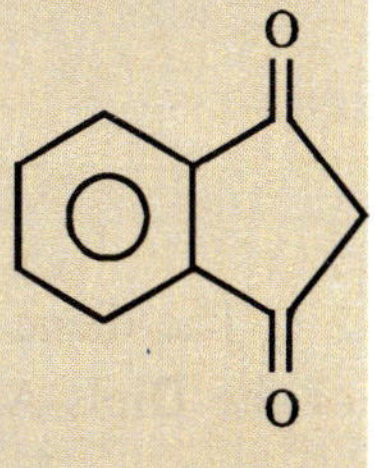

10. 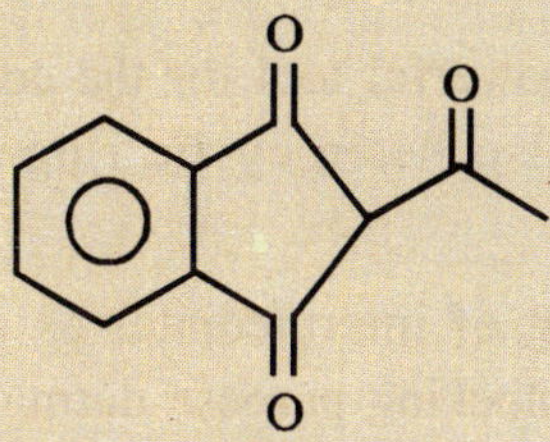 11. 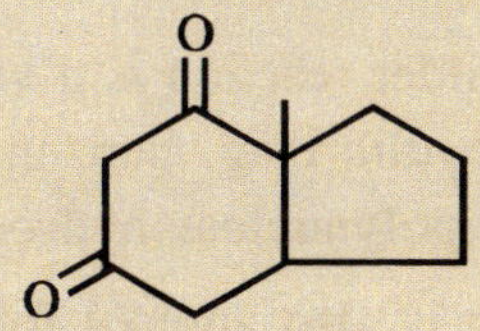 12. 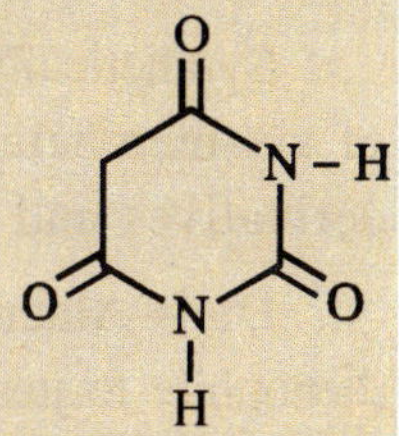 N–H N H

13. 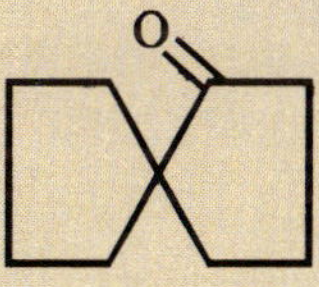 14. 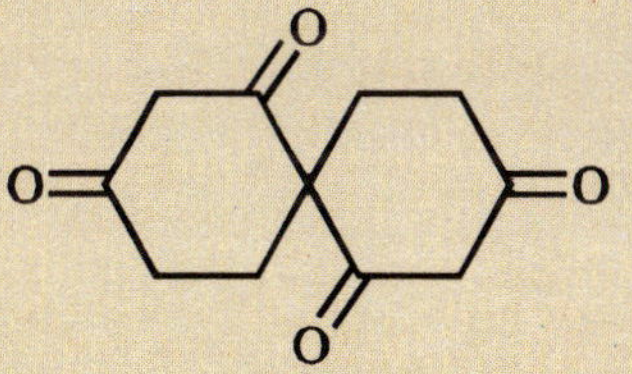 15.

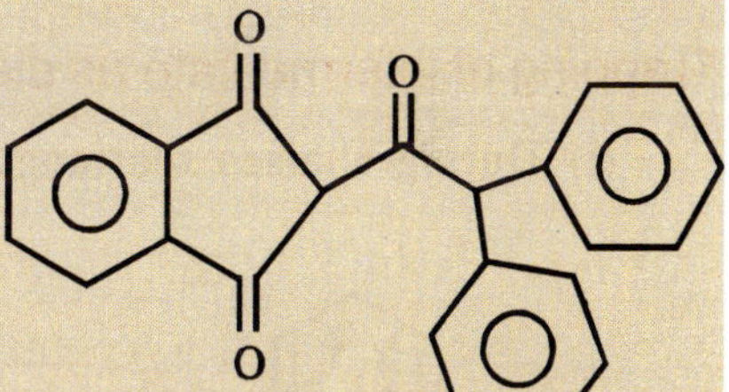

16. 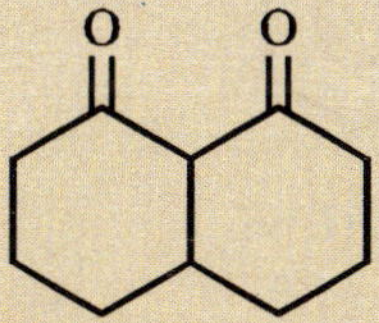 17. 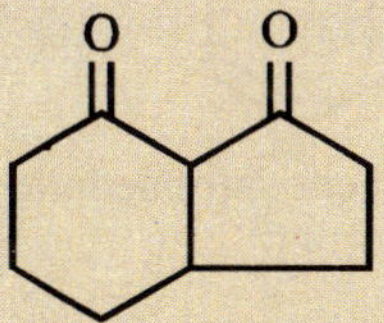 18.

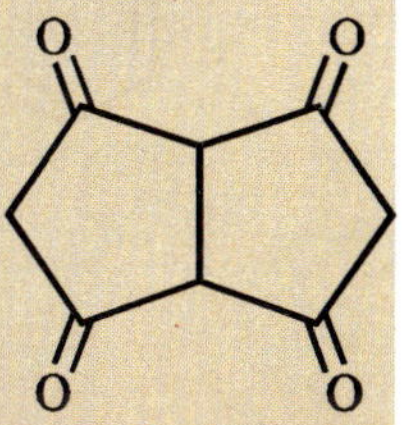

19. 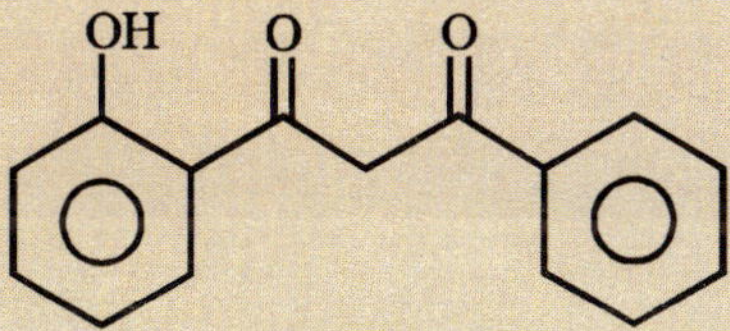 OH 20. 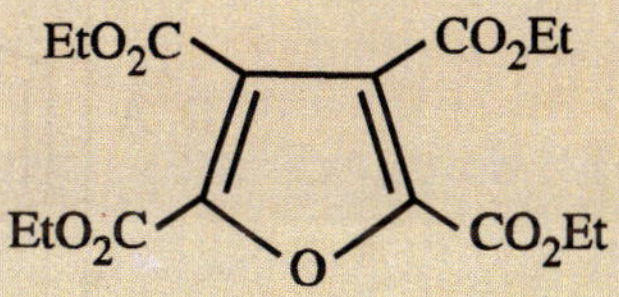 EtO_2C CO_2Et EtO_2C O CO_2Et

3.3 Diels-Alder reaction

3.3.1 What is Diels-Alder reaction?

Diels-Alder reaction is a pericyclic cyclo-addition reaction between '1, 3-diene' (S-cis form) having 4πe system and an 'ene' containing 2πe system under thermal condition. The 4πe system is termed as 'Diene' and the 2πe system is termed as 'Dienophile'. During the diene and dienophile reaction, two new σ-bonds are formed at the cost of 2π bonds.

4πe 2πe Δ Six membered cyclic Diels-Alder adduct

The diene and dienophile may be acyclic or cyclic and these are synthetic equivalent of the Diels-Alder reaction.

3.3.2 Scope of the Diels-Alder reaction

Synthetically Diels-Alder reaction is a versatile and powerful tool for the construction of six-membered carbocyclic ring, benzene ring and also heterocyclic ring and their derivatives and also for polynuclear hydrocarbons.

Diels-Alder reaction is also applied for the trapping of intermediates as diene and dienophile to justify the reaction mechanism and also for blocking purpose during synthesis or conversion, as for example.

Trapping of intermediate as diene

(i) During Claisen rearrangement

O Δ O OH

diene intermediate

D. A O O O

$\rightarrow$ ≡

D. A. adduct

(ii) During Elb's reaction

300°C

diene intermediate

D. A $EtO_2C-C\equiv C-CO_2Et$

$-H_2O$

D. A. adduct

$-H_2O$

Anthracene

Trapping of intermediate as dienophile

(i) During 'benzyne intermediate' via reaction

OMe Br $\overset{\ominus}{NH_2}$ / NH_3 → OMe dienophile intermediate NH_3 → OMe NH_2

O OMe D. A adduct

(ii) During Favorskii reaction

O H_5C_6 C_6H_5 Cl H $\overset{\ominus}{OH}$ → O H_5C_6 C_6H_5 Cl

O H_5C_6 C_6H_5 dienophile intermediate (i) $\overset{\ominus}{OH}$ (ii) $H_3\overset{\oplus}{O}$ → H_5C_6 CO_2H C_6H_5

O C_6H_5 O D. A adduct C_6H_5 O

As blocking purpose during synthesis or conversion

O O → O O O

(1 : 1)

D. A

$\overset{\ominus}{OH}/H_2O_2$

Δ

Retro D. A.

blocked

deblocked

3.3.3 Nature of the diene and dienophile in the Diels-Alder reaction

(i) The presence of electron donating groups ($-CH_3$, $-OMe$, $-NMe_2$, etc.) on the diene part and electron withdrawing groups ($-CN$, $-CO_2Et$, $-CHO$, $-NO_2$, $-NO$, etc.) on the dienophile part favour the Diels-Alder reaction. Cyclic diene is more favoured for Diels-Alder reaction than acyclic diene because of 100% S-cis form in the cyclic diene.

3.3.4 Stereochemistry of Diels-Alder reaction

(a) Since DA reaction is a pericyclic reaction and therefore it is one step concerted process. DA reaction is thermally allowed process and both the components interact suprafacially in the TS and hence it has got cis-stereochemistry. The importance of DA reaction synthetically, is because of its high stereo specificity.

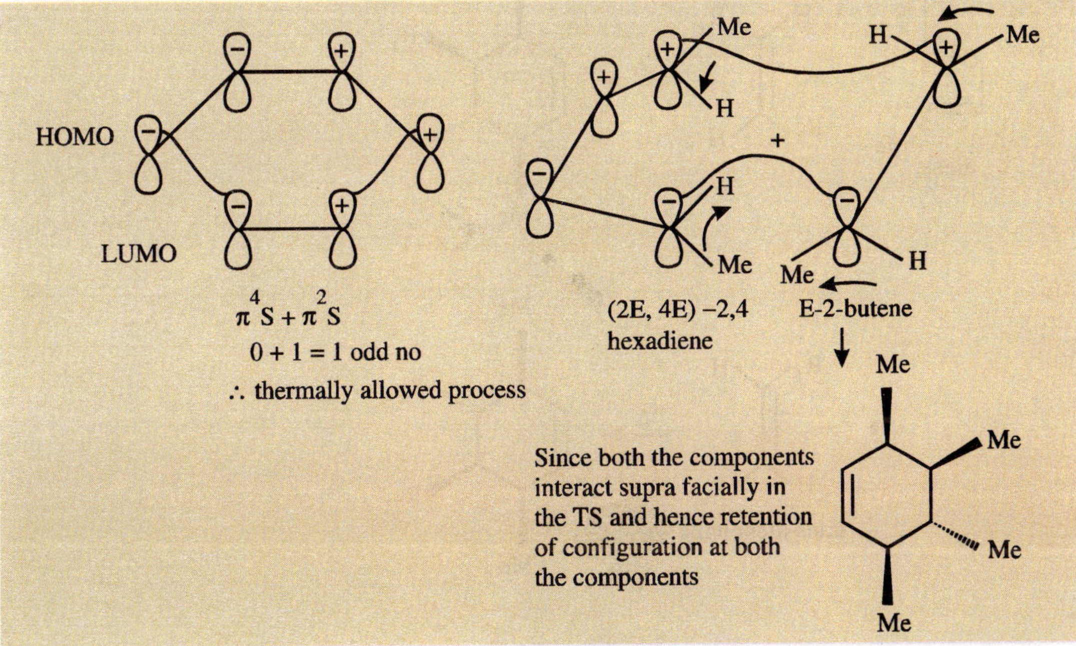

Substituated dienophile may add to the diene in two ways and the stereoselectivity is controlled by steric crowding in the TS.

Me
R_1 H
+ Δ →
R_2 H
Me
diene
Me R_1 R_2 Me
Major (Sterically less crowded TS)

H R_1
+ Δ →
H R_2
cis-dienophile
Me R_1 R_2 Me
Minor (Sterically highly crowded TS since four alkyl groups are in the same phase in the TS)

Me
H R_1
+ Δ →
R_2 H
Me
diene
Me R_1 R_2 Me

R_1 H
+ Δ →
H R_2
trans-dienophile
Me R_1 R_2 Me

(b) Diels-Alder reaction gives kinetically versus thermodynamically controlled product.

furan + maleimide → (25°C) endo product (KCP) (Major) → (90°C (via reactant)) exo product (TCP) (Major); furan + maleimide → (90°C) exo product (TCP) (Major)

In the Diels-Alder reaction by appropriate choice of starting materials, it is often possible in one step to assemble a product with several groupings in the required stereo chemical configuration.

When furan is treated with maleic anhydride, exo product is exclusive even at low temperature.

furan + maleic anhydride → (35°C) exo exclusive

Endo product is formed 500 times faster than the exo-product but it reverts to the reactants about 10,000 times faster and exo-product accumulates.

3.3.5 Regioselectivity of the Diels-Alder reaction

Regio chemistry is a very important part in the Diels-Alder reaction to the synthetic procedure. If the diene and dienophile are unsymmetrical then they may interact in two ways to give two structurally isomeric products (constitutional isomers) which are regioisomers. It is found in practice that formation of one of the isomers is strongly favoured. To understand the orientations, the electron rich and deficient centres in the diene and dienophile are to be located which will interact preferably. Some examples are given below.

(a)

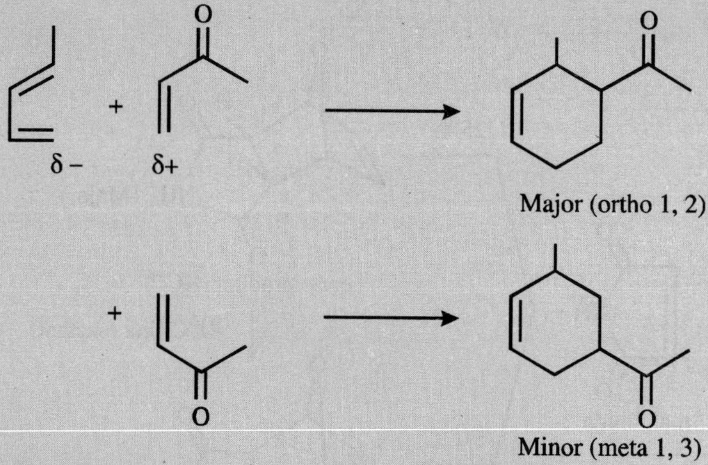

δ− δ+

Major (ortho 1, 2)

Minor (meta 1, 3)

(b)

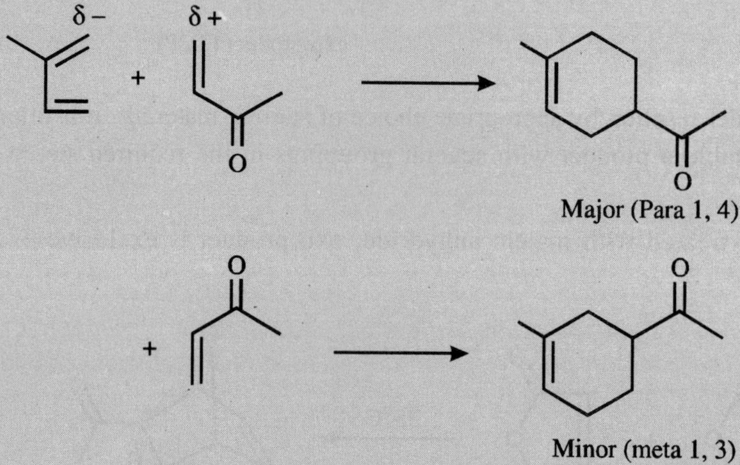

δ− δ+

Major (Para 1, 4)

Minor (meta 1, 3)

(c)

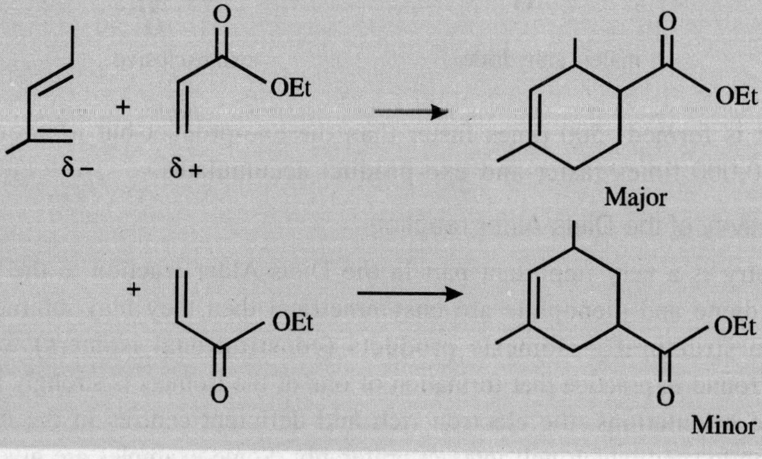

δ− δ+ OEt

Major

Minor

(d)

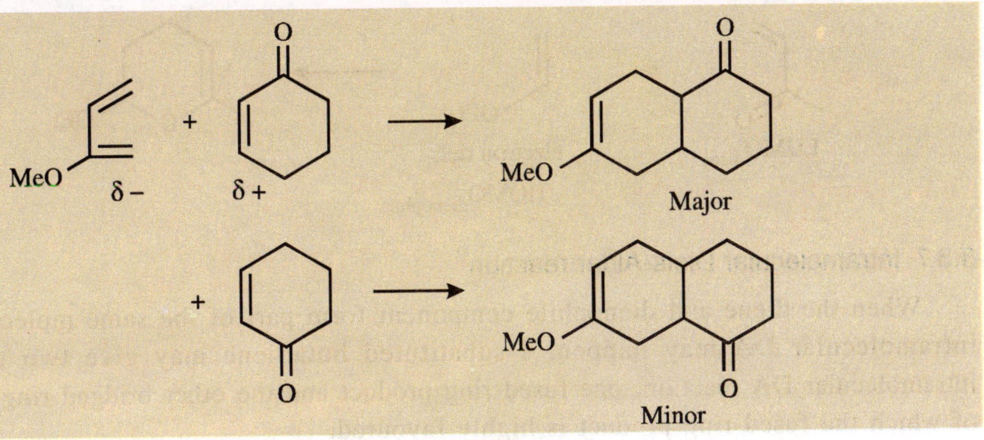

MeO δ− + δ+ → MeO Major

+ → MeO Minor

3.3.6 Inverse electron demand Diels-Alder reaction

Usually E_D group in the diene (which is HOMO $_{diene}$) and E_W group in the dienophile (which is LUMO $_{dienophile}$) interact preferentially in the Diels-Alder reaction. The opposite effect i.e. the LUMO $_{diene}$—HOMO $_{dienophile}$ may interact preferably with dienes bearing an E_W group and dienophile bearing an E_D group—this is the inverse electron demand DA reaction. Two examples are cited below.

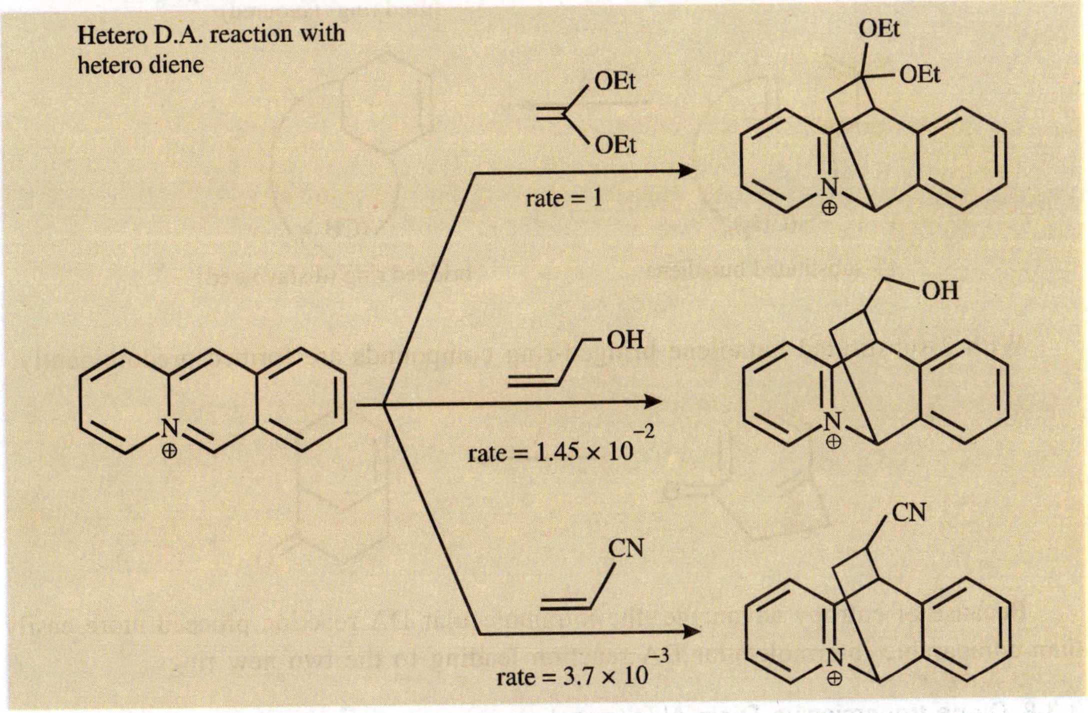

Hetero D.A. reaction with hetero diene

rate = 1

rate = 1.45×10^{-2}

rate = 3.7×10^{-3}

Controlled by LUMO of the diene and HOMO of the dienophile. An electron rich dienophile is reacting faster than electron poor dienophile.

+ OEt → O OEt
O
$LUMO_{diene}$ electron rich $HOMO_{dienophile}$

3.3.7 Intramolecular Diels-Alder reaction

When the diene and dienophile component form part of the same molecule then intramolecular DA may happen. 1-substituted butadiene may give two types of intramolecular DA reaction, one fused ring product and the other bridged ring product of which the fused ring product is highly favoured.

$(CH_2)_n$ Δ → $(CH_2)_n$
fused ring (favoured)

$(CH_2)_n$ Δ → $(CH_2)_n$
1-substituted butadiene bridged ring (disfavoured)

With 2-substituted butadiene bridged-ring compounds are formed predominantly.

O Δ → O

Because of entropy advantage, the intramolecular DA reaction proceed more easily than comparable intermolecular DA reaction leading to the two new rings.

3.3.8 Diene-transmissive Diels-Alder reaction

When a diene reacts with two dienophiles (similar or different) to form a cyclic product in the two stages or simply consecutive double Diels-Alder reaction is the diene-transmissive Diels-Alder reaction. Examples are given below.

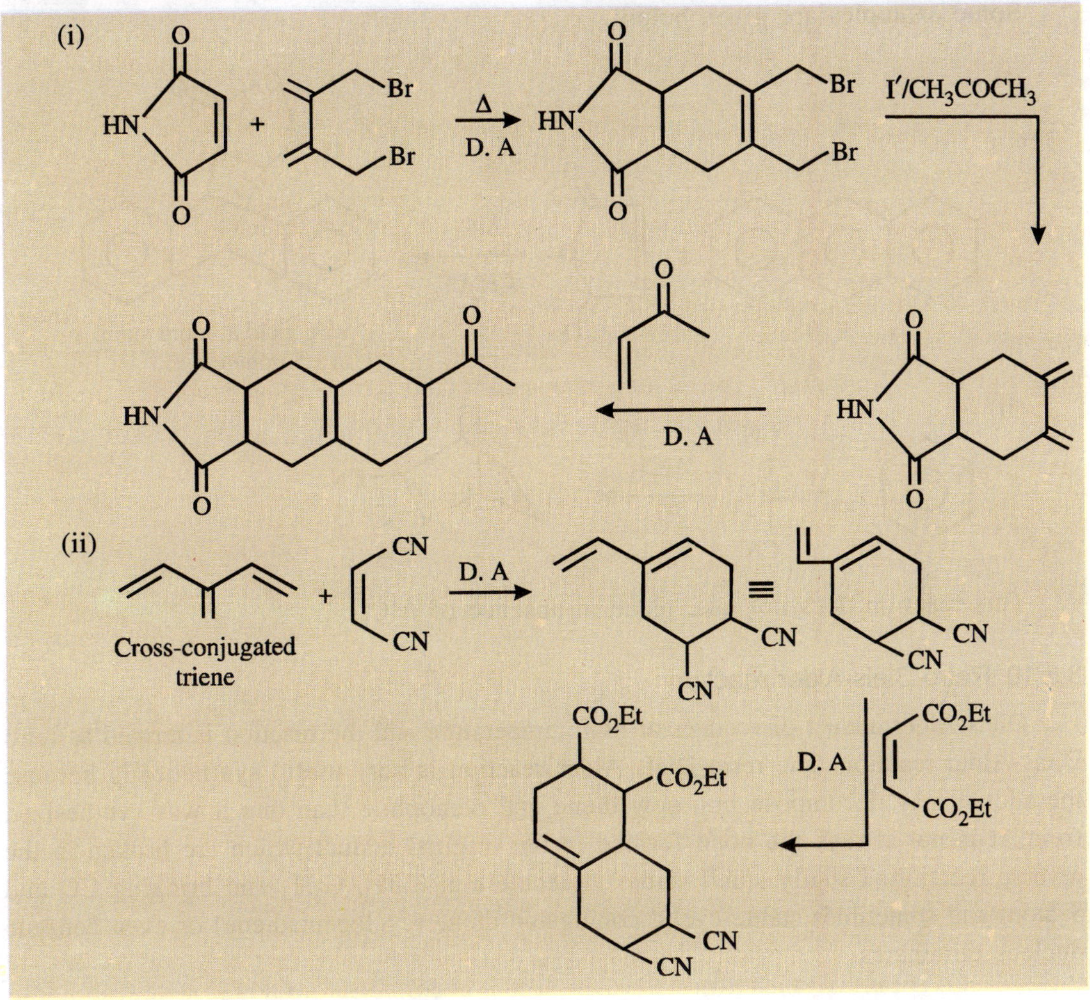

(i)

(ii)

Cross-conjugated triene

3.3.9 Catalysis of Diels-Alder reaction

Diels-Alder reaction may be catalysed with dienophiles particularly containing oxygen functional groups with the molar equivalent of anhydrous $AlCl_3$. The reason for catalysis is as follows.

(i) to accelerate the rate of the DA reaction or the reaction which is very sluggish in the absence of catalyst.

(ii) to run the reaction at a low temperature.

(iii) to increase the yield of the DA adduct.

The reason for acceleration of the rate of the DA reaction is may be due to, dienophile becomes complexed with Lewis acid catalyst which is more electrophilic and thus more reactive.

According to Frontier molecular orbital (FMO) theory, the energy of complexed dienophile comes down and consequently energy gap between $HOMO_{diene}$ and $LUMO_{dienophile}$ decreased and consequently rate increases.

Some examples are given below:

(i)

$AlCl_3$, CH_2Cl_2

95% yield at room temp. in 1.5 minutes

(ii)

CN, CN, $AlCl_3$, CN, CN

This reaction does not take place in absence of $AlCl_3$.

3.3.10 Retro Diels-Alder reaction

Diels-Alder adduct dissociates at high temperature and the reaction is termed as retro Diels-Alder reaction. The retro Diels-Alder reaction is very useful synthetically because the adduct may decompose into new diene and dienophile than one it was synthesized from. It is not always the bond formed in the original adduct which are broken in the reverse reaction. Usually small stable molecule e.g. C_2H_2, C_2H_4 and bridging CO and S atom and sometimes stable cyclic conjugated diene (cyclopentadiene) or even benzene nucleus eliminates.

(i)

CO_2Et, CO_2Et; Δ D. A; CO_2Et, CO_2Et; Δ retro D. A; CO_2Et, CO_2Et; + C_2H_2

(ii)

CO_2Et, CO_2Et; Δ D. A; CO_2Et, CO_2Et; Δ retro D. A; CO_2Et, CO_2Et; + S

Out of the three hetero atoms O, N, S which remain as bridging atom, only 'S' atom eliminates during retro process but O and N bonds are very stable and they do not break during retro process and this technique is taken as advantage for the synthesis of various hetero cyclic ring derivatives.

3.3.11 Some examples of Diels-Alder dienes

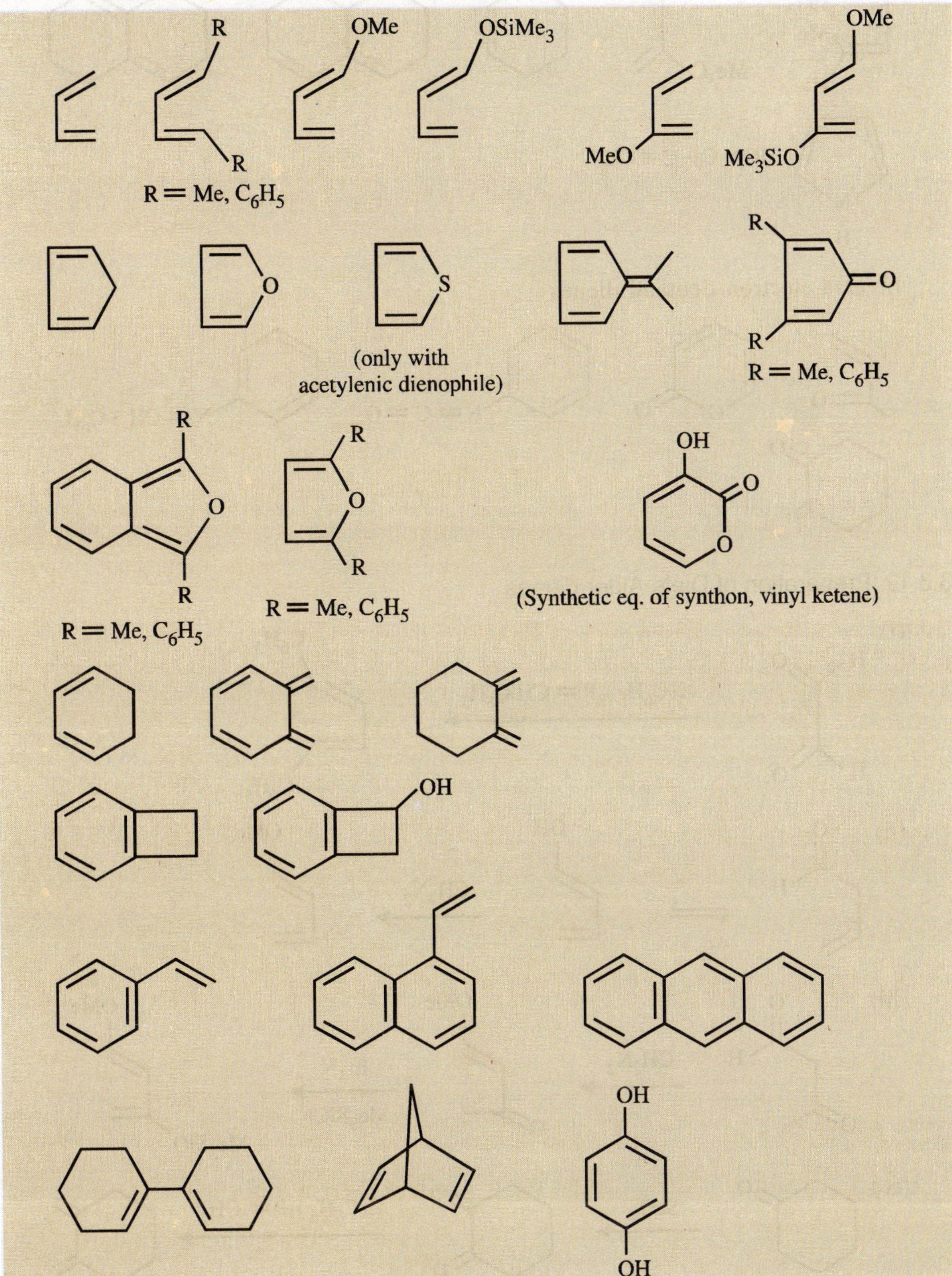

A few examples of Diels-Alder 'negative' dienes (which don't act as D.A. dienes).

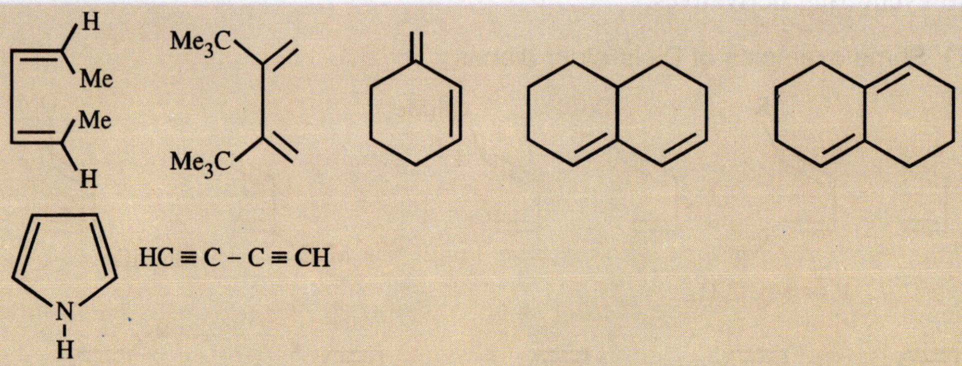

H Me Me H Me_3C Me_3C

N H   HC ≡ C – C ≡ CH

Inverse electron demand dienes

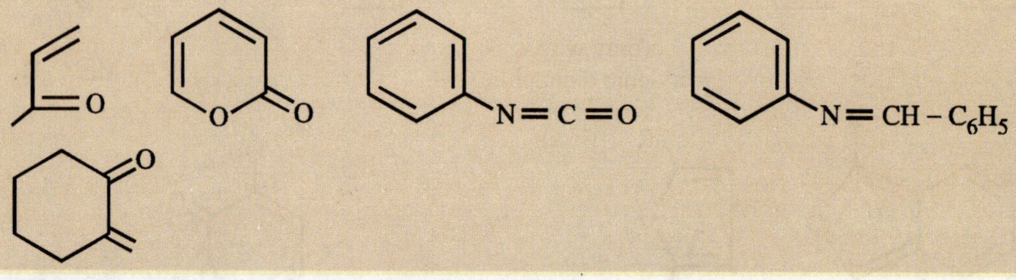

O   O O   N = C = O   N = CH – C_6H_5

O

3.3.12 Preparation of Diels-Alder dienes

(i) H O H O $2(C_6H_5)_3P = CH\,C_6H_5$ → C_6H_5 C_6H_5

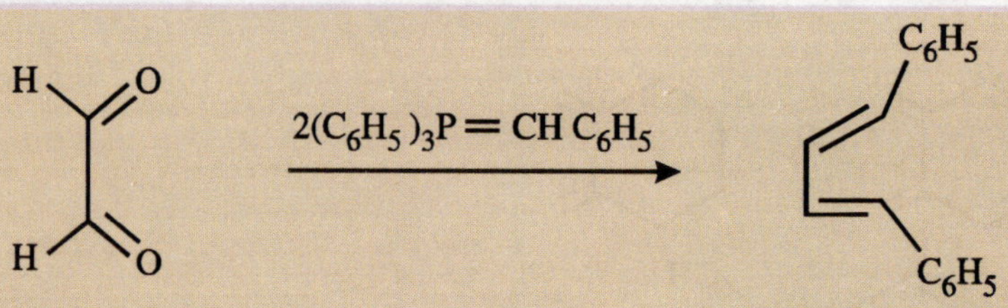

(ii) O H ⇌ OH CH_2N_2 →   OMe

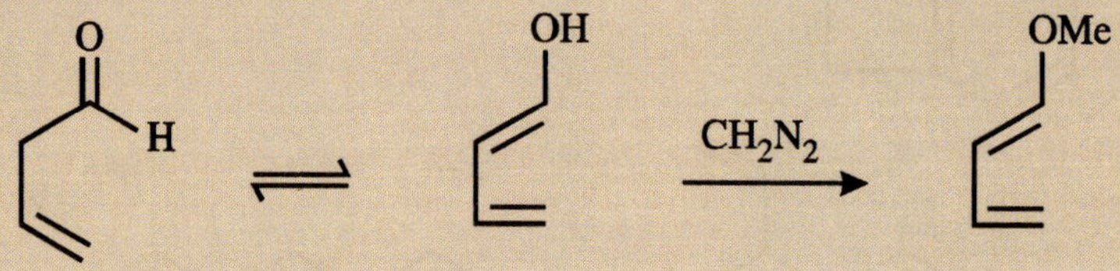

(iii) O H O CH_2N_2 →   OMe O Et_3N / Me_3SiCl →   OMe Me_3SiO

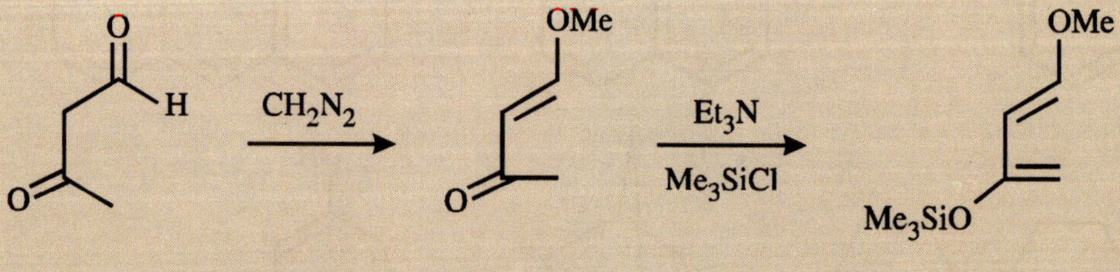

(iv) O SeO_2 →   O O $2(C_6H_5)_3P = CH_2$ →

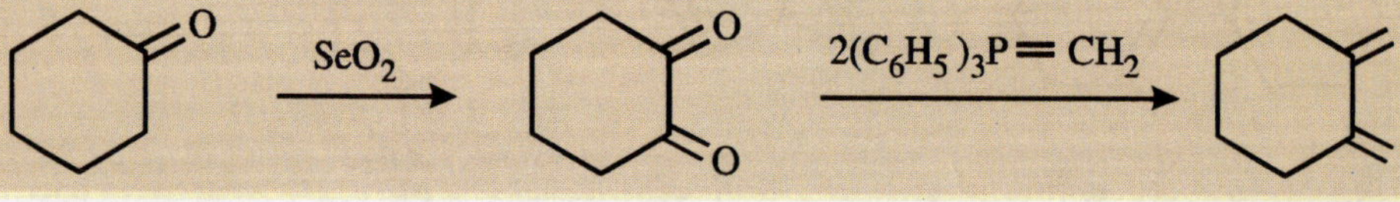

(v)

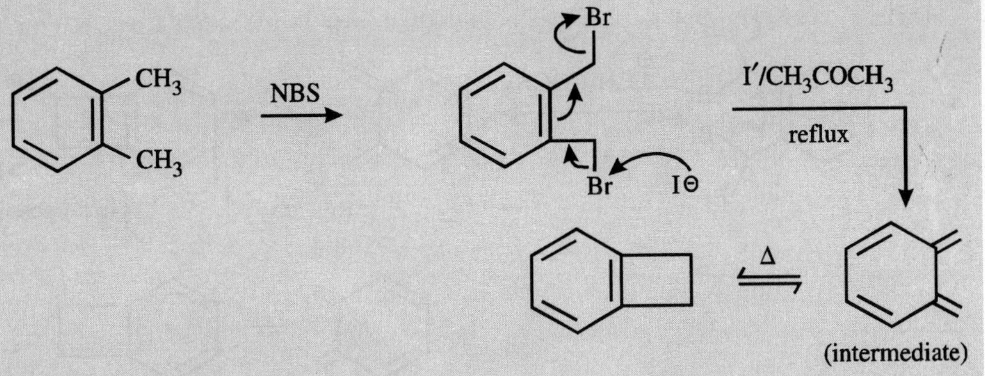

CH_3 / CH_3 —NBS→ Br / Br, I⊖ —I^-/CH_3COCH_3, reflux→ (intermediate) ⇌ Δ

(vi)

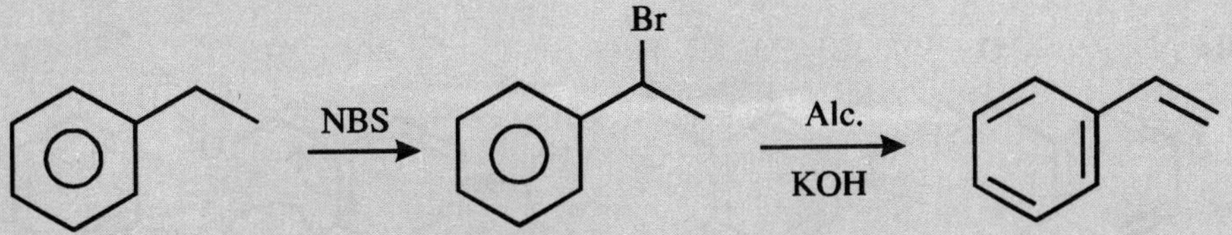

—NBS→ Br —Alc. KOH→

(vii)

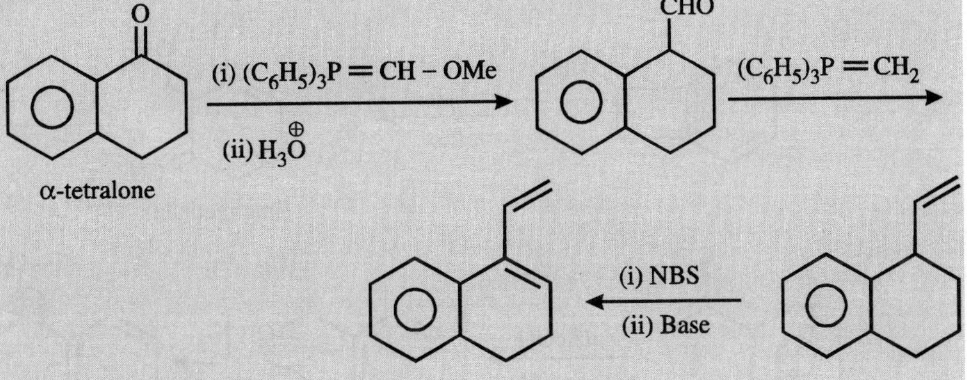

α-tetralone —(i) $(C_6H_5)_3P{=}CH-OMe$ (ii) $H_3\overset{\oplus}{O}$→ CHO —$(C_6H_5)_3P{=}CH_2$→ —(i) NBS (ii) Base→

(viii)

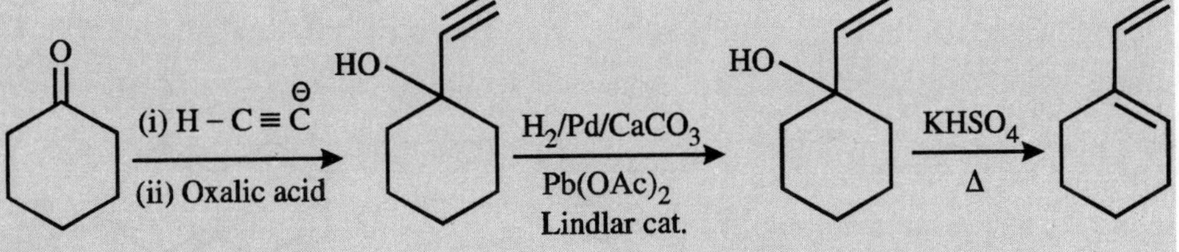

O —(i) $H-C{\equiv}\overset{\ominus}{C}$ (ii) Oxalic acid→ HO —$H_2/Pd/CaCO_3$, $Pb(OAc)_2$, Lindlar cat.→ HO —$KHSO_4$, Δ→

(ix)

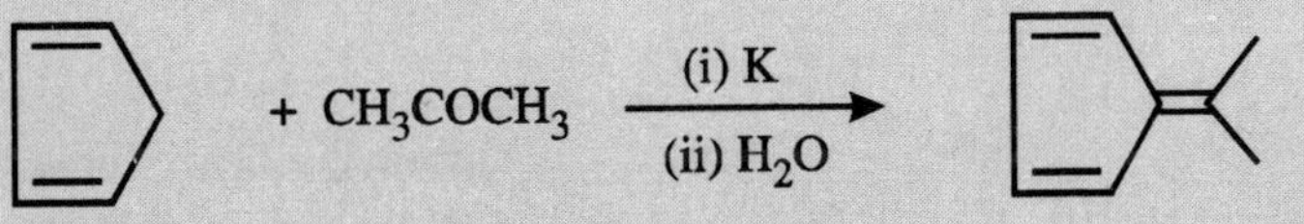

+ CH_3COCH_3 —(i) K (ii) H_2O→

(x)

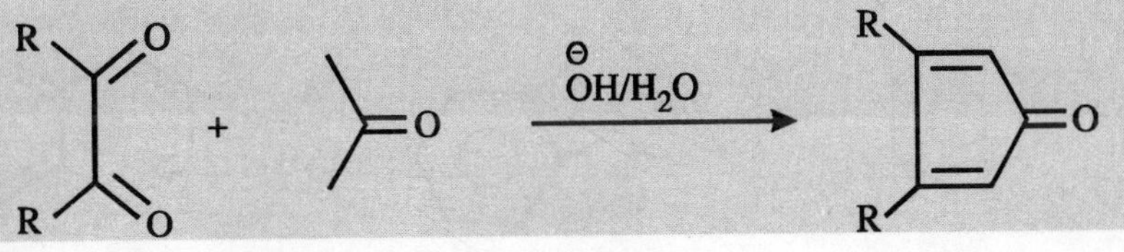

R, O, R, O + O —$\overset{\ominus}{O}H/H_2O$→ R, R, O

(xi)

Br Br Br Br

$\overset{\ominus}{I}/CH_3COCH_3$

I⊖

Br Br

Δ

Br Br

hydrogenolysis

≡

Benzocyclobutene

(xii)

O H CH_3 hν

H O· $CH_2 - H$

H H O CH_2

OH

OH benzene reflux

OH CH_2

(intermediate)

Δ

(xiii)

C_6H_5 O O C_6H_5

Zn/AcOH

Partial (H)

C_6H_5 OH O C_6H_5

C_6H_5 H O OH C_6H_5

$- H_2O$

C_6H_5 O C_6H_5

diphenyl isobenzo furan

O + → O

Δ

$-C_2H_2$

O

O Cl Cl O

$AlCl_3$

Ph O O Ph

P_2O_5

Ph O Ph

O Ph Ph

Retro DA

$-C_2H_2$

Ph O Ph

(xiv) NH_2 O

KOH/Br_2

$CHCl_3$

$N=C=O$

(intermediate)

(xv) NH_2 + H O

$ZnCl_2$

$-H_2O$

$N=CH\ C_6H_5$

(xvi) OEt O

HBr

Br OEt O

Zn/ether

BrZn OEt O

(i) CH_2O/ether (ii) $H_3\overset{\oplus}{O}/\Delta$

OEt O

(i) KOH

(ii) $H_3\overset{\oplus}{O}$

H HO O

$CH_2=\overset{\oplus}{O}$ H

$CH_2O/H_2SO_4/\Delta$

HO :OH O

$-H_2O$

O O

NBS

Br O O

Et_3N

O O

(xvii) R O O R

P_2O_5

R O R

3.3.13 Some examples of Diels-Alder dienophiles

CO_2H
COOH
CO_2H
HO_2C
CO_2Et
CO_2Et
NC
CN
NC
CN
CO_2Et
CO_2Et
C_6H_5
C_6H_5
Me
Me
H
H
CO_2Et
CO_2Et
R
R = –CN, – NO_2, – CHO
– CO_2Et, – $COCH_3$, – $CONH_2$
CH_2
C_6H_5
C_6H_5
hetero dienophiles
R
R = H, CO_2Et, alkyl
$(CH_2)_6$
$P(C_6H_5)_3$
R
R
R= –Me, – C_6H_5
NH

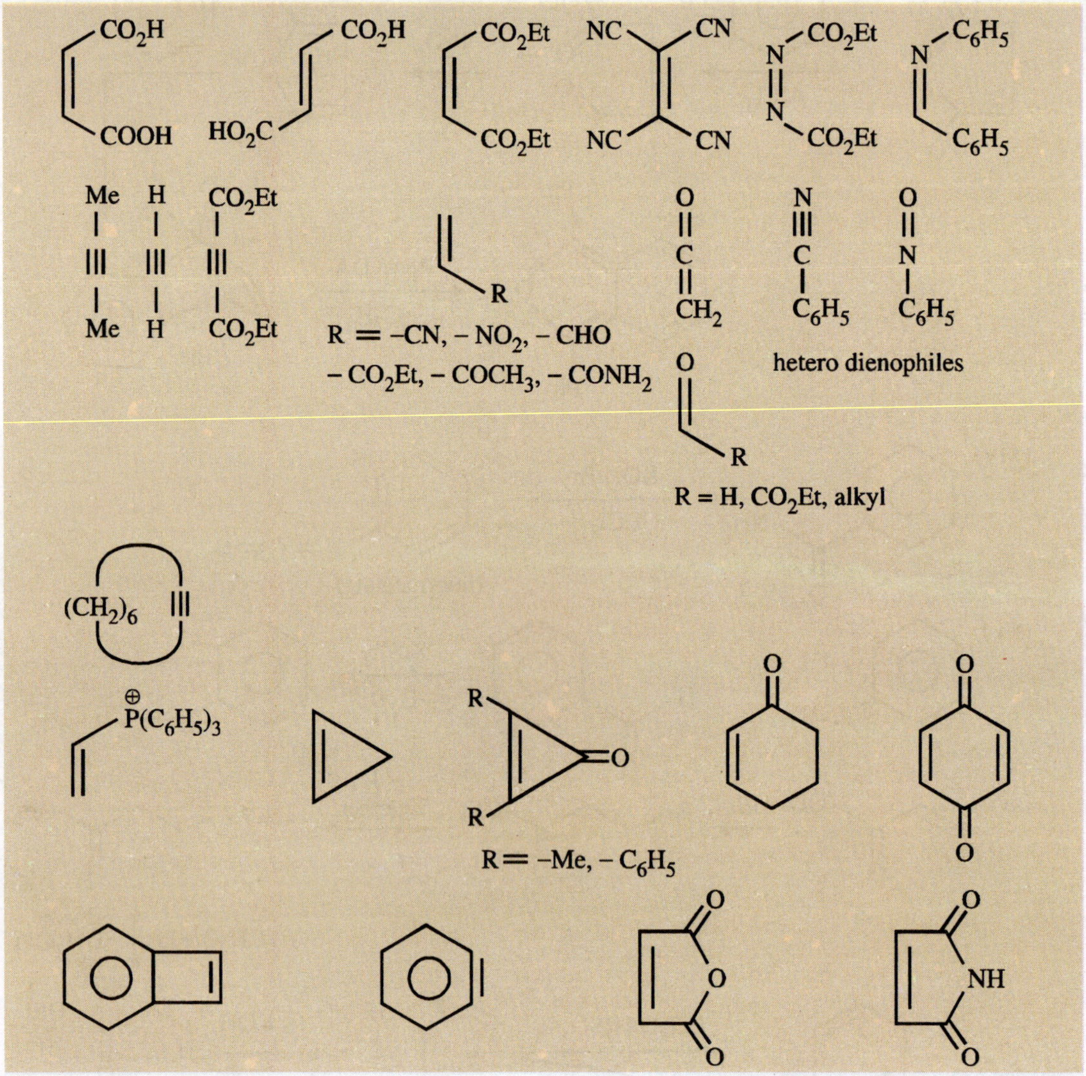

Inverse electron-demand dienophile

OEt
OEt
OEt
NMe_2
Me

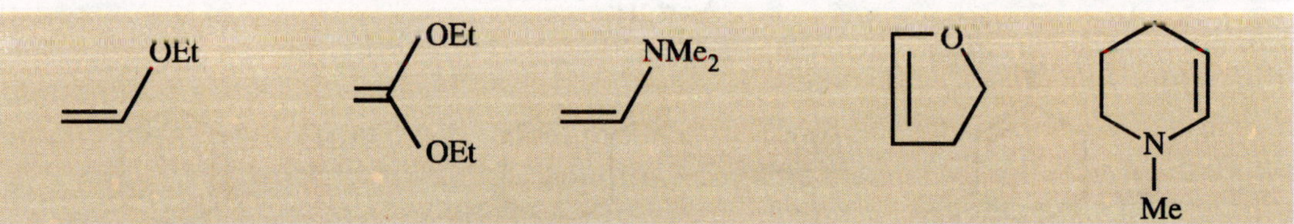

3.3.14 Preparation of Diels-Alder dienophiles

(i) $(C_6H_5)_3\overset{\oplus}{P}-\overset{\ominus}{C}H_2$
Cl
Me
$(C_6H_5)_3\overset{\oplus}{P}$
H
OMe
Δ
$(C_6H_5)_3\overset{\oplus}{P}$

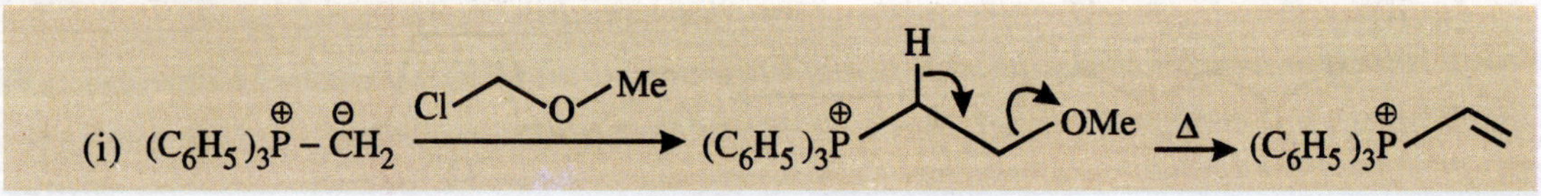

(ii) $C_6H_5CH(Cl)C(=O)CH_2C_6H_5$ $\xrightarrow{\overset{\ominus}{O}H/H_2O}$ 2,3-di(C_6H_5)cyclopropanone (intermediate)

(iii) $CH_2{=}C{=}O \xrightarrow{2EtOH} CH_2{=}C(OEt)_2$

(iv) $CH_3CHO \xrightarrow{Cl_2/CH_3COOH} ClCH_2CHO \xrightarrow[HCl]{EtOH} ClCH_2CH(OEt)_2 \xrightarrow{\overset{\ominus}{N}H_2} H{-}C{\equiv}C{-}OEt$

(v) $(CH_2)_6$ diester (OEt, CO_2Et) $\xrightarrow[(ii)\ H_3\overset{\oplus}{O}]{(i)\ Na/xylene}$ $(CH_2)_6$ cyclic acyloin (C=O, CH–OH) $\xrightarrow{CrO_3/Py}$ $(CH_2)_6$ cyclic 1,2-dione $\xrightarrow[(ii)\ Pb(OAc)_4]{(i)\ N_2H_4}$ $(CH_2)_6$ cycloalkyne ($\equiv$)

3.3.15 Synthesis of target molecules using Diels-Alder reaction with retrosynthetic analysis

1. TM (cyclohexane with COOH, H / H, COOH) $\xRightarrow{FGA}$ cyclohexene with CO_2H, H / H, CO_2H $\Longrightarrow$ butadiene (SE) + $HO_2C(H)C{=}C(H)CO_2H$ (SE)

Forward Synthesis

butadiene + $HO_2C{-}CH{=}CH{-}CO_2H$ $\xrightarrow{\Delta}$ cyclohexene (COOH H / H COOH) $\xrightarrow{H_2/Pd}$ cyclohexane (COOH H / H CO_2H)

2.

TM $\xRightarrow{FGI}$ CO_2H, CO_2H $\Rightarrow$ + CO_2H, CO_2H

Forward Synthesis

+ CO_2H, CO_2H $\xrightarrow{\Delta}$ CO_2H, CO_2H $\xrightarrow{Pb(OAc)_4}$

3.

TM $\xRightarrow{FGA}$ CO_2H, CO_2H $\xRightarrow{FGI}$ CO_2Et, CO_2Et $\Rightarrow$ SE + CO_2Et ≡ CO_2Et SE +

Forward Synthesis

+ CO_2Et ≡ CO_2Et + $\xrightarrow{\Delta}$ CO_2Et, CO_2Et $\xrightarrow{\text{(i) KOH}}$ (ii) $H_3\overset{\oplus}{O}$ (iii) $Pb(OAc)_4$

diene : dienophile = 2 : 1

4.

TM CO_2Et, CO_2Et $\xRightarrow{FGI}$ + CO_2Et, CO_2Et

$\Downarrow$ FGI

Br, Br $\xRightarrow{FGI}$ OH, OH

$\Downarrow$ FGI

CO_2Et, CO_2Et $\Rightarrow$ SE + CO_2Et ≡ CO_2Et SE

Forward Synthesis

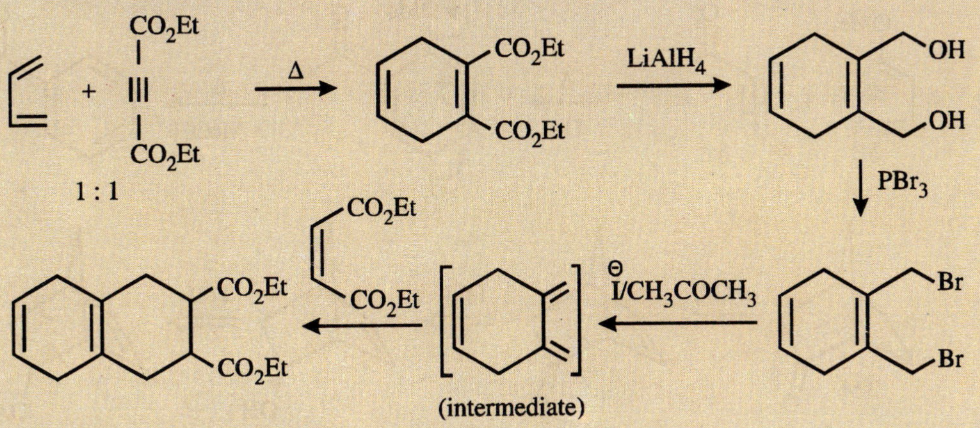

CO_2Et + CO_2Et (1 : 1) —Δ→ CO_2Et, CO_2Et —$LiAlH_4$→ OH, OH —PBr_3→ Br, Br —$\overset{\ominus}{I}/CH_3COCH_3$→ (intermediate) —$CO_2Et$, CO_2Et→ CO_2Et, CO_2Et

5. O, O —FGI⇒ OH, O ⇒ OH + O, SE ⇒ $OSiMe_3$, SE ⇒ O + Et_3N/Me_3SiCl

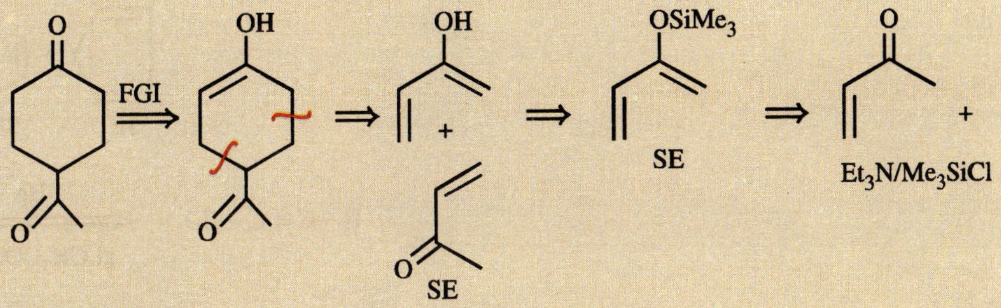

Forward Synthesis

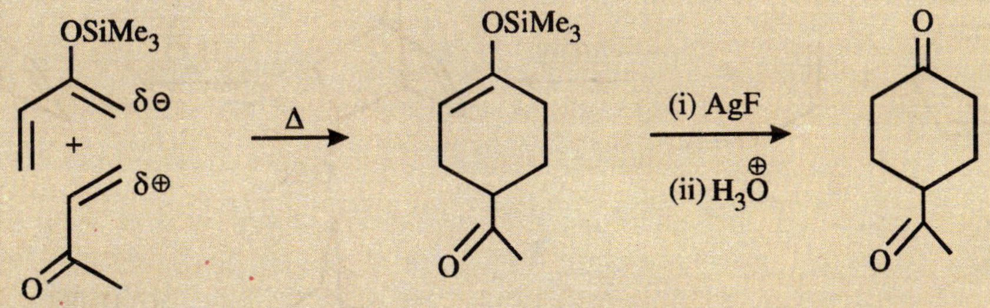

$OSiMe_3$ (δ⊖) + O (δ⊕) —Δ→ $OSiMe_3$, O —(i) AgF, (ii) $H_3\overset{\oplus}{O}$→ O, O

6. O, TM —FGI⇒ O ⇒ + O, SE

⇓FGA

OMe, SE —CH_2N_2⇒ H, O

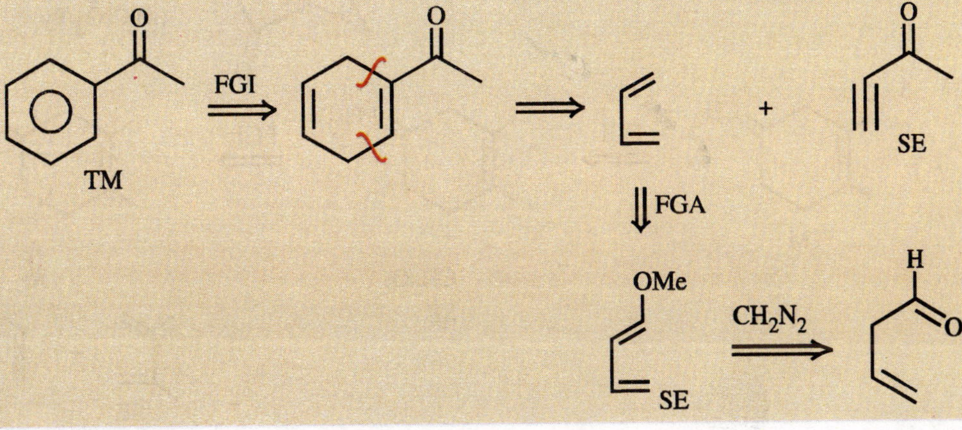

Forward Synthesis

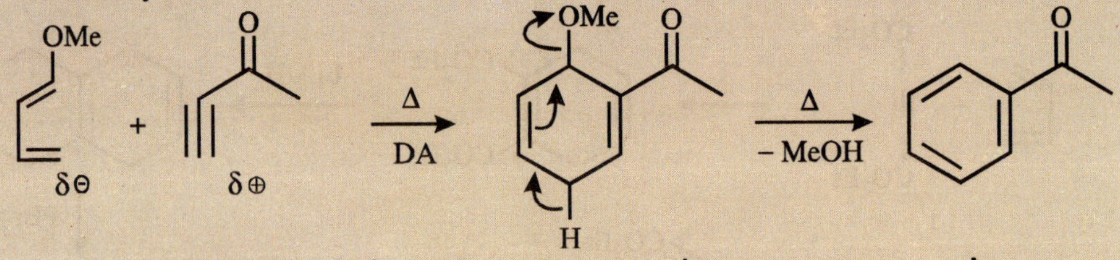

OMe; δ⊖ + δ⊕ $\xrightarrow[\text{DA}]{\Delta}$ OMe, H $\xrightarrow[-\text{MeOH}]{\Delta}$

7. TM, O $\xRightarrow{\text{FGA}}$ O $\xRightarrow{\text{FGI}}$ OH $\xRightarrow{\text{FGI}}$ OAc

⇓

SE + SE OAc

H − C ≡ C − H $\Leftarrow$ Hg^{+2}, gl.CH_3COOH

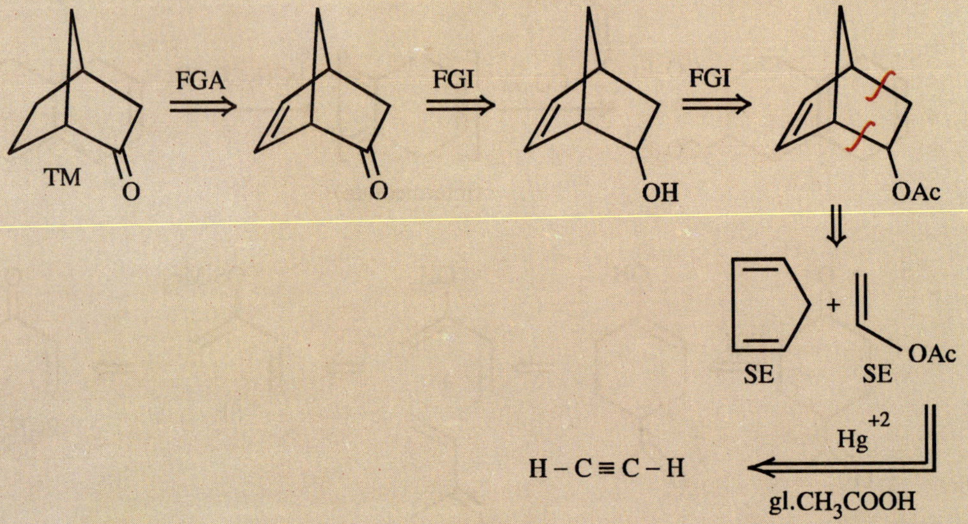

Forward Synthesis

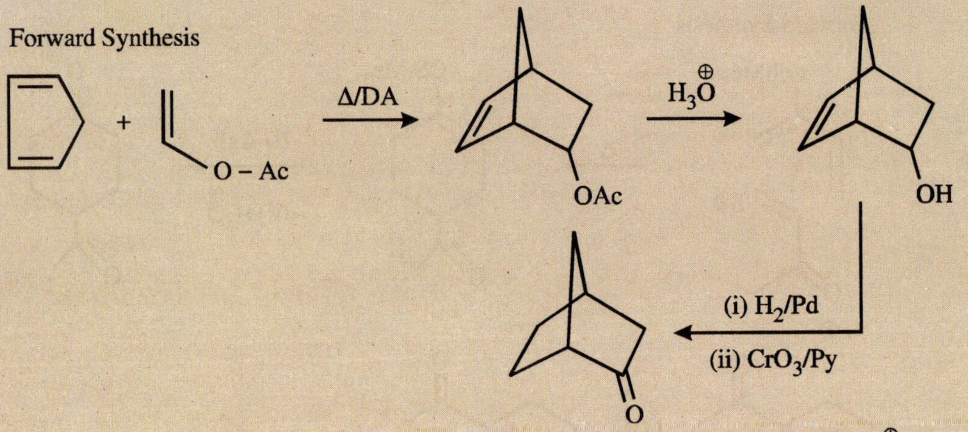

+ O − Ac $\xrightarrow{\Delta/\text{DA}}$ OAc $\xrightarrow{H_3\overset{\oplus}{O}}$ OH

$\xrightarrow[\text{(ii) } CrO_3/Py]{\text{(i) } H_2/Pd}$ O

8. CH_2 TM $\xRightarrow{\text{FGI}}$ PPh_3 + CH_2O SE $\xRightarrow{\text{FGI}}$ $\overset{\oplus}{P}Ph_3$

⇓

SE + $\overset{\oplus}{P}Ph_3$ SE

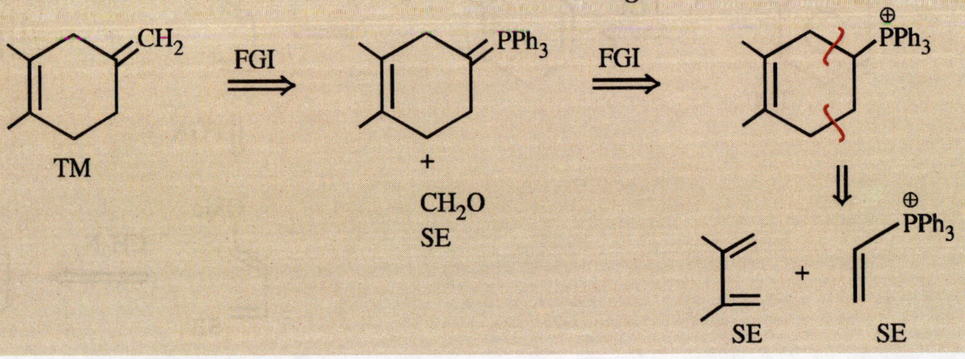

Forward Synthesis

$\oplus$ PPh$_3$ + Δ / DA → PPh$_3$ LDA → ⊖ PPh$_3$ ↓ CH$_2$O

9.

TM FGI ⟹ FGI ⟹

SE ⟸ SN + SE ⟸

Forward Synthesis

Δ / DA → HCl gas / Or HBr → $\oplus$ O–H, H ↓ HO H ← $-H_2O$

10.

FGI

CO_2Et

CO_2Et

CO_2Et

CO_2Et

Route-3 FGI

FGI

FGA

Route-1

TM

(Naphthalene)

FGA Route-2

SE SE

Retro

DA

SE SE

Forward Synthesis

Route-1

Δ

H_2/Pd

$H_3\overset{\oplus}{O}$

$\oplus O-H$

OH

H

$-H_2O$

H

Route-2

Δ

DA

Δ

Retro DA

$-C_2H_2$

Route-3

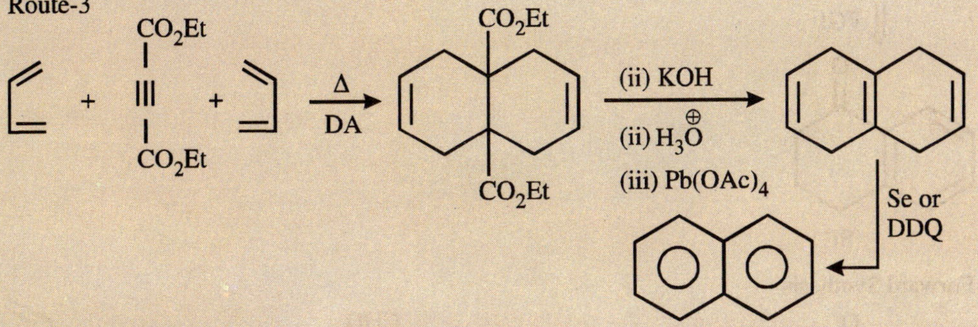

CO_2Et + CO_2Et ... Δ DA → CO_2Et / CO_2Et (ii) KOH (ii) $H_3\overset{\oplus}{O}$ (iii) $Pb(OAc)_4$ → Se or DDQ

11.

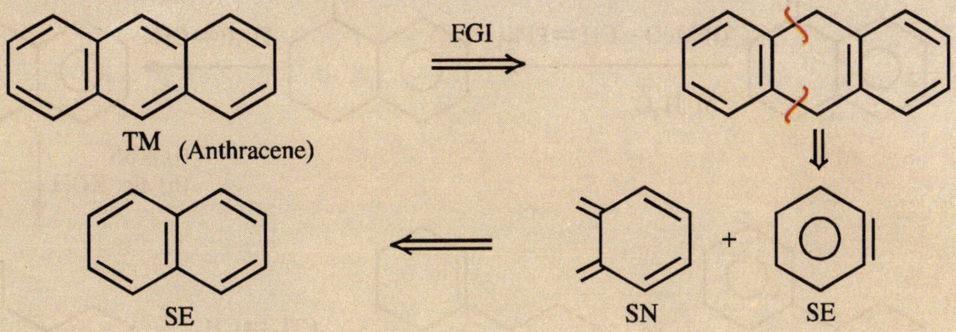

TM (Anthracene) FGI ⟹ ⟹ SN + SE ⟸ SE

Forward Synthesis

+ Δ DA → Δ Retro DA $- C_2H_2$

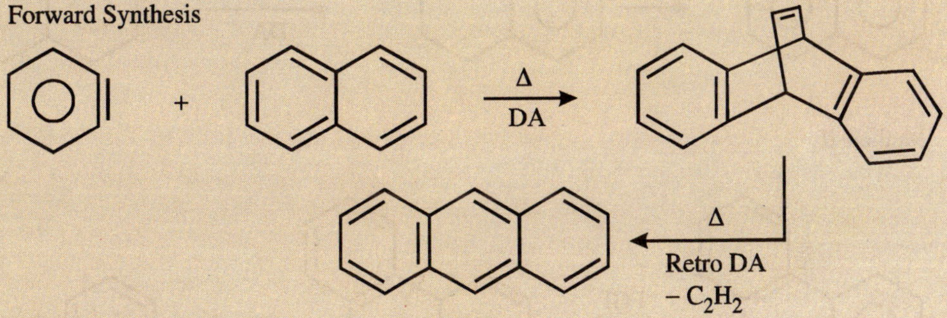

12. Method-I

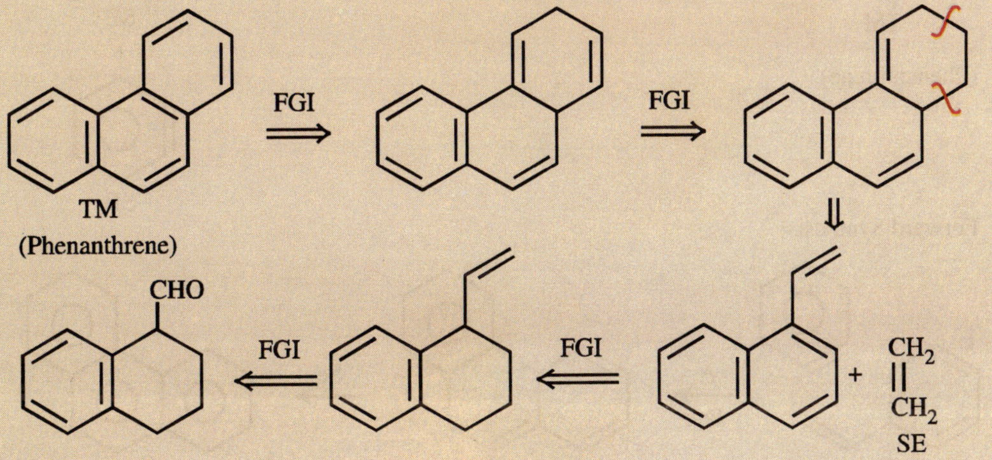

TM (Phenanthrene) FGI ⟹ FGI ⟹ ⟹ + $CH_2 = CH_2$ SE FGI ⟸ FGI ⟸ CHO

⇓ FGI

SE

Forward Synthesis

(i) MeO – CH=PPh_3
(ii) $H_3\overset{\oplus}{O}$

CHO

$Ph_3P=CH_2$

(i) NBS
(ii) alc. KOH

$CH_2=CH_2$
DA

Se

Method-II

FGI

TM

(Phenanthrene)

SE
+
SE

Forward Synthesis

Δ
DA

Se

13.

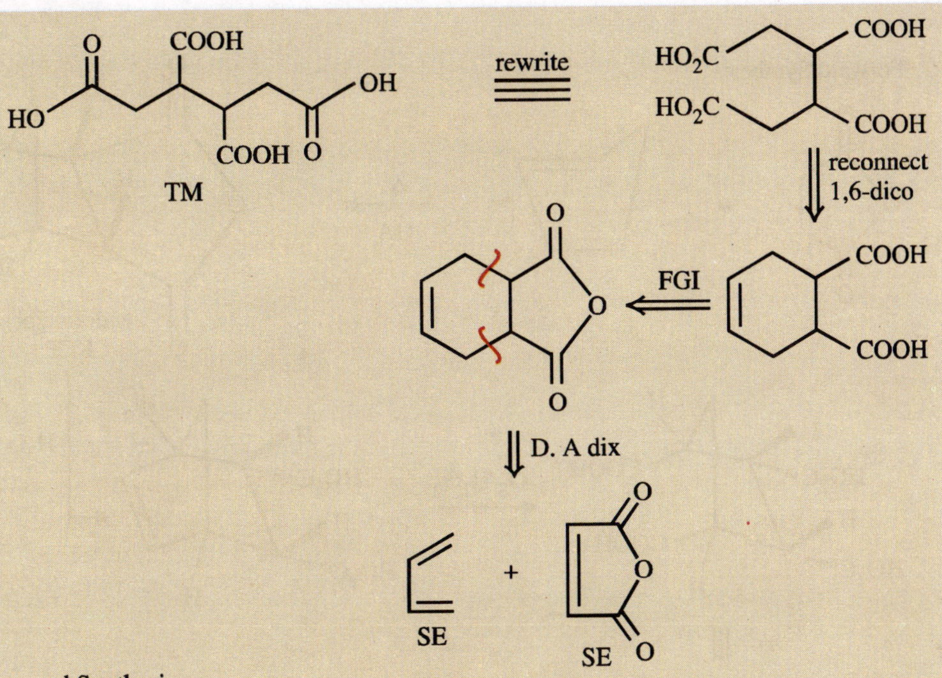

TM

rewrite

reconnect 1,6-dico

FGI

D. A dix

SE + SE

Forward Synthesis

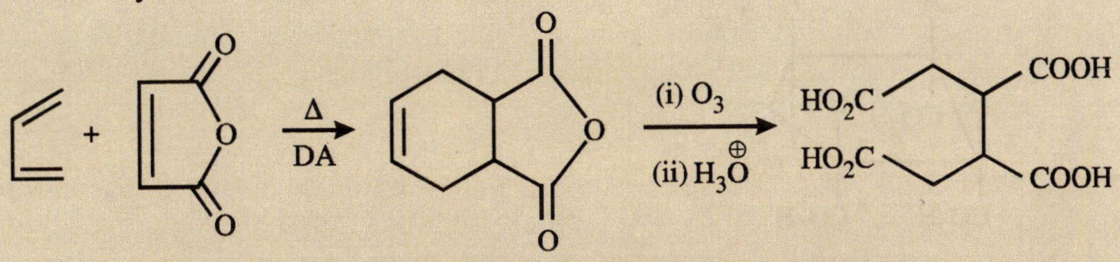

Δ / DA

(i) O_3
(ii) $H_3\overset{\oplus}{O}$

14.

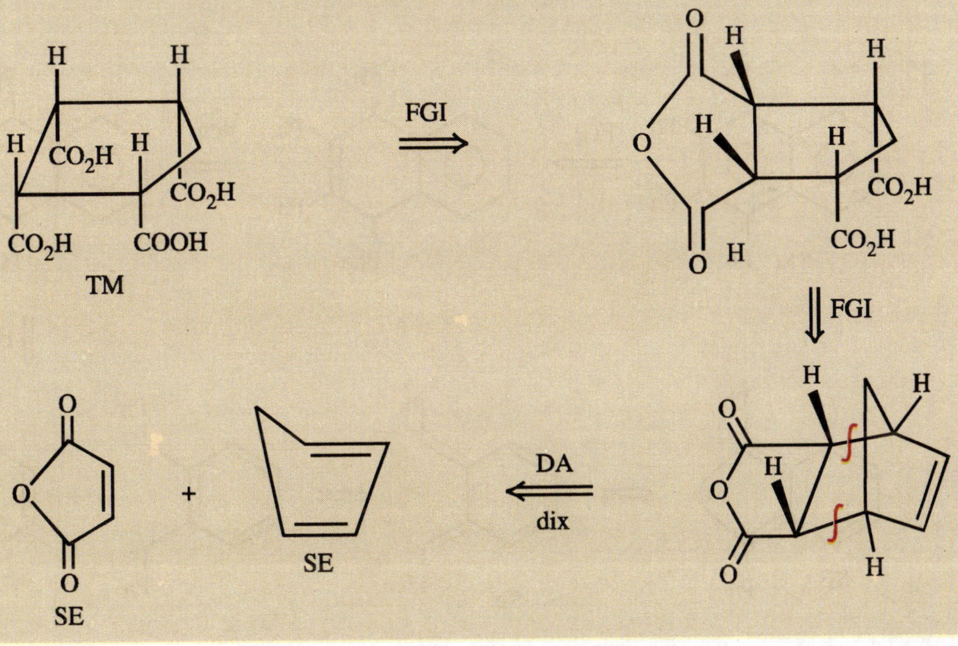

TM

FGI

FGI

DA dix

SE + SE

Forward Synthesis

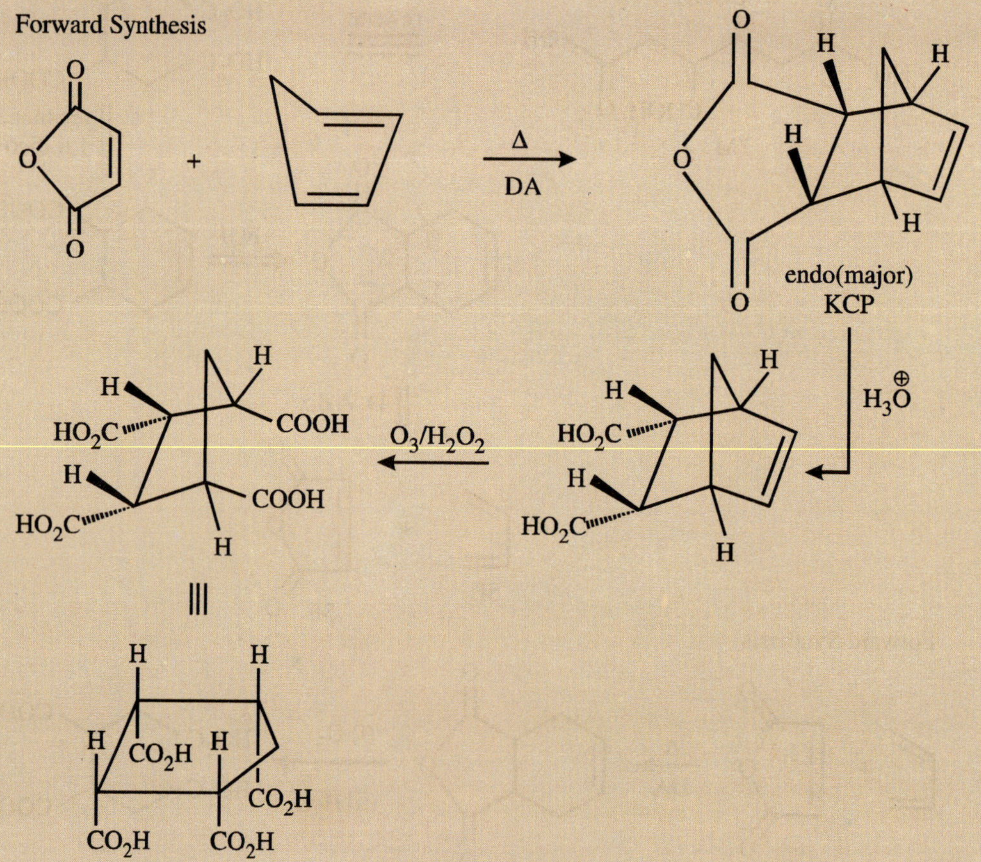

O O O + Δ DA → endo(major) KCP

$H_3\overset{\oplus}{O}$

O_3/H_2O_2

H H HO_2C COOH H HO_2C COOH H H

|||

H H H CO_2H H CO_2H CO_2H CO_2H

15. Ph Ph Ph TM Ph $\xRightarrow{FGI}$ Ph Ph Ph Ph $\xRightarrow{FGI}$ Ph Ph Ph Ph

⇓ FGI

Ph O Ph SE ⇐ Ph Ph SN ⇐ Ph ⊕ ⊖ Ph + Ph Ph SE

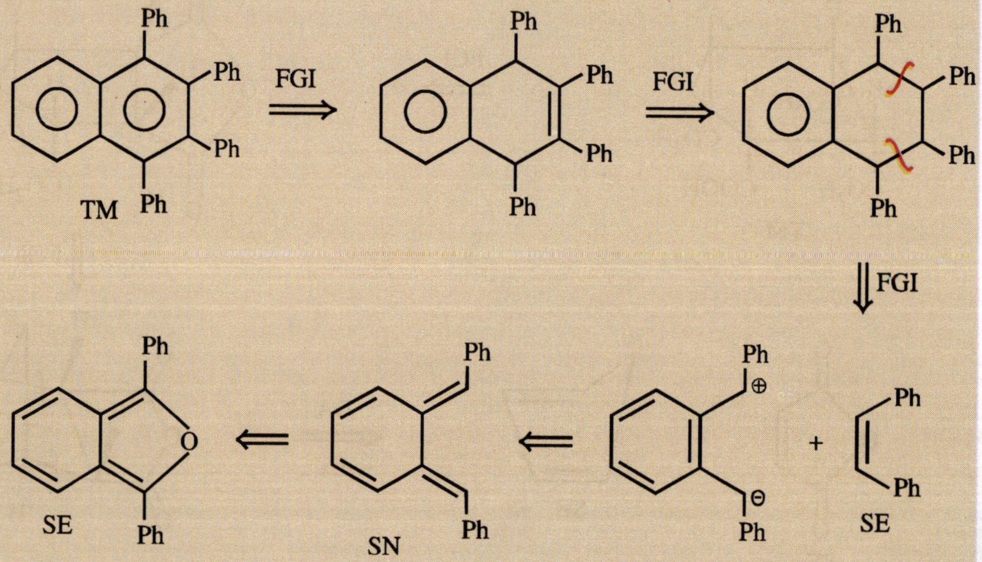

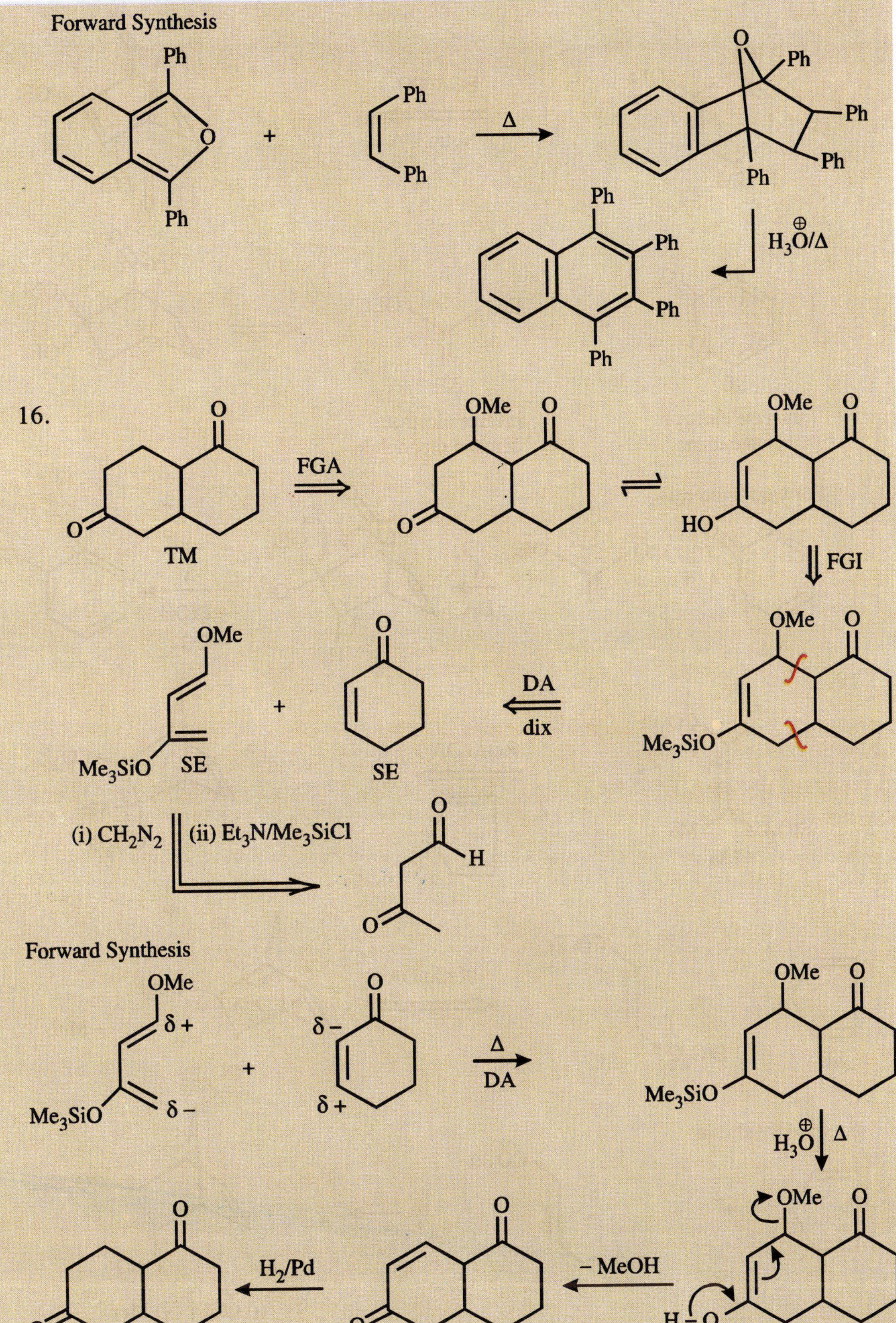

Forward Synthesis
Ph
O
Ph
+
Ph
Ph
Δ
O
Ph
Ph
Ph
Ph
H₃O⊕/Δ
Ph
Ph
Ph
Ph
16.
O
O
TM
FGA
OMe
O
O
OMe
O
HO
FGI
OMe
O
Me₃SiO
DA
dix
OMe
Me₃SiO
SE
+
O
SE
(i) CH₂N₂
(ii) Et₃N/Me₃SiCl
O
H
O
Forward Synthesis
OMe
δ +
Me₃SiO
δ −
+
O
δ −
δ +
Δ
DA
OMe
O
Me₃SiO
H₃O⊕
Δ
OMe
O
H − O
− MeOH
O
O
H₂/Pd
O
O

17.

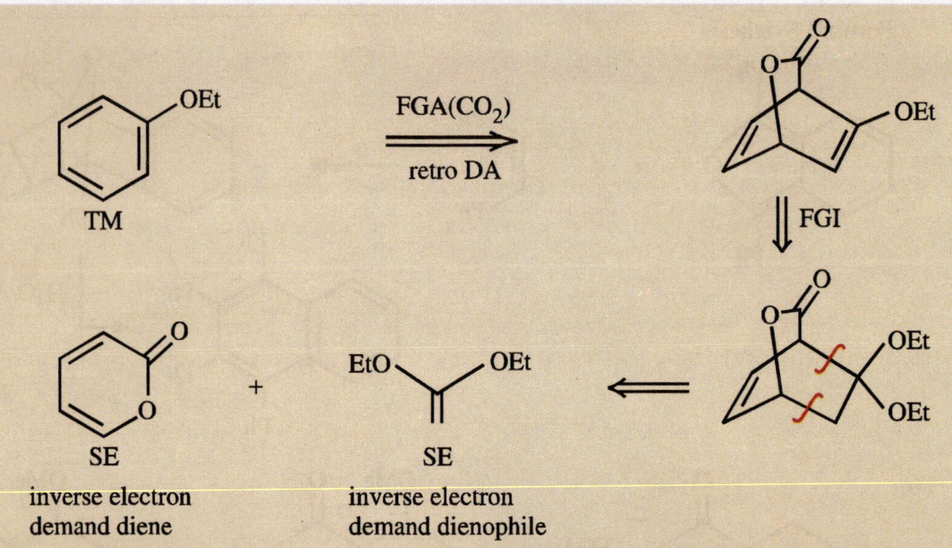

OEt

TM

FGA(CO_2)

retro DA

OEt

FGI

O

O

EtO OEt

OEt

OEt

SE

inverse electron demand diene

+

SE

inverse electron demand dienophile

Forward Synthesis

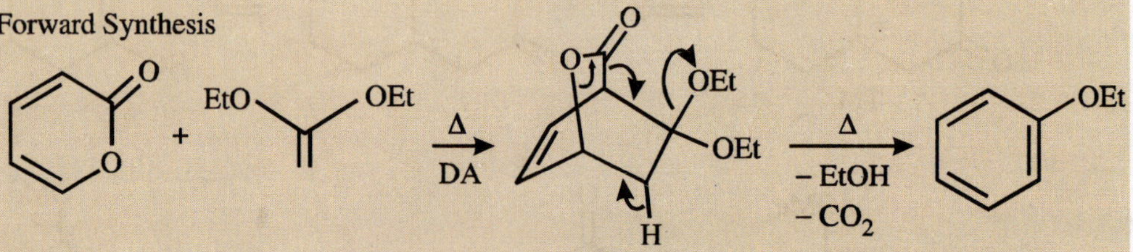

EtO OEt

Δ

DA

OEt

OEt

H

Δ

– EtOH

– CO_2

OEt

18.

H CO_2Et

EtO_2C Me

TM

Retro DA

+

CO_2Et

Me

CO_2Et

CO_2Et

+

EtO_2C

SE SE

Retro DA

CO_2Et

CO_2Et

+ MeI

SE

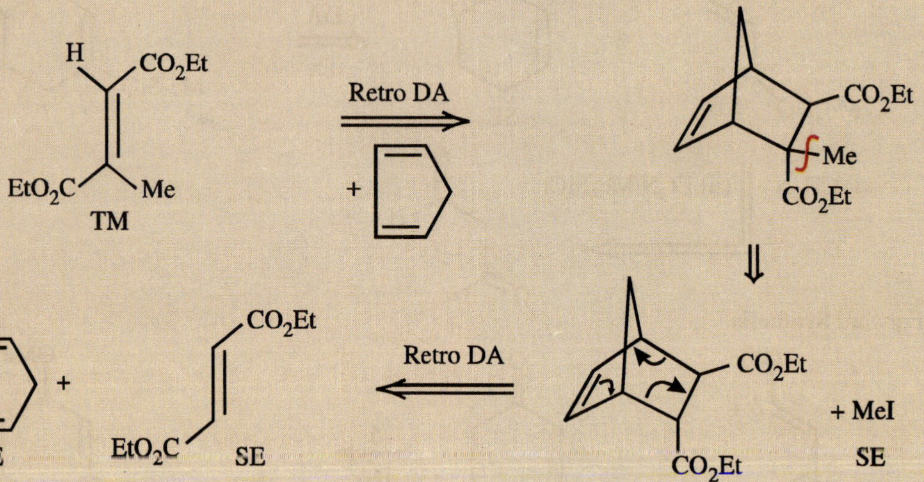

Forward Synthesis

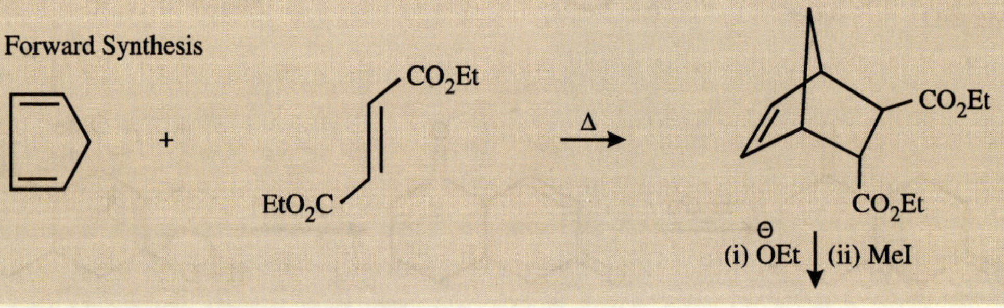

+

CO_2Et

EtO_2C

Δ

CO_2Et

CO_2Et

(i) $\overset{\ominus}{O}Et$ (ii) MeI

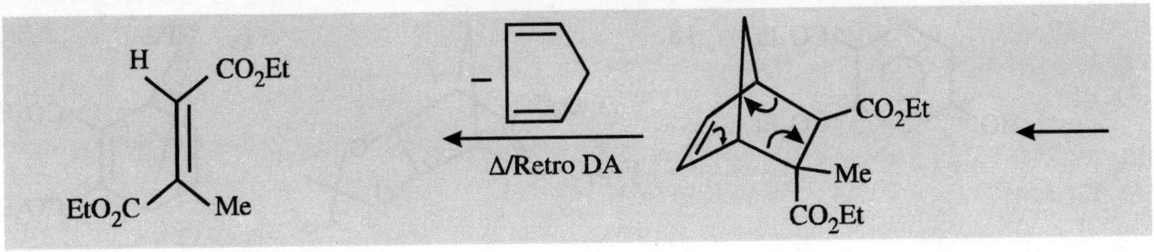

H, CO_2Et, EtO_2C, Me ← Δ/Retro DA ← CO_2Et, Me, CO_2Et ←

3.3.16 Review synthetic problems

Find out the synthetic equivalent after proper retrosynthetic analysis of the following target molecules and then show the forward synthesis using Diels-Alder reaction.

1. 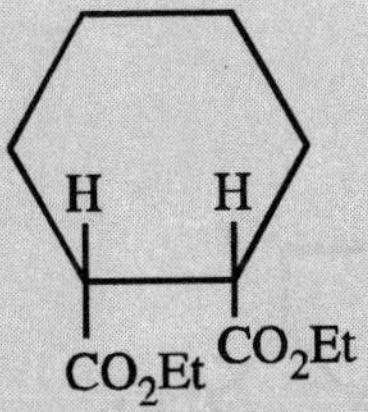 H, H, CO_2Et, CO_2Et
2. 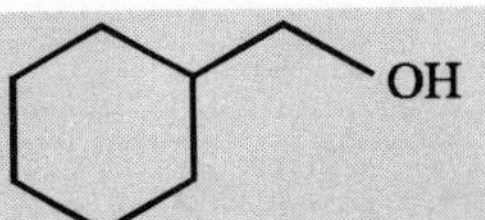 OH
3. 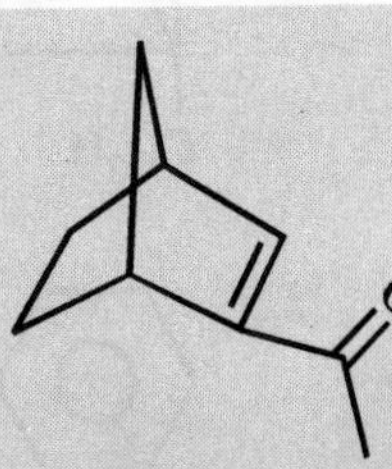
4. 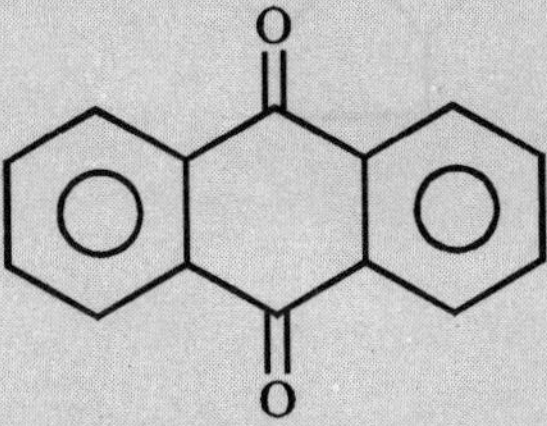 O, O
5. 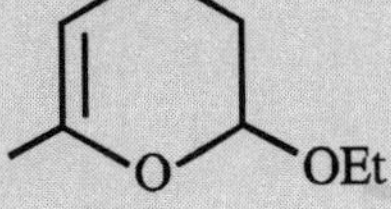 O, OEt
6. 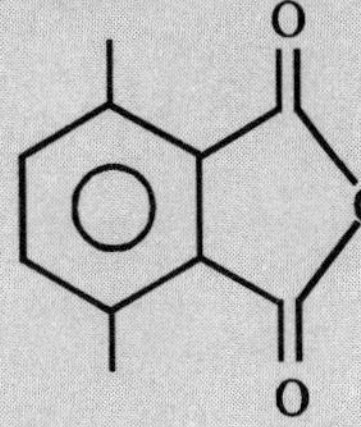 O, O, O
7. 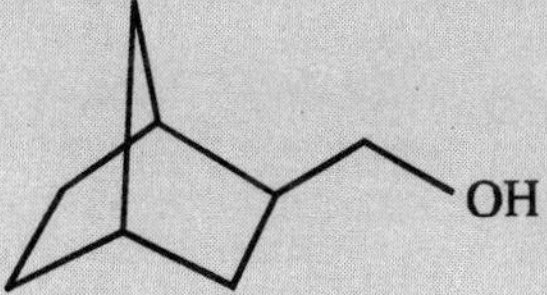 OH
8. 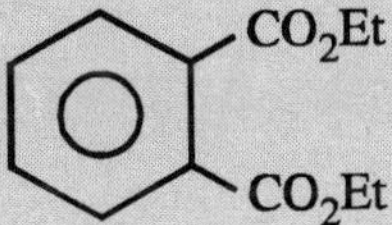 CO_2Et, CO_2Et
9. 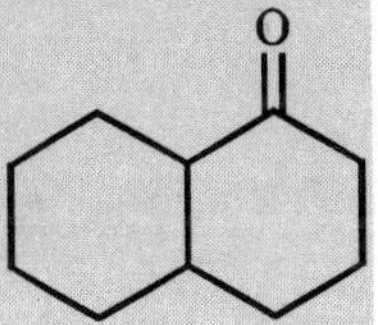 O
10. 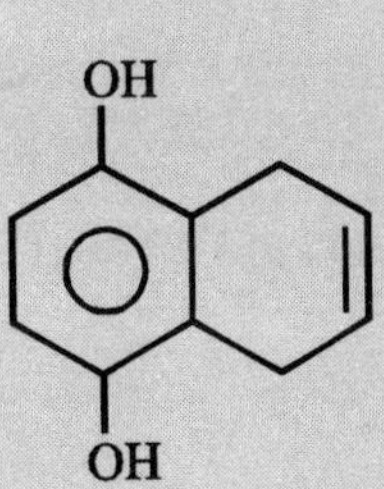 OH, OH
11. 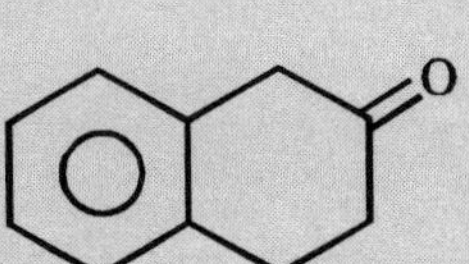 O
12. 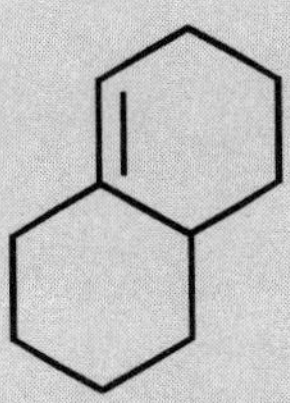
13. 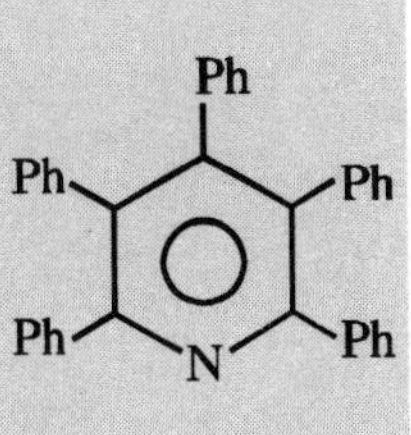 Ph, Ph, Ph, Ph, Ph, N
14. 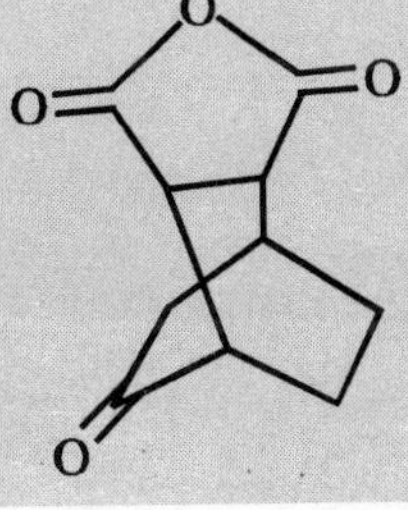 O, O, O, O
15. 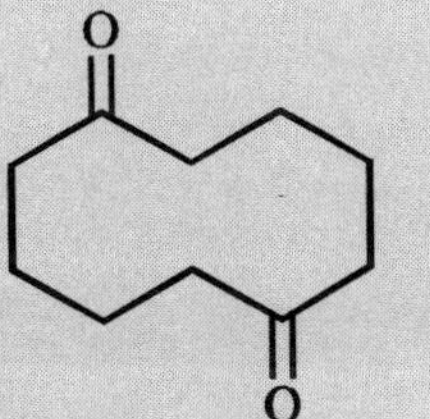 O, O
16. 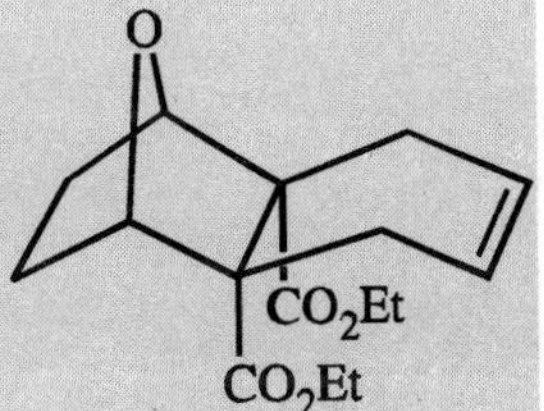 O, CO_2Et, CO_2Et

17. HO, CO_2Et, CO_2Et

18. EtO_2C, EtO_2C, O, O

19. NPh, CO_2Et, CO_2Et

20. H, CO_2H, H, CO_2H, CO_2H, H, CO_2H, H

21. OEt, N, OH

22.

23. O

4

Synthetic Application of Ethyl Aceto Acetate, Diethyl Malonate and Ethyl Cyano Acetate

4.1 Ethyl aceto acetate (EAA)

4.1.1 What is ethyl aceto acetate and what are the reactions of EAA?

Ethyl aceto acetate is a versatile starting material or synthetic equivalent for a number of various classes of organic target molecules aliphatic, alicyclic, carboaromatic, heteroaromatic and for various functionality molecules monoketones, diketones, keto-acids mono and dicarboxylic acids and various heterocyclic molecules furan, pyrrole, pyridine, Quinoline etc. and their derivatives.

Ethyl aceto acetate is a very important compound for knoevenagel reaction and Michael condensation.

Before going into the synthesis of organic target molecules using ethyl aceto acetate as starting material, one should know the answer of two questions:

What is ethyl aceto acetate?

The answer of this question will help one to understand versatility, speciality and utility of its 'structure' for synthetic purpose.

What are the reactions of ethyl aceto acetate with different reagents?

The answer of this question will help one to understand the differences of reactivity of four basic carbon's of the ethyl aceto acetate with different reagents.

Ethyl aceto acetate (carbons numbered 4, 3, 2, 1; OEt)

What is ethyl aceto acetate?

(Structures: carbons 4, 3, 2, 1 with OEt; γ, β, α with OEt)

1. Ethyl aceto acetate is a β-keto ester. It is a bifunctional compound. The C–H bond of methylene (– CH_2 –)is adjacent to two – M group and is highly acidic and hence may be alkylated in milder condition.

$$\text{EAA} \xrightarrow[\text{(ii) } CH_3-Br]{\text{(i) } \overset{\ominus}{O}Et} \text{methylated EAA}$$

80% yield without heat

2. It is the acylated product of ethyl acetate which is formed by Claisen ester condensation.

$$H-CH_2CO_2Et + CH_3CO_2Et \xrightarrow{\overset{\ominus}{O}Et} CH_3COCH_2CO_2Et$$

Two other methods of EAA Synthesis

(a)

$$H-C\equiv C-H \xrightarrow[CH_3I]{\overset{\ominus}{N}H_2\text{(1eq)}} CH_3-C\equiv C-H \xrightarrow[\text{(iii) } H_3\overset{\oplus}{O}]{\text{(i) } \overset{\ominus}{N}H_2 \text{ (ii) } CO_2} CH_3-C\equiv C-CO_2H$$

$$\xrightarrow{EtOH/H^{\oplus}} CH_3-C\equiv C-CO_2Et \xrightarrow[H_2SO_4]{Hg^{+2}} CH_3COCH_2CO_2Et$$

(b) Dimer of ketene in acetone when treated with EtOH gives EAA

$$CH_2=C=O + CH_2=C=O \longrightarrow \text{Dimer} \xrightarrow{EtOH} CH_3COCH_2CO_2Et$$

Ketene

Dimer ($CH_2=C$—O, CH_2—C=O)

3. α-H of acetone is activated since it can take part in enolisation. Now it may be thought that acetone is further activated by adding – CO_2Et in place of α-H of acetone which is ethyl aceto acetate.

4. It is the protection of – COOH of β-keto acid by – CO_2Et

5. It is the activation of carboxyl (– COOH) carbonyl (C=O) of β-keto acid by – CO_2Et so that carboxyl carbon becomes electrophilic in synthetic reaction.

6. β-keto ester may be easily decarboxylated by 'dil HCl/heat' due to cyclic TS.

dil HCl, Δ; – CO_2

6-membered cyclic TS

NaCl | Wet DMSO

+ CO_2 + EtCl
Neutral decarboxylation

7. With weak base it gives stabilised carbanion which may be used as Michael donor.

8. Since it gives stabilised carbanion so it will stop the base catalysed self-condensation (aldol or Claisen) and so exclusive use of alkylation or acylation of methylene carbon in the synthesis.

9. The stabilised carbanion or enolate is an ambident nucleophile, so it may be alkylated or acylated either on carbon or on oxygen.

(i) $\overset{\ominus}{O}Et$/EtOH
(ii) CH_3I

(i) $\overset{\ominus}{O}Et/Me_2SO$
(ii) CH_3OCH_2Cl

10. Ethyl aceto acetate is the synthetic equivalent of the synthon.

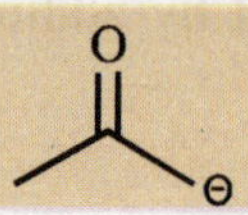

11. EAA is a molecule having 1, 3-electrophilic centres which may be utilised in the organic synthesis.

δ+ δ+ OEt

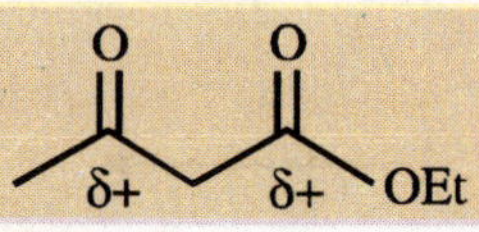

12. EAA is a molecule having 1, 3-nucleophilic centres which may be utilised in the organic synthesis.

OEt; $\ominus$OEt; 2 eq. $\overset{\ominus}{N}H_2$ or LDA; OEt; OEt; Less nucleophilic; More nucleophilic

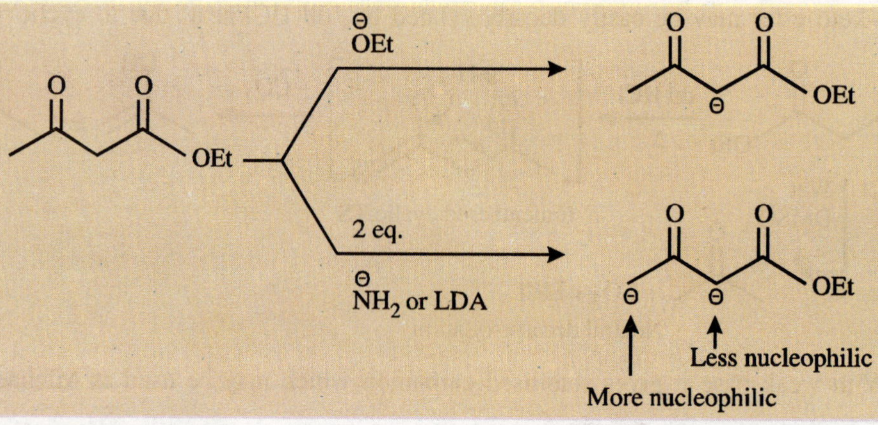

13. After the synthesis of TM starting from EAA, the residual functional group(s) may be modified as desired.

What are the reactions of EAA with different reagents which are synthetically useful?

1. 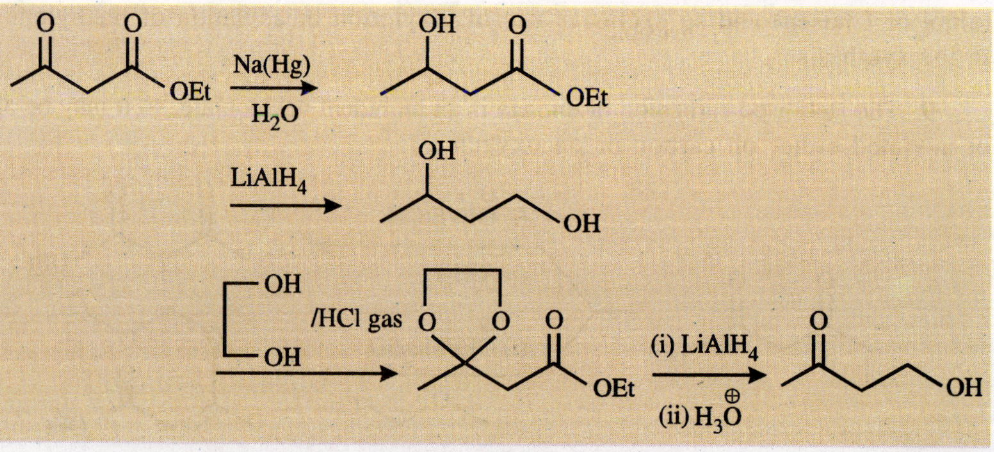 OEt; Na(Hg) / H_2O; OH, OEt; $LiAlH_4$; OH, OH; OH, OH /HCl gas; O O O, OEt; (i) $LiAlH_4$ (ii) $H_3\overset{\oplus}{O}$; O, OH

$\xrightarrow[\text{1eq.}]{CH_3MgBr}$ (OMgBr, O, OEt)

$\xrightarrow{Br_2}$ (O, O, OEt, Br)

(Neither acid nor alkali is necessary as catalyst)

2. (O, O, OEt) $\xrightarrow{CH_2N_2}$ (OMe, O, OEt)

enol ether (C = O protection)

$\xrightarrow[\text{EtOH}]{\overset{\ominus}{O}Et}$ ($O^\ominus$, O, OEt)

It is a Michael donor and is used for α-alkylation or acylation

$\xrightarrow[\overset{\ominus}{N}H_2 \text{ or LDA}]{\text{2eq.}}$ ($O^\ominus$, $O^\ominus$, OEt)

Diene-diolate dianion

It is used for γ-alkylation or acylation

$\xrightarrow[\text{Mannich reaction}]{CH_2O + Me_2NH}$ (O, O, OEt)

It is a good Michael acceptor

$\xrightarrow[\text{Reformatsky reaction}]{BrZn\text{—}CH_2\text{—}CO_2Et}$ (CO_2Et, O, OEt, OH)

Very important synthetic equivalent for various TM

$\xrightarrow[K_2CO_3]{CH_2O}$ (O, CO_2Et, OH) Synthetic equivalent for a number of TM

3. (O, O, OEt) $\xrightarrow[\text{Knoevenagel reaction}]{\text{Py/RCHO}}$ (O, R, CO_2Et) R = aryl

Very important derivative since it is a good Michael acceptor

4.

PCl5

vinylogous substitution

PCl5

Me2CuLi

Corey-House

Vinylogous substitution is a very important reaction in the synthesis of Coumarin derivative (see Coumarin and its derivative synthesis) and alkylation at the carbonyl carbon indirectly by Corey-House reagent.

5.

dil HCl/Δ

Both (I) and (II) are synthetic equivalents for various TM e.g. for Claisen and aldol synthesis with (I) and various hetero cyclic nucleus furan, thiophene and pyrrole may be synthesised starting with 1, 4-dicarbonyl (II).

6. EAA is a very important starting material for the synthesis of γ-δ unsaturated carbonyl compound.

100°C

EtOH +

ester exchange

3,3 -sigma tropic Claisen rearrangement

Δ, $-CO_2$

α β γ δ

Similarly with OH →

OH →

7. Ethyl aceto acetate does not form iodoform with $\overset{\ominus}{O}H/I_2$ although it contains ketomethyl group because of greater acidity of methylene hydrogen than methyl hydrogen.

$\xrightarrow{\overset{\ominus}{O}H/I_2}$ $CHI_2-\overset{\ominus}{CO_2}+CH_3CO\overset{\ominus}{O}+EtOH$

$\overset{\ominus}{O}H$, I_2, $\overset{\ominus}{O}H/I_2$

$\overset{\ominus}{O}H$ cannot attack on ester C=O since it is deactivated compare to keto carbonyl

$OH + \overset{\ominus}{C}I_2CO_2Et \rightleftharpoons O^{\ominus} + I_2HC$ OEt

$CHI_2\overset{\ominus}{CO_2} + EtOH \leftrightharpoons CHI_2COOH + \overset{\ominus}{O}Et \leftarrow I_2HC$ OH OEt

CHI_2CO_2H is a SE for a number of TM

$\overset{\ominus}{O}Et$ is a better leaving group than $\overset{\ominus}{C}HI_2$

8.

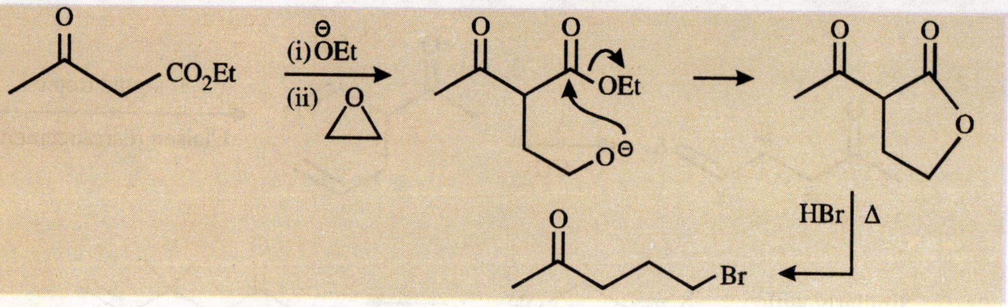
CO2Et
(i) ⊖OEt
(ii)
OEt
O⊖
HBr | Δ
Br

9. **C—N bond formation with EAA**

(i)

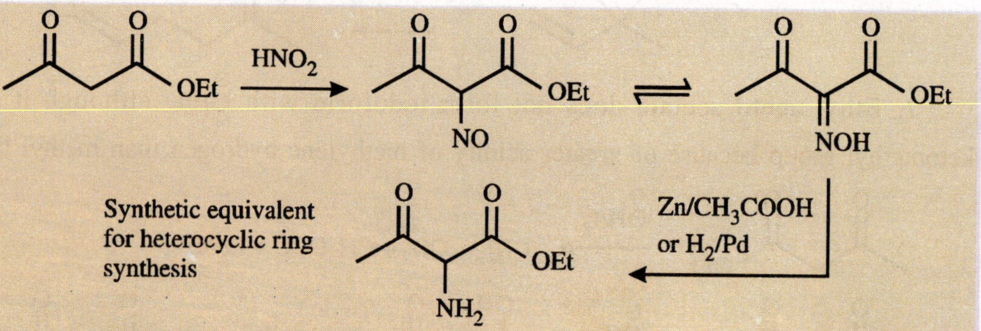
OEt
HNO2
NO
NOH
Synthetic equivalent for heterocyclic ring synthesis
Zn/CH3COOH or H2/Pd
NH2

(ii)

OEt
NH3/Δ
NH2

enamine (ambident nucleophile) β-amino α, β unsaturated ester. It is a good Michael donor. Synthetic equivalent for various heterocyclic ring synthesis

(iii)

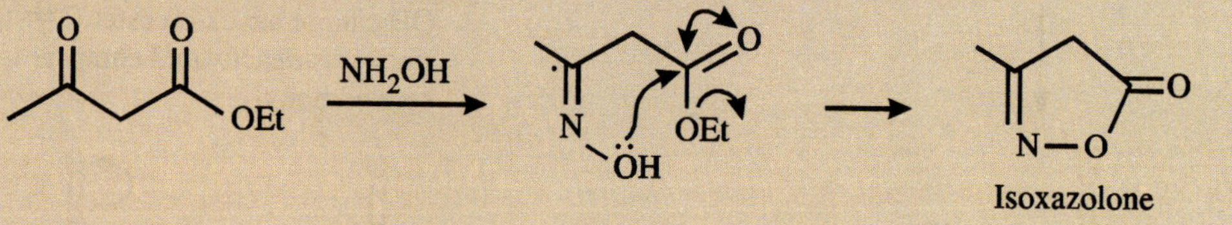
OEt
NH2OH
N
ÖH
N—O
Isoxazolone

(iv)

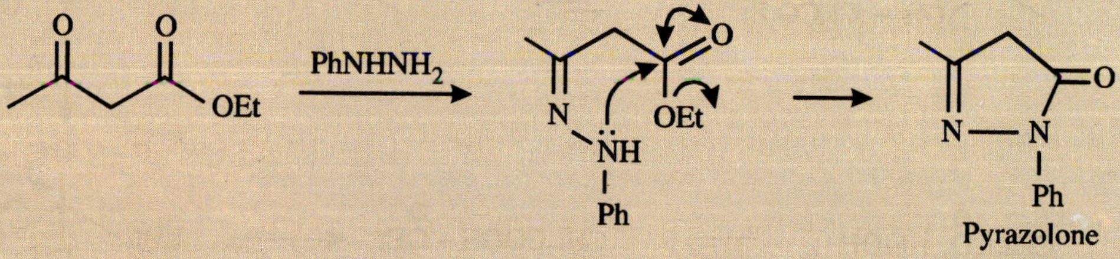
OEt
PhNHNH2
N
NH
Ph
N—N
Pyrazolone

(v)

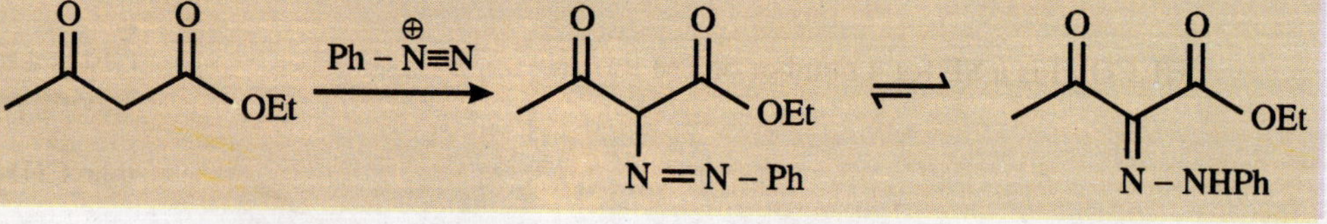
OEt
Ph – N⊕≡N
N=N – Ph
N – NHPh

10. (i) When ethyl aceto acetate is heated under reflux (200°C) with a trace of $NaHCO_3$, EtOH is eliminated and Dehydro acetic acid is formed.

NaHCO$_3$ 200°C; Claisen condensation (1); NaHCO$_3$; (2)

Dehydro acetic acid
(α or 2-Pyrone)

(1) Claisen condensation takes place rather than aldol condensation since the Claisen condensation product is removed from equilibrium as a strongly resonance stabilized anion.

(2) Enolate anion condensation through oxygen and gives 6-membered ring and not through enolate carbon which would give 4-membered ring which is unstable.

(ii) When Dehydro acetic acid is refluxed with EtOH in presence of acid, γ or 4-Pyrone is formed via hemiacetal and dehydration.

Dehydro acetic acid $\xrightarrow[\text{ethanolysis}]{EtOH/H^{\oplus}}$ → $\xrightarrow{H^{\oplus}}$ (acid catalysed hemiacetal formation) $\xrightarrow{-H_2O}$ γ – or 4-Pyrone

11. When ethyl aceto acetate is allowed to stand in conc. H_2SO_4 (5-6 days) then Isodehydro acetic acid and its ester are formed.

EtO_2C... $\xrightarrow[\text{5-6 days, allowed to stand}]{\text{Conc. } H_2SO_4}$ Isodehydro acetic acid

Acid catalysed self-condensation on the carbonyl carbon since it is more electrophilic than ester C=O group

(i) hydrolysis
(ii) – H_2O

4.1.2 Alkylation and hydrolysis of EAA

EAA

There are two positions for alkylation of EAA molecule, methylene hydrogen (α) and methyl hydrogen (γ).

a-alkylation

Since methylene C—H is adjacent to two – M group and hence highly acidic and may be alkylated in milder condition.

EAA may be mono and di-alkylated by primary and secondary halides in presence of base. Now-a-days tosyl derivative (ROTs) is employed.

The products undergo two types of hydrolysis depending on conditions.

(i) Acid medium hydrolysis of ester group and the resulting β-keto acid loses CO_2.

$$CH_3COC(Me)_2CO_2Et \xrightarrow[HCl + H_2O]{dil\ HCl/\Delta} CH_3COC(Me)_2COOH \xrightarrow{-CO_2} CH_3COCHMe_2$$

Ketone derivative

(ii) Concentrated aq. ethanolic NaOH induces cleavage of a C—C bond in the reverse manner of Claisen condensation.

$$CH_3COC(Me)_2CO_2Et \xrightarrow[\Delta]{\overset{\ominus}{O}H + H_2O + EtOH} CH_3C(O^{\ominus})(OH)C(Me)_2CO_2Et \longrightarrow CH_3COOH + \overset{\ominus}{C}(Me)_2CO_2Et \rightleftharpoons CH_3\overset{\ominus}{CO_2} + Me_2CHCO_2Et$$

acid derivative

The equilibria favours to the products because of the stability of $CH_3\overset{\ominus}{CO_2}$ anion.

It is, therefore, possible to synthesize both acetone derivative $CH_3COCHMe_2$ [process (i)] and acetic acid derivative Me_2CHCO_2H [process (ii)]. The former i.e. acetone derivative synthesis is more important since the process (ii) hydrolysis may be contaminated with by-products derived from ketone breakage. Acetic acid derivative is best synthesized from malonic ester.

Introduction of tertiary butyl (Me_3C-) group into methylene carbon (α-position)

Tertiary alkyl halides like Me_3C—Br cannot be used for alkylation of methylene carbon of EAA under basic condition beause of El elimination takes place.

$$CH_3COCH_2CO_2Et + Me_3C-Br \xrightarrow{\overset{\ominus}{O}Et} CH_3CO\overset{\ominus}{C}HCO_2Et + (CH_3)_2C{=}CH_2 + EtOH$$

Both $\overset{\ominus}{O}Et$ and $CH_3CO\overset{\ominus}{C}HCO_2Et$ can act as bases to give El elimination.

Tertiary butyl group may be introduced using acidic condition and tertiary halides or isobutene as reagent. In this case the enol form of EAA and not the enolate ion that takes part in the reaction. Usually Lewis acid BF_3 and strong acid like $HClO_4$ ($ClO_4^{\ominus}$ is non-nucleophilic and least basic) are used as acidic reagent.

(i) Using BF_3

$Me_3C-Br + BF_3 \xrightarrow{\text{ether}} Me_3\overset{\oplus}{C}\ \overset{\ominus}{B}F_3Br$ tight ion-pair

$CH_3COCH_2CO_2Et \xrightarrow[0-10°C]{BF_3/\text{ether}/Me_3CBr} CH_3COCH(CMe_3)CO_2Et$

$CH_3COCH_2CO_2Et \xrightarrow{BF_3} CH_3C(=\overset{\oplus}{O}-\overset{\ominus}{B}F_3)CH_2CO_2Et \longrightarrow CH_3C(O-\overset{\ominus}{B}F_3)=CHCO_2Et \xrightarrow{Me_3\overset{\oplus}{C}\ \overset{\ominus}{B}F_3Br} CH_3COCH(CMe_3)CO_2Et$

(ii) Using $HClO_4$

$CH_3COCH_2CO_2Et \xrightarrow[\text{(ii) } (CH_3)_2C=CH_2/NaHCO_3]{\text{(i) } HClO_4,\ CH_3NO_2\ (\text{solv})/25°C} CH_3COCH(CMe_3)CO_2Et$

$CH_3COCH_2CO_2Et \xrightarrow{HClO_4} CH_3C(OH)=CHCO_2Et \xrightarrow[NaHCO_3]{Me_3\overset{\oplus}{C}\ Cl\overset{\ominus}{O}_4} CH_3COCH(CMe_3)CO_2Et$

$Me_2C=CH_2 \xrightarrow{HClO_4 \mid CH_3NO_2} Me_3\overset{\oplus}{C}\ Cl\overset{\ominus}{O}_4$ tight ion-pair

(iii) EAA has about 5-10% enol-content. Enol can act as a nucleophile. Alkylation under acidic (or neutral) conditions always involves enol as nucleophile.

$$Me_3C-Br \xrightarrow{AgBF_4} Me_3\overset{\oplus}{C}\ \overset{\ominus}{BF_4} + AgBr$$

$Me_3\overset{\oplus}{C}\ \overset{\ominus}{BF_4}$; CO_2Et; CMe_3

(As diethyl malonate does not have detectable enol content, it cannot be alkylated through mechanism involving enol as nucleophile).

Introduction of phenyl group into methylene carbon of EAA

Phenyl group cannot be introduced into the methylene carbon under similar usual conditions. This is because of the fact that aryl halides do not participate in the S_N2 reaction.

Phenylation of EAA is done using benzyne as intermediate. EAA is treated with 2eq. of $NaNH_2$.

$\overset{\ominus}{NH_2}$/liq.NH_3, 2eq.; NH_3

Phenylated EAA can be converted into a ketone or acid by hydrolysis in different condition.

dil HCl, Δ

(i) Conc.aq. ethanolic NaOH

(ii) $H_3\overset{\oplus}{O}$/Δ

COOH

γ-Alkylation: alkylation at the terminal methyl group

When EAA is treated with a very strong base then in addition to methylene carbanion, terminal methyl carbanion will also form in which methyl carbanion is more nucleophilic since methylene carbanion is more resonance stabilised by adjacent two carbonyl groups.

$NaNH_2$/liq.NH_3 1eq.; $NaNH_2$ / (second eq.); CH_3Br 1eq.; aq. NH_4Cl or $CH_3COOH + CH_3CO_2K$ (strong acid may hydrolyse ester); CO_2Et; OEt

Synthesis of isomeric 1, 3-diketone (β-diketone)

The method of α and γ-acylation may be used to synthesize isomeric 1, 3-diketones from EAA. In the following examples where I and II are isomeric 1, 3-diketones.

α-acylation

(i) NaH/C_6H_6 (ii) PhCOCl; (i) $\overset{\ominus}{O}Et$ (ii) CH_3Br; dil HCl Δ; (I)

$\overset{\ominus}{O}Et$/EtOH combination should not be used since it will consume acid chloride PhCOCl

γ-acylation

LDA/THF 2eq.

Ph, Cl, O, O, O, Ph, CO$_2$Et, dil HCl, Δ, (II)

γ-acylation through cyclisation

(i) $\overset{\ominus}{O}$Et, (ii) OEt, CO$_2$Et, EtO, $\overset{\ominus}{O}$Et/EtOH (Claisen cond.) γ-acylation

4.1.3 Synthetic application of EAA

EAA may be used to synthesize the following classes of target molecules.

1. Monoketones 2. Diketones (1, 3 or 1, 4 or 1, 5 or 1, 6) 3. Monocarboxylic acids 4. Dicarboxylic acids 5. Keto acids 6. Amino acids 7. Coumarin derivative 8. Chromone derivative 9. Uracil derivative 10. Antipyrine derivative 11. Pyrone derivative 12. Heterocyclic ring (i) Furan derivative, (ii) Pyrrole derivative, (iii) Pyridine derivative, (iv) Quinoline derivative etc.

4.1.4 Synthesis of target molecules starting from EAA with retrosynthetic analysis

1. TM (1, 3-diketo) ⟹ SN + $\oplus$ SN ⟹ Cl (SE); SN ⟹ CO$_2$Et (SE)

Forward Synthesis

O ‖ CO_2Et —(i) NaH/C_6H_6, (ii) acetyl chloride (O, Cl)→ O O, CO_2Et —dil HCl / Δ→ O O

2. O, TM (1, 4-diketo), O ⟹ O, SN ⊖ + ⊕ O, SN ⟹ Cl, O, SE

⟹ O, CO_2Et, SE

Cl_2 ⟹ AcOH, O

Forward Synthesis

O, CO_2Et —(i) $\overset{\ominus}{O}Et/EtOH$, (ii) Cl, O→ O, CO_2Et, O —dil HCl/Δ→ O, O

3. O O, TM (1, 5-diketo) ⟹ O, ⊖ SN + ⊕ H SN, O

⟹ O, SE, CO_2Et ⟹ O, SE

⟹

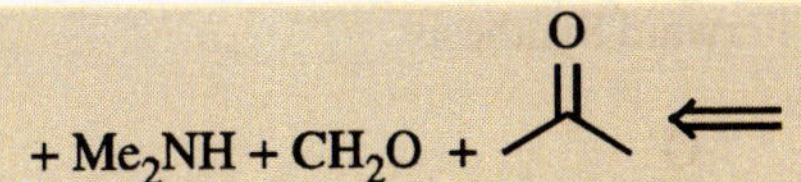

$+ Me_2NH + CH_2O +$ O ⟸

Forward Synthesis

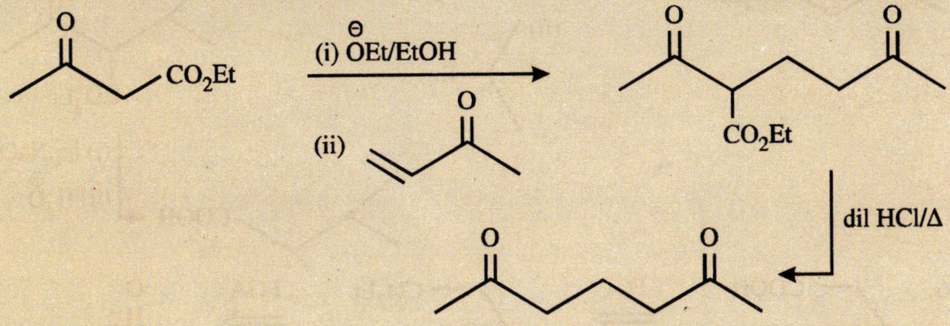

O CO_2Et (i) $\overset{\ominus}{O}Et/EtOH$ (ii) O → O O CO_2Et

dil HCl/Δ

O O

4. O TM (1, 6-diketo) O FGA ⟹ O CO_2Et CO_2Et O

⇓

O 2 CO_2Et + Br Br
SE SE

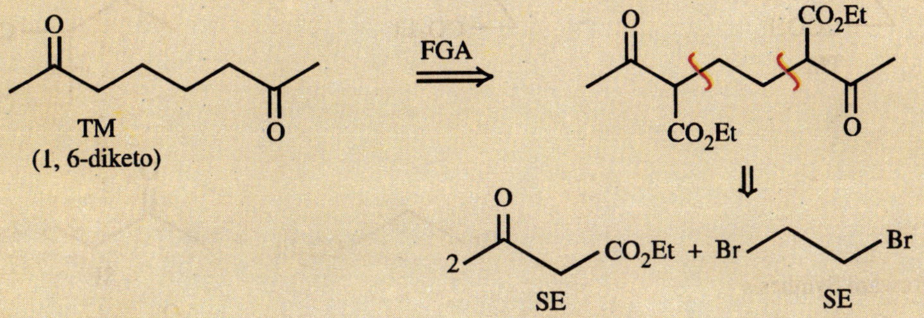

Forward Synthesis

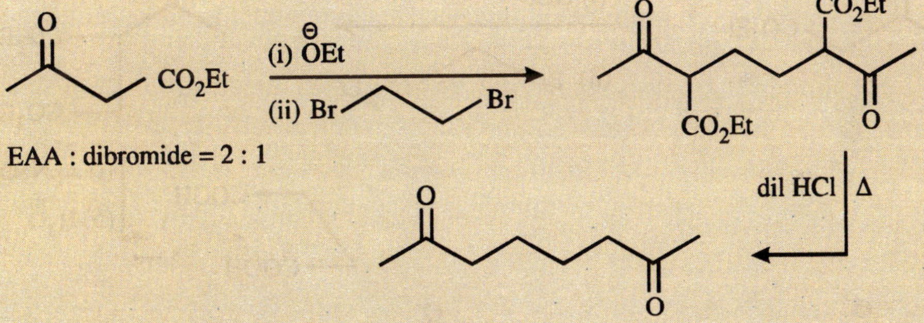

O CO_2Et (i) $\overset{\ominus}{O}Et$ (ii) Br Br → O CO_2Et CO_2Et O

EAA : dibromide = 2 : 1

dil HCl Δ

O O

5. CO_2H TM FGI ⟹ CO_2Et FGA ⟹ CO_2Et O

⇓

Br + CO_2Et O
SE SE

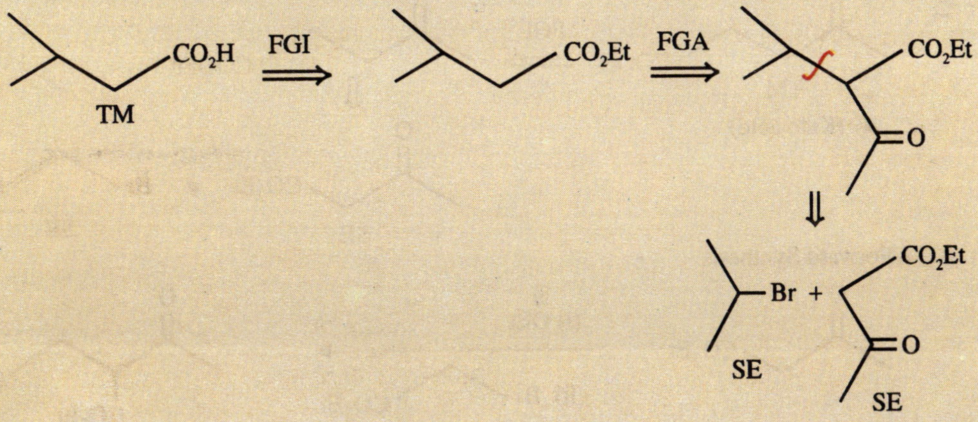

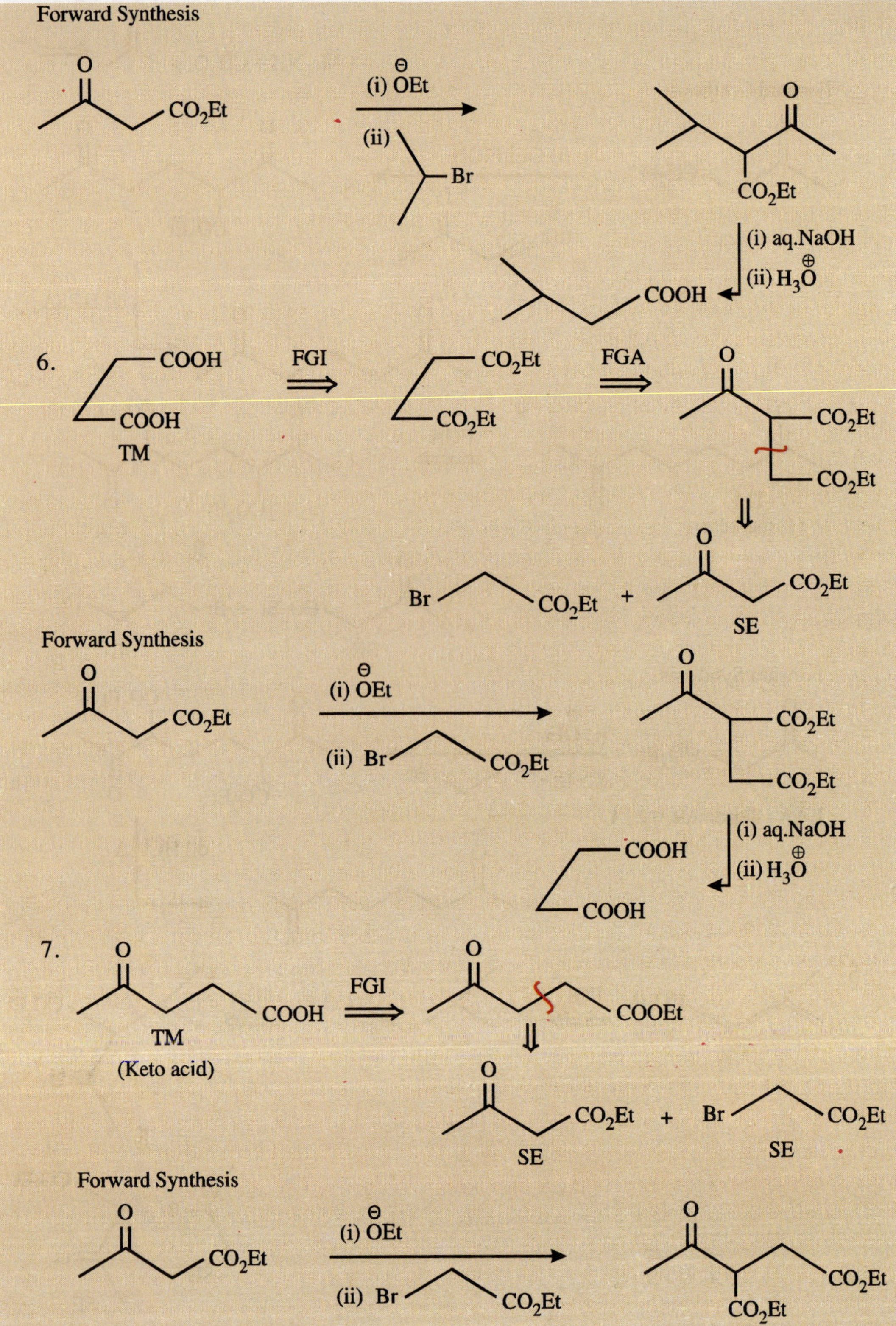

Forward Synthesis
O
CO2Et
(i) OEt
(ii) Br
O
CO2Et
(i) aq.NaOH
(ii) H3O
COOH
6.
COOH
COOH
TM
FGI
CO2Et
CO2Et
FGA
O
CO2Et
CO2Et
O
Br
CO2Et
+
CO2Et
SE
Forward Synthesis
O
CO2Et
(i) OEt
(ii) Br
CO2Et
O
CO2Et
CO2Et
(i) aq.NaOH
(ii) H3O
COOH
COOH
7.
O
COOH
TM
(Keto acid)
FGI
O
COOEt
O
CO2Et
+
Br
CO2Et
SE
SE
Forward Synthesis
O
CO2Et
(i) OEt
(ii) Br
CO2Et
O
CO2Et
CO2Et

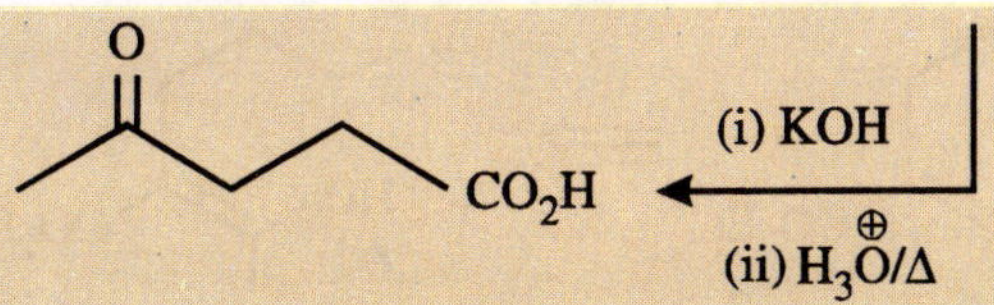

(i) KOH

(ii) $H_3\overset{\oplus}{O}/\Delta$

CO_2H

8. O COOH CO_2H TM FGI ⟹ O COOEt CO_2Et

O CO_2Et SE + CO_2Et CO_2Et SE

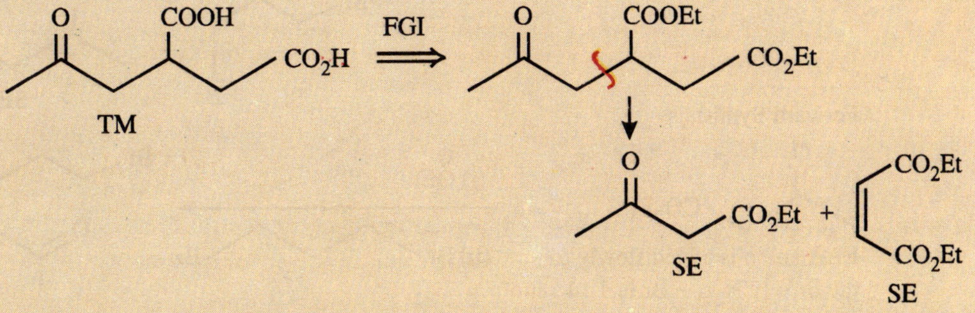

Forward Synthesis

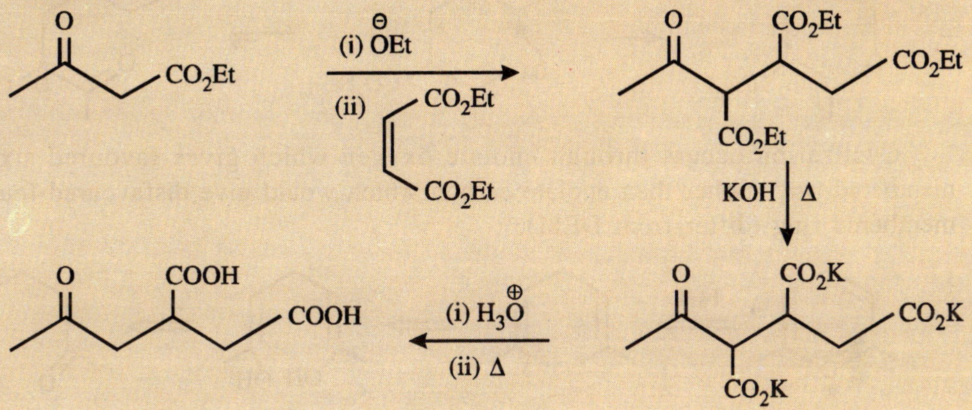

O CO_2Et (i) $\overset{\ominus}{O}Et$ (ii) CO_2Et CO_2Et → O CO_2Et CO_2Et CO_2Et

KOH Δ

O CO_2K CO_2K CO_2K

(i) $H_3\overset{\oplus}{O}$ (ii) Δ

O COOH COOH

9. O CO_2Et TM α, β unsat. C = O aldol dix ⟹ O O CO_2Et Michael dix ⟹ O SE + O CO_2Et SE

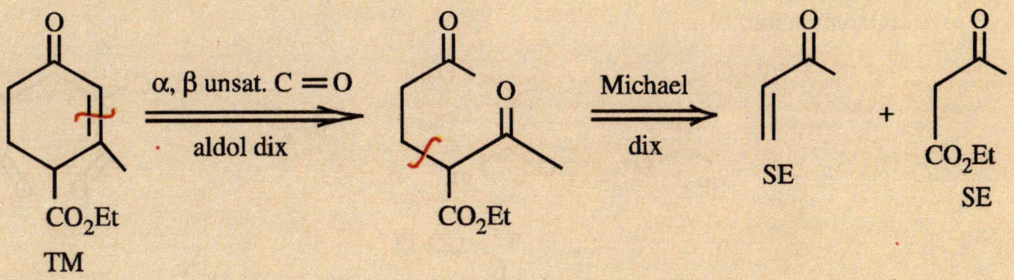

Forward Synthesis

O CO_2Et (i) $\overset{\ominus}{O}Et$ (ii) O → O O CO_2Et $Me_2NH/AcOH$ → O CO_2Et

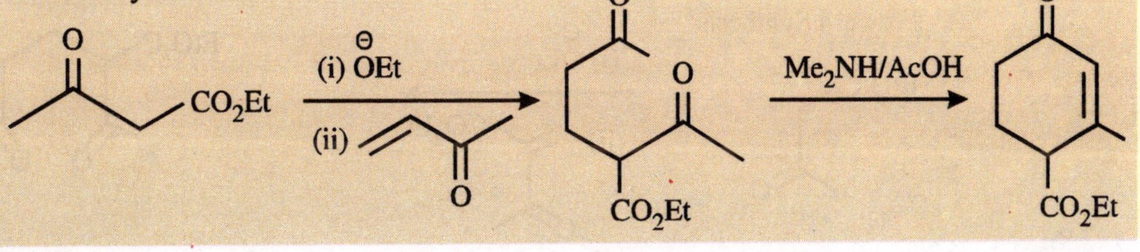

10.

TM

SE SE

Forward Synthesis

(i) $\overset{\ominus}{O}Et$

(ii) Br–$(CH_2)_3$–Br

EAA anion is added slowly to the Br – $(CH_2)_3$ – Br in 1 : 1 ratio

$\overset{\ominus}{O}Et$

Cyclization occurs through enolate oxygen which gives favoured six-membered ring rather than enolate carbon which would give disfavoured four membered ring (differ from DEM).

11.

FGI

TM

(Oxonium ion)

FGA

SE SE

+ Me_2NH + CH_2O

Forward Synthesis

(i) $\overset{\ominus}{O}Et$

(ii)

dil HCl Δ

$H_3O^{\oplus}$

hemiacetal

$- H_2O$

DDQ

Oxonium ion

12.

TM

+ PhCOCl

SE SE

Forward Synthesis

LDA 2eq.

PhCOCl

13.

TM

SE SE

Forward Synthesis

LDA, 2eq.

AcOH + AcOK

14.

TM

FGA

$\Rightarrow$ + PhCOCl

+ SE

$Me-O-CH=\overset{\oplus}{N}Me_2/MeS\overset{\ominus}{O_4}$

SE

$H-\underset{\underset{O}{\|}}{C}-NMe_2 + Me_2SO_4$

Forward Synthesis

$Et_3N/MeO-CH=\overset{\oplus}{N}Me_2$

room temperature

– MeOH

(i) LDA/–70°C

(ii) PhCOCl

$H_3\overset{\oplus}{O}$

vinylogous substitution

$-Me_2NH$

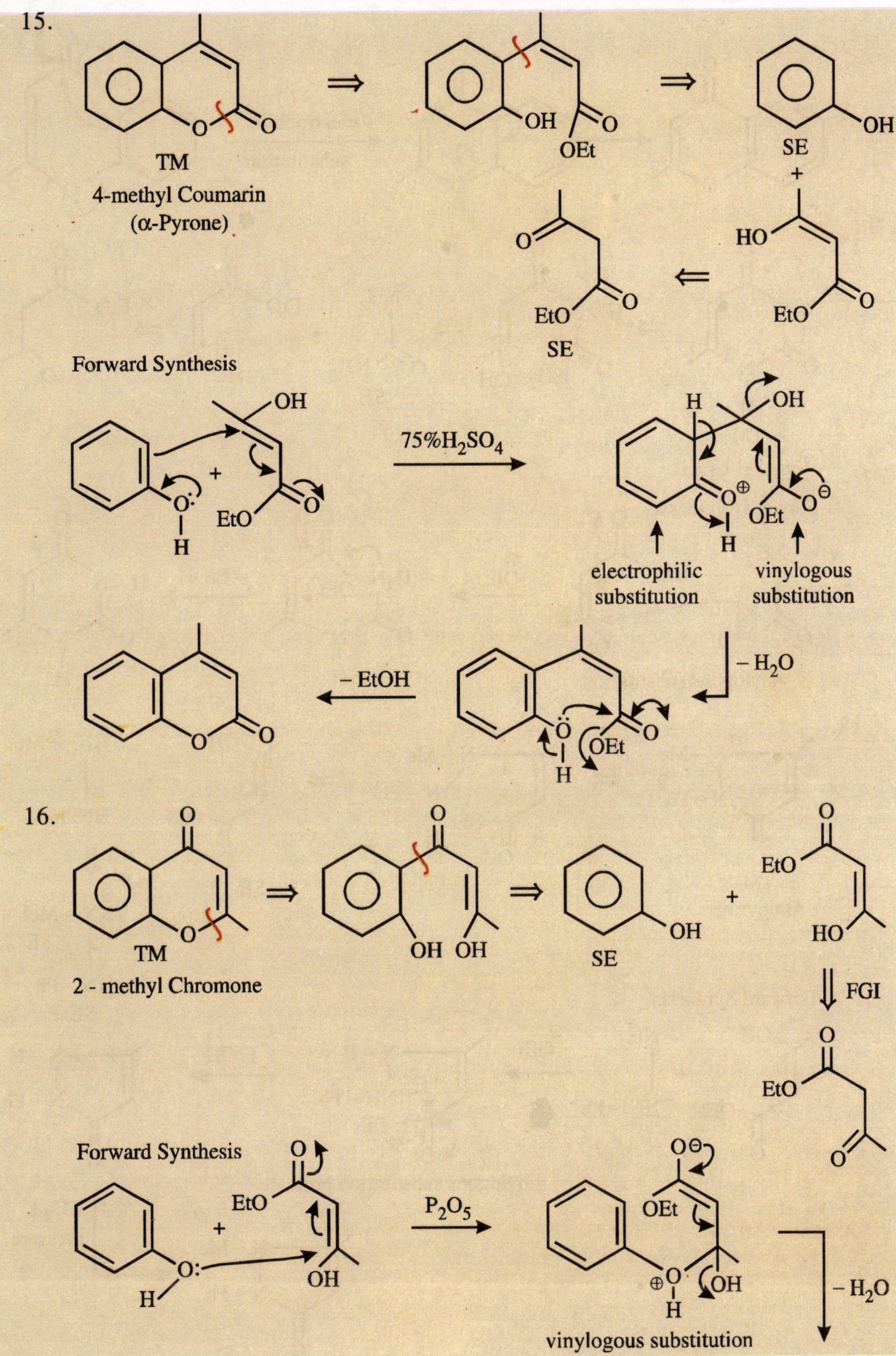
15.
TM
4-methyl Coumarin
(α-Pyrone)
OH
OEt
SE
+
HO
EtO
SE
Forward Synthesis
OH
EtO
H
75%H_2SO_4
electrophilic substitution
vinylogous substitution
OEt
− H_2O
− EtOH
16.
TM
2 - methyl Chromone
OH OH
SE
EtO
HO
FGI
Forward Synthesis
P_2O_5
vinylogous substitution
− H_2O

O H O EtO O

electrophilic substitution

17.

O O NH2 O O

HN H – N EtO EtO

⇒ ⇒ + ⇒

O N NH2 OH O NH2 HO O

H SE

TM

4-methyl Uracil

Forward Synthesis

NH2 EtO O ⊖OEt O H2N OEt – EtOH O H – N

O NH2 OH O N H O N H

vinylogous substitution

18.

N – Me N – Me OH HN–Me

⇒ N – Ph ⇒ N – Ph + HNPh

O OEt H O OEt ⇓

TM O SE H_2N + MeI

Antipyrine SE

HN – Ph

SE

Forward Synthesis

OH NH2 ⊖OEt N – H – EtOH N – H

+ :NH – Ph N – Ph

OEt HN – Ph OEt O

O O

vinylogous substitution product

CH_3 I pressure

N – Me

N – Ph

O

19.

CO_2H NH_2 TM Isoleucine $\xrightarrow{FGI}$ CO_2Et NH_2 $\Rightarrow$ Br SE + $\ominus$ CO_2Et NH_2 activation requires

O CO_2Et NH_2 $\xleftarrow{FGI}$ O CO_2Et NOH $\Leftarrow$ O CO_2Et NO

O CO_2Et

SE
+
HNO_2
SE

Forward Synthesis

O CO_2Et $\xrightarrow{\ominus OEt/EtOH}$ O CO_2Et $\ominus$ $\xrightarrow[\text{alkyl halide}]{\text{Br}}$ CO_2Et =O

$\xrightarrow{EtONO + \overset{\ominus}{O}Et}$ CO_2Et N=O =O $\xrightarrow{\overset{\ominus}{O}Et}$ CO_2Et N=O OEt O$\ominus$

Nitrosyl group blockage the ionization of β-keto ester and reverse Claisen ester condensation cleavage the C – C bond

$\xrightarrow{-CH_3CO_2Et}$ CO_2Et N – $\overset{\ominus}{O}$

$\xrightarrow{H_3\overset{\oplus}{O}}$ COOH N – OH $\xrightarrow{Zn/HCl}$ CO_2H NH_2

With the appropriate choice of alkyl halide or any appropriate Michael acceptor one can synthesize the desired α-amino acid starting from EAA.

20.

CO_2H … O … =O

TM

Terpenylic acid

$\Rightarrow$ CO_2H, OH OH, =O

FGI $\Rightarrow$ CO_2Et, CO_2Et, OH

$\Downarrow$

CO_2Et, CO_2Et, O

+ CH_3MgBr

SE

$\Leftarrow$

$\ominus$ CO_2Et, O

activation requires

Br CO_2Et +

SE

$\Downarrow$

EtO_2C, CO_2Et, O $\Rightarrow$ O, CO_2Et + Br CO_2Et

SE SE

Forward Synthesis

EtO_2C, O $\xrightarrow[\text{Br} \frown CO_2Et]{\ominus OEt}$ CO_2Et, CO_2Et, O $\xrightarrow[\text{Br} \frown CO_2Et]{\ominus OEt}$ EtO_2C, CO_2Et, CO_2Et, O

(i) KOH $\downarrow$ (ii) $H_3\overset{\oplus}{O}/\Delta$

CO_2H, COOH, O $\xleftarrow{EtOH + H^{\oplus}}$ CO_2Et, CO_2Et, O $\xleftarrow[\text{(ii) } H_3\overset{\oplus}{O}]{\text{(i) } CH_3MgBr \text{ 1eq.}}$ CO_2H, OH OH, =O

$\Delta \downarrow - H_2O$

CO_2H, O, =O

21. **Synthesis of isomeric furan derivative starting from EAA.**

CO_2Et EtO_2C (I) (II)

TM FGI ⟹ OH ⟹ CO_2Et OH OH Cl ⟹ FGI

EtO_2C + O Cl SE SE ⟸ EtO_2C OH O Cl

⟹ O + Cl_2 + AcOH

(i) Feist-Benary Synthesis

First step aldol condensation followed by second step intramolecular halogen displacement, medium is water and the reagent is NaOH.

EtO_2C O —aq. NaOH, room temp.→ CO_2Et O Cl → H CO_2Et OH O Cl

Usually α-halo ketone with enolate of EAA gives S_N2 reaction preferentially but here since the medium is H_2O and therefore S_N2 is arrested and aldol condensation occurs preferentially

– HCl

CO_2Et (I) ← $-H_2O$ — CO_2Et OH H O (isolable)

intramolecular halogen displacement occurs in base medium through enolate oxygen to give five-membered ring due to entropy advantage

(ii) Ring closure via 1, 4-dicarbonyl

First step is intermolecular halogen displacement followed by second step which is ring closure, medium is acetone and the reagents are NaOH and catalytic amount of NaI.

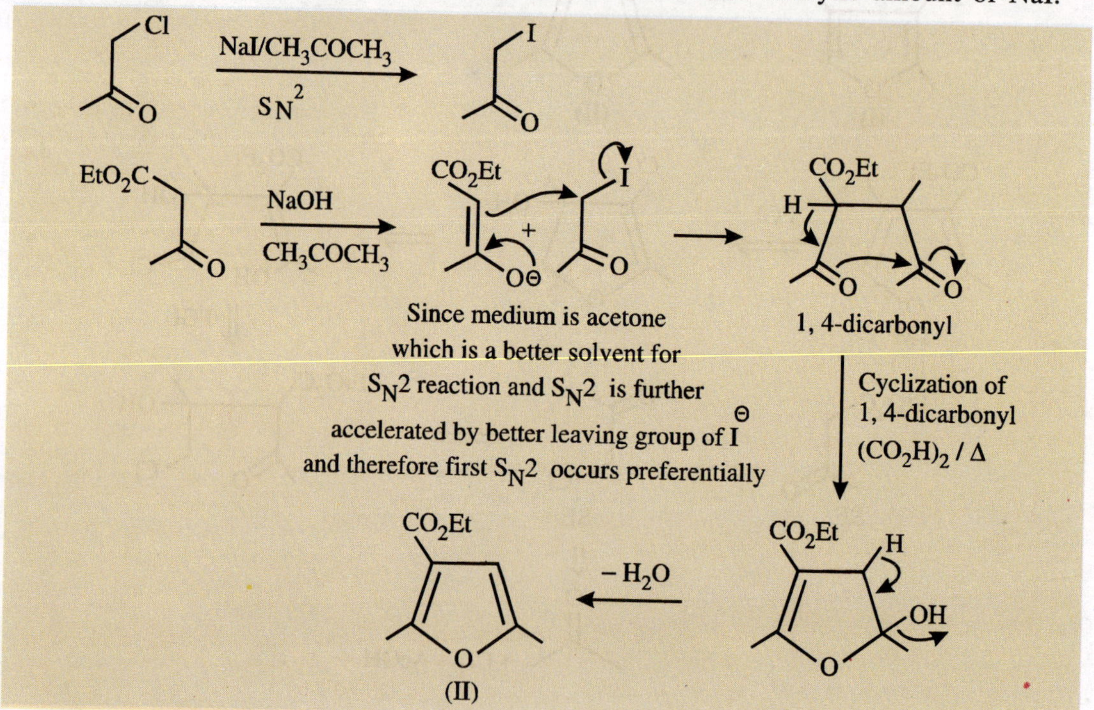

Cl ... NaI/CH_3COCH_3, S_N2 → I ... O

EtO_2C ... O → NaOH, CH_3COCH_3 → CO_2Et ... $O^{\ominus}$ + I ... O → CO_2Et, H ... O ... O

1, 4-dicarbonyl

Since medium is acetone which is a better solvent for S_N2 reaction and S_N2 is further accelerated by better leaving group of $I^{\ominus}$ and therefore first S_N2 occurs preferentially

Cyclization of 1, 4-dicarbonyl $(CO_2H)_2$ / Δ

CO_2Et, H, OH, O → $-H_2O$ → CO_2Et, O

(II)

22. Synthesis of heterocyclic ring pyrrole derivative starting from EAA.

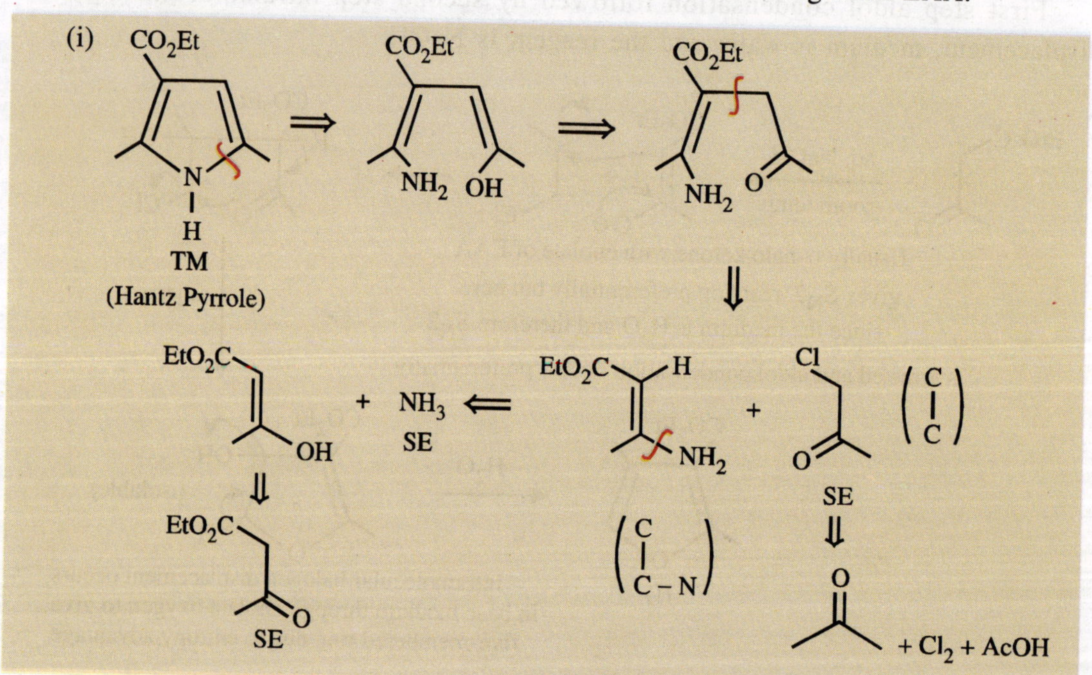

(i) CO_2Et, N, H

TM

(Hantz Pyrrole)

⇒ CO_2Et, NH_2 OH ⇒ CO_2Et, NH_2, O

⇓

EtO_2C, H, NH_2 (C–C–N) + Cl, O (C–C) SE

⇓

O + Cl_2 + AcOH

EtO_2C, OH + NH_3 SE ⇐

⇓

EtO_2C, O SE

Synthesis of Hantz Pyrrole

EtO$_2$C + NH$_3$ → OH NH$_2$ EtO$_2$C Δ → EtO$_2$C NH$_2$

β-amino α, β-unsaturated ester, ambident nucleophile, enamine (activation of ene)

EtO$_2$C NH$_2$ + Cl O → CO$_2$Et H NH$_2$ ⊕ O → CO$_2$Et NH$_2$ O

CO$_2$Et N H ← $-$ H$_2$O ← CO$_2$Et H OH N H

(ii) CO$_2$Et EtO$_2$C N H TM (Knorr Pyrrole) ⟹ CO$_2$Et EtO$_2$C NH$_2$ OH ⟹ CO$_2$Et EtO$_2$C NH$_2$ O

⟹

O EtO$_2$C NOH ⟸ FGI <H> O EtO$_2$C NH$_2$ (C | C − N) ⟸ OH EtO$_2$C NH$_2$ + CO$_2$Et O SE (C | C)

⟹

O EtO$_2$C NO ⟹ O EtO$_2$C SE + HNO$_2$ SE

Synthesis of Knorr Pyrrole

$NaNO_2$ / AcOH; EtO_2C, NO; NOH; Zn | gl.CH_3COOH; CO_2Et; NH_2; OH; $-H_2O$; enamine; $-H_2O$

23. **Synthesis of heterocyclic ring pyridine derivative starting from EAA.**

TM (Hantz Pyridine); FGI; NH_2 OH; (C, C–N); (C, C–C); FGI; $+NH_3$ SE; SE (C); SE; (C–C)

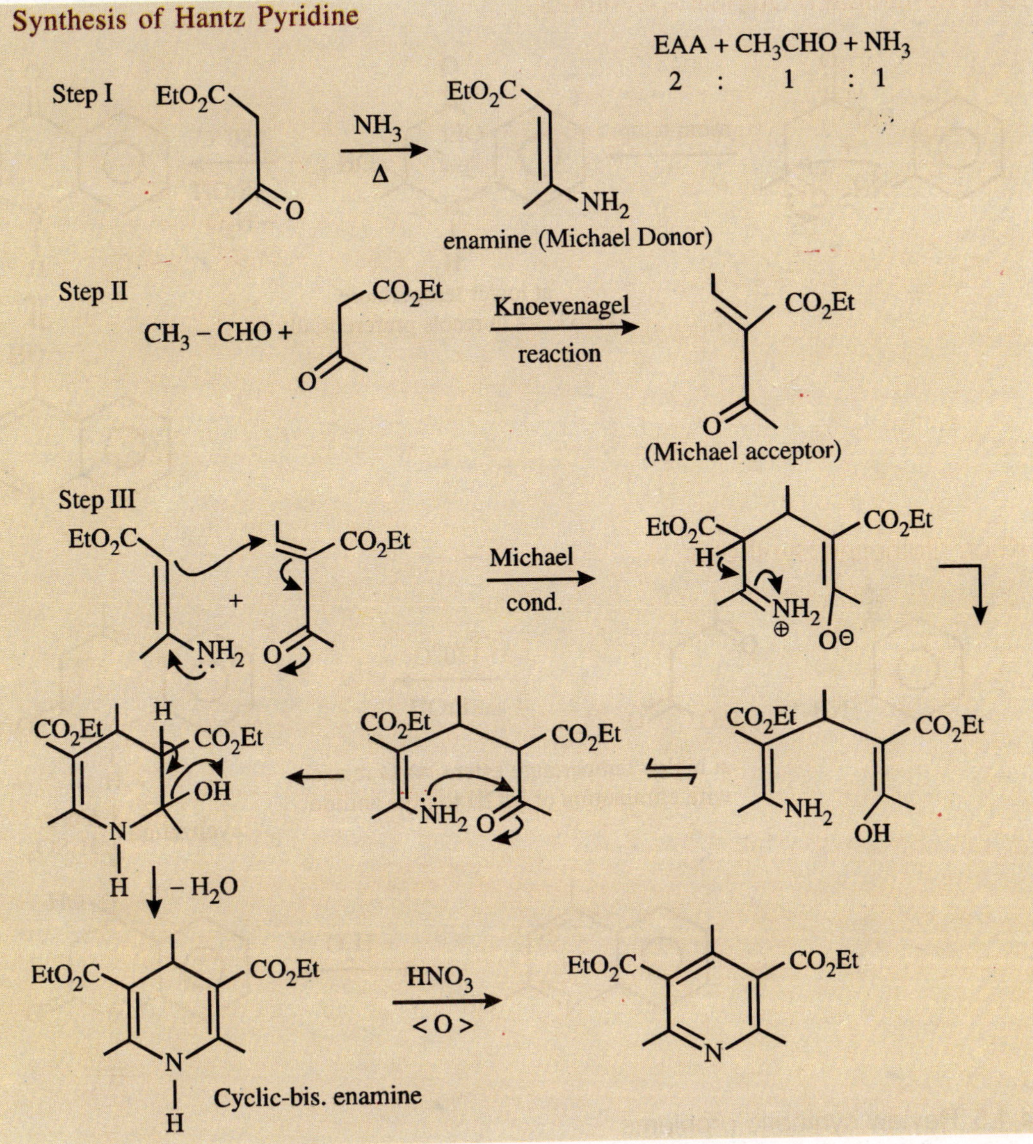

24. **Synthesis of isomeric heterocyclic ring Quinoline derivative starting from EAA.**

FGI

TM

OH

N

H

NH2 OEt

HO

OEt

SE

NH2

Conrad-Limpach Quinoline Synthesis

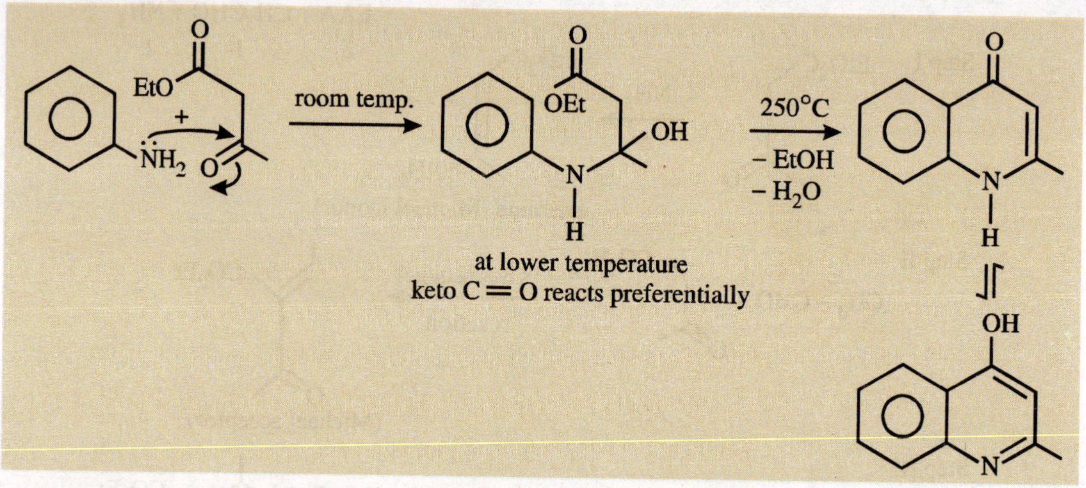

EtO, +, $\ddot{N}H_2$, O — room temp. → OEt, OH, N–H

at lower temperature
keto C=O reacts preferentially

250°C
– EtOH
– H_2O

Knorr Quinoline Synthesis

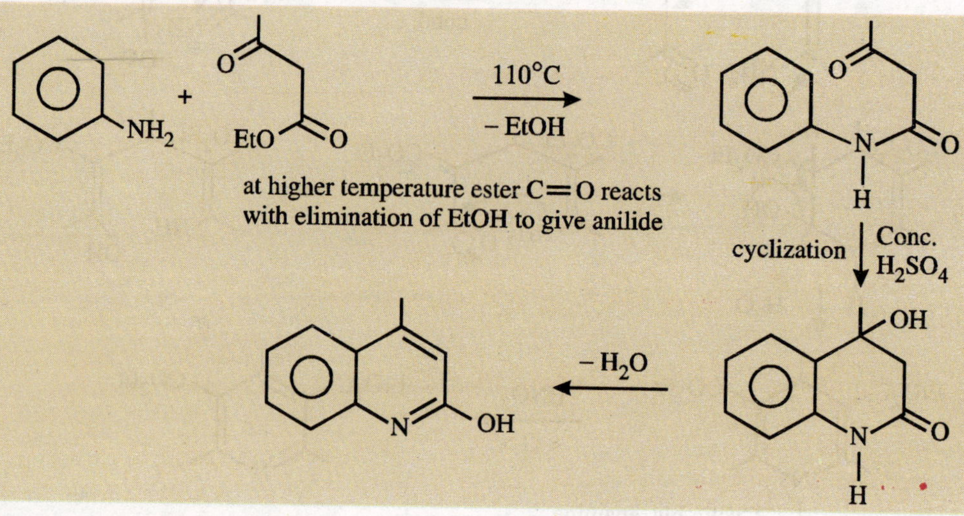

NH_2 + O, EtO, O — 110°C, – EtOH →

at higher temperature ester C=O reacts with elimination of EtOH to give anilide

cyclization, Conc. H_2SO_4

– H_2O

OH

4.1.5 Review synthetic problems

Give the retrosynthetic analysis of the following target molecules in which one of the synthetic equivalent is EAA and then show the forward synthesis with reasonable mechanism.

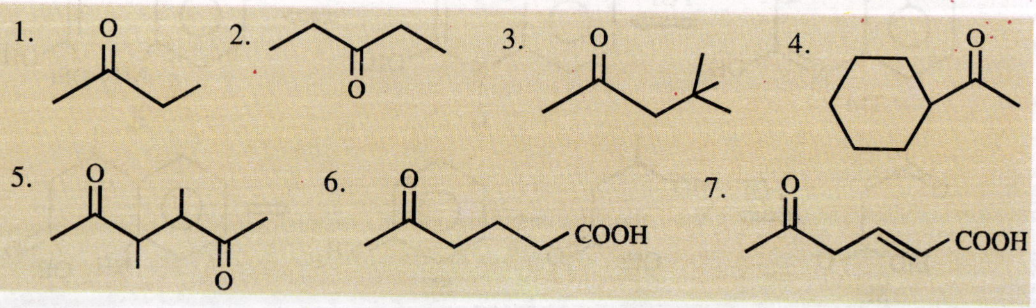

1. 2. 3. 4.

5. 6. COOH 7. COOH

8. HO O O

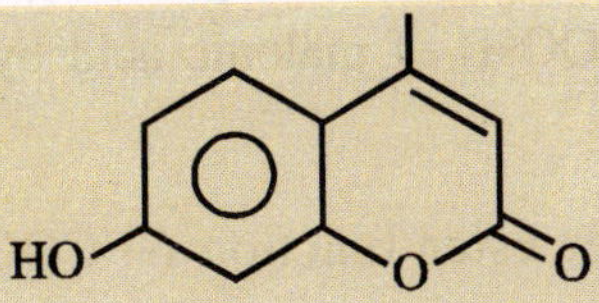

9. OH HO O O

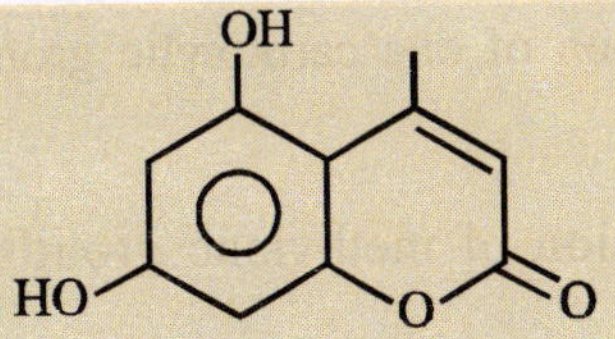

10. O Ph O O

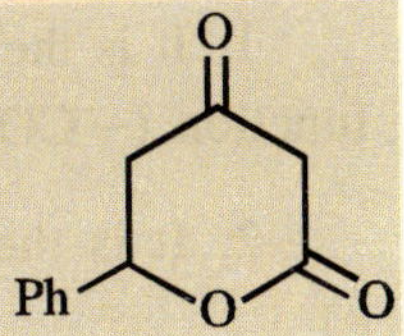

11. O O O

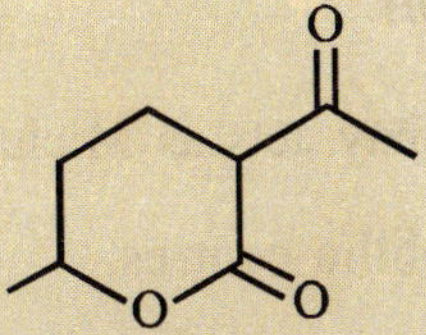

12. O

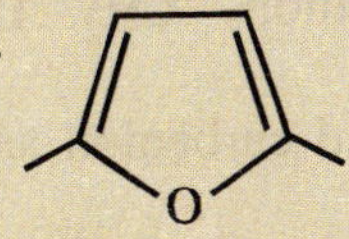

13. CO_2Et O

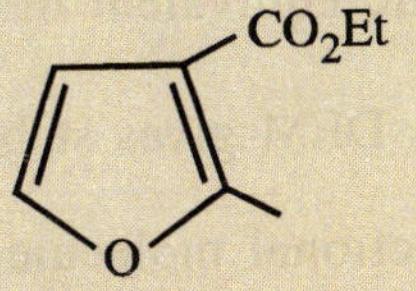

14. S H_2N N CO_2Et NOH

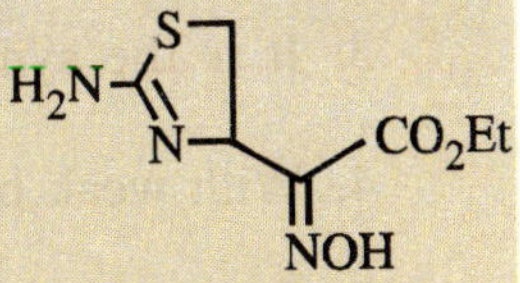

15. O O

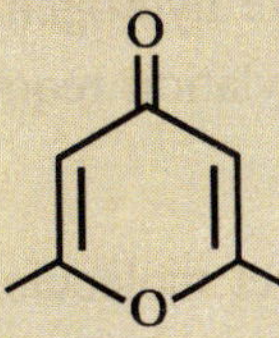

16. OH N N

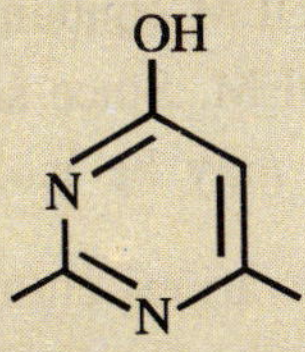

17. COOH O O

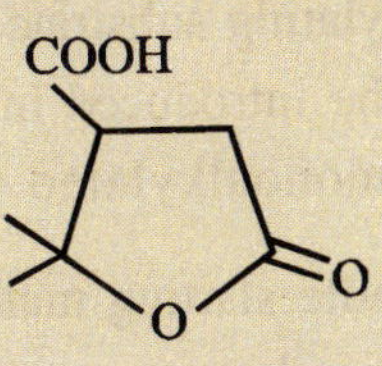

18. O O

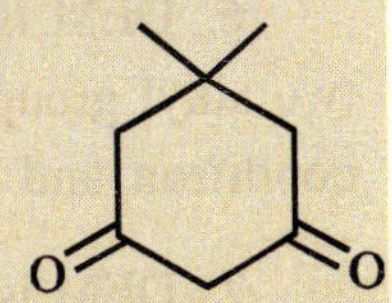

19. CO_2Et N

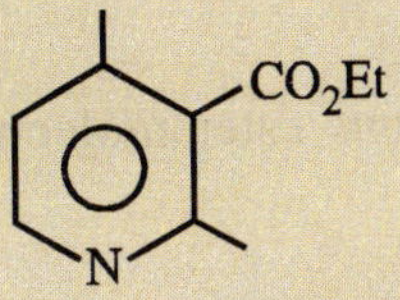

20. CN HO N OH

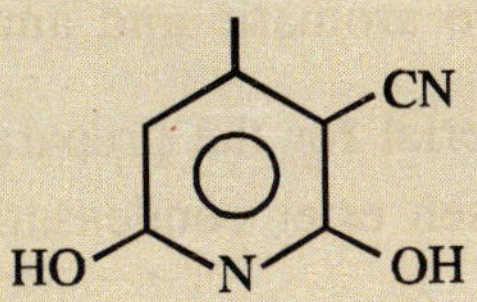

21. Ph EtO_2C CO_2Et N

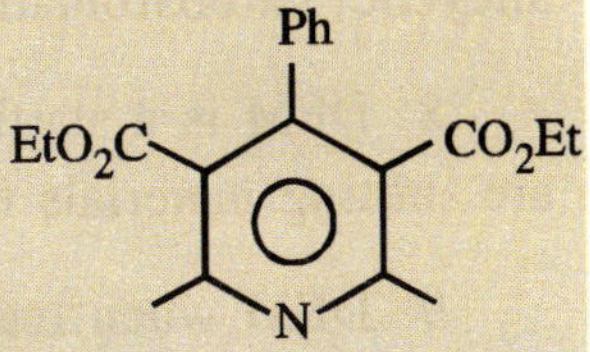

22. O CO_2Et O

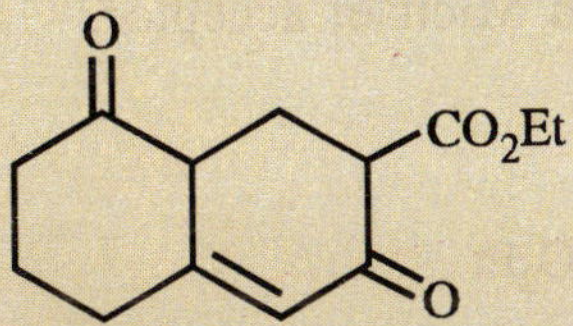

23. EtO_2C CO_2Et N

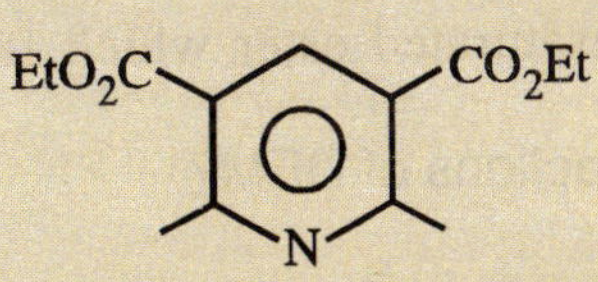

24. O CO_2Et

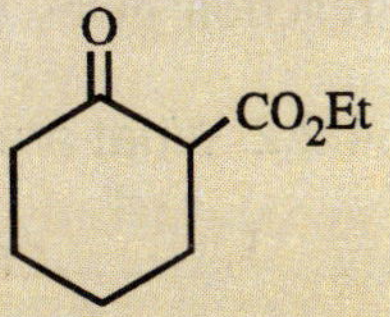

25. HO_2C COOH NH_2

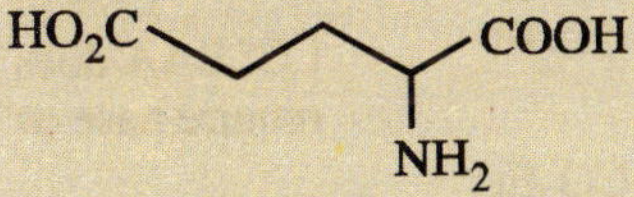

26. COOH NH_2

27. O

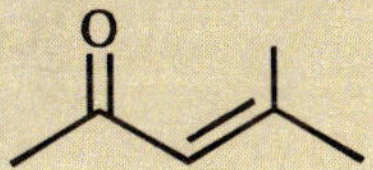

28. OH Me CO_2Et Me

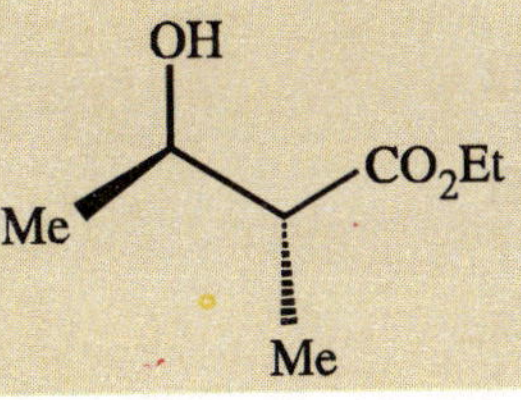

29. O CHO

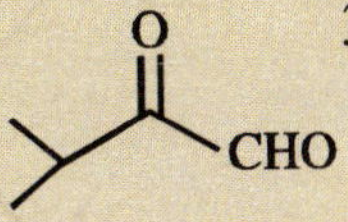

30. O CO_2Et

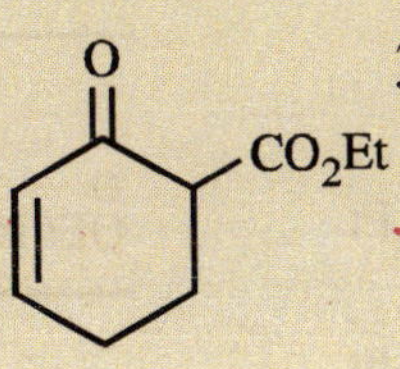

31. O MeN O N H

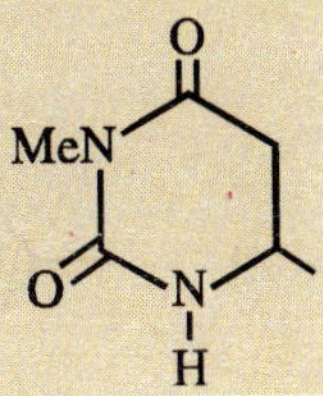

4.2 Diethyl malonate (DEM)

4.2.1 What is diethyl malonate?

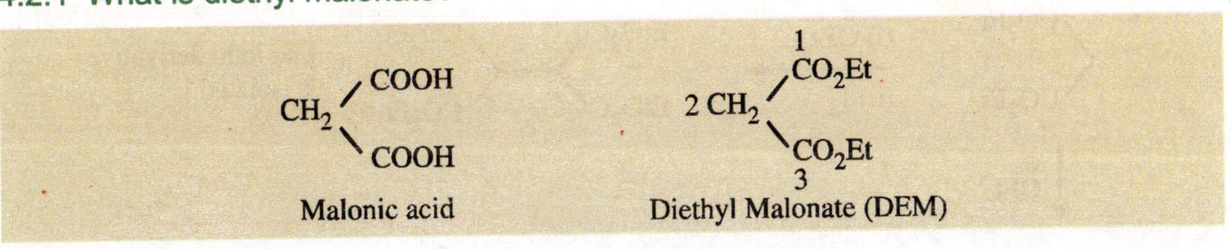

Malonic acid Diethyl Malonate (DEM)

1. It is the protection of one carboxylic group (– COOH) of malonic acid by an ester function (– CO_2Et).

2. It is the activation of methylene group by other ester function (– COOEt).

3. It is the synthetic equivalent of synthon, $\overset{\ominus}{C}H_2 - COOH$.

4. With weak base, DEM gives stabilised carbanion and may acts as Michael donor.

5. DEM is a bifunctional molecule having 1, 3-electrophilic centres.

6. DEM is an alkylating substrate. Since DEM contain two active hydrogen and hence two alkyl groups may be introduced into the DEM. Since second alkylation requires drastic condition and hence monoalkylated product may be isolated.

7. DEM is a versatile starting material for various classes of target molecules aliphatic, alicyclic, carboaromatic, hetero aromatic and amino acids.

8. DEM is a starting material for the preparation of various ester molecules which are starting materials for Claisen ester condensation.

9. DEM when reacts with aliphatic or aromatic aldehyde which is a knoevenagel reaction gives α, β unsaturated ester which is a good Michael acceptor.

4.2.2 What are the reactions of DEM?

1. $CH_2(CO_2Et)_2 \xrightarrow[\text{(ii) } Br_2]{\text{(i) } \overset{\ominus}{O}Et} BrCH(CO_2Et)_2$ [Cf EAA does not require base or acid]

$CH_2(CO_2Et)_2 \xrightarrow{\overset{\ominus}{O}Et/Cl_2} ClCH(CO_2Et)_2$

$CH_3CH(CO_2Et)_2 \xrightarrow{\overset{\ominus}{O}Et/Br_2} CH_3C(Br)(CO_2Et)_2$

This halogenated DEM is an important synthetic equivalent for the synthesis of α-amino acids.

2. $CH_2(CO_2Et)_2 \xrightarrow[\text{(ii) } I_2]{\text{(i) } \overset{\ominus}{O}Et} (EtO_2C)_2CH-CH(CO_2Et)_2$ [no iodo derivative is isolated]

$\downarrow \overset{\ominus}{O}Et$

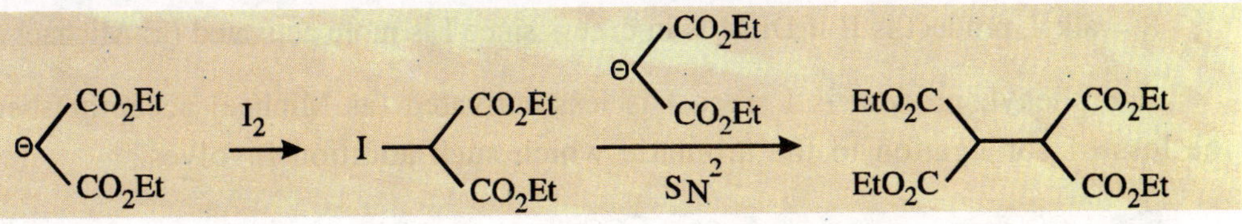

This dimeric product of DEM is a very important derivative for the synthesis of dicarboxylic acid.

3\.

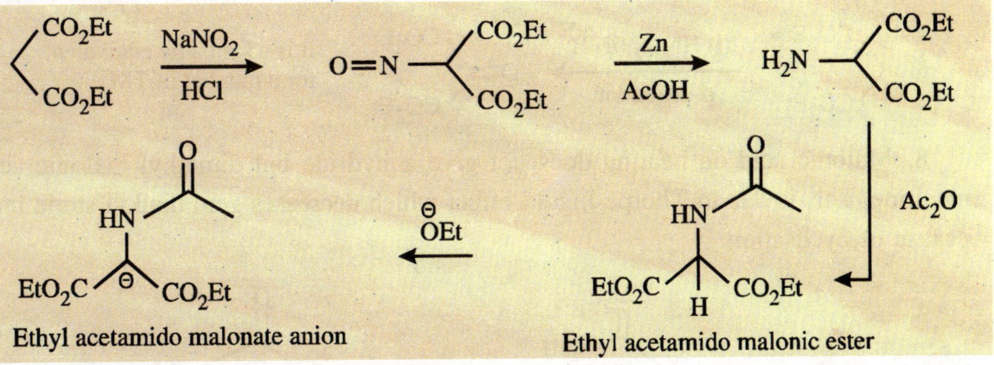

Ethyl acetamido malonate anion

Ethyl acetamido malonic ester

This is a very important derivative for the synthesis of α-amino acid and this is the synthetic equivalent of the synthon $H_2N\overset{\ominus}{C}H\,CO_2H$.

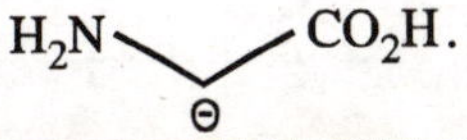

4\.

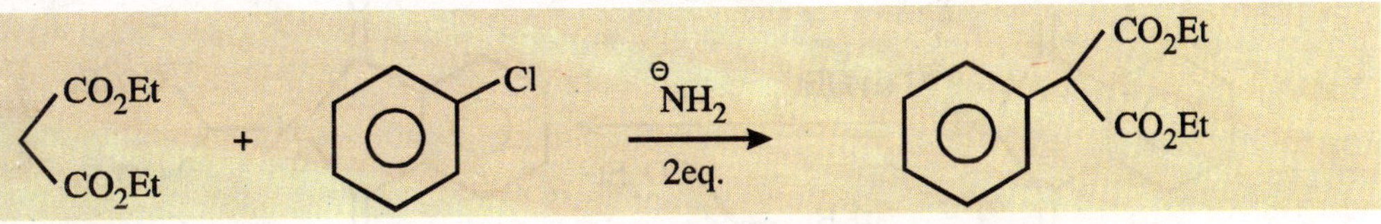

Halobenzene does not take part in S_N2 reaction with malonate anion. Therefore, benzene derivative of DEM is prepared via benzyne intermediate.

5\. Ketones do not undergo knoevenagel reaction with malonic acid or ester but they give reactions with aldehydes.

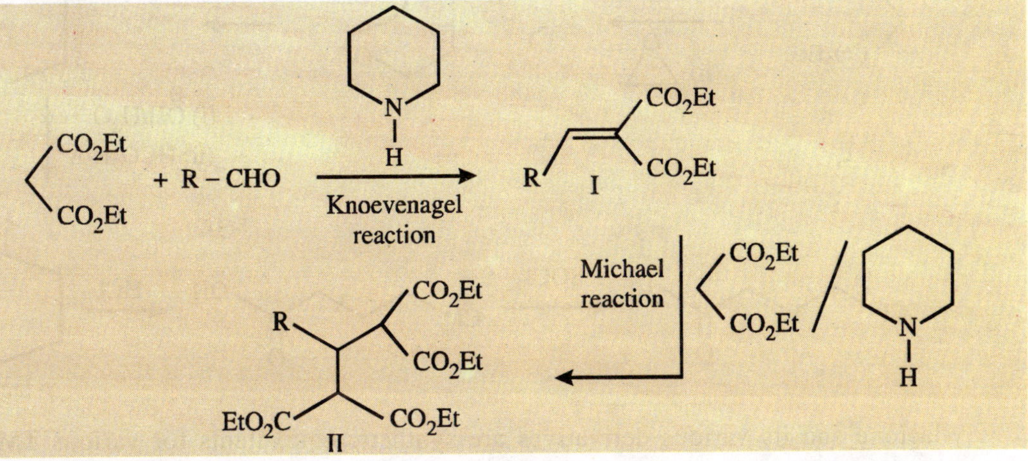

R = alkyl, product is II if DEM is in excess since I is more activated (as Michael acceptor)

R = phenyl, product is I since I is less activated (as Michael acceptor) because of the loss of conjugation to the aromatic which such addition involves.

6. $CH_2(CO_2Et)_2$ —KOH, 1eq.→ CO_2K/CO_2Et Very important derivative for α-amino acid synthesis

7. $CH_2(CO_2Et)_2$ —(i) HNO_2 (ii) $H^{\oplus}$, α-oxidation→ $O{=}C(CO_2Et)_2$ It is a synthetic equivalent for a number of TM

8. Malonic acid on heating does not give anhydride but dimethyl malonic acid gives anhydride. This is due to Thorpe-Ingold effect which decreases gem dialkyl steric interaction because of cyclisation.

OH, OH —Δ, $-H_2O$→ O

9. NH —(i) $\overset{\ominus}{O}Et$ (ii) Br–CH(CO_2Et)$_2$→ N–CH(CO_2Et)$_2$

It is a reaction to synthesize α-amino acid

10. $CH_2(CO_2Et)_2$ —(i) $\overset{\ominus}{O}Et$ (ii) ethylene oxide→ EtO_2C … OEt, $O^{\ominus}$ → EtO_2C lactone —(i) $\overset{\ominus}{O}H/H_2O$ (ii) HCO_2H/Δ→ γ- lactone —HCl→ Cl…OH —$SOCl_2$→ Cl…Cl

γ-lactone and its various derivatives are synthetic equivalents for various TM.

11. DEM does not take part in enolisation and therefore cannot be brominated only with Br_2 but in presence of base it can be brominated. Malonic acid, on the other hand, enolises significantly (unlike CH_3COOH) and bromination occurs readily with bromine alone.

$$CH_2(COOH)_2 \rightleftharpoons CH(COOH)=C(OH)OH \xrightarrow{Br_2} BrCH(COOH)_2$$

4.2.3 Alkylation and hydrolysis of Diethyl malonate.

$$H-CH(CO_2Et)_2$$

DEM

Since methylene C—H is adjacent to two – M groups (– CO_2Et) and hence acidic and may be alkylated in milder condition. DEM may be monoalkylated in the presence of ethoxide ion.

$$CH_2(CO_2Et)_2 \xrightarrow{\ominus OEt} {}^{\ominus}CH(CO_2Et)_2 \xrightarrow{Me-Br} Me-CH(CO_2Et)_2$$

The mono alkylated product still contains an activated α-hydrogen and a second alkyl group may be introduced.

$$Me-CH(CO_2Et)_2 \xrightarrow[\text{(ii) Me-Br}]{\text{(i) } \ominus OEt} Me_2C(CO_2Et)_2$$

Slightly more vigorous conditions are usually required for the second alkylation and it is, therefore, possible to isolate the monoalkylated product in good yield by using only one equivalent of the alkyl halide without external heating in about 80% yield.

Tertiary butylation of DEM

(i)

$$CH_2(CO_2Et)C(=O)OEt \xrightarrow[\text{buffer}]{AcOH + Me_2NH} CH(CO_2Et)=C(OH)OEt \xrightarrow{(CH_3)_2C=CH_2} (CH_3)_3C-CH(CO_2Et)_2$$

Specific enol — Kinetic control

(ii)

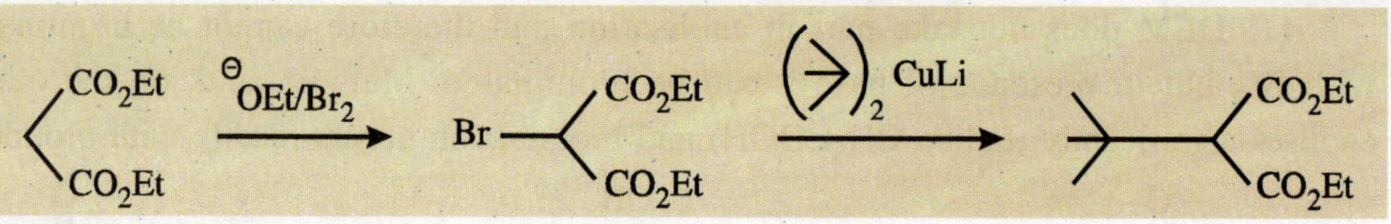

The substituted malonic ester may be hydrolysed to the acid, first by aq. KOH and then acidified with mineral acid.

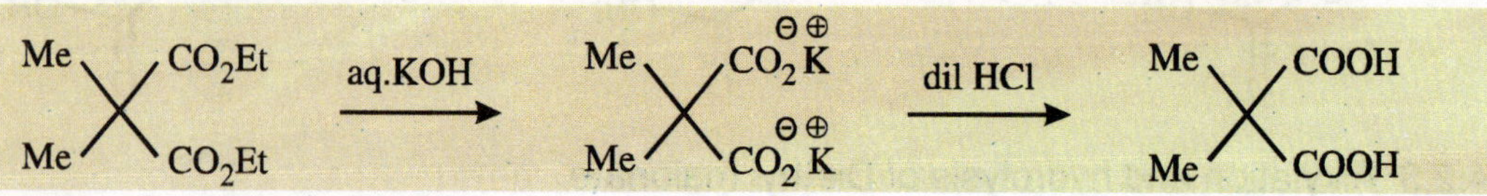

The gem dicarboxylic group may be easily decarboxylated on being heated to 250°C since it seems to be β-keto acid and can give six-membered cyclic TS.

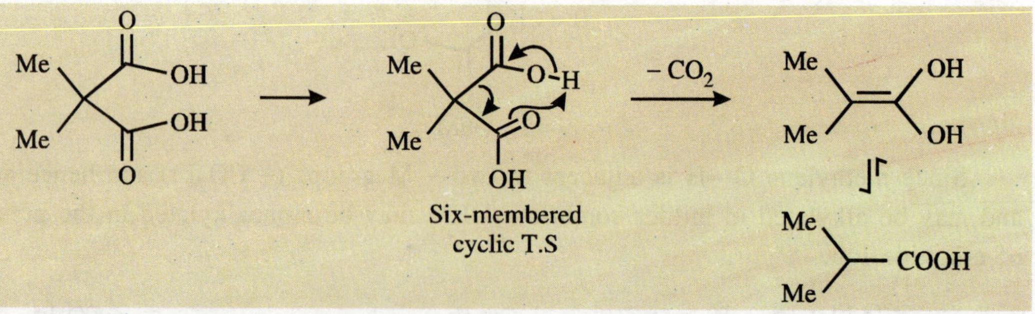

Six-membered cyclic T.S

4.2.4 Synthetic application of DEM

Diethyl malonate may be used to synthesize the following classes of target molecules.

1. Mono and dicarboxylic acid (aliphatic or alicyclic)
2. Tricarboxylic acids
3. α, β unsaturated acids
4. β-keto acids
5. γ-δ unsaturated acids
6. Aliphatic or Alicyclic ketones
7. Barbituric acids and its derivatives.
8. Amino acids.

4.2.5 Synthesis of target molecules starting from DEM with retrosynthetic analysis

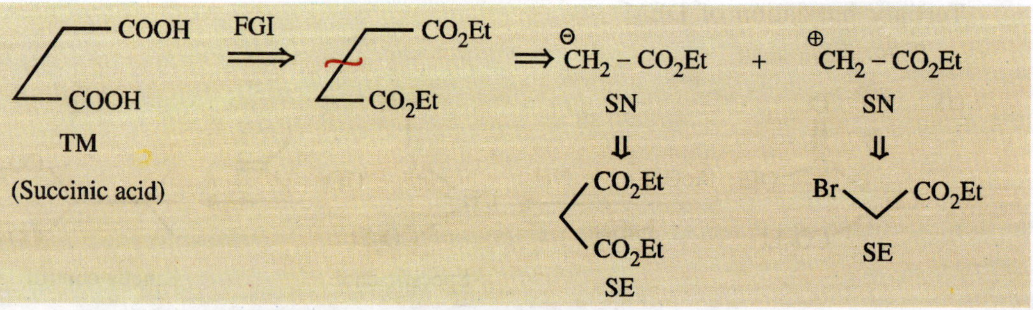

TM

(Succinic acid)

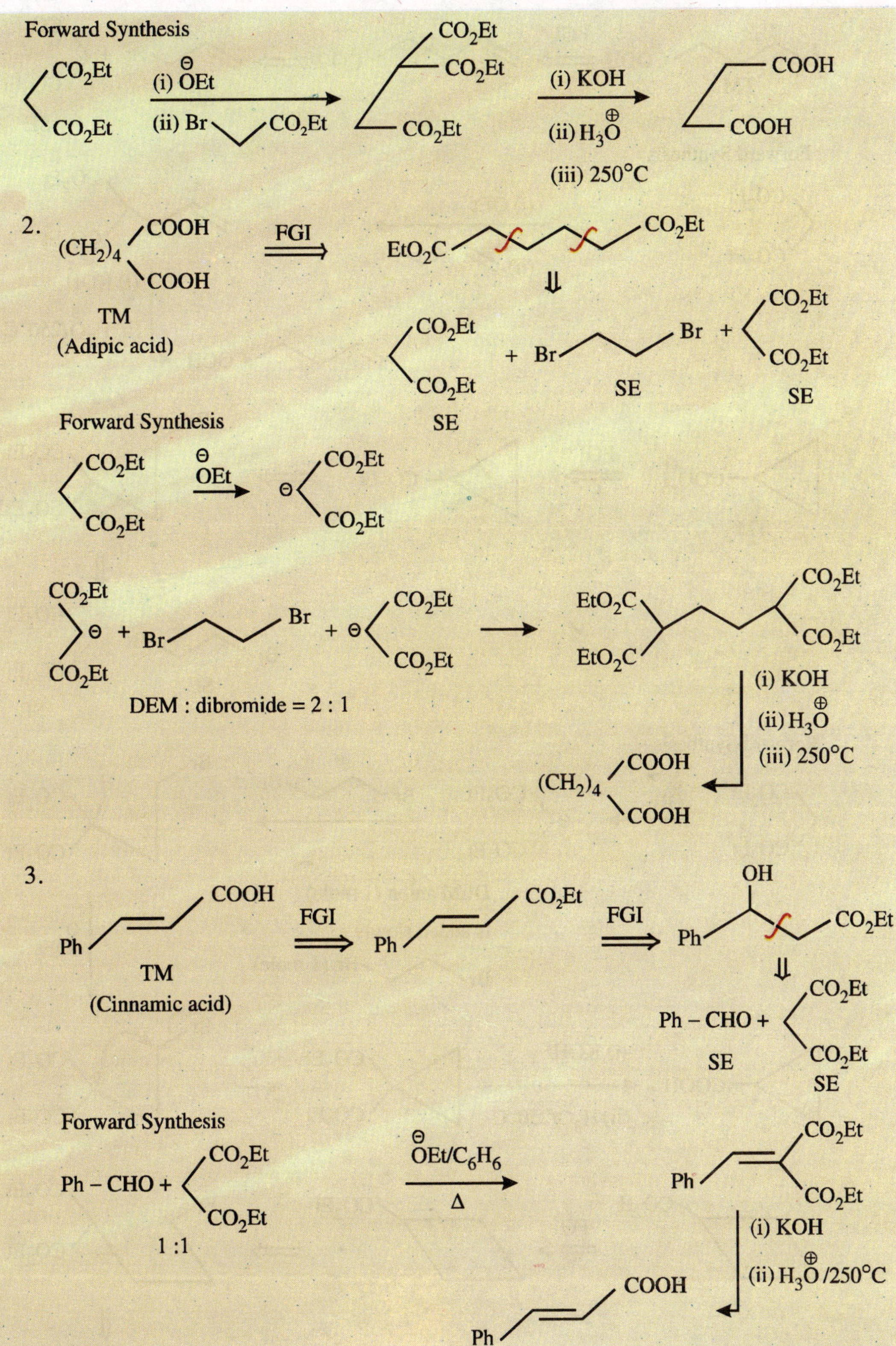

Forward Synthesis
CO2Et
CO2Et
(i) OEt
(ii) Br CO2Et
CO2Et
CO2Et
CO2Et
(i) KOH
(ii) H3O
(iii) 250°C
COOH
COOH
2.
(CH2)4
COOH
COOH
TM
(Adipic acid)
FGI
EtO2C
CO2Et
CO2Et
CO2Et
SE
+ Br
Br
SE
+
CO2Et
CO2Et
SE
Forward Synthesis
CO2Et
CO2Et
OEt
CO2Et
CO2Et
CO2Et
CO2Et
+ Br
Br
+
CO2Et
CO2Et
DEM : dibromide = 2 : 1
EtO2C
EtO2C
CO2Et
CO2Et
(i) KOH
(ii) H3O
(iii) 250°C
(CH2)4
COOH
COOH
3.
Ph
COOH
TM
(Cinnamic acid)
FGI
Ph
CO2Et
FGI
OH
Ph
CO2Et
Ph – CHO +
SE
CO2Et
CO2Et
SE
Forward Synthesis
Ph – CHO +
CO2Et
CO2Et
1 :1
OEt/C6H6
Δ
Ph
CO2Et
CO2Et
(i) KOH
(ii) H3O /250°C
COOH
Ph

4.

δ γ COOH TM $\xRightarrow{FGI}$ CO_2Et $\Rightarrow$ Br + CO_2Et CO_2Et

SE SE

Forward Synthesis

CO_2Et CO_2Et (i) $\overset{\ominus}{O}Et$ (ii) Br → CO_2Et CO_2Et (i) KOH (ii) $H_3\overset{\oplus}{O}$/250°C → COOH

5.

COOH TM $\xRightarrow{FGI}$ CO_2Et $\Rightarrow$ Br CO_2Et CO_2Et $\Downarrow$ Br Br + CO_2Et CO_2Et

SE SE

Forward Synthesis

CO_2Et CO_2Et $\xrightarrow{\overset{\ominus}{O}Et}$ $\ominus$ CO_2Et CO_2Et $\xrightarrow{Br \quad Br}$ Br H CO_2Et CO_2Et

DEM anion (1 mole)
is added slowly to
Br Br (1 mole)

$\overset{\ominus}{O}Et$ ↓ Br $\ominus$ CO_2Et CO_2Et → CO_2Et CO_2Et (i) KOH (ii) $H_3\overset{\oplus}{O}$/250°C → COOH

6.

CO_2H $\xRightarrow{FGI}$ CO_2Et $\Rightarrow$ Br CO_2Et CO_2Et

$\Downarrow$

Br–$CH_2CH_2CH_2CH_2$–Br (SE) + $CH_2(CO_2Et)_2$ (SE)

Forward Synthesis

$CH_2(CO_2Et)_2 \xrightarrow{\ominus OEt} \ominus CH(CO_2Et)_2$ $\xrightarrow{Br(CH_2)_4Br}$ $Br(CH_2)_4CH(CO_2Et)_2$

DEM anion (1 mole) is added slowly over Br–$(CH_2)_4$–Br (1 mole)

$\xrightarrow{\ominus OEt}$ $Br(CH_2)_4\overset{\ominus}{C}(CO_2Et)_2$ → cyclopentane-1,1-dicarboxylate (CO_2Et, CO_2Et) $\xrightarrow[\text{(ii) } H_3\overset{\oplus}{O}/250°C]{\text{(i) KOH}}$ cyclopentane carboxylic acid (CO_2H)

7. TM: cyclobutane-1,2-dicarboxylic acid (COOH, COOH) $\xRightarrow{FGI}$ diethyl cyclobutane-1,2-dicarboxylate (CO_2Et, CO_2Et) $\xRightarrow{FGA}$ tetraethyl cyclobutane-1,1,2,2-tetracarboxylate (CO_2Et ×4)

⇓

$(EtO_2C)_2CH–CH_2CH_2–CH(CO_2Et)_2$ ⟸ Br–CH_2CH_2–Br (SE) + 2 $CH_2(CO_2Et)_2$ (SE)

Forward Synthesis

$2\,CH_2(CO_2Et)_2 \xrightarrow{\ominus OEt} 2\,\ominus CH(CO_2Et)_2 \xrightarrow{Br–CH_2CH_2–Br}$ $(CO_2Et)_2CH–CH_2CH_2–CH(CO_2Et)_2$

$\xrightarrow[\text{(ii) } I_2/\Delta]{\text{(i) NaOEt}}$ cyclobutane with $(CO_2Et)_2$, $(CO_2Et)_2$ $\xrightarrow[\text{(ii) } H_3\overset{\oplus}{O}/250°C]{\text{(i) KOH}}$ cyclobutane-1,2-dicarboxylic acid (COOH, COOH)

8. HOOC ... CO_2H (TM) $\xrightarrow{FGI}$ EtO_2C ... CO_2Et ⟹ EtO_2C ... CO_2Et, CO_2Et, Br

⟸ 2 $CH_2(CO_2Et)_2$ (SE) + CH_2I_2 (SE) ⟸ $(EtO_2C)_2CH-CH_2-CH(CO_2Et)_2$ + Br⁄⁄Br (SE)

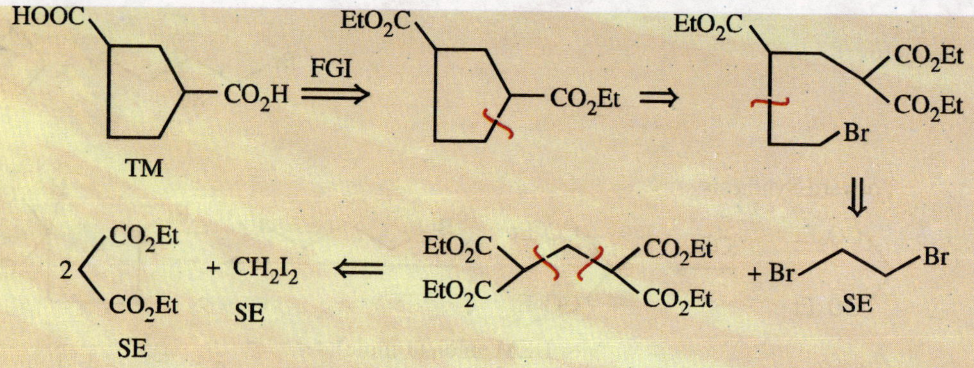

Forward Synthesis

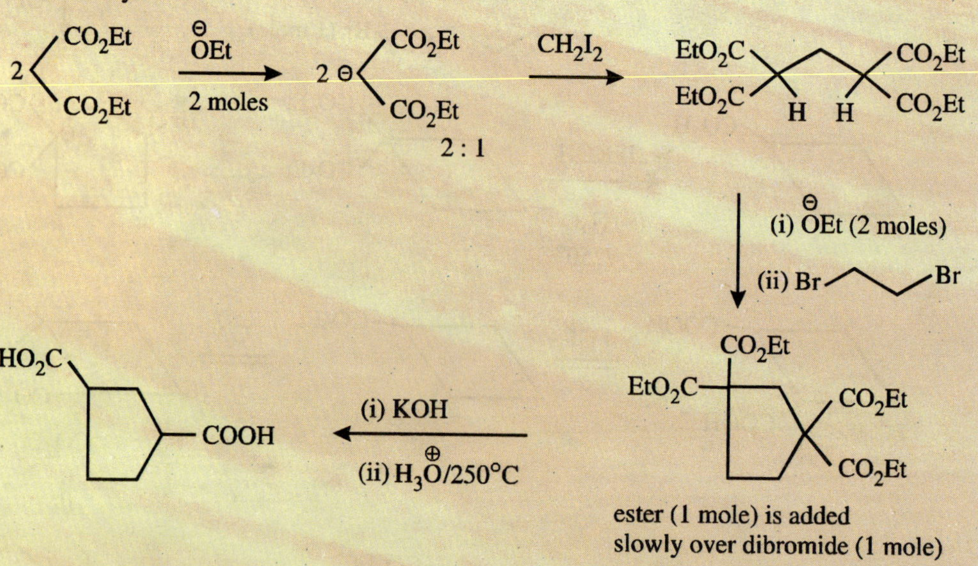

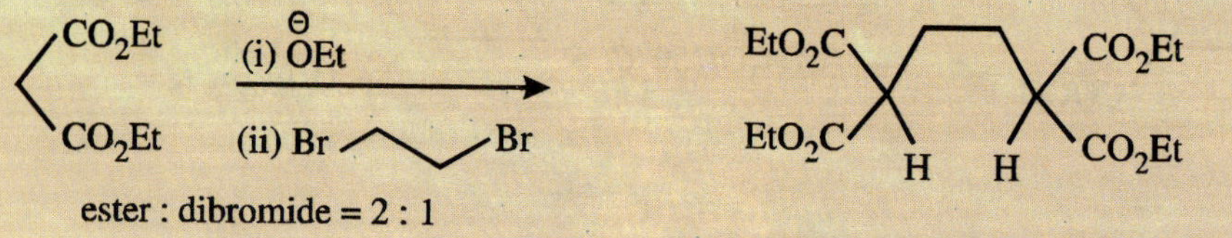

$2\,CH_2(CO_2Et)_2 \xrightarrow[\text{2 moles}]{\overset{\ominus}{O}Et} 2\,\ominus CH(CO_2Et)_2$ (2 : 1) $\xrightarrow{CH_2I_2}$ $(EtO_2C)_2CH-CH_2-CH(CO_2Et)_2$ (H H)

(i) $\overset{\ominus}{O}Et$ (2 moles)
(ii) Br⁄⁄Br

→ EtO_2C, CO_2Et, CO_2Et, CO_2Et (cyclopentane tetraester)

ester (1 mole) is added slowly over dibromide (1 mole)

$\xrightarrow[\text{(ii) } H_3\overset{\oplus}{O}/250°C]{\text{(i) KOH}}$ HO_2C ... COOH

9. HO_2C ... CO_2H (TM) $\xrightarrow{FGI}$ EtO_2C ... CO_2Et

⟹ EtO_2C ... CO_2Et (with CO_2Et, CO_2Et) + Br⁄⁄Br (SE)

⟸ $(EtO_2C)_2CH_2$ (SE) + Br⁄⁄Br (SE) + $CH_2(CO_2Et)_2$ (SE)

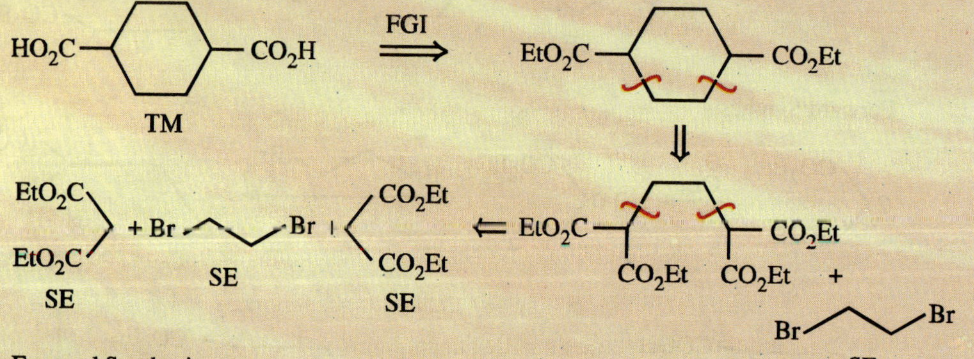

Forward Synthesis

$CH_2(CO_2Et)_2 \xrightarrow[\text{(ii) Br⁄⁄Br}]{\text{(i) } \overset{\ominus}{O}Et}$ $(EtO_2C)_2CH-CH_2CH_2-CH(CO_2Et)_2$ (H H)

ester : dibromide = 2 : 1

(i) $\overset{\ominus}{O}Et$ (2 moles)
(ii) Br⁀⁀Br

EtO_2C, CO_2Et, EtO_2C, CO_2Et

(i) KOH
(ii) $H_3\overset{\oplus}{O}$/250°C

HO_2C — CO_2H

10. TM (Dimedone)

Claisen dix

OEt

FGA

EtO_2C OEt

Michael dix

EtO_2C EtO_2C + O
SE SE

Forward Synthesis

CO_2Et CO_2Et (i) $\overset{\ominus}{O}Et$ (ii) O

EtO_2C OEt

$\overset{\ominus}{O}Et$

EtO_2C O OEt $O^{\ominus}$

EtO_2C O O

dil HCl
Δ

O O

11. CO_2Et O O TM

CO_2Et OH O OEt

FGI

CHO OH SE + CO_2Et CO_2Et SE ⟸ OH CO_2Et OH CO_2Et

Forward Synthesis

CO_2Et CO_2Et $\xrightarrow{\ominus OEt}$ $\ominus$ CO_2Et CO_2Et $\xrightarrow[\Delta]{\text{CHO, OH}}$ CO_2Et O–H O OEt → $\ominus$ OEt → CO_2Et O O

12. TM O O $\xRightarrow[\text{dix.}]{C-O}$ OH OH + CH_2O SE

⇓ FGI

EtO_2C CO_2Et ⟹ CH_3Br SE + CO_2Et CO_2Et SE

Forward Synthesis

CO_2Et CO_2Et $\xrightarrow[\text{(ii) } CH_3-Br]{\text{(i) } \ominus OEt}$ EtO_2C CO_2Et $\xrightarrow{LiAlH_4}$ OH OH $\xrightarrow[\text{HCl gas}]{CH_2O}$ O O

13. O N O H TM $\xRightarrow[\text{imide}]{C-N}$ O NH_2 OH O $\xRightarrow[\text{amide}]{C-N}$ NH_3 + O OH OH O

⇓ FGI

CN, CO_2Et (SE) + CO_2Et ⇐ CO_2Et CO_2Et

⇓ FGI

CH_3CHO + CN, CO_2Et ⇐ HO, CO_2Et

SE SE

Forward Synthesis

CN, CO_2Et —(i) $\overset{\ominus}{O}Et$ (ii) CH_3CHO→ CN, CO_2Et —CN, CO_2Et / $\overset{\ominus}{O}Et$→ CO_2Et, CN, CN, CO_2Et

—(i) KOH (ii) $H_3\overset{\oplus}{O}$/250°C→ O, OH, OH, O —Ac_2O→ O, O, O

NH_3 → [O, $\ddot{N}H_2$, O, OH] —$-H_2O$→ O, N, O, H

14. COOH, N (TM) —FGI⇒ CO_2Et, N ⇒ $\overset{\oplus}{N}$ + $\overset{\ominus}{C}H_2-CO_2Et$ (SN)

⇓ FGI ⇓

CH_3CHO + H–N (SE) ⇐ OH, N ; CO_2Et, CO_2Et (SE)

SE

Forward Synthesis

CH_3CHO + piperidine → N-(1-hydroxyethyl)piperidine —($^{\ominus}$OH, – OH)→ N-ethylidenepiperidinium ion

$CH_2(CO_2Et)_2$ / piperidine (H–N) → $^{\ominus}CH(CO_2Et)_2$ / piperidinium ($H_2N^{\oplus}$)

Iminium ion + malonate anion → $(EtO_2C)_2CH-CH(CH_3)-N$(piperidine) —(i) KOH, (ii) $H_3\overset{\oplus}{O}$/250°C→ $CH_3-CH(N$(piperidine)$)-CH_2-COOH$

15. Target Molecule H_2N-CH_2-COOH glycine

Method 1

$H_2N\diagup\diagdown CO_2H$ (TM) $\overset{FGI}{\Longrightarrow}$ $O=C=N\diagup\diagdown COOH$

$\Downarrow$ Curtius rearrangement

$N\equiv\overset{\oplus}{N}-\overset{\ominus}{N}-C(=O)-CH_2-CO_2H$

$\overset{FGI}{\Longleftarrow}$ $H_2N-\underset{H}{N}-C(=O)-CH_2-CO_2H$

$\Downarrow$ FGI

$H_2N-\underset{H}{N}-C(=O)-CH_2-CO_2^{\ominus}\overset{\oplus}{K}$ $\Longrightarrow$ $EtO-C(=O)-CH_2-CO_2^{\ominus}\overset{\oplus}{K}$ + NH_2-NH_2 (SE) $\overset{FGI}{\Longrightarrow}$ $EtO-C(=O)-CH_2-C(=O)-OEt$ (SE)

Forward Synthesis

$CH_2(CO_2Et)_2$ $\xrightarrow[\text{1eq.}]{KOH}$ $EtO-C(=O)-CH_2-CO_2^{\ominus}\overset{\oplus}{K}$ $\xrightarrow{NH_2-NH_2}$ $NH_2-NH-C(=O)-CH_2-CO_2^{\ominus}\overset{\oplus}{K}$

$\downarrow HNO_2$

$H_2N\text{–}CH_2\text{–}COOH \xleftarrow[-CO_2]{H_2O}$ (COOH, N=C=O) $\xleftarrow[\text{rearrangement}]{\text{Curtius}}$ (COOH, C=O, $\ominus N\text{–}\overset{\oplus}{N}\equiv N$)

This is the curtius method of α-amino acid synthesis. With the choice of appropriate DEM derivative (first alkylation of DEM with required alkyl group) one can synthesize the various desired α-amino acid starting from DEM.

Method 2

$H_2N\text{–}CH_2\text{–}COOH$ (TM) $\xRightarrow{FGI}$ $H_2N\text{–}CH_2\text{–}CO_2Et$ $\xRightarrow{FGA}$ $H_2N\text{–}CH(CO_2Et)_2$

$\Downarrow$ FGI

$HO\text{–}N{=}C(CO_2Et)_2 \xLeftarrow{FGI} O{=}N\text{–}CH(CO_2Et)_2 \Leftarrow \overset{\oplus}{N}O + CH_2(CO_2Et)_2$ (SE)

FGA $\Downarrow$ $HO\text{–}N{=}O$

Forward Synthesis

$CH_2(CO_2Et)_2 \xrightarrow[HCl]{NaNO_2} O{=}N\text{–}CH(CO_2Et)_2 \rightleftharpoons HO\text{–}N{=}C(CO_2Et)_2$

Zn $\downarrow$ CH_3COOH

$H_2N\text{–}CH_2\text{–}COOH \xleftarrow[\text{(ii) } H_3\overset{\oplus}{O}/250°C]{\text{(i) KOH}} H_2N\text{–}CH(CO_2Et)_2$

Method 3

$H_2N\text{–}CH_2\text{–}COOH$ (TM) $\xRightarrow{FGI}$ $H_2N\text{–}CH_2\text{–}CO_2Et$ $\xRightarrow{FGA}$ $H_2N\text{–}CH(CO_2Et)_2$

$\Downarrow$

Phthalimide (NH) $\Leftarrow$ $H_2\overset{\ominus}{N}$ (SN) + $\overset{\oplus}{C}H(CO_2Et)_2$ (SN)

SE (Phthalimide) $\Downarrow$

Br_2 + $CH_2(CO_2Et)_2$ ⟸ $BrCH(CO_2Et)_2$
SE SE

$CH_2(CO_2Et)_2$ —($^{\ominus}$OEt/Br_2)→ $BrCH(CO_2Et)_2$

Phthalimide N–H —(i) $^{\ominus}$OEt; (ii) $BrCH(CO_2Et)_2$→ N-phthalimido-$CH(CO_2Et)_2$

hydrolysis and decarboxylation ↓ (i) KOH (ii) $H_3O^{\oplus}/\Delta$

H_2N–CH_2–COOH

This is the Gabriel method of α-amino acid synthesis. With the introduction of appropriate alkyl group before hydrolysis and decarboxylation (last step) one can synthesize the desired α-amino acid starting from DEM.

16. TM (Leucine) ⟹ $(CH_3)_2CHCH_2CH(Br)CO_2H$ + NH_3 (SE)

⇓ FGA

$(CH_3)_2CHCH_2C(Br)(CO_2H)_2$ ⟹ $(CH_3)_2CHCH_2CH(CO_2H)_2$ + Br_2 (SE)

⇓ FGI

$(CH_3)_2CHCH_2Br$ (SE) + $CH_2(CO_2Et)_2$ (SE) ⟸ $(CH_3)_2CHCH_2CH(CO_2Et)_2$

Forward Synthesis

$CH_2(CO_2Et)_2$ —(i) $^{\ominus}$OEt; (ii) $(CH_3)_2CHCH_2Br$ (alkyl halide)→ $(CH_3)_2CHCH_2CH(CO_2Et)_2$ —(i) KOH; (ii) $H_3O^{\oplus}$, not heated→ $(CH_3)_2CHCH_2CH(CO_2H)_2$

↓ Br_2

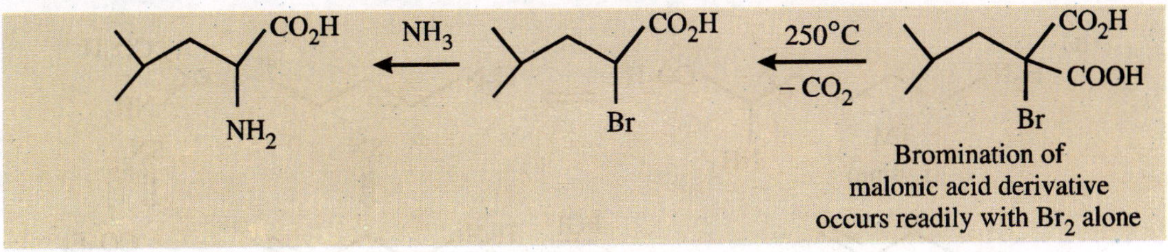
CO2H
NH2
NH3
CO2H
Br
250°C
– CO2
CO2H
COOH
Br
Bromination of malonic acid derivative occurs readily with Br2 alone

With the appropriate choice of alkyl halide one can synthesize the desired α-amino acid starting from DEM by this procedure.

17.

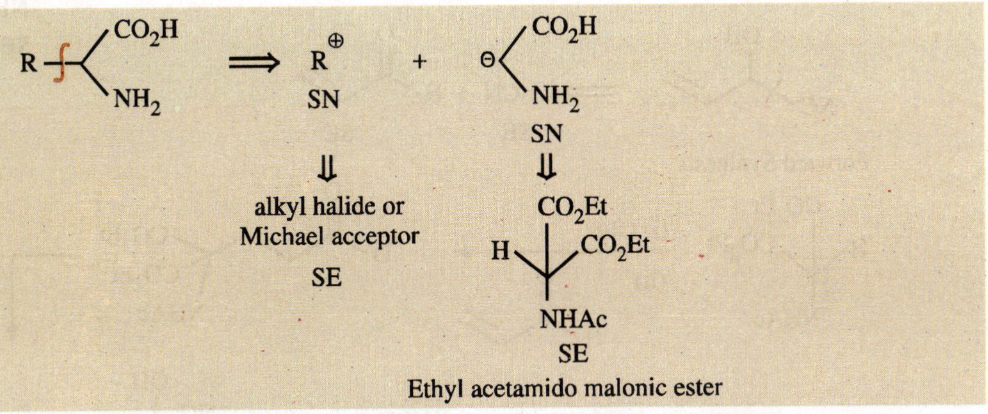
R
CO2H
NH2
R ⊕
SN
+
⊖
CO2H
NH2
SN
alkyl halide or Michael acceptor
SE
CO2Et
H
CO2Et
NHAc
SE
Ethyl acetamido malonic ester

The following syntheses of Glutamic acid, lysine and Tryptophan are illustrative.

(i)

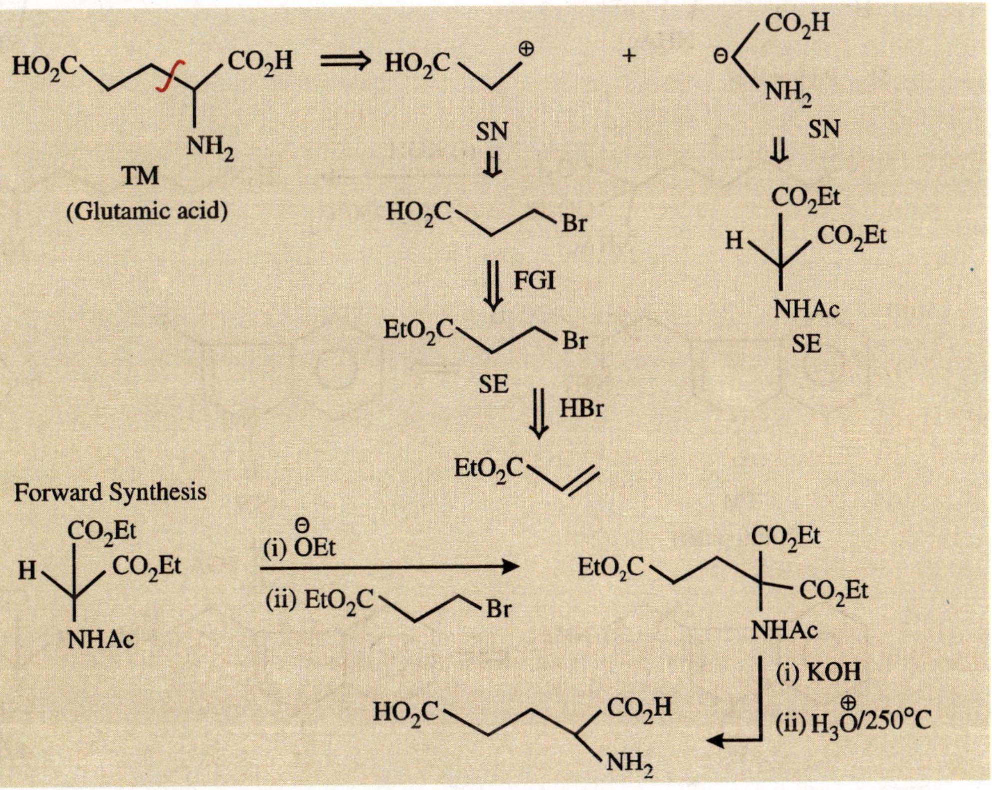
HO2C
CO2H
NH2
TM
(Glutamic acid)
HO2C
⊕
SN
HO2C
Br
FGI
EtO2C
Br
SE
HBr
EtO2C
+
⊖
CO2H
NH2
SN
CO2Et
H
CO2Et
NHAc
SE
Forward Synthesis
CO2Et
H
CO2Et
NHAc
(i) ⊖OEt
(ii) EtO2C
Br
EtO2C
CO2Et
CO2Et
NHAc
(i) KOH
(ii) H3O⊕/250°C
HO2C
CO2H
NH2

(ii)

H_2N ... CO_2H (NH_2) ⟹ H_2N ... $\oplus$ + $\ominus$ CO_2H NH_2

TM (Lysine)

SN ⇓ SN ⇓

NC ⟸ FGI H_2N

CO_2Et, H, CO_2Et, NHAc SE

⇓ FGA

OH, NC ⟹ HCN + H–C(=O)–CH=CH$_2$

SE SE

Forward Synthesis

CO_2Et, H, CO_2Et, NHAc —(i) $\ominus$OEt (ii) H–C(=O)–CH=CH$_2$→ H–C(=O)–... CO_2Et CO_2Et NHAc —HCN→

OH, NC, H, CO_2Et, CO_2Et, NHAc —$KHSO_4$→ NC, H, CO_2Et, CO_2Et, NHAc

H_2 / Pd ↓ < H >

H_2N ... CO_2Et CO_2Et NHAc —(i) KOH (ii) $H_3\overset{\oplus}{O}$/250°C→ H_2N ... CO_2H NH_2

(iii)

TM (Triptophan): indole, N–H, CO_2H, NH_2 ⟹ indole $\oplus$ (N–H) + $\ominus$ CO_2Et NH_2

SN SN

⇓ FGA ⇓

indole NMe_2 (N–H) ⟸ indole $\overset{\oplus}{N}Me_3$ (N–H) CO_2Et, H, CO_2Et, NHAc

SE

⇓

Indole (SE) + $CH_2{=}\overset{\oplus}{N}Me_2$ (SE) ⟹ CH_2O + Me_2NH

Forward Synthesis

Indole $\xrightarrow{CH_2=\overset{\oplus}{N}Me_2}$ 3-(CH_2NMe_2)indole $\xrightarrow{Me_2SO_4}$ 3-($CH_2\overset{\oplus}{N}Me_3$)indole

$\xrightarrow[\text{H–C(CO}_2\text{Et)}_2\text{NHAc}/\overset{\ominus}{O}\text{Et}]{-\,Me_3N}$ 3-[$CH_2C(CO_2Et)_2NHAc$]indole

(i) KOH (ii) $H_3\overset{\oplus}{O}$/250°C → 3-[$CH_2CH(NH_2)CO_2H$]indole

18. TM (Hygrinic acid) [N-Me pyrrolidine-2-CO_2H] $\overset{FGI}{\underset{FGA}{\Longrightarrow}}$ N-Me pyrrolidine-2,2-(CO_2Et)$_2$ ⟹ $MeNH–(CH_2)_3–CBr(CO_2Et)_2$

⇓

$Br(CH_2)_3CBr(CO_2Et)_2$ + $MeNH_2$

⟸

$Br(CH_2)_3CH(CO_2Et)_2$ + Br_2 (SE)

⇓

$Br(CH_2)_3Br$ (SE) + $CH_2(CO_2Et)_2$ (SE)

Forward Synthesis

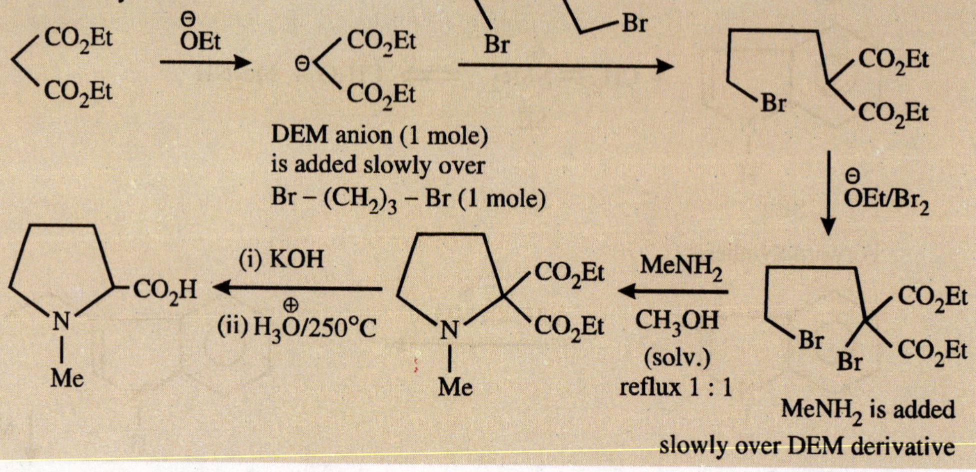

$\overset{\ominus}{O}Et$; $Br-(CH_2)_3-Br$; $\overset{\ominus}{O}Et/Br_2$; CO_2Et ; CO_2H ; Br ; N ; Me

DEM anion (1 mole) is added slowly over $Br-(CH_2)_3-Br$ (1 mole)

$MeNH_2$ / CH_3OH (solv.) reflux 1 : 1

$MeNH_2$ is added slowly over DEM derivative

(i) KOH
(ii) $H_3\overset{\oplus}{O}$/250°C

4.2.6 Review synthetic problems

Show the retrosynthetic analysis of the following target molecules in which one of the synthetic equivalent should be DEM and then give the forward synthesis with reasonable mechanism.

1. 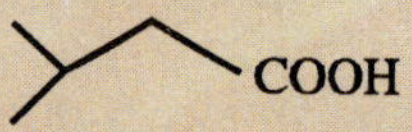 COOH
2. 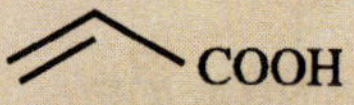 COOH
3. 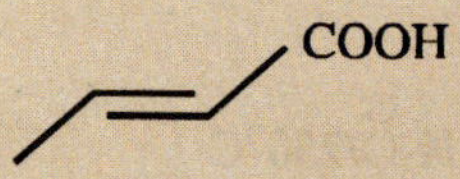 COOH
4. 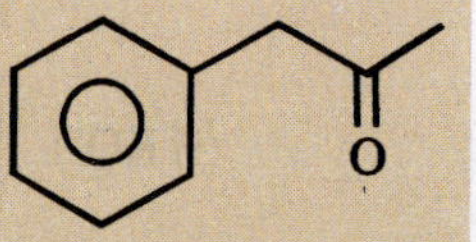 O
5. 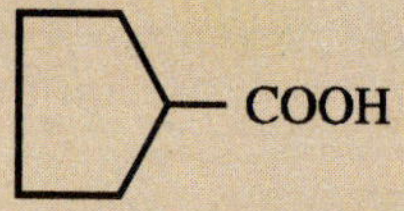 COOH
6. 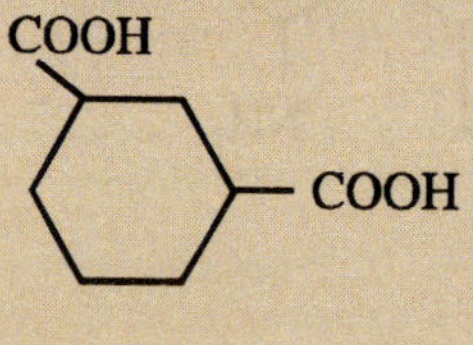 COOH, COOH
7. 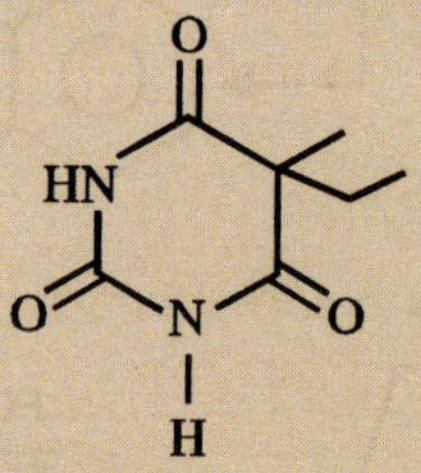 O, HN, O, N, O, H
8. 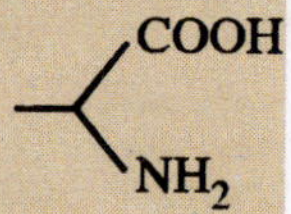 COOH, NH_2
9. 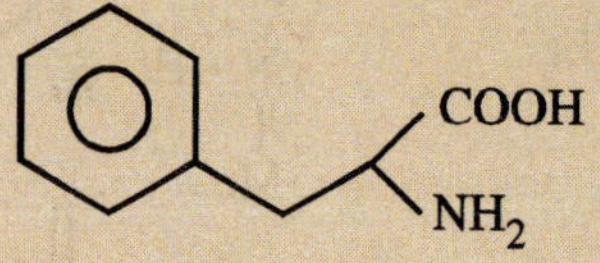 COOH, NH_2
10. 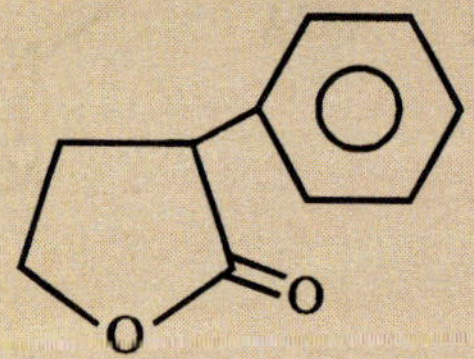 O, O
11. 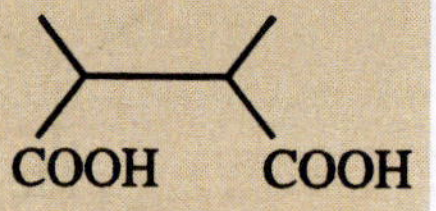 COOH, COOH
12. 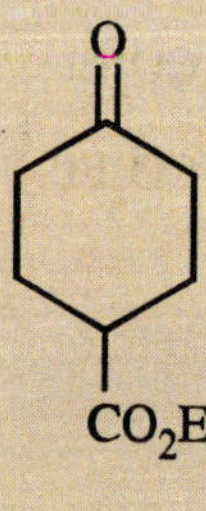 O, CO_2Et
13. 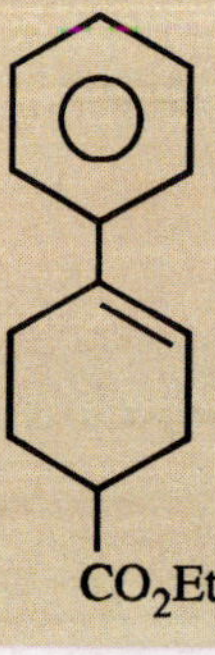 CO_2Et
14. 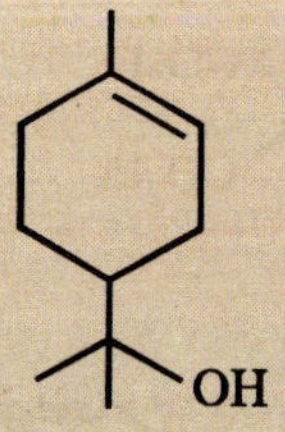 OH
15. 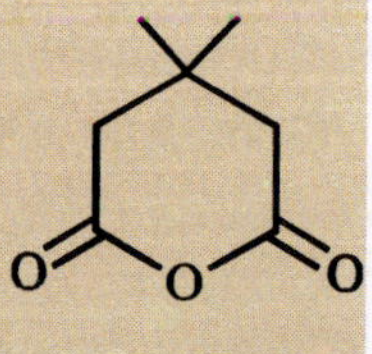 O, O, O

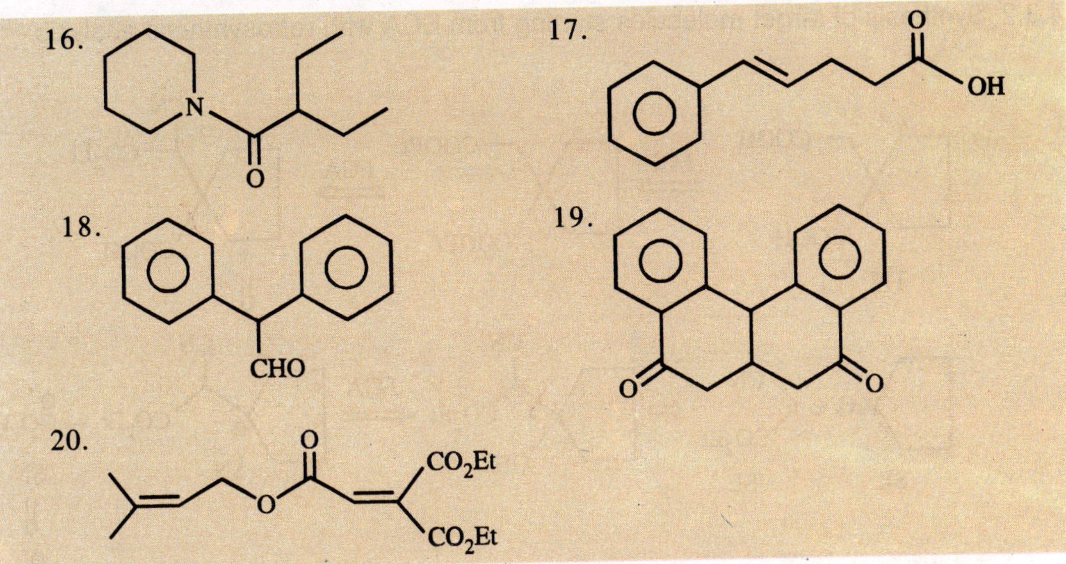

4.3 Ethyl cyano acetate (ECA)

4.3.1 What is ethyl cyano acetate?

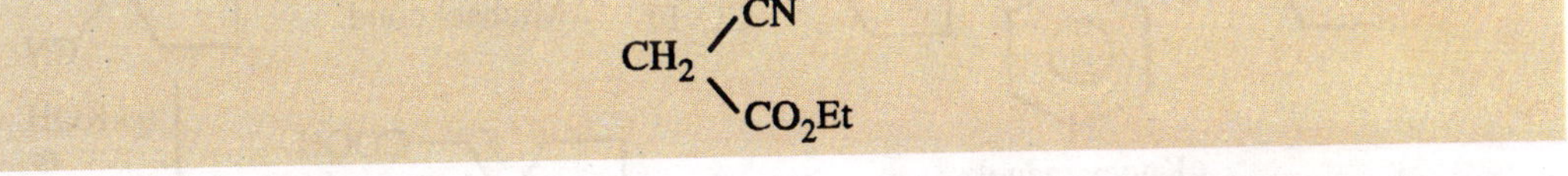

1. It is the protection of carboxylic group (–COOH) by the ester function ($-CO_2Et$).

2. It is the activation of methylene group ($-CH_2-$) by the cyanide group (–CN)

3. With the weak base it gives stabilised carbanion and may acts as a Michael donor.

4. Methylene group ($-CH_2-$) of it is more activated than Diethyl malonate and consequently carbanion of it is more nucleophilic than Diethyl malonate.

Diethyl malonate has no reaction with keto carbonyl but Ethylcyano acetate has.

5. It is the synthetic equivalent of the synthon $\overset{\ominus}{C}H_2-COOH$.

6. It is a bifunctional molecule having 1, 3-electrophilic centres.

7. It is an alkylating substrate.

8. It is a starting material for various classes of target molecules aliphatic, alicyclic, carboaromatic, heteroaromatic and especially pyrimidine derivative.

4.3.2 Synthesis of target molecules starting from ECA with retrosynthetic analysis

1.

COOH, COOH, TM $\overset{FGI}{\Longrightarrow}$ COOEt, COOEt $\overset{FGA}{\Longrightarrow}$ CN, CO_2Et, COOH

$\Downarrow$

=O + CN, CO_2Et (SE, SE) $\Longleftarrow$ CN, CO_2Et, OH $\overset{FGA}{\Longleftarrow}$ CN, $\oplus$, CO_2Et + $\overset{\ominus}{C}O_2H$ (SN, SN)

$\Downarrow$

HCN $\Longleftarrow$ $\overset{\ominus}{C}N$ (SE)

Forward Synthesis

=O $\xrightarrow{\text{CN, } CO_2Et \text{; pyridine (N)}}$ CN, CO_2Et $\xrightarrow[\text{Michael cond.}]{HCN/AlEt_3}$ CN, CO_2Et, CN

Knoevenagel reaction

(i) KOH (ii) $H_3\overset{\oplus}{O}$/250°C → COOH, COOH

2.

COOH, COOH, TM $\overset{FGI}{\Longrightarrow}$ CO_2Et, CO_2Et $\overset{FGA}{\Longrightarrow}$ CN, CO_2Et, CO_2Et, CN

$\Downarrow$

=O + CN, CO_2Et (SE, SE) $\Longleftarrow$ CN, CO_2Et, OH + CN, CO_2Et (SE)

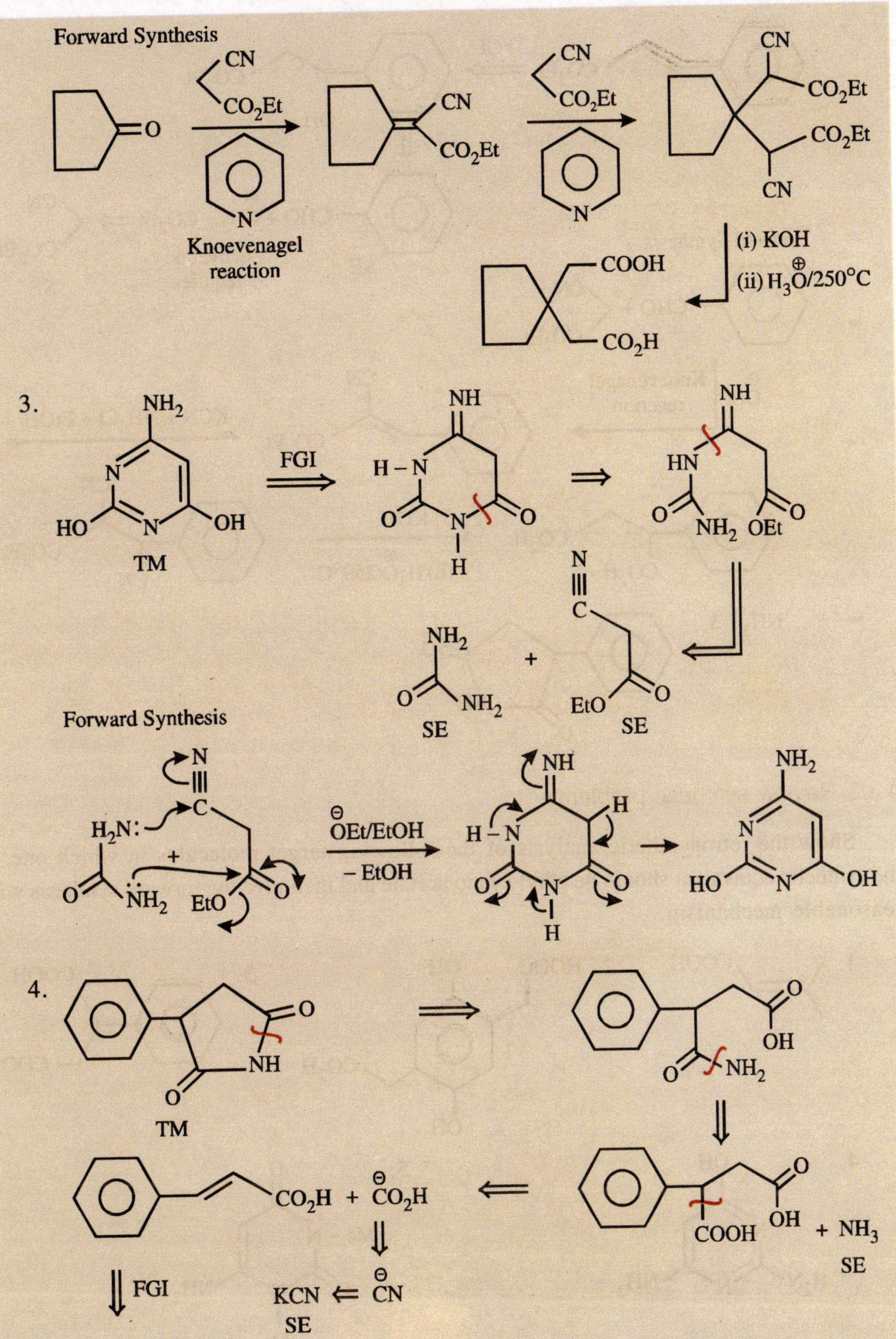
Forward Synthesis
CN
CO_2Et
O
N
Knoevenagel reaction
CN
CO_2Et
(i) KOH
(ii) $H_3\overset{\oplus}{O}$/250°C
COOH
CO_2H
3.
NH_2
N
HO
N
OH
TM
FGI
NH
H – N
O
N
H
HN
NH_2
OEt
N
C
NH_2
+
NH_2
EtO
SE
SE
Forward Synthesis
H_2N:
+
$\overset{\ominus}{O}$Et/EtOH
– EtOH
H
4.
NH
TM
OH
NH_2
CO_2H + $\overset{\ominus}{C}O_2H$
COOH
+ NH_3
SE
FGI
KCN ⇐ $\overset{\ominus}{C}N$
SE

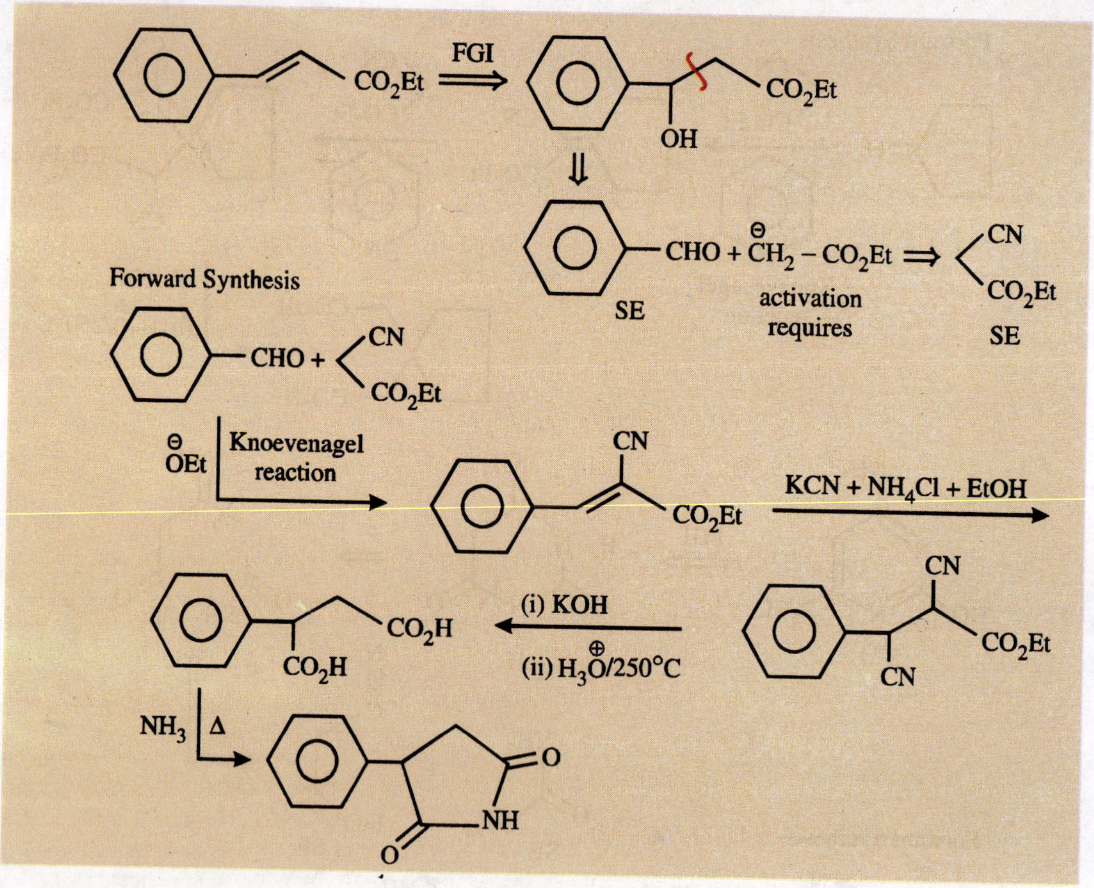

CO_2Et $\xrightarrow{FGI}$ CO_2Et, OH

$\Downarrow$

CHO + $\overset{\ominus}{C}H_2 - CO_2Et$ ⇒ CN, CO_2Et

SE activation requires SE

Forward Synthesis

CHO + CN, CO_2Et

$\overset{\ominus}{O}Et$ Knoevenagel reaction → CN, CO_2Et $\xrightarrow{KCN + NH_4Cl + EtOH}$ CN, CO_2Et, CN

CO_2H, CO_2H ← (i) KOH (ii) $H_3\overset{\oplus}{O}$/250°C

NH_3 Δ → O, NH, O

4.3.3 Review synthetic problems

Show the retrosynthetic analysis of the following target molecules in which one of the synthetic equivalent should be ethyl cyano acetate and then give the forward synthesis with reasonable mechanism.

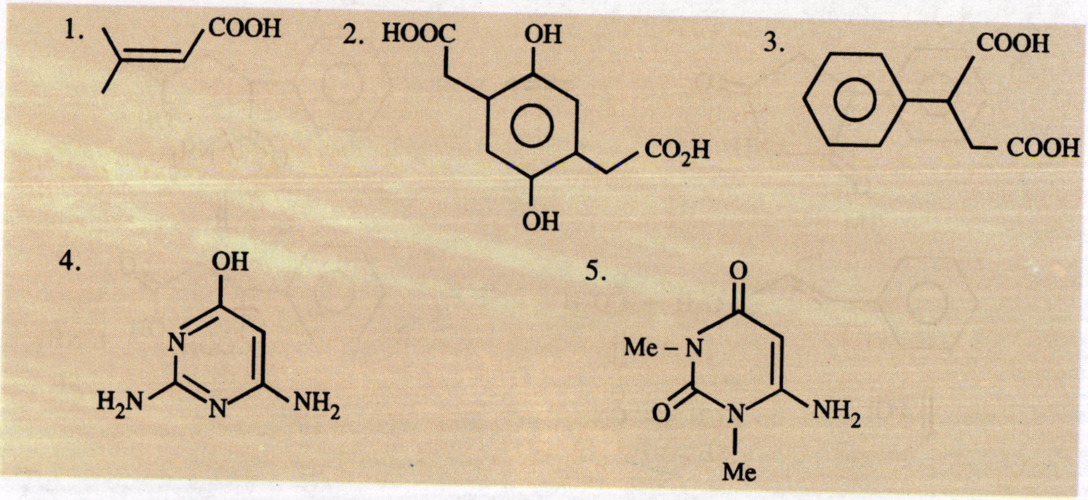

1. COOH
2. HOOC, OH, CO_2H, OH
3. COOH, COOH
4. OH, N, H_2N, N, NH_2
5. O, Me – N, O, N, NH_2, Me

5

Synthesis of Alkylated Cycloalkanone, Fused Bicyclic and Spirocyclic Ring System

5.1 Method of synthesis of alkylated cycloalkanone with retrosynthetic analysis

1. 3,3-dimethylcyclohexanone (TM)

Method I

TM $\Longrightarrow$ 3-methylcyclohexanone cation (SN) + $Me^{\ominus}$ (SN) $\Longrightarrow$ Me_2CuLi

SN $\Downarrow$ 3-methylcyclohex-2-enone

3-methylcyclohex-2-enone $\xLeftarrow[\text{dix}]{\text{aldol}}$ diketone (Me ketone) $\xLeftarrow[\text{dix}]{\text{Michael}}$

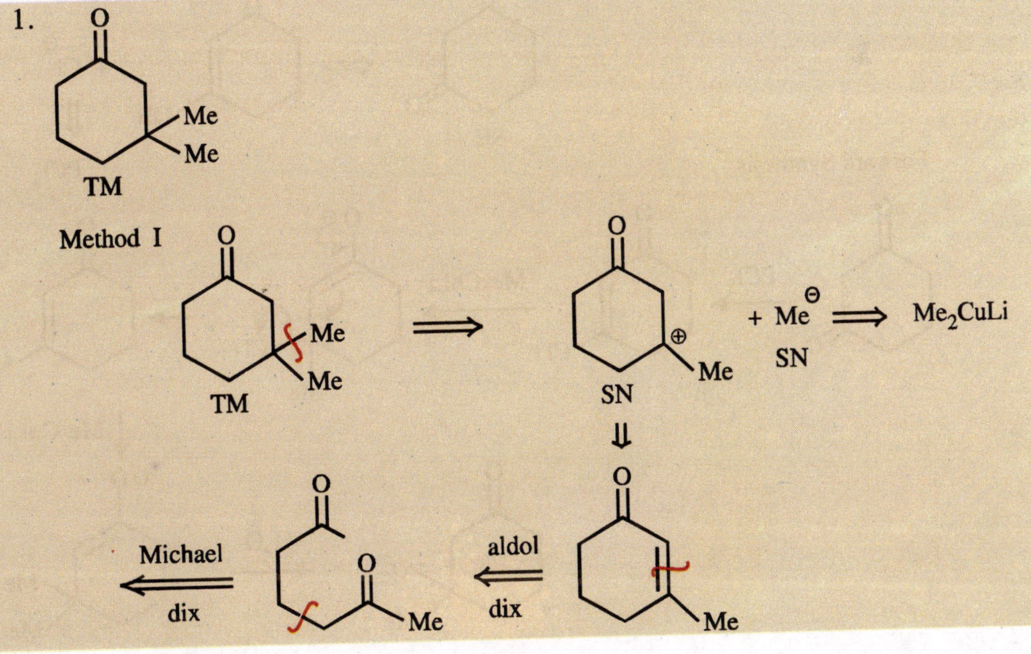

⟹ SE + Me CO_2Et SE

Forward Synthesis

Me CO_2Et (i) $\overset{\ominus}{O}Et$ (ii) (iii) dil HCl/Δ → Me; NH/AcOH Δ → Me

(i) Me_2CuLi ether (ii) $H_3\overset{\oplus}{O}$ → Me Me

Method II

TM Me Me ⟹ Me_2CuLi SE + Me ⟹ Cl + Me_2CuLi SE

⇓

SE ⟸ OH + $Cl^{\ominus}$ ⇓ PCl_5

Forward Synthesis

PCl_5 → Cl; Me_2CuLi → $O^{\ominus}$ Me Cl → Me

↓ Me_2CuLi

$^{\ominus}O$ Me Me → $H_3\overset{\oplus}{O}$ → Me Me

2.

TM

FGA

aldol

dix

+ O

SE

SE

Forward Synthesis

$\overset{\ominus}{OH}/H_2O$

Δ

H_2/Pd

$\overset{\ominus}{OH}$

$- H_2O$ Δ

O

OH

3.

TM

+ ⊕

SN

Cl

SE

$OSiMe_3$

$+ Me_3SiCl$

SE

Forward Synthesis

Et_3N

Me_3SiCl

$OSiMe_3$

thermodynamic
or specific enolate

Cl / $TiCl_4$

$CH_2Cl_2/0°C$

$\overset{\oplus}{O} - SiMe_3$

CH_3Li (1eq.)

Tertiary butylation is not possible in base medium via enolate since tertiary butyl gives isobutene via E1 elimination. Silylated enol ether is neutral, CH_2Cl_2 is inert solvent and $TiCl_4$ is Lewis acid.

4.

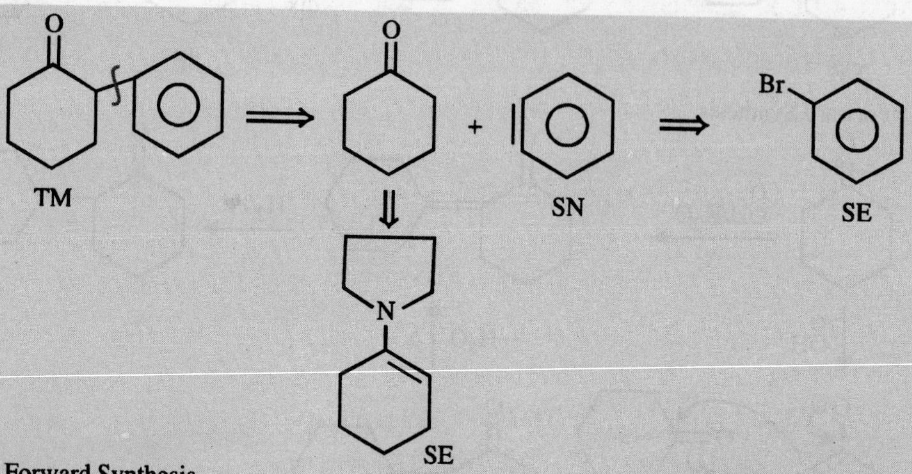

TM ⟹ + SN ⟹ Br SE

⇓

SE

Forward Synthesis

/TsOH

N–H

$\overset{\ominus}{N}H_2$/liq.NH_3

Br

⊕N

$H_3\overset{\oplus}{O}$

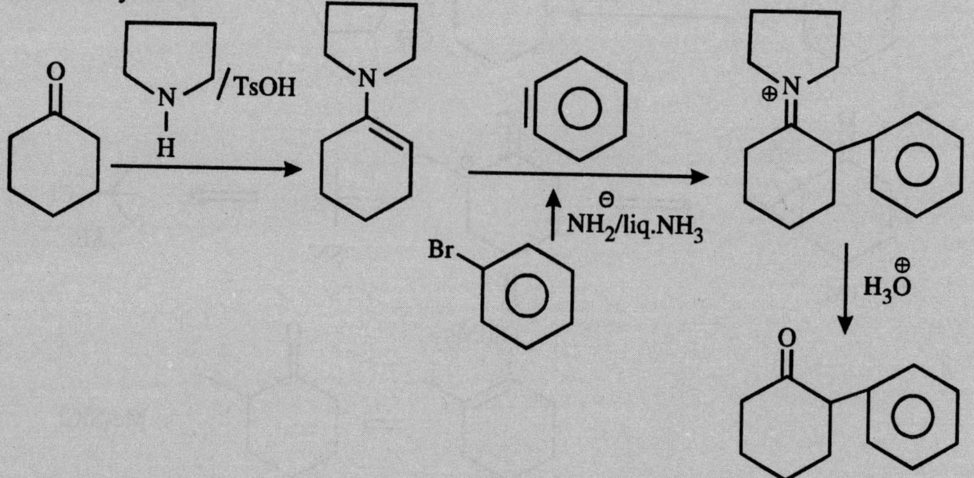

5. Method I

TM

FGA

CHO

CHO

+ Cl SE

⇓

⊕

$Ph_3P = CH - OMe$ ⟸ $\overset{\ominus}{C}HO$ +

SE SN

SN

⇓

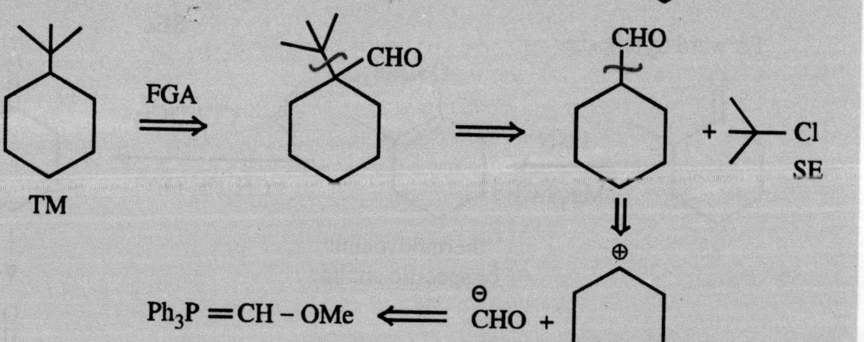

OH ⟹ O

SE

Forward Synthesis

O (i) $Ph_3P=CH-OMe$ (ii) $H_3\overset{\oplus}{O}$ → CHO ⇌ OH

Et_3N/Me_3SiCl ↓ $OSiMe_3$

$\overset{\oplus}{O}-SiMe_3$ $\overset{\ominus}{CH_3}$ ← Cl / $TiCl_4$, CH_2Cl_2/– 22°C

CH_3Li (1 eq.) ↓

$SiMe_4$ + CHO $\xrightarrow{KOH/H_2O/\Delta}$ $-HCO_2^{\ominus}$

Method II

TM FGA ⟹ ⟹ Br + $(\succ)_2CuLi$

SE SE

Forward Synthesis

Br $\xrightarrow{(\succ)_2CuLi}$ $\xrightarrow{Raney\ Ni/H_2}$

6.

TM ⟹ + Me_2CuLi

⇓ aldol dix.

1, 5-dico dix ⟹ SE + SE (CO_2Et)

Forward Synthesis

CO_2Et —(i) $\overset{\ominus}{O}Et$ (ii) [enone]→ CO_2Et —dil. HCl, Δ→ —$Me_2NH/AcOH$→ —Me_2CuLi→ $O^{\ominus}$ —$H_3\overset{\oplus}{O}$→

7.

TM ⟹ $O^{\ominus}$ + MeI SE

⇓

Me_2CuLi + SE; $O^{\ominus}$ ⊕ SN ⟹

⇓ aldol dix.

SE + SE ⟸ 1, 5-dico

Forward Synthesis

LDA/ether

kinetically controlled enolate

Michael cond.

NH/AcOH aldol cond.

Me_2CuLi

CH_3I

Since under this condition aldol condensation occurs via more substituted enol

8.

TM

FGA

TM

aldol dix.

EtO_2C SE + Cl SE

Forward Synthesis

EtO_2C

(i) $\overset{\ominus}{O}Et$

(ii) Cl SE

EtO_2C

(i) $\overset{\ominus}{O}Et$ (ii) $H_3\overset{\oplus}{O}/\Delta$

H_2/Pd

9.

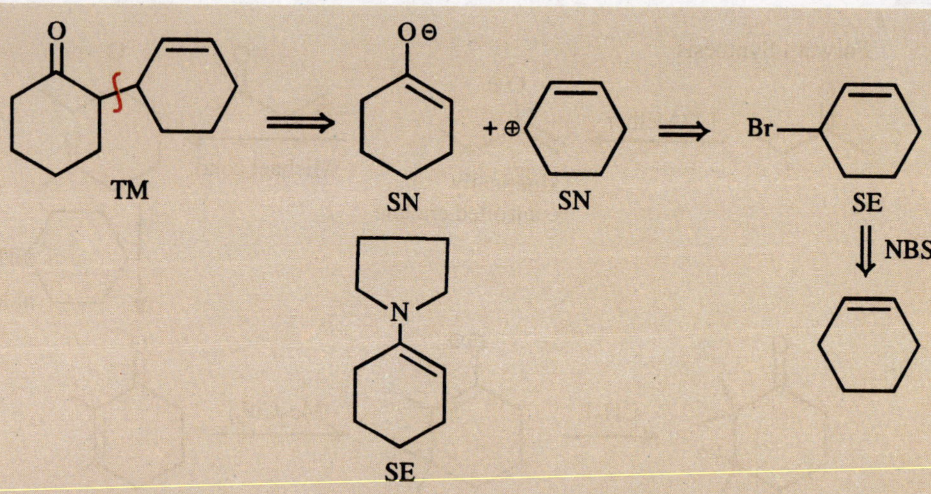
O
TM
O⊖
SN
+⊕
SN
Br
SE
NBS
N
SE

Forward Synthesis

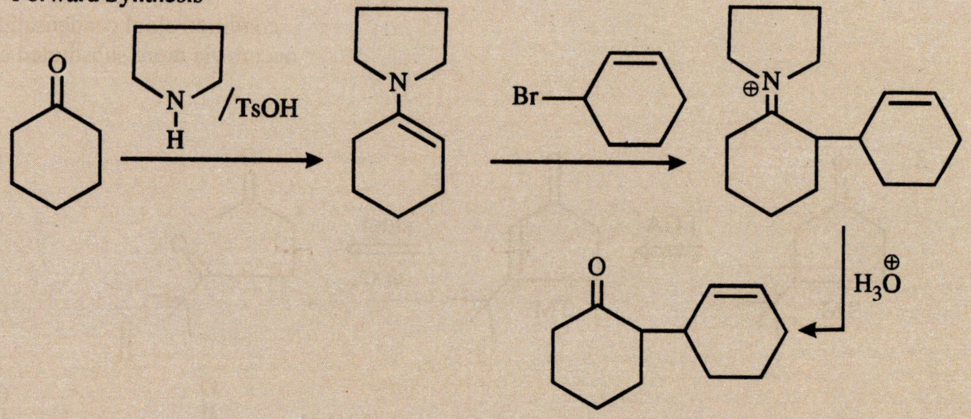
O
N
H
/TsOH
N
Br
⊕N
H3O⊕
O

10.

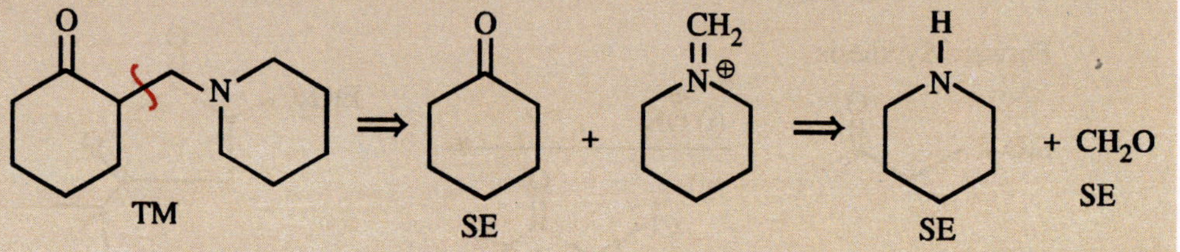
O
N
TM
O
SE
+
CH2
N⊕
H
N
SE
+ CH2O
SE

Forward Synthesis

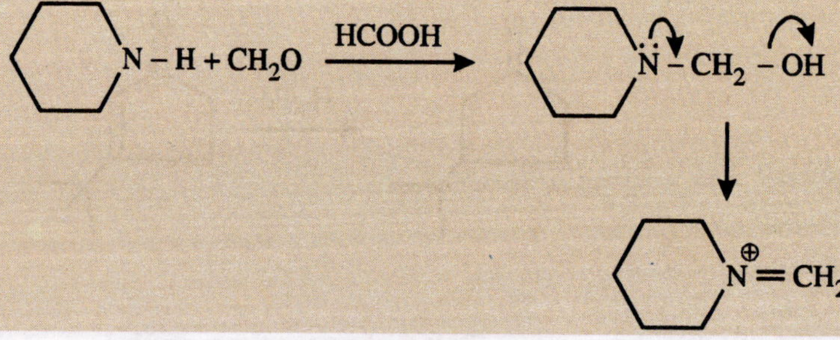
N – H + CH2O
HCOOH
N – CH2 – OH
⊕
N = CH2

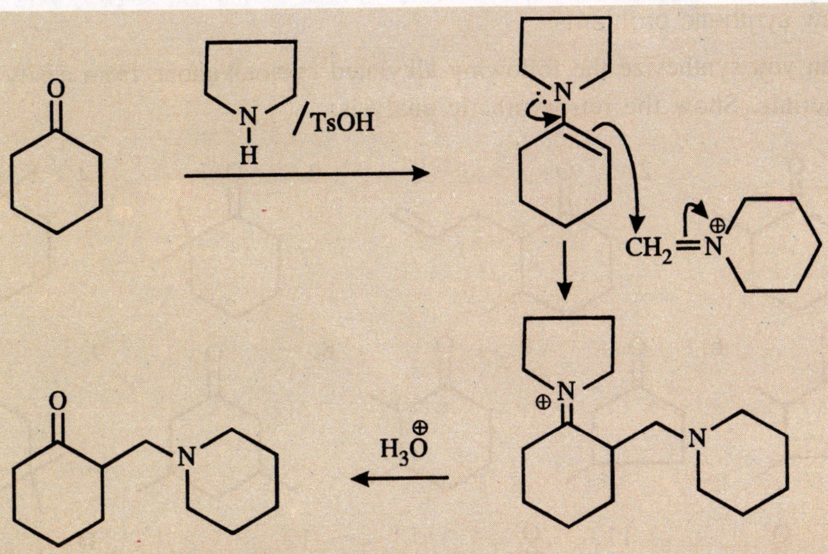

/TsOH

$CH_2=\overset{\oplus}{N}$

$H_3\overset{\oplus}{O}$

11.

TM

$\Rightarrow$

+ Cl

SE

SE

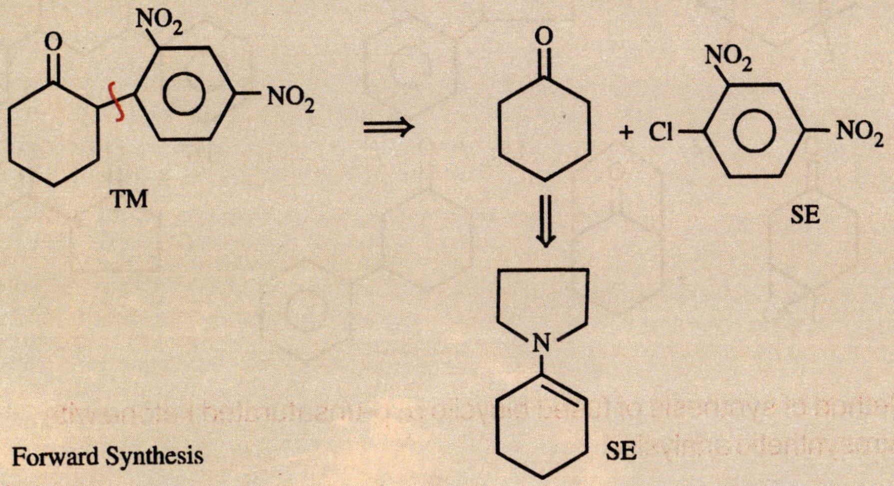

Forward Synthesis

$H_3\overset{\oplus}{O}$

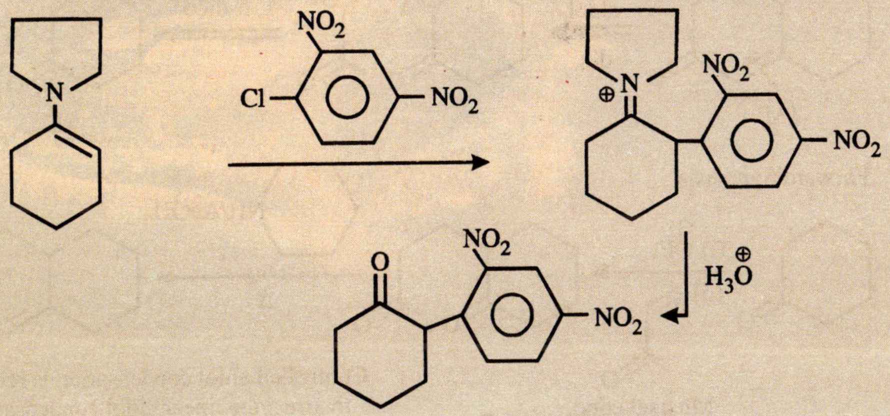

5.1.1 Review synthetic problems

How can you synthesize the following alkylated cycloalkanone from easily available starting materials. Show the retrosynthetic analysis.

1. O
2. O
3. O
4. 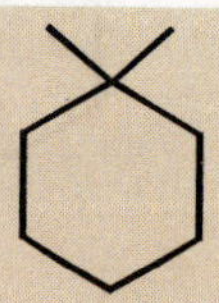
5. O
6. O
7. O
8. O
9. O
10. O
11. O
12. O
13. O O
14. O
15. O
16. O O

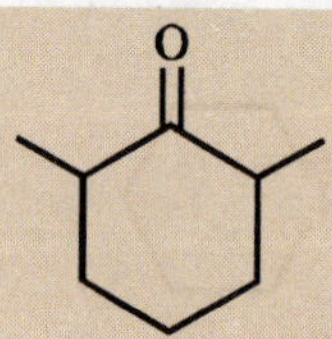

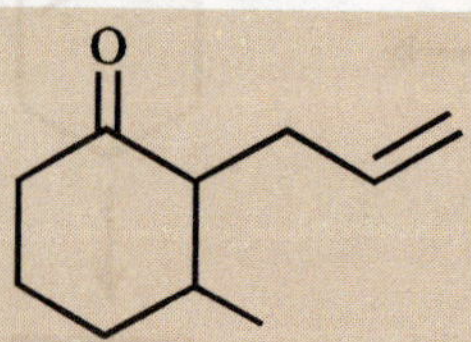

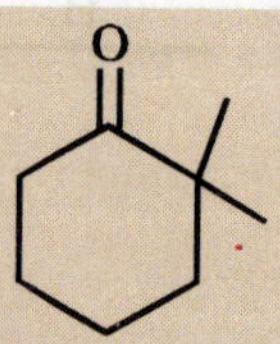

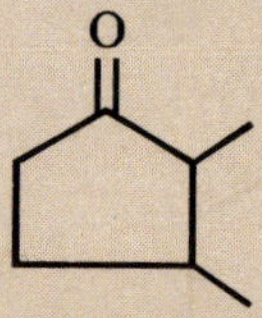

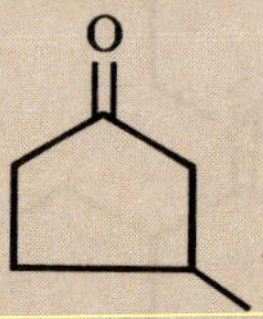

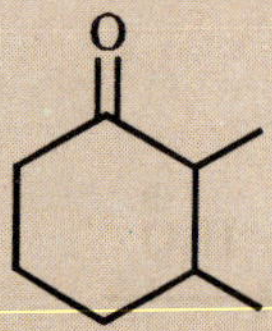

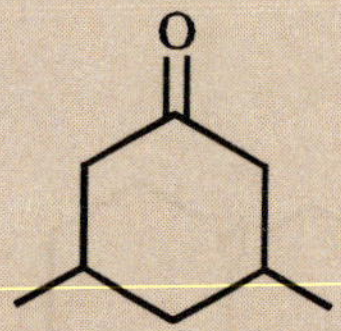

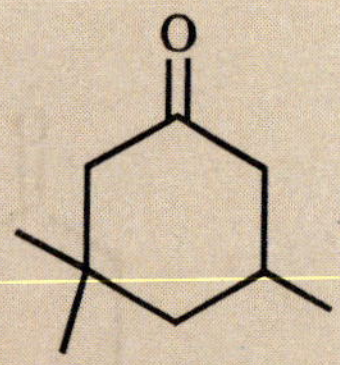

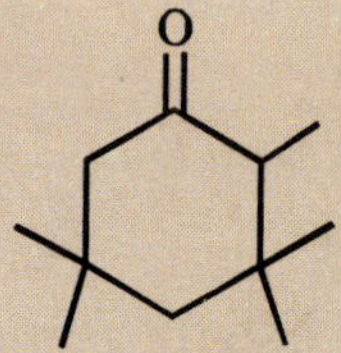

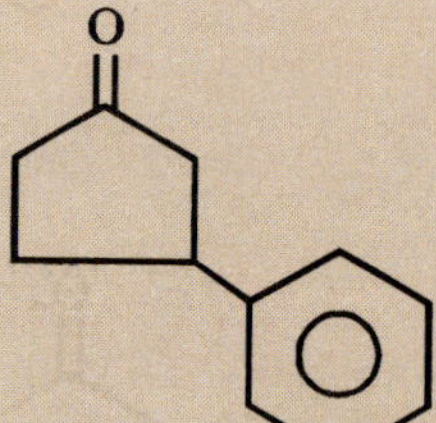

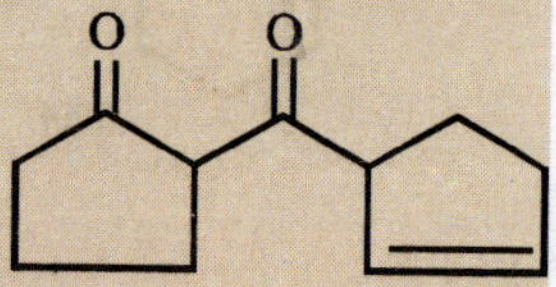

5.2 Method of synthesis of fused bicyclic α, β-unsaturated ketone with retrosynthetic analysis

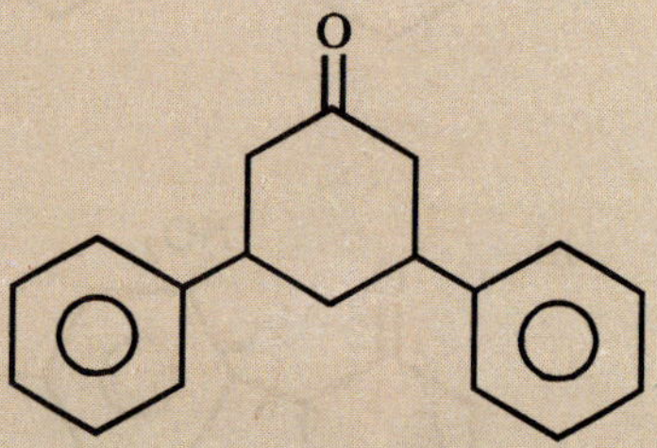

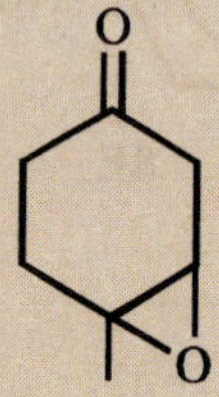

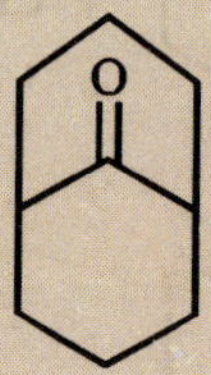

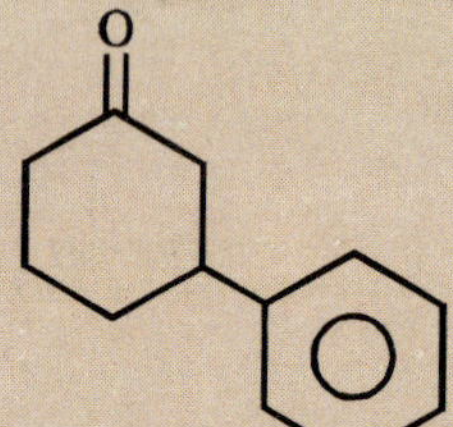

1. TM $\xRightarrow{\text{aldol dix.}}$ O O $\xRightarrow{\text{1, 5-dico}}$ O (SE) + O (SE)

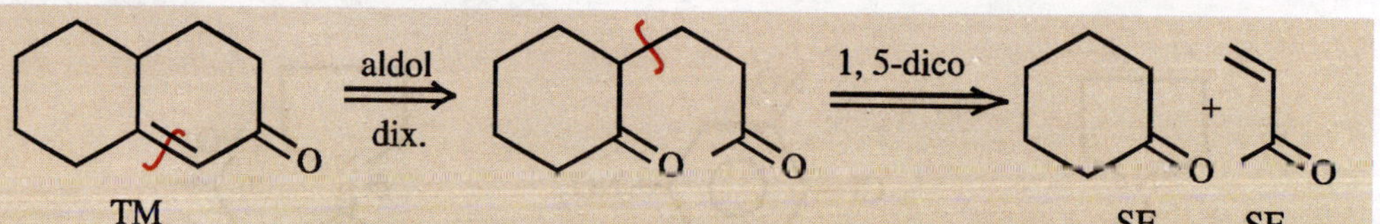

Forward Synthesis

O $\xrightarrow{\text{(i) } \overset{\ominus}{\text{O}}\text{Et} \quad \text{(ii) }}$ O O $\xrightarrow[\Delta]{\text{NH/AcOH}}$ O

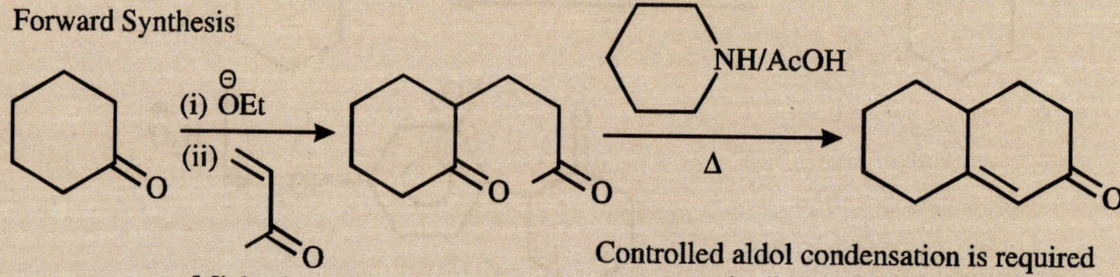

Michael cond.

Controlled aldol condensation is required to arrest the linear aldol condensation.

2.
TM
aldol dix.
5 4 3 2 1
1, 5-dico dix. at 3, 4
SE + SE

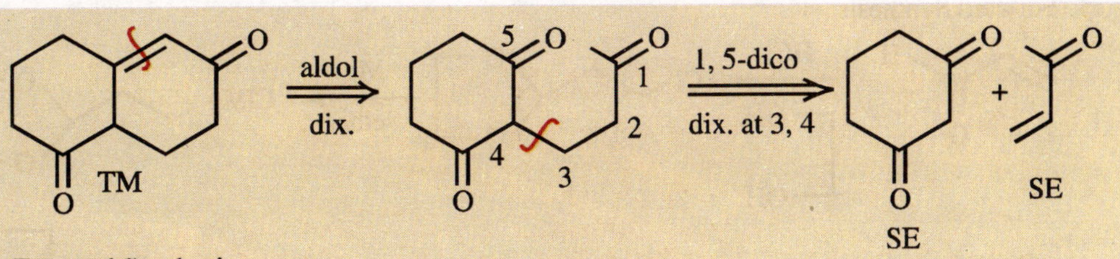

Forward Synthesis

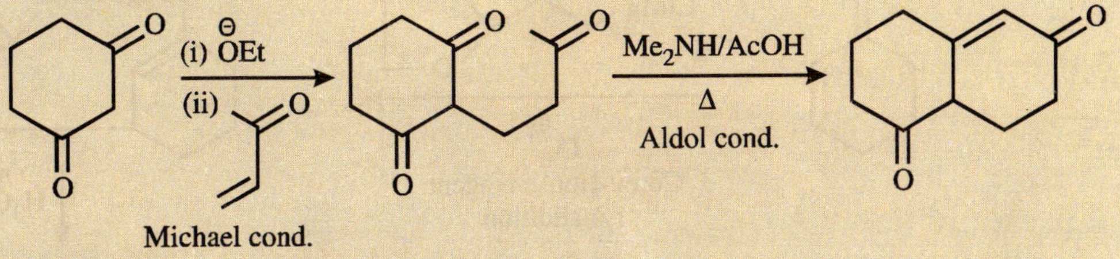

(i) $\overset{\ominus}{O}Et$
(ii)
Michael cond.
$Me_2NH/AcOH$
Δ
Aldol cond.

3.
TM
aldol dix.
SE + SE

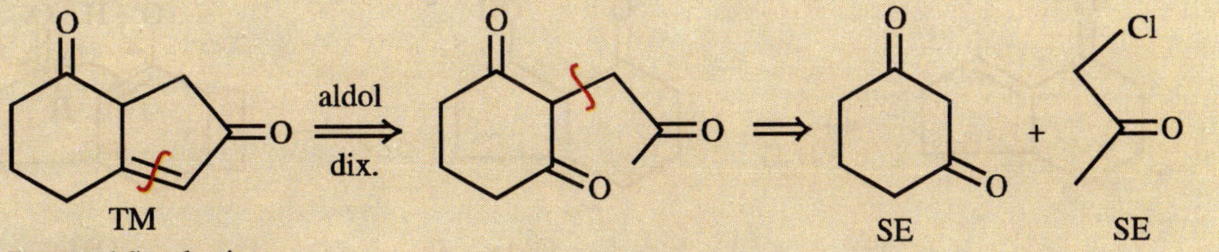

Forward Synthesis

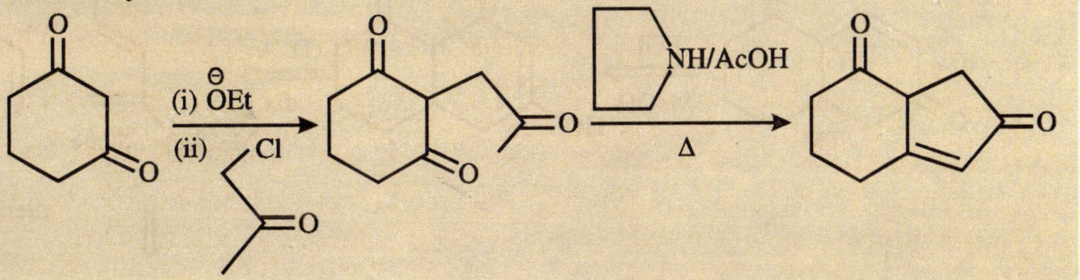

(i) $\overset{\ominus}{O}Et$
(ii)
NH/AcOH
Δ

4.
TM
FGI
SE + ClMg
HCl
SE

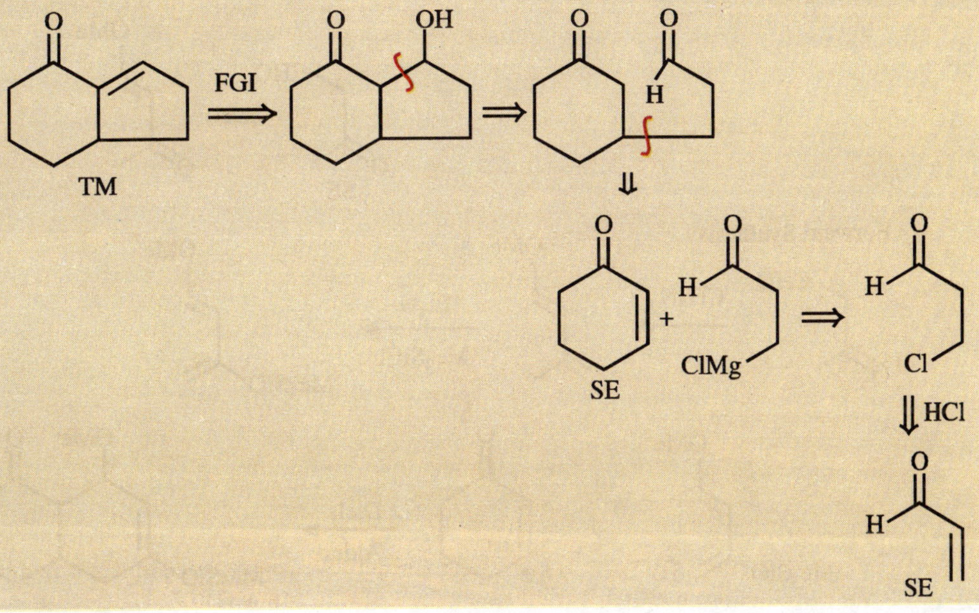

Forward Synthesis

H HCl OH OH O Cl O O Mg ether ClMg O O

O ClMg O O Cu$^{\oplus}$ Corey-House reagent 1,4-addition OMgCl O O

$H_3O^{\oplus}$

O O-H H

O OH

O Δ $-H_2O$

5. O O TM FGI MeOH OMe O HO D. A dix. OMe Me_3SiO diene + O SE dienophile

⇓

CHO O SE ⇐ OMe O

Forward Synthesis

CHO O CH_2N_2 OMe O Et_3N Me_3SiCl OMe Me_3SiO

OMe Me_3SiO δ⊖ + O δ⊕ Diels Alder OMe O Me_3SiO

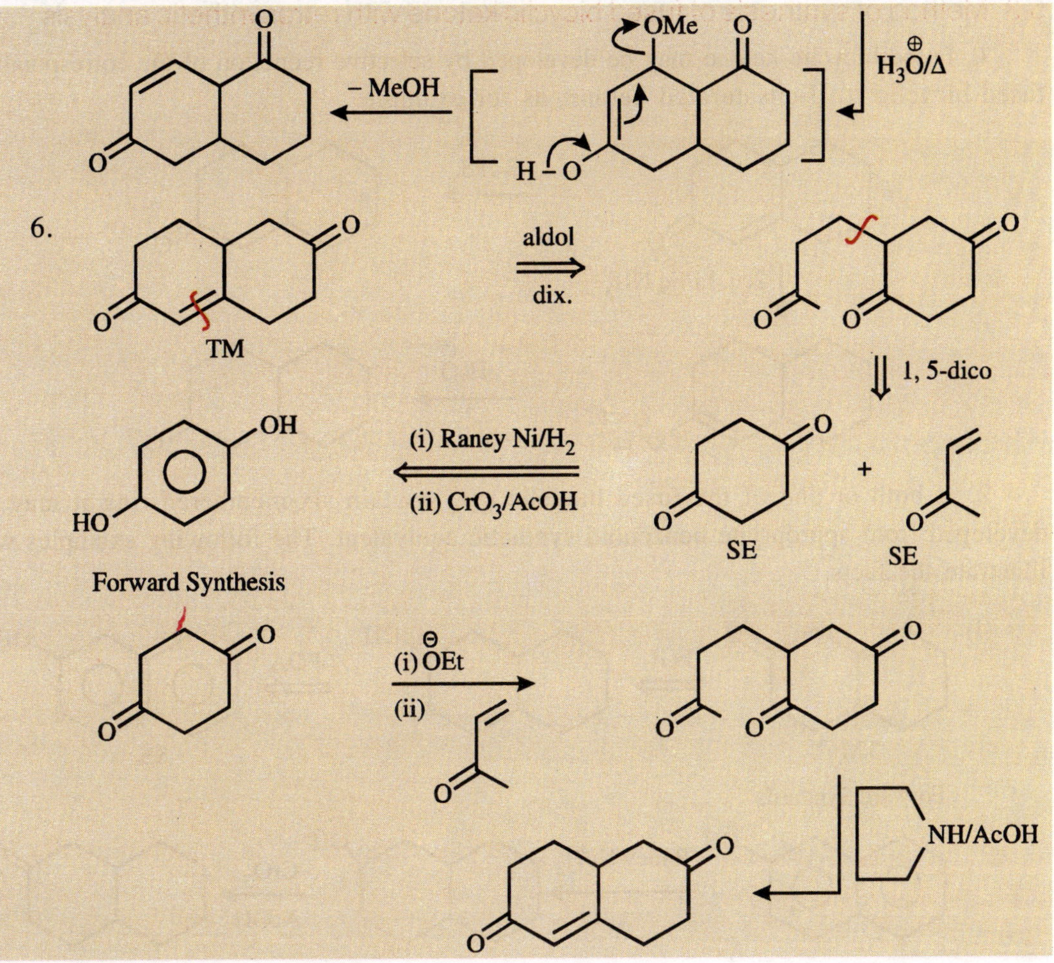

– MeOH

OMe O

H–O

$\overset{\oplus}{H_3O}/\Delta$

6.

TM

aldol
dix.

1, 5-dico

(i) Raney Ni/H_2
(ii) CrO_3/AcOH

OH
HO

SE + SE

Forward Synthesis

(i) $\overset{\ominus}{O}Et$
(ii)

NH/AcOH

5.2.1 Review synthetic problems

After proper retrosynthetic analysis show the forward synthesis for the following fused bicyclic α, β-unsaturated ketone.

1. 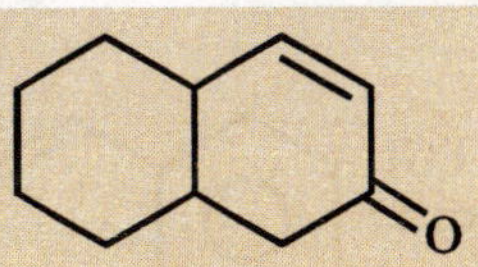 2. 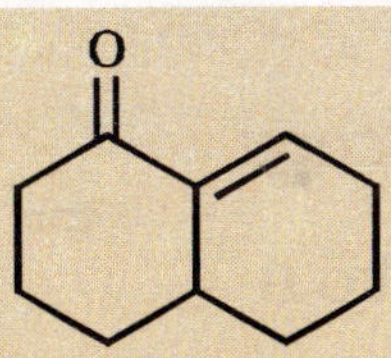 3. 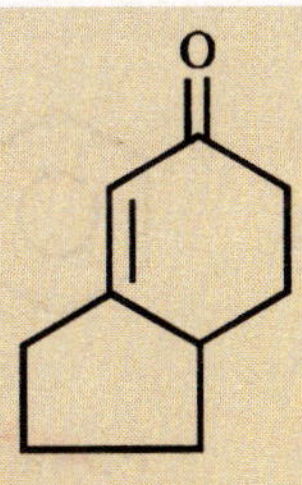 4. 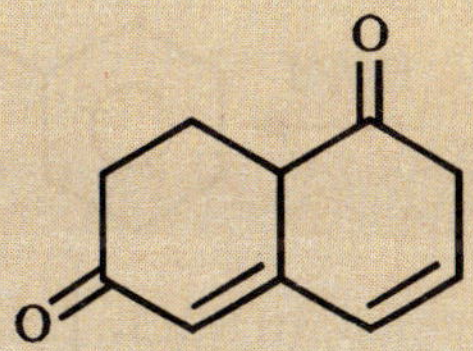 5.

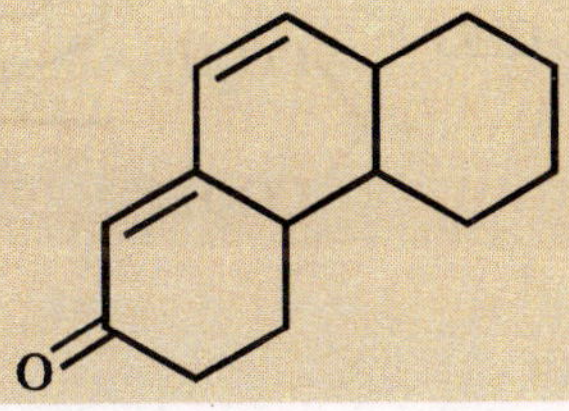

5.3 Method of synthesis of fused bicyclic ketone with retrosynthetic analysis

1. Fused bicyclic ketone may be developed by selective reduction of the corresponding fused bicyclic α, β-unsaturated ketone, as for example.

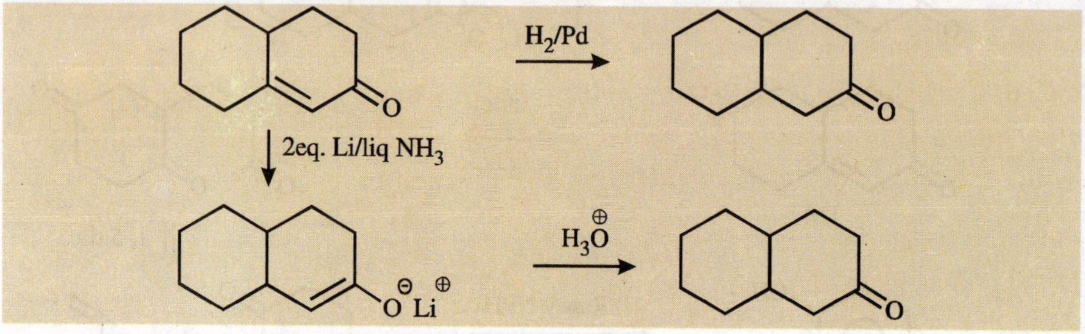

H_2/Pd

2eq. Li/liq NH_3

$H_3\overset{\oplus}{O}$

$\overset{\ominus}{O}\ \overset{\oplus}{Li}$

2. If both or one of the fused bicyclic ring contain six-membered ring it may be developed from appropriate benzenoid synthetic equivalent. The following examples will illustrate the facts.

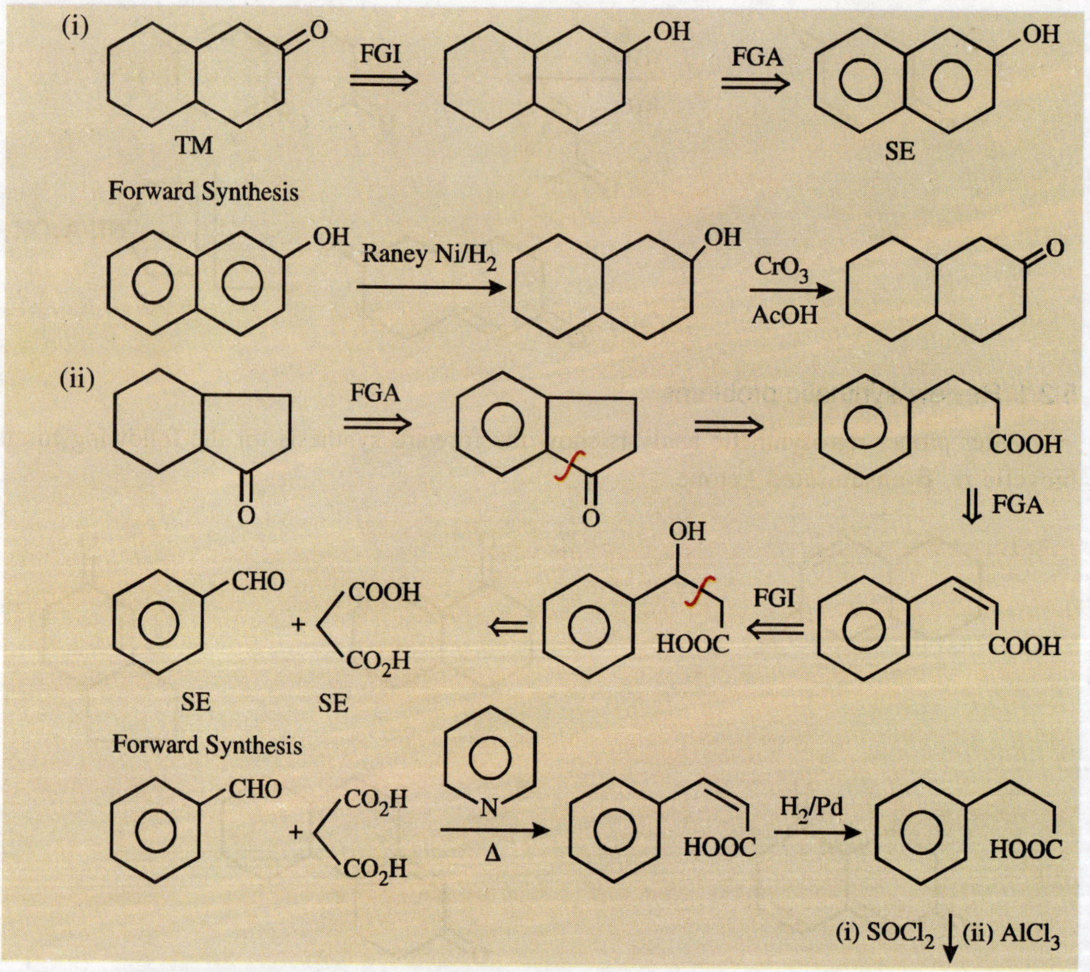

(i) O FGI ⟹ OH FGA ⟹ OH

TM SE

Forward Synthesis

OH Raney Ni/H_2 → OH CrO_3 / AcOH → O

(ii) FGA ⟹ ⟹ COOH

O O

⇓ FGA

CHO + COOH / CO_2H ⟸ OH / HOOC ⟸ FGI COOH

SE SE

Forward Synthesis

CHO + CO_2H / CO_2H —(pyridine, N) Δ→ HOOC —H_2/Pd→ HOOC

(i) $SOCl_2$ ↓ (ii) $AlCl_3$

CrO_3 / AcOH; Raney Ni / H_2

OH O O

(iii)

O O O CO_2Et O COOH

FGI

+ EtOH

TM

SE

FGI, reduction

OH COOH OH COOH OH COOH

FGI FGA

OH

OH

CO_2H

+

CHO COOH

SE SE

Forward Synthesis

OH CO_2H OH COOH

+ N

CHO CO_2H

Raney Ni/H_2

O O O OH

EtO O EtOH HO_2C CrO_3 CO_2H

$H^{\oplus}$ AcOH

$\overset{\ominus}{O}Et$/EtOH

Claisen ester cond.

O O

5.3.1 Review synthetic problems

With the retrosynthetic analysis, show the forward synthesis for the following fused bicyclic ketone.

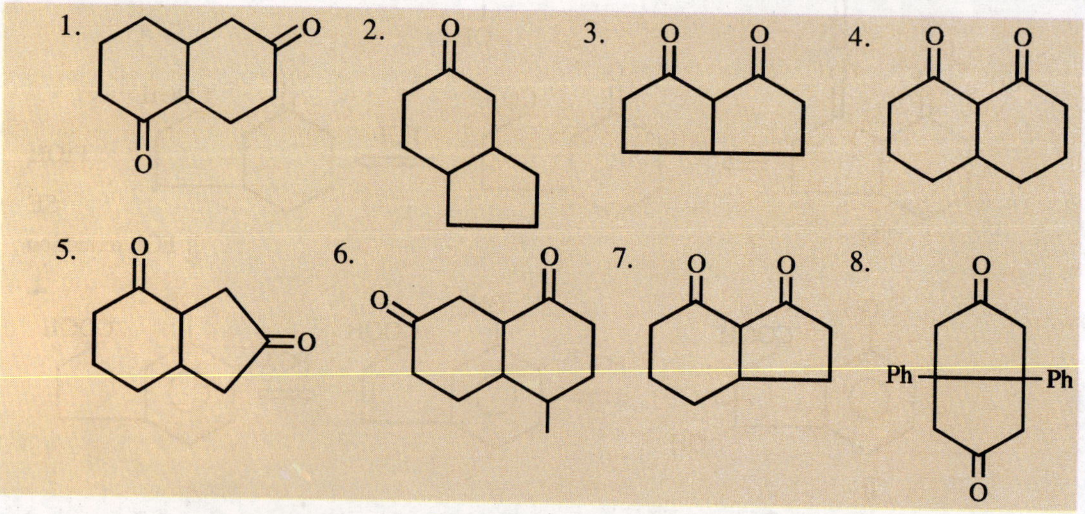

1. 2. 3. 4.

5. 6. 7. 8.

Ph Ph

5.4 Method of synthesis of angular alkylated fused bicyclic ketone with retrosynthetic analysis

1. Angular alkylation may be done with the selective alkylating agent (for example Corey-House reagent) to the appropriate choice of fused bicyclic α, β-unsaturated ketone, as for example.

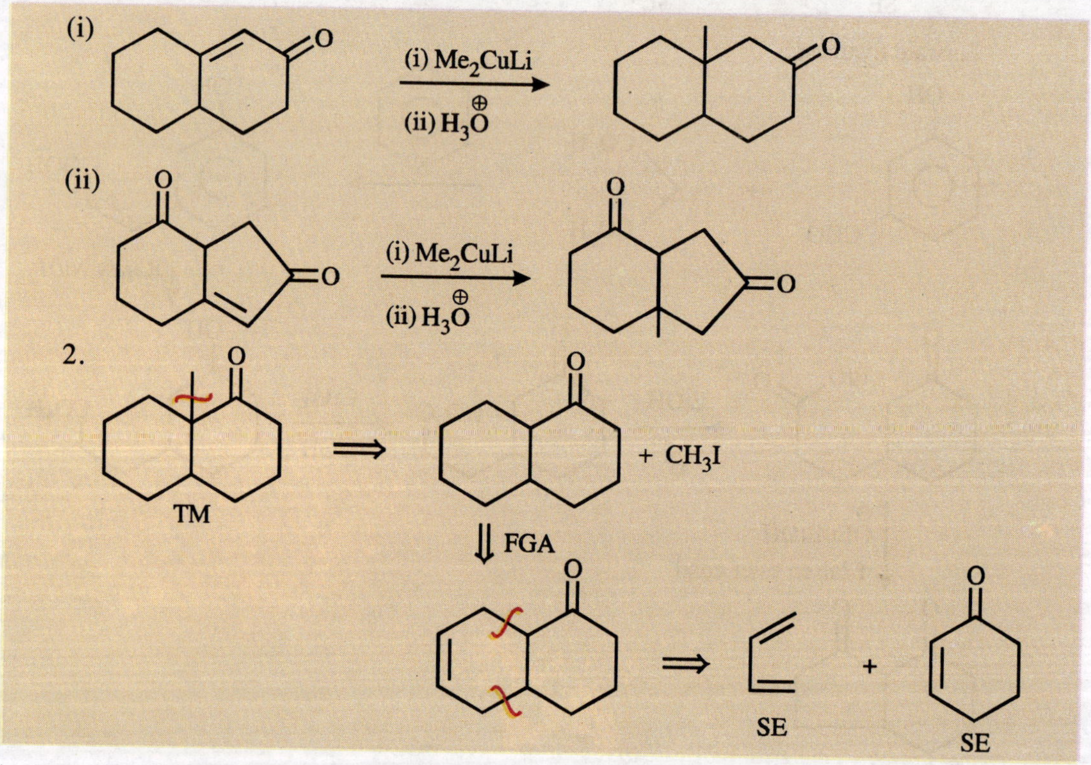

(i) (i) Me_2CuLi (ii) $H_3\overset{\oplus}{O}$

(ii) (i) Me_2CuLi (ii) $H_3\overset{\oplus}{O}$

2. TM ⟹ + CH_3I

⇓ FGA

⟹ SE + SE

Forward Synthesis

D. A, Δ

H_2/Pd

$(CH_3)_3C\overset{\ominus}{O}/HCO_2Et$

CHO

C_4H_9SH

$CHSC_4H_9$

(i) $(CH_3)_3C\overset{\ominus}{O}$

(ii) CH_3I

$CHSC_4H_9$

aq.KOH

HO⌒⌒OH

reflux

3.

TM

FGI ⟹

HO

FGI ⟹

Me_3SiO

⟹ D. A dix.

FGI ⟸

SE

Me_3SiO

⟹ aldol dix.

1, 5-dico ⟸

SE

CO_2Et

SE

Forward Synthesis

Et_3N

Me_3SiCl

Me_3SiO

O, CO_2Et (i) $\overset{\ominus}{O}Et$ (ii) O (iii) NaCl/DMSO/Δ O O Me_2NH AcOH O

Me_3SiO + O D. A O Me_3SiO

$H_3\overset{\oplus}{O}$

O O ⇌ O HO

4. TM ⇒ O + Me_2CuLi aldol dix. ⇒ O O

⇓ 1, 5-dico

CH_3I + O SE ⇐ O + O SE

Forward Synthesis

O $\overset{\ominus}{O}Et/CH_3I$ O NaH $O^{\ominus}$ thermodynamic enolate

O

O (i) Me_2CuLi (ii) $H_3\overset{\oplus}{O}$ O Me_2NH AcOH O O

5.

TM

FGA

FGI

hydrogenolysis

$CHBr_2$

Reimer-Tiemann reaction

FGI

OH

OMe

FGA

OMe

OMe

+

SE

Br

OMe

SE

Forward Synthesis

Br

OMe

$\overset{\ominus}{N}H_2$

liq.NH_3

OMe

D. A

OMe

H_2/Pd

OMe

HI

OH

$\overset{\ominus}{O}H/CHBr_3$

$CHBr_2$

$NaBH_3CN$

(i) Raney Ni/H_2

(ii) CrO_3/AcOH

5.4.1 Review synthetic problems

How can you synthesize the following angular methylated fused bicyclic ketone? Show the retro-synthetic analysis.

1. O
2. =O
3. O
4. O
5. O
6. O

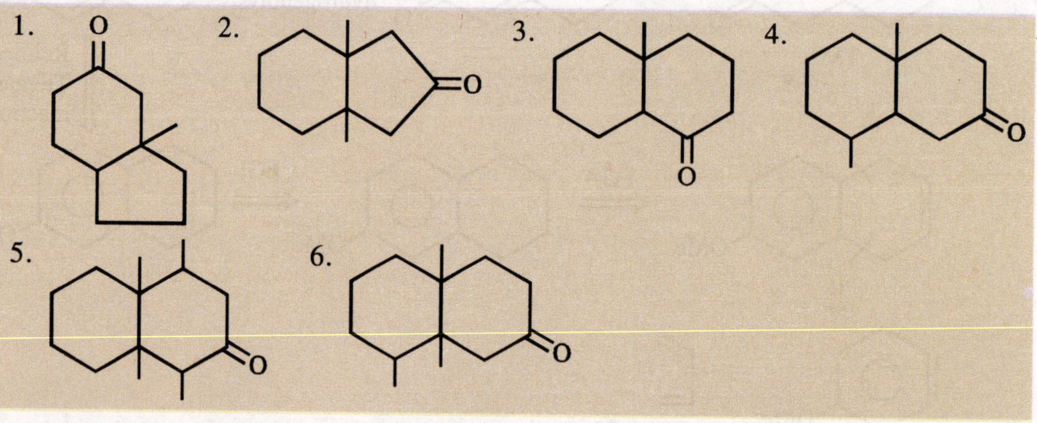

5.5 Method of synthesis of spirocyclic ring system with retrosynthetic analysis

1. TM O — Retro Pinacol ⟹ OH OH ⟹ O O Mg ⇓ O O Mg ⟸ =O (SE) + O= (SE) +Mg

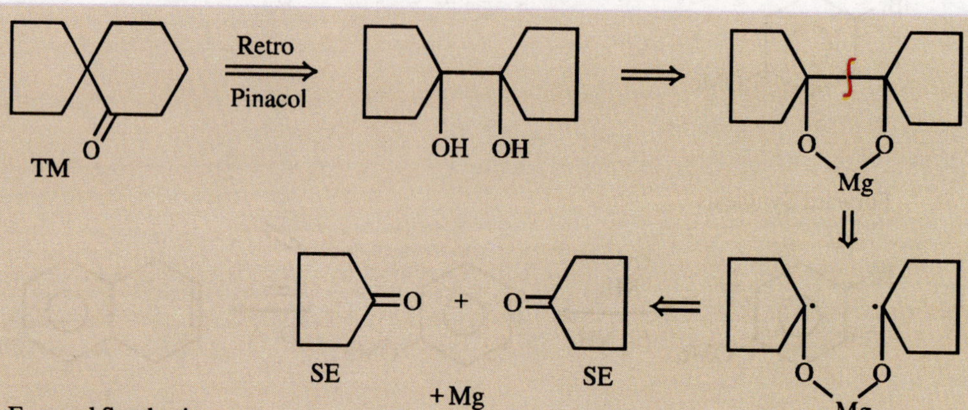

Forward Synthesis

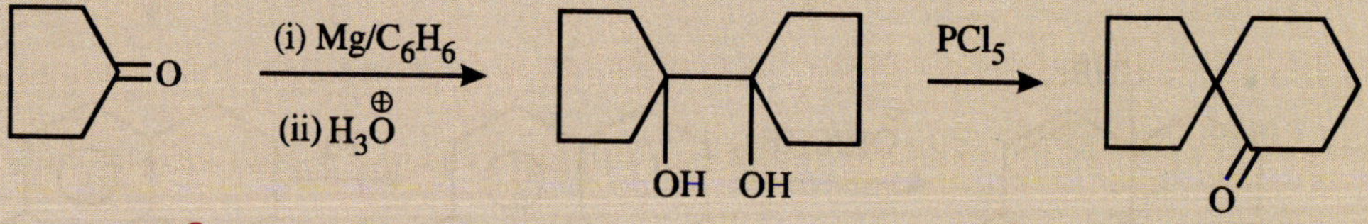

=O —(i) Mg/C_6H_6 (ii) $H_3\overset{\oplus}{O}$→ OH OH —PCl_5→ O

2. TM =O — aldol dix ⟹ CHO =O — 1, 5-dico ⟹ CHO + =O (SE) ⟹ ⊕ (SN) + $\overset{\ominus}{C}HO$ (SN)

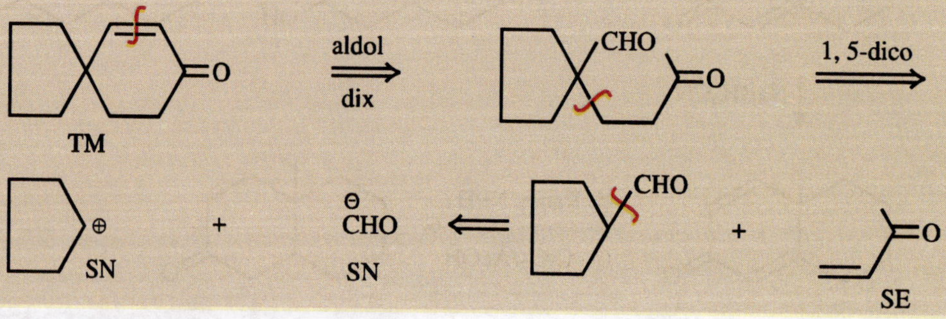

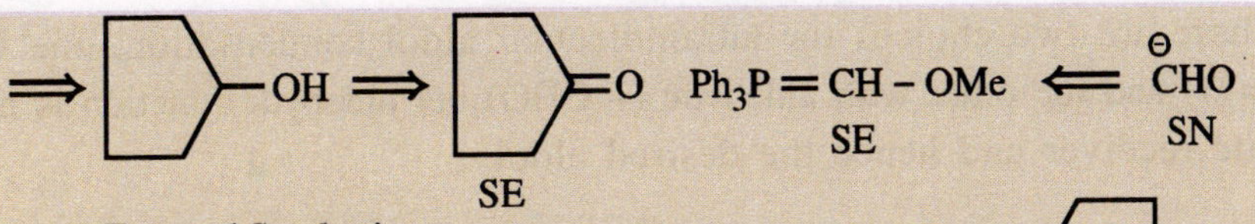

⟹ cyclopentyl–OH ⟹ cyclopentanone (SE) + $Ph_3P=CH-OMe$ (SE) ⟸ $\overset{\ominus}{C}HO$ (SN)

Forward Synthesis

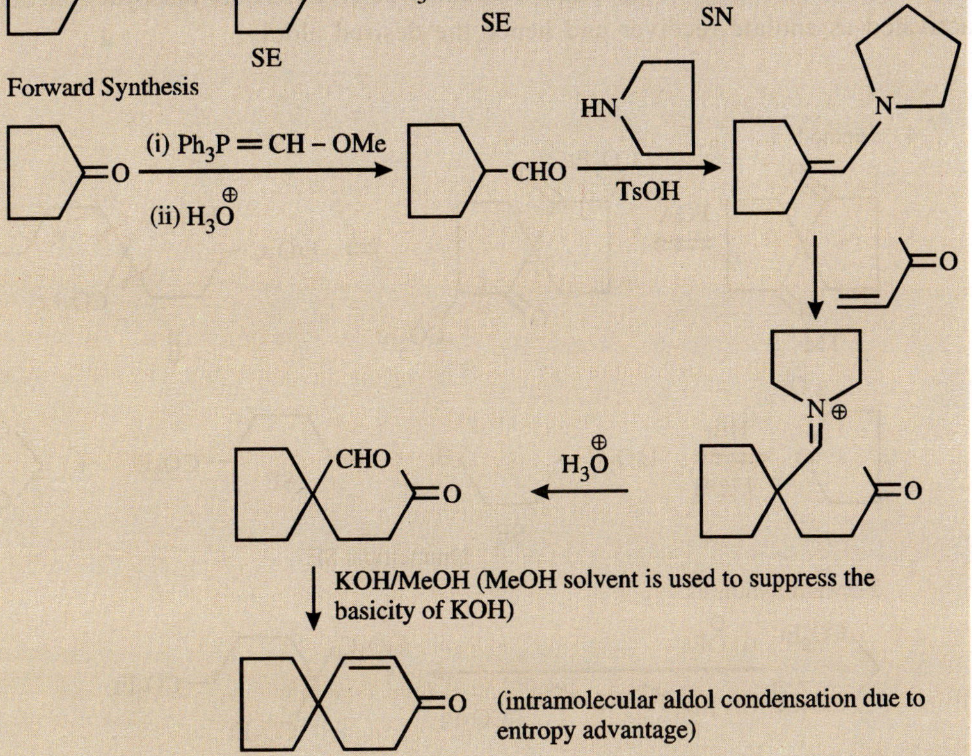

Cyclopentanone $\xrightarrow[\text{(ii) } H_3\overset{\oplus}{O}]{\text{(i) } Ph_3P=CH-OMe}$ cyclopentane–CHO $\xrightarrow[\text{TsOH}]{\text{pyrrolidine (HN)}}$ enamine (N) ⟶ (methyl vinyl ketone) ⟶ iminium ($N^{\oplus}$) spiro intermediate $\xrightarrow{H_3\overset{\oplus}{O}}$ keto-aldehyde (CHO, =O)

↓ KOH/MeOH (MeOH solvent is used to suppress the basicity of KOH)

Spiro enone (=O) (intramolecular aldol condensation due to entropy advantage)

3. Spiro diketone ⟹ (aldol, dix.) keto-aldehyde (CHO, =O) ⟹ (1, 5-dico, dix.) methyl vinyl ketone (SE) + 2-formylcyclohexanone (CHO, =O)

2-formylcyclohexanone ⟹ cyclohexanone (SE) + HCO_2Et (SE)

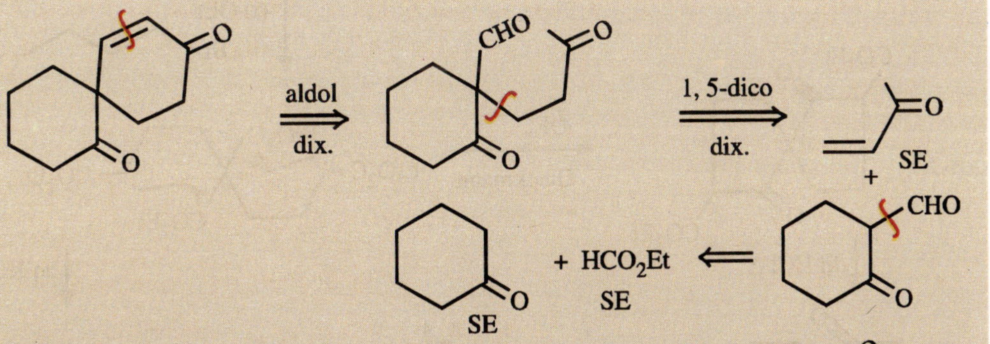

Forward Synthesis

Cyclohexanone $\xrightarrow{\overset{\ominus}{O}Et/HCO_2Et}$ 2-formylcyclohexanone (CHO, =O) $\xrightarrow[\text{methyl vinyl ketone}]{\overset{\ominus}{O}Et}$ keto-aldehyde (CHO, =O) $\xrightarrow[\text{(ii) } H_3\overset{\oplus}{O}/\Delta]{\text{(i) } \overset{\ominus}{O}Et/EtOH}$ spiro enone diketone (=O, =O)

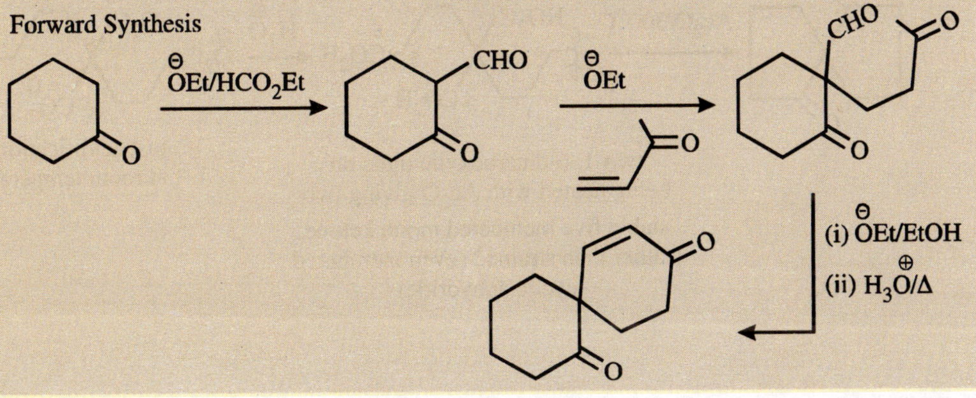

In this case there are two choices for intramolecular aldol condensation, one with ring carbonyl (C=O) and the other with aldehyde (– CHO) but aldehyde function is more activated as enolate receiver and hence the desired aldol.

4. Method I

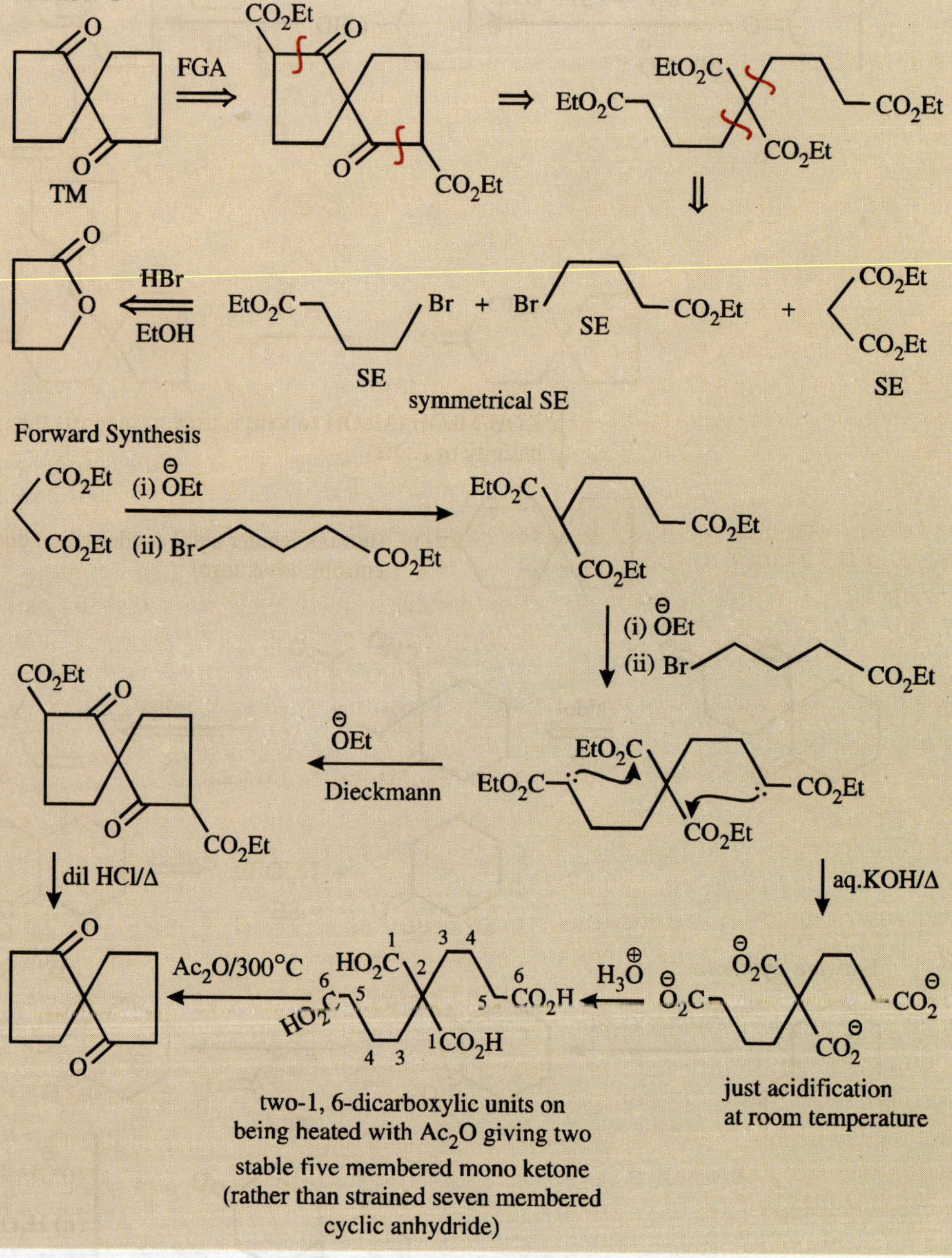

TM FGA ⟹ CO_2Et ... CO_2Et ⟹ EtO_2C ... CO_2Et

HBr / EtOH ⟸ EtO_2C–(CH$_2$)$_3$–Br (SE) + Br–(CH$_2$)$_3$–CO_2Et (SE) + $CH_2(CO_2Et)_2$ (SE)

symmetrical SE

Forward Synthesis

$CH_2(CO_2Et)_2$ (i) $\overset{\ominus}{O}Et$ (ii) Br–(CH$_2$)$_3$–CO_2Et ⟶ EtO_2C, CO_2Et, CO_2Et

(i) $\overset{\ominus}{O}Et$ (ii) Br–(CH$_2$)$_3$–CO_2Et

$\overset{\ominus}{O}Et$ Dieckmann

dil HCl/Δ

aq.KOH/Δ

$H_3\overset{\oplus}{O}$

just acidification at room temperature

$Ac_2O/300°C$

two-1, 6-dicarboxylic units on being heated with Ac_2O giving two stable five membered mono ketone (rather than strained seven membered cyclic anhydride)

5. Method II

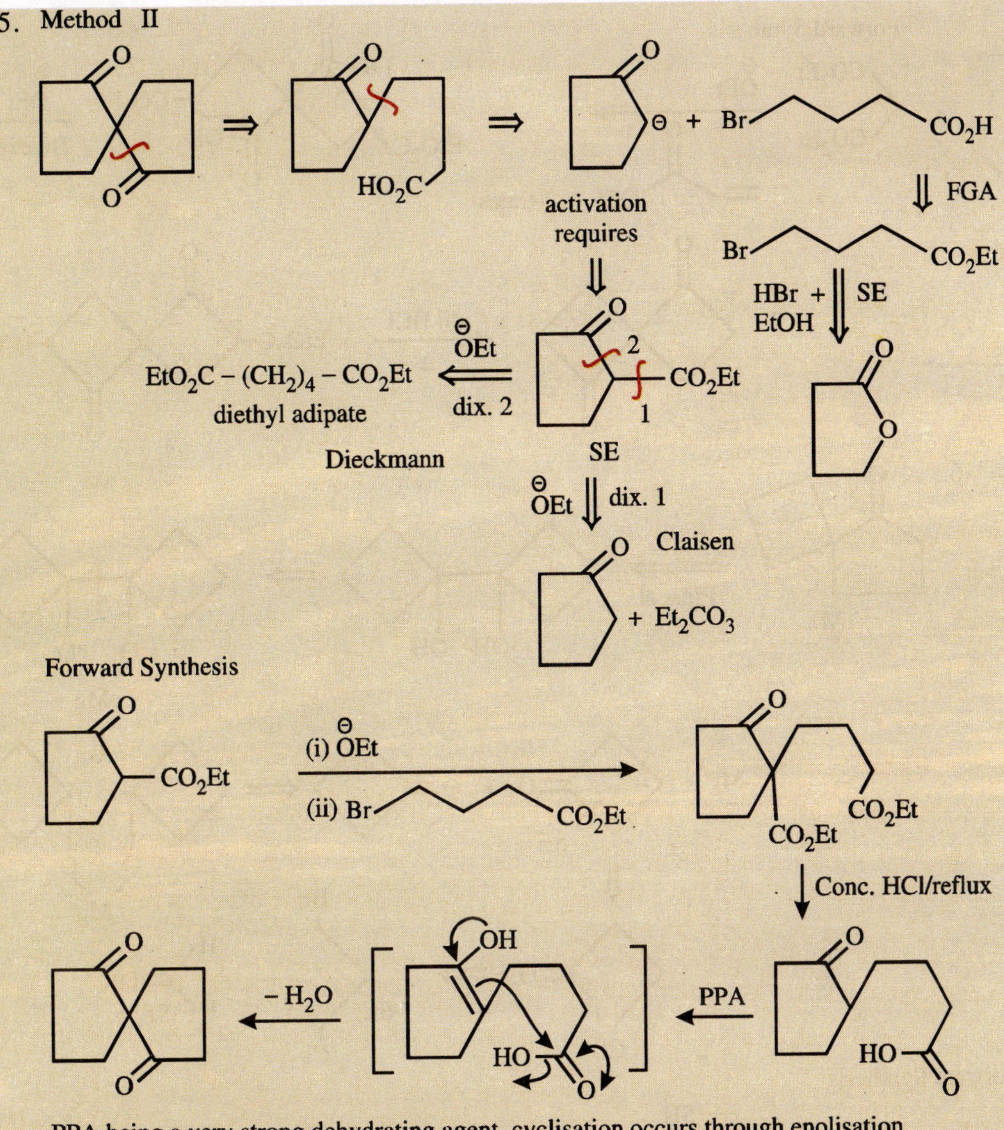

PPA being a very strong dehydrating agent, cyclisation occurs through enolisation

6.

O

TM

O

FGA ⟹

O

EtO$_2$C– –CO$_2$Et

O

⇓

EtO$_2$C CO$_2$Et

EtO$_2$C CO$_2$Et

⟸

O

OEt

SE

+

CO$_2$Et

CO$_2$Et

SE

Forward Synthesis

CO_2Et / CO_2Et ; $\overset{\ominus}{O}Et$; OEt 2-times ; EtO ; EtO_2C ; CO_2Et ; OEt ; $\overset{\ominus}{O}Et$ Dieckmann

dil HCl Δ ; EtO_2C ; CO_2Et

6. TM ; Retro Pinacol ; OH OH ; O O Mg

Mg + =O + O= ; SE ; O O Mg

=O ; H ; Cl ; Br ; SE ; Cl ; + ; H ; =O ; H ; SE

Forward Synthesis

(a) H ; C=O ; H ; SH ; SH ; $/BF_3$; H ; S ; H ; S ; (i) nBuLi ; (ii) Br ; Cl ; H ; S ; Cl ; S ; nBuLi

(b) =O ; (i) Mg/C_6H_6 ; (ii) $H_3\overset{\oplus}{O}$; HO OH ; =O ; Hg^{+2} ; $H_3\overset{\oplus}{O}$; S ; S

(c) HO OH

Conc. H_2SO_4

7. TM

Ketal ⟹ dix

SE + HO OH HO OH + SE

cannizaro ⇓ FGI reduction

HO CHO HO OH

FGA ⟸ aldol

HO CHO HO H + CH_2O

FGA ⟸ aldol

HO CHO H H + CH_2O

⇓ FGA aldol

$CH_3CHO + CH_2O$
SE

Forward Synthesis

CH_3CHO $\xrightarrow[\text{3 times three aldol}]{\overset{\ominus}{O}H/CH_2O}$ HO CHO HO OH $\xrightarrow[\text{one cannizaro}]{\overset{\ominus}{O}H/CH_2O(X'S)}$ HO OH HO OH

2 (acetone), dry HCl gas ↓

5.5.1 Review synthetic problems (Spirocyclic ring system)

After proper retrosynthetic analysis show the synthetic route with mechanism for the following spirocyclic target molecules.

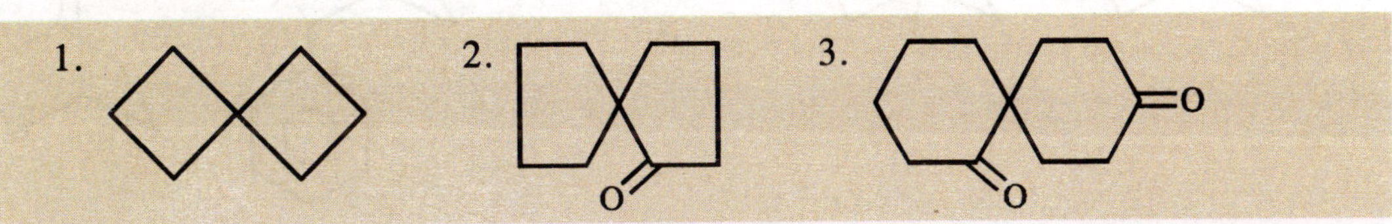

4. 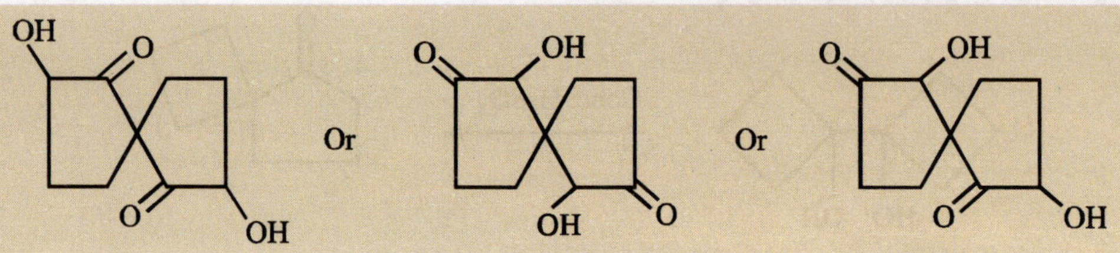 OH O O OH Or O OH O OH Or O OH O OH

5. 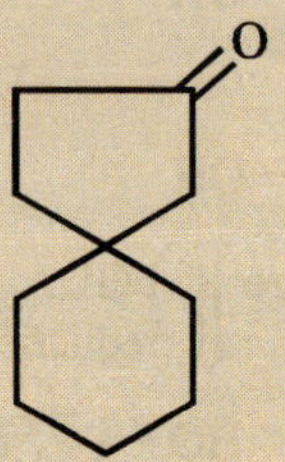 O 6. 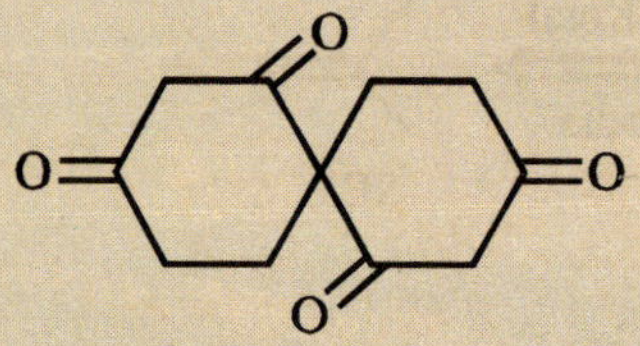 O O O O 7.

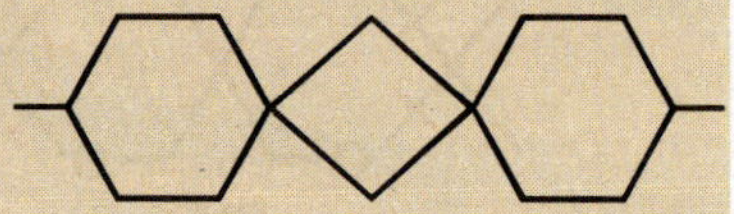

8. 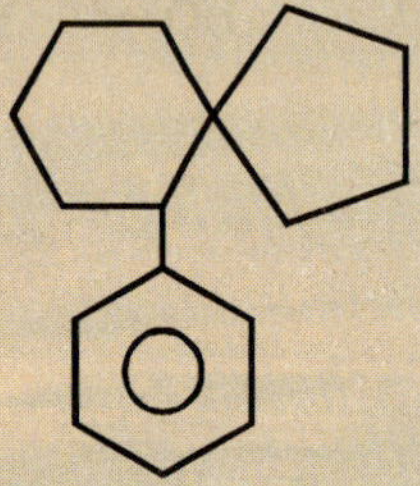 9. 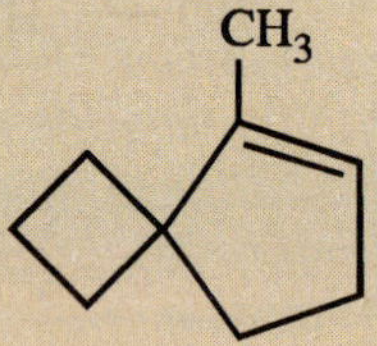 CH_3 10. 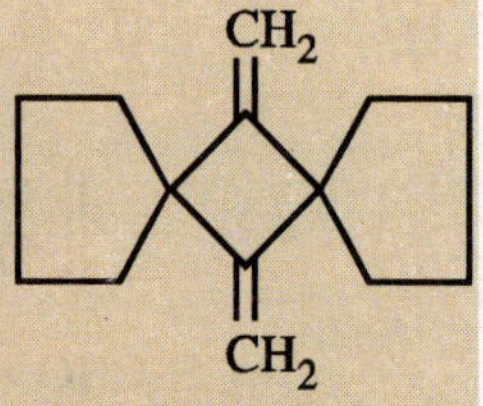 CH_2 CH_2

11. 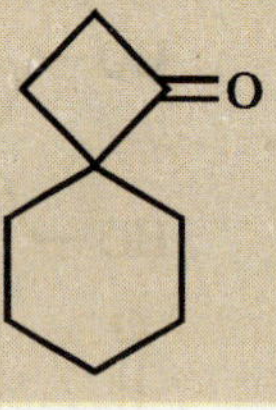 O 12. 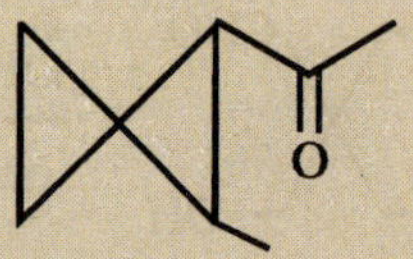 O 13. 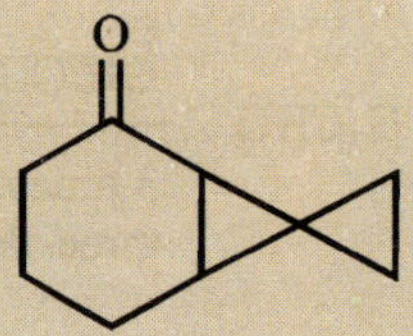 O

5.5.2 Review synthetic problems (Bridged ring bicyclic system)

After proper retrosynthetic analysis show the synthetic route with mechanism for the following bridged ring bicyclic target molecules.

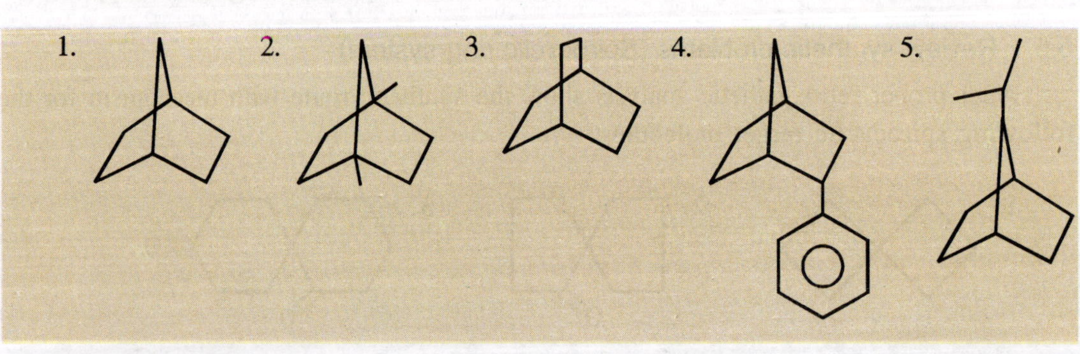

6

Umpolung Synthesis

6.1 What is umpolung synthesis?

The transformation of an electrophilic carbon into a nucleophilic one i.e. dipole reversal which is umpolung and the reverse process i.e. addition of electrophile to the dipole reversed (umpoled) nucleophile are said to involve umpolung synthesis.

All dipole reversed nucleophiles are actually illogical nucleophilic synthons or synthetic equivalents.

The opposite is also true i.e. transformation of a nucleophilic carbon into an electrophilic one and the addition of a nucleophile to the umpoled electrophile is also umpolung synthesis.

All dipole reversed electrophiles are actually illogical electrophilic synthons or synthetic equivalents and all illogical nucleophiles and electrophiles are umpoled reagents.

Synthesis from the reaction of an illogical nucleophile and illogical electrophile is the double umpolung synthesis.

6.2 Synthesis of umpoled synthons and synthetic equivalents

6.2.1 Conversion of electrophilic carbon to the nucleophilic carbon

(i) $Me-\overset{\delta+}{CH_2}-Br \xrightarrow[DMF]{KNO_2} Me-CH_2-NO_2 \xrightarrow{\overset{\ominus}{O}Et} Me-\overset{\ominus}{CH}-NO_2 \equiv Me\overset{\ominus}{C}{=}O$

deprotonated nitroethane
acyl anion equivalent

(ii)

H(δ+)HC=O + HS–CH₂CH₂–SH, BF_3 → H_2C(S–CH₂CH₂CH₂–S) (1,3-dithiane) → nBuLi → H–C⊖ dithiane ≡ H–C⊖=O

deprotonated thiane formyl anion eq.

(iii) Ph(δ+)HC=O → KCN, EtOH → Ph–C⊖(CN)(OH) ≡ Ph–C⊖=O

deprotonated cyanohydrin, benzoyl anion eq.

(iv) Ph(δ+)HC=O → KCN, Me_3SiCN → Ph–C(H)($OSiMe_3$)(CN) → LDA/THF, – 78°C → Ph–C⊖($OSiMe_3$)(CN) ≡ Ph–C⊖=O

deprotonated silylated cyanohydrin, benzoyl anion eq.

(v) H_2C=C(δ+)H–O–Me → tBuLi → H_2C=C⊖–O–Me ≡ Me–C⊖=O

acyl anion eq.

(vi) (δ+) CH_2=CH–C(=O)Me → MPV<H> → CH_2=CH–C(H)(OH)Me → $Me_3O^{\oplus}BF_4^{\ominus}$ → CH_2=CH–C(H)(OMe)Me → nBuLi → CH_2=CH–C⊖(OMe)Me ↔

$\overset{\ominus}{C}H_2 - CH_2 - C(=O) - Me \equiv \overset{\ominus}{C}H_2 - CH{=}C(OMe)Me$

β-acyl carbanion eq.

(vii) CH_2=CH–CH₂–C(=O)H (δ+) → CH_2N_2 → CH_2=CH–CH=CH(H)–OMe → tBuLi → CH_2=CH–CH=C⊖–OMe ≡ CH_3–CH=CH–C⊖=O

Crotonyl anion eq.

(viii) $\overset{\delta+}{CH_2}{=}CH{-}CO{-}OEt \xrightarrow{HBr} Br{-}CH_2CH_2{-}CO{-}OEt \xrightarrow[\text{ether}]{Zn} BrZn{-}\overset{\delta-}{CH_2}CH_2{-}CO{-}OEt$

(ix) In any alkyl cyanide (e.g. $Me-\overset{\delta+}{C}\equiv N$), cyanide carbon is electrophilic in nature but $\overset{\ominus}{C}\equiv N$ is umpoled reagent in which carbon bears negative charge.

$$H-\overset{\delta+}{C}\equiv N \xrightarrow{\overset{\ominus}{O}H} \overset{\ominus}{C}\equiv N$$

Carboxyl anion eq. ($\overset{\ominus}{C}OOH$)

(x) Grignard reagent is umpoled reagent. Carbon is electrophilic in alkyl halide is converted to alkyl magnesium halide in which alkyl carbon is nucleophilic.

(a) $\overset{\delta+}{Me}-Br \xrightarrow[\text{ether}]{Mg} \overset{\delta-}{Me}MgBr$

(b) $CH_3{-}O{-}\overset{\delta+}{CH_2}{-}Cl \xrightarrow[\text{ether}]{Mg} CH_3{-}O{-}\overset{\delta-}{CH_2}{-}MgCl$

(xi) Preparation of Reformatsky reagent is a dipole reversed.

$$Br{-}\overset{\delta+}{CH_2}{-}CO{-}OEt \xrightarrow[\text{ether}]{Zn} BrZn{-}\overset{\delta-}{CH_2}{-}CO{-}OEt$$

(xii) Wittig reagent is actually umpoled reagent.

(a) $\overset{\delta+}{Me}-Br \xrightarrow[\text{(ii) nBuLi}]{\text{(i) } Ph_3P} Ph_3P=\overset{\delta-}{CH_2}$

(b) $Me{-}O{-}\overset{\delta+}{CH_2}{-}Cl \xrightarrow[\text{(ii) nBuLi}]{\text{(i) } Ph_3P} Ph_3P=\overset{\delta-}{C}(H)(OMe)$

(xiii) Simmon's-Smith reagent is actually umpoled reagent.

$$\overset{\delta+}{CH_2I_2} \xrightarrow[\text{ether}]{Zn-Cu} \overset{\delta-}{CH_2}(ZnI)(I)$$

6.2.2 Conversion of nucleophilic carbon to the electrophilic carbon

(i) Ketomethyl or methylene group in base medium gives enolate which is a versatile nucleophilic carbon (and even in acid medium it behaves as a nucleophile through enol) but when ketomethyl or methylene is monohalogenated then α-carbon becomes highly electrophilic (illogical electrophile) in base medium. It is to be noted that halocarbon is more electrophilic than carbonyl carbon (halocarbon positive charge (δ +) is stabilised by adjacent HOMO of C = O, π during S_N2 attack)

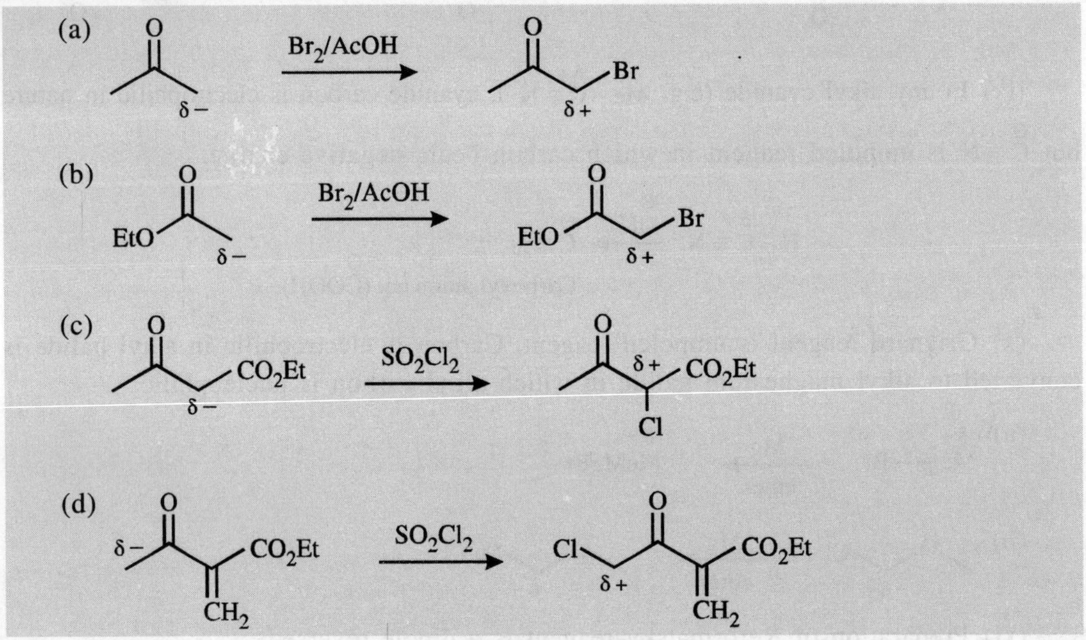

(a) $Br_2/AcOH$

(b) $Br_2/AcOH$

(c) SO_2Cl_2

(d) SO_2Cl_2

(ii) Alkene 'ene' is nucleophilic. If –M group (e.g. –CN, $-COCH_3$, –CHO) is added to 'ene' than 'ene' nucleophilicity is reversed and that's why in all Michael acceptors enone, ene is electrophilic

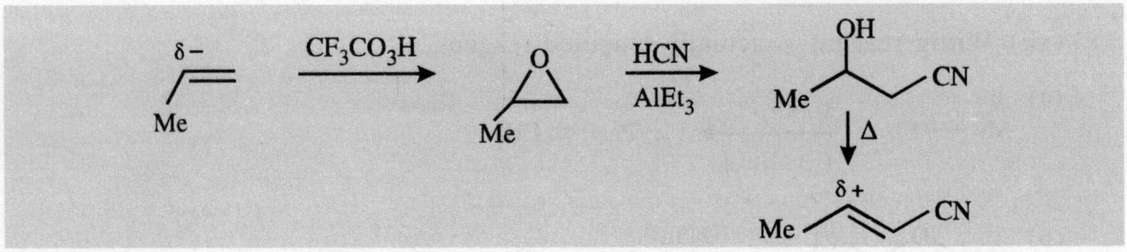

(iii) When alkene 'ene' is converted to epoxide then dipole is reversed

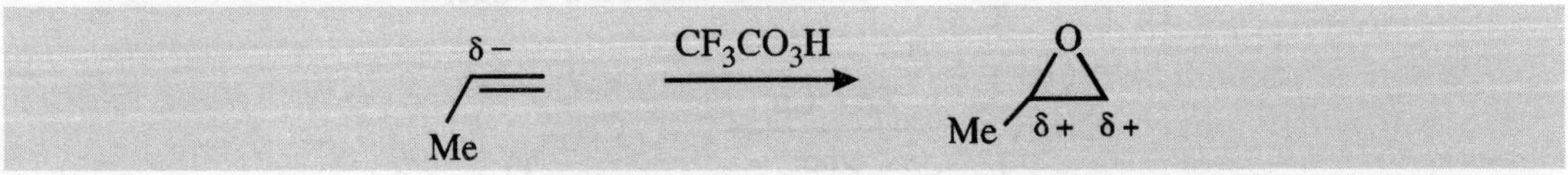

(iv) Sometimes structural feature may also reversed 'ene' nucleophilicity

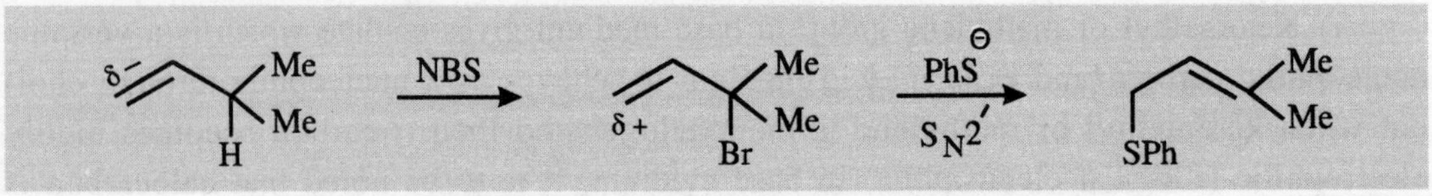

6.3 Umpoled Synthons: Latent polarity approach during retrosynthetic analysis (disconnection) of target molecules

Latent polarity concept (during disconnection) is the imaginary pattern of alternating (+) and (–) charges used to help in the identification of the disconnection and synthons of a target molecule.

The polarities are arranged in such a way that the (+) charge is initially placed on the carbon adjacent to oxygen, nitrogen and halogen atom whether linked to single or double bond. For example.

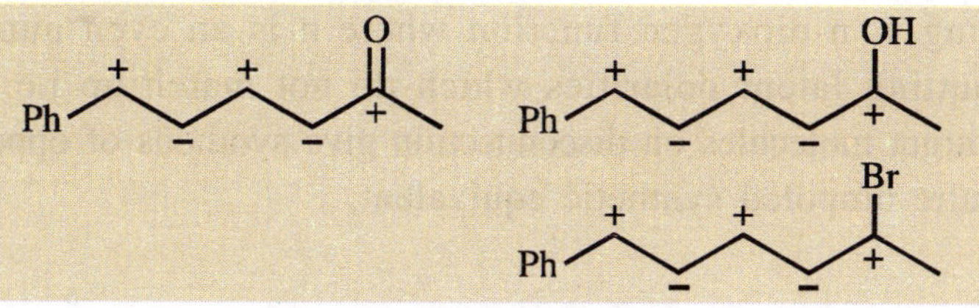

The best disconnection leads to synthons which have an alternate pattern of charges as we move away from the carbonyl (C = O) or hydroxyl (– OH) starting with (+) charge on the carbon adjacent to oxygen.

(i) Any system containing a 1, n-dioxygen function where 'n' is an odd number (e.g. 1, 3 or 1, 5) follows this alternate charges with right choice [i.e. C = O, carbon becomes (+)]. If the latent polarities arising from each of the functional group are added, starting from either left hand side C = O or from right hand side C = O, we see that the relationship of latent polarities are overlapping i.e. superimposable. Such type of target molecules on disconnection gives synthons of normal polarities and forward synthesis can be done with normal synthetic equivalent.

Thus for example,

(a)

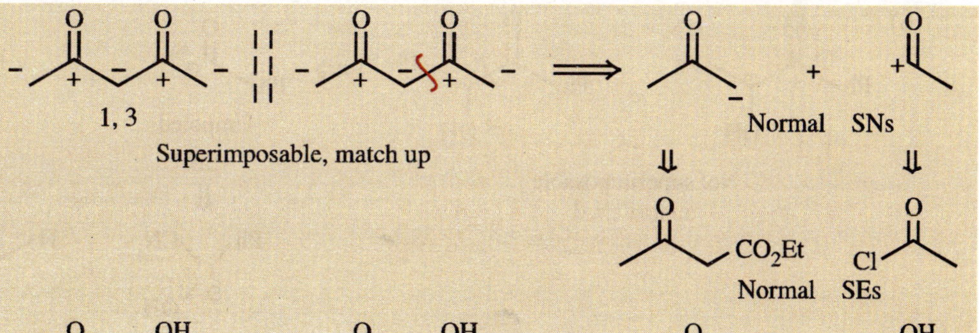

1, 3

Superimposable, match up

Normal SNs

CO_2Et Cl

Normal SEs

(b)

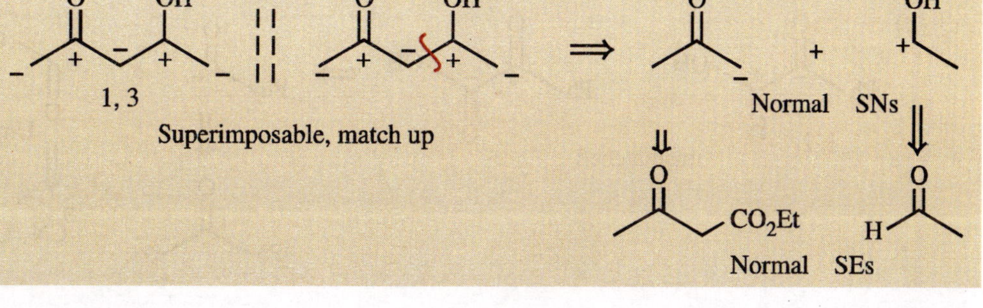

1, 3

Superimposable, match up

Normal SNs

CO_2Et H

Normal SEs

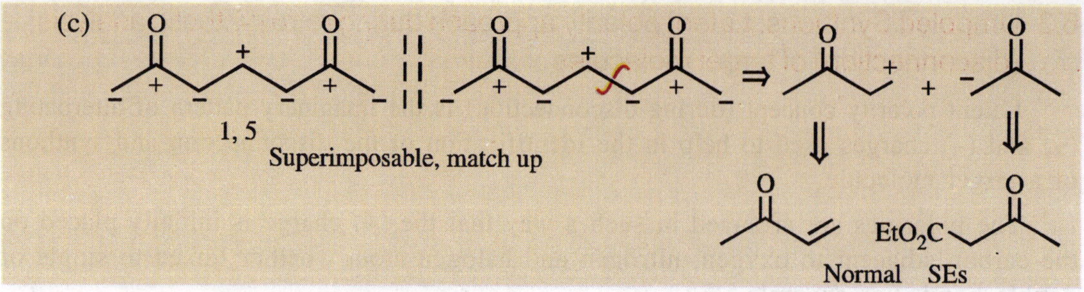

(c) 1, 5 Superimposable, match up ⇒ Normal SEs EtO_2C

(ii) Any system containing 1, n-dioxygen function where n is an even number (e.g. 1, 2 or 1, 4) then on putting latent polarities which do not match up i.e. not superimposable. Such type of target molecules on disconnection give synthons of opposite polarity and forward will require umpoled synthetic equivalent.

Thus for example,

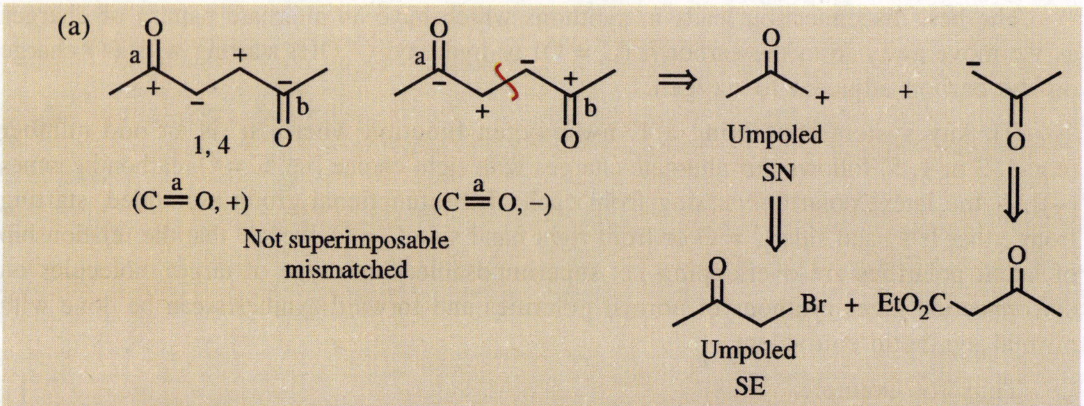

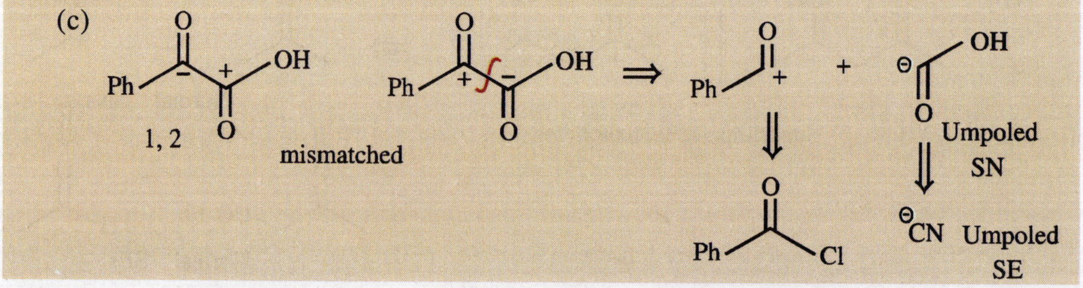

(a) 1, 4 (C $\overset{a}{=}$ O, +) (C $\overset{a}{=}$ O, −) Not superimposable mismatched ⇒ Umpoled SN ⇓ Br + EtO_2C Umpoled SE

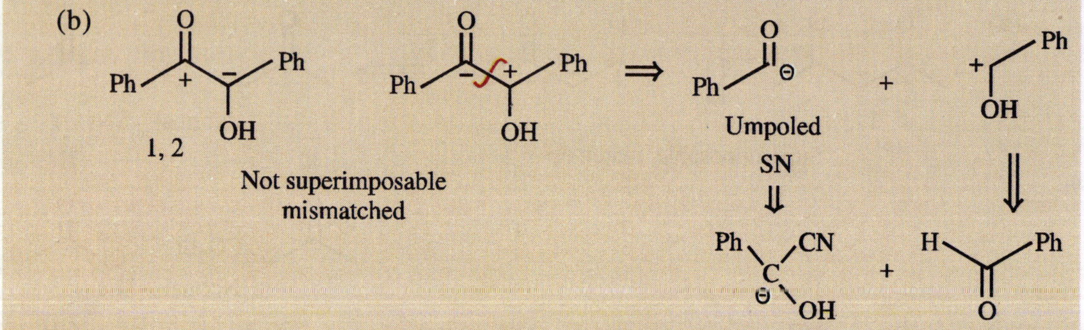

(b) Ph Ph OH 1, 2 Not superimposable mismatched ⇒ Ph Umpoled SN ⇓ Ph CN C OH + H Ph O

(c) Ph OH 1, 2 mismatched ⇒ Ph + OH Umpoled SN ⇓ Ph Cl $^{\ominus}CN$ Umpoled SE

Any system which do not follow latent polarity rule i.e. does not follow alternate (+) and (–) charges with right choice (as for examole, if C = O, carbon bears (–) charge, it is not right choice) are either umpoled synthetic equivalents (reagents) e.g. (C) and (D) below or whose synthesis is to be done with umpoled synthetic equivalents (reagents) e.g. 1, 2 and 1, 4 dioxygen system.

(C) (D)

6.4 Correlation of umpoled synthons with umpoled and normal synthetic equivalents with FGI

Some correlated umpoled (illogical) synthons with their corresponding umpoled and normal (logical) synthetic equivalents (reagents) are given below:

Umpoled Synthon	Umpoled Synthetic eq. (reagent)	Umpoled Synthon	Normal Synthetic eq.
$H-\overset{\ominus}{C}=O$	$^{\ominus}C\equiv N$	$H-\overset{\ominus}{C}=O$	$^{\ominus}CH_2-NO_2$
$Me-\overset{\ominus}{C}=O$	Me-dithiane anion	$Me-\overset{\ominus}{C}=O$	$H-C\equiv C^{\ominus}$
Me, OH, ⊕	Me-epoxide	Me, ⊖, NH_2	Me, NO_2
$^{\ominus}CH_2-NH_2$	$^{\ominus}C\equiv N$	O, Me, ⊕ CH_2	$HC\equiv C-CH_2-Br$
$H-O-CH_2^{\ominus}$	$Me-O-CH_2MgCl$	$^{\ominus}CH_2-NH_2$	$CH_3-N=CPh_2$
$H_2C^{\ominus}$, O	⊖, $Li^{\oplus}$, OMe	⊕ CH_2, O, Me	Me, O (cyclopropyl ketone)

Some illustrations

(a) R–CHO (TM) ⟹ $R-\overset{\oplus}{C}H_2$ + $^{\ominus}CHO$

⇓ ⇓

R–Cl $^{\ominus}CN$

Forward Synthesis

$$R\text{–Cl} \xrightarrow[\text{acetone}]{\text{KCN}} R\text{–CN} \xrightarrow{\text{LiAlH(OEt)}_3} R\text{–CHO}$$

(b) R–CH$_2$–CH$_2$–NH$_2$ (TM) ⟹ $R-\overset{\oplus}{C}H_2$ + $\overset{\ominus}{C}H_2-NH_2$

$R-CH_2-Cl$ ⇓ ; $\overset{\ominus}{C}H_2-NO_2$ ⇓ ⟹ CH_3NO_2

Forward Synthesis

CH_3NO_2 $\xrightarrow[2eq.]{nBuLi}$ $\overset{\ominus}{C}H=\overset{\oplus}{N}(O^{\ominus})_2$ $\xrightarrow{R-CH_2-Cl}$ $R-CH_2-CH=\overset{\oplus}{N}(O^{\ominus})_2$ $\xrightarrow{H_3\overset{\oplus}{O}}$ $R-CH_2-CH_2-NO_2$ $\xrightarrow{H_2/Pd}$ $R-CH_2-CH_2-NH_2$

(c) R–CH$_2$–CH$_2$–CHO (TM) ⟹ $R-\overset{\oplus}{C}H_2$ + $\overset{\ominus}{C}HO$

$R-CH_2-Cl$ ⇓ ; $\overset{\ominus}{C}H_2-NO_2$ ⇓ ⟹ CH_3NO_2

Forward Synthesis

CH_3NO_2 $\xrightarrow[(ii)\ R-CH_2-Cl]{(i)\ nBuLi\ (2eq)}$ $R-CH_2-CH=\overset{\oplus}{N}(O^{\ominus})_2$ $\xrightarrow[H_2O]{TiCl_3}$ $R-CH_2-CHO$

(d) R–CH(OH)–C(=O)–CH$_3$ (TM) ⟹ $R-\overset{\oplus}{C}H-OH$ + $\overset{\ominus}{C}(=O)-CH_3$ ⟹ $\overset{\ominus}{C}\equiv C-H$

⇓ R–CHO

Forward Synthesis

$H-C\equiv C-H$ $\xrightarrow[(ii)\ CH_3CHO]{(i)\ \overset{\ominus}{N}H_2}$ $CH_3-CH(O^{\ominus})-C\equiv C-H$ $\xrightarrow[H_2SO_4]{Hg^{+2}}$ $CH_3-CH(OH)-C(=O)-CH_3$

6.5 Umpolung synthesis with retrosynthetic analysis

1. Me–C(=O)–CH$_2$–OH (TM) ⟹ $Me-\overset{\ominus}{C}=O$ + $\overset{\oplus}{C}H_2OH$ ⟹ HCHO (SE)

⇓ 2-Me-1,3-dithiane anion (Me–$\overset{\ominus}{C}$(S–(CH$_2$)$_3$–S)) (SE)

Forward Synthesis

Me, S, C, ⊖, S $\xrightarrow{HCHO}$ Me, ⊖O, C, S, S $\xrightarrow[H_3\overset{\oplus}{O}]{Hg^{+2}}$ O, Me, OH

2. O, Ph, Ph, OH, Me, TM ⟹ O, ⊖, Ph + ⊕, Ph, C, OH, Me ⟹ O, Ph, Me, SE

⇓

Ph, $OSiMe_3$, C, ⊖, CN, SE

Forward Synthesis

Ph, $OSiMe_3$, C, ⊖, CN → (O, Ph, Me) Ph, $OSiMe_3$, Me, C, CN, Ph, $O^{\ominus}$ → (i) CH_3Li (1eq) (ii) H_2O/Δ

Ph, OH, Me, C, CN, Ph, OH ← – HCN O, Ph, Ph, OH, Me

3. O, OH, Me, Me, TM ⟹ O, Me, ⊖, Umpoled SN + OH, ⊕, Me, Umpoled SN

⇓ ⇓

Me, S, ⊖, S, Umpoled SE; O, Me, Umpoled SE

Forward Synthesis

Me, S, ⊖, S + O, Me → ⊖O, Me, C, S, S $\xrightarrow[H_3\overset{\oplus}{O}]{Hg^{+2}}$ O, OH, Me, Me

double umpolung synthesis

4. NH_2, Me, CO_2Et, TM ⟹ ⊕, Me, CO_2Et + $\overset{\ominus}{C}H_2-NH_2$, SN ⟹ $\overset{\ominus}{C}N$ umpoled SE

⇓

Me, CO_2Et, SE

Forward Synthesis

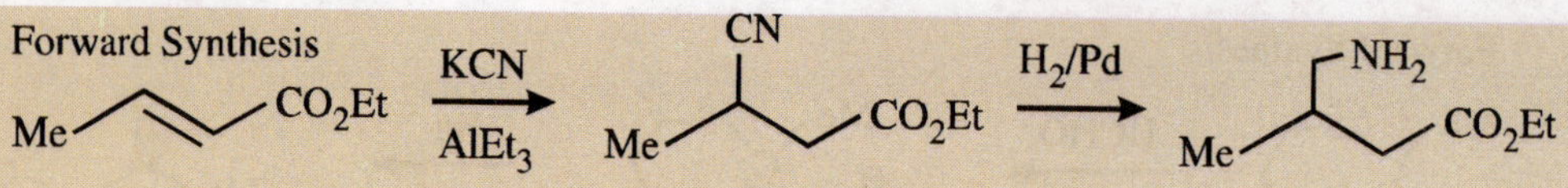

KCN / $AlEt_3$; H_2/Pd

Me CO$_2$Et, CN, NH$_2$

5.

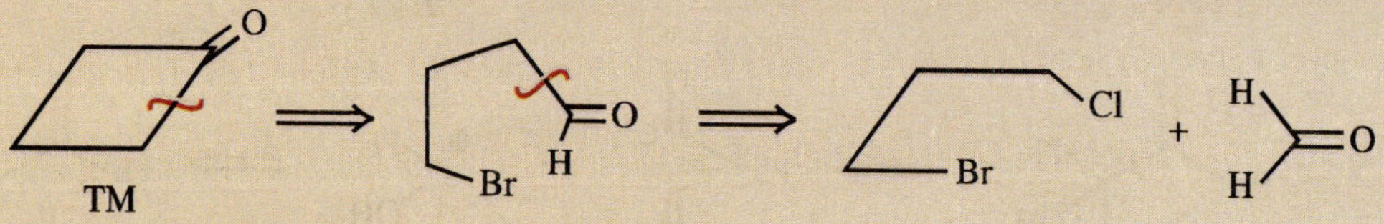

TM

Br, H, O, Cl, H, H

Forward Synthesis

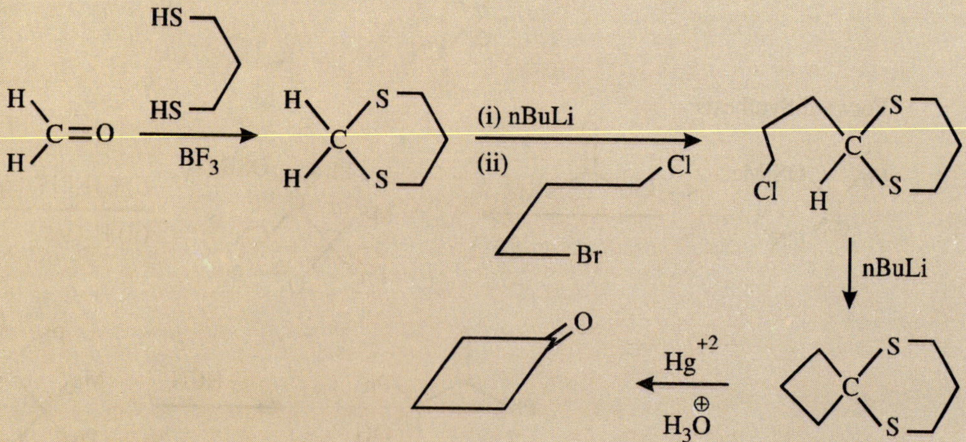

HS, HS, BF_3, (i) nBuLi, (ii) Cl, Br, nBuLi, Hg^{+2}, $H_3O^{\oplus}$

6.

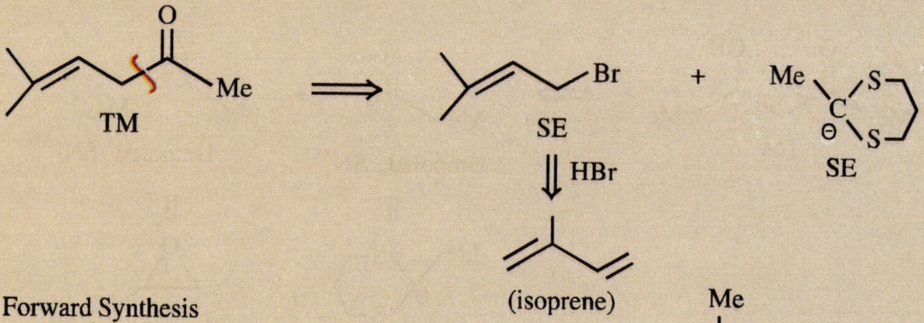

TM, Me, Br, SE, HBr, (isoprene), SE

Forward Synthesis

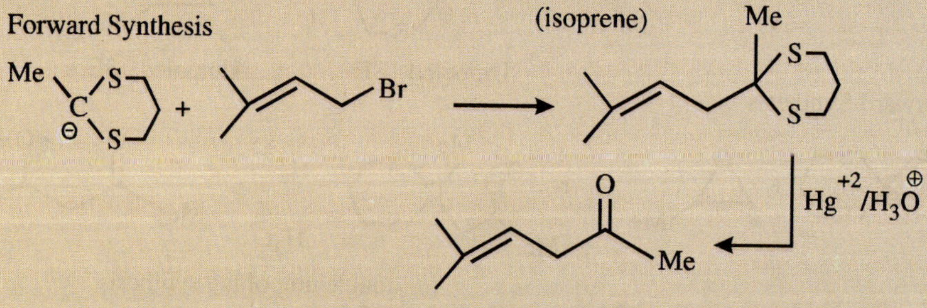

Me, Br, $Hg^{+2}/H_3O^{\oplus}$

7.

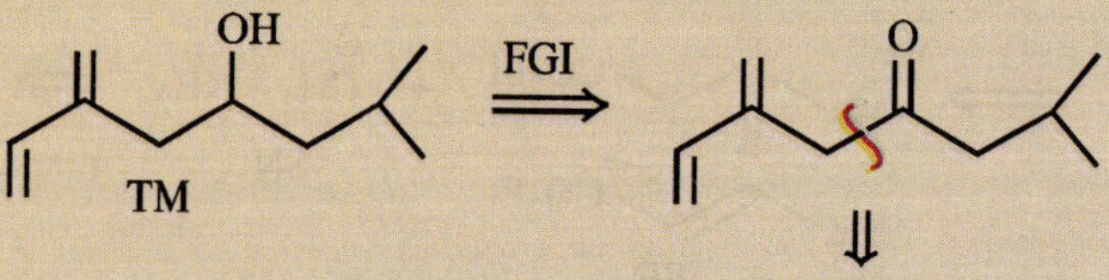

OH, TM, FGI, O

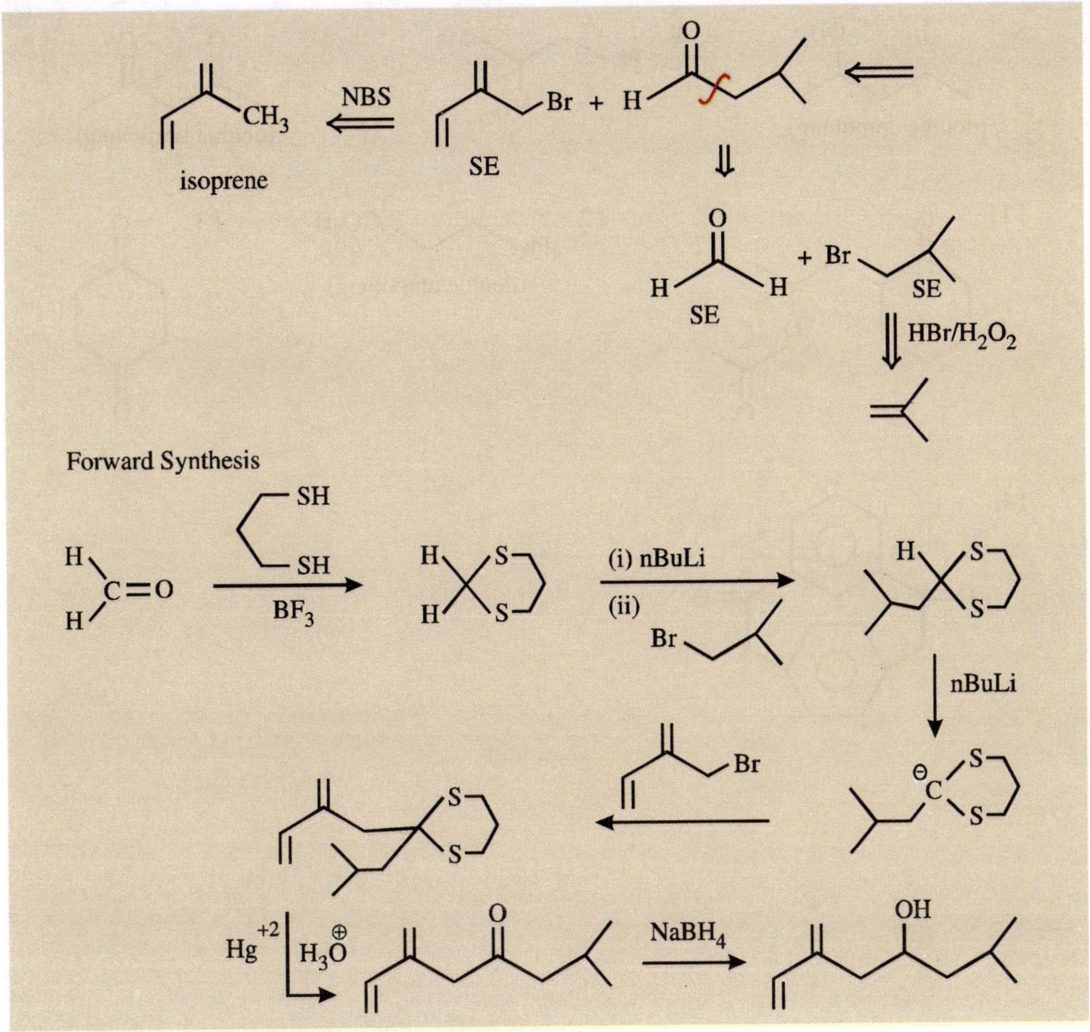

isoprene ⟸(NBS) SE (allyl bromide) + H–C(=O)– ⟸
⇓
H–C(=O)–H (SE) + Br– (SE)
⇓ HBr/H_2O_2

Forward Synthesis

H₂C=O —(HS–CH₂CH₂CH₂–SH, BF_3)→ 1,3-dithiane —((i) nBuLi (ii) Br–CH₂CH(CH₃)₂)→ 2-alkyl-1,3-dithiane —nBuLi→ dithiane anion —(allyl bromide)→ dialkylated dithiane —(Hg^{+2}, $H_3O^{\oplus}$)→ ketone —$NaBH_4$→ alcohol (OH)

6.6 Review synthetic problems

Show the retrosynthetic analysis of the following target molecules and then show the forward umpolung synthesis with at least one umpoled synthetic equivalent.

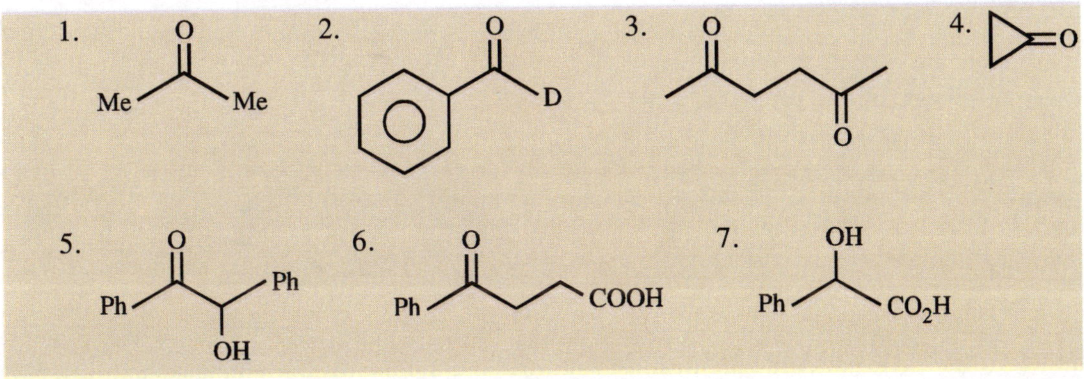

1. Me–C(=O)–Me
2. Ph–C(=O)–D
3. CH₃C(=O)CH₂CH₂C(=O)CH₃
4. cyclopropanone (C=O)
5. Ph–C(=O)–CH(OH)–Ph
6. Ph–C(=O)–CH₂CH₂–COOH
7. Ph–CH(OH)–CO_2H

8. OH

(double umpolung)

9. Ph Me NH_2

10. O O

(double umpolung)

11. O O O

12. Ph CO_2H

(double umpolung)

13. O O

14. O O

7

Synthesis of Medium and Large Ring Compounds

7.1 Synthesis by high dilution principle

7.1.1 What is high dilution principle?

Intramolecular cyclisation is an essential part to synthesize medium (8 to 11-membered) and large ring (12-membered and over) from an acyclic precursor. Several factors influence during the intramolecular ring closure.

(i) 'Distance factor'—For the formation of n-membered ring, the new bond must be formed between two atoms which are separated by (n–2) other atoms and it follows that, as 'n' increases, there is a decreasing probability of the molecule adopting a conformation in which the 'reactive' atoms are sufficiently close for bond formation to occur.

(ii) There are various kinds of 'Strain factors'. Angle strain and cis-double bonds nature in the cyclised compound may destabilise it relative to its acyclic precursor.

(iii) Unfavourable steric interactions in the cyclised product (e.g. 1, 3-syn-diaxial repulsion between the substituents and transannular interactions).

(iv) Angle strain and unfavourable steric interactions in the transition state for the ring-closure step—and if the ring-closure step is reversible, the equilibrium may lie in favour of the acyclic precursor.

Medium and large rings may be synthesised by high dilution principle.

According to this principle, by using sufficiently dilute solution of the acyclic precursor (diester, dicyanide, bromo acid and acyl chloride of benzene derivative), the intermolecular distance can be made greater than the intramolecular reactive centres. Thus, the cyclic compound is formed instead of linear condensation product. In order to apply high dilution principle it is necessary to have all the reactants in solution in a suitable solvent.

7.1.2 High dilution technique

The acyclic precursor is introduced very slowly into the reaction medium so that its concentration is always very low 10^{-3} (M) or less—at this concentration the probability of intermolecular reaction is greatly reduced. Under such high dilution conditions, Dieckmann or related reactions lead to acceptable yields of medium and large rings.

7.1.3 Synthesis of medium and large ring compounds with high dilution principle

(i) Monoketone from diester

(a) $(CH_2)_6$ (CO_2Et, CO_2Et) $\xrightarrow[\text{Xylene}]{\text{NaH}}$ $(CH_2)_6$ (O, CO_2Et) (48%)

dil HCl/Δ

$(CH_2)_6$ (O)

The diester (1M) solution is added dropwise over nine days to a stirred suspension of the hydride (2.5 fold excess ~ 1M).

(b) $(CH_2)_{13}$ (CO_2Et, CO_2Et) $\xrightarrow[\text{Xylene}]{t\text{-BuO}^{\ominus}\,K^{\oplus}}$ $(CH_2)_{13}$ (O, CO_2Et) (48%)

dil HCl/Δ

$(CH_2)_{13}$ (O)

The diester (4M solution) is added to base (4.8 fold excess ~ 4M) dropwise over 24 hours.

(ii) Monoketone from dicyanide

$(CH_2)_n$(CN)(CN) $\xrightarrow[\text{ether}]{Ph(Et)\overset{\ominus}{N}\overset{\oplus}{Li}}$ $(CH_2)_n$(C=NH)(CN) ⇌ $(CH_2)_n$(C–NH_2)(CN) $\xrightarrow[H_2O]{\overset{\oplus}{H}}$ $(CH_2)_n$(C=O)(CH_2)

7-membered ring	95% yield
8-membered ring	88% yield
9-13-membered ring	< 15% yield
14 or more membered ring	60-80% yield

Ph (Et) $\overset{\ominus}{N}\overset{\oplus}{Li}$—soluble in ether and it is highly hindered base so as to minimise its reaction as a nucleophile with nitrile group.

(iii) Lactone from bromoacid

$(CH_2)_9$(CH$_2$Br)(COOH) $\xrightarrow[\text{MeCOEt}]{K_2CO_3}$ $(CH_2)_9$ lactone (C(=O)–O) 85%

Bromoacid (0.15M solution) is added over 2-days to the base (~ 0.25M).

(iv) Aromatic ketone from acylchloride

$C_6H_5-(CH_2)_n-COCl$ $\xrightarrow[PhNO_2]{AlCl_3}$ cyclic aromatic ketone bridged by $(CH_2)_n$, n = 8 – 10

Substrate acylchloride is added dropwise in a mixture of lewis acid $AlCl_3$ and nitro benzene solvent.

7.2 Synthesis by acyloin condensation method

7.2.1 Synthesis of medium and large ring compounds

Long chain diester gives cyclic acyloin, since the reaction is heterogeneous, taking place on the surface of the metal, no need to employ high dilution method.

$(CH_2)_8(CO_2Et)_2$ —(i) Na/xylene/N_2 – atm.; (ii) $H_3\overset{\oplus}{O}$→ $(CH_2)_8$ cyclic acyloin (C=O, CH–OH)

Diester (undiluted) added over 3 hours to a suspension of sodium (4-fold excess) in xylene.

7.2.2 What are the usual complications in acyloin condensation in the preparation of four to seven membered rings? How are these difficulties overcome?

Complications:

(a) In the four to seven membered rings there is a steric strain (angle strain, conformational strain) compare to medium and large rings.

(b)

Diester (CO_2Et, CO_2Et) —Na / Xylene→ cyclopentane-1,2-dione + 2 $Et\overset{\ominus}{O}$ (per cond. 2 moles of ethoxide released) —Na→ bis-enolate —$H_3\overset{\oplus}{O}$→ ene-diol ⇌ acyloin

The ethoxide ion produced during the acyloin condensation gives competitive reactions or wasteful base catalysed side reactions as for example β-eliminations, Dieckmann condensation to give cyclic β-keto ester or Claisen ester condensation (linear condensation) to give acyclic β-keto ester.

(c) The bis-enolate or ene-diol produced in the acyloin condensation is oxygen sensitive and hence impurity in the products.

(d) Throughout the acyloin condensation reaction, the medium should be neutral but here basic in nature (due to $Et\overset{\ominus}{O}$).

All these complications may be avoided by the treatment of trimethyl silyl chloride Me_3SiCl.

+ $2EtO^{\ominus}$ $\xrightarrow{4Me_3SiCl}$ + $2EtO\text{-}SiMe_3$

1. Me_3SiCl acts as a scavanger (remover) for the ethoxide ion.

2. Bis-enolate is protected as bis-trimethyl silyl enol ether which may be isolated and purified before hydrolysis to get acyloin.

3. No competitive or side reactions since ethoxide ion is removed.

4. Medium is kept neutral.

Thus for example,

(i) CO_2Et CO_2Et $\xrightarrow[Me_3SiCl]{Na/xylene}$ $OSiMe_3$ $OSiMe_3$ $\xrightarrow{H_3\overset{\oplus}{O}}$ O OH

(ii) $(CH_2)_5$ CO_2Et CO_2Et $\xrightarrow[Me_3SiCl]{Na/xylene}$ $(CH_2)_5$ $OSiMe_3$ $OSiMe_3$ $\xrightarrow{H_3\overset{\oplus}{O}}$ $(CH_2)_5$ O OH

7.2.3 Synthesis of catenane with acyloin condensation method

Catenane which contain large ring may be synthesized by the following steps using acyloin condensation.

(a) An acyloin condensation on the diester of the C_{36}-dicarboxylic acid give the cyclic acyloin. This is reduced under clemmensen reduction condition and the C_{36}-cycloalkane is formed.

$(CH_2)_{34}$ CO_2Et CO_2Et $\xrightarrow[(ii)\ H_3\overset{\oplus}{O}]{(i)\ Na/xylene}$ $(CH_2)_{34}$ O OH $\xrightarrow[HCl]{Zn(Hg)}$ $(CH_2)_{36}$

(b) A repetition of the reaction in the presence of C_{36}-cycloalkane (A) is anticipated to give (C) since there is a chance that some molecules of diester will get threaded through (A) before cyclisation of (B) and indeed the catenane (D) was isolated.

(A) $(CH_2)_{36}$ + $(CH_2)_{34}(CO_2Et)_2$ diester → (B) —(i) Na (ii) $H_3O^{\oplus}$→ (C) —Zn(Hg)/HCl→ (D)

This type of synthesis is known as statistical synthesis.

The ready formation of large rings where the two ends of the chain must approach each other is thought due to the two ends becoming attached to the nearby sites on the surface of the sodium.

7.3 Construction of medium and large rings taking advantage of restricted rotation

If a long chain of atoms contains one or more rigid sections, in which free rotation about bonds is not possible, there may be an increased chance of cyclisation to form a medium or large rings.

Thus for example,

(a) 2 [o-xylylene dibromide] —2PhLi→ ... → (10-membered ring)

(b)

3 [o-iodophenyl]–C≡C Cu $\xrightarrow[\text{Py}]{\Delta}$ [intermediate: –I CuC≡C–] ↓

(12-membered ring)

7.4 Synthesis of medium and large ring compounds by pericyclic reaction

7.4.1 Via electrocyclic ring-opening

(a) The thermal electrocyclic ring-opening of cyclobutene is a conrotatory process. This stereospecific conrotatory ring-opening of cyclobutene has got important synthetic consequence.

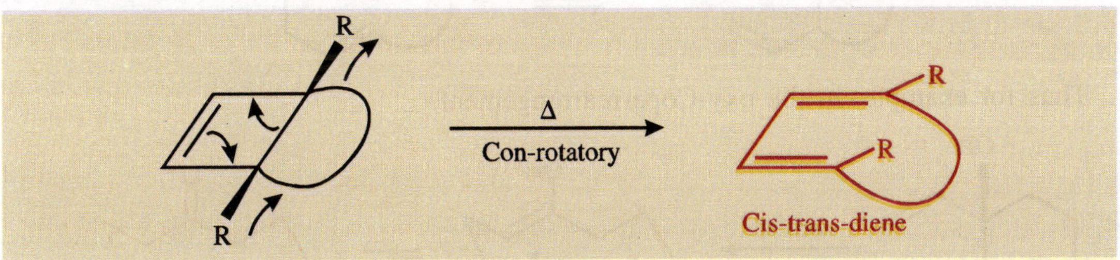

If the R-groups are part of a ring system and are cis-fused then conrotatory ring-opening generates a cis-trans-diene which cannot be accomodated within a normal sized ring but it is possible for medium and large rings.

Thus for example,

H, H $\xrightarrow[\text{Con-rotatory}]{200^\circ C}$ H, H 95%

(10-membered ring)

(b) The thermal ring-opening of cyclohexadiene is a disrotatory process. If the two R-groups are part of a ring system and is cis-fused—the disrotatory ring-opening presents no problem.

7.4.2 Via Cope-rearrangement

In the Cope-rearrangement which involves a six-membered cyclic chair transition state, a 1, 5-diene is rearranged to another 1, 5-diene by concerted formation of a C_1—C_6 single bond, breaking of C_3—C_4 single bond and migration of both the double bonds.

It follows now that if the original 3, 4 -bond is part of a ring system, the method may be used for a type of ring-opening and cyclisation simultaneously.

Thus for example, in the oxy-Cope-rearrangement.

In the normal Cope (not oxy-Cope) rearrangement, the equilibrium is shifted largely to the ring-closure reactant.

7.5 Synthesis of large ring Crown-ether

Crown ethers are large ring polyethers. The following two examples are given for Crown-ether synthesis.

(i) Synthesis of 18-Crown-6

The 18-Crown-6 is a cyclic polymer of ethylene glycol. The $PhCH_2\overset{\oplus}{N}Et_3\overset{\ominus}{Br}$ is a quaternary ammonium salt and which is a phase-transfer catalyst whose addition to a two-phase organic-water system helps to transfer a water soluble reactant (alkoxide) across the interface to the organic phase (haloether) where a homogeneous reaction can take place, thus enhancing the rate of reaction. The phase-transfer cation as bromide or alkoxide is soluble in both water and organic solvent.

(ii) Synthesis of Benzo-Crown ether

The $Cr(CO)_3$ fragment is electron-withdrawing and activates the aromatic compound to nucleophiles (activated aromatic nucleophilic substitution). The $Cr(CO)_3$ fragment is released from the complex by mild oxidation, usually with cerium (IV).

Note

8

Chiral Synthesis: Diastereoselective and Enantioselective Synthesis

8.1 The need of the stereo and regioselectivity in the design of organic synthesis

The biological and physical properties of organic molecules used as drugs, insecticides, plant growth regulators, perfumes, polymers, pharmaceuticals, dyestuffs, food colourings and flavourings etc. depend to a large extent on the stereochemistry of the substituents and functional groups. A synthetic route to the target compound must be designed to enable the required pattern of carbon bonds to be constructed with all the functional groups and substituents in their correct position (regiochemistry) and with the control of relative orientation of the substituents (Stereochemistry: Controlled by Stereospecific and Stereoselective reaction). Stereochemistry of the substituents and functional groups also have an important effect on the reactivity and biological or physical properties.

Therefore, the knowledge of stereoselectivity, stereospecificity and regioselectivity are of great importance in the design of organic synthesis.

The following examples will illustrate the need of these selectivities.

Example 1

The compound (1) which is *cis*-hydroxy aldehyde has a strong and pleasant smell and is used as a perfumes whereas the *trans*-isomer (2) is odourless.

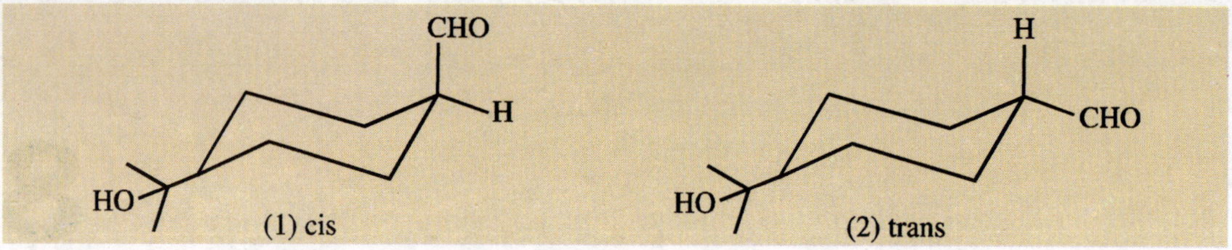

(1) cis (2) trans

This *trans*-isomer (2) is thermodynamically more stable than the *cis*-isomer, compound (1). At equilibrium there is 92% of (2). Therefore, any synthesis should and must be stereoselective and kinetic controlled.

Example 2

The compound (3) which is siglure and is trans-isomer has a strong attractant power for fruit fly used as a bait for preservation of fruits.

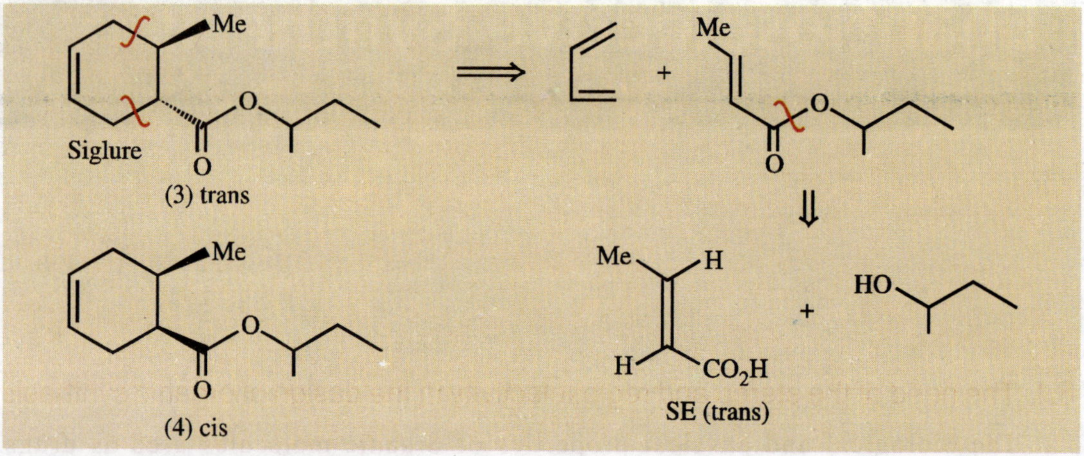

Siglure (3) trans

(4) cis

SE (trans)

The *cis*-isomer (4) has no any attractant power for fruit-fly. Since it is a Diels-Alder product and *trans*-dienophile synthetic equivalent (SE) must be used to get the desired stereochemistry. This synthesis should and must be stereospecific.

Example 3

The compound (5) is a weedkiller propanil used in rice fields whereas the other isomer (6) which is regiomer has no any killer property.

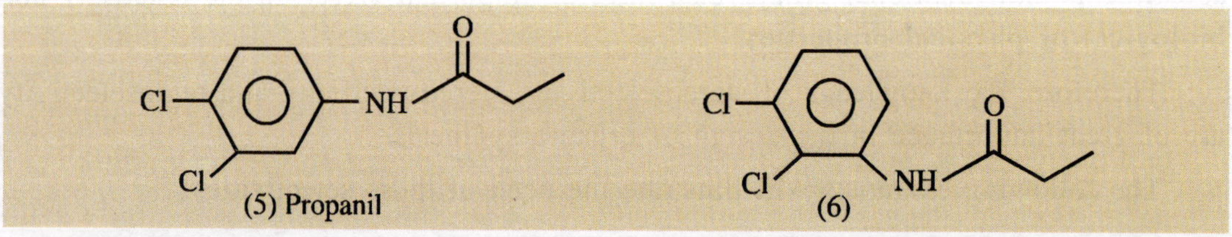

(5) Propanil (6)

Therefore, any synthesis must give pure (5) and not a mixture of (5) with other regiomer (6). The synthesis must be regioselective.

8.1.1 What is stereospecific reaction?

1. Stereospecific reaction is a reaction in which a reactant can exist as different stereoisomers (diastereomer or enantiomer) and each stereoisomeric reactant leads to a different stereoisomeric product (diastereomer or enantiomer) under identical conditions.

As for example, Iodide-induced debromination of meso and active 2, 3-dibromo butane (E2 reaction) which are diastereomeric gives *trans*-2-butene and *cis*-2-butene respectively which are diastereomeric.

Meso-2, 3-dibromo butane → Less crowded TS → ($I^{\ominus}$, RDS, Relative rate = 2.3) → Trans-2-butene

Active-2, 3-dibromo butane → More crowded TS → ($I^{\ominus}$, RDS, Relative rate = 1) → Cis-2-butene

If a reaction is carried out on a reactant that has no stereoisomers, it cannot be stereospecific. As for example, addition of bromine to methyl acetylene produces trans 1, 2-dibromo propene predominantly and this is not a stereospecific reaction (but stereoselective since some amount of *cis* product formed) although product is a stereoisomer.

$$\mathrm{Me-C\equiv C-H} \xrightarrow{Br_2} \text{(Me)(Br)C=C(Br)(H)}$$

Stereospecificity is not related with the chirality or optical activity.

2. In a stereospecific reaction different stereoisomers (diastereomer or enantiomer) produces different products stereochemically at different rates under identical conditions. As for example, during the debromination of the meso and active 2, 3-dibromo butane, the meso isomer reacts 2.3 times faster than the active isomer since TS correspond to active is more crowded and one gauche-butane interaction is greater than the TS correspond to the meso isomer.

3. Stereospecific reaction may also be defined as a reaction in which a reactant can exist as different stereoisomers and one single stereoisomeric reactant leads to different products under different conditions.

Cis-3, 4-dimethyl cyclobutene ← hν — (2E, 4E)-2, 4-hexadiene — Δ → Trans-3, 4-dimethyl cyclobutene

2, 4-hexadiene may exist in different stereoisomeric forms (2E, 4E or 2Z, 4Z or 2E, 4Z) and on giving thermal and photoenergy to the (2E, 4E) –2, 4-hexadiene produces *trans* and *cis*-isomer respectively and stereospecifically.

4. Stereospecific reaction may be thought in terms of transition state (TS) geometry.

A. If any stereoisomer gives reaction which goes via one and only one single TS to give another stereoisomer exclusively then it is obviously stereospecific reaction provided basic condition is fulfilled i.e. other stereoisomer should give opposite stereo product via single TS. As for example, any pure S_N2 reaction where nucleophile and leaving groups are different (if nucleophile and leaving groups are same then both the stereoisomers will give dl product which is also stereospecific because of common single TS) goes via one single TS (involving pentavalent SP^2 carbon).

(R) I — NaOH / Me_2SO → (S) OH

(S) I — NaOH / Me_2SO → (R) OH

B. If any stereoisomer gives reaction which may go via two TS and which are homomeric then the reaction will be stereospecific.

Example (i)

Bromination of *trans*-2-butene gives meso-2, 3-dibromo butane.

Trans-2-butene has got two enantiotopic faces and on reaction with bromine to both the faces of it gives two chiral cyclic bromonium ion intermediates which are homomeric.

Trans-2-butene

β-face attack

α-face attack

180° in plane

Meso-2, 3-dibromo butane

Chiral homomeric intermediates.

The two intermediates arise due to α-and β-face attack of trans-2-butene with bromine are chiral (since the two intermediates are non-superposable mirror images) and homomeric (since each intermediate has got C_2-symmetry). Attack with bromide ion either to the front or back carbon of any intermediate (produces homomeric TS) giving homomers (single product).

Example (ii)

Reaction of singlet methylene to the *cis*-2-butene gives *cis*-1, 2-dimethyl cyclopropane (Meso).

Addition of singlet methylene to the *cis*-2-butene goes via cyclic concerted and one step process. Attack of singlet methylene from the front or backside of *cis*-2-butene gives two TS which are superposable to each other (since on C_2-operation in any one TS mutual conversion occurs) and gives single product.

cis-2-butene

back side attack

front side attack

cis-1, 2-dimethyl cyclopropane

two homomeric TS (presence of C_2-vertical)

C. If any stereoisomer gives reaction which may go via two TS and which are enantiomeric then the reaction will be stereospecific.

Example (i)

Bromination of *cis*-2-butene gives active (±) 2, 3-dibromobutane.

Cis-2-butene has got two homotopic faces and on reaction with bromine to both the faces of it gives two achiral (presence of σ-plane) cyclic bromonium ion intermediates.

cis-2-butene

Br_2 β-face attack

Br_2 α-face attack

180° in plane

σ-plane

$Br^{\ominus}$

2, 3-dibromo butane dl pair

achiral intermediates with σ-plane

The two intermediates arise due to α-and β-face attack of *cis*-2-butene with bromine, are achiral since the two intermediates have got σ-plane. Therefore, attack with bromide ion either to the front or back carbon of any intermediate (produces enantiomeric TS) giving enantiomers (dl pair).

Example (ii)

Reaction of singlet methylene to the *trans*-2-butene gives active (±) 1, 2-dimethyl cyclopropane.

trans-2-butene

back side attack

front side attack

(±) trans-1, 2-dimethyl cyclopropane

two-non-superposable enantiomeric TS

Addition of singlet methylene to the *trans*-2-butene goes via cyclic concerted and one step process. Attack from the front side and backside gives two enantiomeric transition states since they are non-superposable to each other and consequently dl pair will form with equal facility (due to same energy of activation)—enatiospecificity.

Example (iii)

Addition of bromine to cyclohexene gives (±) 1, 2-dibromo cyclohexane. This is a stereospecific reaction.

⊕ Br ≡ H H σ-plane Br_2 Br⊕ 1 2 H H attack of Br⊖ to the 1C from α-face H Br Br H attack of Br⊖ to the 2C from α-face Br H H Br dl pair

By definition this reaction should not be stereospecific since cyclohexene has no other existable stereoisomer (cyclohexene is chiral but inseparable dl pair). But if we consider the transition state geometry which is achiral (due to the presence of σ-plane of the intermediate) and produces two enantiomeric TS then it should be stereospecific.

5. Stereospecific reaction depends on structure of the substrate and dielectric constant of the medium.

Example (i)

Addition of bromine to the *cis* or *trans*-2-butene in CCl_4 which are going via completely cyclic bromonium ion intermediate (no free rotation occurs) are highly stereospecific but addition of bromine to the *cis* or *trans* 1-phenyl propene in CCl_4 are going via completely open cation (free rotation occurs) due to the formation of stable benzylic cation is non-stereospecific.

Ph Me H H or Ph H H Me Br_2 Ph ⊕ H Me H Br

Example (ii)

Addition of bromine to any stereoisomer of stilbene (Ph CH=CH Ph) in very low dielectric constant is 90 to 100% stereospecific (due to exclusive cyclic bromonium cation) but in high dielectric constant it is non-stereospecific due to exclusive open cation.

6. All stereocontrol reactions for a given stereoisomer are stereoselective but when it is 100% stereoselective (to give either one stereoisomer exclusively e.g. bromination of *trans*-2-butene to give meso product or bromination of *cis*-2-butene to give dl product exclusively) then it is termed as stereospecific. Therefore, all stereospecific reactions are stereoselective reactions but all stereoselective reactions are not necessarily stereospecific.

7. Stereospecificity depends on mode of addition. If two substrates are stereoisomeric then in the following condition the reaction will be stereospecific.

Example (i)

In case of 100% SYN addition

(a) Cyclohexene $\xrightarrow{OsO_4}$ cyclic osmate ester (OsO_2) $\xrightarrow{H_3\overset{\oplus}{O}}$ cis-diol (OH, OH)

(b) cis (Me, Me / H, H) $\xrightarrow{CF_3CO_3H}$ meso (epoxide)

trans (Me, H / H, Me) $\xrightarrow{CF_3CO_3H}$ dl pair (epoxide)

(c) cis (Ph, Ph / Me, Me) $\xrightarrow{H_2/Pd}$ meso (Ph; Me—H; Me—H; Ph)

trans (Ph, Me / Me, Ph) $\xrightarrow{H_2/Pd}$ dl pair (Ph; H—Me; Me—H; Ph)

Example (ii)

In case of 100% *trans* anti addition

mCPBA; $H_3\overset{\oplus}{O}$; trans-diol

Example (iii)

In case of 100% *anti* substitution.

Any pure S_N2 reaction

8. All stereospecific reactions are substrate selectivity i.e. stereochemistry developed on the product, controlled by the stereostructure of the substrate. As for example, Diels-Alder reaction which is highly stereospecific reaction occurs in one step so that neither diene nor dienophile has time to rotate and stereochemistry of each must be faithfully reproduced in the product. *Cis* dienophile gives *cis* product and *trans* dienophile gives *trans* product.

CO_2Et, CO_2Et, cis — DA → cis

CO_2Et, EtO_2C, trans — DA → trans

Substrate selectivity are of two types:

(i) Substrate diastereoselectivity: Symmetry and stereoelectronic controlled. As for example, during the bromination of *cis* and *trans*-2-butene, *cis* nature (stereo) of the substrate is controlled by the σ-symmetry and *trans* nature (stereo) of the substrate is controlled by the C_2-symmetry in the TS and addition of bromine via *trans* anti is electronic controlled in both the cases and accordingly stereostructures (diastereomers, active and meso) developed in the products—this is substrate diastereoselectivity.

(ii) Substrate enantioselectivity: Enzyme which very often reacts with one enantiomer only and this shows total substrate enantioselectivity. Ordinary chemical reactions using chiral reagents are substrate enantioselective reactions (see ee synthesis with chiral borolane).

9. Rules for stereospecific addition to alkene

(i) $R-CH=CH-R$

Reagents	Stereochemistry of the alkene	Addition mode	Stereochemistry of the products
Br_2 (Bromination)	*cis* *trans*	trans addition trans addition	threo dl pair meso
$KMnO_4$ (Hydroxylation)	*cis* *trans*	cis addition cis addition	meso threo dl pair
(ii) $R_1-CH=CH-R_2$			
Br_2 (Bromination)	*cis* *trans*	trans addition trans addition	threo dl pair erythro dl pair
$KMnO_4$ (Hydroxylation)	*cis* *trans*	cis addition cis addition	erythro dl pair threo dl pair

8.1.2 What is stereoselective reaction?

1. Stereoselective reactions are those reactions in which a given structure (may or may not be stereoisomeric) produce diastereomeric products in unequal amounts.

Stereoselectivity is not related with chirality or optical activity.

In all stereoselective reactions since products are diastereomeric and unequal amount and therefore the two transition states are diastereomeric and have got different free energy and consequently energy of activation (Ea) are different and therefore reaction pathway has a choice. The pathway of lower energy of activation is preferred and kinetically controlled and forms predominantly.

In the stereoselective reactions stereoselectivity is controlled by several factors stereoelectronic factor, steric factor, electronic effects (including participation of chclate and hydrogen bonds), bond rotation and spin inversion etc.

Example (i)

Ph–CH(OH)–CH$_2$–Me $\xrightarrow{H^{\oplus}}$ (H)(Ph)C=C(Me)(H) + (H)(Ph)C=C(H)(Me)

major / minor

controlled by stereoelectronic factor

Example (ii)

(S) 2-methylcyclohexanone (Me, H) —$LiAlH_4$→ (SS) H, OH, Me, H major + (SR) OH, H, Me, H minor

controlled by stereoelectronic factor and thermodynamic stability of the product

Example (iii)

$HO_2C-C\equiv C-CO_2H$ —Br_2→ CO_2H, Br / Br, CO_2H (major) + Br, Br / CO_2H, CO_2H (minor)

controlled by stereoelectronic factor

2. Stereoselective reactions are those reactions in which a given structure or structures (stereoisomeric) produces dl pair (or active) and meso in unequal amount.

Example (i)

H, Me / Me, H (trans) or Me, Me / H, H (cis) —$:CH_2$ triplate→ Me, H / H, Me (dl pair, major) + Me, Me / H, H (meso, minor)

controlled by steric and bond rotation versus spin inversion

Example (ii)

$Ph-\underset{\underset{O}{\|}}{C}-\underset{\underset{OH}{|}}{CH}-Ph$ (S) —$LiAlH_4$→ $Ph-\underset{\underset{OH}{|}}{CH}-\underset{\underset{OH}{|}}{CH}-Ph$ (R) (S) meso, major + $Ph-\underset{\underset{OH}{|}}{CH}-\underset{\underset{OH}{|}}{CH}-Ph$ (S) (S) active, minor

controlled by steric factor in the TS

3. Stereoselective reactions are those reactions in which a given structure produces one diastereomeric dl pair in considerable predominance over the other diastereomeric dl pair.

Ph – C – C – Me (O, NMe) $\xrightarrow{LiAlH_4}$ (±)-ephidrine major + (±)-ψ-ephidrine minor

4. Diastereomeric conformers react through diastereomeric transition states to give either diastereomeric products or constitutional products.

diastereomers minor

Constitutional isomer major

5. All stereoselective reactions are product selectivity. Product selectivity are of two types.

(a) Product diastereoselectivity which is again are of two types.

(i) Diastereoselectivity with kinetic control in which case the stereoselectivity depends on the free energies of the respective transition states.

(i) Me_2CuLi (ii) $H_3\overset{\oplus}{O}$

trans 98% cis 2%

Me_2CuLi (more crowded TS)

Me_2CuLi (less crowded TS), kinetically controlled

This is a diastereomeric excess (de) synthesis.

(ii) Diastereoselectivity with thermodynamic control in which case, the stereoselectivity depends on the difference in the free energies of the products.

$LiAlH_4$

trans
major (thermodynamic controlled)

cis
minor (kinetic controlled)

Attack of $\overset{\ominus}{A}lH_4$ to the carbonyl either from equatorial or from axial side experience same TS interaction and therefore the reaction is controlled by thermodyanamic stability of the product. Trans isomer is more stable than cis where two 1, 3-syn diaxial interaction occurs between OH and two axial Hs.

However, kinetic controlled product may be developed by using highly hindered metal hydride e.g. with $LiAlH(OBu\text{-}t)_3$.

(b) Product enantioselectivity which is controlled by the extent of asymmetric induction in the TS. (See ee synthesis e.g. Prelogs' synthesis).

6. One very important point regarding stereoselective reaction is that two stereoisomeric reactants may react through two diastereomeric transition states but if the product is homomeric or single (same) product then it is not diastereoselective reaction although the two transition states are diastereomeric and the rate of the reaction happens to be different.

Cis

CrO_3
acetic acid
Relative rate = 3.23

CrO_3 acetic acid
Relative rate = 1

trans

7. All enantiomeric excess (ee) and diastereomeric excess (de) synthesis are stereoselective reactions when they are going through diastereomeric transition states.

8. Sequential double diastereoselection.

When 4-methyl cyclohexenone is reacted with Lithium dimethylcuprate which results in conjugate addition of a methyl group to the β-position of the enone. This stereoselective addition gives *trans* isomer as the major product since approach of the bulky curprate reagent occurs predominantly on the less sterically hindered face of the enone.

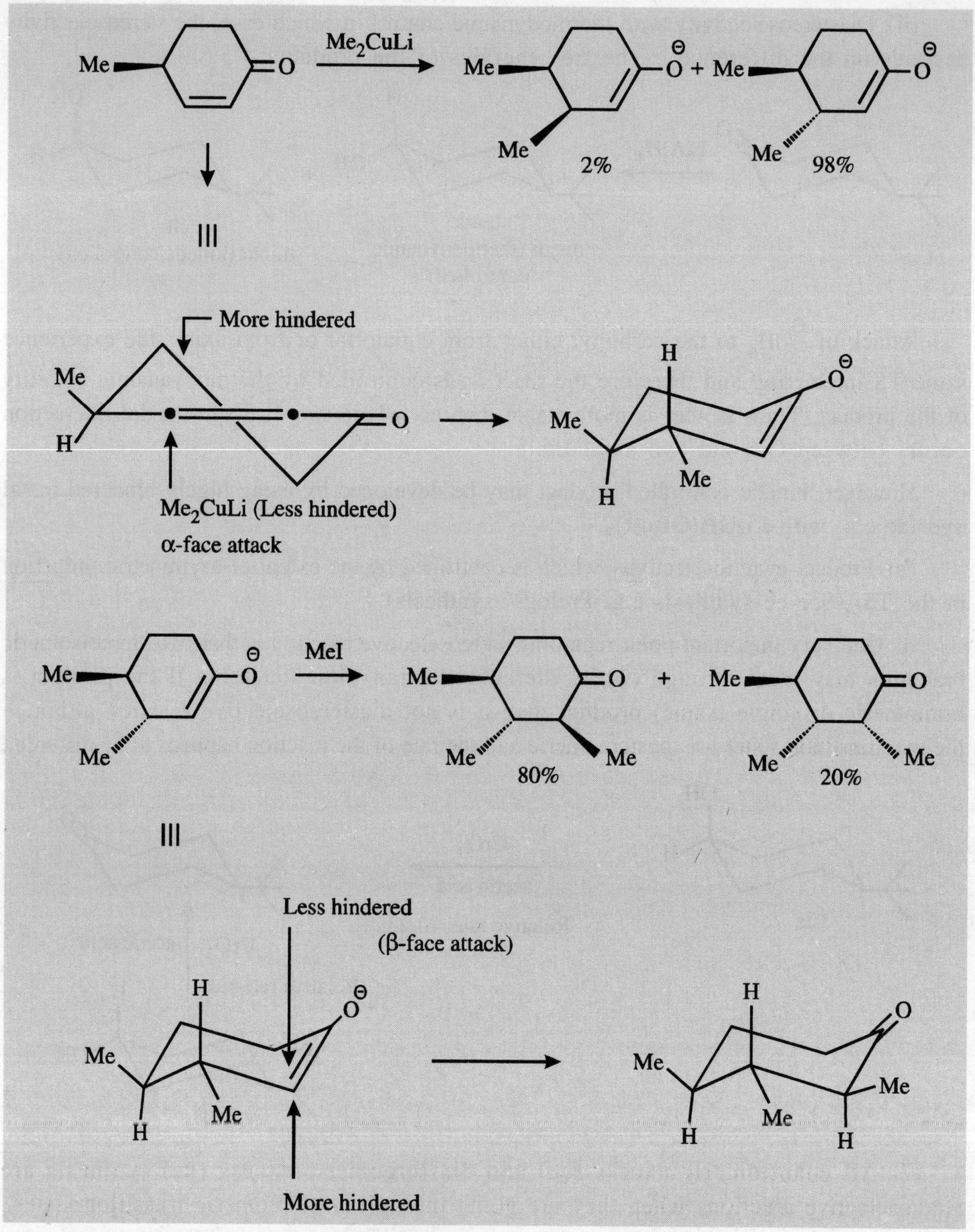

Now this enolate which is formed in the conjugate reaction may be further alkylated with methyl iodide. Steric factor once again control the stereochemistry. The preferred product is the one in which the substituents α and β to the carbonyl group are *trans* to each other. This is the sequential control of stereochemistry with double diastereoselectivity.

8.1.3 Controlling of diastereoselectivity

There are several factors which control the diastereoselectivity.

1. Selectivity is controlled by stereoelectronic factor.

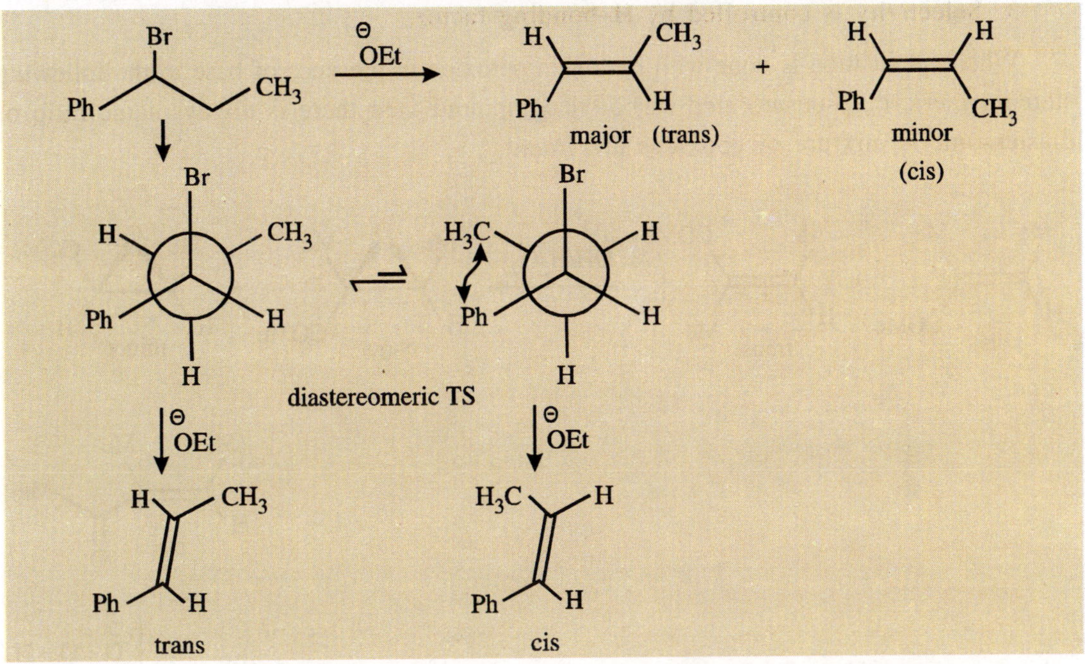

Stereoelectronic factor is a combination of steric and electronic factor. In both the transition states hydrogen and bromine are in the antiparallel (electronic factor) but the diastereomeric conformers correspond to *trans* is more favoured since phenyl and methyl are anti to each other than the diastereomeric conformer correspond to *cis* where bulky groups phenyl and methyl are in gauche interaction (steric factor).

2. Selectivity is controlled by electronic factor only.

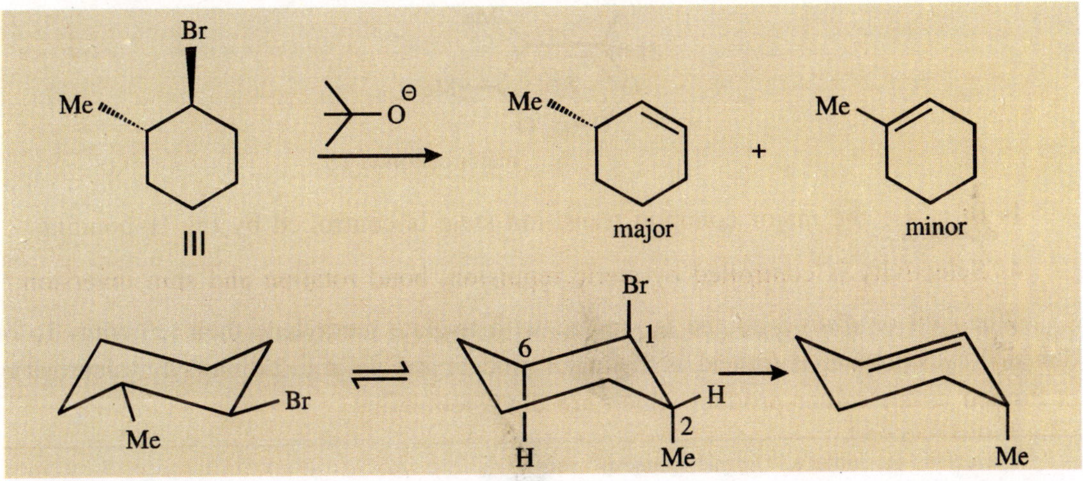

In this case the equatorial hydrogen (2) and axial bromine (1) are not in antiparallel but axial hydrogen (6) and axial bromine (1) are in anti parallel and hence major product. Not steric but electronic factor only controlling the selectivity.

3. Selectivity is controlled by H–bonding factor.

When epoxidation is done with hydrogen peroxide in presence of base to the following stereoisomeric α, β-unsaturated carbonyl compound then there is always same ratio of diastereomeric mixture of epoxides are found.

Me Me H COMe cis + Me COMe H Me trans $\xrightarrow{\ominus OH/H_2O_2}$ Me O Me H COMe major + Me O COMe H H minor

Me Me Me H cis O

$\ominus$ O–O–H

Me O Me H Me trans $\xrightarrow{\ominus O-O-H}$ Me O$^\ominus$ Me H O O–H Me → Me Me Me H O O–H···O δ⊖

Me Me H O δ⊖ Me O–H O

major common TS

In this case the major common transition state is controlled by the H–bonding.

4. Selectivity is controlled by steric repulsion, bond rotation and spin inversion.*

When *cis* or *trans*-2-butene is reacted with triplate methylene then (±) *trans* 1, 2-dimethyl cyclopropane is formed as the major product and *meso* 1, 2-dimethyl cyclopropane is formed as the minor product which are diastereomeric.

---

*For details see author's "Organic Synthesis" vol. 1

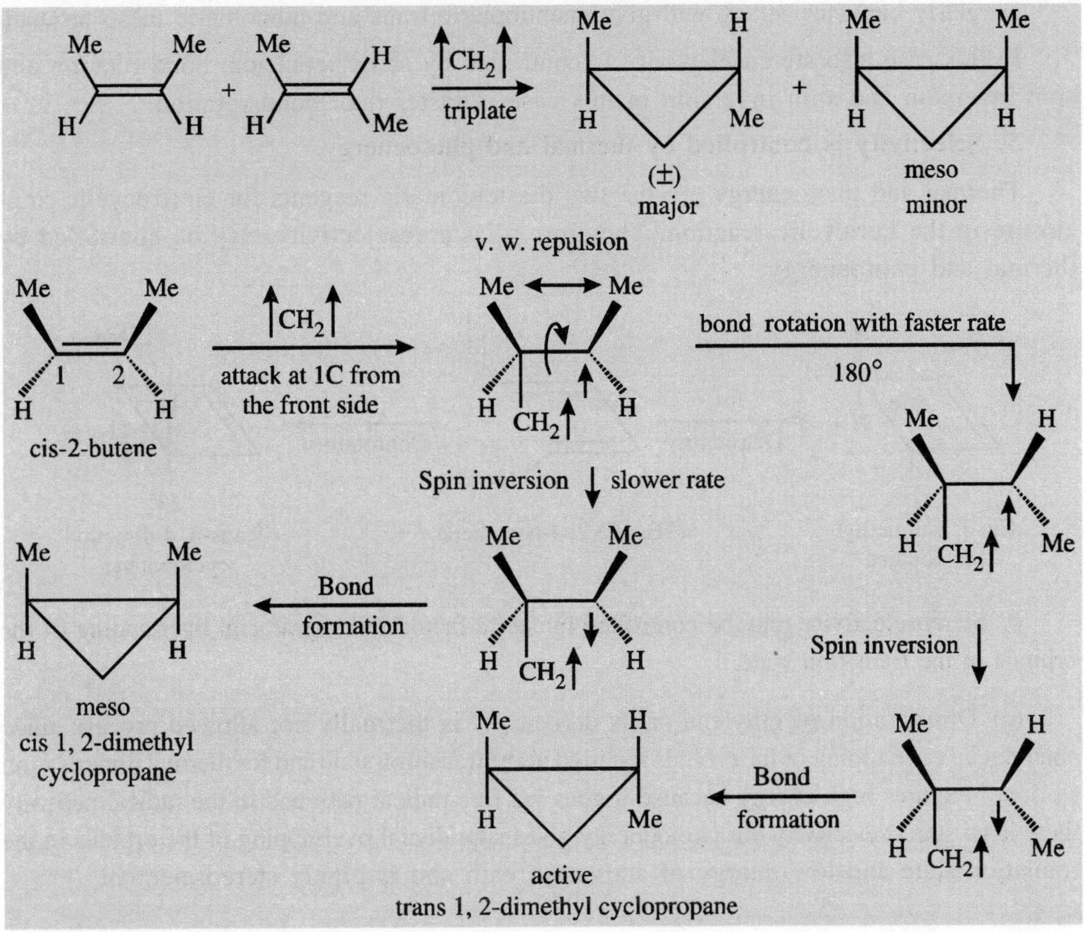

Me Me + Me H
H H H Me
$\uparrow CH_2 \uparrow$
triplate
Me H + Me Me
H Me H H
(±)
major
meso
minor

Me Me
1 2
H H
cis-2-butene
$\uparrow CH_2 \uparrow$
attack at 1C from
the front side
v. w. repulsion
Me ⟷ Me
H H
$CH_2\uparrow$
bond rotation with faster rate
180°
Me H
H Me
$CH_2\uparrow$
Spin inversion ↓ slower rate
Me Me
H H
$CH_2\uparrow$
Bond
formation
Me Me
H H
meso
cis 1, 2-dimethyl
cyclopropane
Spin inversion ↓
Me H
H Me
$CH_2\uparrow$
Bond
formation
Me H
H Me
active
trans 1, 2-dimethyl cyclopropane

Similarly backface attack at 1C of alkene will give enantiomeric *trans*-1, 2-dimethyl cyclopropane with equal facility and homomeric meso product.

In this reaction stereoselectivity is controlled by steric repulsion, bond rotation and spin inversion. Methyl methyl trans is stereically less crowded in the transition state (in case of *trans*-2-butene) and spin inversion is faster than bond rotation.

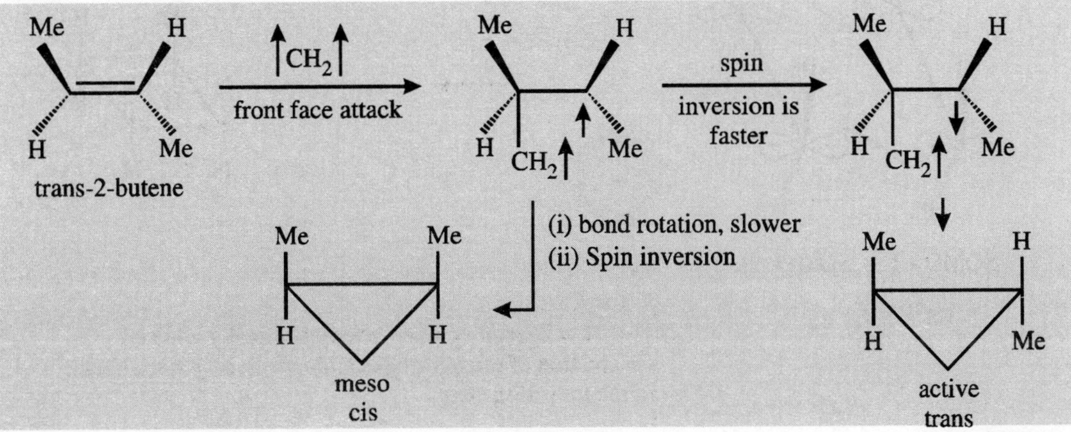

Similarly backface attack will give enantiomeric trans and homomeric meso product.

In this case also stereoselectivity is controlled by steric repulsion, bond rotation and spin inversion and spin inversion in this case is faster than bond rotation.

5. Selectivity is controlled by thermal and photoenergy.

Thermal and photoenergy are the two diastereomeric reagents for electrocyclic ring-closure in the Pericyclic reaction. Therefore, diastereoselectivity may be controlled by thermal and photoenergy.

Cis-3, 4-dimethyl cyclobutene ← hν, Disrotatory — (2E, 4E)-2, 4-hexadiene — Δ, Conrotatory → Trans-3, 4-dimethyl cyclobutane

6. Stereoselectivity may be controlled by steric factor and suprafacial overlapping of the orbitals in the transition state.

(a) Dimerisation of ethylene or its derivatives is thermally not allowed process since antarafacial overlapping of the orbitals required in the transition state and for thermal dimerisation (if done) requires high energy because it goes via free radical path and in the radical pathway there is no stereoselectivity but photoenergy gives suprafacial overlapping of the orbitals in the transition state and low energy of activation path and is highly stereoselective.

SOMO - 1 SOMO - 2

$\pi^2 s + \pi^2 s$

$\pi^2 s + \pi^2 s$

1 + 1 = 2, even number, photochemically allowed
Orientation of phenyl rings are controlled by steric factor in the transition state.

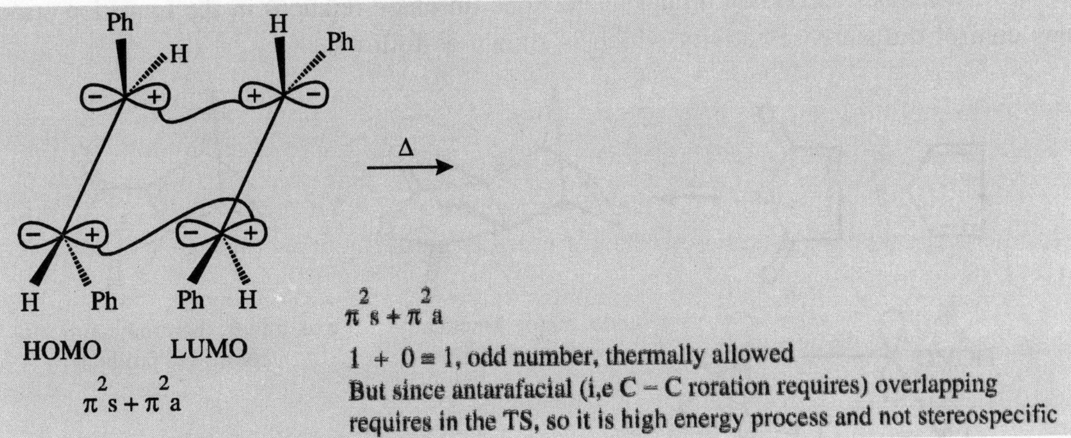

(b) In the Diels-Alder reaction stereochemistry of the substituents in the joining carbon of the product is as in the *Diene* and *Dienophile.* This is because of suprafacial overlapping in the transition state.

HOMO + LUMO

$\pi^4 s + \pi^2 s$

0 + 1 = 1, odd number, thermally allowed

(c) Sometimes secondary orbital interactions (in phase relation) in the transition state may control the stereoselectivity which is shown as follows.

endo, major, kinetically controlled product

exo, minor, thermodynamically controlled product

HOMO

LUMO

(endo approach)

.... Secondary orbital in phase (matching phase) interaction controls the stereoselectivity

7. Stereoselectivity is controlled by favoured diequatorial chair-like transition state over the diaxial for six-membered cyclic transition state.

When active threo 3, 4-dimethyl hexa 1, 5-diene is heated which is a Cope rearrangement gives a mixture of diastereomeric products EE (90%) and ZZ (10%)-2, 6-octadienes. In this rearrangement the diastereoselectivity is controlled by diequatorial chair-like transition state which is energetically favoured.

Threo

180°C

dieq.chair-like TS

EE (90%)

two reactants are in conformational diastereomers

diaxial chair-like TS

ZZ (10%)

8.1.4 What is regioselective reaction?

For substrates which are capable of reacting at more than one centres but react at one centre with a higher rate than at the others to give constitutional or positional isomers is the regioselective reaction.

Reaction of a particular functional group or position in the substrate molecule, in a desired way (control of the reactivity of functional groups or position) may be done either by selective reagent or by protection of comparable undesired functional group/position or by activating a desired reacting centre—these are all control of regioselectivity.

During the addition of unsymmetrical reagent (as for example H_2O) to the unsymmetrical alkene there is a regioselective problems, this may be controlled by selective reagents (e.g. Markownikov hydration by oxymercuration-demercuration process and anti Markownikov hydration by hydroboration process).

Control of stereochemistry is the control of geometry and the control of regiochemistry is the control, as for example, where the double bond is formed if more than one choices of eliminations are possible for a given structure in an elimination reaction i.e. of fixing the double bond predominantly in our desired choice. Oppositely when unsymmetrical reagent is added to the unsymmetrical double bond then more than one choices are possible to give constitutional products. Therefore, any preferential reaction product over the other to give constitutional or positional isomers is the regioselectivity.

The following examples are some illustrations of regioselective reactions.

1. Regioselective hydration of unsymmetrical alkene*

(a) Markownikov hydration via oxymercuration-demercuration process

$=CH_2$, H; (i) Hg $(OAc)_2$, THF, H_2O; (ii) $NaBH_4/\overset{\ominus}{O}H$; OH, Me

Hg $(OAc)_2$

$:OH_2$; $\oplus$Hg; OAc; acet (oxymercuration)

OH; Hg – OAc; $\overset{\ominus}{H}$; – Hg; – $Ac\overset{\ominus}{O}$; OH; Me; demercuration

This reaction is creditful in the sense it does not give descrete carbonium ion (as in the case of dil H_2SO_4 treatment) and hence no rearrangement occurs. Solvent nucleophile H_2O attacks preferentially to the more electrophilic carbon adjacent to the tertiary butyl group.

---

*For details see author's "Organic Synthesis" vol. 1

(b) Anti Markownikov hydration via hydroboration

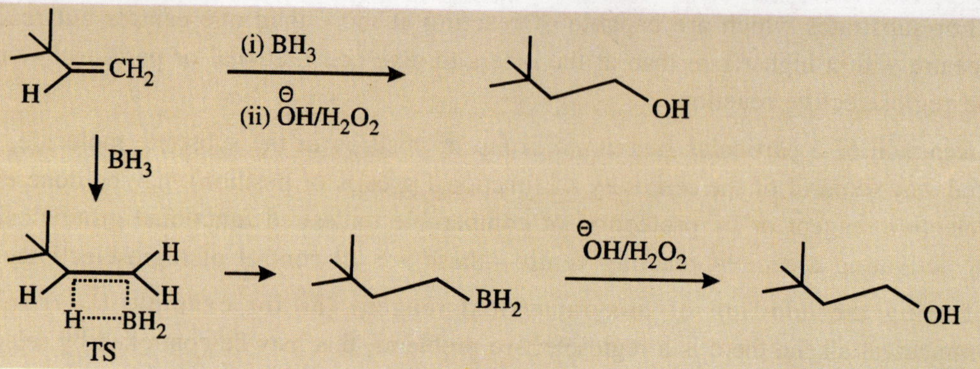

The first step of hydroboration involves addition of borane across the double bond of the substrate, with the formation of carbon-boron bond at the less sterically hindered end. This is then converted to the alcohol with hydrogen peroxide in base medium. The net result of hydroboration and oxidation is the anti Markownikov hydration of the double bond.

In this case regioselectivity is controlled by steric factor (cyclic cis addition of BH_3).

2. Regioselective oxidation of ketones to ester—the Baeyer-Villiger reaction.*

Baeyer-Villiger oxidation which transforms acyclic ketones to ester and cyclic ketones to lactones. The reaction proceeds involving a 1, 2-migration of an alkyl group to the electron deficient oxygen atom.

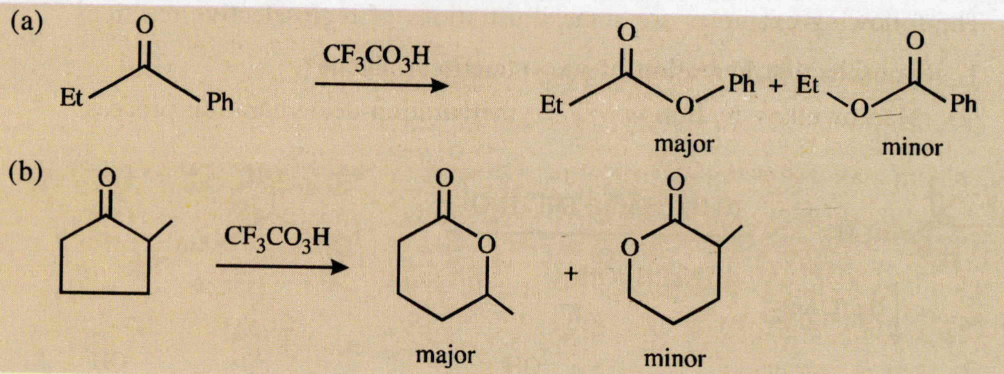

In this case regioselectivity is controlled by preferential migration of the alkyl groups Migratory aptitude: tertiary alkyl > secondary alkyl > phenyl > primary alkyl.

3. Regioselective Diels-Alder reaction

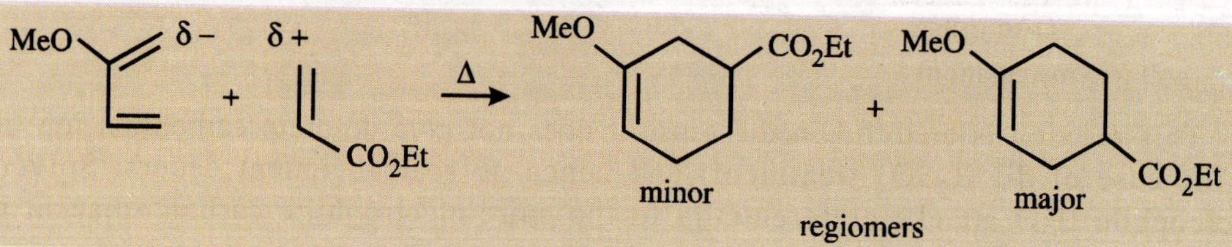

*For details see author's "Organic Synthesis" vol. 1

In this Diels-Alder reaction regioselectivity is controlled by electronic factor which is guided by electron donating group (OMe) and electron withdrawing group (– CO_2Et).

4. Regioselective alkylation

(a) Alkylation of α, β-unsaturated carbonyl

α, β-unsaturated carbonyl compound has two sites for alkylation, one is carbonyl carbon alkylation and the other is conjugate alkylation.

(i) Me_2CuLi

(ii) $H_3\overset{\oplus}{O}$

In this case regioselectivity is controlled by the selective alkylating reagent Lithium dimethyl cuprate which is a good Michael donor and α, β unsaturated carbonyl is a good Michael acceptor and therefore conjugate addition and alkylation occurs in almost exclusively.

(b) Alkylation of unsymmetrical ketone

Two types of enolates (which are specific enolates) may be generated from an unsymmetrical ketone at different conditions. With hindered strong base at low temperature and in presence of trimethyl silyl chloride, kinetic specific enolate is formed almost exclusively and in presence of weak base at higher temperature in presence of trimethyl silyl chloride thermodynamic specific enolate is formed.

LDA, – 78°C / THF, Me_3SiCl → $OSiMe_3$ (kinetic enolate) → MeI → $\overset{\oplus}{O}SiMe_3$ → (i) $F^{\ominus}$ (ii) $H_3\overset{\oplus}{O}$

Et_3N, 130°C / Me_3SiCl → $OSiMe_3$ (thermodynamic enolate) → MeI → $\overset{\oplus}{O}SiMe_3$ → (i) $F^{\ominus}$ (ii) $H_3\overset{\oplus}{O}$

5. Regioselective O-alkylation versus C-alkylation

Enolate is an ambident nucleophile. Alkylation of oxygen versus carbon is a regioselective reaction which may be controlled by various ways as for example withappropriate solvent, by protection, by activation and with selective alkylating agent.

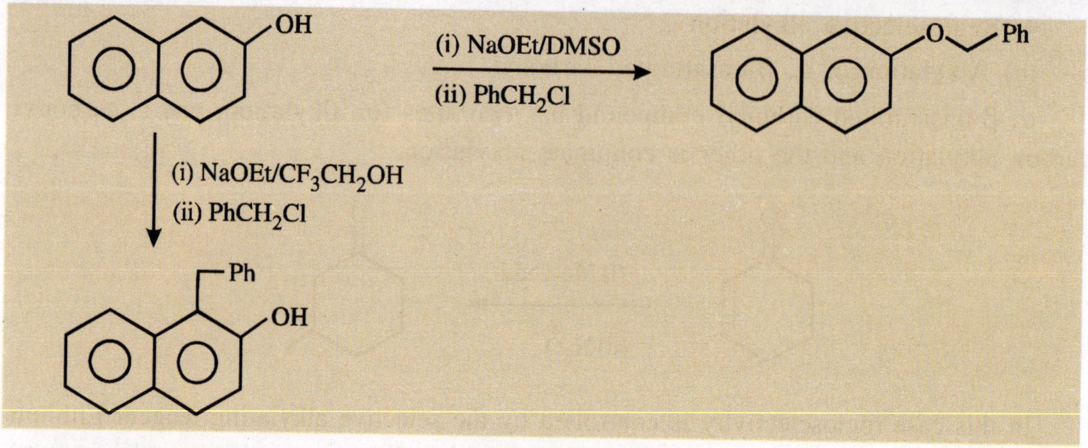

OH

(i) NaOEt/DMSO

(ii) $PhCH_2Cl$

O Ph

(i) $NaOEt/CF_3CH_2OH$

(ii) $PhCH_2Cl$

Ph

OH

In this case selectivity is controlled by solvents.

6. Regioselective electrophilic aromatic substitution

The position of substitution of the electrophile on the benzene ring depends on the substituent(s) already present in the ring.

If the aromatic compound has a substituent which will stabilise an adjacent positive charge then the electrophile adds predominantly at the ortho or para position e.g.,

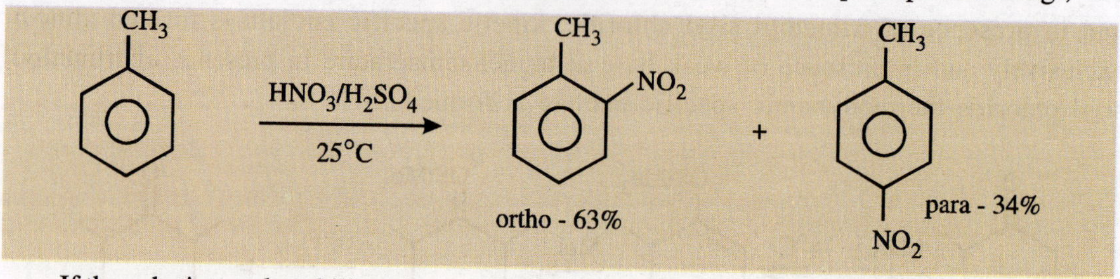

CH_3

HNO_3/H_2SO_4

25°C

CH_3

NO_2

ortho - 63%

+

CH_3

NO_2

para - 34%

If the substituent destabilises the pentadienyl cation, regioselective attack occurs at the meta position of the ring e.g.,

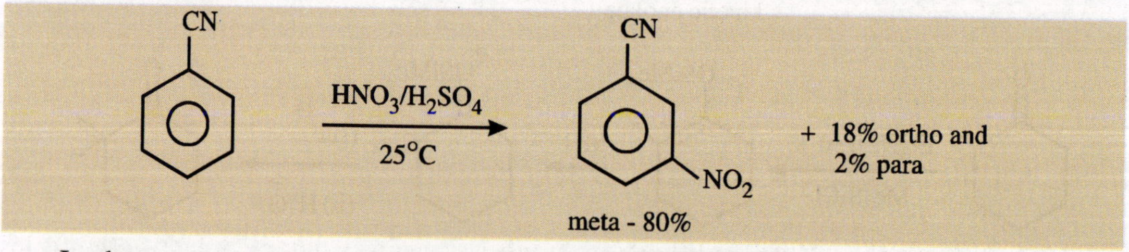

CN

HNO_3/H_2SO_4

25°C

CN

NO_2

meta - 80%

+ 18% ortho and 2% para

In these cases regioselectivity is controlled by the stability of the σ-complex which is controlled by the group already present in the ring and temperature.

7. Regioselectivity in the polynuclear hydrocarbons, Anthracene and Phenanthrene

Anthracene and Phenanthrene contain two unsymmetrical benzenoid rings, one is middle ring and the other is side ring (two side rings are equivalent). In these cases regioselectivity is controlled by the sacrifice of less resonance energy. That pathway is favoured which goes via sacrifice of less resonance energy. Usually middle ring is more activated because reaction of middle ring, sacrifice of less resonance energy occurs.

(a) anthracene $\xrightarrow[\text{regio-selective reduction}]{Na/EtOH}$ major + minor

major: Loss of RE = 84 – 2 × 36 = 12 KCal

minor: Loss of RE = 84 – 61 = 23 KCal

$\xrightarrow[\text{regioselective reduction}]{Na/EtOH}$ major + minor

major: Loss of RE = 92 – 2 × 36 = 20 KCal

minor: Loss of RE = 92 – 61 = 31 KCal

8. Regioselectivity in the elimination reaction

(a) H_3C, OH (1-methylcyclohexanol) $\xrightarrow[120°C]{H_3PO_4}$ major (1-methylcyclohexene) + minor (CH_2 methylenecyclohexane)

In this case regioselectivity is controlled by E1-elimination and stability of the product.

(b) H–CH_2, Cl, H (cyclohexane) $\xrightarrow{Et_3C-\overset{\ominus}{O}\overset{\oplus}{K}}$ major (CH_2 methylenecyclohexane) + minor (1-methylcyclohexene)

The base $Et_3C-\overset{\ominus}{O}\overset{\oplus}{K}$ is bulky and the ring hydrogen is hindered and therefore less substituated alkene is the major product by E2-elimination.

In this case regioselectivity is controlled by steric factor in the transition state.

(c) Hofmann and saytzeff E2 elimination reactions are regioselective reactions which are controlled by stereoelectronic factor which is a combination of steric factor which is mainly controlled by size of the leaving group and attacking base and electronic factor which is antiperiplanarity of the leaving neutral molecule.

$$CH_3-\overset{H_2}{CH}-\underset{X}{CH}-\overset{H_1}{CH_2} \xrightarrow{\text{Base}} \underset{CH_3-CH_2-CH{=}CH_2}{\text{Hofmann \%}} + \underset{CH_3-CH{=}CH-CH_3}{\text{Saytzeff \%}}$$

		Hofmann %	Saytzeff %
Base = $\overset{\ominus}{O}Et$	X = Br	20	80
	X = $\overset{\oplus}{N}Me_3$	95	5
X = Br	Base = $\overset{\ominus}{O}Et$	30	70
	Base = $Et_3C\overset{\ominus}{O}$	80	20

Hofmann TS (H_1, Et, H, H, H, X)

Saytzeff TS (H_2, Me, H, H, Me, X)

9. In certain cases kinetically controlled product (KCP) versus thermodynamically controlled product (TCP) are actually regioselective reactions. Some examples are given below.

(i) Bromination of 1, 3-butadiene to give 1, 2-and 1, 4-dibromo derivative.

$$\text{1,3-butadiene} \xrightarrow{Br_2} \text{3,4-dibromo-1-butene} + \text{1,4-dibromo-2-butene}$$

	KCP	TCP
– 80°C	major	minor
+ 40°C	minor	major

(ii) Fries rearrangement to give ortho and para derivative.

$$\text{phenyl acetate} \xrightarrow[\Delta]{AlCl_3} \text{p-hydroxyacetophenone} + \text{o-hydroxyacetophenone}$$

	KCP	TCP
25°C	major	minor
165°C	minor	major

(iii) Kolbe-Schmidt reaction of sodium phenoxide with carbon dioxide to give ortho and para derivative.

(iv) Sulphonation of Naphthalene to give 1 and 2-sulphonic acid derivative.

8.1.5 Stereoselectivity and regioselectivity either at a time or sequential

There are some organic compounds (target molecules) where substituents and functional groups are placed in a definite position which is regiochemistry and in a definite orientation which is stereochemistry. Therefore, to synthesis such types of organic compounds, knowledge of application of regioselectivity and stereoselectivity at a time for a given structure or sequential regioselectivity followed by stereo control or oppositely sequential stereoselectivity followed by regio control for a given structure are necessary. For this purpose the following examples may be taken as illustrations.

1. Stereospecificity and regioselectivity at a time.

1-methylcyclohexene has got two enantiotopic faces and therefore boron hydride may add either of the faces α or β with equal facility and give two enantiomeric transition states and BH_3 is added to the unsymmetrical alkene with a *cis* mode because of cyclic transition state (stereospecificity) and hydrogen is added to the more substituated carbon of the alkene which is regioselective.

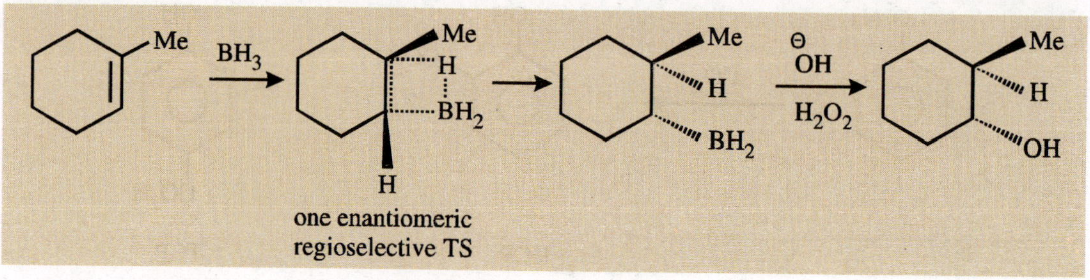

Me BH_3 Me H BH_2 H Me H BH_2 $\ominus$OH H_2O_2 Me H OH

one enantiomeric
regioselective TS

Therefore, hydroboration oxidation of 1-methylcyclohexene is a stereospecific *SYN* addition (enantiospecific) reaction with a regioselective control (H_2O is added to the unsymmetrical double bond with definite pattern).

(ii) Diels-Alder reaction is highly stereospecific because neither diene nor dienophile has time to rotate and the stereochemistry of each must be faithfully reproduced in the product.

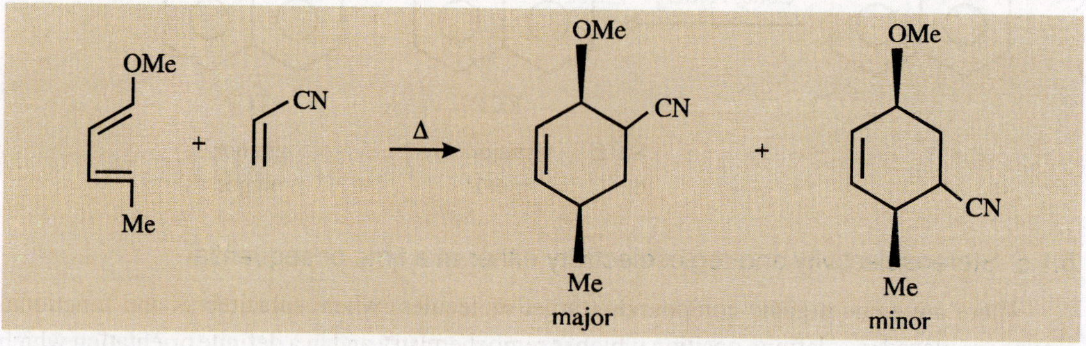

OMe Me + CN Δ OMe CN Me + OMe CN Me

major

minor

This reaction is stereospecific since the two substituents (OMe and Me) in both the two regio products are in the same *(cis)* stereochemistry as in the reactant diene and regioselective since dienophile and diene are added in two different modes to give two constitutional products.

2. Stereoselectivity and regioselectivity at a time.

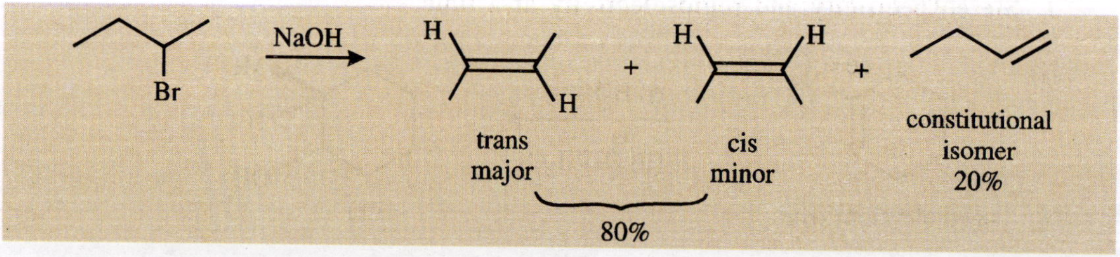

Br NaOH H H + H H +

trans
major

cis
minor

80%

constitutional
isomer
20%

This reaction is stereoselective since diastereomers are formed in unequal amount and regioselective since constitutional isomer is formed at a time.

3. Sequential regioselectivity followed by stereoselectivity

(i) $H-C\equiv C-H \xrightarrow{^{\ominus}NH_2} {}^{\ominus}C\equiv C-H \xrightarrow{EtBr}$ ⁄—≡C−H $\xrightarrow{^{\ominus}NH_2}$ ⁄—≡C$^{\ominus}$ $\xrightarrow{nPrBr}$ ⁄—≡—⁄⁄ $\xrightarrow[\text{Lindlar catalyst}]{H_2/Pd/BaSO_4}$ (cis-hept-3-ene, carbons numbered 1–7, H and H on C3 and C4)

This alkene synthesis is regioselective since double bond is fixing at the 3, 4 position (position of the double bond may be fixed by the length of alkyl groups) and then *cis* stereoselectivity is controlled by selective reducing agent, lindlar catayst.

(ii) When Lithium dimethyl cuprate is reacted with cyclohexenone then regioselective conjugate addtion of a methyl group to the β-position of the enone occurs.

Cyclohexenone $\xrightarrow[\text{regioselective alkylation}]{Me_2CuLi}$ enolate ($O^{\ominus}$, 3-Me) $\xrightarrow{MeI}$ trans-2,3-dimethylcyclohexanone (80%) + cis-2,3-dimethylcyclohexanone (20%)

stereoselective alkylation

The enolate which is formed in the conjugate addition reaction may be alkylated further by MeI, steric factors now control the stereochemistry. The preferred product is the one in which the substituents α and β to the carbonyl group are *trans* to each other.

8.2 Stereochemistry in the design of chiral synthesis

8.2.1 Definition of chiral (asymmetric) synthesis

Chiral or asymmetric synthesis is a reaction in which an achiral unit in the substrate molecule (having either enantiotopic or diastereotopic groups or faces) is converted into a chiral unit in such a manner that unequal amounts of chiral stereoisomers either diastereomeric excess (de) or enantiomeric excess (ee) are produced.

enantiomeric excess (ee)

It is the excess of one enantiomer over the dl pair in a dl mixture. It is also optical purity. Optical purity is equal to zero for racemic modification (dl mixture).

$$ee = \frac{\text{difference of mole - fraction of d and l}}{\text{total number of moles of the d and l}} \times 100$$

diastereomeric excess (de)

It is the excess of one diastereomer over the other diastereomer in a diastereomeric mixture.

$$de = \frac{\text{difference of mole - fraction of the two diastereomers}}{\text{total number of diastereomers}} \times 100$$

8.2.2 Criteria of chiral synthesis

In the enantioselective synthesis the product is enantiomeric excess (ee) and optically active.

In the chiral diastereoselective synthesis the product is either diastereomeric excess (de) and optically active or racemic and optically inactive.

For example,

Ph – C(=O) – CH(OH) – Ph (S)-Benzoin $\xrightarrow{LiAlH_4}$ Ph – CH(OH) – CH(OH) – Ph (R) (S) 92% meso + Ph – CH(OH) – CH(OH) – Ph (S) (S) 8% active

diastereomeric excess

Ph – C(=O) – CH(OH) – Ph (R)-Benzoin $\xrightarrow{LiAlH_4}$ Ph – CH(OH) – CH(OH) – Ph (S) (R) 92% meso + Ph – CH(OH) – CH(OH) – Ph (R) (R) 8% active

diastereomeric excess

If single enantiomer either (S) or (R)-Benzoin is taken, the product is (de) and optically active but if the racemic mixture of reactants Benzoin are taken then the product is racemic and optically inactive although the individual reaction is chiral diastereoselective.

8.2.3 Conditions for chiral or asymmetric synthesis

(i) The products must be chiral (one of the products may be *meso*).

(ii) The reaction must go through two diastereomeric transition states.

(iii) The product must be unequal amount either diastereomeric excess (de) or enantiomeric excess (ee).

The following reactions are not chiral synthesis:

(a)

4-tert-butylcyclohexanone $\xrightarrow{LiAlH_4}$ (H axial, OH equatorial) 90% + (OH axial, H equatorial) 10%

This reaction goes via diastereomeric transition states and is giving unequal amount of diastereomeric products but this reaction is not chiral synthesis because products are not chiral.

(b) [cyclohexanol (OH axial) → CrO_3/AcOH → cyclohexanone ← CrO_3/AcOH ← cyclohexanol (OH equatorial)]

This oxidation reaction goes via two diastereomeric transition states but the product is single achiral molecule which is stereodestructive reaction. Therefore, it is not a chiral synthesis.

(c) PhCOMe $\xrightarrow{LiAlH_4}$ Ph – CH(OH) – Me

racemic product

enantiomeric transition states

R-alcohol S-alcohol

Although this reaction is giving chiral products but this reaction is not a chiral synthesis because the transition states are not diastereomeric (rather enantiomeric) and the products are not unequal amount since enantiomeric transition states will form with equal facility because of same energy of activation.

The following reactions are chiral synthesis.

(a) (S) $\xrightarrow{LiAlH_4}$ (SS) major + (SR) minor

This reaction is chiral synthesis since products are chiral, involve diastereomeric TS and products are unequal amount (de).

(b) If a racemic mixture of (S)-2-methyl and (R)-2-methyl cyclohexanone are reduced by $LiAlH_4$ then in that condition this is also chiral synthesis.

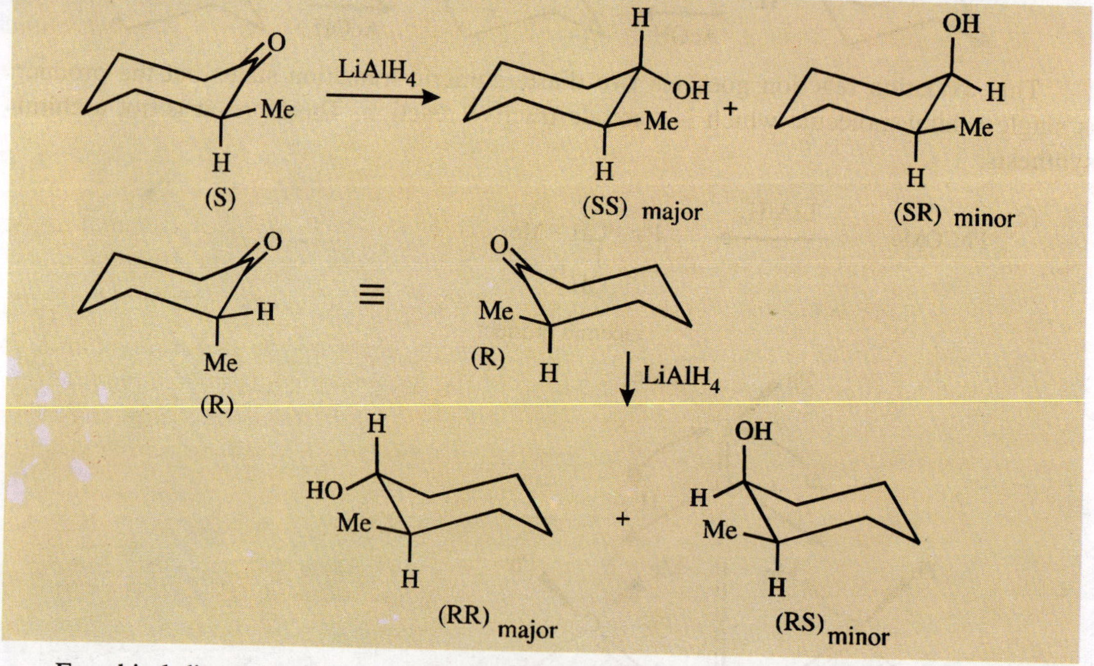

For chiral diastereoselective synthesis overall racemic product is not a factor but chiral diastereoselectivity (two diastereomeric TS) in the individual reaction to give unequal amount of diastereomers (or enantiomers) is the factor.

8.2.4 Controlling of diastereoselectivity in the chiral synthetis

In any chiral or asymmetric synthesis, the two transition states are diastereomeric and therefore they have different free energy and the two diastereomeric transition states must be unequal amount and consequently they have different energy of activation (Ea). The diastereomeric transition state having low Ea will give predominant product which is kinetically controlled product (KCP) and the kinetically controlled product is always associated with irreversible reaction.

The product stability controlled product under equilibrium condition is the thermodynamic controlled product (TCP) and the thermodynamic controlled product is always associated with reversible reaction.

Stereoselectivity of the products depends on difference in free energy of the two TS and consequently difference in energy of activation (Ea). The more the difference in energy of activation of the two diastereomeric TS, the more will be the stereoselectivity.

Stereoselectivity of the two diastereomeric TS may be increased by various factors during the chiral synthesis as shown in the following examples.

(i) Stereoselectivity may be controlled by only steric factor in the transition state.

Example: See chiral synthesis with chiral cyclic secondary amine (enamine) and with chiral borolane.

(ii) Stereoselectivity may be controlled by both steric factor and chelating in the transition state.

Example: See chiral synthesis by Cram's Chelate cyclic model.

(iii) Stereoselectivity may be controlled by both steric factor and chiral reagents.

Example: See chiral synthesis by Meerwein-Poundorf-Verley (MPV) reduction and with chiral Grignard reagent.

(iv) Stereoselectivity may be controlled by both steric factor and asymmetric induction in the transition state.

Example: See chiral synthesis by Cram's and Prelog's models.

(v) If there is a choice of chair and boat like transition states for six-membered cyclic ring then chair conformation (also steric factor) controls the stereoselectivity.

Example: See chiral synthesis by Cope and Claisen rearrangement.

(vi) When the addition of a nucleophile to the diastereotopic face is reversible, it is likely that the thermodynamically more stable of the two possible diastereomeric products will be formed predominantly. In any first order asymmetric synthesis where diastereomeric products are in equilibrium then in that condition thermodynamic stability of the product controls the stereoselectivity.

Example: See first order asymmetric synthesis—mutarotation of glucose.

8.2.5 Conditions for chirality imposition

In the chiral or asymmetric synthesis where products are enantiomeric excess (ee) or diastereomeric excess (de) the conditions are:

1. Chiral reagents and achiral substrate.

Example: See chiral synthesis by chiral aluminium alkoxide (MPV reduction), chiral grignard reagent, chiral borolane and chiral cyclic secondary amine (enamines). In all the cases enantiomeric excess (ee) product results.

2. Achiral reagents and chiral part in the substrate molecule having pro chiral centre or face.

Example: Cram's synthesis for diastereomeric excess (de) product, Prelog's synthesis for enantiomeric excess (ee) product.

3. Two achiral units, one achiral aldehyde and the other achiral enolate having enantiotopic faces in each case can combine in two modes (like and unlike mode) to give two diastereomeric transition states controlled by six-membered cyclic chair conformation and diastereomeric excess (de) product results.

Example: See Aldol-SYN Diastereoselective synthesis.

4. 1, 5-diene with two enantiotopic faces may rearrange (interact) under thermal condition controlled by six-membered cyclic chair conformation to give almost total enantioselective product.

Example: See chiral synthesis by Cope and Claisen rearrangement.

8.2.6 Asymmetric induction

It is the asymmetry induced at a prochiral centre of the substrate either by the chirality of the reagent (e.g. chiral Al-alkoxide, Grignard or enanmine) or by one or more chiral centres present in the substrate molecule itself (Cram's and Prelog's synthesis).

In the enantioselective synthesis asymmetric induction generates the ee product and in the diastereoselective synthesis (giving to a new chiral centre) it generates the de product.

1, 2-asymmetric induction—Cram's synthesis of de

1, 4-asymmetric induction—Prelog's synthesis of ee

ee synthesis with chiral enamine

Although any chiral synthesis involves diastereomeric transition states then how enantiomeric excess product developed?

The answer of this question is, if the chiral part (chiral Al-alkoxide, Grignard, borane and enamine) or chiral part of the substrate molecule (Prelog's) is detouched from the product at the end of the reaction—the reaction is ee.

On the otherhand, if the chiral part remains attached to the product—the reaction is de as in the Cram's synthesis.

8.3 Chiral synthesis

8.3.1 Diastereoselective Synthesis (de)

8.3.1.1 Cram's Synthesis

When a carbonyl group attached to a chiral centre $R-\underset{\underset{O}{\|}}{C}-CR_LR_MR_S$ in which L, M and S stand for large, medium and small groups respectively undergoes nucleophilic addition with CH_3MgBr / CH_3Li (nucleophile = $\overset{\ominus}{C}H_3$) or with $LiAlH_4$ / $NaBH_4$ (nucleophile = $\overset{\ominus}{H}$) and the two diastereomeric products results erythro and threo of which one predominates.

The relative configuration of the predominant stereoisomer is predicted by Cram's rule based on three models.

(i) **Open chain model**

$CH_3MgBr/CH_3Li/LiAlH_4/NaBH_4$

Disposition of R_L, R_M and R_S in the Newman Projection must be according to the configuration as in the given Fisher Projection

$M = \overset{\oplus}{M}gBr/Li^+/Na^+$

New chiral carbon

Predominant stereoisomer

The carbonyl group is flanked by two smaller groups (R_S and R_M) with the large group (R_L) nearly eclipsed with R (carbonyl part of alkyl or hydrogen).

The metallic part of the CH_3MgBr / CH_3Li / $LiAlH_4$ / $NaBH_4$ get complexed with carbonyl oxygen and the carbanion equivalent or $\overset{\ominus}{H}$ is transferred to the trigonal carbon from the side of smaller group in preference to that of medium group. It is reasoned that >C=O being complexed with the reagent becomes effectively the bulkiest group is thus better placed between small and medium group.

(ii) Dipolar model

Dipoles of the C=O and C—X oppose each other and so they are placed anti as in the model. The dipole repulsion is minimised and the electrophilicity of carbonyl carbon increases.

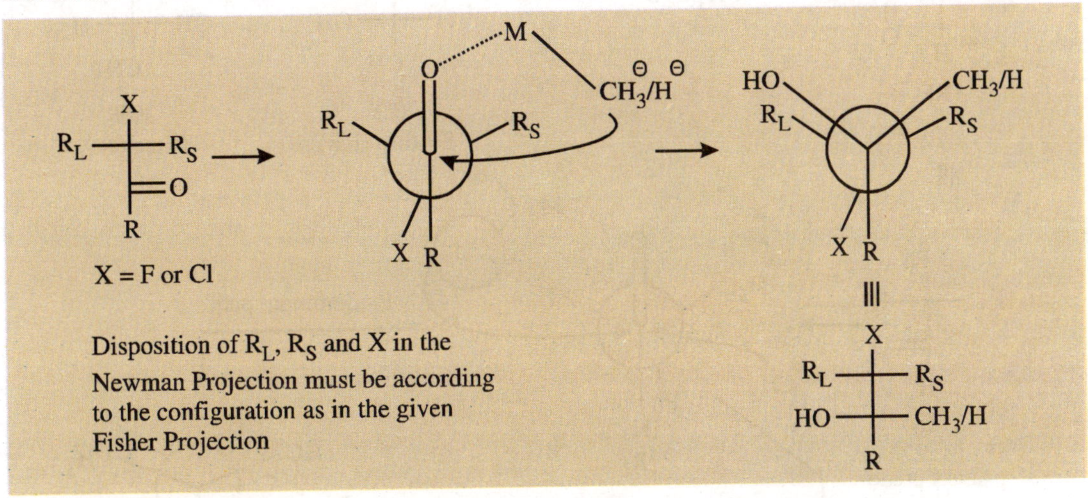

X = F or Cl

Disposition of R_L, R_S and X in the Newman Projection must be according to the configuration as in the given Fisher Projection

(iii) Rigid (chelate) cyclic model

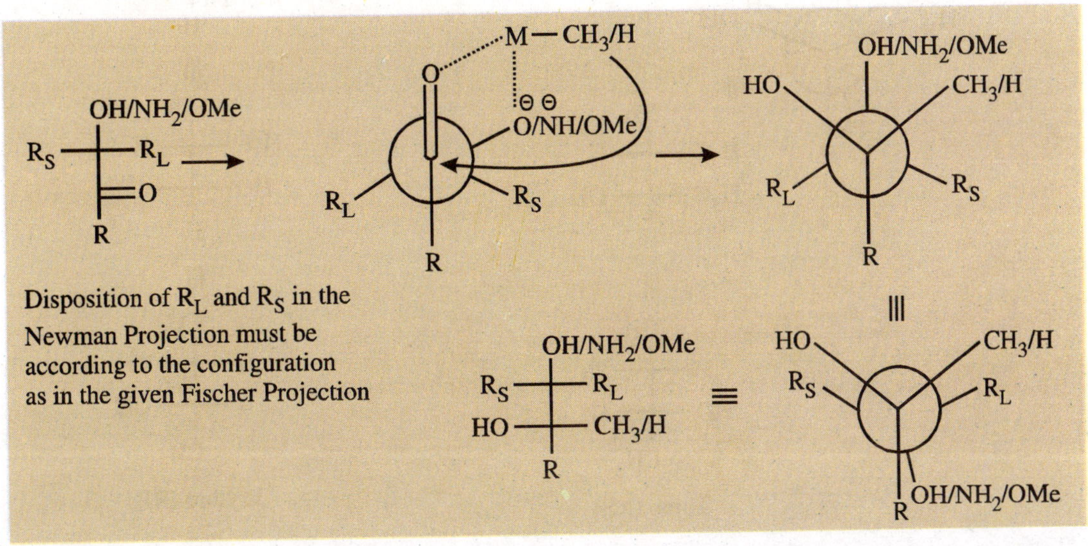

Disposition of R_L and R_S in the Newman Projection must be according to the configuration as in the given Fischer Projection

Metallic part is doubly coordinated to form a 5-membered ring—stereoselectivity is high of the model.

During the synthetic representation according to Cram's models one should be careful about the configuration of the chiral carbon in the given substrate molecule.

(a) the configuration of the old chiral carbon in the product should remain the same as in the given substrate molecule.

(b) during the transfer of Fischer projection to the Newman projection the configuration should remain the same.

The following examples are some illustrations of Cram's diastereoselective synthesis.

Open Chain Model

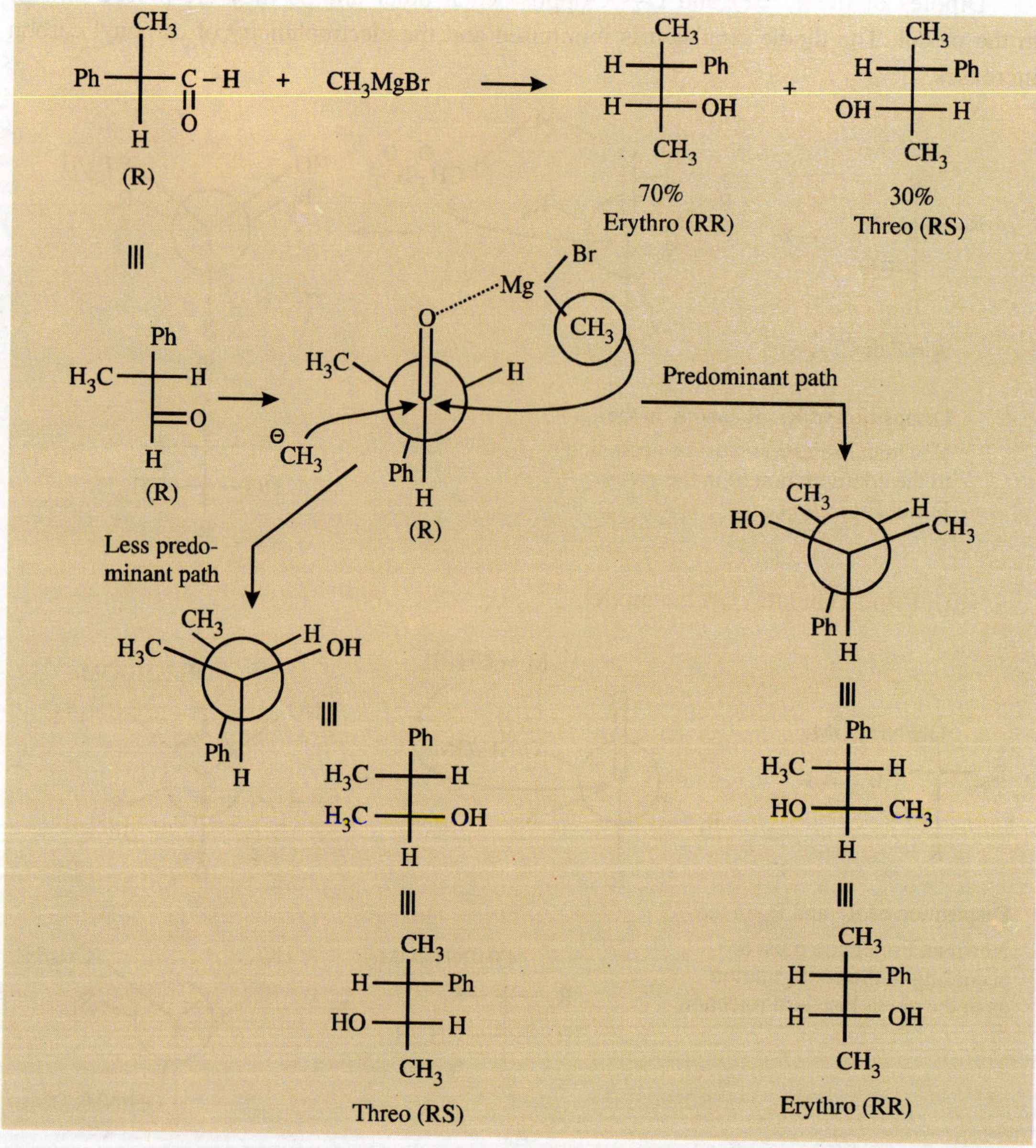

CH$_3$, Ph, C–H, O, H (R) + CH_3MgBr ⟶ 70% Erythro (RR) + 30% Threo (RS)

Mg, Br, O, CH$_3$, H$_3$C, H, Ph, (R)

Predominant path

Less predominant path

Threo (RS)

Erythro (RR)

Ph–C(CH$_3$)(H)–C(=O)–CH$_3$ + NaBH$_4$ ⟶ Threo (RS) + Erythro (RR)

(R) | 70% Threo (RS) | 30% Erythro (RR)

Pred. path

Less pred-path

Erythro (RR)

Threo (RS)

Dipolar Model

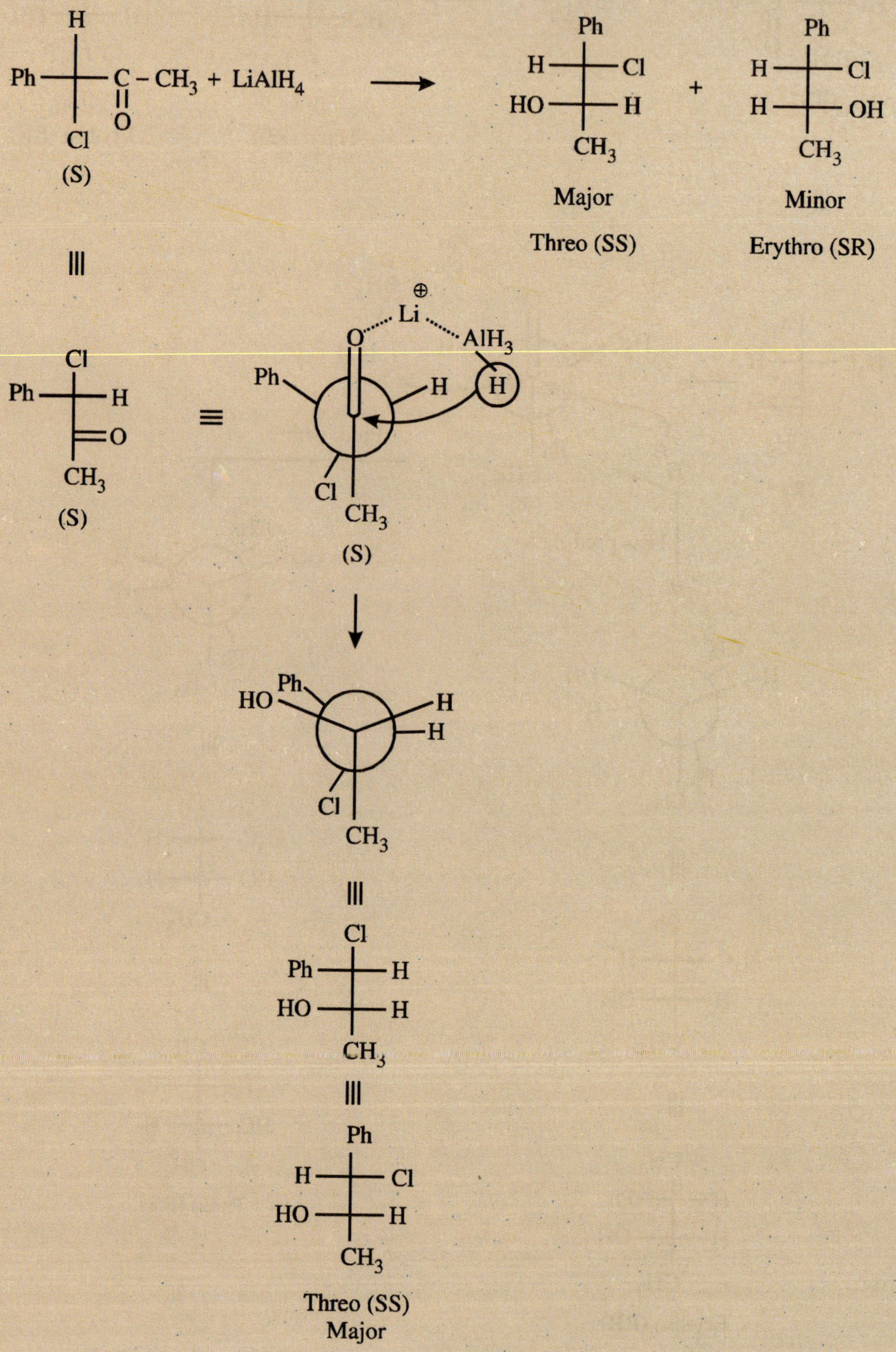

H
Ph
C – CH3 + LiAlH4
O
Cl
(S)
Ph
H
Cl
HO
H
CH3
Major
Threo (SS)
+
Ph
H
Cl
H
OH
CH3
Minor
Erythro (SR)
Cl
Ph
H
O
CH3
(S)
⊕
Li
O
AlH3
Ph
H
H
Cl
CH3
(S)
HO
Ph
H
H
Cl
CH3
Cl
Ph
H
HO
H
CH3
Ph
H
Cl
HO
H
CH3
Threo (SS)
Major

Rigid (chelate) Cyclic Model

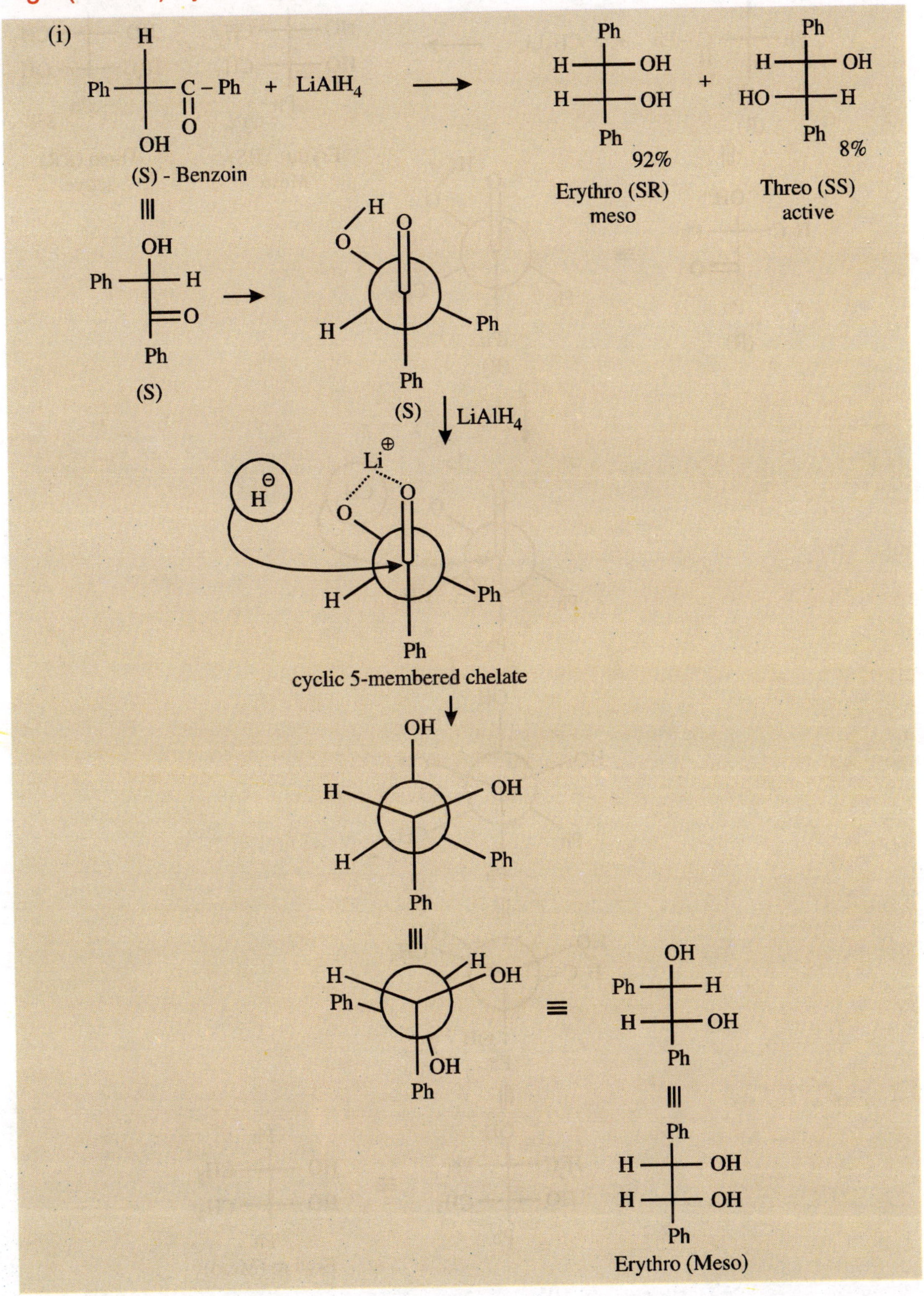

(i) (S) - Benzoin + $LiAlH_4$ → Erythro (SR) meso 92% + Threo (SS) active 8%

(S) → (S) ↓ $LiAlH_4$

cyclic 5-membered chelate

≡ Erythro (Meso)

(ii) (R) + CH_3Li ⟶ Erythro (RS) Meso 92% + Threo (RR) active 8%

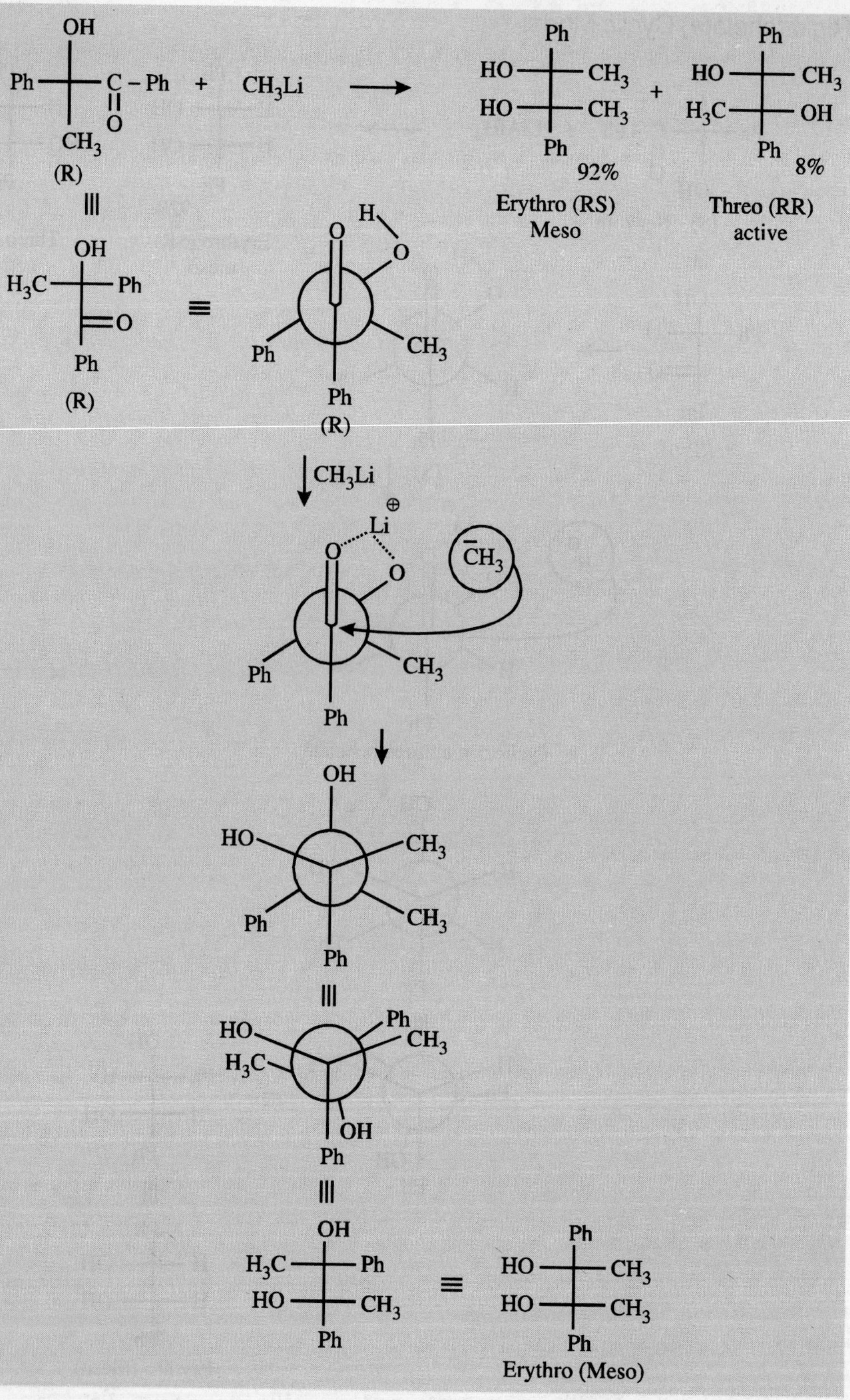

(R) ≡ (R)

↓ CH_3Li

↓

≡

≡ Erythro (Meso)

8.3.1.2 Aldol-SYN-Diastereoselective Synthesis

An enolate derived from a carbonyl compound undegoes nucleophilic addition to another carbonyl compound usually an aldehyde to give a β-hydroxy aldehyde or ketone, known as an aldol condensation.

Depending on the symmetry of the two reactants an achiral aldehyde and an achiral enolate each with two enantiotopic faces may combine in two modes: an unlike mode (ul) in which combination takes place between Si and Re (or Re and Si) faces of the reactants and a like mode (lk) in which combination takes place between Re and Re (or Si and Si) faces.

Combination of unlike (ul) mode goes via lower energy TS (lower Ea) which is kinetically controlled product (KCP) whereas combination of like mode (lk) goes via higher energy TS (higher Ea) which is thermodynamically controlled product (TCP), controlled by product stability under equilibrating condition.

As for example, condensation between 2, 2-dimethyl-3-pentanone (ethyl t-butyl ketone) and benzaldehyde in the presence of LDA gives a preponderance (> 98%) of the SYN-diastereomeric product. This experimental findings may be explained as shown below.

Ph – CHO + Me (ethyl t-butyl ketone, O) —LDA→ Ph – C*H (OH) – C*H(Me) – C(=O) – t-Bu

R	S (SYN) erythro or SR
S	S (ANTI) threo or RR

When benzaldehyde and Lithium enolate of ethyl t-butyl ketone reacts may give four stereosiomers i.e. two enantiomeric pairs (±) of diastereomers are possible but only RS (SYN) erythro forms in 98% yield and SS (ANTI) threo forms in 2% yield.

Benzaldehyde has got two enantiotopic faces.

Ph, O, H — Si-face   H, O, Ph — Re-face

Ethyl t-butyl ketone with LDA gives two enolates of which only z-enolate is favoured under equilibrium condition.

Me (ethyl t-butyl ketone, O) —LDA→ Li⊕ O⊖ (Me, H) Z-enolate ⇌ Li⊕ O⊖ (H, Me) E-enolate unstable due to methyl t-butyl steric repulsion   Li⊕ O⊖ (Me, H) Re-face

Reaction scheme for favoured SYN aldol diastereoselective product: combination via ul mode.

Si + Re

ul mode

Zimmerman and Traxler Model

Chair-like six-centre low energy TS where only Ph and Me gauche interaction occurs and hence low Ea

$Li^{\oplus}$ co-ordinated to both the carbonyl and enolate oxygen in the TS

taking long carbon chain σ-frame-work in the same plane

SYN (RS)

erythro

Kinetically controlled product, less stable since all bulkier groups are in the same side

Reaction scheme for disfavoured ANTI aldol diastereoselective product: Combination via like (lk) mode.

Re + Re

lk mode

Zimmerman and Traxler Model

Chair-like six-centre high energy TS where severe 1, 3-SYN diaxial interaction occurs between Ph and Bu^t in the TS and hence high energy of activation.

$Li^{\oplus}$ co-ordinated to both the carbonyl and enolate oxygen in the TS

taking long carbon chain σ-frame work in the same plane

ANTI (SS)

threo

Thermodynamically controlled product, more stable since bulkier groups are separated by two anti H's

Unlike and like mode interactions are giving two diastereomeric TS and for the former low energy TS and therefore Ea low and the corresponding SYN diastereoselective product is preponderance under kinetic control.

If the reaction is effected under equilibrating condition usually leads to ANTI diastereoselective product because of higher stability.

Kinetically controlled product is guided by Ea whereas thermodynamically controlled product is guided by product stability under equilibrium condition.

8.3.2 Enantioselective synthesis (ee)

What is enantioselective synthesis?

Enantioselective synthesis is a reaction that converts an achiral unit with enantiotopic face or ligands (with chiral reagent) in the starting material to chiral products in which one enantiomer is present in excess of the other.

Conversion of an achiral unit with diastereotopic face with detouchable chiral unit in the starting material to chiral products (with unequal ammount of enantiomers) is also enantioselective synthesis if the chiral part is detouched at the end of the reaction.

8.3.2.1 Prelog's Synthesis

When α-keto acid, phenyl glyoxylic acid (PhCOCOOH) [or pyruvic acid (MeCOCOOH)] is esterified with optically active alcohol e.g. R (–)-2-Octanol [C_6H_{13} CH(OH)CH_3] and when the ester is reacted with grignard reagent (or hydride reagent, $NaBH_4$) the product diastereomeric alcohols (since keto carbonyl is in diastereotopic face in the ester) are formed in unequal amount (this is a reaction of diastereomeric transition states with enantiomeric excess product) and on complete hydrolysis (the step where chiral part is detouched) the enantiomeric products of atrolactic acid (or mandelic acid) are formed in unequal amount. This is a case of 1, 4-asymmetric induction.

The predominant enantiomer can be predicted by Prelog's rule by conformational analysis of the ester which is shown as follows.

(i) When the α-keto acid reacts with optically active alcohol it will give an ester. Now this ester has two parts, one is chiral part and the other is achiral part. The achiral part lies in one plane in such a way that the oxo groups are planar and trans-anti to each other (S-trans).

$$R{-}CO{-}CO{-}O{-}H \;+\; H{-}O{-}C(R_L R_M R_S) \longrightarrow R{-}CO{-}CO{-}O{-}C(R_L R_M R_S)$$

H—OC ($R_L R_M R_S$) is taking as chiral alcohol where R_L, R_M and R_S are large, medium and small groups and let according to the CIP rule the priority order is $R_L > R_M > R_S$.

(ii) The O—C ($R_L R_M R_S$) single bond is also lie in the same plane.

$$R{-}CO{-}CO{-}O{-}C(R_L R_M R_S)$$

(iii) The rotation of the single bond O—C ($R_LR_MR_S$) can give two energetically favoured conformations.

The large group R_L at the chiral centre of the alcohol moiety is on the same side of the ketonic carbonyl and lies in the same plane with the achiral part. R_M and R_S should be placed in the back and front of the plane according to the configuration of the alcohol. This is the most populated conformer.

The small group R_S at the chiral centre of the alcohol moiety is on the same side of the ketonic carbonyl and lies in the same plane with the achiral part. R_L and R_M should be placed in the back and front of the plane according to the configuration of the alcohol.

If the chiral alcohol is R-configurated then the conformers may be either.

(R)
Conformer - I

(R)
Conformer - II

(iv) Methyl magnesium bromide (or $NaBH_4$) approaches the electrophilic carbon (of the ketonic $C=O$) from the side of the R_S-α-face for conformer-I and from the side of the R_M-α-face for conformer-II more easily than from the side of R_M which is large group for conformer-I or from the side of R_L which is a large group for conformer-II. In both the cases the same stereochemistry will result.

Stereostructures of α-ketoacids, hydroxy acids and R (–)-2-Octanol.

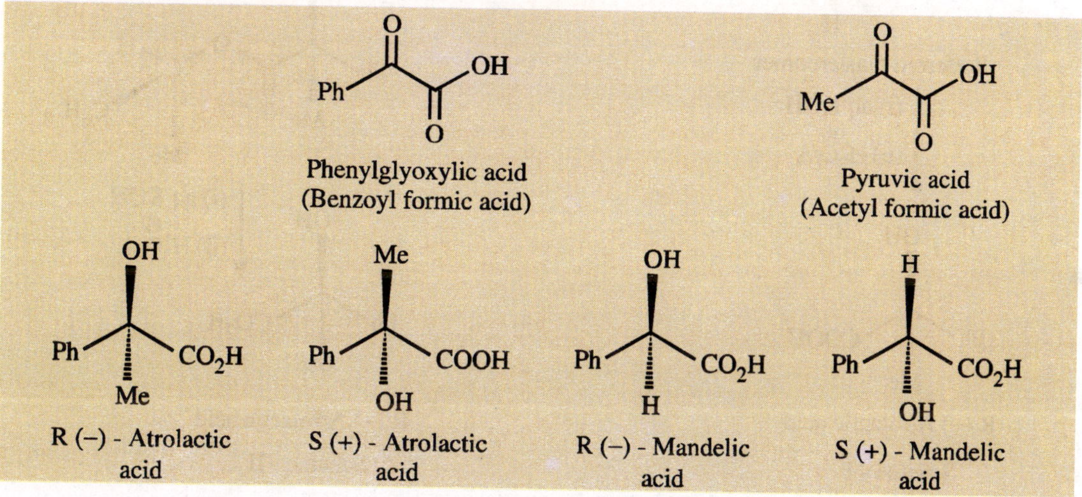

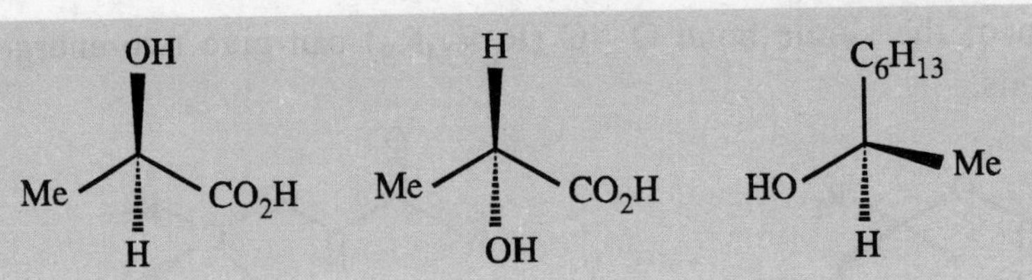

OH
Me CO$_2$H
H
R (–) Lactic acid

H
Me CO$_2$H
OH
S (+) Lactic acid

C_6H_{13}
HO Me
H
R (–) – 2-Octanol

The following four schemes will illustrate the application of Prelog's synthesis of α-hydroxyacids with enantiomeric excess (ee).

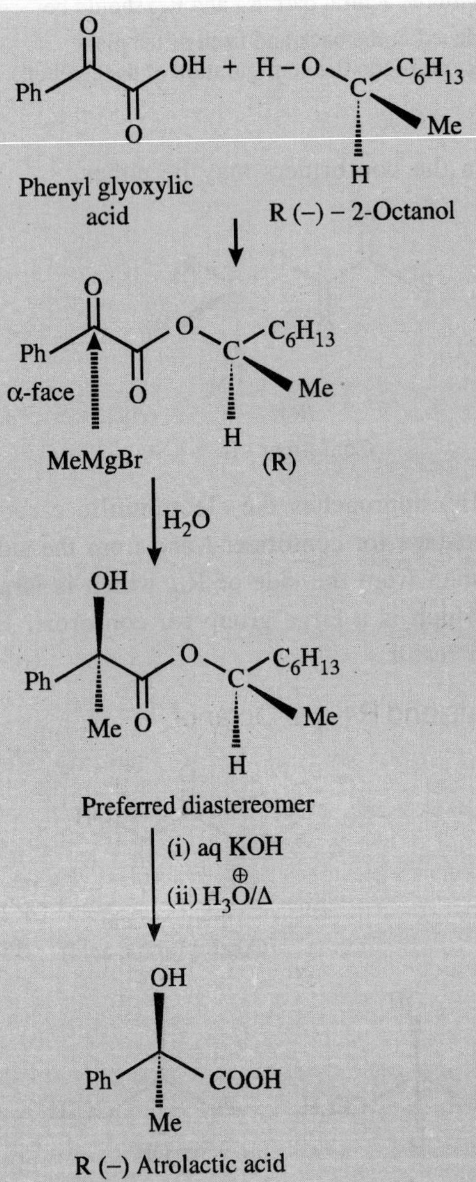

Phenyl glyoxylic acid + R (–) – 2-Octanol → (R) ester; α-face MeMgBr; H_2O → Preferred diastereomer → (i) aq KOH (ii) $H_3\overset{\oplus}{O}/\Delta$ → R (–) Atrolactic acid

Scheme - I

Phenyl glyoxylic acid + R (–) – 2-Octanol → (R) ester → (R); α-face MeMgBr; H_2O → (i) aq KOH (ii) $H_3\overset{\oplus}{O}/\Delta$ → R (–) Atrolactic acid

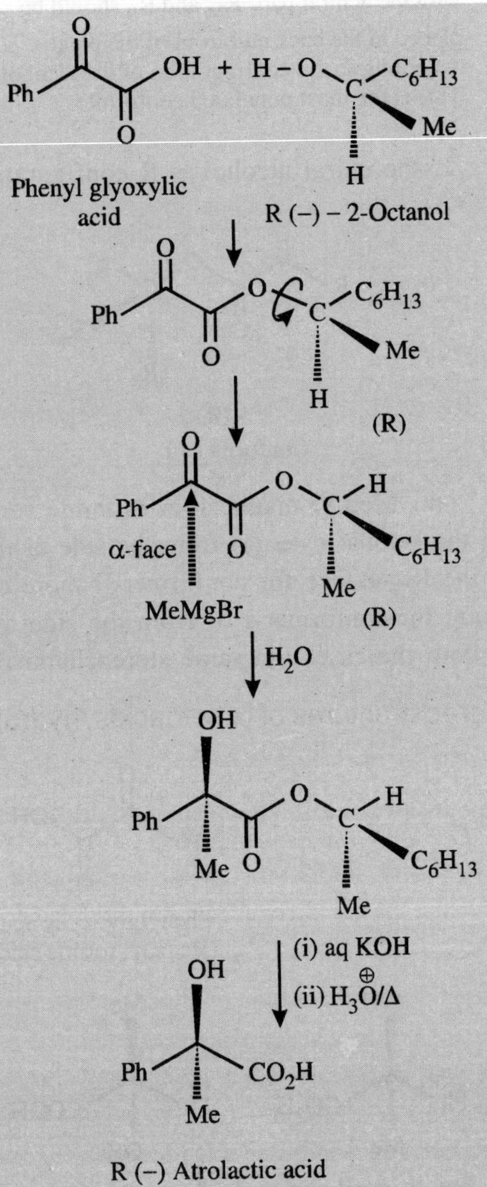

Scheme - II

Phenyl glyoxylic acid + S (+) – 2-Octanol

↓

β-face MeMgBr (S)

↓ H_2O

Preferred diastereomer

↓ (i) aq KOH (ii) $H_3\overset{\oplus}{O}/\Delta$

S (+) Atrolactic acid

Scheme - III

Pyruvic acid + R (–) – 2-Octanol

↓

α-face PhMgBr (R)

↓ H_2O

Preferred diastereomer

↓ (i) aq KOH (ii) $H_3\overset{\oplus}{O}/\Delta$

S (+) Atrolactic acid

Scheme - IV

(i) In the above schemes I, II and III if we use $NaBH_4$ in place of grignard reagent then corresponding configurated Mandelic acid will form and in scheme IV, R (–) lactic acid will form with $NaBH_4$.

(ii) Considering the scheme I and II, the large group ($R_L = C_6H_{13}$) and the small group (R_S = H) are in the same plane of the achiral part of the molecule respectively but the configuration (R) remain the same giving the same configuration of α-hydroxyacid [R(–) Atrolactic acid]. The definite configurated alcohol controls the approach of grignard reagent similarly. For scheme-I, hydrogen is smaller group whereas in scheme-II the smaller group is methyl.

(iii) Considering the scheme I and III, for a given α-keto acid (phenylglyoxylic acid) different configurated alcohols [R(–)-2-Octanol and S(+)-2-Octanol] are giving different α-hydroxyacids [R(–) Atrolactic acid and S(+) Atrolactic acid] with the same grignard reagent.

(iv) Considering the scheme I and IV, it is seen that if the α-keto acid is Pyruvic acid and the grignard reagent is phenylmagnesium bromide then for a given chiral alcohol [R(–)-2-Octanol], the Atrolactic acid is produced to have the opposite configuration of that produced by phenylglyoxylic acid and methyl magnesium bromide. The phenyl and methyl groups are now interchanged.

(v) Considering the scheme I and II, it is clear that if the configuration of the chiral alcohol is known, it is possible to deduce the configuration of the enantiomer of Atrolactic acid produced in excess. The opposite is also true i.e. if the configuration of the Atrolactic acid is known, it is possible to deduce the configuration of the alcohol used in the synthesis. This method of correlating the configurations has been called the 'Atrolactic acid method'.

Furthermore, since the absolute configuration of (–) Atrolactic acid is (R) and (+) Atrolactic acid is (S) and therefore the absolute configuration of the chiral alcohols can be correlated with the sign of rotation of the Atrolactic acid produced in excess.

8.3.2.2 Synthesis with chiral enamine

The enamine derived by the reaction of (2S, 5S)-2, 5-dimethyl pyrrolidine with cyclohexanone is chiral enamine.

O + Me H H Me N H

(2S, 5S) - 2, 5-dimethyl pyrrolidine

Me H H N Me

Chiral enamine - I

or

Me H H N Me

Chiral enamine - II

$R-CH_2-Cl$

$R-CH_2-Cl$

major diastereomer | major diastereomer

$\downarrow H_3\overset{\oplus}{O}$

major product (S - configuration) | major product (S - configuration)

Now alkylation of chiral enamine-I (which has got a diastereotopic face and it is a case of 1, 4-asymmetric induction and this synthesis involve diastereomeric transition states with enantiomeric excess product) from the bottom is severely hindered by the α-placed methyl group. Thus alkylation takes place much more rapidly from the top face. Similarly, for alkylation of chiral enamine-II from the top face is severely hindered by the β-placed methyl group. Thus alkylation takes place more rapidly from the bottom face. After hydrolysis (the step where chiral part is detouched) in both the cases the predominant alkylated product is S-configurated enantiomer which is enantiomeric excess (ee) product.

Group R	Yield	Optical purity
H–	50%	83%
$CH_3—CH_2$ –	57%	93%
$CH_2 = CH$ –	80%	82%

8.3.2.3 Synthesis with chiral borolane

When hydroboration reaction is performed with (2R, 5R)-2, 5-dimethyl borolane to the *cis*-2-butene and the product borane is treated with alkaline hydrogen peroxide, mostly a single enantiomer (S-configurated) of the product alcohol is formed. This is an example of diastereomeric excess transition states with enantiomeric excess (ee) product.

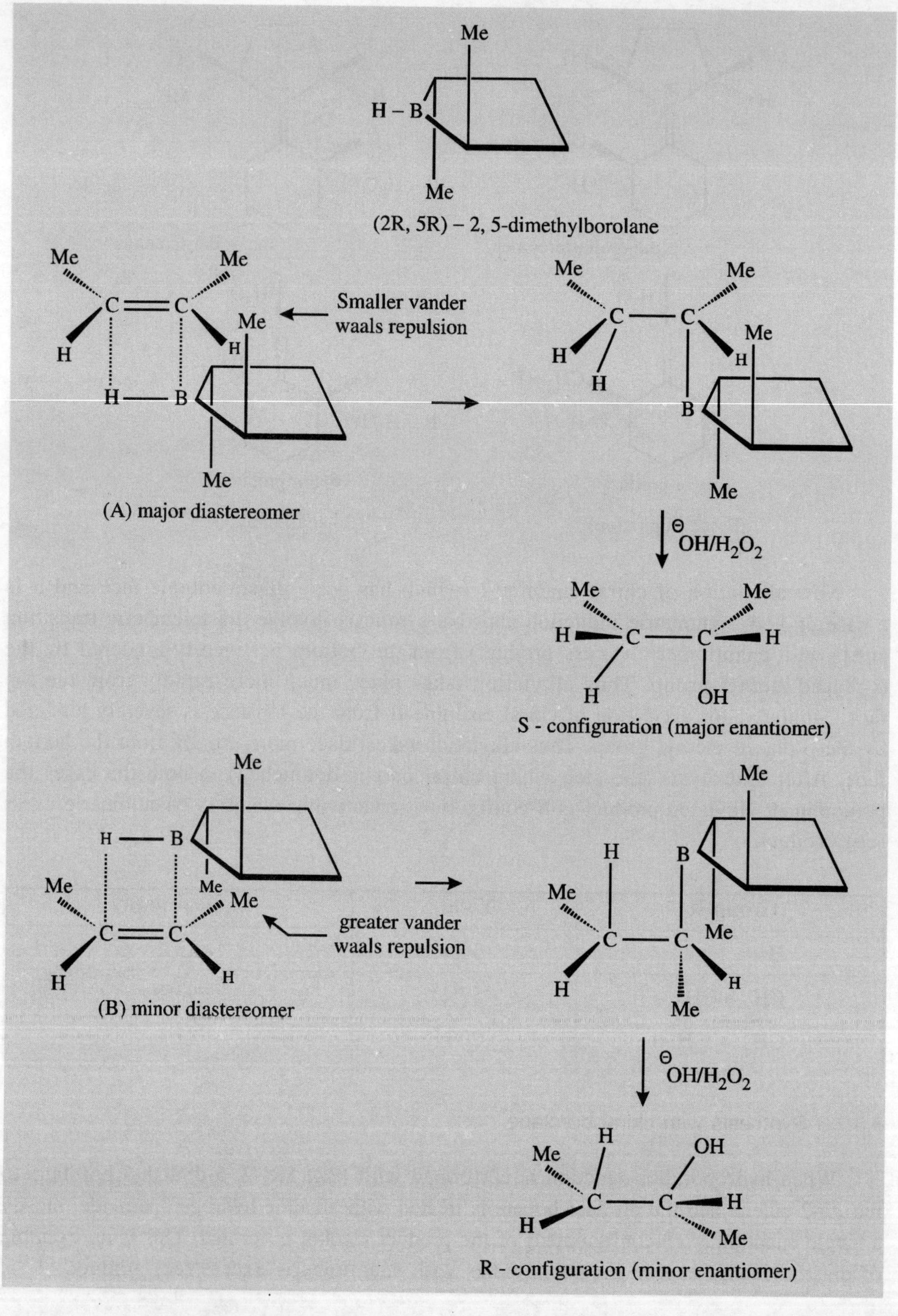

Me

H – B

Me

(2R, 5R) – 2, 5-dimethylborolane

Smaller vander waals repulsion

(A) major diastereomer

$\overset{\ominus}{O}H/H_2O_2$

S - configuration (major enantiomer)

greater vander waals repulsion

(B) minor diastereomer

$\overset{\ominus}{O}H/H_2O_2$

R - configuration (minor enantiomer)

The organoborane can approaches the π bond of a cis-2-butene molecule in two distinguishable ways. Approach from one face gives a transition state in which there is severe van der waals repusion between *methyl* group of the alkene and *methyl* group on the organoborane (Structure B). Approach from the opposite face gives a transition state in which van der waals repulsion is in between *hydrogen* of the alkene and *methyl* group of the borane (Structure A). Therefore, the transition state (A) has fewer van der waals repulsion and the corresponding product borane is in major quantity which when oxidised with alkaline hydrogen peroxide (the step where chiral part is detouched) gives enantiomeric alcohol in excess with S-configuration.

8.3.2.4 Synthesis using Meerwein-Ponndorf-Verley (MPV) reduction with chiral aluminium alkoxide

Reduction of acyclic enantiotopic ketone, as for example, methyl n-hexyl ketone ($MeCOC_6H_{13}$) with S-butane-2-ol as solvent in presence of chiral aluminium alkoxide, as for example Al(S-2-butoxide)$_3$ gives S-octane-2-ol as major enantiomeric excess product. The MPV reduction is a kinetically controlled reaction. This chiral synthesis is an example of diastereomeric transition states with enantiomeric excess product.

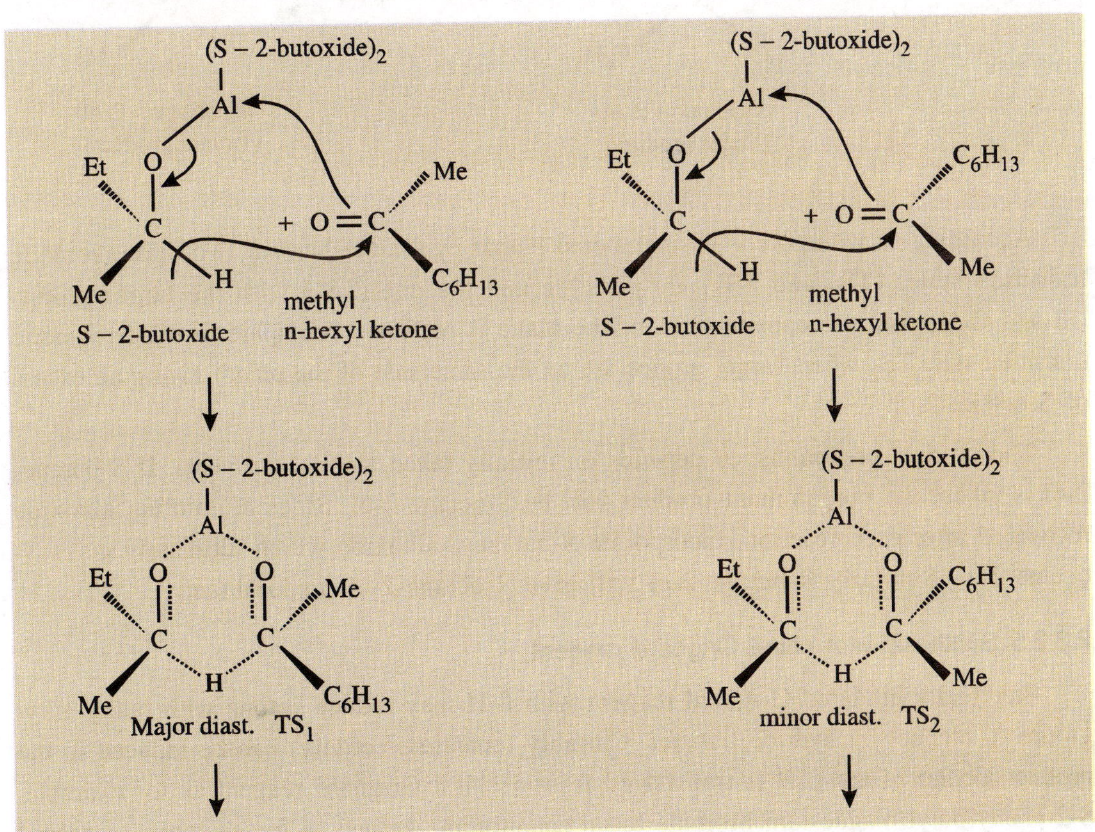

(S – 2-butoxide)$_2$

Et, Me >C=O + Al

H—O: ; C; H; Me; Et

O; C; H; Me; C_6H_{13}

chiral part is detouched

S-butane – 2-ol (Solvent)

(S – 2-butoxide)$_2$

Et, Me >C=O + Al

H—O: ; C; H; Me; Et

O; C; H; C_6H_{13}; Me

chiral part is detouched

S-butane – 2-ol (Solvent)

Al(S – 2-butoxide)$_3$ (recovered) + HO–C(H)(Me)(C_6H_{13})

S-Octane – 2-ol (major product)

Al(S – 2-butoxide)$_3$ (recovered) + HO–C(H)(C_6H_{13})(Me)

R-Octane – 2-ol (minor product)

According to whitmore's six-membered planar cyclic mechanism two diastereomeric transition states (TS_1 and TS_2) are possible and the one (TS_1) with the larger groups (Et and C_6H_{13}) on the oppsite sides of the plane is preferred (compare to diastereomeric transition state TS_2 where larger groups are on the same side of the plane) giving an excess of S-octane-2-ol.

The product predominance depends on initially taken alcohol (solvent). If S-butane-2-ol is taken, the predominant product will be S-octane-2-ol. Since aluminium alkoxide recovered after each reaction, incorporate S-butane-2-alkoxide which ultimately gives S-octane-2-ol. Similarly R-butane-2-ol will give R-octane-2-ol predominantly.

8.3.2.5 Synthesis with chiral Grignard reagent

Stereically hindered Grignard reagent with β-H may reduce ketone with bulky alkyl groups to alcohol by hydride transfer. Chirality (enantioselectivity) can be induced in the product alcohol if the β-H is transferred from a chiral Grignard reagent as for example, S-2-phenylbutyl magnesium bromide to an enantiotopic ketone as for example, isopropyl phenyl ketone. This reduction is kinetically controlled and is an example of diastereomeric transition states with enantiomeric excess product.

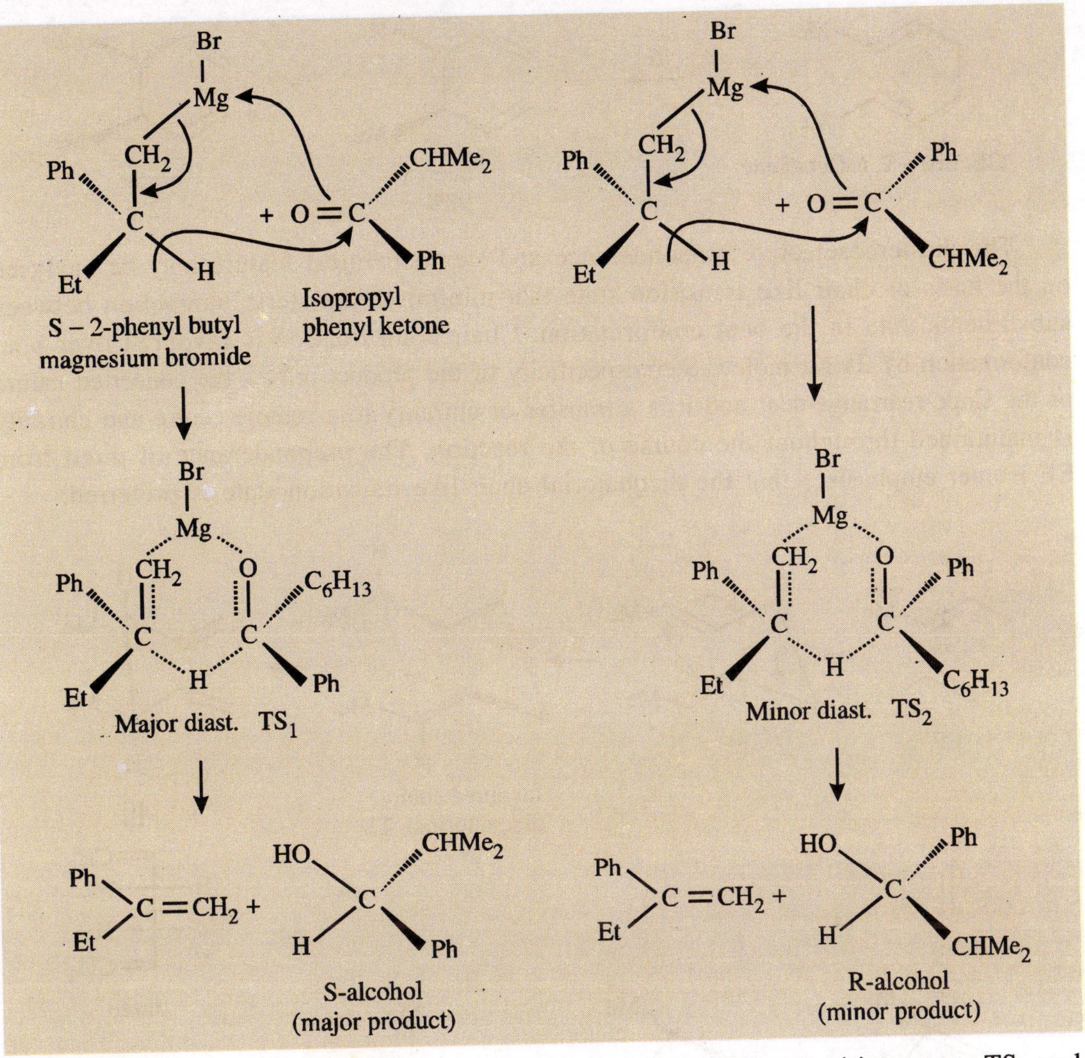
Br
Mg
CH2
Ph
C
Et
H
+ O=C
CHMe2
Ph
S – 2-phenyl butyl magnesium bromide
Isopropyl phenyl ketone
Major diast. TS1
Minor diast. TS2
C6H13
HO
C=CH2 +
S-alcohol (major product)
R-alcohol (minor product)

There are two diastereomeric six-membered planar cyclic transition states TS_1 and TS_2 and the one (TS_1) with larger groups (phenyl and phenyl) on the opposite sides is preferred compare to diastereomeric TS_2 where larger groups are on the same side giving an excess of S-alcohol.

8.4 Diastereoselectivity cum high enantiospecificity in the Cope and Claisen rearrangement.

The thermal rearrangement of 1, 5-diene by [3, 3] sigmatropic shift is called Cope and Claisen rearrangement. The reaction proceeds in the thermodynamic favoured direction. These are pericyclic reaction.

It is found that in the Cope-rearrangement of (2E, 6E)-2, 6-Octadiene on heating produces exclusively active threo-3, 4-dimethyl hexa-1, 5-diene in 98% yield and inactive erythro (*meso*)-3, 4-dimethyl hexa-1, 5-diene in 2% yield which is a diastereoselective cum enantiospecific synthesis.

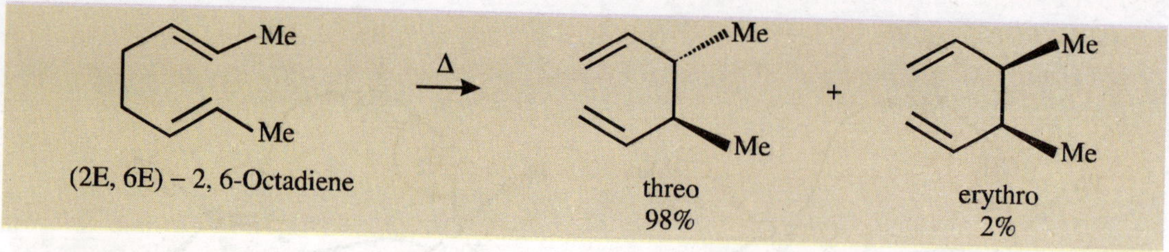

(2E, 6E) – 2, 6-Octadiene $\xrightarrow{\Delta}$ threo 98% + erythro 2%

The diastereoselective preponderance and stereochemical feature may be analysed on the basis of chair like transition state that minimises the steric interaction between substituents than in the boat conformation. Chair conformation is favoured over boat conformation by 25 KJ mole^{-1}. Stereospecificity of the product proves the concerted nature of the Cope-rearrangement and it is a transfer of chirality to a remote centre and chirality is maintained throughout the course of the reaction. The preponderance of threo from EE isomer emphasises that the diequatorial chair-like transition state is preferred.

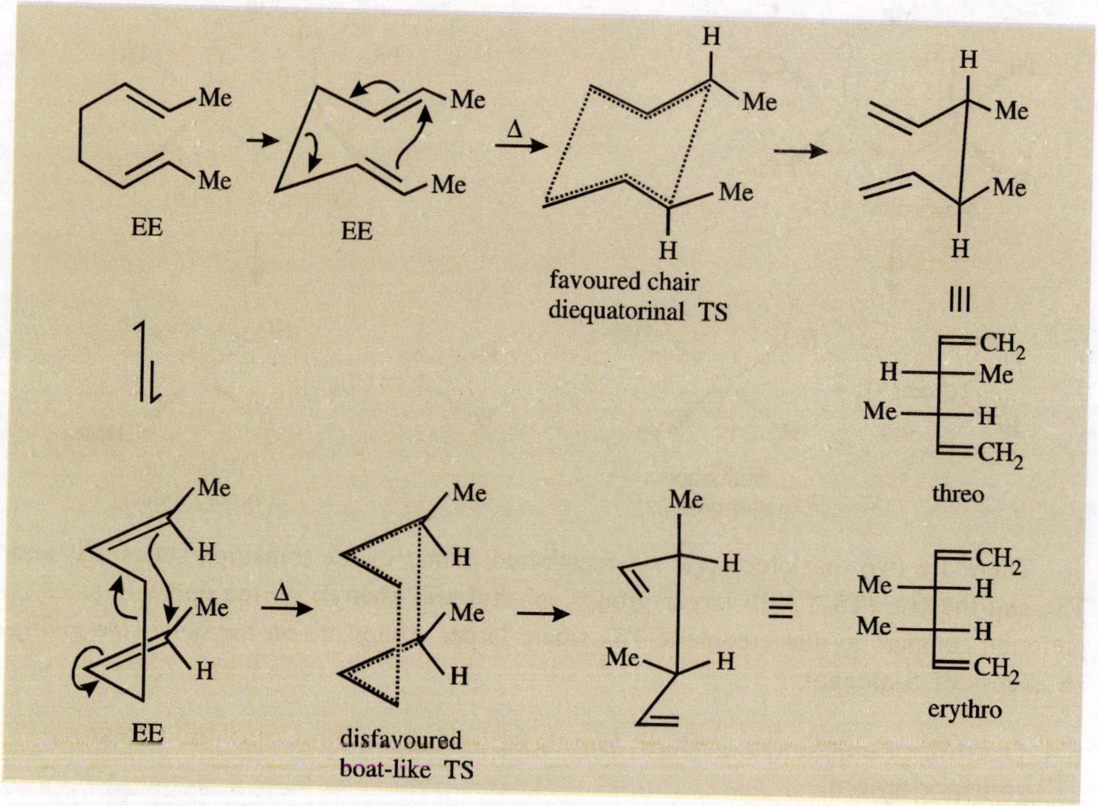

EE → EE $\xrightarrow{\Delta}$ favoured chair diequatorinal TS → threo

EE $\xrightarrow{\Delta}$ disfavoured boat-like TS → erythro

Frontier molecular orbital (FMO) approach

The transition state for the Cope-rearrangement may be seen as the interaction of two allylic systems, one is cation component and the other is anion component, the frontier MO involved are the HOMO of the allyl anion and the LUMO of the allyl cation which are same orbital Ψ_2 of the allyl system. The two FMO involved achieved bonding overlap when [3, 3]-sigmatropic rearrangement occur suprafacially on both the components.

Me Me Me Me Me Me

conrotatory movement

$\ominus$ Me — 4πe component

+

$\oplus$ Me — 2πe component

ψ_2(HOMO of allyl anion)

ψ_2(LUMO of allyl cation)

Suprafacial overlap of both the components

$\pi^4 S + \pi^2 S$

O + 1 = 1, odd number, thermally allowed

diequatorial chair-like TS

active threo

Conrotatory movement gives in phase relation (matching phases) of the two FMO in the transition state and the two methyl groups are in the equatorial relation which is favoured process because of avoiding 1, 3-SYN-diaxial interaction of the two methyl groups with hydrogens if they are in axial orientation during the movement in the transition state.

Claisen rearrangement gives similar stereofeature, as for example, when (E, E)-Crotyl propenyl ether is heated at about 160°C, it gives 97% threo aldehyde and 3% erythro aldehyde.

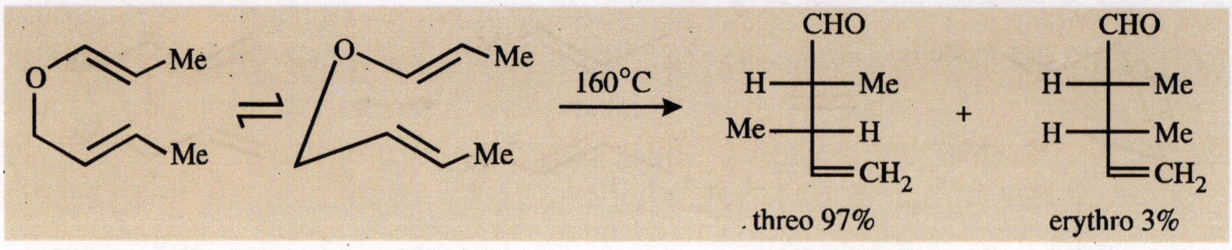

threo 97% erythro 3%

8.5 First order asymmetric synthesis (transformation)

A compound having a labile chiral centre may undergo spontaneous configurational change in solution. If there is a second chiral centre (centres) in the molecule itself, one of the diastereomers predominates in equilibrium which is actually spontaneous epimerisation in solution (same phase). This is known as first order (first kind) asymmetric transformation and is accompanied with a change of optical rotation to a steady equilibrium value—a phenomenon known as mutarotation. The configurational change is known as asymmetric transformation while the change in optical rotation is known as mutarotation.

The following first order asymmetric transformation of glucose may be taken as illustration.

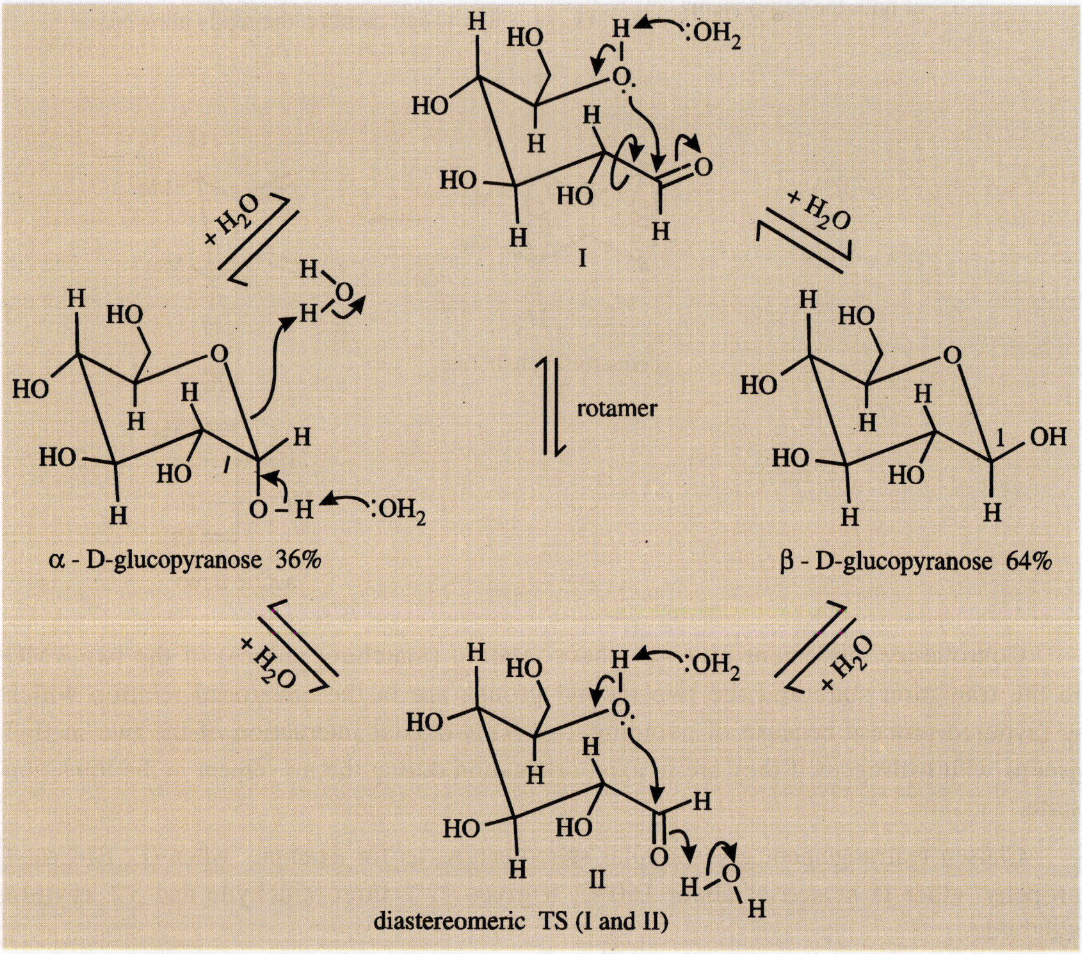

α - D-glucopyranose 36%

β - D-glucopyranose 64%

diastereomeric TS (I and II)

A freshly prepared solution of glucose in water is α-D-glucopyranose gradually changes its specific rotation from the initial value of + 112° to an equilibrium value of +52.5°. This is mutarotation. In this case glucose changes its configuration at the anomeric centre (C—1) under equilibrium condition between α-and β-forms of glucose through the intermediate aldehyde form which has got diastereotopic face because of the chiral part present adjacent to it in the rest of the glucose molecule.

This transformation is ultimately to a mixture of unequal amount of two chiral diastereomers (involving a single phase) where β-form predominates over the α-form (de) without isolation which is first order asymmetric transformation. This transformation is equilibrium controlled as well as thermodynamic controlled since major diastereomer β-form is more stable thermodynamically than α-form where it experiences 1, 3-SYN diaxial interaction due to anomeric axial hydroxyl group.

8.6 Second order asymmetric transformation

After the first order asymmetric transformation where two diastereomers are in equilibrium then if one of the equilibrating diastereomeric species precipitates out of the solution under some specific condition and as a result the entire amount of the substrate (i.e. both the diastereomers) may come out as a single diastereomer. This process is known as second order (second kind) asymmetric transformation.

In the second order asymmetric transformation one of the diastereomers actually crosses the phase boundary i.e. comes out of the solution as crystals. Therefore, transformation of the first kind may be turned into those second kind by changing the condition e.g. with the solvent.

For example, when a solution of glucose in ethanol is concentrated, the less soluble α-form crystalises first and only α-form of glucose is obtained. In the similar way, crystalisation of glucose from pyridine gives the β-form only.

8.7 Review synthetic problems

1. Answer the following questions with proper stereo justification.

(i) Predict the major product of the following and justify

MeCOCOC(Et)(Me)(H) $\xrightarrow[\text{(ii) } H_3O^{\oplus}]{\text{(i) PhMgBr}}$

(ii) How would you obtain Ph (Me)C$=$CH CH_2 CH_2 CHO applying Claisen rearrangement? Draw the preferred stereochemistry of the transition state.

(iii) What is the major stereoisomeric product of the following reaction?

(R) Ph CO CH(Me) CH_2 CH_3 $\xrightarrow[\text{(ii) } H_3O^{\oplus}]{\text{(i) } LiAlH_4}$

(iv) In an enantioselective reaction, the transition state leading to the enantiomers bears diastereoisomeric relationship. Justify or Criticise.

(v) Predict the nature of the product formed when (S)-α-amino benzyl phenyl ketone as its hydrochloride is reacted with excess of para bromophenyl magnesium bromide. Give the configuration of the newly generated chiral centre as R or S.

(vi) How do you correlate the configuration of a chiral α-hydroxy acid with a homochiral alcohol by means of Prelog's rule?

(vii) Predict the major product of the following reaction:

$$\text{(S) } Me_3CH(OMe)-COCH_3 \xrightarrow[\text{(ii) } H_3\overset{\oplus}{O}]{\text{(i) PhMgBr}}$$

(viii) Explain the atrolactic acid rule using (R)-2-butanol as reagent giving outlines of the reactions.

(ix) Predict the major product of the following reaction:

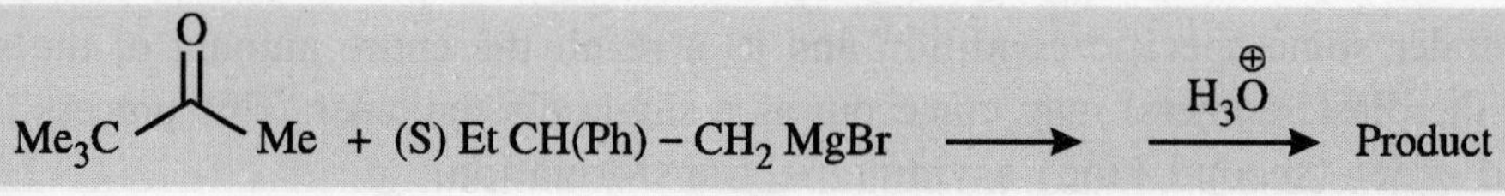

$$Me_3C-CO-Me + \text{(S) } Et\,CH(Ph)-CH_2\,MgBr \longrightarrow \xrightarrow{H_3\overset{\oplus}{O}} \text{Product}$$

(x) Predict the major product of the following reaction showing proper stereochemistry with the Zimmerman and Traxler model.

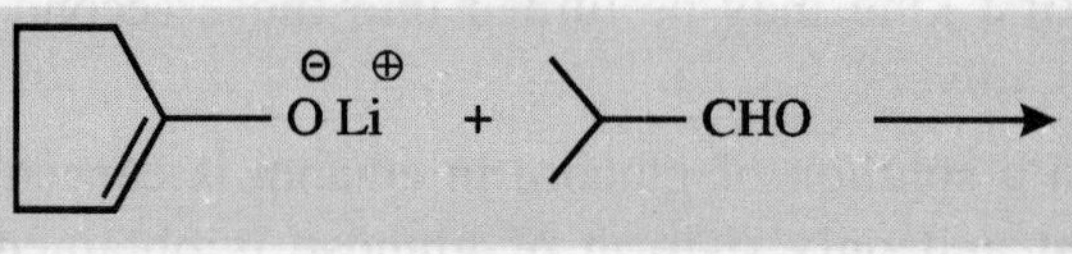

$$\text{(cyclopentenyl)}-\overset{\ominus}{O}\overset{\oplus}{Li} + (CH_3)_2CH-CHO \longrightarrow$$

2. Devise synthesis of the following target molecules below from the starting materials indicated and bearing in mind the need to control regiochemistry.

(i) OH, H, H (structure) from Ethyne

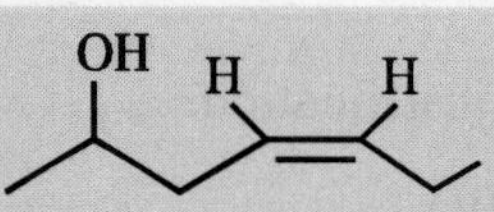

(ii) CH_2 (structure) from O (structure)

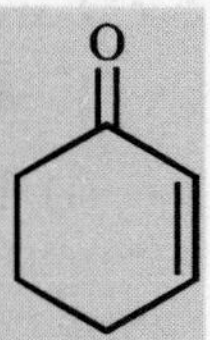

(iii) O (structure) from O (structure)

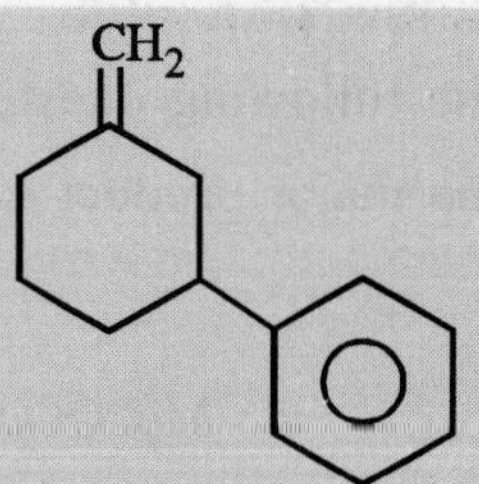

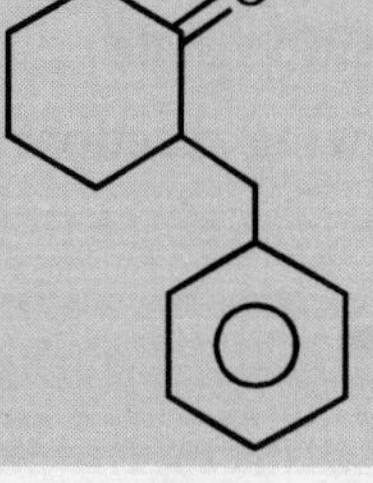

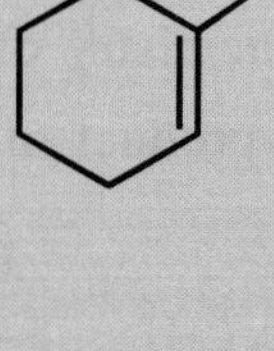

(iv) 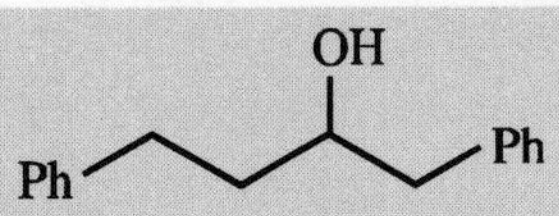 from

(v) 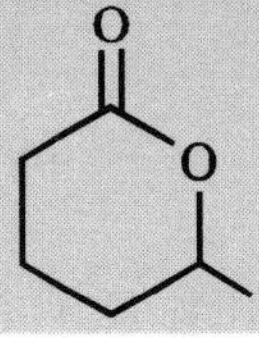 from Adipic acid

3. Using retrosynthetic analysis how can you synthesis the following target molecules taking care of stereocontrol in the synthetic manipulations.

(i) 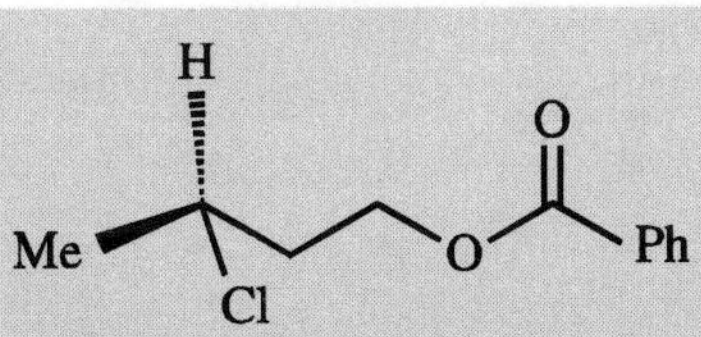 (ii) 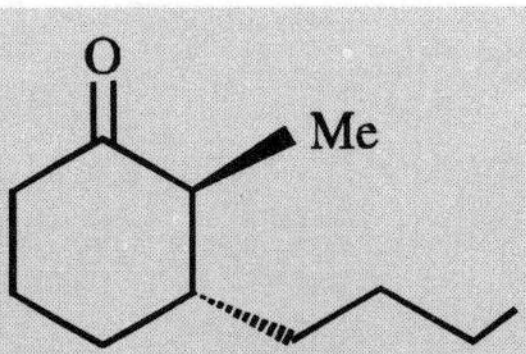 (iii)

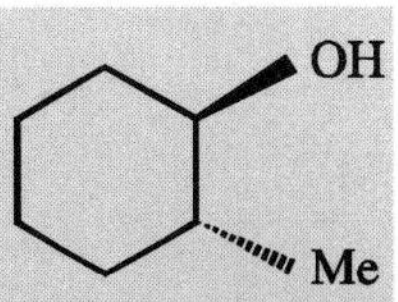

(iv) 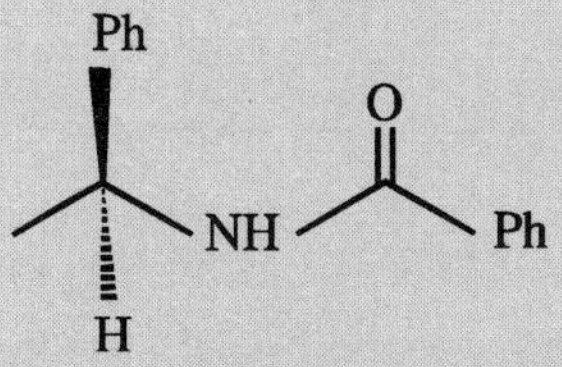 (v) 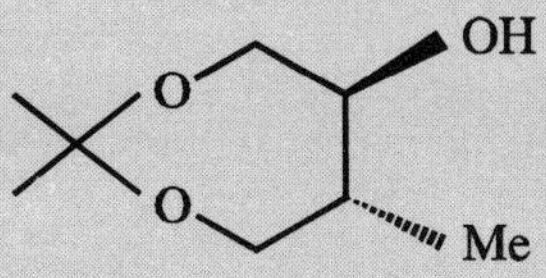 (vi)

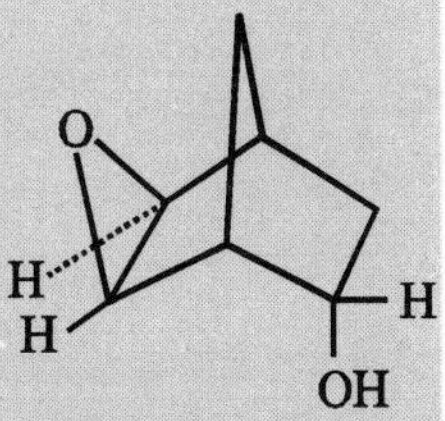

(vii) 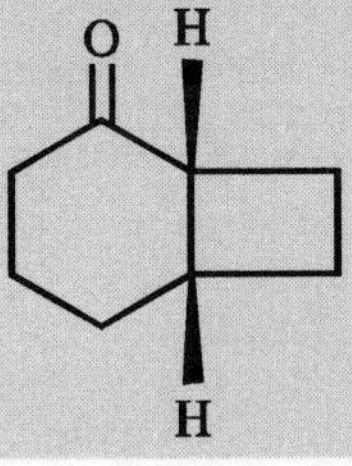 (viii) 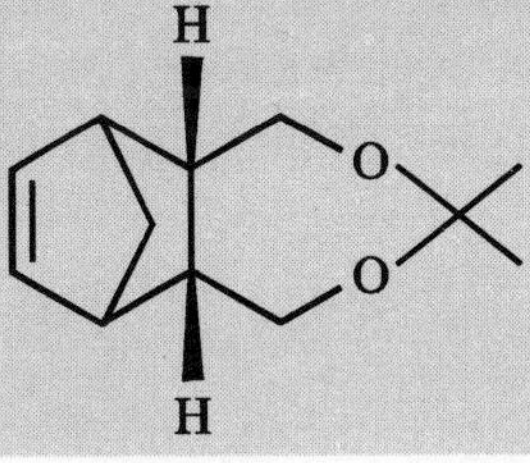 (ix)

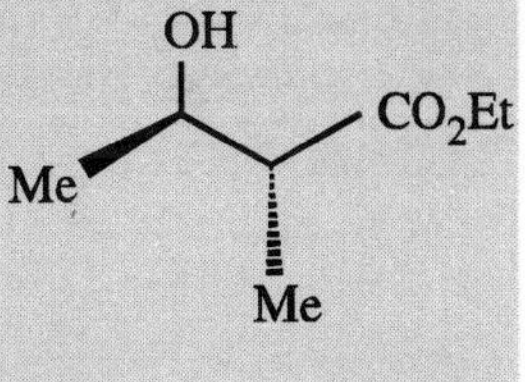

4. What is double asymmetric synthesis? Give predominant diastereomer produced (out of four possible diastereomers) showing the transition state by the reaction of chiral borolane (trans-2, 5-dimethyl borolane, RR or SS) with MeEtC = CHnPr (*Cis* or *trans*) followed by oxidation with alkaline hydrogen peroxide.

Note

9

Synthesis of Dyes

9.1 What is a dye?

A dye is an organic coloured compound, normally used in solution, which is capable of being fixed to a fabric.

For a substance to act as a dye, certain conditions must be fulfilled.

(i) It must have a suitable colour.

(ii) It must, be able to 'fix' itself or be capable of being 'fixed' to the fabric.

(iii) When fixed, it must be fast to light and it must be resistant to the action of water and to a certain extent (the more the better), to dilute acids and alkalis, particularly the latter, on account of the alkaline nature of 'washing soda'.

9.2 Chromophores in dyes

The colour of organic compounds is due to the presence of certain multiple bonded groups called chromophores.

Quinonoid

$-N{=}N-$
Azo

$-\overset{O}{\overset{\|}{C}}-\left[\underset{|}{C}{=}\underset{|}{C}\right]_n-\overset{O}{\overset{\|}{C}}-$
n = 0 or more

It has been noted that the presence of a chromophore is not necessarily sufficient for colour. To make a substance coloured, the chromophore has to be conjugated with an extensive system of alternate single and double bonds as exists in aromatic ring.

9.3 Auxochromes in dyes

Certain groups, which do not itself act as a chromophore but when present along with chromophores in an organic substance, intensify the colour. Such colour assisting groups are called auxochromes.

Acidic auxochromes	– OH	– SO_3H	– COOH
Basic auxochromes	– NH_2	– NHR	– NR_2

9.4 Classification of dyes

Dyes are classified on the basis of:

(i) Structure: The type of chromophores present in their structure which is a theoretical value to the chemist.

(ii) Application to the fibre: The dyer is concerned mainly with the application of dyes to the fabrics i.e. the reaction of dyes towards the fibre being dyed. This depends on the nature of both the dye and the fibre to be dyed. This classification is on the basis of technique employed for their application.

9.5 Structure of different dyes

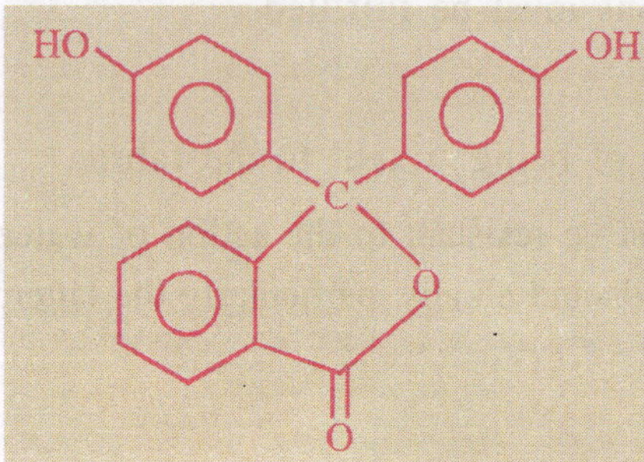

HO OH C O O

Phenolphthalein: Triarymethane dye

Chromophore: 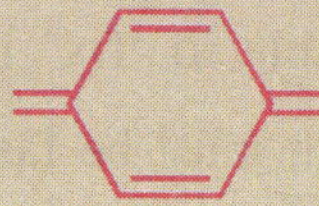 , quinonoid

auxochrome: – OH

HO O OH C O O

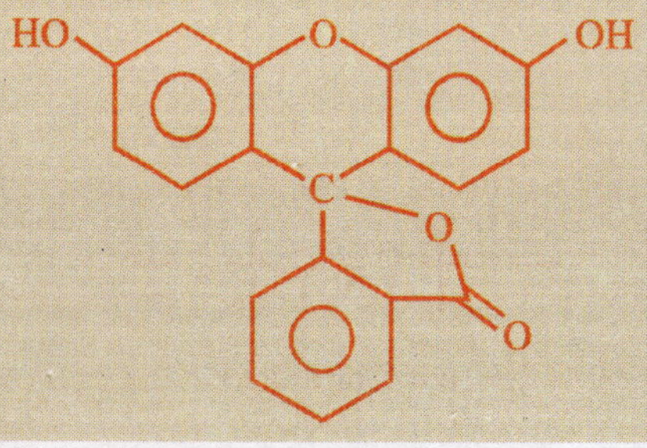

Fluorescein: Xanthen dye

Chromophore: 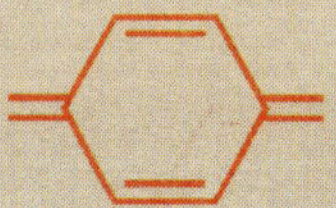 , quinonoid

auxochrome: – OH

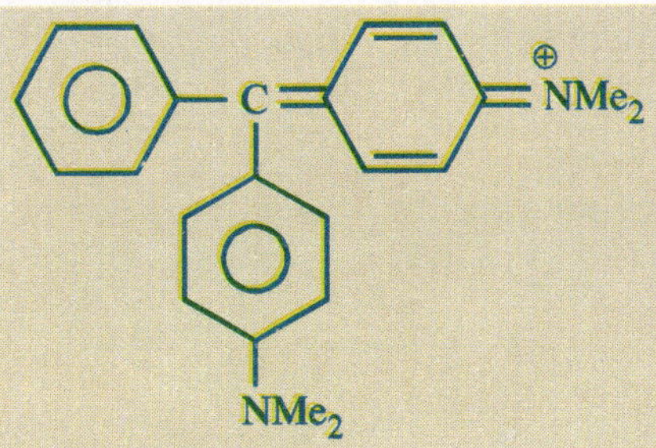

$\oplus$ NMe$_2$

NMe$_2$

Malachite green: Triarylmethane dye

Chromophore: =⟨ ⟩=, quinonoid

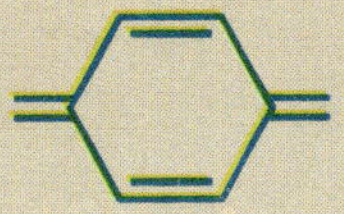

auxochrome: $-NMe_2$

Me_2N — C= — $\oplus$ NMe$_2$

NMe$_2$

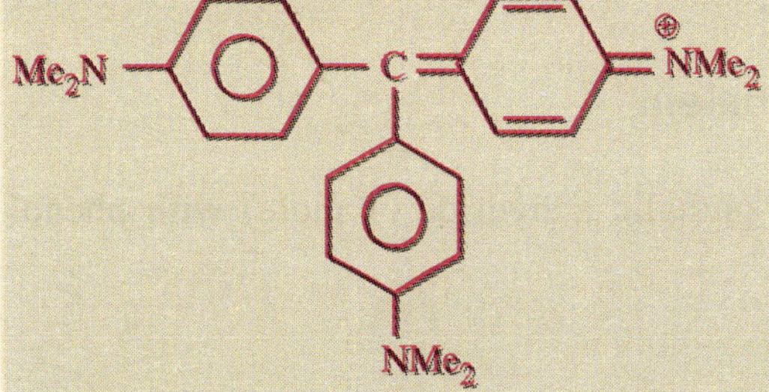

Crystal Violet: Triarylmethane dye

Chromophore: =⟨ ⟩=, quinonoid

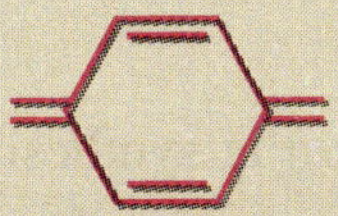

auxochrome: $-NMe_2$

O OH OH

O

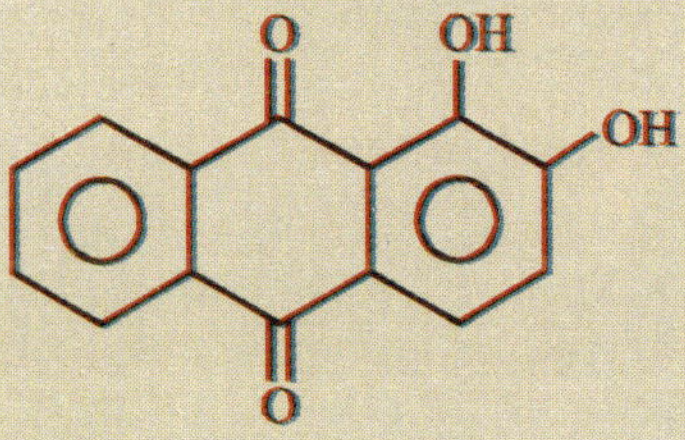

Alizarin: Anthraquinone dye

Chromophore: =⟨ ⟩=, quinonoid

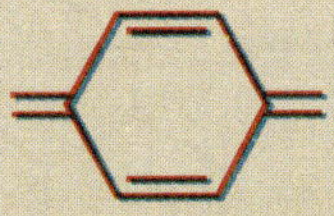

auxochrome: $-OH$

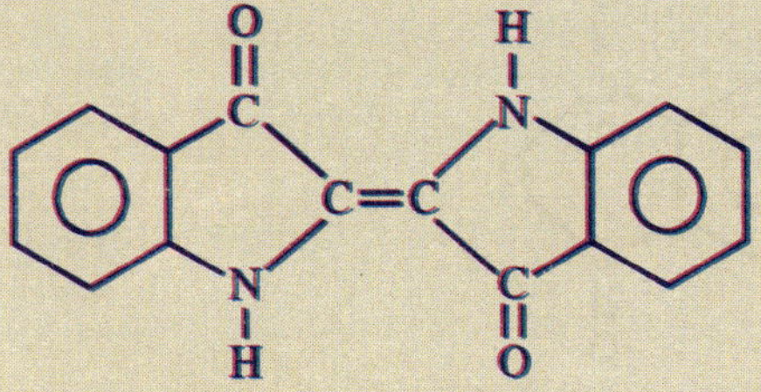

Indigo: Carbonyl chromophore dye

Chromophore: $-\overset{O}{\overset{\|}{C}}-\underset{|}{C}=\overset{|}{C}-\underset{\underset{O}{\|}}{C}-$

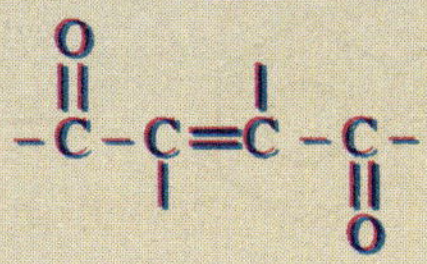

auxochrome: $-NHPh$

HO_3S — N=N — NMe$_2$

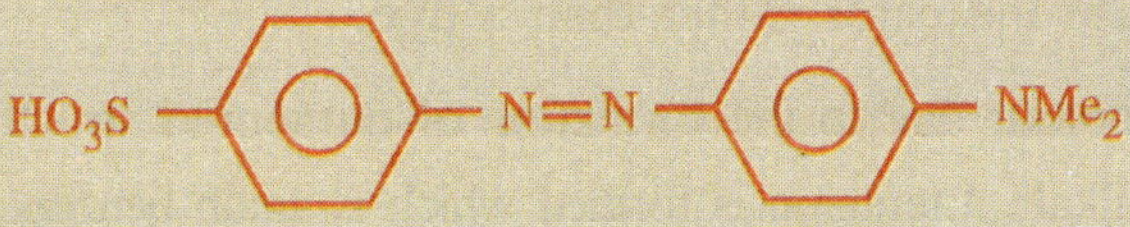

Methyl Orange: Azo dye

Chromophore: $-N{=}N-$

auxochrome: $-NMe_2$, $-SO_3H$

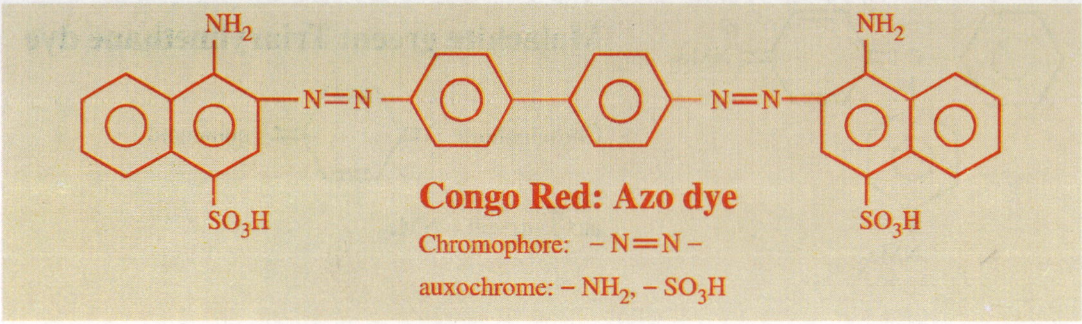

NH_2 N=N N=N NH_2 SO_3H SO_3H

Congo Red: Azo dye

Chromophore: – N=N –

auxochrome: – NH_2, – SO_3H

9.6 Synthesis of dyes and their uses

Phenolphthalein

Phenolphthalein is synthesized by heating phthalic anhydride (1 mole) with phenol (2 moles) in the presence of conc. H_2SO_4.

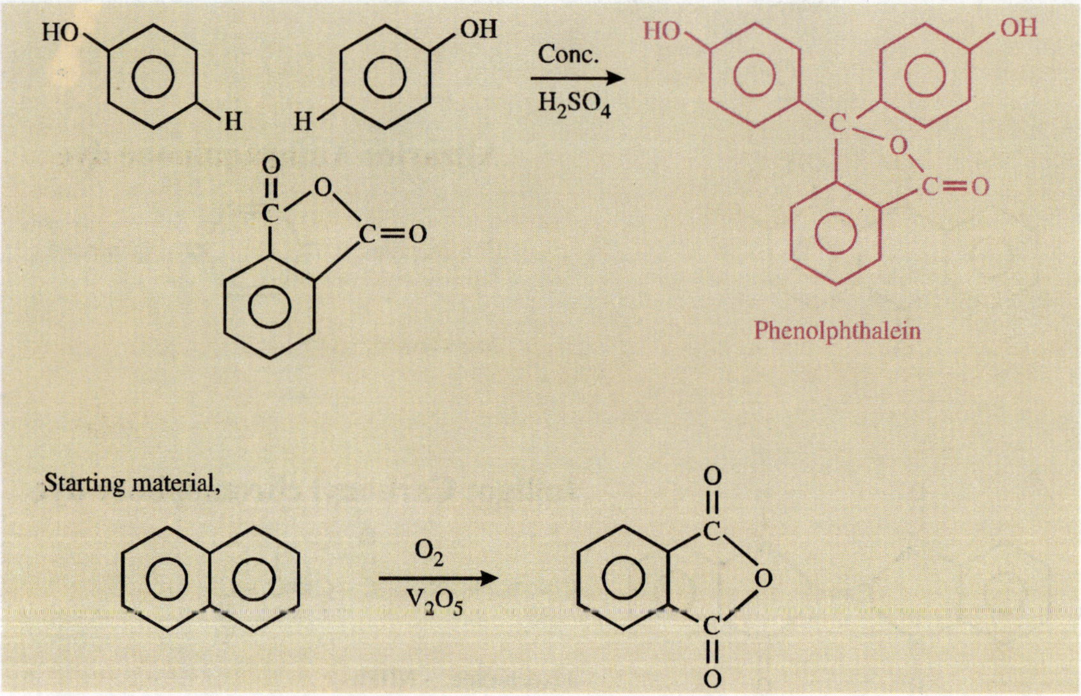

HO H H OH + phthalic anhydride $\xrightarrow[H_2SO_4]{Conc.}$ Phenolphthalein

Starting material,

Naphthalene $\xrightarrow[V_2O_5]{O_2}$ phthalic anhydride

Phenolphthalein is a white crystalline solid, insoluble in water but soluble in alkali and in alcohol. It is used as an acid-base indicator rather than a dye.

In alkali phenolpthalein gives a red colour due to the formation of quinonoid structure (dianion). On addition of a large excess of alkali, a teranion is formed which is again benzenoid. The coloured dianion may easily revert back to colourless benzenoid form in acid. On addition of conc. HCl, the colourless teranion miraculously, gives yellow colour while the dilute acids simply protonate the teranion. The following reactions explain the whole sequence.

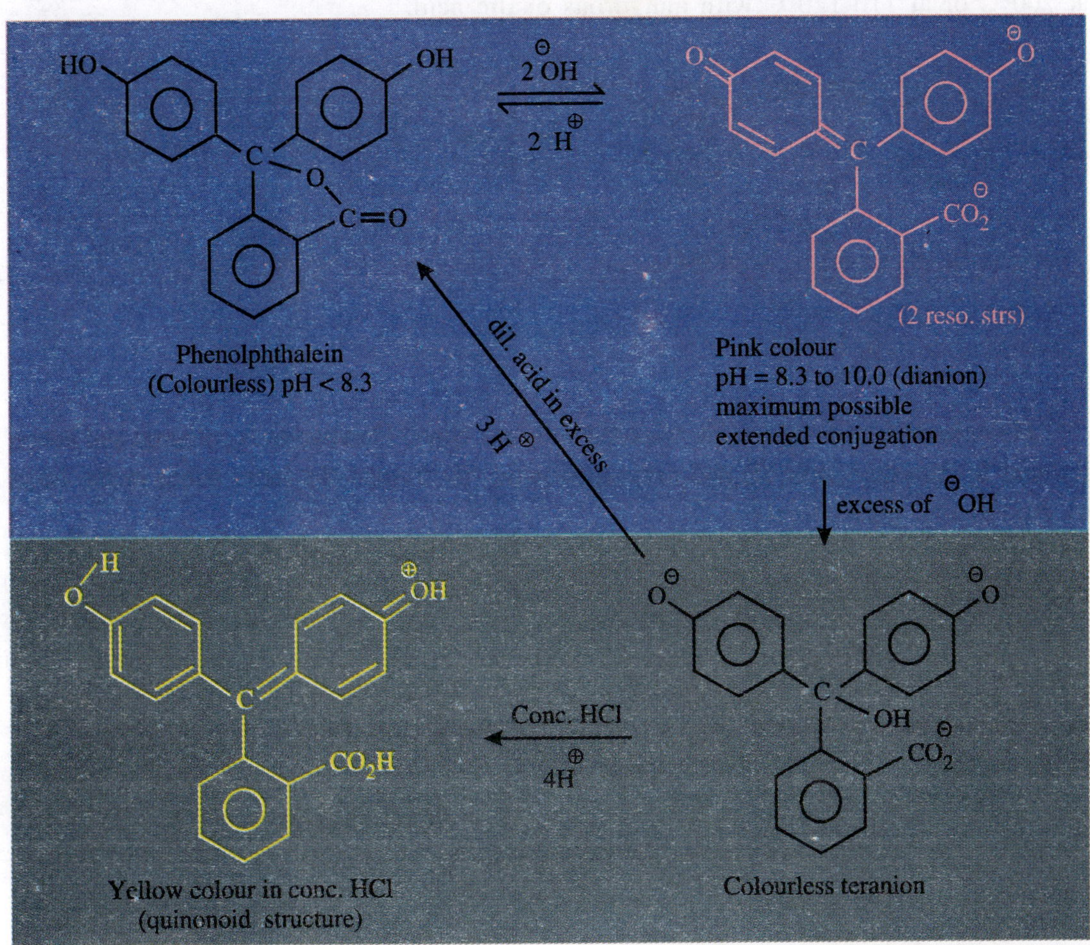

	Change of colour		
$Pk_{Ind.}$	acid	base	pH-range ($Pk_{Ind} \pm 1$)
9.6	colourless	Red	8.3-10.0

Phenolphthalein indicator can be used during the titration of (i) strong acid-strong base, at the equivalence point pH = 7, pH jumps sharply from 4.0 to 10.0, Red to colourless (ii) strong base-weak acid, at the equivalence point, pH > 7, Red to colourless.

This indicator cannot be used during the titration of strong acid-weak base e.g. H_2SO_4 versus Na_2CO_3. At the equivalence point pH < 7. This is not within the range of $Pk_{Ind} \pm 1$.

Fluorescein

Fluorescein is prepared by heating phthalic anhydride (1 mole) and resorcinol (2 moles) at 200°C or at 110-120°C with anhydrous oxalic acid.

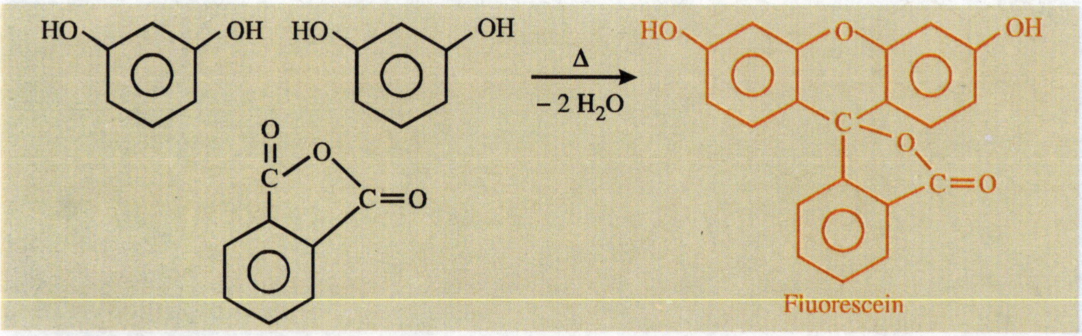

$\xrightarrow[-2\,H_2O]{\Delta}$

Fluorescein

Fluorescein is an orange solid, insoluble in water. Since Fluorescein is coloured and hence it exists in para-quinonoid (I) and ortho-quinonoid (II) forms—the latter being in the form of dipolar-ion, predominates.

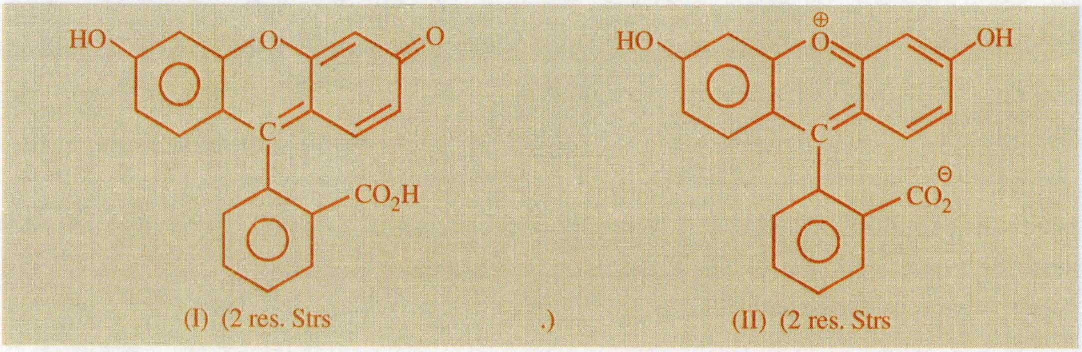

(I) (2 res. Strs .) (II) (2 res. Strs

It dissolves in alkali to a reddish brown solution which on dilution exhibits a strong yellowish-green fluorescence (persists up to 4 × 10^6 times dilution). It dyes wool and silk yellow from the acid bath.

Malachite green

Malachite green is synthesized by condensing N, N-dimethyl aniline (2 moles) with benzaldehyde (1 mole) at 100°C in the presence of conc. H_2SO_4. The leuco-base produced is oxidised with PbO_2 in a solution of acetic acid with a little hydrochloric acid, the resulting colour base (colourless) gives Malachite green with excess hydrochloric acid.

H—C₆H₄—NMe$_2$

C$_6$H$_5$—C(H)=O + H—C$_6$H$_4$—NMe$_2$ $\xrightarrow[H_2SO_4]{Conc.}$ Leuco-base

Benzaldehyde

N, N-dimethyl aniline

Leuco-base

$\langle O \rangle$ $\downarrow$ PbO$_2$/AcOH + a little dil. HCl

Colour base (Colourless)

$\xleftarrow[-\,H_2O]{excess\ HCl}$

$\overset{\oplus}{N}Me_2\overset{\ominus}{Cl}$

Malachite green

Malachite green is a deep green-blue colour. It dyes wool and silk directly and cotton mordanted with tannic acid to deep green. The colour is not too fast in light and fades rapidly in light and because of this reason it is rarely used in textiles. This dye is commonly used for colouring paper, type writer ribbons and for those articles where fastness to light is not of much importance.

It is a powerful antiseptic.

Crystal violet

Crystal violet is synthesized by heating N, N-dimethyl aniline with carbonyl chloride.

(i) O=CCl$_2$ + 2 C$_6$H$_5$—NMe$_2$ $\xrightarrow[-\,2HCl]{\Delta}$ O=C(C$_6$H$_4$NMe$_2$)$_2$

Michler's ketone

(ii)

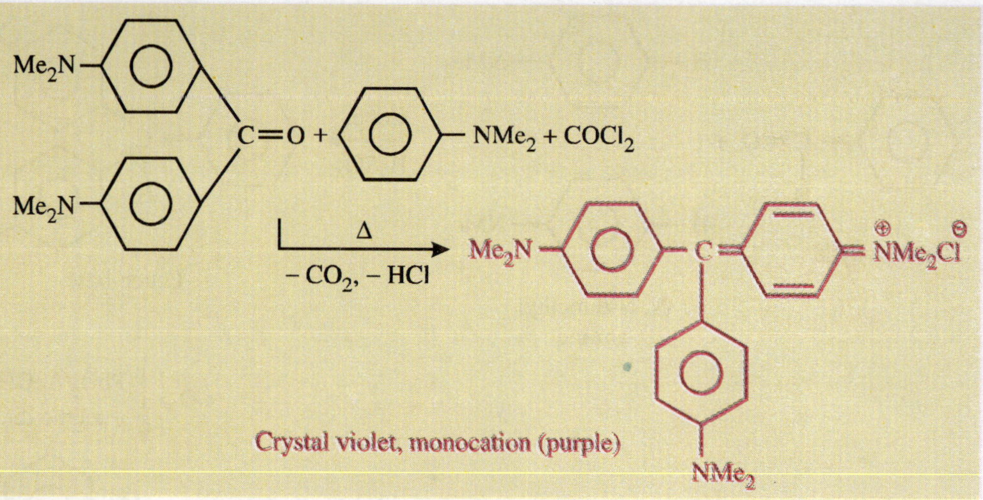

$(Me_2N\text{–}C_6H_4)_2C{=}O + C_6H_5\text{–}NMe_2 + COCl_2 \xrightarrow[-CO_2,\ -HCl]{\Delta}$

Crystal violet, monocation (purple)

Colour of the crystal violet depends on pH, length of the chromophore and charge involved.

(a) In weak acid solution crystal violet is purple and is due to monocation nature and length of the chromophore involving three aryl groups and 75% of the charge is in horizontal direction and 25% is in vertical direction.

(b) In strong acid solution crystal violet is green and is due to dication nature and in this state total charge is in horizontal direction and consequently colour deepens and vertical direction of resonance is inhibited due to lone-pair on nitrogen is protonated.

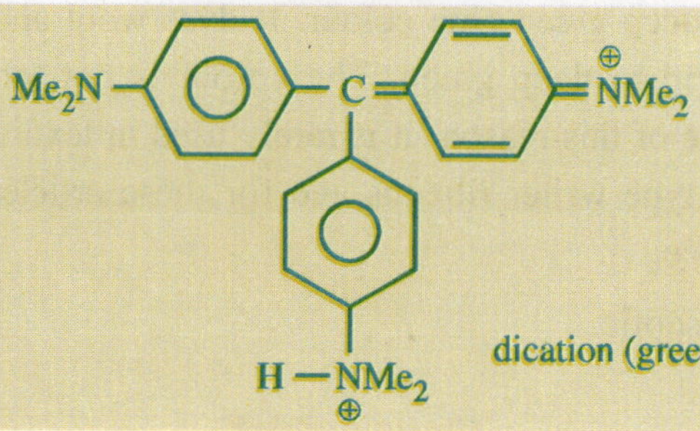

dication (green)

(c) In more strong acid solution crystal violet is yellow and is due to ter-cation nature and in this state no resonance is possible and consequently shortens the length of the chromophore and the colour lightens.

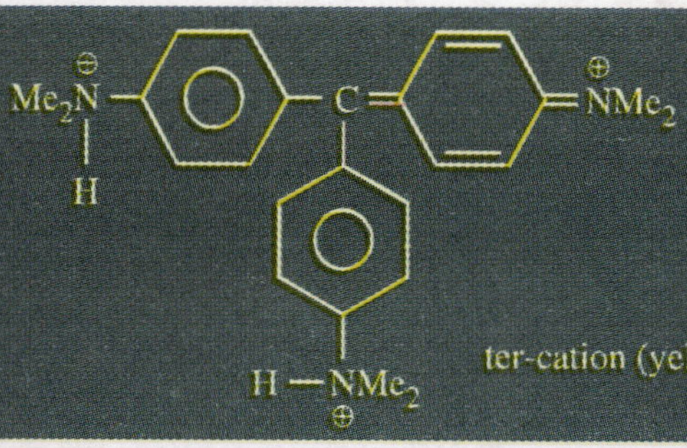

ter-cation (yellow)

Alizarin

Alizarin can be synthesized by the following sequence of reactions.

Phthalic anhydride + benzene $\xrightarrow[\text{Friedel Crafts acylation}]{AlCl_3/\Delta}$ o-benzoylbenzoic acid (COOH) $\xrightarrow{\text{Conc. } H_2SO_4}$ Anthraquinone

Anthracene $\xrightarrow[<O>]{V_2O_5/\text{air}/400^\circ}$ Anthraquinone

Anthraquinone $\xrightarrow{\text{Oleum } 160^\circ}$ Anthraquinone - 2-sulphonic acid (SO_3H) $\xrightarrow[\text{fused}]{NaOH}$ sodium anthraquinone-2-sulphonate (SO_3Na)

$\xrightarrow{\text{Fused with NaOH in presence of calculated } KClO_3 \text{ at } 200^\circ C \text{ under pressure}}$ 1,2-di(ONa) anthraquinone $\xrightarrow{\text{acid}}$ Alizarin (1, 2-dihydroxy anthraquinone)

The function of $KClO_3$ for nucleophilic substitution of hydrogen (to leave as hydride) at 1 – C by nucleophilic $\overset{\ominus}{O}H$ which is facilitated by the presence of an oxidiser $KClO_3$ (to remove hydride).

Alizarin is a ruby-red crystal. It is a mordant dye—the common mordants being metallic hydroxides

Mordants	Colour
Al^{+3}	Turkey red
Fe^{+2}	Violet
Fe^{+3}	Violet black
Cr^{+3}	Brown-violet

Aluminium and iron lakes are usually employed for cotton dyeing and for printing and aluminium and chromium lakes for wood dyeing.

Indigo

Indigo can be synthesized by any of the following sequence of reactions.

1. Anthranilic acid $\xrightarrow[-\,HCl]{ClCH_2COOH/\Delta}$ Phenylglycine – O – carboxylic acid ($NH-CH_2-COOH$, COOH) $\xrightarrow[-\,H_2O]{KOH/NaNH_2/\Delta}$ [Indoxylic acid, unstable] $\xrightarrow{-\,CO_2}$ Indoxyl $\xrightarrow[-\,2H_2O]{O_2(air)}$ Indigo

2. Aniline $\xrightarrow[-HCl]{ClCH_2COOH/\Delta}$ N-phenylglycine (C_6H_5–NH–CH_2–COOH) $\xrightarrow[\text{fused}]{NaOH + KOH/\Delta}$ C_6H_5–N(H)–CH_2–$CO_2^{\ominus}$ {$Na^{\oplus}$, $K^{\oplus}$} $\xrightarrow[\text{NaOH + KOH fused, 240°C}]{NaNH_2}$ Indoxyl $\xrightarrow[-2H_2O]{O_2(\text{air})}$ Indigo

Equimolecular mixture of Na and K-salts of phenyl glycine - the mixed salts requires to lower the temperature and the yield of Indoxyl is better.

At one time it was obtained in large quantities in India from the Indigo plant (Indigofera). Indigofera plants contain Indican which on hydrolysis produces, Indoxyl which on air-oxidation gives Indigo. Indican is the glycoside of Indoxyl.

Indican $\xrightarrow[HCl]{\text{dilute}}$ Indoxyl + α-D-glucose

Indoxyl $\xrightarrow{\text{air, } O_2}$ Indigo

Indigo is a dark-blue amorphous powder with a copper lustre.

Methyl orange

Methyl orange is synthesized by coupling diazotised sulphanilic acid with N, N-dimethyl aniline.

$HO_3S-C_6H_4-NH_2$ (Sulphanilic acid) $\xrightarrow[\text{at } 0^\circ C]{NaNO_2 + HCl}$ $HO_3S-C_6H_4-\overset{\oplus}{N_2}\overset{\ominus}{Cl}$ $\xrightarrow[NaOH]{C_6H_5-NMe_2}$ $\overset{\ominus}{O_3S}-C_6H_4-N{=}N-C_6H_4-NMe_2$

Methyl orange

Synthesis of starting materials.

$C_6H_5-NH_2 + \text{Conc. } H_2SO_4$ (1 :1 by weight) $\xrightarrow[200^\circ C]{\text{baking process}}$ $[C_6H_5-\overset{\oplus}{N}H_3\ H\overset{\ominus}{SO_4}]$ $\rightarrow$ $HO_3S-C_6H_4-NH_2$

$C_6H_5-NH_2 + 2\,CH_3I \xrightarrow[\text{Pressure } CH_3OH, \text{ large excess}]{\text{Conc. } H_2SO_4/230^\circ C} C_6H_5-NMe_2$

Methyl orange imparts orange colour to wool and silk but the colour is not fast to sunlight or washing. That is why it is not used as a dye.

It is a valuable indicator for acid-base titration because it gives yellow colour in basic solution and red colour in acid solution. The change in colour is due to the change in the chromophore of the ion.

$\overset{\ominus}{O_3S}-C_6H_4-N{=}N-C_6H_4-NMe_2$

Azo (yellow)

base ⇌ acid

$\overset{\ominus}{O_3S}-C_6H_4-NH-N{=}C_6H_4{=}\overset{\oplus}{N}Me_2$

Quinonoid (Red)

Pk_{Ind}	Change of colour		pH-range
	acid	base	$Pk_{Ind} \pm 1$
3.7	Red	Yellow	2.7 - 4.7

Methyl orange indicator can be used during the titration of (i) strong acid-strong base e.g. HCl versus NaOH, at the equivalence point pH = 7 and pH jumps sharply from 4 to 10, yellow to Red. (ii) Strong acid-weak base e.g. HCl versus Na_2CO_3 at the equivalence point pH < 7. Yellow to Red.

This indicator cannot be used during the titration of strong base-weak acid e.g. NaOH versus CH_3COOH, at the equivalence point pH > 7. This is not within the range of $Pk_{Ind} \pm 1$ of Methyl orange.

Congo red

Congo red is synthesized by coupling, bis-diazotised benzidine (1 mole) with Naphthionic acid (1-naphthyl-amine-4-sulphonic acid) (2 moles).

NH_2, SO_3H — Naphthionic acid + $\overset{\ominus}{Cl}\,\overset{\oplus}{N_2}$–C₆H₄–C₆H₄–$\overset{\oplus}{N_2}\,\overset{\ominus}{Cl}$ (bis-diazotised benzidine) + Naphthionic acid (NH_2, SO_3H)

↓ coupling

Congo red (NH_2, SO_3H naphthyl–N=N–biphenyl–N=N–naphthyl NH_2, SO_3H)

Synthesis of starting materials

Naphthalene $\xrightarrow[\text{(ii) Sn/HCl, (iii) NaOH}]{\text{(i) } HNO_3 + H_2SO_4}$ 1-naphthylamine (NH_2) $\xrightarrow[200^\circ C]{\text{Conc. } H_2SO_4}$ Naphthionic acid (NH_2, SO_3H)

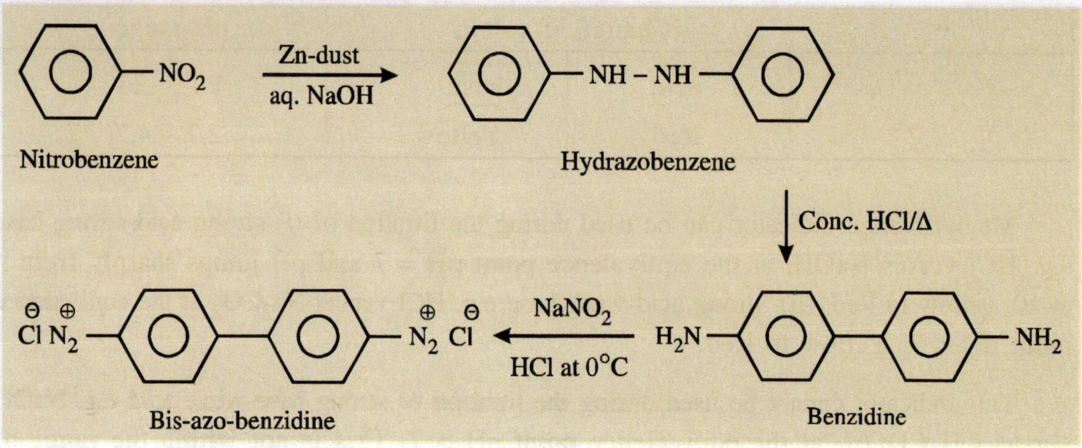

Nitrobenzene $\xrightarrow[\text{aq. NaOH}]{\text{Zn-dust}}$ Hydrazobenzene $\xrightarrow{\text{Conc. HCl}/\Delta}$ Benzidine $\xrightarrow[\text{HCl at } 0^\circ\text{C}]{NaNO_2}$ Bis-azo-benzidine

Congo red is a direct dye and its sodium salt dyes cotton a full red from aqueous solution.

Congo red is also used as an indicator being red in alkali and blue in acid solution.

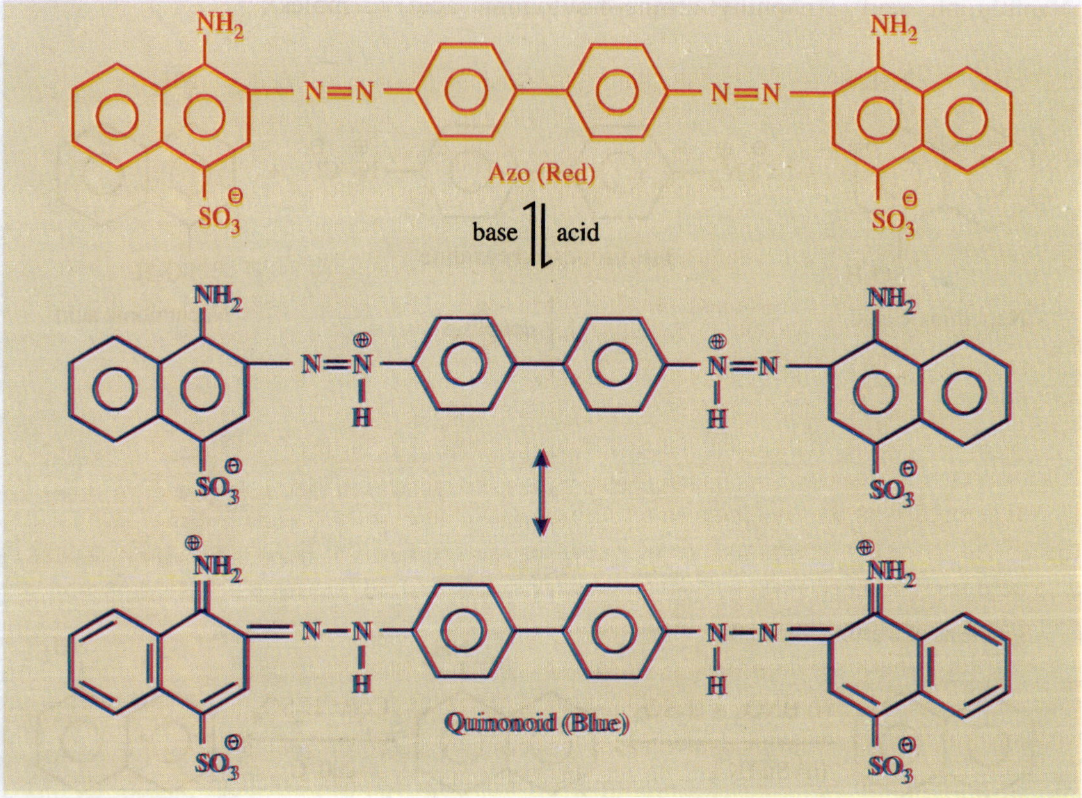

Azo (Red) $\underset{\text{base}}{\overset{\text{acid}}{\rightleftharpoons}}$ Quinonoid (Blue)

10

Synthetic Drugs and Pharmaceuticals

10.1 Structures and uses of drugs

COOH, O, Me, O (structure of aspirin)

Aspirin (Acetyl salicylic acid)

Trade name: Aspro, Anacin, Disprin, Coldarin, Codopyrin.

It is used both as analgesic (pain killer) and antipyretic (fever reducing) agent. It has the anticoagulant property and hence now-a-days used in the heart-attack patients. It is also used for rheumetic fever and in a variety of conditions ranging from headache, discomfort and fever associated with cold and muscular pain.

Aspirin in the gastrointestinal tract, is hydrolysed into salicylic acid and absorb into blood stream. Salicylic acid is a metabolite of aspirin and inhibits the enzymetic activity of prostaglandin. This results in the controlled product of prostaglandin which is responsible for the sensation of pain. The result is that the impulses responsible for pain are supressed. Salicylic acid cannot be used as a drug directly. It is sour and causes severe intestinal bleeding, when taken orally. Aspirin does not cause such bleeding. This is the advantage of aspirin over salicylic acid as a drug. Aspirin is a pro-drug.

Drugs, in the form of plant products have been used from times immemorial. Primitive man discovered appropriate cures through trial and error by chewing plant products such as roots, bark, leaves and fruits, and this information was passed on to subsequent generations by word of mouth.

The bark of the tree "willow" has been used from time immemorial to reduce fever. It contains "Salicin" which is the glycoside of salicylic acid (aglycon part).

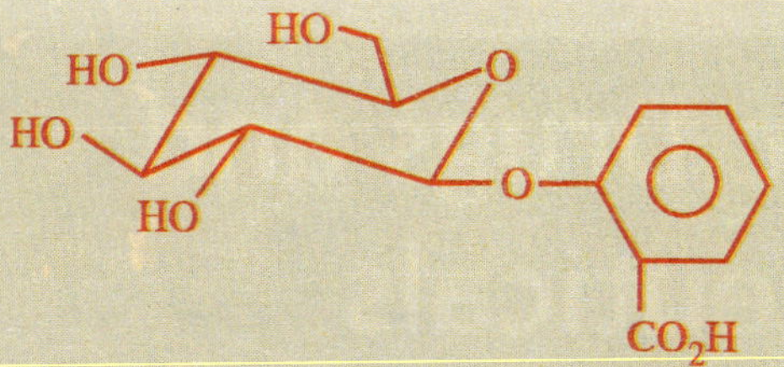

HO HO HO HO O O CO_2H

Salicin (glycoside of salicylic acid)

In the body, salicin is hydrolyzed into salicylic acid, which is responsible for its antipyretic and analgesic activity.

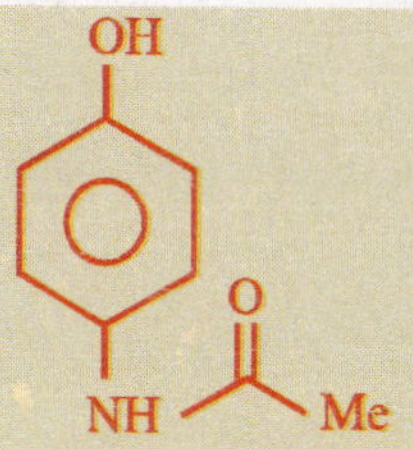

OH O NH Me

Paracitamol (4-hydroxy acetanilide)

Trade name: Crocin, Metacin, Calpal.

Paracitamol is used both as an analgesic and antipyretic agent. It acts on the hypothalamus to reduce temperature—heat dissipation increases as a result of vasodilation and increased peripherial blood flow.

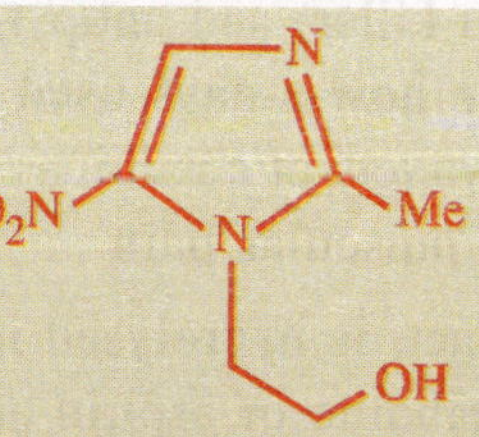

N O_2N N Me OH

Metronidazole

Trade name: Flagyl, Unimezol, Metrogyl.

Metronidazole is an antibiotic. It is used as a drug to cure amoebic dysentery and diarrhoea in the case of bacterial infections.

Metronidazole is targeted at the DNA of the bacteria present inside the human body. This drug breaks the DNA strands (which controls the bacterial cell-wall synthesis, is inhibited, leading to the death of bacteria) of the bacteria, causing bactericidal action.

$$H_2N-C_6H_4-SO_2-NH_2$$

Sulphanilamide (Para aminobenzene sulphonamide)

Trade name: Sulphanilam, Rhinamid.

This drug is used for the cure of bacterial infections in human such as pneumonia and diphtharia. This is a bacterio static drug (bacteria cannot multiply).

$$H_2N-C_6H_4-SO_2-NH-C(=NH)-NH_2$$

Sulphaguanidine

Trade name: Sulphaguanid.

This drug is used in the treatment of bacillary dysentery for the treatment of local gastrointestinal infections.

$$H_2N-C_6H_4-SO_2-NH-C_4H_3N_2$$

Sulphadiazine

Trade name: Codiazine, Coco-diazine, Diazyl.

This drug is used for the cure of bacterial infections.

$$H_2N-C_6H_4-SO_2-NH-C_3H_2NS$$

Sulphathiazole

Trade name: Cibazole, Cerazole, Sulfex.

This drug is used against staphylococcal infections and in bubunic plague.

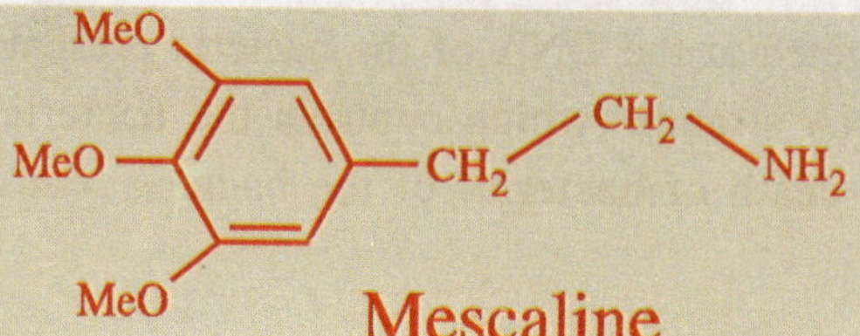

Mescaline

Trade name: Peyote.

This is a Psychotomimetic drug and is used for mental illness. This drug is habit-forming if used repeatedly.

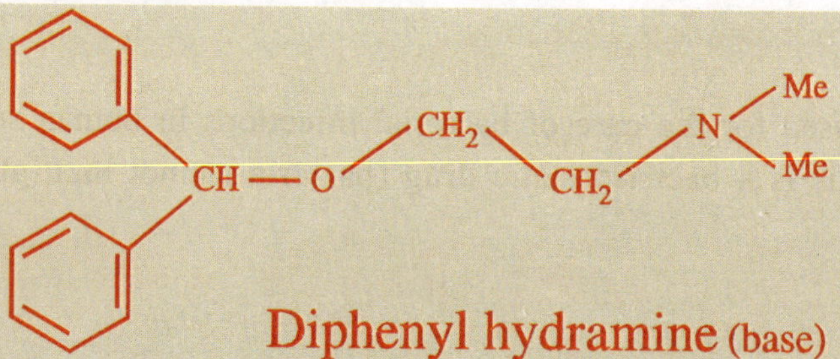

Diphenyl hydramine (base)

Trade name: Benadryl.

It is used as hydrochloride. It possesses sedative and antiemetic properties. It is used as seasonal anti allergic drug. This drug controls cough and is used as cough mixture (syrup).

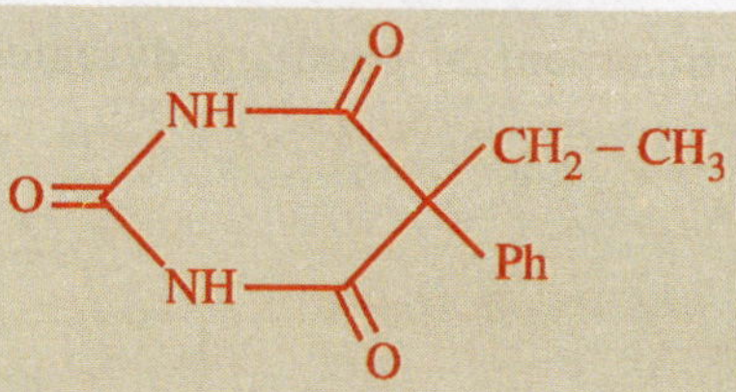

Phenobarbital

Trade name: Luminal, Gardenal.

It is used both as a sedative (low dose) and hypnotic (high dose). It is a sleep producing drug.

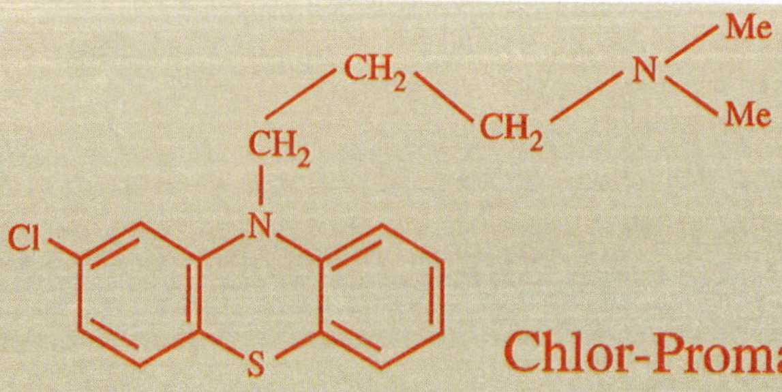

Chlor-Promazine

Trade name: Thorazine.

This drug reduces mental tension, relieve anxiety and mental disorder such as schizophrenia, psychotic depression and mania. It is also used for the prevension of vomiting.

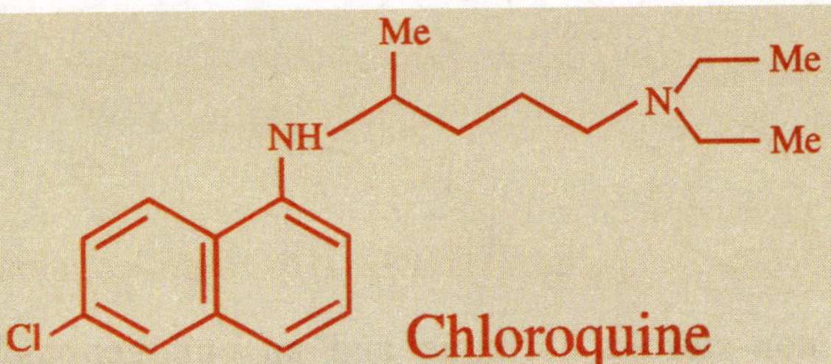

Me NH N Me Me Cl

Chloroquine

It is used as an antimalarial drug. Hydrochloride as injection and sulphate (Nivaquine) as tablets.

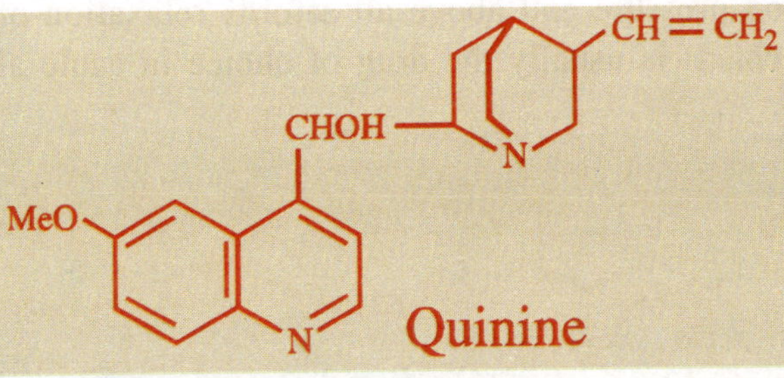

$CH=CH_2$ CHOH N MeO N

Quinine

It is a natural alkaloid found in Cinchona bark. It is used as an antimalarial drug.

O_2N—C₆H₄—CH—CH(NH—CO—$CHCl_2$)—CH_2OH, OH

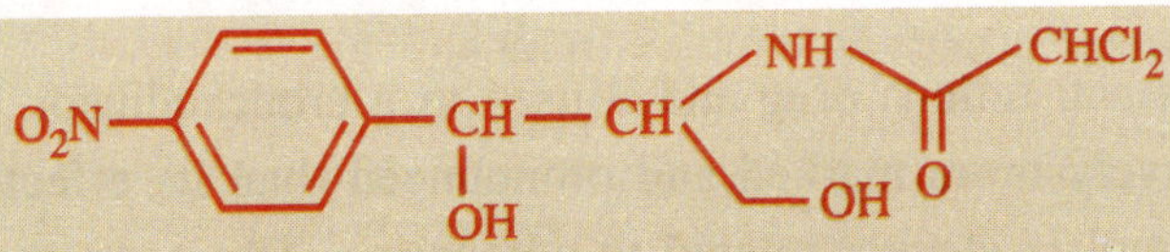

Chloramphenicol [D (–) Threo]

Trade name: Chloromycetin.

It is a broad spectrum antibiotic. It is used in the treatment of Typhoid fever. It also find its enormous applications in topical infections of eye and skin.

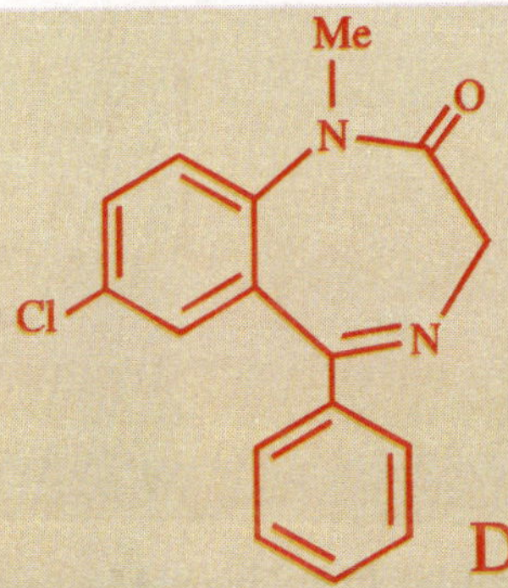

Me N O Cl N

Diazepam/Benzodiazepine

Trade name: Valium.

It is used as an anti anxiety and anti tension drug. It also possesses muscle relaxant properties.

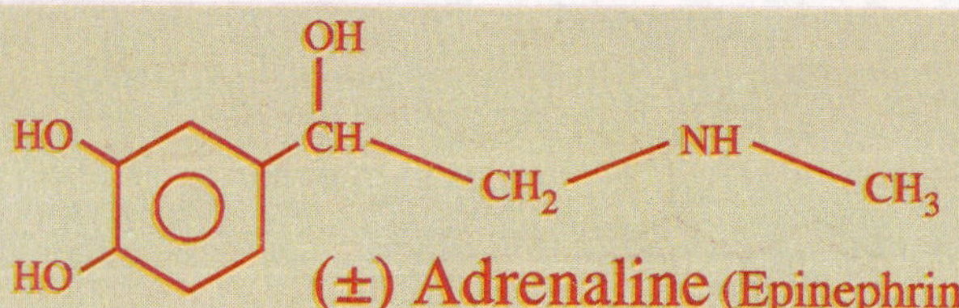

(±) Adrenaline (Epinephrine)

Adrenaline is a non-steroid hormone and an endogenous chemical produces bronchodilation. If there is less adrenaline in the body, it leads to symptoms of asthma. Epinephrine, the sympathomimetic adrenal hormone is found to act on smooth muscles, heart and gland cells. Hence, it invariably used to stimulate the heart, enhances the heart rate, tones up the blood pressure and above all affords relaxation of the musculature of the intestine and bronchi. It is usually the drug of choice in acute allergic disorders and histamine reactions.

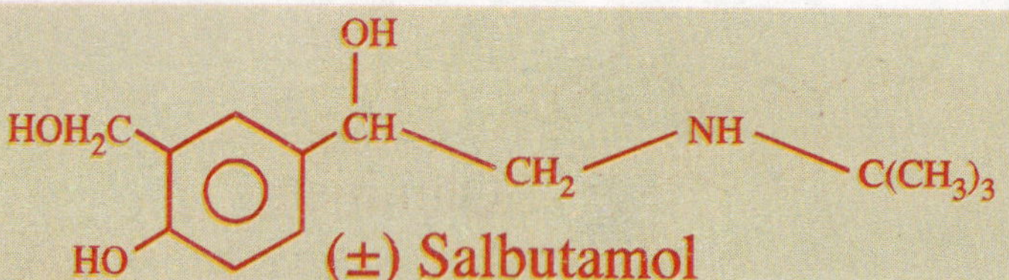

(±) Salbutamol

Trade name: Ventolin.

Salbutamol is an anti asthma drug and is used as a bronchodilator. Its bronchodilating property being relatively more marked and pronounced than its effect on the heart. The asthma drug salbutamol is closely related to adrenaline. The extra carbon atom (benzyl alcohol carbon) prevents dangerous side effects on the heart and the tertiary butyl group makes the drug longer lasting.

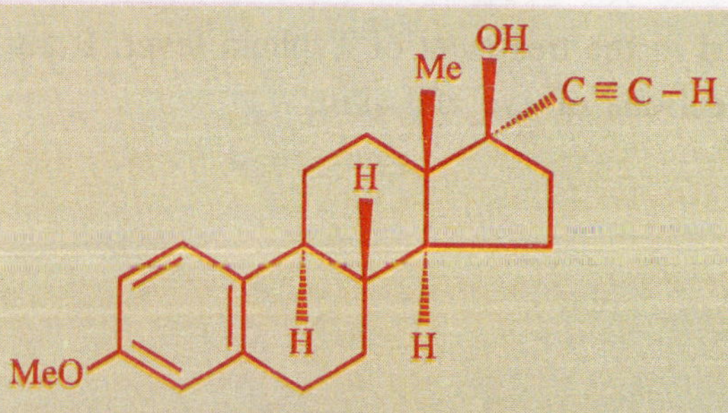

Mestranol (Estrogen) 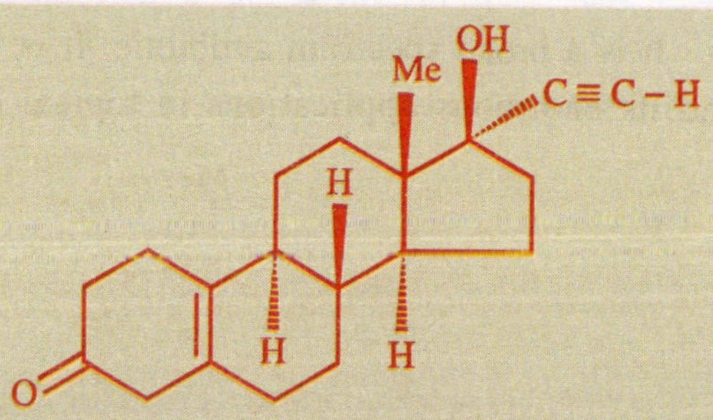 Norethynodrel (Progestrin)

Enovid

Enovid is a combination of two synthetic steroidal hormones, Mestranol (Estrogen) and Norethynodrel (Progestrin). They are used as oral contraceptive or Pills.

Estrogen and Progesterone both are female sex hormones which are secreted under the influence of pituitary gland, mostly by the overies. Estrogen is known to induce reproductive activity during the period.

Estrogen (Mestranol) and Progestrin (Norethynodrel) together control the development of secondary sex characteristic in females. These two combined steroids control the growth and development of vagina, uterus and fallopian tube and also control the female sex characteristics, menstrual cycle and pregnancy. That is why they are used in the prevention of abortion and uterus bleeding.

10.2 Synthesis of drugs

Aspirin

COOH, OH — Salicylic acid $\xrightarrow[50-60^{\circ}C]{Ac_2O\ +\ Conc.\ H_2SO_4}$ COOH, O, O, Me — Aspirin

Synthesis of starting material

$\overset{\ominus}{O}\ \overset{\oplus}{Na}$ $\xrightarrow[120-140^{\circ}C]{CO_2}$ OH, $\overset{\ominus}{CO_2}\ \overset{\oplus}{Na}$ $\xrightarrow{H_3\overset{\oplus}{O}}$ OH, CO_2H

Kolbe-Schmidt reaction

Paracitamol

Method A (starting from parachloro nitrobenzene)

Cl, NO_2 — p-chloronitrobenzene $\xrightarrow[(ii)\ H_3\overset{\oplus}{O}]{(i)\ aq.\ NaOH/\Delta}$ OH, NO_2 $\xrightarrow{H_2/Pd\text{-}C}$ OH, NH_2 $\xrightarrow[0^{\circ}C]{Ac_2O\ +\ H_2SO_4}$ OH, NH, O, Me — Paracitamol

First step is activated aromatic nucleophilic substitution and last step is chemoselective acetylation of amino group. Amino group is more reactive and nucleophilic in acid medium.

Method B (starting from phenol)

Phenol → ($NaNO_2 + H_2SO_4/0°C$, Nitrosation) → p-nitrosophenol (OH, NO) → (aq. NaHS, <H>) → p-aminophenol (OH, NH_2) → (Ac_2O + Conc. H_2SO_4, 0°C) → Paracitamol (OH, NH–C(=O)Me)

Paracitamol

Metronidazole

glyoxal + ammonia (NH_3, NH_3) + acetaldehyde → ($Zn(OH)_2$, Δ) → 2-methylimidazole (N, N–H, Me) → (Conc. HNO_3 + Conc. H_2SO_4) → 2-methyl-5-nitroimidazole (O_2N, N, N–H, Me) → ($ClCH_2CH_2-OH$, Δ, − HCl) → Metronidazole (O_2N, N, Me, N–CH_2CH_2–OH)

Metronidazole

Sulphanilamide

Method A (starting from acetanilide)

$$AcHN{-}C_6H_4 \xrightarrow{ClSO_3H} AcHN{-}C_6H_4{-}SO_2Cl \xrightarrow[40-58^\circ C]{NH_3} AcHN{-}C_6H_4{-}SO_2NH_2 \xrightarrow{H_3\overset{\oplus}{O}} H_2N{-}C_6H_4{-}SO_2NH_2$$

Acetanilide

Sulphanilamide

Method B (starting from sulphanilic acid)

$$H_2N{-}C_6H_4{-}SO_3H \xrightarrow{Ac_2O} AcHN{-}C_6H_4{-}SO_3H \xrightarrow{PCl_5} AcHN{-}C_6H_4{-}SO_2Cl \xrightarrow[40-58^\circ C]{NH_3} AcHN{-}C_6H_4{-}SO_2NH_2 \xrightarrow{H_3\overset{\oplus}{O}} H_2N{-}C_6H_4{-}SO_2NH_2$$

Sulphanilamide

Method C (starting from benzene)

$$C_6H_6 \xrightarrow[Conc.\ H_2SO_4]{Conc.\ HNO_3} O_2N{-}C_6H_5 \xrightarrow{Sn/HCl} H_2N{-}C_6H_5 \xrightarrow[\Delta]{Conc.\ H_2SO_4} H_2N{-}C_6H_4{-}SO_3H \xrightarrow{PCl_5} H_2N{-}C_6H_4{-}SO_2Cl \xrightarrow[40-58^\circ C]{NH_3} H_2N{-}C_6H_4{-}SO_2NH_2$$

Sulphanilamide

Sulphaguanidine

AcHN—C_6H_4 (Acetanilide) $\xrightarrow{ClSO_3H}$ AcHN—C_6H_4—SO_2Cl $\xrightarrow[\text{Guanidine} \; -HCl]{H_2N-C(=NH)NH_2}$ AcHN—C_6H_4—$SO_2-NH-C(=NH)NH_2$

$\xrightarrow[\text{hydrolysis, work up}]{\text{aq. NaOH}}$ H_2N—C_6H_4—$SO_2-NH-C(=NH)NH_2$

Acetanilide

Sulphaguanidine

Sulphadiazine

AcHN—C_6H_4 (Acetanilide) $\xrightarrow{ClSO_3H}$ AcHN—C_6H_4—$SO_2Cl + H_2N$—(2-pyrimidinyl) (2-amino pyrimidine)

$\xrightarrow[\text{Py/– HCl}]{\text{Condensation}}$ AcHN—C_6H_4—SO_2-NH—(2-pyrimidinyl)

$\xrightarrow[\text{work up}]{\text{aq. NaOH}}$ H_2N—C_6H_4—SO_2-NH—(2-pyrimidinyl)

Acetanilide

2-amino pyrimidine

Sulphadiazine

Preparation of starting material 2-amino pyrimidine

Malic acid → (Conc. H_2SO_4, $-CO, -H_2O$) → formyl acetic acid (in situ)

$H_2N-C(=NH)NH_2$ + formyl acetic acid → (Condensation) → 2-amino-4-pyrimidinone ⇌ 2-amino-4-hydroxypyrimidine → ($POCl_3$) → 2-amino-4-chloropyrimidine → (Zn/NH_4OH) → 2-amino pyrimidine

Sulphathiazole

Acetanilide → ($ClSO_3H$) → AcHN–C_6H_4–SO_2Cl + 2-amino thiazole → (Py/– HCl Condensation) → AcHN–C_6H_4–SO_2–NH–(thiazolyl) → (aq. NaOH hydrolysis work up) → H_2N–C_6H_4–SO_2–NH–(thiazolyl) Sulphathiazole

Preparation of starting material, 2-amino thiazole

$CH \equiv CH \xrightarrow[\text{gl. } CH_3COOH]{Hg^{+2}} CH_2{=}CH-OCOCH_3 \xrightarrow[\text{low temp.}]{Cl_2} CH_2(Cl)-CH(Cl)-OCOCH_3$

1, 2-dichloro ethyl acetate

thiourea $\rightleftharpoons$ $H_2N-C(=N-H)-S-H$ + 1, 2-dichloro ethyl acetate

$\xrightarrow{Py/-2HCl}$ (intermediate, H, $OCOCH_3$) $\xrightarrow{-AcOH}$ 2-amino thiazole

Mescaline

3, 4, 5-trimethoxy benzoic acid $\xrightarrow{PCl_5}$ $(MeO)_3C_6H_2-C(=O)-Cl \xrightarrow{CH_2N_2} (MeO)_3C_6H_2-C(=O)-CHN_2 \xrightarrow{NH_3 \text{ in aq. } AgNO_3} (MeO)_3C_6H_2-CH_2-C(=O)-NH_2 \xrightarrow{LiAlH_4} (MeO)_3C_6H_2-CH_2-CH_2-NH_2$

Mescaline

Preparation of starting material, 3, 4, 5-trimethoxy benzoic acid.

HO–C$_6$H$_4$–CO$_2$H (parahydroxy benzoic acid) $\xrightarrow[\text{H}_2\text{SO}_4]{\text{Conc.}}$ HO$_3$S, HO–C$_6$H$_3$–CO$_2$H $\xrightarrow[\text{H}_2\text{SO}_4]{\text{Conc.}}$ HO$_3$S, HO, HO$_3$S–C$_6$H$_2$–CO$_2$H

$\xrightarrow[\text{(ii) } H_3\overset{\oplus}{O}]{\text{(i) KOH fused}}$ HO, HO, HO–C$_6$H$_2$–CO$_2$H (gallic acid)

$\xrightarrow[\text{(ii) } H_3\overset{\oplus}{O} \text{ work up}]{\text{(i) MeI/NaOH}}$ MeO, MeO, MeO–C$_6$H$_2$–CO$_2$H

Diphenyl hydramine

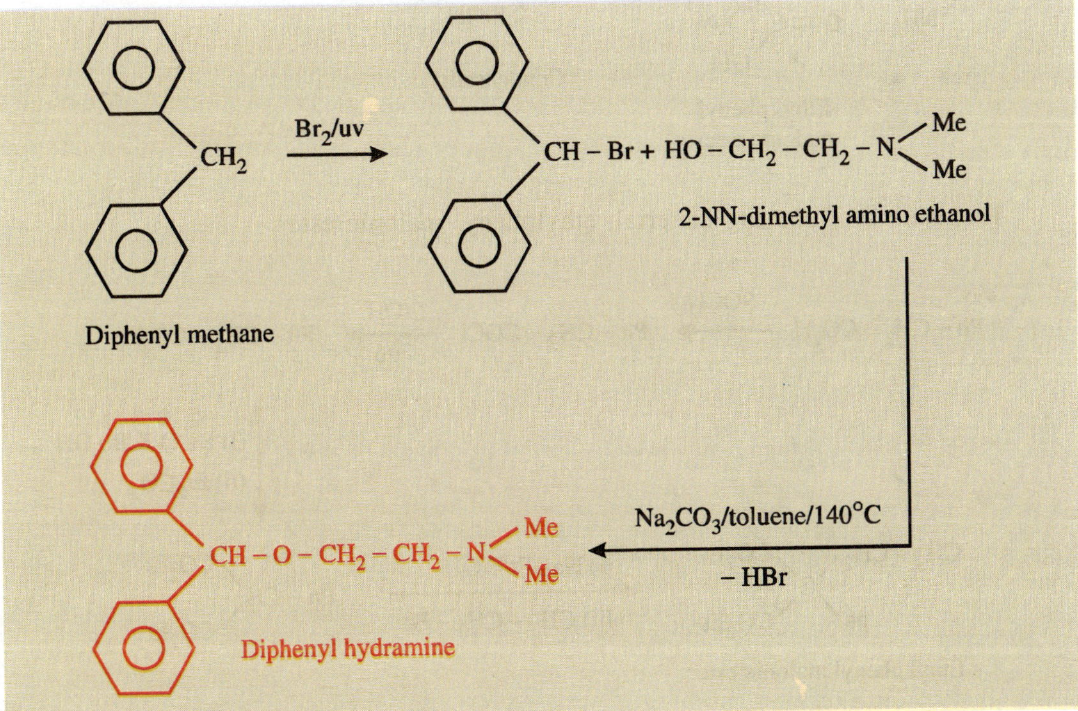

Preparation of starting material, 2-NN-dimethylamino ethanol and Diphenyl methane.

$$\text{(ethylene oxide)} \xrightarrow[\text{MeOH}]{HN(Me)_2} HO-CH_2-CH_2-N(Me)_2$$

2-NN-dimethyl amino ethanol

$$C_6H_6 + C_6H_5-CH_2-Cl \xrightarrow{AlCl_3} C_6H_5-CH_2-C_6H_5$$

Diphenyl methane

Phenobarbital

$$O{=}C(NH_2)_2 + (EtO-CO)_2C(CH_2-CH_3)(Ph) \xrightarrow[-2\,EtOH]{NaOEt/EtOH} \text{Phenobarbital}$$

Urea; Ethyl phenyl malonic ester; Condensation; Phenobarbital

Preparation of starting material, ethylphenyl malonic ester.

$$Ph-CH_2-CO_2H \xrightarrow{SOCl_2} Ph-CH_2-COCl \xrightarrow[Py]{EtOH} Ph-CH_2-CO_2Et$$

$$Ph-CH_2-CO_2Et \xrightarrow{\text{(i) }{}^tBuO^{\ominus}K^{\oplus}/{}^tBuOH\text{ (ii) }Et_2CO_3} Ph-CH(CO_2Et)_2$$

$$Ph-CH(CO_2Et)_2 \xrightarrow{\text{(i) NaOEt/EtOH (ii) }CH_3-CH_2-Br} CH_3-CH_2-C(Ph)(CO_2Et)_2$$

Ethyl phenyl malonic ester

Chlor-Promazine

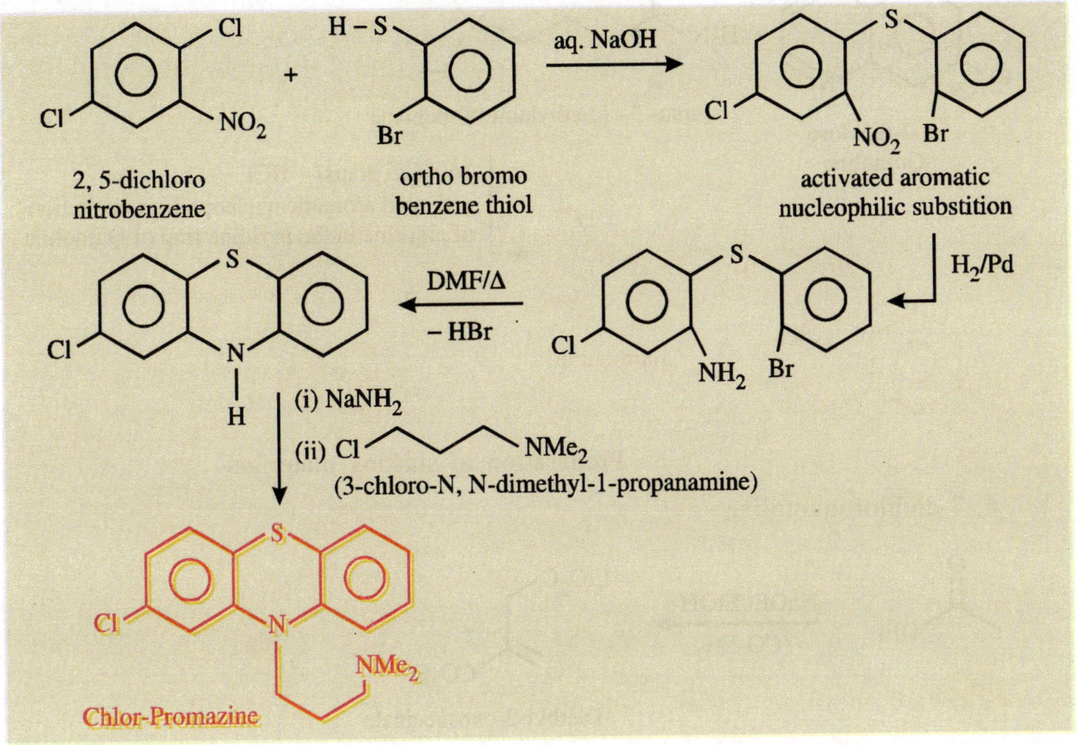

2, 5-dichloro nitrobenzene + ortho bromo benzene thiol —aq. NaOH→ activated aromatic nucleophilic substition

—H_2/Pd→ —DMF/Δ, – HBr→

(i) $NaNH_2$
(ii) Cl–$CH_2CH_2CH_2$–NMe_2
(3-chloro-N, N-dimethyl-1-propanamine)

Chlor-Promazine

Preparation of starting material, 3-chloro-N, N-dimethyl-1-propanamine.

$CH_2=CH-CHO$ —Me_2NH→ $Me_2N-CH_2CH_2-CHO$ —$LiAlH_4$→ $Me_2N-CH_2CH_2CH_2-OH$ —PCl_5→ $Me_2N-CH_2CH_2CH_2-Cl$

OR

$CH_2=CH-CN$ —Me_2NH→ $Me_2N-CH_2CH_2-CN$ —(i) H_2/Pd, (ii) HNO_2→ $Me_2N-CH_2CH_2CH_2-OH$ —$SOCl_2$→ $Me_2N-CH_2CH_2CH_2-Cl$

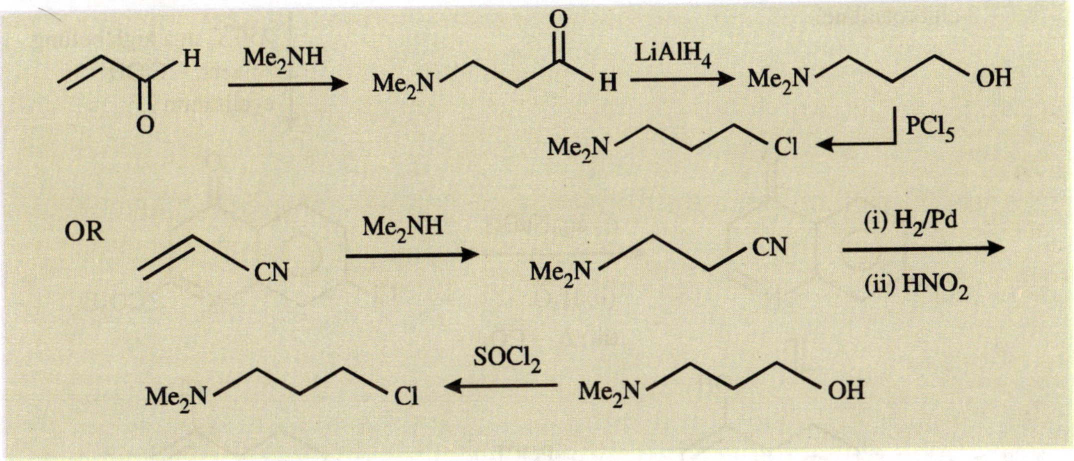

Chloroquine

Cl

Cl N

+ H_2N N

4, 7-dichloro Quinoline

2-amino- 5 - (diethylamino)-pentane

160°C/5 hours/ – HCl
activated aromatic nucleophilic substitution of chlorine in the pyridine ring of Quinoline

HN N

Cl N

Chloroquine

Preparation of starting materials

4, 7-dichloroquinoline

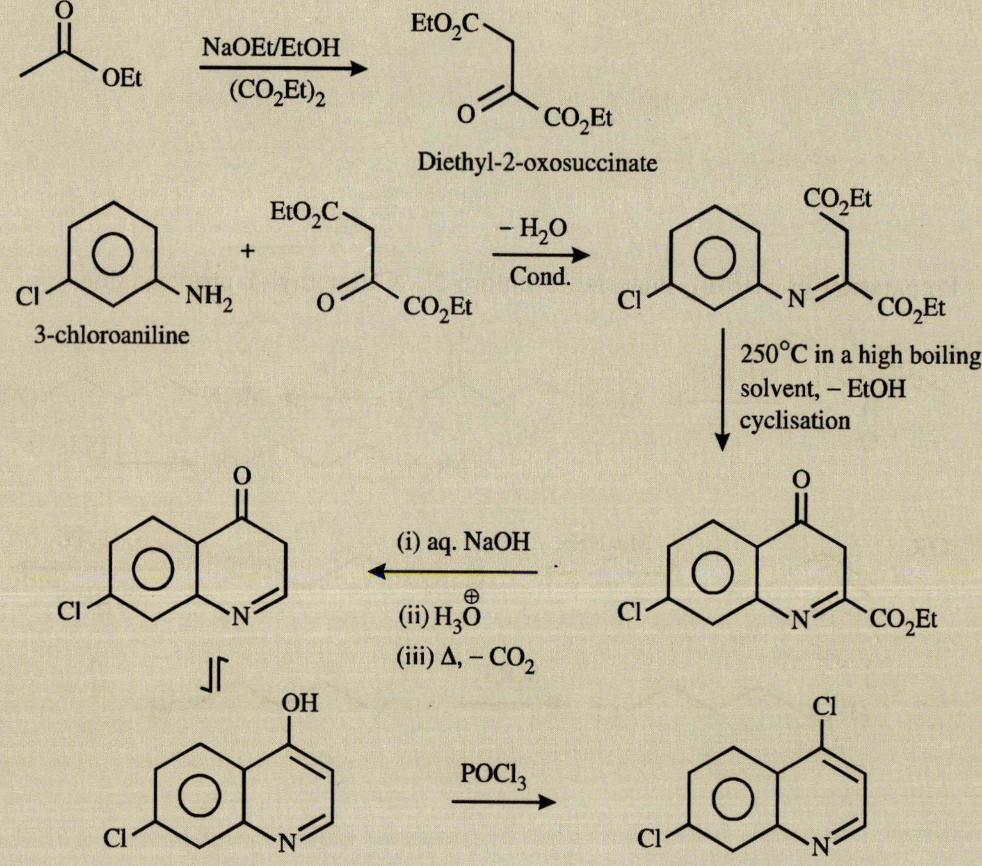

2-amino-5-(diethylamino)-pentane

HN
O
MeOH
HO
N
$SOCl_2$
Cl
N
O
$CO_2Et/NaOEt$
O
N
(i) dil HCl
(ii) 250°C
O
N
CO_2Et
reductive amination
NH_3
H_2 - Raney Ni
H_2N
N

2-amino -5-(diethylamino)-pentane

Chloramphenicol [D (–) threo]

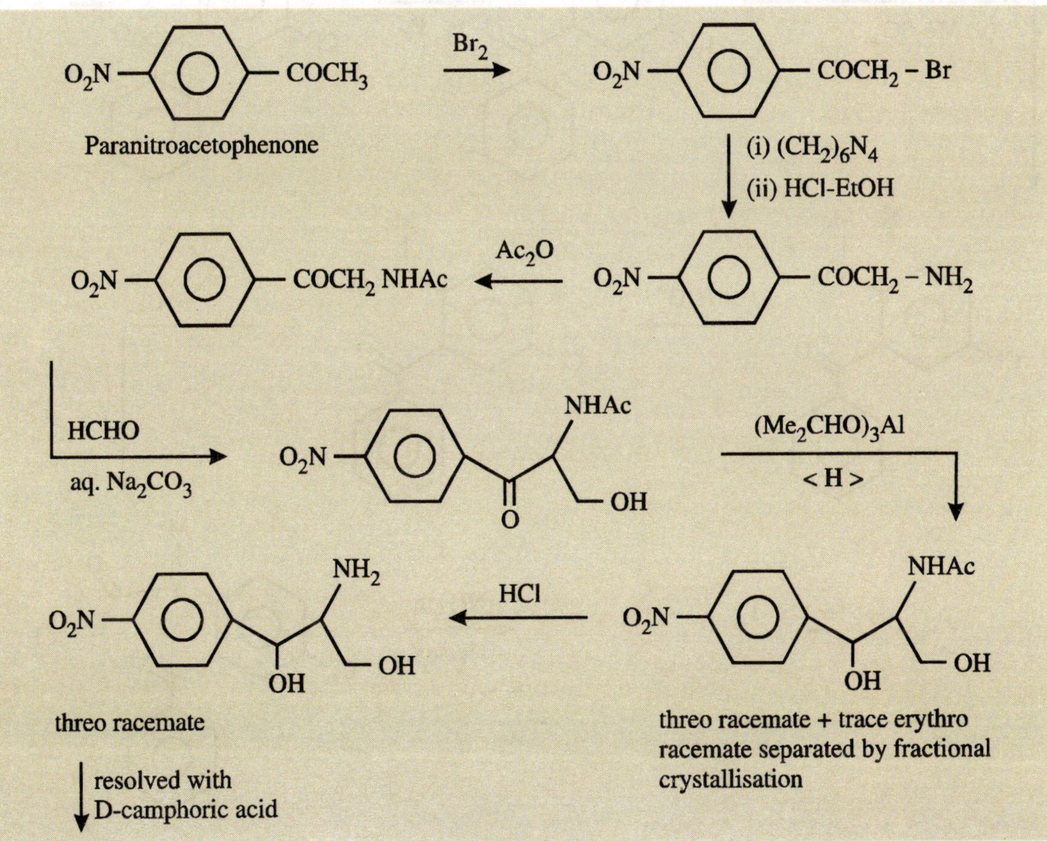

O_2N NH$_2$ OH OH

D(–) threo

OEt O CHCl$_2$

– EtOH

O_2N NH CHCl$_2$ O OH OH

Chloramphenicol/D(–) threo

Diazepam

NHAc Cl

Para chloro acetanilide

\+ O Cl

$AlCl_3$ or $ZnCl_2$

Friedel-Crafts reaction

NHAc Cl O

$H_3O^{\oplus}$

NH$_2$ Cl O

TsCl

Py

NHTs Cl O

(i) 40% NaOH in C_6H_6

(ii) Me_2SO_4

Me NTs Cl O

$H_3O^{\oplus}$

Me N – H Cl O

O Cl Cl

Me N O Cl Cl O

NH_3 in alcohol

Me N O Cl N

Diazepam

Adrenaline

HO, HO catechol + Cl–CH_2–COOH $\xrightarrow[-H_2O]{POCl_3}$ HO, HO C(=O) CH_2 Cl

– HCl | Ph–CH_2–NH–CH_3 /Py

HO, HO C(=O) CH_2 N(CH_2 – Ph)(CH_3)

$LiAlH_4$

HO, HO CH(OH) CH_2 N(CH_2 – Ph)(CH_3) (±)

H_2 / Pd -C

HO, HO CH(OH) CH_2 NH CH_3 (±)

Adrenaline

Salbutamol

Method A

HO_2C, HO Salicylic acid $\xrightarrow{Ac_2O}$ HO_2C, AcO Aspirin $\xrightarrow[\text{Fries rearr.}]{AlCl_3/\Delta}$ HO_2C, HO C(=O) CH_3

$Br_2/CHCl_3$

HO_2C, HO C(=O) CH_2 Br

Ph–CH_2–NH – $C(CH_3)_3$/Py

– HBr

(i) $LiAlH_4$ (ii) $H_3\overset{\oplus}{O}$

(±)

$H_2/Pd-C$

(±)

Salbutamol

Method B

Para hydroxy benzaldehyde

(i) $\overset{\ominus}{O}H/CH_2O$

(ii) $H_3\overset{\oplus}{O}$

HCl

$Me_3\overset{\oplus}{S}\overset{\ominus}{I}$ + NaH | $Me_2S = CH_2$/DMSO

(±)

(i) $H_2N-C(CH_3)_3$

(ii) $H_3\overset{\oplus}{O}$

(±)

Salbutamol

Enovid (Mestranol and Norethynodrel)

Methyl ether of Estrone $\xrightarrow[H-C\equiv C-H]{K/liq\ NH_3}$ Mestranol

Methyl ether of Estrone $\xrightarrow{LiAlH_4}$ $\xrightarrow{Li/liq\ NH_3}$ $\xrightarrow{CrO_3}$ $\xrightarrow[H-C\equiv C-H]{K/liq\ NH_3}$ $\xrightarrow{HCl}$ Norethynodrel

Me, O, OMe, MeO, OH, H, $C\equiv C-H$

Methyl ether of Estrone

Mestranol

Norethynodrel

Preparation of starting material, Methyl ether of Estrone.

MeO, CH_2-Br
$H-C\equiv CNa$
DMF
H C ≡ C
MeO
Et_2NH CH_2O
NEt_2
MeO
Hg^{+2}
H_2SO_4
O
MeO
Me O O
$^{\ominus}OH$
Me O
O O
MeO
TsOH
Me O
MeO
H_2-Ni
Me O
H
MeO
$K/liq\ NH_3/NH_4Cl$
OH Me
H H H
MeO
CrO_3
Me O
H H H
MeO
Methyl ether of Estrone

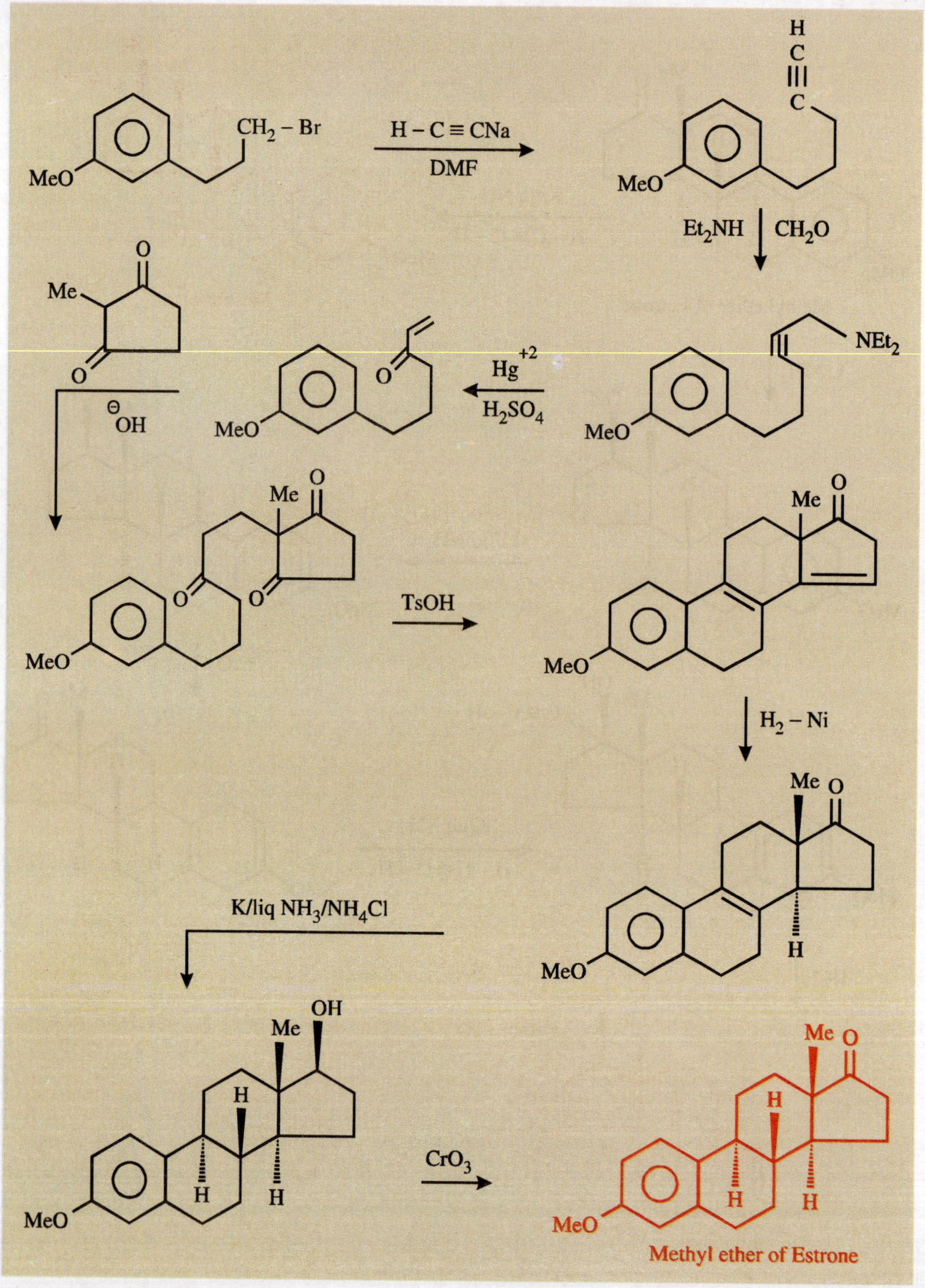

11

Synthesis of Polymers

11.1 What is a polymer?

A polymer is a large molecule made up of a repeating sequence of smaller units called monomers.

Polymers which are synthesized from one kind of monomers are called homopolymers and the polymers which are synthesized from more than one kind of monomers to give a single polymer are called co-polymers.

Synthetic polymers are classified as chain-growth polymers (addition polymers) as for examples polyethene, polyvinyl chloride (PVC), polystyrene, orlon, teflon and poly (methyl methacrylate) and step-growth polymers (condensation polymers) as for example, Terylene (polyester), Nylon (polyamide) or thermosetting polymers (resins), e.g., Bakelite which is a phenol-formaldehyde resin.

Polymers are long chain, large surface area and many conformations. The polymer rubber (elastomer) has a flexible chain, low cohesive force and moderate ease of packing into a crystal lattice and the polymer fibre has a rigid chain, strong cohesive force and ease of packing into crystal lattice and the polymer plastic has a flexible to tough chain.

Synthetic polymers have found a variety of important applications as for example, synthetic fibre for clothing, synthetic rubber for automobile tyres and plastics for making utencils, toys, toothbrushes and many other applications e.g. adhesive, paints, etc.

Natural polymer, rubber is a (Z)-polyisoprene and Gutta-percha is another polymer of (E)-polyisoprene.

Natural fibres, Cotton (consists of Cellulose), jute (consists of ligno Cellulose) i.e. cellulose is the main constituent of most natural fibres. The cellulose again is a condensed high polymer of glucose molecules. Starch is also a polymer of glucose molecules.

Silk on the otherhand contain fibrion which is made up of various proteins—all proteins again are condensed high polymers of amino acids (polypeptides). Wool is also a polymerised amino acids.

DNA and RNA are natural polymers made up of sugars, heterocyclic nitrogen bases and phosphate groups.

11.2 Synthesis of polymers

11.2.1 Addition polymers (chain-growth polymers)

Addition polymerisation are those obtained from monomers containing carbon-carbon double bond. The process is referred to a vinyl polymerisation. A few examples are cited below.

$$nCH_2{=}CH_2 \longrightarrow \left[CH_2-CH_2\right]_n$$

Polythene

$$nCF_2{=}CF(F) \longrightarrow \left[CF_2-CF(F)\right]_n$$

Teflon

$$nCH_2{=}CH(Ph) \longrightarrow \left[CH_2-CH(Ph)\right]_n$$

Polystyrene

$$nCH_2{=}CH(CN) \longrightarrow \left[CH_2-CH(CN)\right]_n$$

Orlon

$$nCH_2{=}CH(Cl) \longrightarrow \left[CH_2-CH(Cl)\right]_n$$

PVC

$$nCH_2{=}CH(OCOCH_3) \longrightarrow \left[CH_2-CH(OCOCH_3)\right]_n$$

Polyvinyl acetate (PVA)

$$nCH_2{=}C(CH_3)(CO_2Me) \longrightarrow \left[CH_2-C(CH_3)(CO_2Me)\right]_n$$

Poly(methyl methacrylate)

$$nCH_2{=}CH(Ph) + nCH_2{=}C(Me)(CO_2Me) \longrightarrow \left[CH_2-CH(Ph)-CH_2-C(Me)(CO_2Me)\right]_n$$

Polystyrene co-polymer

The reactions involved in the formation of these polymers may proceed by free radical, cationic or anionic mechanism.

Free radical vinyl polymerisation

Vinyl polymerisation is a chain reaction proceeding in three distinct steps, initiation, propagation and termination.

Initiation: It involves the formation of a free radical from a radical initiator such as benzoyl peroxide (I) or azobisisobutyronitrile (II).

$$\mathrm{Me_2C(CN)-N{=}N-C(CN)Me_2}\ \text{(II)} \xrightarrow{\Delta} \mathrm{2Me_2\dot{C}(CN)} + N_2$$

$$\mathrm{PhC(O)O{-}OC(O)Ph}\ \text{(I)} \xrightarrow{\Delta} \mathrm{2Ph\cdot} + 2CO_2$$

$$\mathrm{R\cdot} + \mathrm{CH_2{=}CHX} \longrightarrow \mathrm{RCH_2\dot{C}HX}$$

Propagation:

$$\mathrm{RCH_2\dot{C}HX} + \mathrm{CH_2{=}CHX} \longrightarrow \mathrm{RCH_2CHXCH_2\dot{C}HX}$$

$$\mathrm{RCH_2CHXCH_2\dot{C}HX} + \mathrm{CH_2{=}CHX} \longrightarrow \mathrm{RCH_2CHXCH_2CHXCH_2\dot{C}HX} \longrightarrow \text{etc.}$$

R = Ph· or $\mathrm{Me_2\dot{C}(CN)}$

X = H	Polyethene	X = CN	Orlon
X = Cl	PVC	X = $OCOCH_3$	PVA
X = Ph	Polystyrene		

Termination: Termination of kinetic chains usually occurs by radical coupling (combination) or disproportionation as shown below.

$$\mathrm{R[CH_2CHX]_m CH_2\dot{C}HX} + \mathrm{R[CH_2CHX]_n CH_2\dot{C}HX} \xrightarrow{\text{coupling}} \mathrm{R[CH_2CHX]_m CH_2CHX{-}CHXCH_2[CHXCH_2]_n R}$$

disproportionation:

$$\longrightarrow \mathrm{R[CH_2CHX]_m CH{=}CHX} + \mathrm{R[CH_2CHX]_n CH_2CH_2X}$$

Termination is occasionally brought about by chain-transfer reactions due to the presence of an impurity in the monomer. A growing polymer radical, abstracts an atom from the impurity to terminate the original polymerisation chain. In some cases, radical inhibitors are added from the external source which reacts with the polymer radical to terminate. Such radical inhibitors are phenol (or phenol derivative) or hydroquinone.

Free radical vinyl polymerisation gives low density polymer.

Cationic vinyl polymerisation

Cationic polymerisation is initiated by strong protic or Lewis acid initiator. A proton for example, adds to an electron rich alkene to form the most stable carbocation. The carbocation adds to another electron rich alkene to build the polymer chain. Monomers with electron releasing substituents such as phenyl group readily undergo this type of polymerisation to give polystryrene.

Initiation: $H^{\oplus}$ + $CH_2=CH-Ph$ → $H-CH_2-\overset{\oplus}{C}H-Ph$

Propagation: $H-CH_2-\overset{\oplus}{C}H-Ph$ + $CH_2=CH-Ph$ → $H-CH_2-CH(Ph)-CH_2-\overset{\oplus}{C}H-Ph$ + $CH_2=CH-Ph$ → $H-CH_2-CH(Ph)-CH_2-CH(Ph)-CH_2-\overset{\oplus}{C}H-Ph$ → etc.

Termination: $H-[CH_2-CH(Ph)]_n-CH_2-\overset{\oplus}{C}H-Ph$ (loss of H) → $H-[CH_2-CH(Ph)]_n-CH=CH-Ph$

Polystyrene

Anionic vinyl polymerisation

Anionic polymerisation is initiated by nucleophiles such as alkyllithium (alkoxide or hydroxide). The monomer must possess an electron withdrawing substituents such as nitrile or phenyl. The nucleophile adds to the alkene to form most stable carbanion (in a Michael type reaction). The carbanion then adds to another alkene to build the polymer chain.

Initiation: $\overset{\ominus}{C_4H_9}\overset{\oplus}{Li}$ + $CH_2=CH-CN$ → $C_4H_9-CH_2-\overset{\ominus}{C}H-CN$

Propagation:

$C_4H_9-CH_2-\overset{\ominus}{C}H-CN$ + $CH_2=CH-CN$ → $C_4H_9-CH_2-CH(CN)-CH_2-\overset{\ominus}{C}H-CN$ + $CH_2=CH-CN$

Vinyl polymerisation may be effected by co-ordination polymerisation by using Zieglar-Natta Catalysts (Mixture of $AlEt_3$—$TiCl_4$) in an inert solvent such as heptane. It gives high density polymer.

Uses of vinyl polymers

Polyethene: High density polyethene is used for molded products in the manufacture of house ware such as buckets and dustbins, toys and carpet backing and low density polyethene is used as a packaging materials and cable insulation etc.

PVC: PVC is used in the manufacture of initation leather, pipes, floor covering, corrugated roofing material and gramophone records etc.

Polystyrene: Polystyrene is used in the manufacture of food containers, cosmetic bottles, television cabinates, plastic cups, packaging and toys.

Teflon: Because of its low chemical reactivity, excellent toughness, electrical and heat resistance, teflon is used as insulation for electrical items and in the manufacture of gaskets and valves. Teflon has an extremely low co-efficient of friction and it is used as lining of frying pans and for coating particularly for cooking utensils.

Orlon: Orlon is used as a synthetic textile fibre in clothing and carpeting.

PVA: It is soft, rubbery polymer and is used as an adhesive and in emulsion paints.

Poly (methyl methacrylate): Clear and semiflexible used as a lenses and windows, fibre optics.

11.2.2 Condensation polymers (step-growth polymers)

Condensation polymers are formed from two monomers containing two functional groups. The formation of a bond usually leads to the elimination of a simple by-product (such as water), and this occurs in descrete steps i.e. not via a chain reaction.

Synthesis of Nylon 6,6 (Polyamide)

Nylon 6,6 is formed by the thermal decomposition of salts of diamines and dicarboxylic acids. The driving force for polymerisation is the elimination of water between an acid and an amine resulting in the formation of an amide linkage.

Nylon 6,6 is produced from hexamethylene diamine and adipic acid in two steps. The first step is the preparation and purification of the salt which is then polymerised by heating at 280°C.

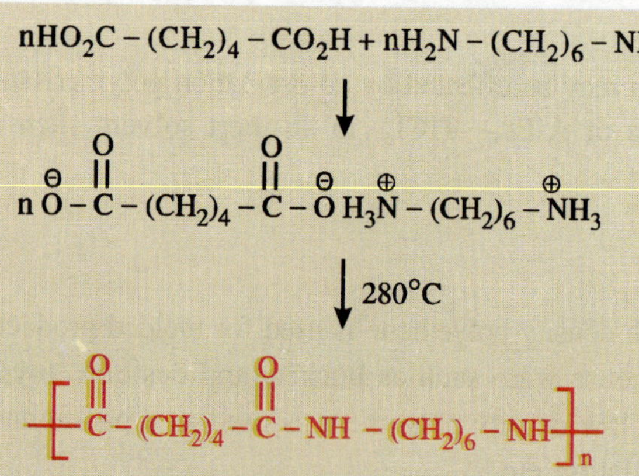

$$nHO_2C-(CH_2)_4-CO_2H + nH_2N-(CH_2)_6-NH_2$$

$$\downarrow$$

$$n\,\overset{\ominus}{O}-\overset{\overset{O}{\|}}{C}-(CH_2)_4-\overset{\overset{O}{\|}}{C}-\overset{\ominus}{O}\;H_3\overset{\oplus}{N}-(CH_2)_6-\overset{\oplus}{N}H_3$$

$$\downarrow 280°C$$

$$\left[\overset{\overset{O}{\|}}{C}-(CH_2)_4-\overset{\overset{O}{\|}}{C}-NH-(CH_2)_6-NH\right]_n$$

Nylon 6,6

System of numbering in Nylon 6,6 indicates from which monomers a particular Nylon is made. In the Nylon 6,6, the first 6 refers to the number of carbon atoms in the amine and the second to the number of carbon atoms in the acid.

In Nylon 6,6, molecular chains are held together by interchain hydrogen bonding between — NH — and —C = O groups.

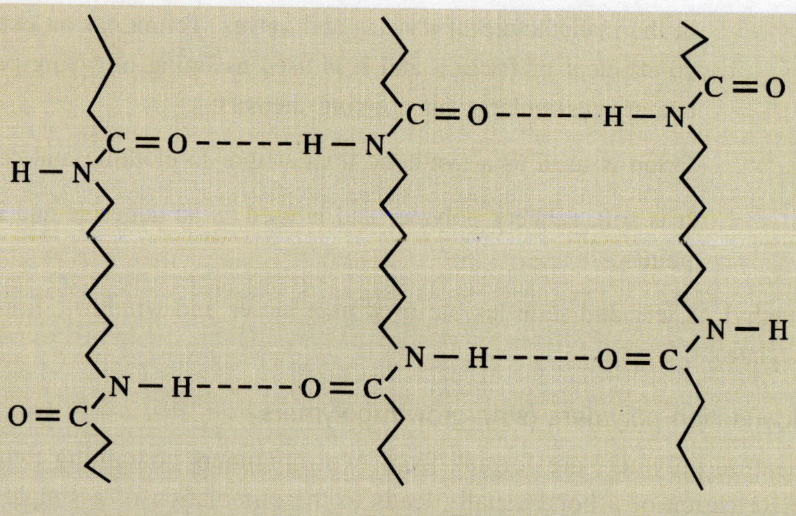

C=O-----H-N
C=O-----H-N
H-N
C=O
N-H
N-H-----O=C
N-H-----O=C
O=C

Nylon 6, 6

Due to such interchain of hydrogen bonding, Nylon 6,6 polymer is long, thin and threadlike and are lined up in the direction of the fibre. This interchain hydrogen bonding also prevent the molecules from returning to random looping and coiling.

Uses of Nylon 6,6

Nylon 6,6 is used as a fibre for clothing, ropes, tire cords and for tooth brushes as well as for many molded products such as gears and bearings.

Synthesis of Nylon 6 (Perlon)

Nylon 6 is made from caprolactam (which is prepared by Beckmann rearrangement of cyclohexanone oxime). The reaction is initiated by the hydrolysis of caprolactam whose amino group attacks the other molecule of caprolactam and the continuation of the process result in the formation of a polymer of Nylon 6.

Cyclohexanone $\xrightarrow{NH_2OH}$ Cyclohexanone oxime ($=N-OH$) $\xrightarrow{H_2SO_4}$ Caprolactam ($N-H$)

Caprolactam

$$\text{Caprolactam} + H_2O \longrightarrow HO_2C-(CH_2)_5-NH_2$$

$$HO_2C-(CH_2)_5-NH_2 + \text{Caprolactam} \longrightarrow HO_2C-(CH_2)_5-NHCO-(CH_2)_5-NH_2$$

$$\downarrow$$
$$\downarrow$$

Nylon 6 $\quad HO_2C-(CH_2)_5-HN\left[CO-(CH_2)_5-NH\right]_n CO(CH_2)_5-NH_2$

In Nylon 6, the number 6 indicates it is made from one monomer containing six carbon atoms.

In Nylon 6, molecular chains are held together by inter chain hydrogen bonding between –NH– and $-C=O$ groups.

$$\text{H}-\text{N}\overset{}{}\text{C}=\text{O}\text{-----}\text{H}-\text{N}\text{C}=\text{O}\text{-----}\text{H}-\text{N}\text{C}=\text{O}$$

$$\text{O}=\text{C}\text{N}-\text{H}\text{-----}\text{O}=\text{C}\text{N}-\text{H}\text{-----}\text{O}=\text{C}\text{N}-\text{H}$$

Nylon 6

Uses of Nylon 6

Nylon 6 polymer is used as a synthetic fibre.

Synthesis of Terylene (Polyester)

Polyesters are formed as a result of reactions between polyols and polybasic acids. The polyester fibre, terylene is prepared by reaction of either terephthalic acid or dimethylterephthalate with ethylene glycol. The reaction takes place in two steps. The first product which is formed at 200°C, is a new monomer that contains two ethylene glycol units, when heated to 280°C, loses ethylene glycol and gives polyester fibre, terylene.

$$HO-CH_2-CH_2-OH + HO_2C-C_6H_4-CO_2H$$

ethylene glycol terephthalic acid

$$\downarrow \; 200°C, \; -2H_2O$$

$$HO-CH_2-CH_2-O-\overset{O}{\overset{\|}{C}}-C_6H_4-\overset{O}{\overset{\|}{C}}-O-CH_2-CH_2-OH$$

$$\downarrow \; 280°C, \; -HOCH_2CH_2OH$$

$$\left[\overset{O}{\overset{\|}{C}}-C_6H_4-\overset{O}{\overset{\|}{C}}-O-CH_2-CH_2-O\right]_n$$

Terylene

There are strong intermolecular (inter chain) attractions which prevent the molecules from returning to random looping and coiling. In the terylene this intermolecular force is provided by the presence of polar carbonyl groups which lead to powerful dipole-dipole attractions.

Uses of Terylene

The polyester, terylene is used as a fabric in the cloth industry. The combination of cotton and terylene yields a fabric (e.g. 60/40 cloth) that still retains the coolness and comfort of 100% cotton.

Note

Model Questions and Answers (2000-2006, C. U. PART-II)

2000

Q.1 Work backward to devise synthetic route to the following target molecules.

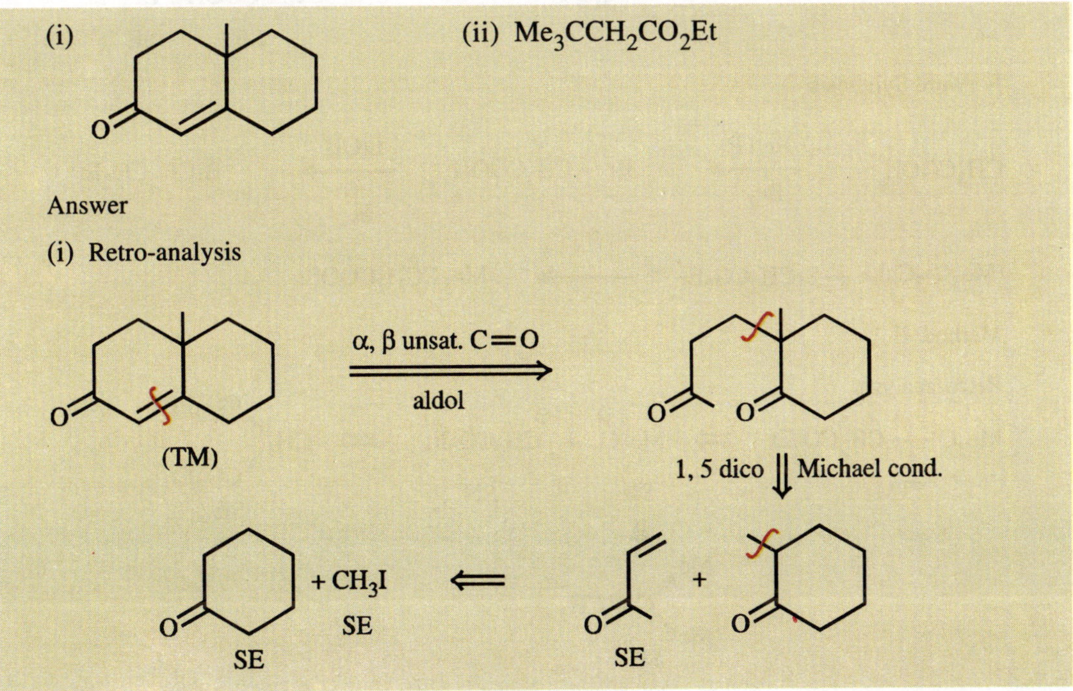

Forward Synthesis

Cyclohexanone $\xrightarrow[\text{(ii) } CH_3I]{\text{(i) } Me_3C-\overset{\ominus}{O}}$ 2-methylcyclohexanone $\xrightarrow[\text{(ii) methyl vinyl ketone}]{\text{(i) } \overset{\ominus}{N}H_2 \text{ (2eq)}}$ enolate adduct $\xrightarrow{\text{aq. } NH_4Cl}$ 2-methyl-2-(3-oxobutyl)cyclohexanone $\xrightarrow[AcOH]{Me_2NH}$ octalone

$Me_2NH/AcOH$ gives intramolecular aldol condensation via enol and which arrest the linear (aldol) condensation.

(ii) Method I

Retro-analysis

$$Me_3C-\!\!\!/\!\!-CH_2CO_2Et \text{ (TM)} \Rightarrow \overset{\ominus}{Me_3C} \text{ (SN)} + \overset{\oplus}{C}H_2CO_2Et \text{ (SN)} \Rightarrow Br-CH_2-CO_2Et$$

$$\overset{\ominus}{Me_3C} \text{ (SN)} \Downarrow (Me_3C)_2CuLi \text{ (SE)}$$

$$Br-CH_2-CO_2Et \Downarrow Br-CH_2COOH \Downarrow CH_3COOH + Br_2 \text{ (SE, SE)}$$

Forward Synthesis

$$CH_3COOH \xrightarrow[Br_2]{\text{Red P}} Br-CH_2COOH \xrightarrow[\overset{\oplus}{H}]{EtOH} BrCH_2CO_2Et$$

$$(Me_3C)_2CuLi + BrCH_2CO_2Et \longrightarrow Me_3CCH_2COOEt$$

Method II

Retro-analysis

$$Me_3C-\!\!\!/\!\!-CH_2CO_2Et \text{ (TM)} \Rightarrow \overset{\oplus}{Me_3C} \text{ (SN)} + \overset{\ominus}{C}H_2CO_2Et \text{ (SN)} \Rightarrow CH_2(CO_2Et)_2 \text{ (SE)}$$

$$\overset{\oplus}{Me_3C} \text{ (SN)} \Downarrow Me_2C=CH_2 \text{ (SE)}$$

Forward Synthesis

$CH_2(CO_2Et)_2$ —[$Me_2NH/AcOH$, buffer]⇌ HO–$C(OEt)=CH$–CO_2Et (enol) + $Me_2C=CH_2$ —[kinetically controlled]→ Me_3C–$CH(CO_2Et)_2$ —[(i) KOH; (ii) $H_3O^{\oplus}$; (ii) 250°]→ Me_3CCH_2COOH —[EtOH, $H^{\oplus}$]→ $Me_3CCH_2CO_2Et$

Q.2 Identify the products A—D in the following reaction sequence.

Acetone (1 mole) + Ethyl cyano acetate (2 moles) —[NH_3, EtOH]→ A —[(i) NaOEt (2 moles); (ii) CH_2I_2 (1 mole)]→ B —[(i) NaOH; (ii) $H_3O^{\oplus}$]→ C —[heat]→ D

Answer

Acetone + Ethylcyano acetate —[NH_3/EtOH, Knoevenagel reaction]→ $Me_2C(OH)$–$CH(CN)CO_2Et$ —[$-H_2O$]→ $Me_2C=C(CN)CO_2Et$ —[$NCCH_2CO_2Et$ /Michael cond.]→ $Me_2C[CH(CN)CO_2Et]_2$ (A)

(A) —[NaOEt, 2 moles]→ $Me_2C[C^{\ominus}(CN)CO_2Et]_2$ —[CH_2I_2, 1 mole]→ 3,3-dimethylcyclobutane-1,2-dicyano-1,2-dicarboxylate ester (B)

(B) —[(i) NaOH; (ii) $H_3O^{\oplus}$]→ tetracarboxylic acid (C): HO_2C, CO_2H, HO_2C, CO_2H on cyclobutane —[heat, $-2CO_2$]→ (D): 3,3-dimethylcyclobutane-1,2-dicarboxylic acid (CO_2H, CO_2H)

2001 (New Syllabus)

Q.1 Describe the synthesis of the following compounds with proper retrosynthetic analysis.

(i) (ii) 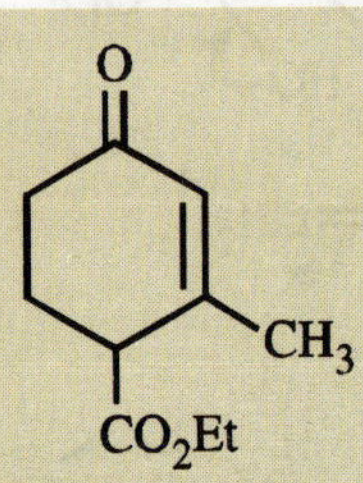 O, CH_3, CO_2Et (iii) O, N – H, O, N, O, H

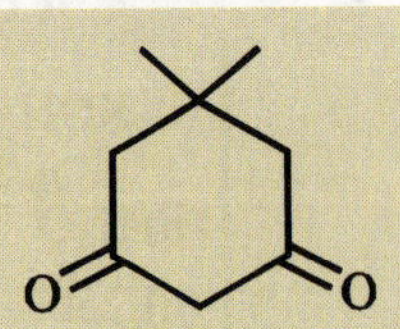

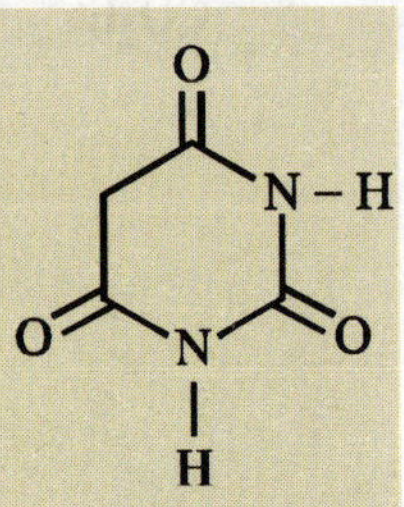

Answer

(i) Method I

Retro-analysis

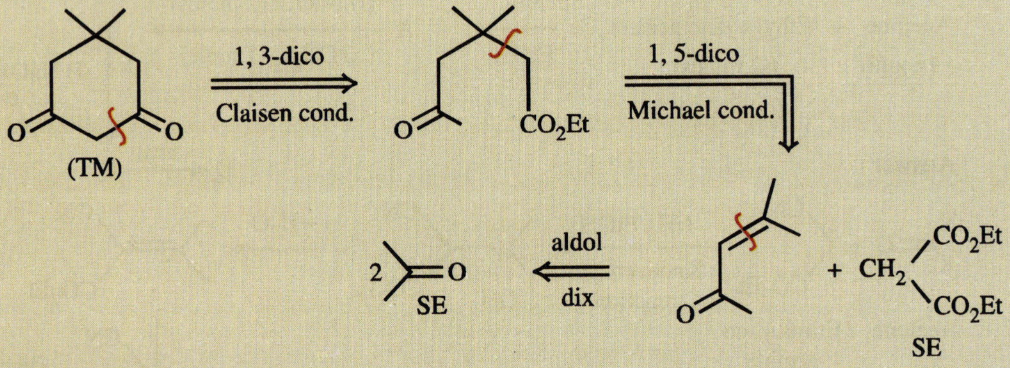

(TM) 1, 3-dico / Claisen cond. ⟹ O, CO_2Et 1, 5-dico / Michael cond. ⟹ O + $CH_2(CO_2Et)_2$ SE

2 =O SE ⟸ aldol dix

Forward Synthesis

=O (SE) $\xrightarrow{H^{\oplus}}$ (O)

$\xrightarrow[\overset{\ominus}{O}Et]{CH_2(CO_2Et)_2}$ O, CO_2Et, CO_2Et $\xrightarrow{\overset{\ominus}{O}Et}$ CO_2Et, O, $\ominus$, OEt, O

↓

O, O $\xleftarrow{\text{dil HCl}/\Delta}$ CO_2Et, O, O

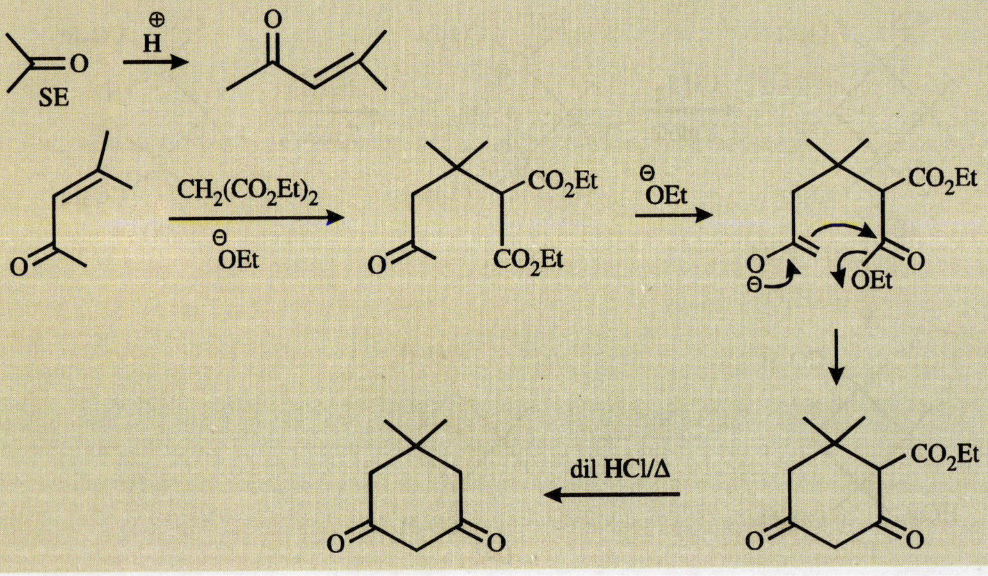

Method II

Retro-analysis

FGA

EtO_2C

1, 3-dico

Claisen cond.

EtO_2C CO_2Et

1, 5-dico

Michael cond.

EtO_2C + CO_2Et

SE

α, β-unsat. ester

Reformatsky

=O + Br – CH_2 – CO_2Et

SE SE

Forward Synthesis

=O + $BrCH_2CO_2Et$ (i) Zn/ether (ii) $H_3O^{\oplus}$ → CO_2Et

EtO_2C + CO_2Et $\overset{\ominus}{O}Et$ → [EtO_2C CO_2Et] $\overset{\ominus}{O}Et$

EtO_2C ⊖O OEt

← EtO_2C

dil HCl Δ

(ii) Retro-analysis

O

CH_3

CO_2Et

(TM)

α, β-unsat. CO

aldol cond.

CH_3 CO_2Et

1, 5-dico

Michael cond.

CH_3 CO_2Et +

SE

α, β-unsat. CO

aldol

CH_2O +

SE SE

Forward Synthesis

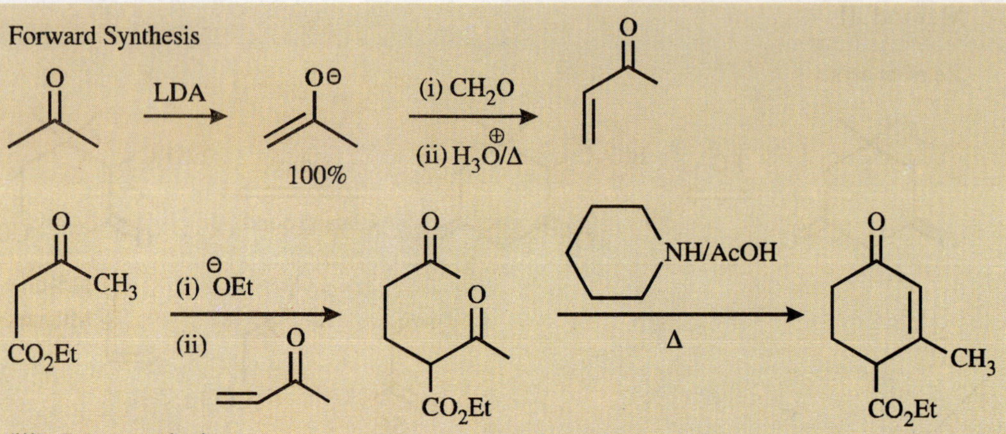

LDA → 100% → (i) CH_2O (ii) $H_3\overset{\oplus}{O}/\Delta$

CH_3 CO_2Et (i) $\overset{\ominus}{O}Et$ (ii) → CO_2Et NH/AcOH Δ → CH_3 CO_2Et

(iii) Retro-analysis

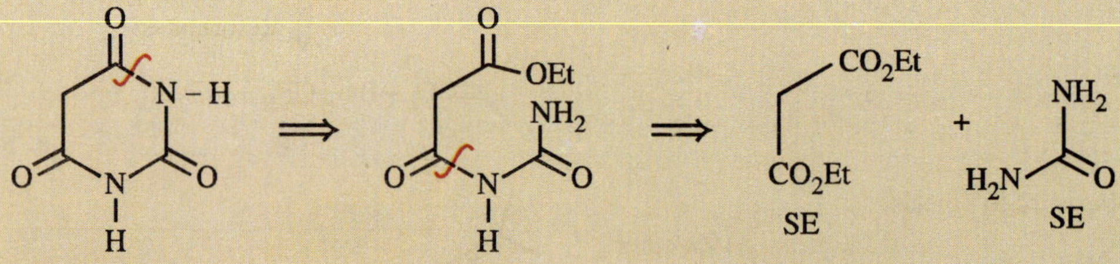

N – H O O N H ⟹ OEt NH_2 O N O H ⟹ CO_2Et CO_2Et SE + NH_2 H_2N O SE

Forward Synthesis

OEt O OEt + H_2N H_2N =O $\overset{\ominus}{O}Et/EtOH$ – 2EtOH → N – H O N O H

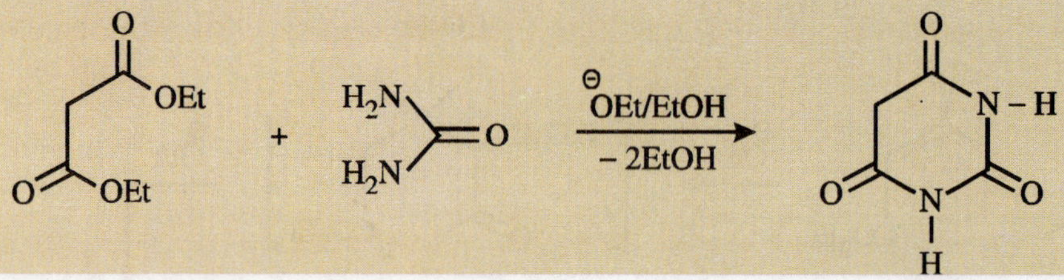

Q.2 Outline the synthesis of the compound (A) from diethyl adipate. Give mechanism of the steps involved.

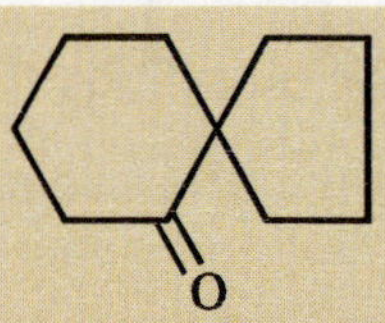

(A)

Answer

Retro-analysis

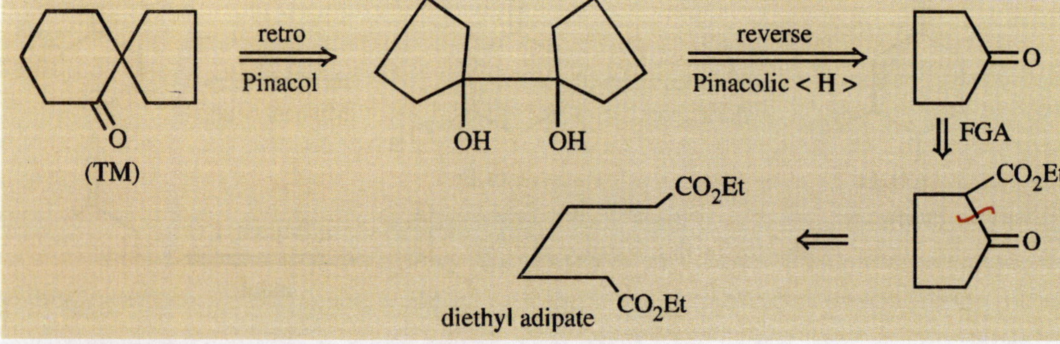

(TM) ⟹ retro Pinacol ⟹ OH OH ⟹ reverse Pinacolic < H > =O ⇓ FGA CO_2Et =O ⟸ CO_2Et CO_2Et diethyl adipate

Forward Synthesis

$(CH_2)_3$ C(=O)OEt / CH_2-CO_2Et diethyladipate $\xrightarrow[\text{Dieckmann cond.}]{NaOEt/EtOH}$ $(CH_2)_3$ C(=O)OEt / $\overset{\ominus}{C}H$ / CO_2Et $\longrightarrow$ 2-(CO_2Et)cyclopentanone $\xrightarrow{\text{dil HCl}/\Delta}$ cyclopentanone

cyclopentanone $\xrightarrow{Mg/xylene}$ (O–Mg–O diradical) $\longrightarrow$ (O–Mg–O pinacolate) $\xrightarrow{H_3O^{\oplus}}$ (OH OH pinacol) $\xrightarrow[\Delta]{H_2SO_4}$ (H–O, $\oplus OH_2$) $\longrightarrow$ spiro ketone (O)

Q.3 Predict the product of the following reaction and justify.

1-vinyl-3H-indene + $CH_2{=}CH{-}CO_2Me$ $\xrightarrow{\Delta}$

Answer

This is a Diels-Alder reaction. Diels-Alder reaction is regioselective if either of the diene or dienophile or both the diene and dienophile are unsymmetrical. In the concerned reaction both the diene and dienophile are unsymmetrical and the reaction is guided by regioselectivity. In the Diels-Alder reaction, regioselectivity in the usual sense (not by FMO approach) is determined by the electrophilicity and nucleophilicity of the carbon centres of the diene and dienophile.

In this reaction in the dienophile, methylacrylate, methylene carbon is electrophilic carbon since it is α, β-unsaturated ester. For the diene part, we know primary carbanion and secondary carbonium ion is more contributing than secondary carbanion and primary carbonium ion.

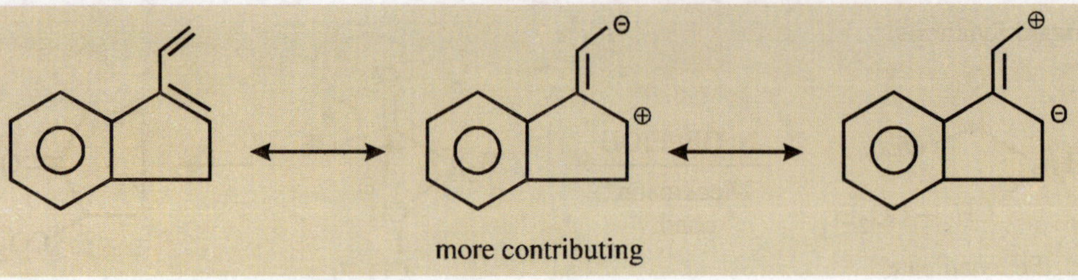

more contributing

Therefore, regioselective products are

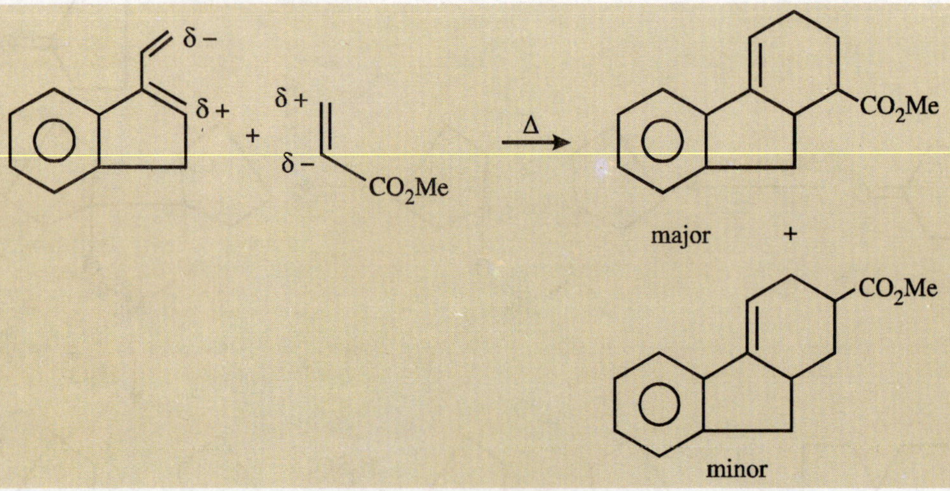

δ − δ + δ + δ − CO_2Me Δ CO_2Me

major +

CO_2Me

minor

2001 (Old Syllabus)

Q.1 Using Claisen ester condensation reaction, how would you prepare the following compounds.

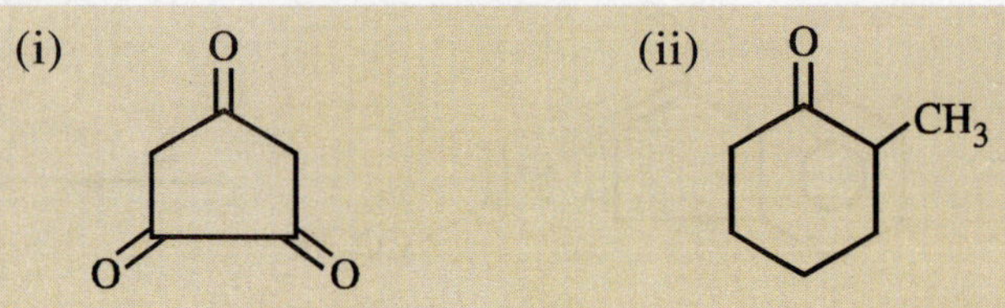

(i) (ii) CH_3

Answer

(i) Method I

Retro-analysis

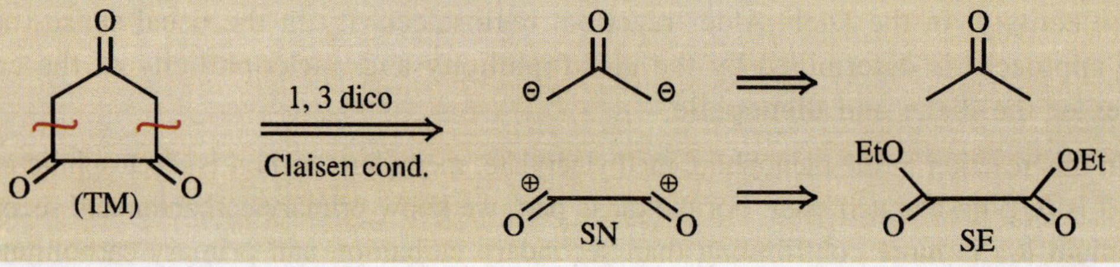

(TM) 1, 3 dico Claisen cond. SN EtO OEt SE

Forward Synthesis

Acetone + diethyl malonate —$\overset{\ominus}{O}Et$/EtOH→ cyclopentane-1,2,4-trione

Method II

Retro-analysis

(TM) —FGA⟹ EtO_2C / CO_2Et substituted trione

1, 3 dico ⟹ Claisen cond.

EtO_2C–CH_2–CO–CH_2–CO_2Et + EtO–CO–CO–OEt (SE)

⟹ FGI

HO_2C–CH_2–CO–CH_2–CO_2H ⟹ HO–C(CO_2H)(CH_2CO_2H)(CH_2CO_2H)

SE

(citric acid)

Forward Synthesis

Citric acid —fuming H_2SO_4, – HCOOH→ O=C(CH_2CO_2H)$_2$ —EtOH/$H^{\oplus}$→ O=C(CH_2CO_2Et)$_2$

EtO_2C–CH–CO–CH–CO_2Et + EtO–CO–CO–OEt —$\overset{\ominus}{O}Et$, EtOH→ EtO_2C / CO_2Et substituted trione —dil HCl, Δ→ cyclopentane-1,2,4-trione

(ii) Retro-analysis

2-Methylcyclohexanone (TM) $\Longrightarrow$ CH_3I + cyclohexanone $\xrightarrow{\text{FGA}}$ 2-(ethoxycarbonyl)cyclohexanone $\xRightarrow[\text{Claisen cond.}]{\text{1, 3 dico / intra}}$

$EtO_2C-H_2C-(CH_2)_3-CH_2-CO_2Et$ $\Longrightarrow$ $Br-(CH_2)_3-Br + 2\overset{\ominus}{C}H_2-CO_2Et$

SE

$\Downarrow$

$2CH_2(CO_2Et)_2$

SE

(DEM)

Forward Synthesis

$2CH_2(CO_2Et)_2 \xrightarrow{2\ \overset{\ominus}{O}Et} (EtO_2C)_2\overset{\ominus}{C}H + Br-(CH_2)_3-Br + \overset{\ominus}{C}H(CO_2Et)_2$

$\downarrow$

$(EtO_2C)_2CH-(CH_2)_3-CH(CO_2Et)_2$

$\downarrow$ (i) KOH (ii) $H_3\overset{\oplus}{O}$ (iii) 250°

$HO_2C-(CH_2)_5-CO_2H$

$\downarrow$ $EtOH/\overset{\oplus}{H}$

$EtO_2C-(CH_2)_5-CO_2Et$

$\downarrow$ $\overset{\ominus}{O}Et/EtOH$, intra Claisen cond.

2-(ethoxycarbonyl)cyclohexanone $\xrightarrow[\text{(ii) } CH_3I]{\text{(i) } \overset{\ominus}{O}Et}$ 2-methyl-2-(ethoxycarbonyl)cyclohexanone $\xrightarrow[\text{HCl}/\Delta]{\text{dil}}$ 2-methylcyclohexanone

Q.2 How can you synthesize the following compounds.

(i) (ii) 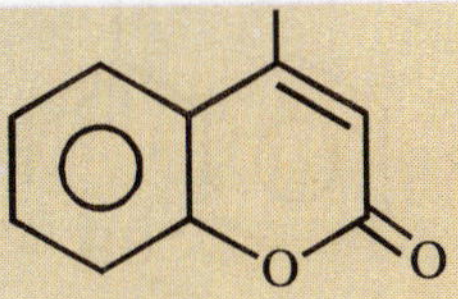 (iii)

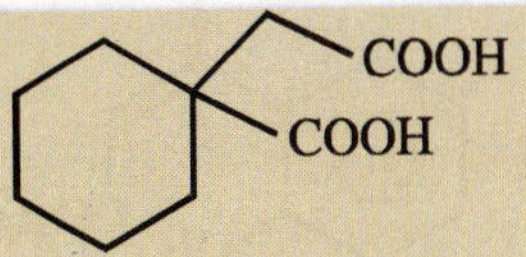

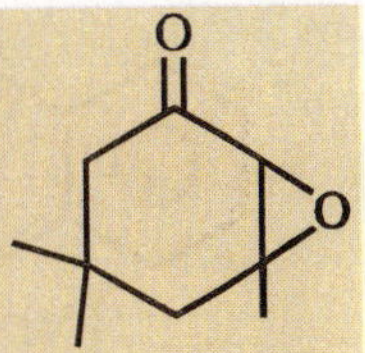

Answer

(i) Retro-analysis

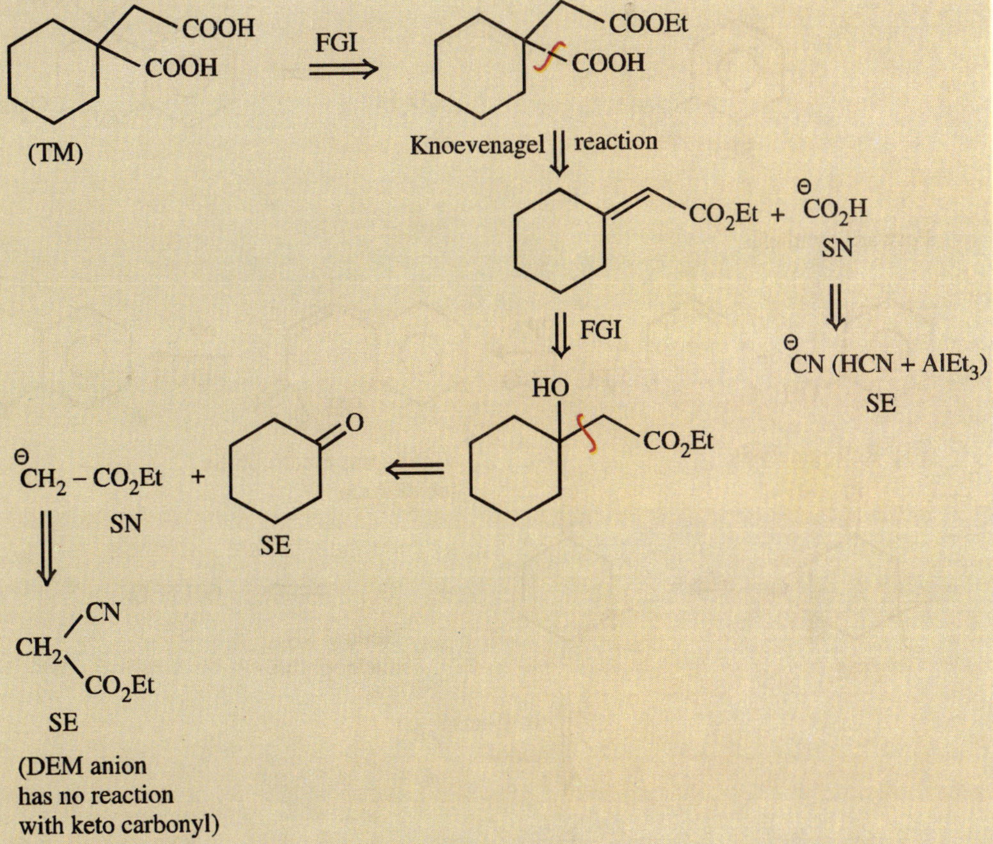

COOH COOH (TM) $\xRightarrow{FGI}$ COOEt COOH

Knoevenagel $\Downarrow$ reaction

CO_2Et + $\overset{\ominus}{C}O_2H$ SN

$\Downarrow$ FGI

$\Downarrow$ $\overset{\ominus}{C}N$ (HCN + $AlEt_3$) SE

HO CO_2Et $\Leftarrow$ O SE + $\overset{\ominus}{C}H_2-CO_2Et$ SN

$\Downarrow$

$CH_2(CN)(CO_2Et)$ SE

(DEM anion has no reaction with keto carbonyl)

Forward Synthesis

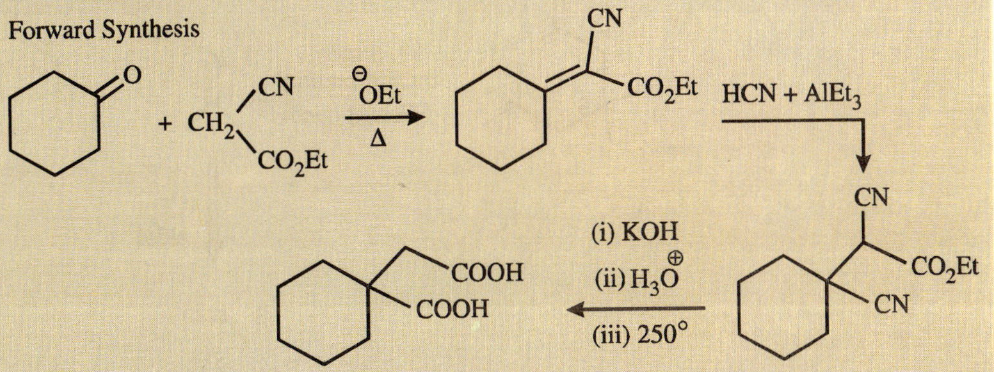

O + $CH_2(CN)(CO_2Et)$ $\xrightarrow[\Delta]{\overset{\ominus}{O}Et}$ CN CO_2Et $\xrightarrow{HCN + AlEt_3}$ CN CO_2Et CN $\xrightarrow{(i)\ KOH\ (ii)\ H_3\overset{\oplus}{O}\ (iii)\ 250°}$ COOH COOH

(ii) Retro-analysis

(TM) ⟹ OH COOH —FGI⟹ OH CO_2Et

↓ FGI

OH, OH CO_2Et ⟸ OH (SE) + O, CO_2Et (SE)

Forward Synthesis

OH + O, CO_2Et —PPA, $-H_2O$→ OH, O, OEt —PPA, $-EtOH$→ O, O

Vinylogous electrophilic substitution

(iii) Retro-analysis

(TM) O, O ⟹ O + 'O' ⟹ $\overset{\ominus}{O}H/H_2O_2$ SE

Neutral SN nucleophilic

⇓ α, β-unsat. co aldol

O, O —1, 5-dico Michael cond.⟹ O + O, CO_2Et SE

⇓ α, β-unsat. CO aldol

2 >=O

SE

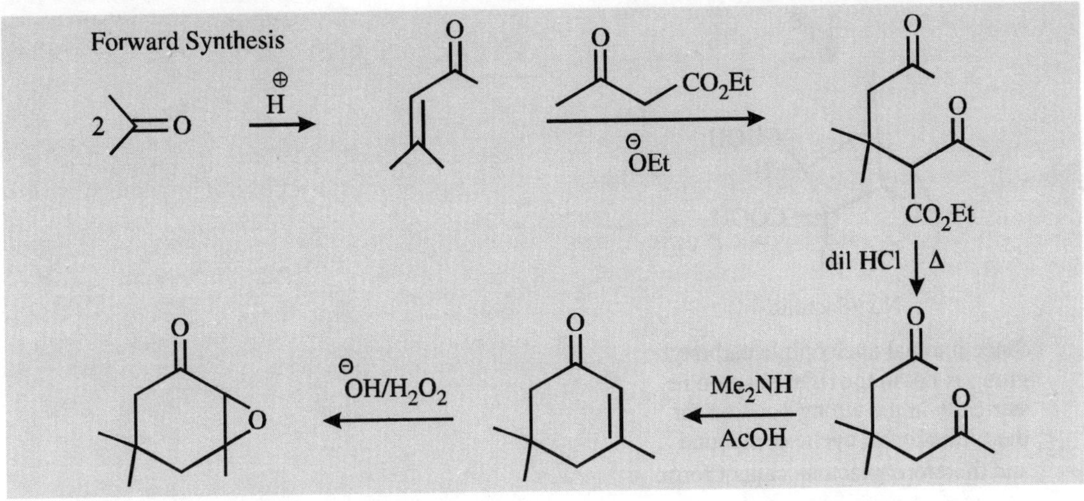

Forward Synthesis

2 acetone —$H^{\oplus}$→ mesityl oxide —(ethyl acetoacetate, CO_2Et; $\overset{\ominus}{O}Et$)→ Michael adduct (CO_2Et) —dil HCl, Δ→ diketone —Me_2NH, AcOH→ 3,5,5-trimethylcyclohex-2-enone —$\overset{\ominus}{O}H/H_2O_2$→ epoxy ketone

Q.3 The products formed by the Diels-Alder reaction of cyclopentadiene and maleic anhydride are distinguished by hydrolysis followed by treatment with iodine. Write down the products with explanations.

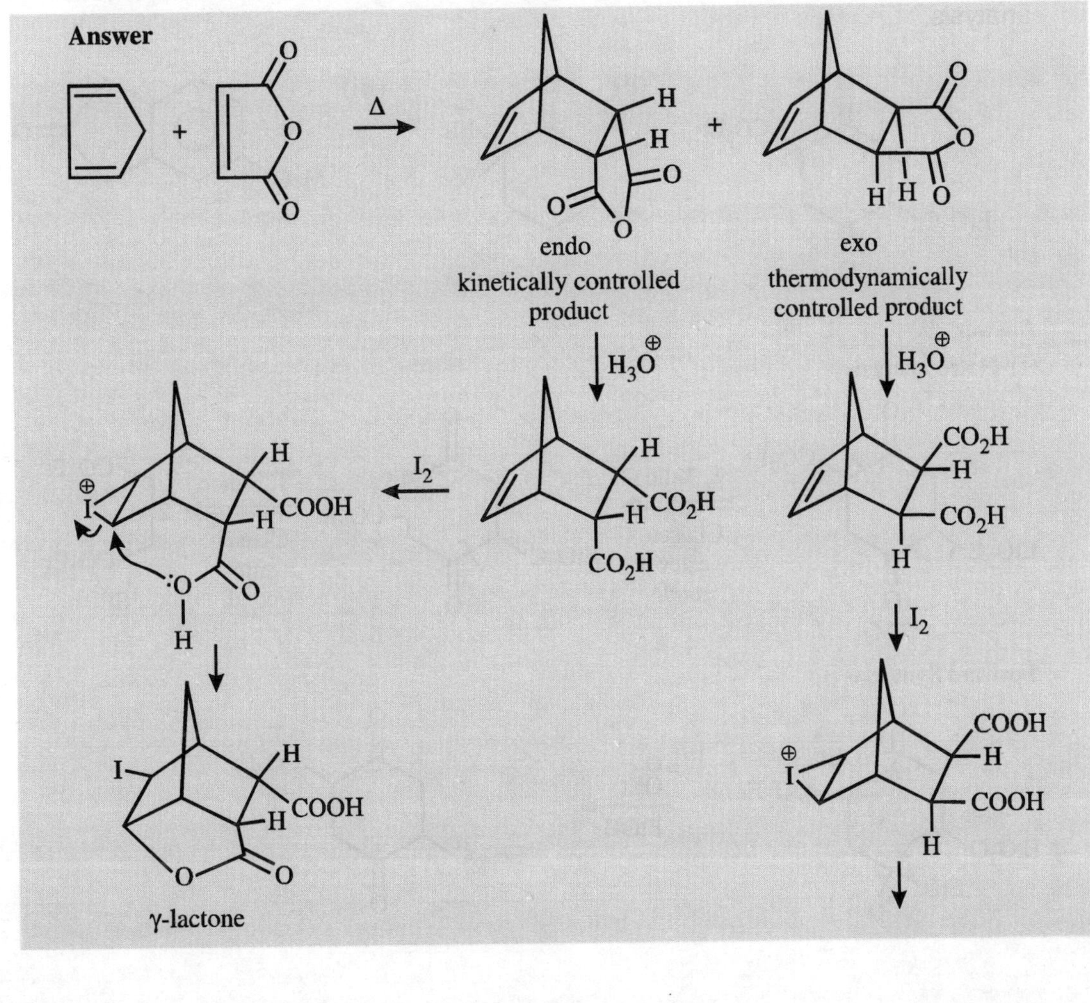

Answer

Cyclopentadiene + maleic anhydride —Δ→ endo (kinetically controlled product) + exo (thermodynamically controlled product)

endo —$H_3\overset{\oplus}{O}$→ endo diacid (CO_2H, CO_2H) —I_2→ iodonium ion; neighbouring COOH (:O–H) attacks → γ-lactone (I, H, COOH)

γ-lactone

exo —$H_3\overset{\oplus}{O}$→ exo diacid (CO_2H, CO_2H) —I_2→ iodonium ion (COOH, COOH) →

$I^{\ominus}$

I, COOH, H, COOH, I, H

No γ-lactone

Since internal nucleophile carboxylic group is not in the right position i.e. sterically in the wrong position for the formation of cyclic compound and therefore γ-lactone cannot form.

2002 (New Syllabus)

Q.1 Describe the synthesis of the following compounds with proper retrosynthetic analysis.

(i) O, CO_2Et, EtO_2C, O

(ii) O, Me, Me

(iii) MeO, O

Answer

(i) Retro-analysis

O, CO_2Et, EtO_2C, O (TM) $\xrightarrow[\text{Claisen ester cond.}]{\text{1, 3-dico}}$ O, OEt, CO_2Et, EtO_2C, O $\xrightarrow[\text{Claisen ester cond.}]{\text{1, 3-dico}}$ 2 CO_2Et, CO_2Et SE

Forward Synthesis

O, OEt, CO_2Et, EtO_2C, EtO, O $\xrightarrow[\text{EtOH}]{\overset{\ominus}{\text{O}}\text{Et}}$ O, CO_2Et, EtO_2C, O

(ii) Retro-analysis

(TM) ⟹ 2-methylcyclohexanone + MeI (SE) ⟹ cyclohexanone (SE) + MeI (SE)

Forward Synthesis

Cyclohexanone —(i) $Me_3C-O^{\ominus}$, (ii) MeI→ 2-methylcyclohexanone (CH_3)

2-Methylcyclohexanone —NaH, MeO⌒⌒OMe→ enolate ($O^{\ominus}$, Me)

thermodynamically controlled enolate

—MeI→ 2,2-dimethylcyclohexanone (Me, Me)

2-Methylcyclohexanone —Me_3SiCl/Et_3N, CH_2Cl_2→ $OSiMe_3$ (Me)

specific enolate silylated enol ether

—MeI→ $\oplus O-SiMe_3$ (Me, Me)

—(i) $(nBu)_4\overset{\oplus}{N}\overset{\ominus}{F}$, (ii) H_2O→ 2,2-dimethylcyclohexanone (Me, Me)

(iii) Method I

Retro-analysis

MeO-indan-2-one (TM) =O —FGI⟹ MeO-benzene with COOH and CO_2H side chains

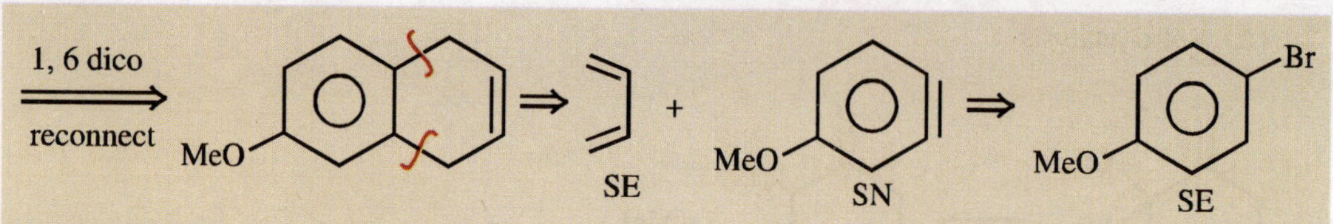

1, 6 dico ⟹ reconnect

MeO ... ⟹ SE + MeO ... SN ⟹ MeO ... Br SE

Forward Synthesis

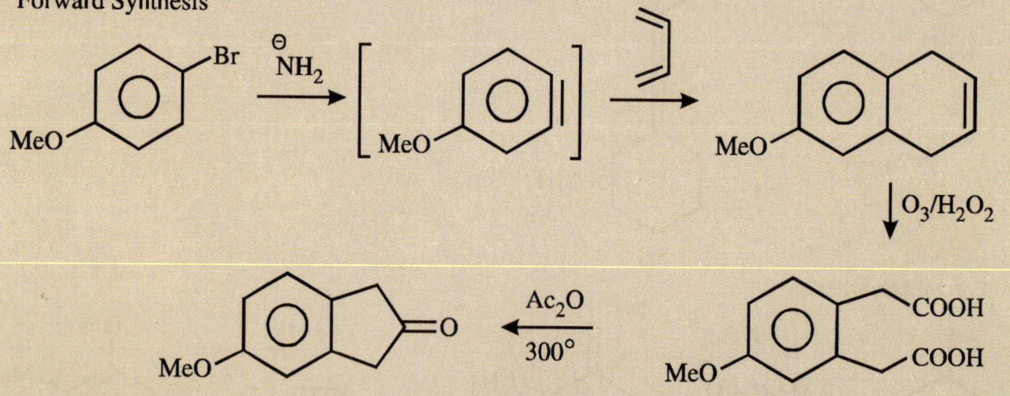

MeO ... Br —($^{\ominus}NH_2$)→ [MeO ...] —→ MeO ...

↓ O_3/H_2O_2

MeO ... =O ←(Ac_2O, 300°)— MeO ... COOH, COOH

Method II
Retro-analysis

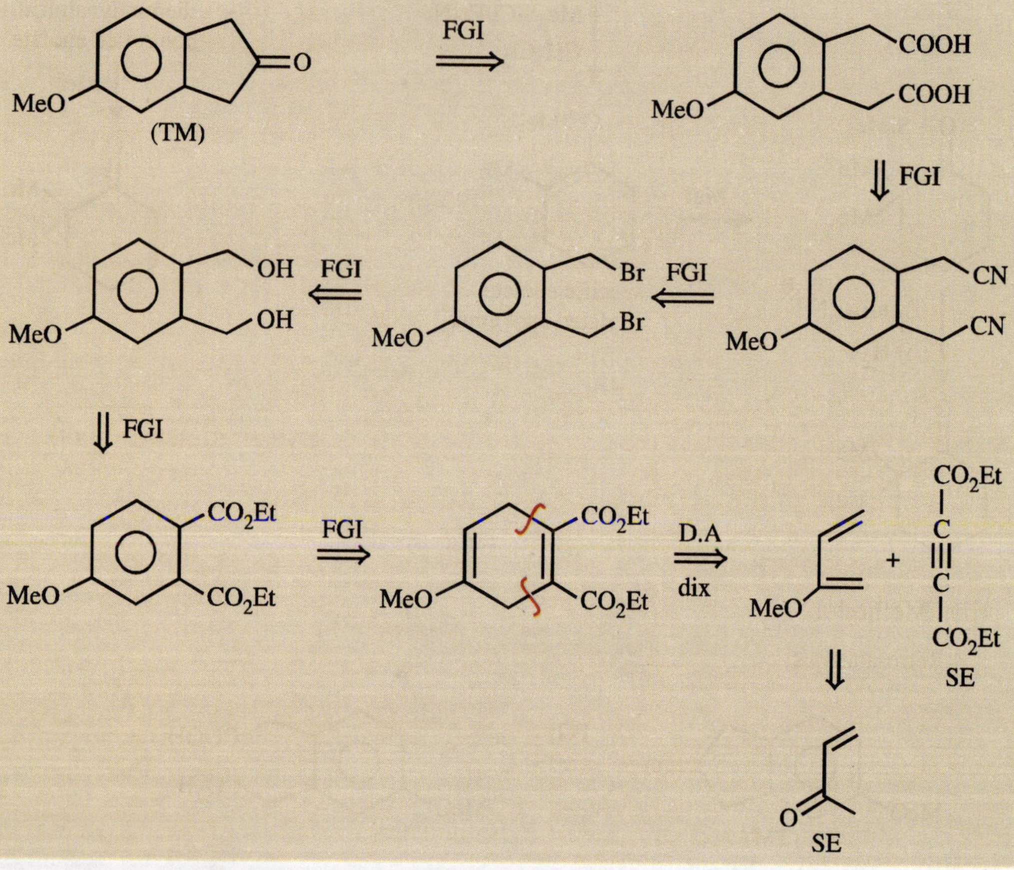

MeO ... =O (TM) ⟹ FGI ⟹ MeO ... COOH, COOH

⇓ FGI

MeO ... OH, OH ⟸ FGI ⟸ MeO ... Br, Br ⟸ FGI ⟸ MeO ... CN, CN

⇓ FGI

MeO ... CO_2Et, CO_2Et ⟹ FGI ⟹ MeO ... CO_2Et, CO_2Et ⟹ D.A dix ⟹ MeO ... + CO_2Et–C≡C–CO_2Et SE

⇓

O ... SE

Forward Synthesis

(Scheme: methyl vinyl ketone → CH_2N_2 → 2-methoxybutadiene + $EtO_2C-C\equiv C-CO_2Et$ → Δ → diethyl 4-methoxycyclohexa-1,4-diene-1,2-dicarboxylate → DDQ → diethyl 4-methoxyphthalate → $LiAlH_4$ → diol (OH, OH) → (i) PBr_3 (ii) KCN (iii) $H_3O^{\oplus}$ → diacid (CO_2H, CO_2H) → Ac_2O, 300° → 5-methoxy-2-indanone)

Method III

Retro-analysis

(TM) $\overset{FGA}{\underset{FGR}{\Longrightarrow}}$ 5-methoxy-1-indanone ⇒ 3-(4-methoxyphenyl)propanoic acid (CO_2H) ⇓ FGA ⇐ 4-methoxycinnamic acid (HO_2C) $\overset{FGI}{\Longleftarrow}$ β-hydroxy acid (OH, HO_2C) ⇐ 4-methoxybenzaldehyde (CHO, MeO) SE + $CH_2(COOH)_2$ SE

Forward Synthesis

4-methoxybenzaldehyde (CHO) $\xrightarrow[Py/\Delta]{CH_2(CO_2H)_2}$ 4-methoxycinnamic acid (CO_2H) → H_2/Pd → 3-(4-methoxyphenyl)propanoic acid (CO_2H) → (i) $SOCl_2$ (ii) $AlCl_3$ → 5-methoxy-1-indanone → (i) $Me_3C-O^{\ominus}$ (ii) PhCHO → 2-benzylidene indanone (=CHPh) $\xrightarrow[AlCl_3]{LiAlH_4}$ 2-benzylidene indane (=CHPh) $\xrightarrow{O_3}$ 5-methoxy-2-indanone (=O)

2002 (Old Syllabus)

Q.1 How can you synthesize the following compounds from readily available starting materials.

(i) HO_2C–cyclopentane–CO_2H (1,3-dicarboxylic acid) (ii) tetraphenylcyclopentadienone (Ph, Ph, Ph, Ph; O) (iii) 3-methylcyclohex-2-enone (O, Me)

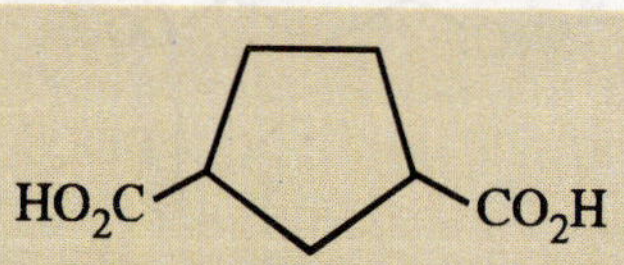

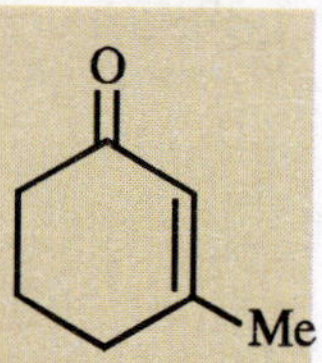

Answer

(i) Retro-analysis

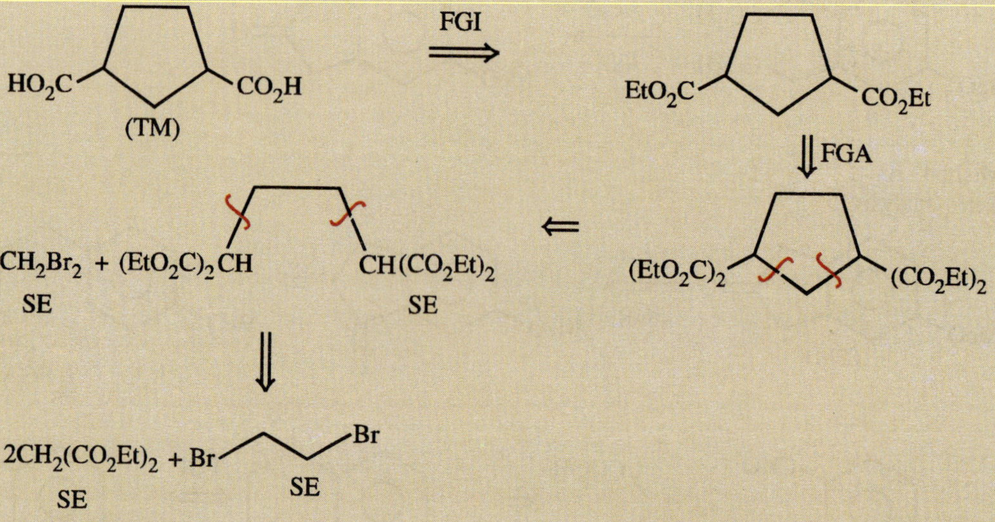

HO_2C–C₅ ring–CO_2H (TM) $\xrightarrow{\text{FGI}}$ EtO_2C–C₅ ring–CO_2Et

⇓ FGA

$(EtO_2C)_2$–C₅ ring–$(CO_2Et)_2$ ⇐ CH_2Br_2 + $(EtO_2C)_2CH$–$CH_2CH_2CH_2$–$CH(CO_2Et)_2$

SE SE

⇓

$2CH_2(CO_2Et)_2$ + $Br\diagup\diagdown\diagup Br$

SE SE

Forward Synthesis

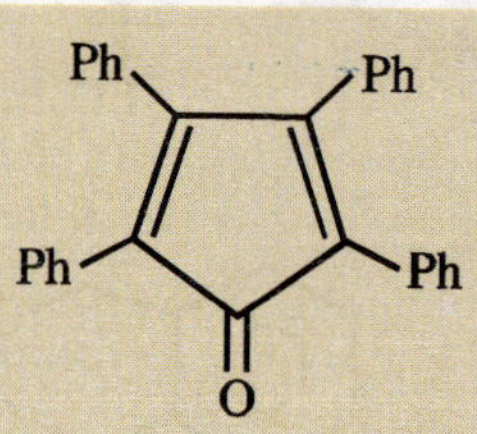

$$2CH_2(CO_2Et)_2 \xrightarrow{2\overset{\ominus}{O}Et} (EtO_2C)_2\overset{\ominus}{C}H + Br\text{–}CH_2CH_2CH_2\text{–}Br + \overset{\ominus}{C}H(CO_2Et)_2$$

↓

$(EtO_2C)_2CH$–$CH_2CH_2CH_2$–$CH(CO_2Et)_2$

↓ (i) $2\overset{\ominus}{O}Et$ (ii) CH_2Br_2

$(EtO_2C)_2$–C₅ ring–$(CO_2Et)_2$ $\xrightarrow{\text{(i) KOH, (ii) } H_3\overset{\oplus}{O}\text{, (iii) 250°C}}$ HO_2C–C₅ ring–CO_2H

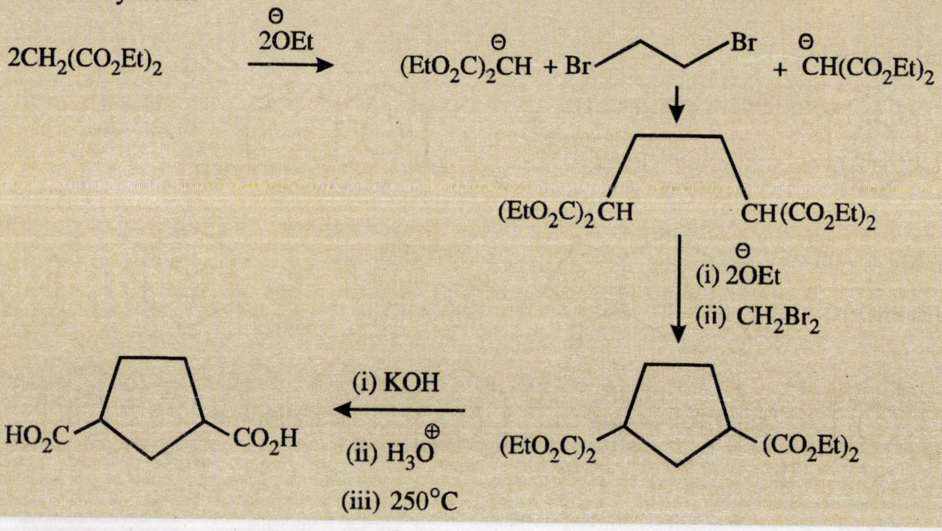

(ii) Ph, Ph, Ph, Ph (tetraphenylcyclopentadienone, O) $\xRightarrow{\text{FGI}}$ Ph Ph HO OH Ph Ph O $\xRightarrow[\text{aldol cond.}]{\text{1, 3-dio}}$

Ph Ph (ketone, O) + Ph Ph (O O) $\xRightarrow{\text{FGA}}$ CO_2Et Ph Ph O

$\Downarrow$ 1, 3-dico Claisen cond.

2Ph CO_2Et
SE

<O> $\Downarrow$ FGI

Ph Ph O OH

benzoin cond. $\Downarrow$ 1, 2-dio

2Ph CHO
SE

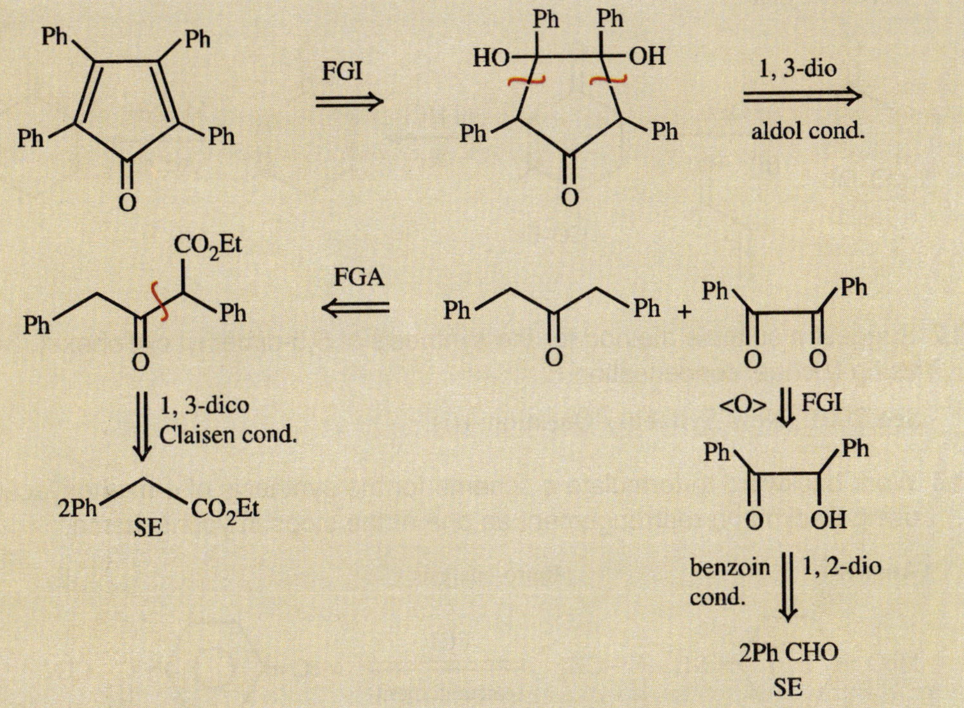

Forward Synthesis

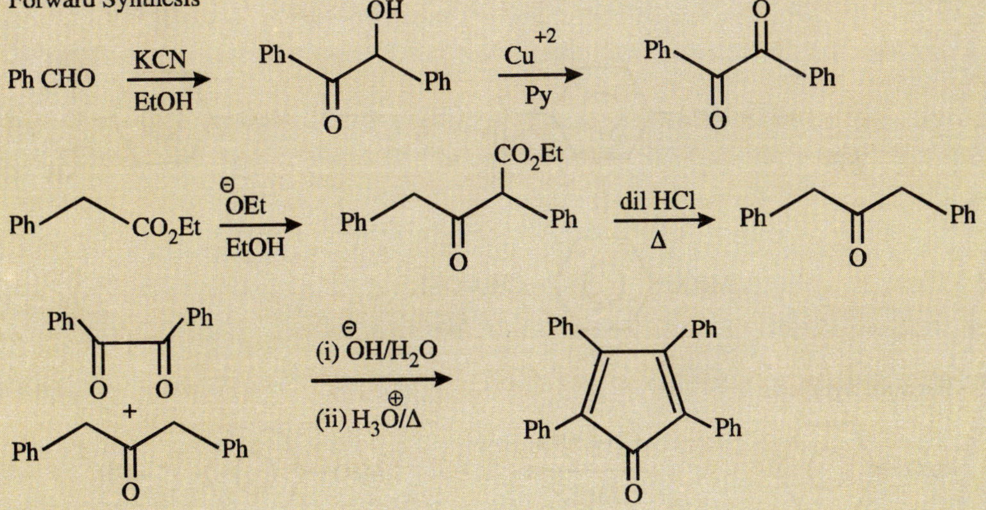

Ph CHO $\xrightarrow[\text{EtOH}]{\text{KCN}}$ Ph OH Ph O $\xrightarrow[\text{Py}]{Cu^{+2}}$ Ph O Ph O

Ph CO_2Et $\xrightarrow[\text{EtOH}]{\overset{\ominus}{O}Et}$ CO_2Et Ph Ph O $\xrightarrow[\Delta]{\text{dil HCl}}$ Ph Ph O

Ph Ph O O + Ph Ph O $\xrightarrow[\text{(ii) } H_3\overset{\oplus}{O}/\Delta]{\text{(i) } \overset{\ominus}{O}H/H_2O}$ Ph Ph Ph Ph O

(iii) O (cyclohexenone) $\xRightarrow[\text{aldol cond.}]{\alpha, \beta\text{-unsat. co}}$ O O $\xRightarrow{\text{FGA}}$ O O CO_2Et $\xRightarrow[\text{Michael cond.}]{\text{1, 5-dico}}$ O SE + O CO_2Et SE

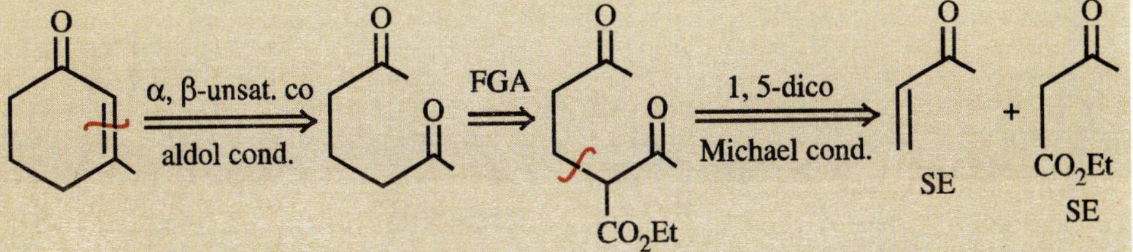

Forward Synthesis

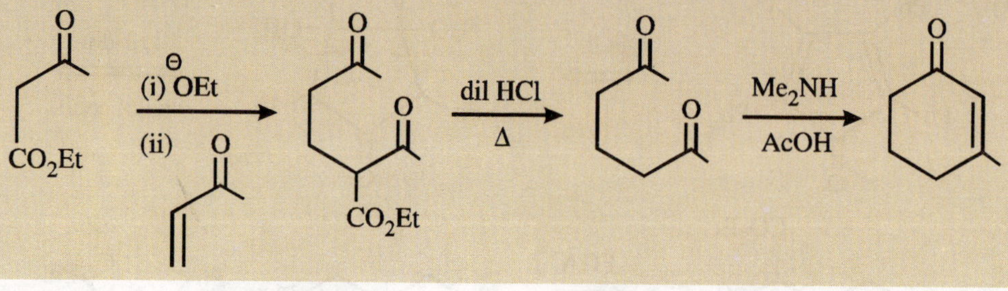

CO_2Et — (i) $\overset{\ominus}{O}Et$ (ii) methyl vinyl ketone → CO_2Et — dil HCl, Δ → — Me_2NH, AcOH →

Q.2 Suggest a suitable method for the synthesis of 5,5-dimethyl cyclohex-1, 3-dione using Michael condensation.

See 2001 (New Syllabus) Question 1(i).

Q.3 Work backward to formulate a scheme for the synthesis of 4-methoxyacetanilide using Beckmann rearrangement as one of the steps in your scheme.

Answer

Retro-analysis

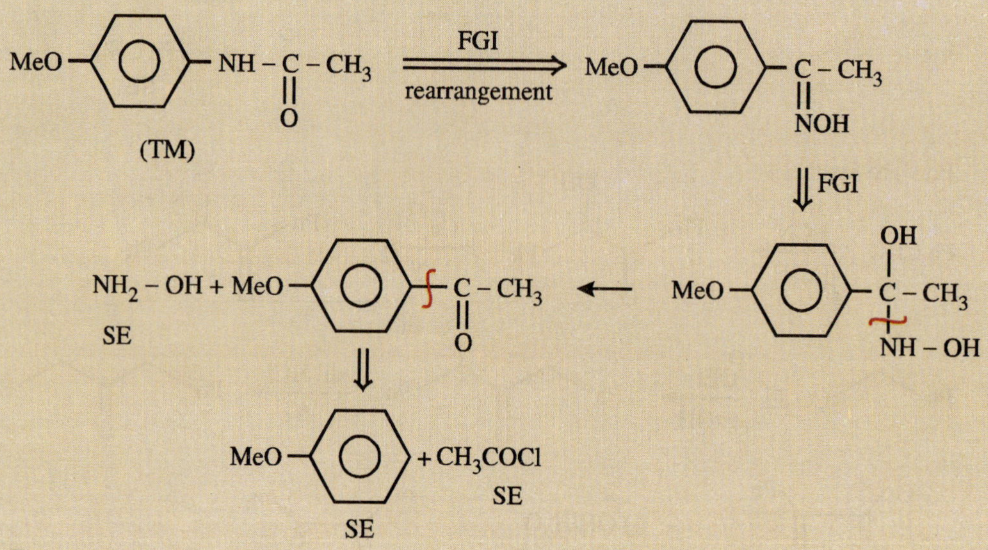

MeO–C₆H₄–NH–C(=O)–CH₃ (TM) $\xRightarrow[\text{rearrangement}]{\text{FGI}}$ MeO–C₆H₄–C(=NOH)–CH₃

⇓ FGI

MeO–C₆H₄–C(OH)(NH–OH)–CH₃ ⟵ NH_2 – OH + MeO–C₆H₄–C(=O)–CH₃
SE

⇓

MeO–C₆H₄ + CH_3COCl
SE SE

Forward Synthesis

MeO–C₆H₄ $\xrightarrow[AlCl_3]{CH_3COCl}$ MeO–C₆H₄–C(=O)–CH₃

↓ NH_2OH

MeO–C₆H₄–C(=NOH)–CH₃ $\xrightarrow[\text{(ii) } H_3\overset{\oplus}{O}]{\text{(i) } PCl_5}$ MeO–C₆H₄–NH–C(=O)–CH₃

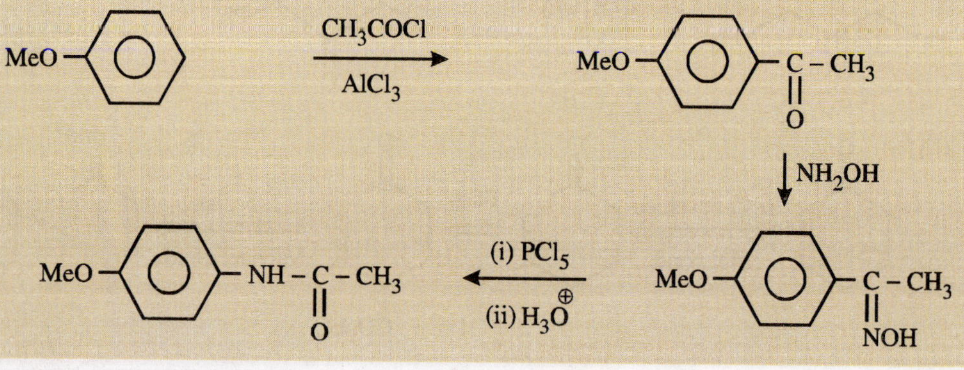

2003 (New Syllabus)

Q.1 Describe the synthesis of the following compounds with proper retrosynthetic analysis.

(i) (ii) 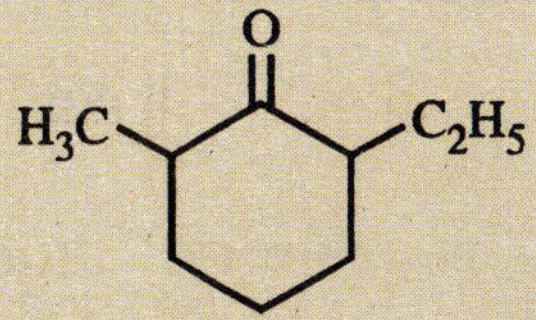 (iii) $Ph-CH=CH-CH_2-C(=O)-Ph$

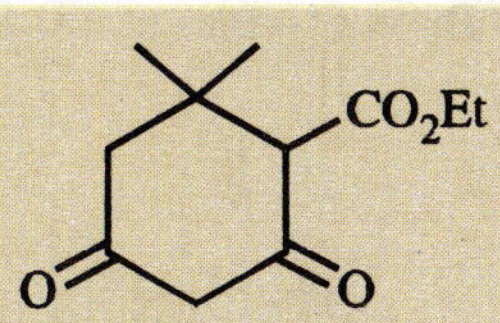

Answer

(i) See 2001 (New Syllabus) Question 1(i).

(ii) Retro-analysis

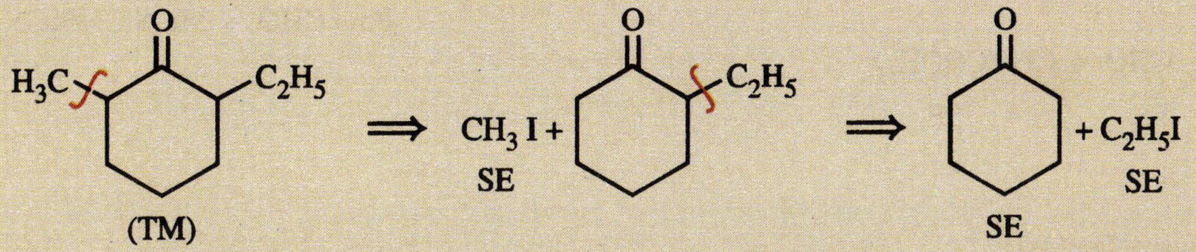

(TM) $\Rightarrow$ CH_3I + SE $\Rightarrow$ SE + C_2H_5I SE

Forward Synthesis

(i) $Me_3C-O^{\ominus}$ (ii) C_2H_5I → TsOH → CH_3I → $H_3O^{\oplus}$

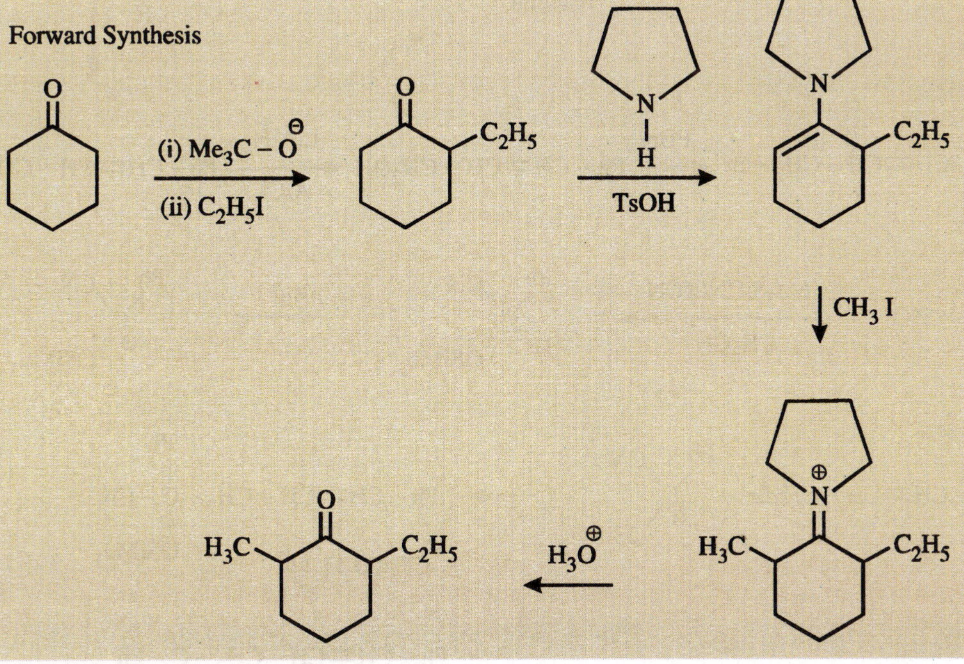

(iii) Method I

Retro-analysis

$$Ph-CH=CH-CH_2 \,/\!\!/\, C(=O)-Ph \Longrightarrow Ph-CH=CH-CH_2-Br + {}^{\ominus}C(=O)-Ph$$

$Ph-CH=CH-CH_2-Br$ $\Downarrow$ FGI $Ph-CH=CH-CH_2OH$

$Ph-CH=CH-CH_2OH$ $\overset{FGI}{\Longleftarrow}$ $Ph-CH=CH-COOH$

$Ph-CH=CH-COOH$ $\Downarrow$ FGI $Ph-CH(OH) \,/\!\!/\, CH_2-CO_2H$

$\Downarrow$

$Ph-CHO + CH_2(COOH)_2$

SE SE

${}^{\ominus}C(=O)-Ph$ $\Downarrow$ $Ph(CN)C^{\ominus}(OSiMe_3)$ (SE)

$\Downarrow$

$Ph-CHO + Me_3SiCN/KCN$

Forward Synthesis

$$Ph-CHO + CH_2(CO_2H)_2 + Py \xrightarrow[\text{reaction}]{\text{Knoevenagel}} Ph-CH=CH-CO_2H \xrightarrow[H^{\oplus}]{EtOH} Ph-CH=CH-CO_2Et$$

$$Ph-CH=CH-CO_2Et \xrightarrow[AlCl_3]{LiAlH_4} Ph-CH=CH-CH_2OH \xrightarrow{PBr_3} Ph-CH=CH-CH_2-Br$$

$$Ph-CHO \xrightarrow[CH_3CN]{Me_3SiCN/KCN} Ph(CN)CH(OSiMe_3) \xrightarrow{nBuLi} Ph(CN)C^{\ominus}(OSiMe_3)$$

$$Ph-CH=CH-CH_2Br + Ph(CN)C^{\ominus}(OSiMe_3) \longrightarrow Ph-CH=CH-CH_2-C(Ph)(OSiMe_3)-CN$$

$$\xrightarrow{CF_3CO_2H,\ \Delta} Ph-CH=CH-CH_2-C(=O)-Ph$$

Method II

Retro-analysis

$$\underset{\text{(TM)}}{Ph-CH{=}CH-CH_2 \nmid \underset{\underset{O}{\|}}{C}-Ph} \Longrightarrow Ph-CH{=}CH-CH_2MgBr + \underset{SE}{PhCN}$$

$$\Downarrow$$

$$\underset{SE}{Ph-CH{=}CH-CH_2Br}$$

Forward Synthesis

$$Ph-CH{=}CH-CH_2-Br \xrightarrow[\text{ether}]{Mg} Ph-CH{=}CH-CH_2-MgBr \xrightarrow{PhCN}$$

$$Ph-CH{=}CH-CH_2-\underset{\underset{O}{\|}}{C}-Ph \xleftarrow{H_3\overset{\oplus}{O}} Ph-CH{=}CH-CH_2-\underset{\underset{NMgBr}{\|}}{C}-Ph$$

Method III

Retro-analysis

$$\underset{\text{(TM)}}{Ph-CH{=}CH-CH_2-\underset{\underset{O}{\|}}{C}-Ph} \overset{FGA}{\Longrightarrow} Ph-C\equiv C \nmid CH_2-\underset{\underset{O}{\|}}{C}-Ph \Longrightarrow \underset{SE}{Ph-C\equiv C-H}$$

$$+$$

$$\underset{SE}{Br_2} + \underset{SE}{H_3C-\underset{\underset{O}{\|}}{C}-Ph} \Longleftarrow Br \nmid CH_2-\underset{\underset{O}{\|}}{C}-Ph$$

Forward Synthesis

$$H_3C-\underset{\underset{O}{\|}}{C}-Ph \overset{Br_2}{\underset{\overset{\oplus}{H}}{\Longrightarrow}} Br-CH_2-\underset{\underset{O}{\|}}{C}-Ph$$

$$Ph-C\equiv C-H \xrightarrow{\overset{\ominus}{N}H_2} Ph-C\equiv \overset{\ominus}{C} \xrightarrow{Br-CH_2-\underset{\underset{O}{\|}}{C}-Ph} Ph-C\equiv C-CH_2-\underset{\underset{O}{\|}}{C}-Ph$$

$$\Big\downarrow \; H_2/Pd/CaCO_3/S \text{ Lindlar cat.}$$

$$Ph-CH{=}CH-CH_2-\underset{\underset{O}{\|}}{C}-Ph$$

2003 (Old Syllabus)

Q.1 How would you prepare the following compounds:

(i) 2-methyl cyclopentane from diethyladipate;

(ii) 2, 2, 6, 6-tetramethyl-4-pipyridone from acetone.

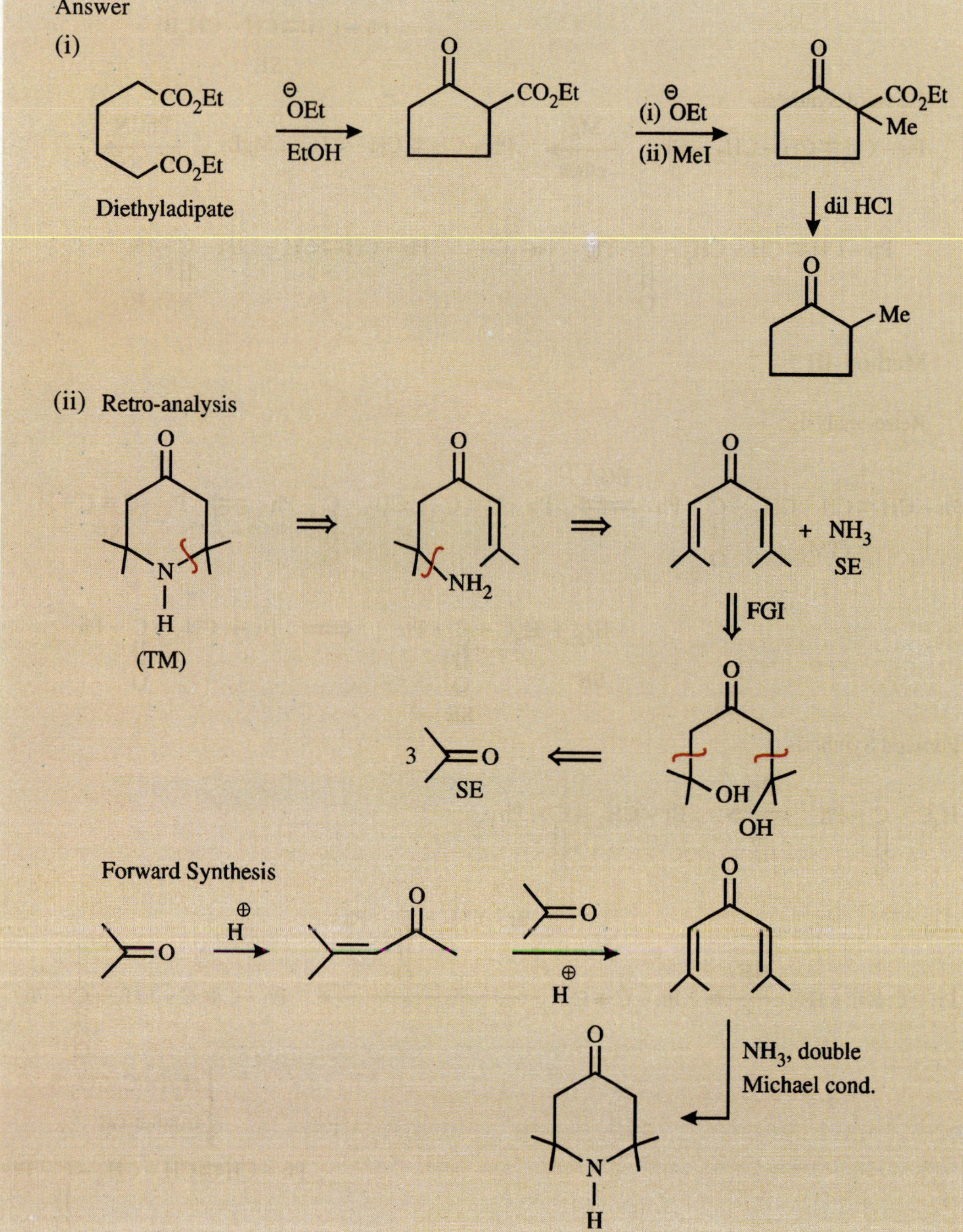

Q.2 Write down the synthesis of the following compounds:

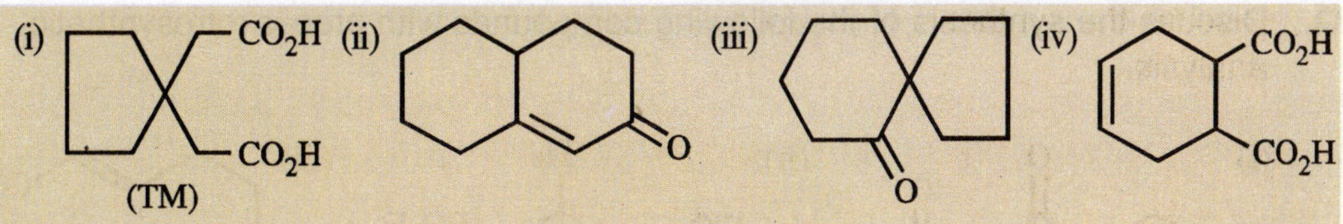

(i) (TM) (ii) (iii) (iv)

Answer

(i) Retro-analysis

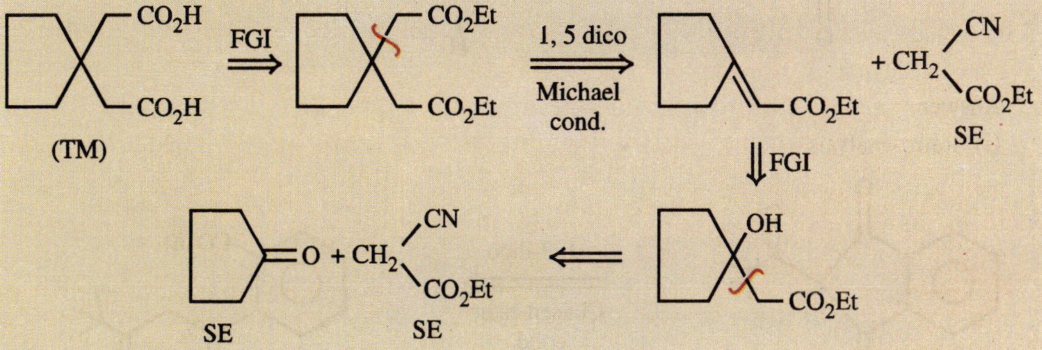

(TM) $\xRightarrow{\text{FGI}}$ $\xRightarrow[\text{Michael cond.}]{\text{1, 5 dico}}$ $+ CH_2(CN)CO_2Et$ SE

$\Downarrow$ FGI

SE $=O + CH_2(CN)CO_2Et$ SE $\Leftarrow$

Forward Synthesis

$=O + CH_2(CN)(CO_2Et) \xrightarrow{\overset{\ominus}{O}Et}$ $\xrightarrow[CH_2(CN)CO_2Et]{\overset{\ominus}{O}Et}$

CO_2H, CO_2H $\xleftarrow[\text{(ii) } H_3\overset{\oplus}{O} \text{ (iii) } 250°C]{\text{(i) KOH}}$ CN, CO_2Et, CO_2Et, CN

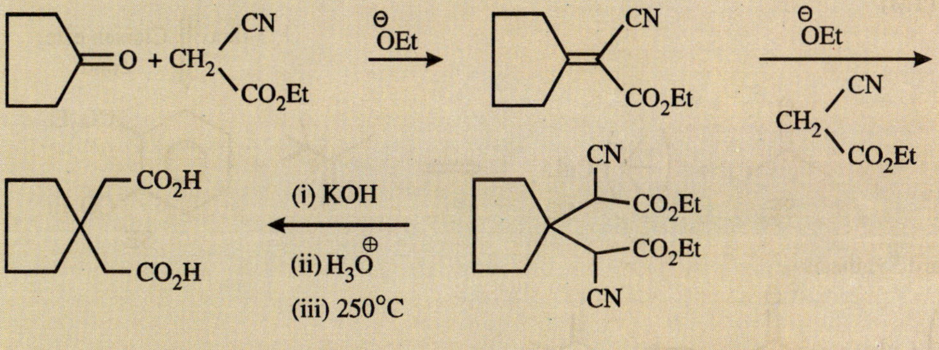

(ii) See 2000, Question 1(i).

(iii) See 2001 (New Syllabus), Question 2.

(iv) Retro-analysis

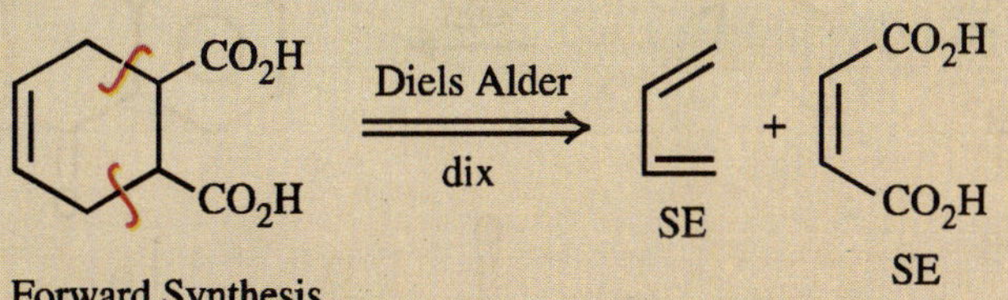

CO_2H, CO_2H $\xRightarrow[\text{dix}]{\text{Diels Alder}}$ SE + CO_2H, CO_2H SE

Forward Synthesis

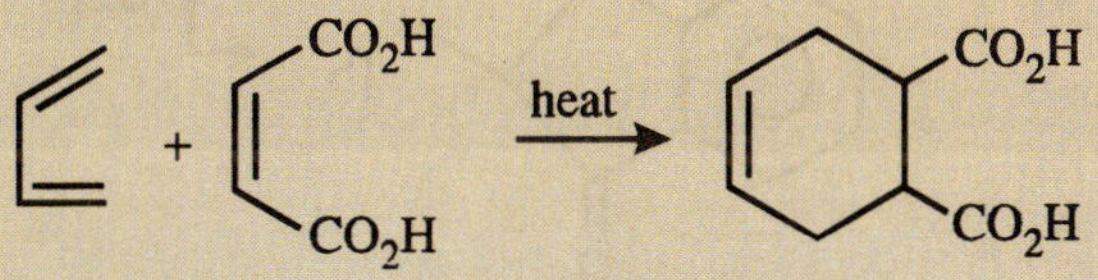

+ CO_2H, CO_2H $\xrightarrow{\text{heat}}$ CO_2H, CO_2H

2004 (New Syllabus)

Q. Discuss the synthesis of the following compounds with proper retrosynthetic analysis.

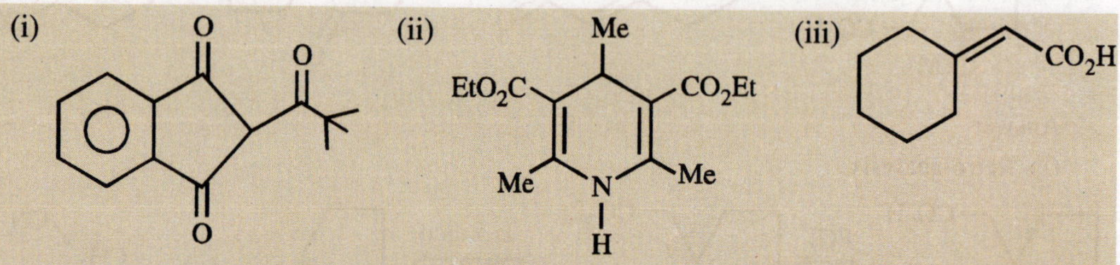

(i) (ii) Me, EtO_2C, CO_2Et, Me, N, Me, H (iii) CO_2H

Answer

(i) Retro-analysis

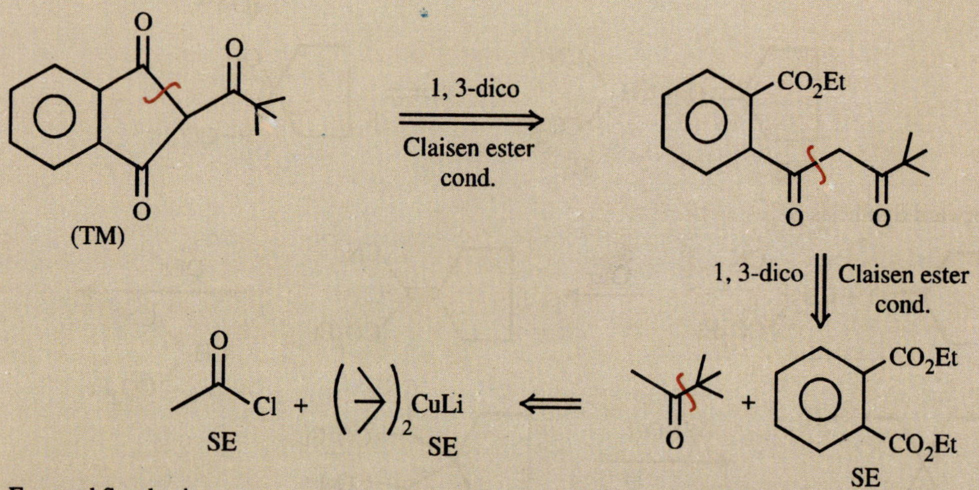

(TM) 1, 3-dico Claisen ester cond. CO_2Et

1, 3-dico Claisen ester cond.

Cl + $CuLi$ SE SE ⟸ O + CO_2Et CO_2Et SE

Forward Synthesis

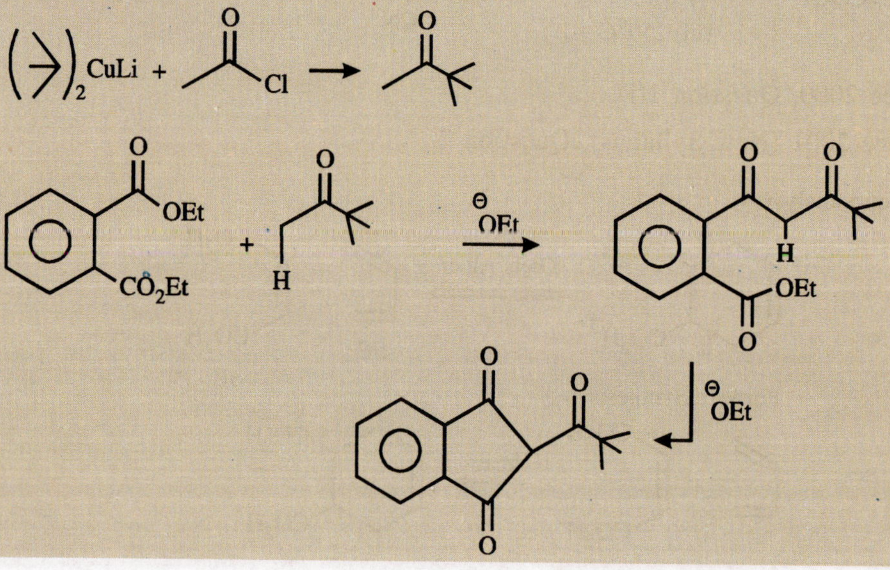

$CuLi$ + Cl → O

OEt CO_2Et + H $\overset{\ominus}{O}Et$ → H OEt

$\overset{\ominus}{O}Et$

(ii) Retro-analysis

(TM) $\overset{FGI}{\Longrightarrow}$ $\Longrightarrow$ $\Downarrow$

+ NH_3 SE

SE

Knoevenagel reaction

MeCHO + SE SE

Forward Synthesis

NH_3 Δ; $-H_2O$

SE

β-amino α, β unsat. ester
enamine - ambident nucleophile
Michael donor

MeCHO + ; NH_3

α, β- unsat. carbonyl/ Michael acceptor

Michael cond.

$-H_2O$

(iii) Retro-analysis

(TM) cyclohexylidene-CH-CO_2H $\xrightarrow{FGI}$ cyclohexylidene-CH-CO_2Et $\xrightarrow{FGI}$ 1-hydroxycyclohexyl-CH_2-CO_2Et (OH)

⇓

Cyclohexanone (SE) + $\overset{\ominus}{C}H_2-CO_2Et$ (SN)

⇓

$CH_2(CN)(CO_2Et)$
SE

$(EtO)_3P{=}CH-CO_2Et$
SE

(modified wittig reagent)

$BnZn-CH_2-CO_2Et$

(Reformatsky reagent)

$\overset{\oplus}{Li}\ \overset{\ominus}{C}H(SiMe_3)(CO_2Et)$
SE
(Peterson reagent)

Forward Synthesis

$$Br-CH_2CO_2Et \xrightarrow[\text{(ii) LDA}]{\text{(i) }(EtO)_3P} (EtO)_3P{=}CH-CO_2Et$$

$$Br-CH_2CO_2Et \xrightarrow{Zn/ether} BrZn-CH_2CO_2Et \xrightarrow{Me_3SiCl} CH_2(SiMe_3)(CO_2Et)$$

$$\downarrow PhLi$$

$$\overset{\oplus}{Li}\ \overset{\ominus}{C}H(SiMe_3)(CO_2Et)$$

$$\text{Cyclohexanone} + (EtO)_3P{=}CH-CO_2Et \rightarrow \text{Cyclohexylidene}{=}CH-CO_2Et \xrightarrow[\text{(ii) } H_3\overset{\oplus}{O}]{\text{(i) KOH}} \text{Cyclohexylidene}{=}CH-CO_2H$$

$$+ CH_2(CN)(CO_2Et) \xrightarrow{\overset{\ominus}{O}Et} \text{1-}O^{\ominus}\text{-cyclohexyl}-CH(CN)(CO_2Et) \xrightarrow[\text{(ii) } H_3\overset{\oplus}{O}/\Delta]{\text{(i) KOH}} \text{Cyclohexylidene}{=}CH-CO_2H$$

$$+ BnZnCH_2CO_2Et \rightarrow \text{1-OZnBr-cyclohexyl}-CH_2-CO_2Et \xrightarrow[\text{(ii) } H_3\overset{\oplus}{O}]{\text{(i) KOH}} \text{Cyclohexylidene}{=}CH-CO_2H$$

$$+ \overset{\oplus}{Li}\,\overset{\ominus}{C}H(SiMe_3)(CO_2Et) \rightarrow \text{1-}\overset{\ominus}{O}\overset{\oplus}{Li}\text{-cyclohexyl}-CH(SiMe_3)(CO_2Et)$$

$$\xrightarrow[\Delta]{-\,Me_3Si\overset{\ominus}{O}\,\overset{\oplus}{Li}} \text{Cyclohexylidene}{=}CH-CO_2Et \xrightarrow[\text{(ii) } H_3\overset{\oplus}{O}]{\text{(i) KOH}} \text{Cyclohexylidene}{=}CH-CO_2H$$

2004 (Old Syllabus)

Q.1 Illustrate the meaning of the following terms with example (i) Regioselective synthesis. (ii) Stereoselective synthesis.

Answer

(i) Regioselective synthesis*

The ultimate goal of any synthesis is to get a pure sample of the desired product by an efficient and convenient procedure. When devising a synthetic route to an organic molecule substituents and functional groups must be placed in the required position in an unsymmetrical situation i.e. with the correct regiochemistry—this is regioselective synthesis.

*For details see author's "Organic Synthesis" vol. 1

Examples:

(a) Regioselectivity of double bond fixing in an unsymmetrical position.

1 2 3 4 5 6 7 —(3, 4 double bond fixing)⟹ 1 2 3 4 5 6 7

FGA ⟹ ⇒ CHO + Ph_3P= (SE, SE)

⇓ FGI

⇒ I + H − C ≡ C − H + I (SE, SE, SE)

Forward Synthesis*

CHO + Ph_3P= —(mixed with / low temp.)→ —((i) PhLi / (ii) Me_3COH / warm)→

H − C ≡ C − H —($\overset{\ominus}{N}H_2$)→ H − C ≡ $\overset{\ominus}{C}$ —(I)→ H − C ≡ C —($\overset{\ominus}{N}H_2$)→ $\overset{\ominus}{C}$ ≡ C —(I)→ C ≡ C —(Li./Liq.NH_3 / Birch < H >)→ (H, H)

(b) Regioselectivity of alkylation to the unsymmetrical ketone.

O —(Me_3SiCl / Et_3N)→ $OSiMe_3$ —((i) Ph Br / (ii) CF_3CO_2H)→ O, Ph

O —((i) LDA/THF/0° / (ii) Ph Br)→ Ph, O

(c) Regioselectivity of hydration to the unsymmetrical double bond.

—((i) $Hg(OAc)_2$/THF / (ii) $NaBH_4/\overset{\ominus}{O}H$)→ OH, H

—((i) Sia_2BH / (ii) $\overset{\ominus}{O}H/H_2O_2$)→ H, OH

*For details see author's "Organic Synthesis" vol. 1

(ii) Stereoselective synthesis

A synthetic route to the target compound must be designed to enable the required pattern of carbon bonds to be constructed with all the functional groups and substituents with the control of relative orientation (stereochemistry).

Stereoisomers have the same carbon frame-work and the substituents have identical regiochemistry—the stereoisomers differ in their three dimensional spatial arrangement of atoms (groups) within the molecule (e.g. enantiomers and 3D, σ-diastereomers) or orientation of atoms (groups) around the rigid part with the molecule (e.g. double bond isomers-π diastereomers).

In any stereoselective synthesis mechanism offer alternative pathways and one is predominant over the other and the selectivity of this type is achieved through the presence of a physical barrier which hinders the formation of one stereoisomer (steric hindrance) and it happens through diastereomeric transition states with different free energy of activation.

Stereoselective synthesis is the synthesis of one enantiomer, one diastereomer or one double bond isomer in preference to others.

Examples:

(a) Stereoselective controlled with temperature.

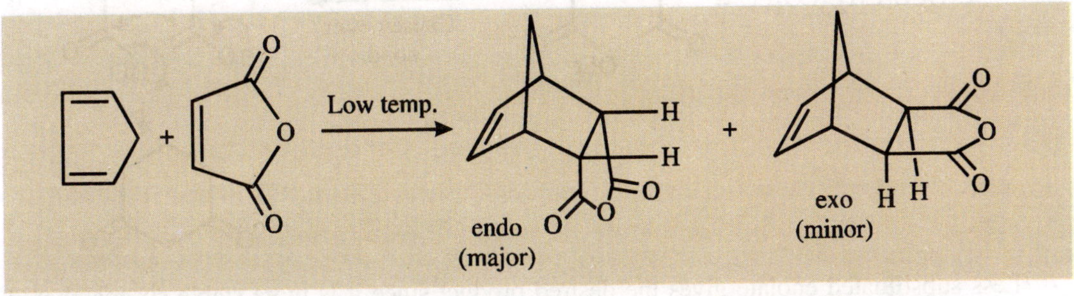

Low temp.

endo (major) exo (minor)

(b) Stereoselective controlled by steric hindrance in the transition state.

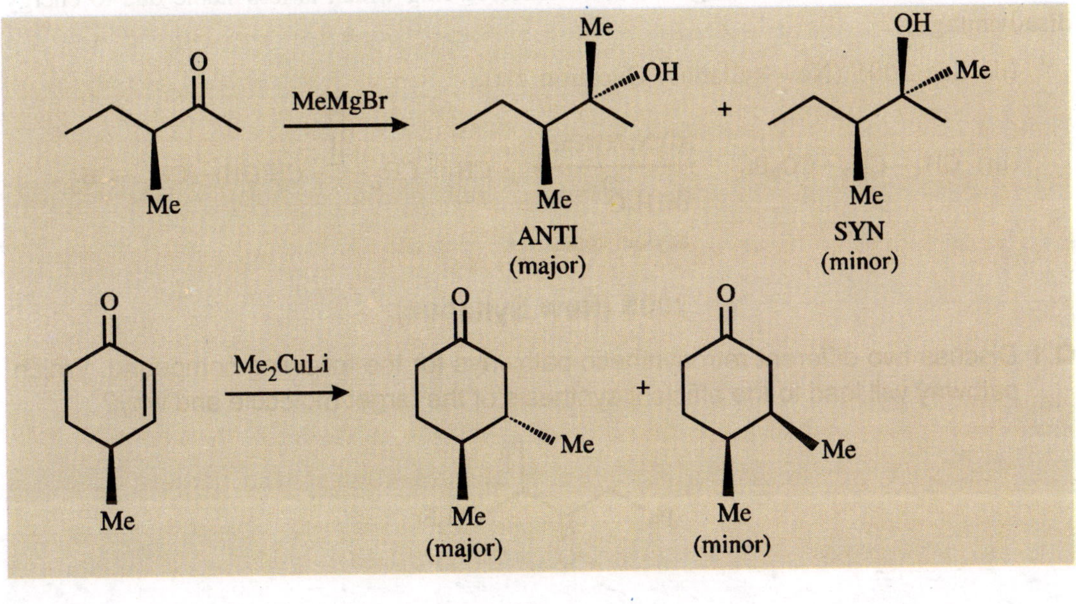

MeMgBr

ANTI (major) SYN (minor)

Me_2CuLi

(major) (minor)

(c) CH_2-COOH / $CH(OH)COOH$ (Malic acid) $\xrightarrow{H^{\oplus}}$ Fumaric acid (major) + Maleic acid (minor)

controlled by steric crowdedness of the ground state diastereomeric conformers

Q.2 How would you prepare the following compounds:

(i) cyclohexane-1,3-dione from $CH_3CO(CH_2)_3CO_2Et$

(ii) 5,5-dimethylcyclohexane-1,3-dione

(iii) $CH_3-CH_2-\overset{O}{\overset{||}{C}}-CH(OH)-CH_2-CH_3$ from $CH_3CH_2CO_2Et$

(i) **Answer**

$CH_3CO(CH_2)_3CO_2Et \equiv$ [structure] $\xrightarrow[\text{Claisen ester cond.}]{\overset{\ominus}{O}Et/EtOH}$ [enolate, cyclization] → cyclohexane-1,3-dione

Less substituted enolate gives the desired product since it is more stable six-membered ring (more substituted enolate gives four-membered ring which is less stable due to energy disadvantage).

(ii) See 2001 (New syllabus) Question 1(i).

(iii) $CH_3-CH_2-CO_2Et \xrightarrow[\text{(ii) } H_3\overset{\oplus}{O}]{\text{(i) Na/xylene}} CH_3-CH_2-\overset{O}{\overset{||}{C}}-CH(OH)-CH_2-CH_3$

acyloin cond.

2005 (New Syllabus)

Q.1 Discuss two different retrosynthetic pathways for the following compound. Which pathway will lead to the efficient synthesis of the target molecule and why?

$PhCH_2-CO-CH(Ph)-CO_2Et$

Answer Retrosynthetic analysis

$$\underset{\text{(TM)}}{PhCH_2COCH(Ph)CO_2Et} \xrightarrow{(1)} \underset{\text{SE}}{PhCH_2CO_2Et} + \underset{\text{SE}}{PhCH_2CO_2Et}$$

$$\underset{\text{(TM)}}{PhCH_2COCH(Ph)CO_2Et} \xrightarrow{(2)} \underset{\text{SE}}{PhCH_2COCH_2Ph} + EtO-CO_2Et \equiv Et_2CO_3$$

The target molecule is a β-keto ester and hence it may be synthesised by Claisen ester condensation. Therefore, retro-analysis (1) is most logical since it is giving two symmetrical (same) synthetic equivalent and therefore Claisen condensation would give only one product (target molecule) without giving any side product.

$$PhCH_2CO_2Et \xrightarrow[\text{EtOH}]{\ominus\text{OEt}} PhCH_2COCH(Ph)CO_2Et$$

The second retro-analysis (2) is not good one, since in addition to the Claisen ester condensation to give the desired target molecule, the keto compound (synthetic equivalent) may give base induced aldol condensation which would minimise the target molecule of β-keto ester giving impurity to the product.

$$PhCH_2COCH_2Ph + Et_2CO_3 \xrightarrow[\text{EtOH}]{\ominus\text{OEt}} PhCH_2COCH(Ph)CO_2Et + \underset{\text{side product}}{PhCH_2COCH(Ph)C(OH)(CH_2Ph)_2}$$

Q.2 Outline the synthesis of the following compound from benzene:

$$\text{4-}(CH_3)_2CHCH_2\text{-}C_6H_4\text{-}CH(CH_3)CO_2H$$

Answer

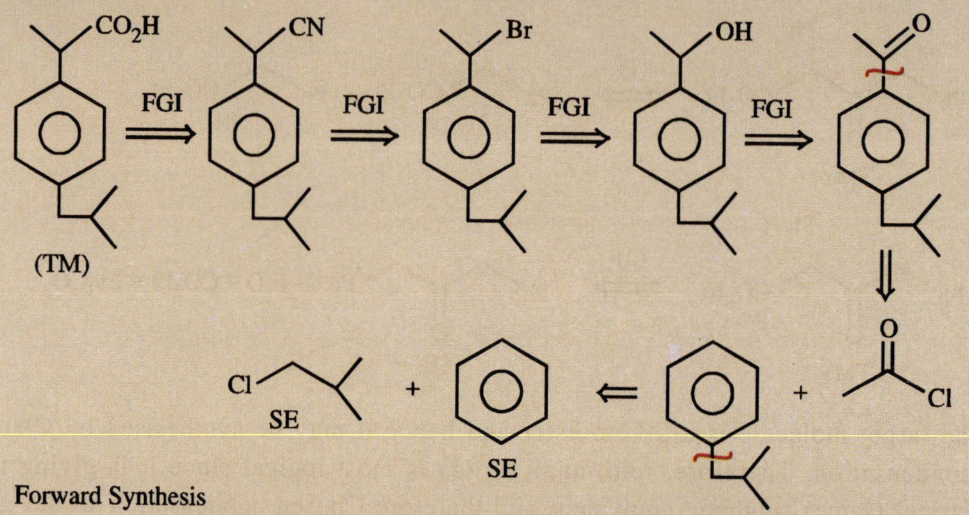

(TM) ⟹ FGI ⟹ FGI ⟹ FGI ⟹ FGI

CO_2H CN Br OH O

SE

Cl + SE ⟸ + Cl

O

Forward Synthesis

Cl

$FeCl_3$

No rearrangement of alkyl chloride with weak Lewis acid catalyst, $FeCl_3$

O Cl

$AlCl_3$

O

$LiAlH_4$

OH

(i) PBr_3

(ii) KCN

CN

$H_3O^{\oplus}$

CO_2H

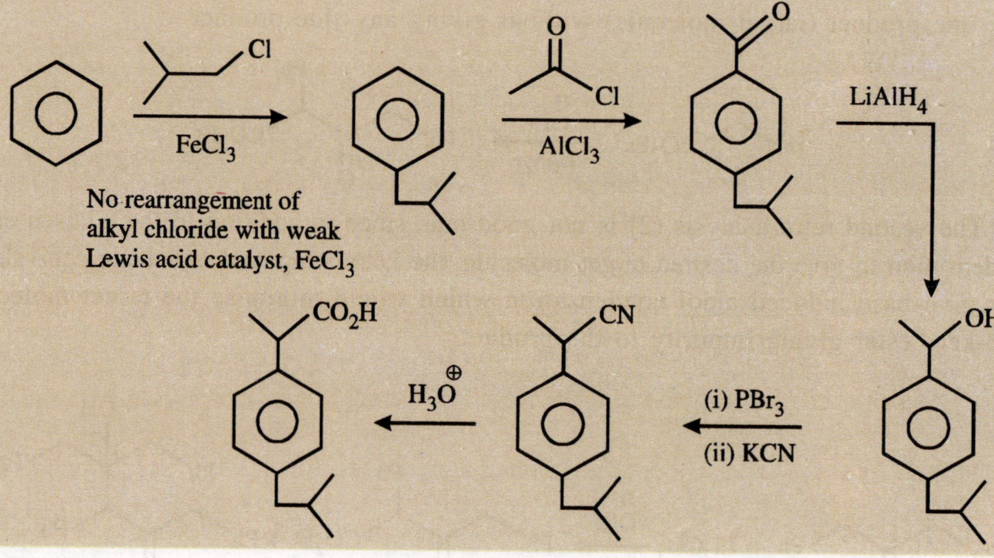

2005 (Old Syllabus)

Q.1 How can you synthesize the following compounds:

(i) O, O

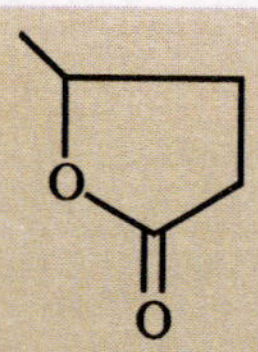

(ii) O, NH, O, NH, O

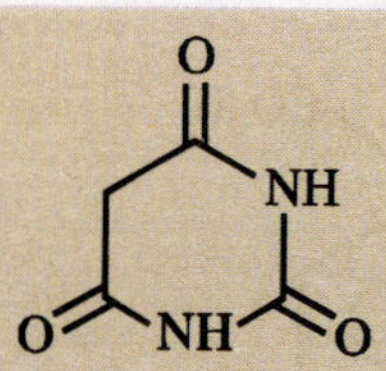

(iii) $CH_3-CH_2-\underset{\underset{O}{||}}{C}-CH_2-CO_2Et$ from Ethylacetoacetate

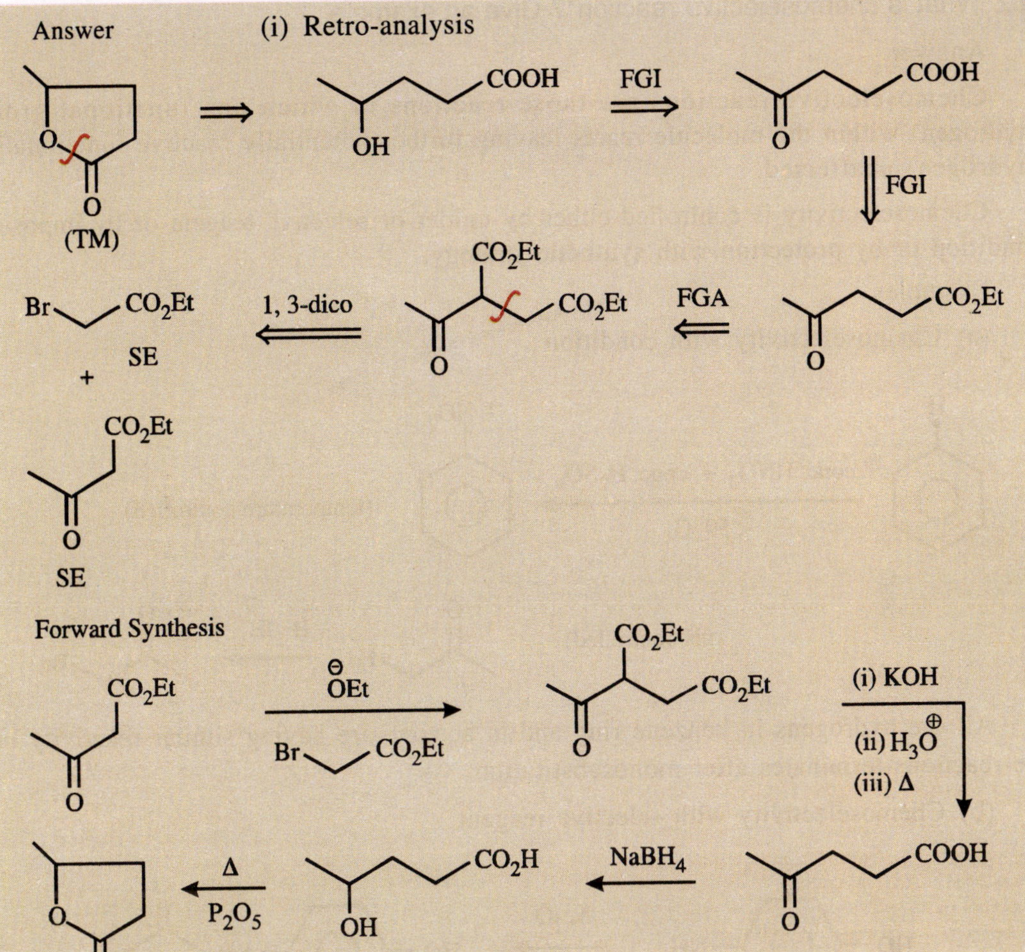

(ii) See 2001 (New Syllabus) Question 1(iii).

(iii) Retro-analysis

$$\underset{\text{(TM)}}{CH_3 - CH_2 - \underset{\underset{O}{\|}}{C} - CH_2 - CO_2Et} \Longrightarrow CH_3I + \underset{\text{Ethylacetoacetate}}{CH_3 - \underset{\underset{O}{\|}}{C} - CH_2 - CO_2Et}$$

Forward Synthesis

$$CH_3 - \underset{\underset{O}{\|}}{C} - CH_2 - CO_2Et \xrightarrow[\text{2eq.}]{\overset{\ominus}{N}H_2} \overset{\ominus}{C}H_2 - \underset{\underset{O}{\|}}{C} - \overset{\ominus}{C}H - CO_2Et \xrightarrow{CH_3I} CH_3 - CH_2 - \underset{\underset{O}{\|}}{C} - \overset{\ominus}{C}H - CO_2Et$$

↑ more nucleophilic

$$\downarrow \text{aq. } NH_4Cl$$

$$CH_3 - CH_2 - \underset{\underset{O}{\|}}{C} - CH_2 - CO_2Et$$

Q.2 What is chemoselective reaction*? Give an example.

Answer

Chemoselective reactions are those reactions in which one functional group (hydrogen) within the molecule reacts leaving further potentially reactive functionality (hydrogen) unaffected.

Chemoselectivity is controlled either by milder or selective reagent or by imposing condition or by protection with synthetic strategy.

Examples

(a) Chemoselectivity with condition

conc. HNO_3 + conc. H_2SO_4, < 50°C (temperature controlled)

(pH controlled) $Br_2/H^{\oplus}$

All the hydrogens in benzene ring and in acetone are having similar reactivity but the reactions terminates after monosubstitution.

(b) Chemoselectivity with selective reagent

HO–C₆H₄–NH_2 → (Ac_2O, H_2SO_4) → HO–C₆H₄–NH–COCH₃

NaHS, MeOH: m-dinitrobenzene → 3-nitroaniline (NO_2 / NH_2, NO_2)

(c) Chemoselectivity with protection

CO_2Et → (HOCH₂CH₂OH, dry HCl) → CO_2Et → ((i) $LiAlH_4$, (ii) $H_3O^{\oplus}$) → OH

*For details see author's "Organic Synthesis" vol. 1

2006 (New Syllabus)

Q.1(a) Write the retrosynthesis of the following compound in two different ways:

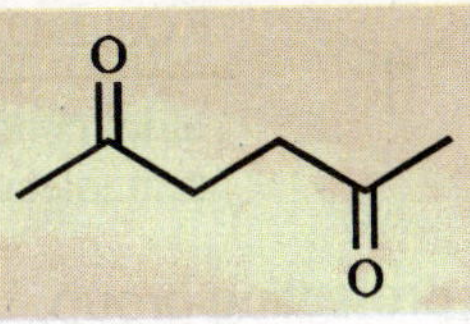

Give one synthetic method for the preparation of above compound. What happens when this compound is treated with ethanolic alkali at an elevated temperature?

Answer

Method I

Retro-analysis

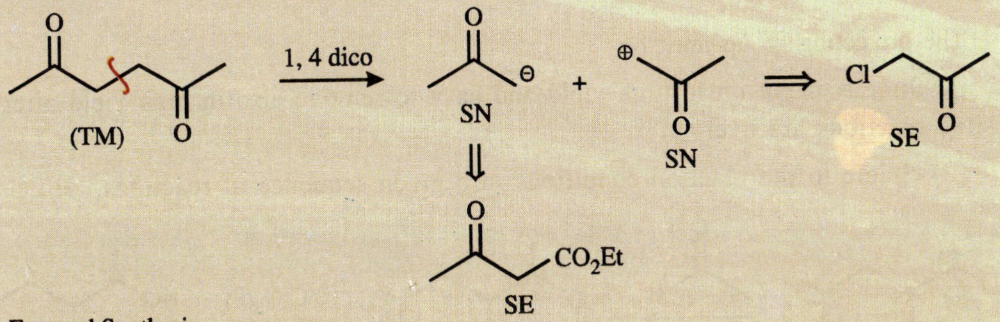

(TM) — 1, 4 dico → SN + SN ⇒ Cl (SE); SN ⇓ CO_2Et (SE)

Forward Synthesis

CO_2Et — (i) $\overset{\ominus}{O}Et$ (ii) Cl → CO_2Et — dil HCl, Δ →

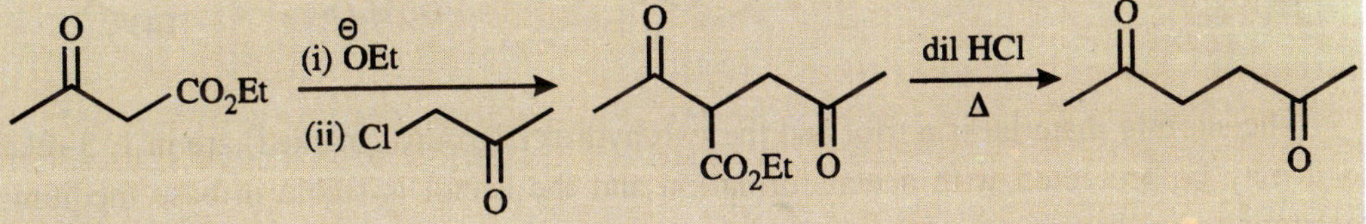

Method II

Retro-analysis

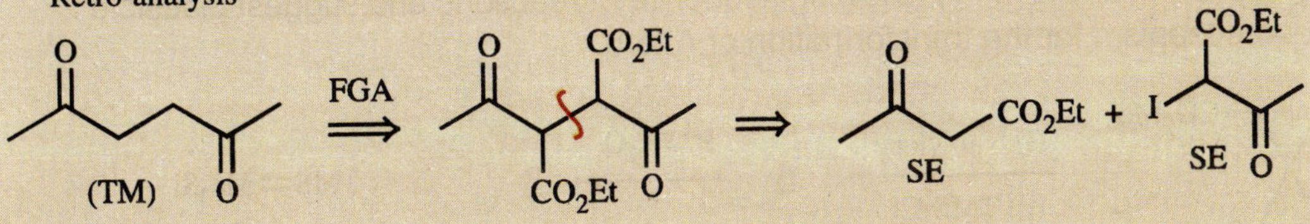

(TM) — FGA ⇒ CO_2Et, CO_2Et ⇒ CO_2Et (SE) + I, CO_2Et (SE)

Forward Synthesis

2 CO_2Et — $\overset{\ominus}{O}Et$ → 2 CO_2Et ⊖ — I_2 → CO_2Et, CO_2Et — dil HCl, Δ →

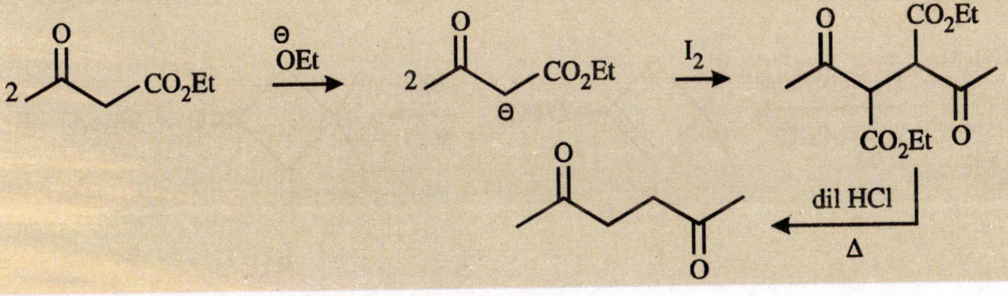

Last part

KOH/EtOH, aldol cond. intramolecular

(b) Mention two criteria for a good protecting group. Using protection/deprotection techniques outline the following transformation.

The protecting group must be

(i) Simple to put on in high yield and easy to remove also in high yield after the desired reactions are over.

(ii) Stable to the reaction conditions in a given sequence of reactions.

HCl gas; acetal; (i) CrO_3/Py (ii) $H_3O^{\oplus}$

The starting material is a triol and the two hydroxyl groups (marked) are in 1, 3-relation, so it may be protected with acetal formation and the acetal is stable in base medium and resistant to oxidising agent but unstable to aqueous acid and hence deprotected.

(c) Identify A and B in the following sequence of reactions and suggest plausible mechanism for the transformation of A to B.

(i) Na/xylene (ii) TMS-Cl → A; Br_2/H_2O → B; TMS $=$ Me_3Si-

Answer

(i) Na/xylene (ii) TMS-Cl → A (OTMS, OTMS); Br_2, H_2O → B + $2Me_3Si-Br$

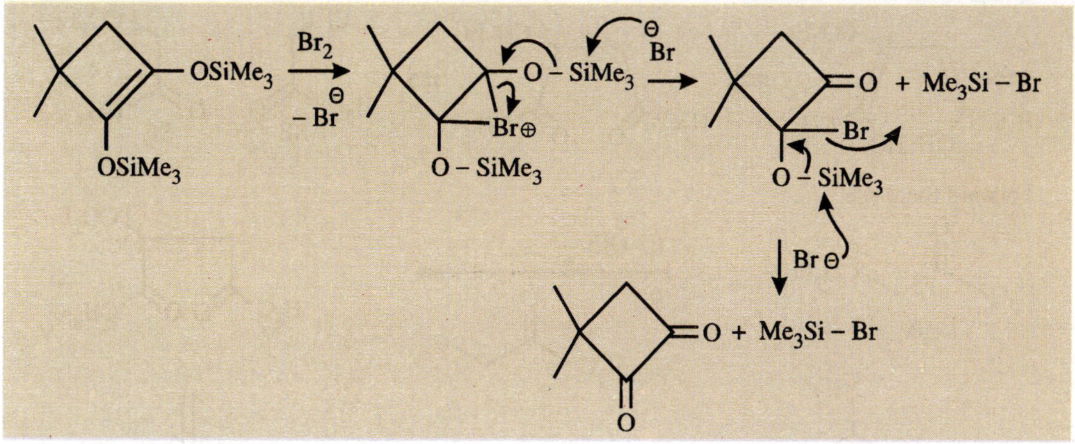

OSiMe$_3$ / OSiMe$_3$ $\xrightarrow[-\,Br^{\ominus}]{Br_2}$ Br⊕ / O – SiMe$_3$ / O – SiMe$_3$ ← $Br^{\ominus}$ → =O + Me$_3$Si – Br ; Br / O – SiMe$_3$ ↓ $Br^{\ominus}$ → =O / O + Me$_3$Si – Br

(d) Carry out the following conversions. Write the reaction conditions and the reagents involved.

(i) EAA ⟶ (1-(4-chlorophenyl)-2,5-dimethylpyrrole-3-carboxylic acid: CO_2H, H_3C, CH_3, N, Cl)

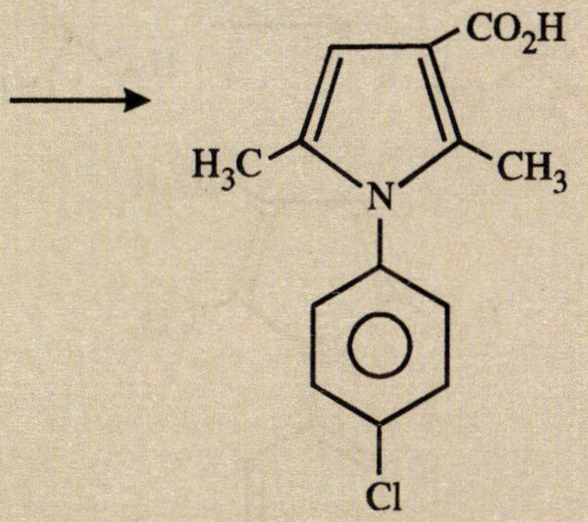

(ii) EAA ⟶ (CH_3, H_3C, N, CH_3)

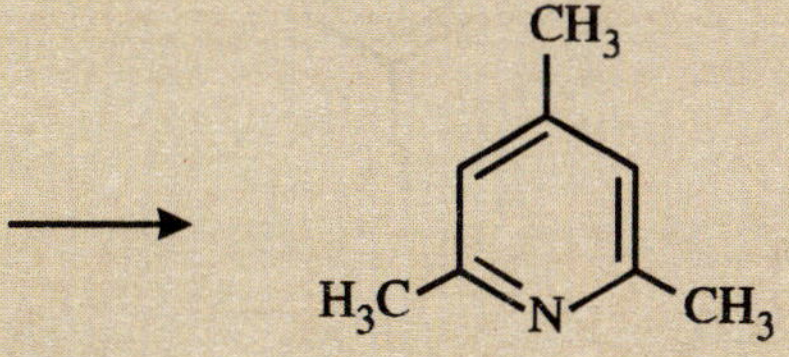

Answer

(i) Retro-analysis

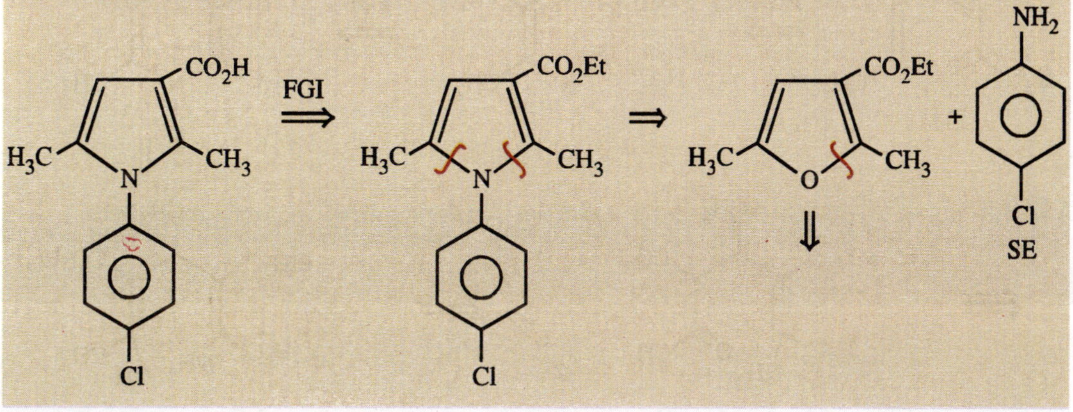

(CO_2H, H_3C, CH_3, N, Cl) $\xRightarrow{FGI}$ (CO_2Et, H_3C, CH_3, N, Cl) ⟹ (CO_2Et, H_3C, CH_3, O) ⇓ + (NH_2, Cl) SE

⇒ H_3C, OH HO, CO_2Et, CH_3 ⇒ H_3C, O O, CO_2Et, CH_3 ⇒ Cl, H_3C, O, SE + CO_2Et, O, CH_3, SE

Forward Synthesis

EAA (H_3C, O, CO_2Et) —(i) $\overset{\ominus}{O}Et$; (ii) H_3C, O, Cl→ H_3C, O O, CO_2Et, CH_3 ⇌ H_3C, OH OH, CO_2Et, CH_3 —P_2O_5→ H_3C, O, CO_2Et, CH_3

NH_2, Cl —reflux→ H_3C, N, CO_2Et, CH_3, Cl —(i) KOH; (ii) $H_3\overset{\oplus}{O}$→ H_3C, N, CO_2H, CH_3, Cl

(ii) Retro-analysis

(TM) CH_3, H_3C, N, CH_3 —FGA⇒ CH_3, EtO_2C, CO_2Et, H_3C, N, CH_3 —FGI⇒ CH_3, EtO_2C, CO_2Et, H_3C, N–H, CH_3 ⇓ CH_3, EtO_2C, CO_2Et, H_3C, NH_2 OH, CH_3 ⇐ CH_3, EtO_2C, CO_2Et, H_3C, NH_2, O, CH_3 ⇐

CH_3CHO + SE EAA

SE EAA + NH_3

Forward Synthesis

Step I

$\xrightarrow{\Delta}$ enamine Michael donor

Step II

(i) $\overset{\ominus}{O}Et$ (ii) CH_3CHO → Michael acecptor

Step III

Michael cond.

$-H_2O$

HNO_3

(i) KOH (ii) CaO/Δ

Q.2(a) Write two different retrosynthesis and one synthesis of the following compound.

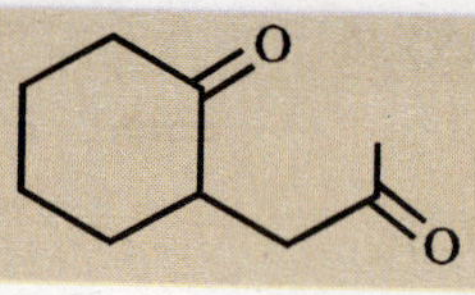

What happens when this compound is heated with acqueous ethanolic KOH?

Answer

Method I

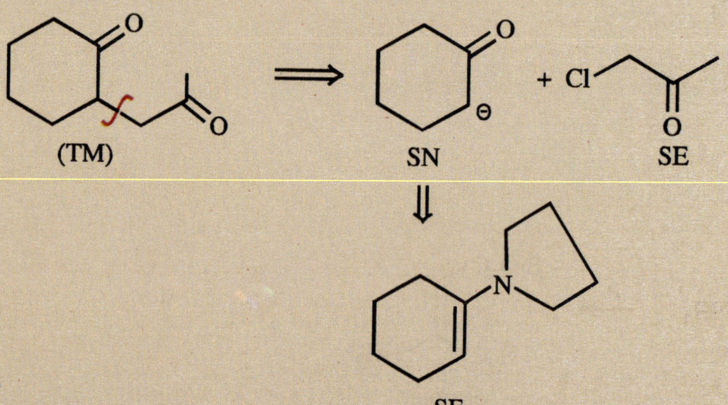

(TM) ⟹ SN + SE

SE

Forward Synthesis

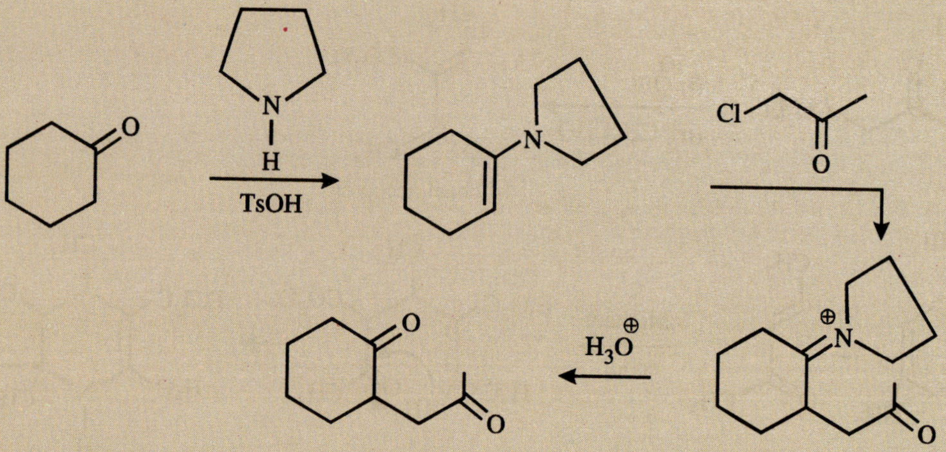

TsOH

$H_3O^{\oplus}$

Method II

Retro-analysis

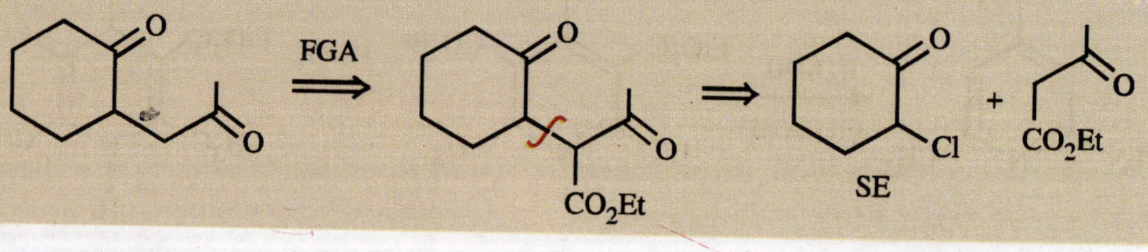

FGA

CO_2Et

SE

CO_2Et

Forward Synthesis

(i) $\overset{\ominus}{O}Et$ (ii) 2-chlorocyclohexanone; CO_2Et; NaCl / wet DMSO; CO_2Et

Last part

KOH / EtOH + H_2O / intra-aldol

(b) Alcoholic functionalities are often protected by making its t-butylether, as R—OH → R—$OCMe_3$.

What advantage does this t-butyl group provide? Mention the protection and deprotection techniques also.

Answer

Alcohol is protected by t-butyl as ether. Since protected function is ether and hence any reaction can survive in ether and in base medium and it is stable to usual redox agent and Grignard reagent which performs its reaction in ether medium.

Tertiary butyl ether may be easily solvolyzed (deprotected) to the original alcohol under mild condition with aqueous strong non-nucleophilic protic acid under S_N1 condition since t-butyl cation is stable due to three + I effect or nine σ-p conjugation.

Example

Br–CH₂CH₂CH₂–OH → (isobutylene, trace H_2SO_4) → Br–CH₂CH₂CH₂–O–t-Bu (protected) → (Mg, ether) → BrMg–CH₂CH₂CH₂–O–t-Bu → (D_2O) → D–CH₂CH₂CH₂–O–t-Bu → (dil H_2SO_4, S_N1) → D–CH₂CH₂CH₂–OH (deprotected)

(c) Identify the product of the following reaction and offer mechanism.

H_3C (epoxide O) ... CH_2-CH_3 (C=O) → BF_3-Et_2O / room temperature

BF_3 here is acting as a lewis acid and condensing agent with dehydration.

(d) Carry out following conversion. Write the reaction condition and reagents involved.

(i) Acetylacetone ⟶ 2, 5-Dimethyl pyrrole

(ii) β-Phenylethylamine ⟶ Isoquinoline

Answer

(i)

(i) aq. NaOH/ reflux (ii) $H_3O^{\oplus}$

(i) K_2CO_3/ acetone (i) Br

Acetylacetone

rewrite

NH_3

Δ $-2H_2O$

2, 5-Dimethyl pyrrole

(ii)

β-phenyl ethyl amine $\xrightarrow[\text{Py}]{CH_3COCl}$ $\xrightarrow{POCl_3}$ → → $\xrightarrow{-HOPOCl_2}$ $\xrightarrow[\Delta]{Pd-C}$ $\xrightarrow[\text{(ii) } H_3\overset{\oplus}{O}/\Delta]{\text{(i) } KMnO_4/\overset{\ominus}{O}H}$ Isoquinoline

N–H, O, Cl, P, Cl, Cl, N–H, ⊕, O – $POCl_2$, ⊕, H, N – H, O – $POCl_2$, N – H, $OPOCl_2$, N, N, N

(e) 2-Aminofuran is unstable but 5-aminofuran-2-carboxylate is relatively stable compound. With this idea in mind predict the product of the following reaction with mechanism.

MeO_2C–(furan)–NH_2 $\xrightarrow[\text{(ii) p–TsOH}/C_6H_6/\Delta]{\text{(i) } CH_2=CHCO\text{–}OCH_3/C_6H_6\text{/reflux}}$

Answer

NH_2, O, δ⊖, CO_2Me + CO_2Me, δ⊕ $\xrightarrow[\text{D.A}]{\Delta/C_6H_6}$ O, NH_2, CO_2Me, CO_2Me $\xrightarrow{p-TsOH}$

⊕O – H, NH_2, H, CO_2Me, CO_2Me ← NH_2, CO_2Me, H, MeO_2C, OH $\xrightarrow{-H_2O}$ NH_2, CO_2Me, CO_2Me

Q.3(i) Predict the product of the following reaction with mechanism.

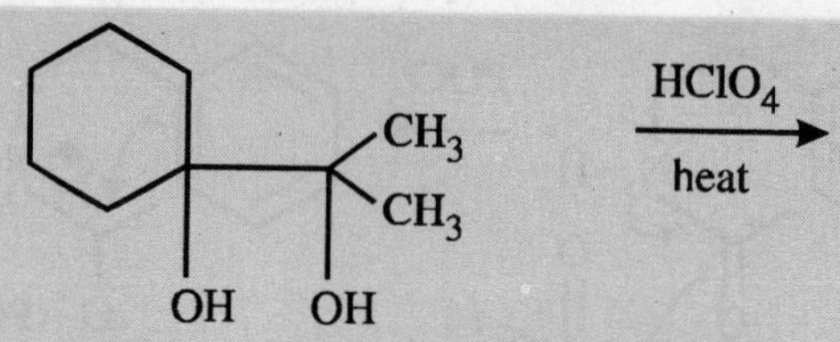

$HClO_4$, heat

Answer

$HClO_4$

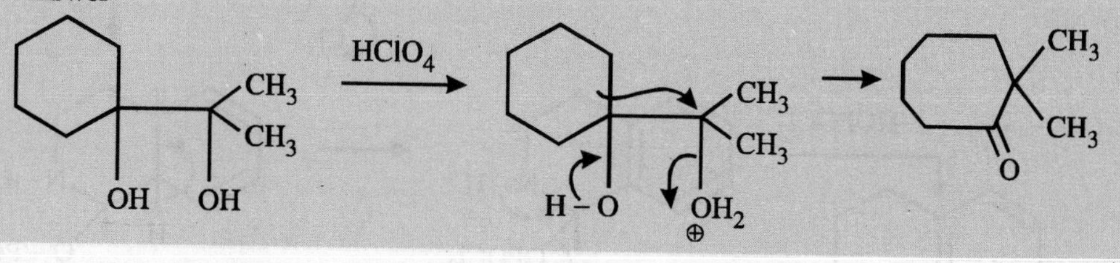

This is an example of Pinacol-Pinacolone rearrangement.

The reactant molecule is a Pinacol and both the alcohols are tertiary and one hydroxyl group is linked with the alicyclic part and the other is linked with the acyclic part. Now protonation occurs preferentially to the acyclic alcohol may be due to the corresponding (potential) acyclic carbonium ion is more stable (6 σ-p) than alicyclic carbonium ion (4 σ-p) and therefore ring-expansion occurs.

(ii) What happens when cis-3, 4-dimethyl cyclobutene is heated with maleic anhydride? Write the product with stereochemistry.

Answer

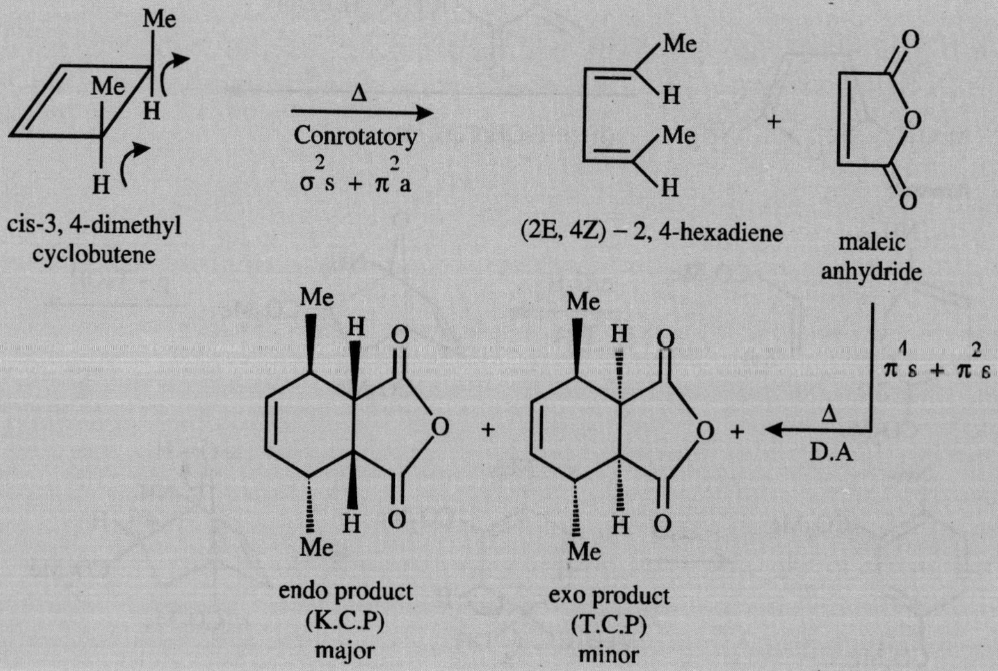

cis-3, 4-dimethyl cyclobutene

Δ, Conrotatory, $\sigma^2 s + \pi^2 a$

(2E, 4Z) – 2, 4-hexadiene + maleic anhydride

Δ, D.A, $\pi^4 s + \pi^2 s$

endo product (K.C.P) major + exo product (T.C.P) minor

Q.4(a) Write the plausible mechanism of the following reactions.

(i) 1-(2-hydroxy-2-propyl)cyclohexanol (OH, OH, CH_3, CH_3) $\xrightarrow[CH_2Cl_2]{BF_3.Et_2O}$ 1-methyl... ketone (CH_3, CH_3, O)

(ii) 1-ethynylcyclohexanol ($C{\equiv}C-H$, OH) $\xrightarrow[\Delta]{90\%\ HCO_2H}$ 1-acetylcyclohexene (CH_3, O)

Answer

(i) CH_3, CH_3, OH, OH $\xrightarrow{BF_3}$ CH_3, CH_3, $F_3\overset{\ominus}{B}-\overset{\oplus}{O}H$, O–H → CH_3, CH_3, O + $H_2O.BF_3$

In this case BF_3 is a weak (Lewis) acid and preferential co-ordination occurs with alicyclic alcohol. This may be because of the fact if acyclic alcohol co-ordinates with BF_3—the corresponding pinacolic carbon becomes highly crowded whereas co-ordination with alicyclic alcohol it is comparatively less crowded since alicyclic ring is tied-back (or held-back) structure.

(ii) $C{\equiv}C-H$, OH $\xrightarrow{HCOOH}$ H, $C{\equiv}C-H$, $\oplus OH_2$ $\xrightarrow{-H_2O}$ $C{\equiv}C-H$ + H–O–CH=O (H) $\xrightarrow{H_2\ddot{O}}$ $C=CH_2$ (O–H) ⇌ CH_3–C=O

(b) Outline the synthesis of Dewar benzene. Comment on its stability from symmetry point of view.

Answer

Synthesis of Dewar benzene

Δ / D.A ; Br_2 ; base

Or

D.A ; 140°C/Retro D.A, $- CO_2$

Retro-DA is effectively irreversible since cleavage product (CO_2) is volatile

hν, Disrotatory electrocyclic ring-closure ; $Pb(OAc)_4$/Py

Dewar benzene

Dewar benzene is less stable by 60 KCal/mole than its isomer benzene. Energy of activation (Ea) for conversion of Dewar benzene to benzene is 37 KCal. It has a half-life at room temperature of two days in pyridine. At 90°C complete conversion into benzene takes one-half hour.

Thermal ring-opening of the cyclobutene unit within the Dewar benzene (like the ring-closure reaction) must be a conrotatory process according to selection rules (symmetry criteria).

In the electrocyclic ring-opening reaction of Dewar benzene, the allowed thermal conrotatory process gives a highly strained molecule containing a trans double bond within a small ring.

Dewar benzene
(open book structure)

Δ conrotatory

HOMO

LUMO

$\sigma^2 s + \pi^2 a$

$1 + 0 = 1$, odd no. thermally allowed

Although the selection rule suggests that the reaction can occur but it does not because of strain (angle strain) in the product.

Ring-opening to the relatively unstrained all-cis-diene may occur with a disrotatory process— a process forbidden by the pericyclic selection rule (symmetry criteria) for a concerted 4ne reaction under thermal condition.

That is why Dewar benzene although a very unstable molecule is not spontaneously transformed to benzene under thermal condition.

Dewar benzene forms benzene when heated and this reaction requires a surprisingly high temperature and is believed not to be concerted (may be via free radical).

Q.5(a) Write any suitable method for the synthesis of L-proline.

Answer

KOH

$Br-CH(CO_2Et)_2$

(i) $\overset{\ominus}{O}Et$

(ii) $Br-(CH_2)_3-Br$

(i) NaOH

(ii) $H_3\overset{\oplus}{O}$/250°C

HCl/Δ

resolve

L -proline

dl -proline

(b) Synthesize the dipeptide L-alanyl-L-phenylalanine using suitable protecting groups

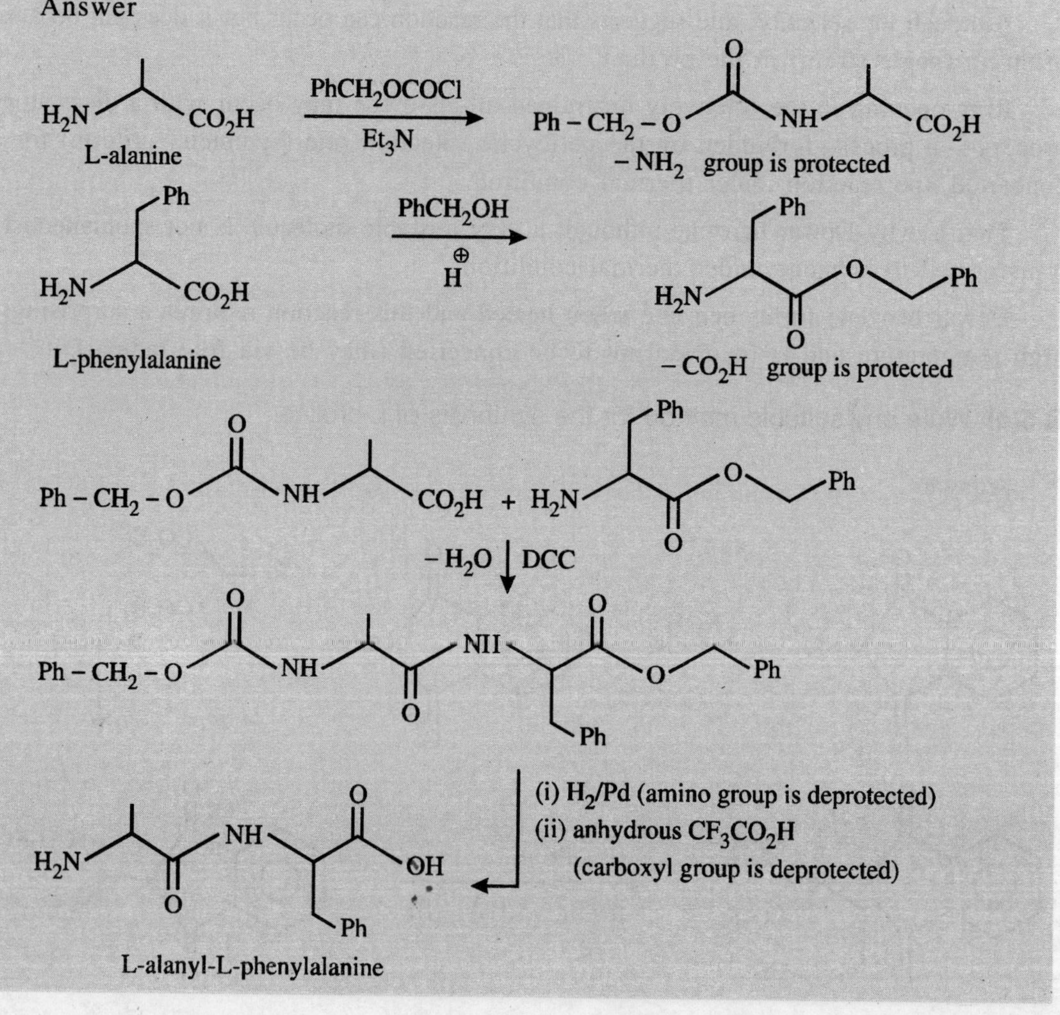

2006 (Old Syllabus)

Q.1(a) Discuss possible retrosynthetic analysis and write one efficient synthesis of the following compound.

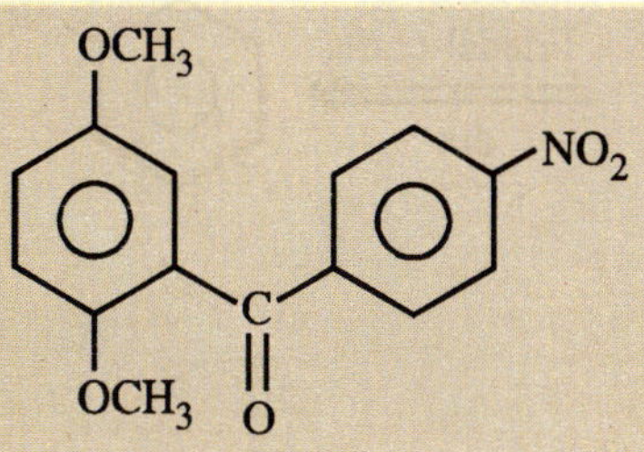

OCH$_3$, NO$_2$, C, OCH$_3$, O

Answer

Retrosynthetic analysis

Method I

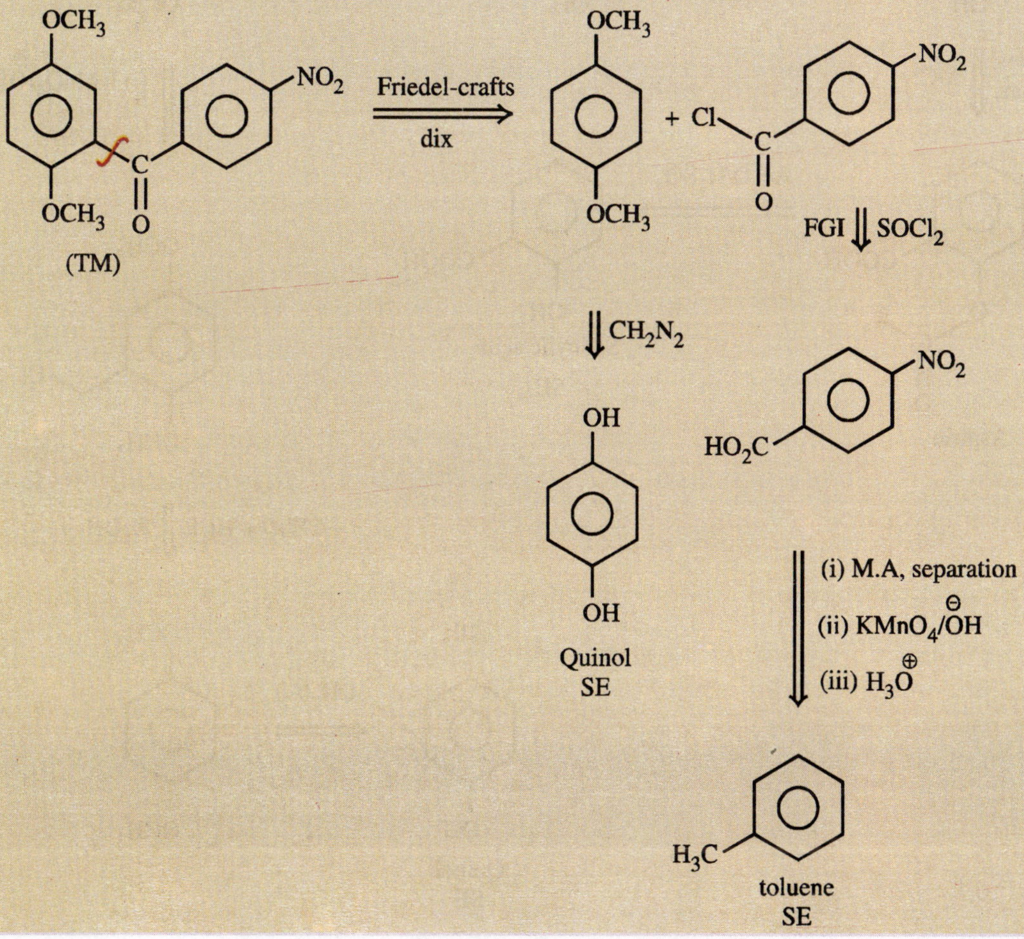

OCH$_3$, NO$_2$, C, OCH$_3$, O, (TM)

Friedel-crafts dix

OCH$_3$, OCH$_3$, + Cl, C, O, NO$_2$

FGI ⇓ SOCl$_2$

⇓ CH$_2$N$_2$

OH, OH, Quinol SE

HO$_2$C, NO$_2$

(i) M.A, separation

(ii) KMnO$_4$/$\overset{\ominus}{O}$H

(iii) H$_3\overset{\oplus}{O}$

H$_3$C, toluene SE

Method II

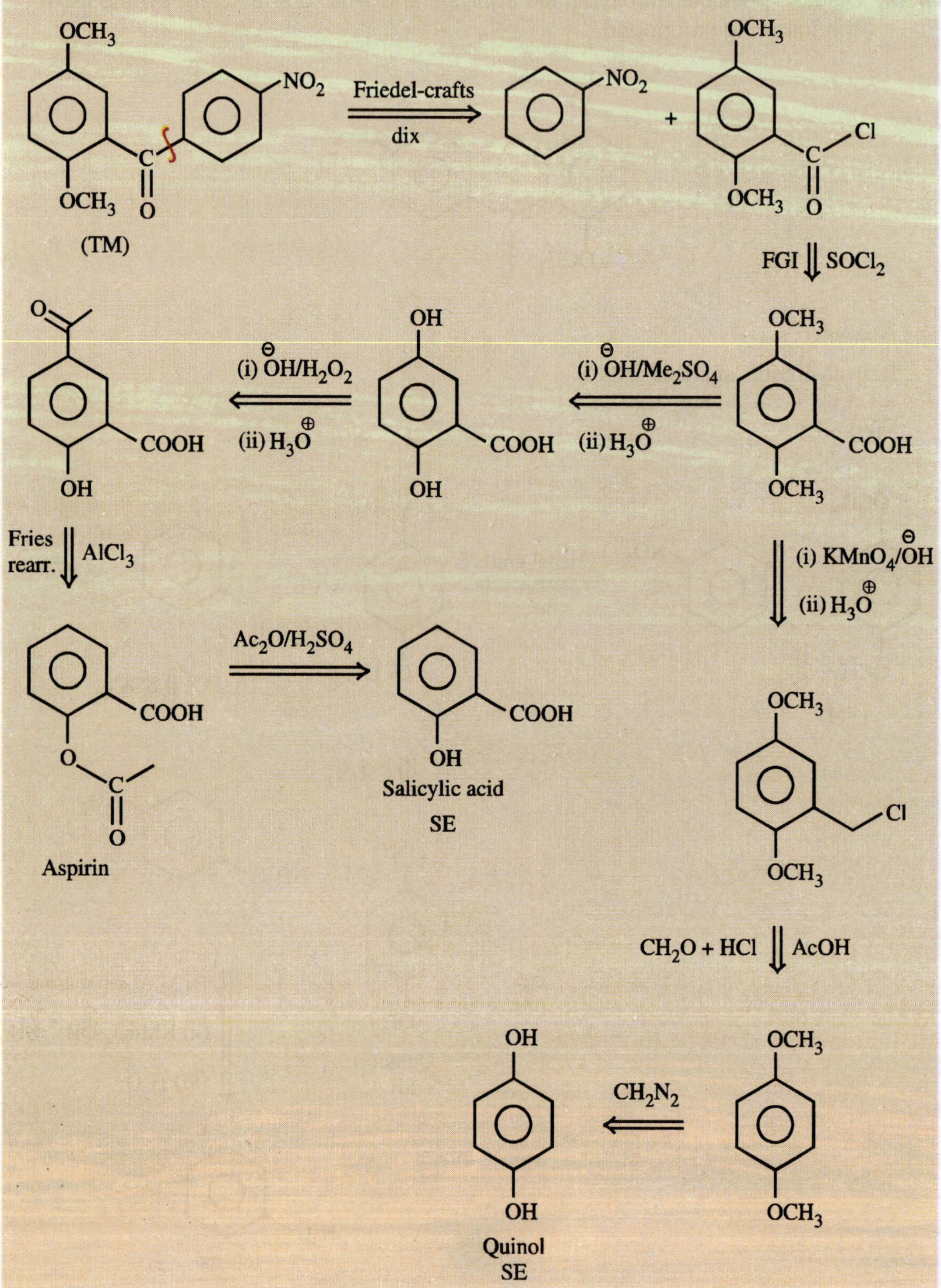

(TM)

Friedel-crafts

dix

$+$

FGI $\Downarrow$ $SOCl_2$

(i) $\overset{\ominus}{O}H/H_2O_2$

(ii) $H_3\overset{\oplus}{O}$

(i) $\overset{\ominus}{O}H/Me_2SO_4$

(ii) $H_3\overset{\oplus}{O}$

Fries rearr. $\Downarrow$ $AlCl_3$

(i) $KMnO_4/\overset{\ominus}{O}H$

(ii) $H_3\overset{\oplus}{O}$

Ac_2O/H_2SO_4

Aspirin

Salicylic acid

SE

$CH_2O + HCl$ $\Downarrow$ AcOH

CH_2N_2

Quinol

SE

Disconnection in method II is not good one because of two reasons (i) nitro group is meta orienting but in the TM it is para isomer (ii) nitro benzene does not react under Friedel-Crafts condition.

Therefore, the efficient synthesis is according to method I.

Cl–C(=O)–C₆H₄–NO_2 + p-dimethoxybenzene (OCH_3, OCH_3) + $AlCl_3$ ⟶ 2,5-dimethoxyphenyl–C(=O)–C₆H₄–NO_2

Method III

Retro-analysis

(TM) 2,5-dimethoxyphenyl–C(=O)–C₆H₄–NO_2 ⟹ 2,5-dimethoxybenzonitrile (OCH_3, CN, OCH_3) + ClMg–C₆H₄–NO_2

ClMg–C₆H₄–NO_2 ⟹ (i) M.A, separation (ii) Mg/ether ⟹ Chlorobenzene SE

2,5-dimethoxybenzonitrile ⟹ $Me_3\overset{\oplus}{O}\overset{\ominus}{BF_4}$ ⟹ 2,5-dihydroxybenzonitrile (OH, CN, OH)

2,5-dihydroxybenzonitrile ⟸ KCN, conc. H_2SO_4 ⟸ Parabenzoquinone SE

(Quinone : KCN = 1 : 1)

Forward Synthesis

OCH_3, CN, OCH_3 + ClMg, NO_2 (1 : 1) → OCH_3, OCH_3, C=NMgCl, NO_2

$H_3O^{\oplus}$ → OCH_3, OCH_3, C=O, NO_2

Method IV

OCH_3, OCH_3, C=O, NO_2 (TM) ⇒ OCH_3, MgCl, OCH_3 + NC, NO_2

⇓ (i) Cl_2 (ii) Mg/ether

OMe, OMe ⇒ (CH_2N_2) OH, OH Quinol SE

NC, NO_2 ⇓ KCN → Me_2N, NO_2 ⇓ HNO_3 → Me_2N, NO ⇓ HNO_2 → Me_2N

N, N-dimethyl aniline SE

Forward Synthesis

(b) How would you prepare both active and meso-variety of butane-2, 3-diol separately from cis-2-butene? Exaplain the reactions.

Answer

Epoxide of cis-2-butene and trans-2-butene on acid hydrolysis give active and meso-butane-2, 3-diol respectively.

Attack by H_2O to the epoxide of front and back carbon to the protonated epoxide occurs with equal facility because of same energy of activation and gives active dl pair.

Conversion of cis-2-butene to trans-2-butene

In place of Ph_3P, one can use $Me_3\overset{\ominus}{Si}(Me_3Si - SiMe_3 + NaOMe \rightarrow Me_3\overset{\ominus}{Si} + Me_3SiOMe)$ which gives trans-2-butene with similar mechanism.

SYN-elimination of $Me_3Si\overset{\ominus}{O}$

trans-2-butene

trans-2-butene $\xrightarrow{CF_3CO_3H}$ active (±) epoxide $\xrightarrow{H_3\overset{\oplus}{O}}$ meso-butane-2, 3-diol

(c) Predict the major product of the following reaction with reason.

$$\text{(R)-EtCOCH(Me) Bu}^t \xrightarrow[\text{(ii) } H_3\overset{\oplus}{O}]{\text{(i) LAH/THF}}$$

Answer

The reaction can be explained by Cram's open-chain model as the following way.

(R) → (R) attack of $\overset{\ominus}{H}$ from smaller (H) side

Me—H (R)

HO—H (S)

Et

erythro (major)

newly chiral centre is 'S'

Q.2 Predict the product of the following reactions explaining the mechanism involved.

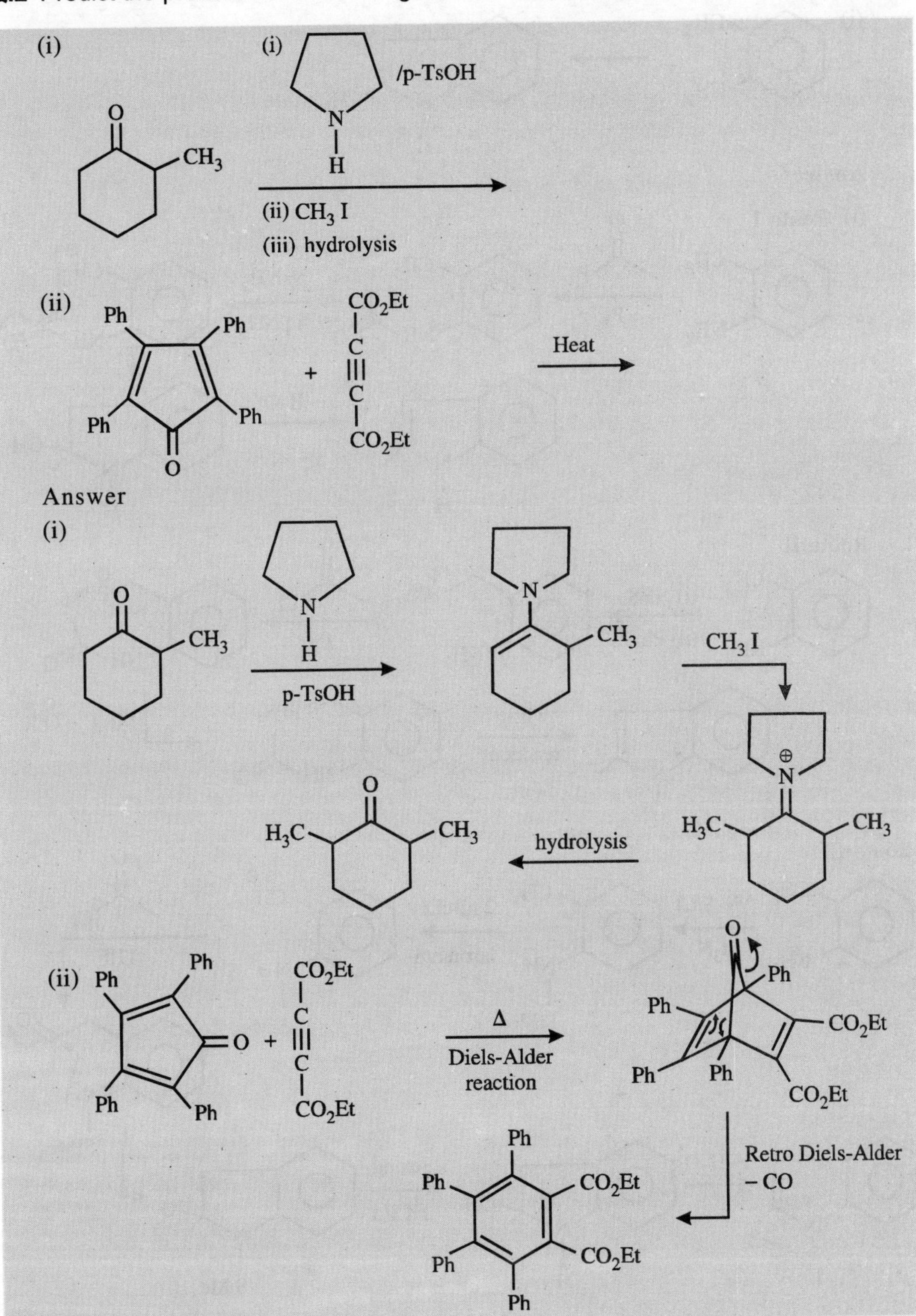

Q.3 Carry out the following conversion.

(i)

Answer

(i) Route I

$NaNH_2$ / liq.NH_3

$-H_2O$

Route II

(i) NBS (ii) Ph_3P

Py

$Me_3C-\overset{\ominus}{O}\overset{\oplus}{K}$

wittig reaction

Route III

Me_3SiCl / Et_3N

2 nBuLi / hexane/Δ

THF

strong HCl/Δ

$-H_2O$

(ii) EAA ⟶ [diethyl 2,4,6-trimethylpyridine-3,5-dicarboxylate: pyridine ring with Me, EtO_2C, CO_2Et, Me, Me]

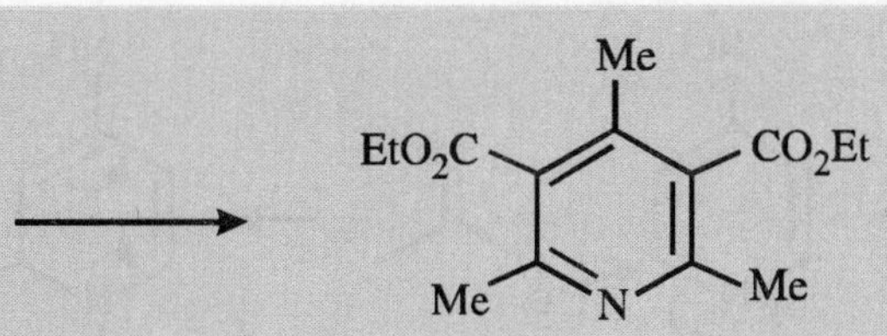

see Question 1d(ii), 2006 (New syllabus).

(iii) Identify the product and offer mechanism

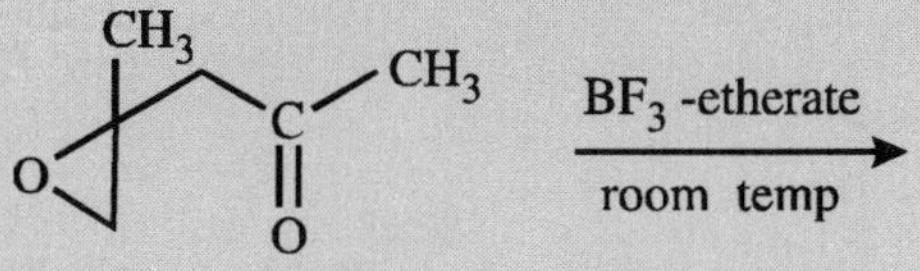

[2-methyl-2-(2-oxopropyl)oxirane: CH_3, CH_3, C=O] $\xrightarrow[\text{room temp}]{BF_3\text{-etherate}}$

see Question 2(c), 2006 (New syllabus).

(iv) Identify the products(s) of the following reaction and offer mechanism

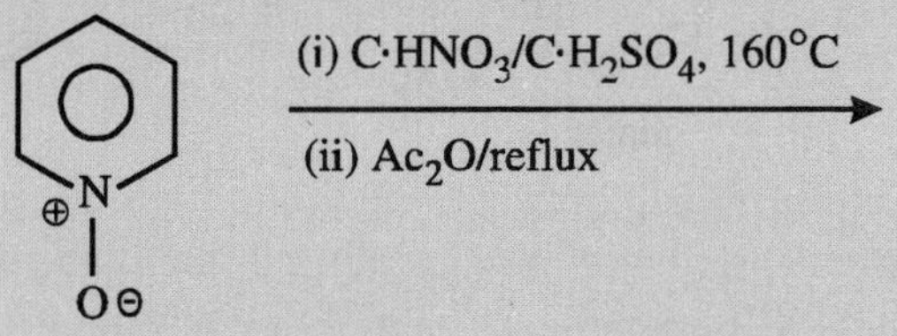

[pyridine N-oxide] $\xrightarrow[\text{(ii) } Ac_2O/\text{reflux}]{\text{(i) C·}HNO_3\text{/C·}H_2SO_4\text{, 160°C}}$

Answer

[pyridine N-oxide] $\xrightarrow[\text{160°C}]{\text{C·}HNO_3\text{/C·}H_2SO_4}$ [4-nitropyridine N-oxide, NO_2]

[4-nitropyridine N-oxide, NO_2] $\xrightarrow[\text{reflux}]{Ac_2O}$ [4-nitro-2-acetoxypyridine, NO_2, OAc] + [4-nitro-3-acetoxypyridine, NO_2, OAc] + [N-(4-nitropyridin-2-yl)-4-nitro-2-pyridone, NO_2, NO_2, O]

major minor minor

[pyridine N-oxide with $O_2N^{\oplus}$, curved arrows] $\xrightarrow{\text{C·}HNO_3\text{/C·}H_2SO_4}$ [intermediate with O_2N, H, N=O, curved arrows] ⟶ [4-nitropyridine N-oxide, NO_2]

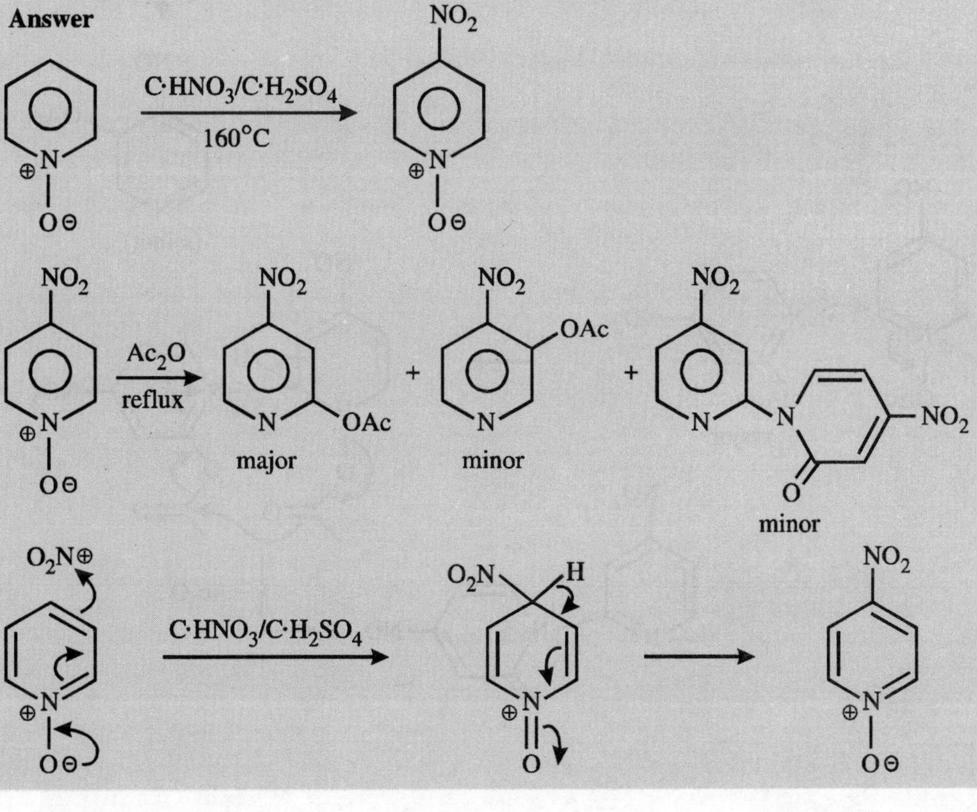

NO$_2$ ⊕N O⊖ O O O NO$_2$ ⊕N OAc O O⊖ NO$_2$ N OAc O O

$-$ AcO$^{\ominus}$

NO$_2$ N OAc major $-$ H$_1^{\oplus}$ O$_2$N less acidic (H$_2$) H$_2$ O ⊕ N O H$_1$ more acidic(H$_1$)

$-$ H$_2^{\oplus}$

NO$_2$ OAc N (minor)

NO$_2$ ⊕N OAc :N NO$_2$ OAc major NO$_2$ N H ⊕N NO$_2$ O O O O

NO$_2$ N N NO$_2$ O $-$ Ac$_2$O

Q.4 (i) Outline a synthesis of L-triptophan.

Answer

Indole

$CH_2{=}\overset{\oplus}{N}Me_2$

$CH_2O + Me_2NH{\cdot}HCl$

NMe_2

$AcNH\overset{\ominus}{C}(CO_2Et)CO_2Et$/toluene/Δ

CO_2Et CO_2Et NHAc

(i) aq. NaOH
(ii) dil H_2SO_4
(ii) 250°C

CO_2H NH_2

H dl pair

resolve

CO_2H NH_2

L-triptophan

(ii) Write the synthesis of the following compound showing its retrosynthesis.

Answer

Retro-analysis

CO_2H O H_3C Cl

(TM) ⇒ OH, H_3C, Cl + $\overset{\oplus}{C}H_2-CO_2H$ ⇒ $Br-CH_2-CO_2H$

SN SE

Cl_2/HCl

Red P/Br_2 | H.V.Z reaction

CH_3CO_2H

OH H_3C

ortho cresol

SE

Forward Synthesis

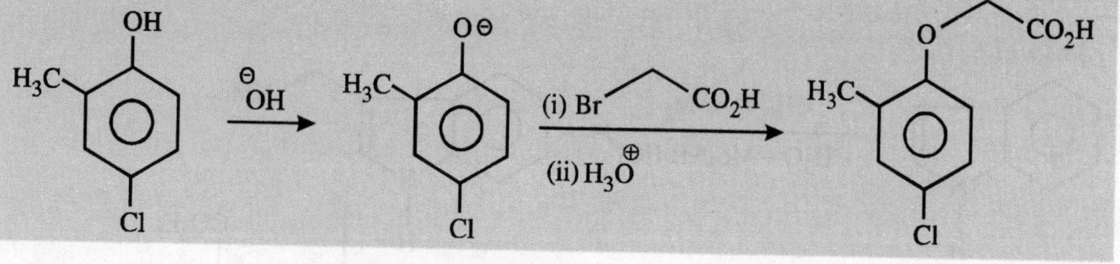

OH, H_3C, Cl → ($^{\ominus}OH$) → $O^{\ominus}$, H_3C, Cl → (i) Br–CH_2–CO_2H (ii) $H_3O^{\oplus}$ → O–CH_2–CO_2H, H_3C, Cl

(iii) Synthesize L-serine from phthalimide.

Answer

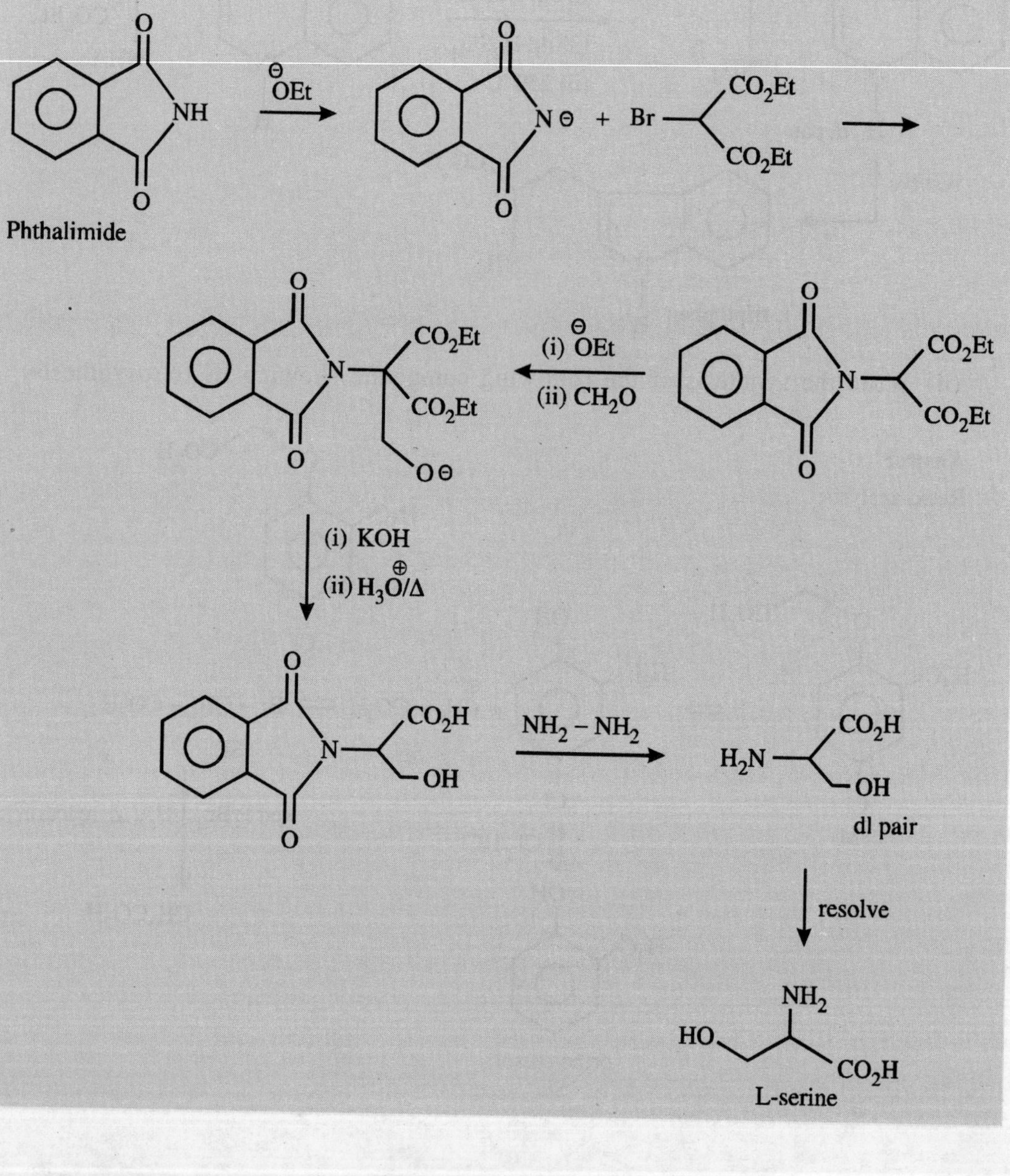

Phthalimide (O, NH, O) → ($^{\ominus}OEt$) → $N^{\ominus}$ + Br–CH(CO_2Et)(CO_2Et) →

N–CH(CO_2Et)(CO_2Et) → (i) $^{\ominus}OEt$ (ii) CH_2O → N–C(CO_2Et)(CO_2Et)–$CH_2O^{\ominus}$

↓ (i) KOH (ii) $H_3O^{\oplus}/\Delta$

N–CH(CO_2H)–OH → (NH_2-NH_2) → H_2N–CH(CO_2H)–OH

dl pair

↓ resolve

NH_2, HO, CO_2H

L-serine

Glossary of Terms

Acyclic Non-cyclic.

Acyl group A carbonyl group bonded to an alkyl or aryl group (RCO –).

Aprotic Normally applied to solvent, meaning devoid of acidic protons e.g. $HCONMe_2$(DMF), Me_2SO(DMSO).

Annulation reaction A ring forming reaction.

Angular alkylation When alkylation is done at the common carbon atom in a fused bi-or polycyclic ring.

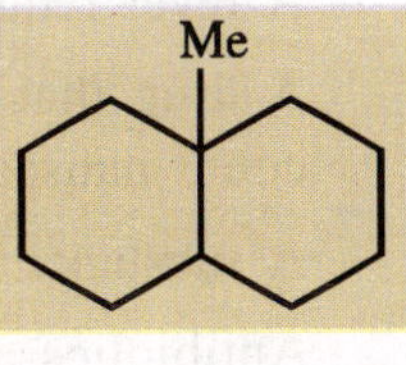

Ambident nucleophile A nucleophile that can react at two different sites such as an enolate, which may react at the carbon and oxygen.

Ambident electrophile An electrophile that can react at two different sites, such as an α, β-unsaturated carbonyl compound, which may react at the carbonyl carbon or at the β-position.

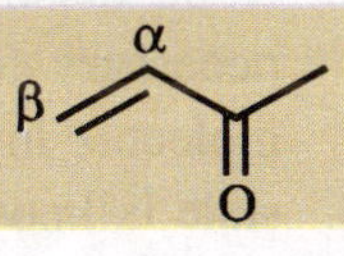

Anti Bonding Molecular orbital — A molecular orbital is antibonding if the number of bonding interactions is fewer than the number of nodes between the nuclei.

Activating substituent — A group that when present in place of hydrogen causes a particular reaction to occur faster. As for example, electron donating substituent methoxy group (in anisol) activating benzene ring towards electrophilic attack and electron withdrawing substituent nitro group (in meta dinitro benzene) activate benzene ring towards nucleophilic attack. When acetone hydrogen is replaced by carbethoxy group (in ethyl aceto acetate), the methylene hydrogen becomes more acidic (activated) and hence preferential enolate and therefore corresponding alkylation.

Asymmetric (chiral) synthesis — Asymmetric synthesis is a reaction in which an achiral unit in the substrate molecule (having either enantiotopic or diastereotopic groups or faces) is converted into a chiral unit in such a manner that unequal amount of chiral stereoisomers (ee or de) are produced.

Asymmetric induction — It is the extent of asymmetry induced at a prochiral centre either by chirality of the reagent (e.g. chiral aluminium alkoxide in the MPV reduction) or by chiral centre present in the substrate molecule itself (e.g. 1, 2-asymmetric induction—Cram's synthesis and 1, 4-asymmetric induction—Prelog's synthesis).

Alkaloids — Alkaloids are compounds obtained from plants having a basic character and containing at least one nitrogen atom in a heterocyclic ring.

Analgesic — A pain reliever.

Allergy — A disorder in which the body becomes hypersensitive to a particular antigen—the antigen is called allergen.

Antipyretic — A fever reducing drug.

Anti-emetic — A drug that prevents vomiting—such drugs are also used to count nausea.

Antibacterial — A substance that prevents the growth of or kills, bacteria.

Antibiotic — Antibiotics are specific chemical substances derived from or produced by living organism, which in small concentration, are capable of inhibiting the life process of micro-organisms.

Antiviral drug — A drug that interferes with DNA or RNA synthesis in order to prevent a virus from replicating.

B.D.E. — Bond dissociation energy. It is the amount of energy that is necessary for the cleavage of a particular covalent bond in a molecule in its gaseous state to produce radicals in the gaseous state is known as B.D.E.

	The B.D.E. (given as ΔH in KJ $mole^{-1}$) is energy needed for the endo thermic homolysis of covalent bond where ΔH is positive.
Bi-functional : molecule	A molecule with two functional groups.
Bio-organic compound	An organic compound that is formed in biological system.
Bond energy	The average value of B.D.E. is known as Bond energy, B.D.E. depends on the molecule for a particular bond but bond energy remain same for a particular bond in different molecules.
Bicyclic compound	A compound that contains two rings that share at least one carbon.
Bridged-ring compound	When two rings joined through non-adjacent atoms so that more than two atoms are common giving rise to bridged-ring compound.

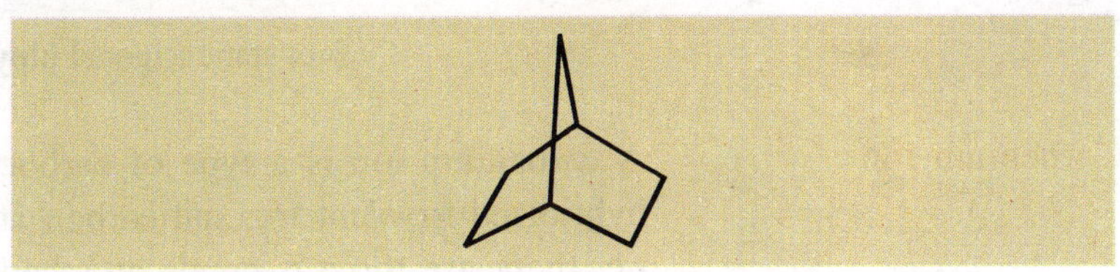

Bond-line formula	Formula in which connecting between carbons are shown but individual carbons and hydrogens are not. The bond-line formula represents the compound $(CH_3)_2CHCH_2CH_3$.
Bonding molecular orbital	A molecular orbital is bonding if the number of bonding interaction is greater than the number of nodes between the nuclei.
Bond order	It is half the difference of number of electrons in the bonding molecular orbital and the number of electrons in the antibonding molecular orbital.
Bacteria	There are a group of microorganisms which are unicellular and surrounded by rigid complex, protein cell-wall. They are highly adaptable and can survive extremes of temperature, atmospheric pressure and pH and are, therefore, found in all natural environment. They are resistant and can withstand in the temperature range,−190°C to +75°C. They are present everywhere in water, air, soil, in all living organisms, animals and plants.

Carbene A divalent carbon species. Carbene carbon has only six electrons in the valence shell instead of normal eight. There are two types of carbene, singlet and triplet carbene with SP^2 and SP hybridization respectively.

Carbanion A species with a negatively charged carbon atom. Carbanion carbon may be SP^3 or SP^2 or SP.

Carbonium ion The term carbonium ion is restricted in modern usage to penta-coordinate carbocation i.e. derivative of protonated CH_4. In such species there are only eight valence electrons involved in bonding the five ligands to the central carbon, so they involve non-classical multicentre bonds.

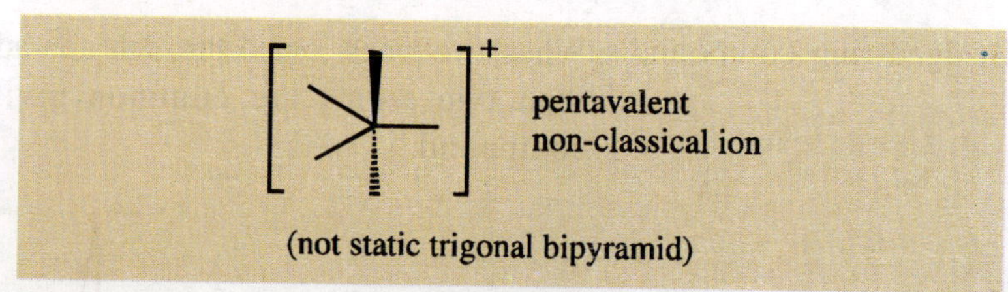

(not static trigonal bipyramid)

Carbenium ion A carbenium ion is a type of carbocation. it has a central SP^2 hybridized trivalent trigonal carbon atom with a vacant p-orbital i.e. there are three σ-bonds attached to the positively charged central carbon which has only six valence electrons. It may be seen as a derivative of a protonated carbene.

trivalent classical ion

Chemoselective reaction A reaction in which one functional group reacts leaving other potentially reactive functional groups unaltered within the same molecules. As for example,

Chemoselective reduction

$NaBH_4$

Chemoselective acetylation

HO–C$_6$H$_4$–NH_2 $\xrightarrow[\text{conc. } H_2SO_4]{Ac_2O}$ HO–C$_6$H$_4$–NHAc

Complex molecule	Large molecules whose total synthesis requires a large number of steps.
Cross conjugation	Nonlinear conjugation.
Combinatorial library	A group of structurally related compounds.
Combinatorial organic synthesis	The synthesis of a library of compounds by covalently connecting sets of building blocks of varying structures.
Convergent synthesis	A synthesis in which pieces of the target compound are individually prepared and then assembled.
Configuration	By 'configuration' is meant, the arrangements of atoms or group of atoms in space around the asymmetric (e.g. lactic acid) or dissymmetric (e.g. trans 1, 2-dimethyl cyclopropane) or rigid part of a molecule (e.g. cis and trans cinnamic acid or properly substituated allene). **Designation of configuration :** 'R' and 'S' (absolute), D and L (relative or genetic), erythro, threo or meso (diastereomeric), cis and trans or 'Z' and 'E' (diastereomeric geometric).
Conformation	The term 'conformation' is used to donate any one of the infinite number of momentary arrangements of the atoms or group of atoms in space that results from rotation about a C—C single bond. The different arrangements differ only in the dihedral angle and different conformations are possible for a single classical structure (configuration). The conformational isomers (rotamers) are actually stereoisomers which can be interconverted by mechanical processes without making or breaking of bonds (chemical processes). Mechanical processes may include rotation about a single bond, twisting or puckering of bonds (e.g. cyclohexane). When a molecule has a number of conformations, the energy minima conformation (or conformations) is called 'conformer'. **Designation of conformers :** eclipsed, gauche and staggered ± sp, ± sc, ± ac and ± ap.
CNS-drugs	The drugs which produce depressing effect on the central nervous system (CNS)

$$\text{Patient} \xrightarrow{\text{CNS-drug}} \text{Sedation} \rightleftharpoons \text{Hypnosis}$$

$$\text{Hypnosis} \rightleftharpoons \text{Anaesthesia}$$

$$\text{Death} \longleftarrow \text{Coma} \rightleftharpoons \text{Anaesthesia}$$

Driving force The thermodynamic push behind a reaction, analogous to gravity being the driving force for apples falling off trees.

Deactivating substituent A group that when present in place of hydrogen causes a particular reaction to occur more slowly. As for example, amino groups of aniline is highly activating to the electrophilic attack on benzene ring but when hydrogen of amino group is replaced by acetyl group, benzene ring becomes deativated to the electrophilic attack.

DNA A polymer of deoxyribonucleotides.

Detergent A salt of sulphonic acid.

Dextrorotatory The enantiomer that rotates polarized light in a clockwise direction.

DBE Double bond equivalent. it is the number of double bonds and/ or rings, in the compound to be ascertained. If the general formula of the compound is $C_aH_bN_cO_d$ then,

$DBE = a + 1 - (b - c)/2$

For example,

(i) Benzene C_6H_6

$DBE = 6 + 1 - {}^6/_2 = 4$

(3 double bonds and one ring)

(ii) Allylamine C_3H_7N

$DBE = 3 + 1 - (7 - 1)/2 = 1$

(one double bond)

Dihedral angle Dihedral angle (possible for 3D molecule and with at least four atoms) is the angle between the two planes $X - C_1 - C_2$ and $Y - C_2 - C_1$ of th type molecule $X - C_1 - C_2 - Y$ where X, C_1, C_2 and Y are in a row

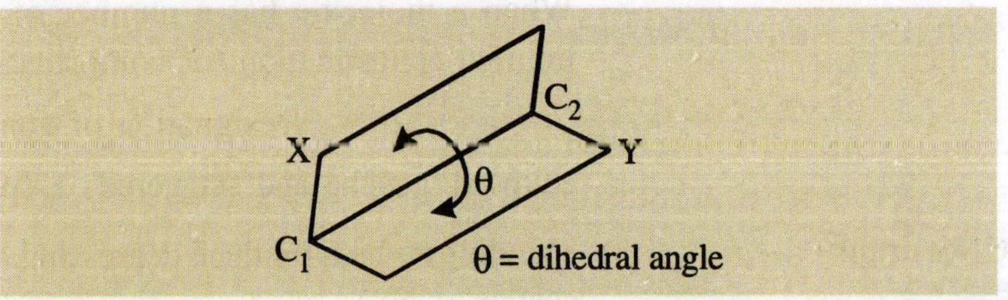

Dihedral angle is also possible when the ligands are branched

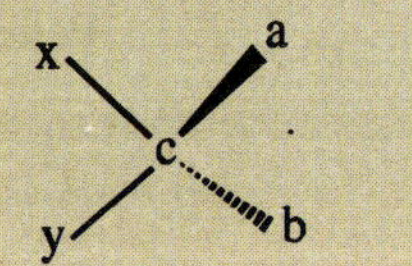

Dihedral angle between the two planes x – c – a and y – c – a.

Dihedral angle only has a magnitude but no direction and therefore it is scalar quantity and dihedral angle ranges from 0 to 360° and it is looked over the two planes.

Drug A drug is a substance used in the prevention, diagnosis, treatment or cure of desease in man or other animals.

Drug reacts with a biomolecules, triggering a physiological effect.

Drug actions Drug actions are involved with (i) absorption, distribution, biotransformation (metabolism) and elimination (excretion) of drugs (ii) the interaction of the drug with the specific biological macromolecules of the human body (such as protein or DNA) or with the bacteria.

Enamine An α, β-unsaturated tertiary amine. Enamine is activation of 'ene' which is used for alkylation or acylation with the protection of carbonyl in the synthetic strategy.

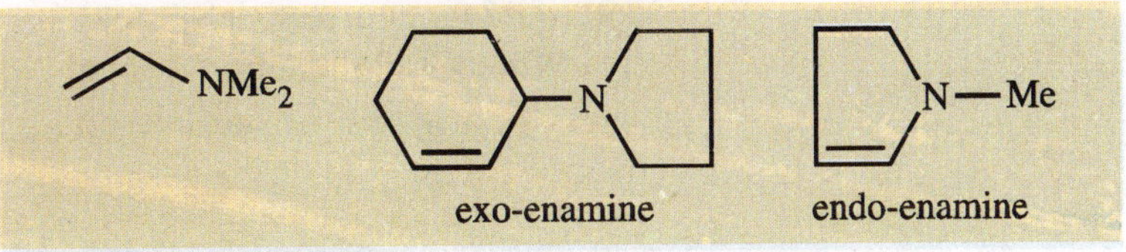

Enamine synthesis Synthesis of organic molecule using enamine for selective alkylation or acylation (regioselective) and also for chiral (ee) synthesis.

Ester exchange reaction (trans esterification) The reaction of an ester with an alcohol to form a different ester.

$$MeCO_2Ph + MeOH/\overset{\oplus}{H} \rightarrow MeCO_2Me + PhOH$$

Electrophile An electron-deficient species either with positive charge or with a vacant orbital capable of accepting electron—an electrophile is an electron sink.

Endo and Exo Endo is the position of a substituent in a bicyclic molecule opposite the main bridge. The endo substituent is considered to be in the 'inside' or more hindered position. The main bridge is determined by applying the priorities (highest first) (a) bridge with hetero atom, (b) bridge with fewest members, (c) saturated bridge, (d) bridge with fewest substituent, (e) bridge with lowest priority substituent.

Exo is the position of a substituent in a bicyclic molecule nearer the main bridge. The exo is considered to be on the outside or less hindered position.

exo endo

exo

endo

O

Energy profile diagram

An energy profile diagram is a plot of energy (molecular potential energy, molecular enthalpy or free energy) against reaction co-ordinate (progress of reaction) for the conversion of reactants into products based upon a particular mechanism.

An energy profile diagram describes (i) energy position of reactants and products, (ii) Whether the reaction is exothermic or endothermic, (iii) the step which is rate determining step (e.g. TS_1 step) (iv) Whether the reaction is one step, two steps or multistep processes.

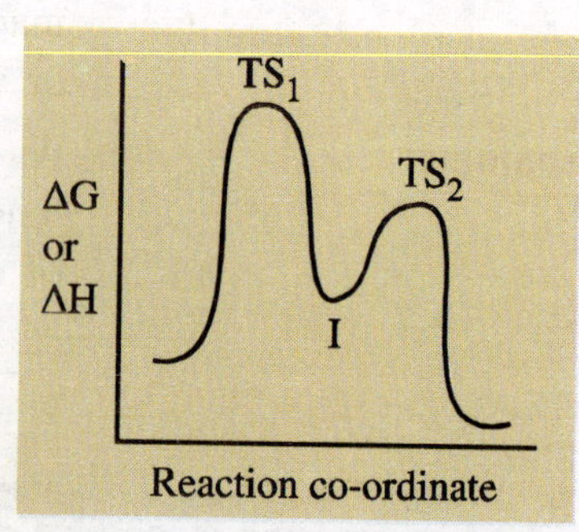

For multistep processes there must be some intermediates. The depth of the intermediate (I) indicates the quality of intermediate—the more the depth, the more stable the intermediate is. For each step, there is a transition state (TS) and per intermediate there are two transition states. For one step process of reaction one transition state and no intermediate (e.g. any S_N2 reaction).

Energy of activation

The barrier of highest energy (transition state) that has to be overcome in converting starting material into products.

Energy of activation is related with the rate of the reaction. The lower the energy of activation—higher will be the rate and vice-versa.

Enantioselective synthesis

Enantioselective synthesis is a reaction that converts an achiral unit with enantiotopic face or ligands (with chiral reagent) in the starting material to a chiral product in which one enantiomer is present in excess of the other.

Conversion of an achiral unit with diastereotopic face with (detouchable) chiral unit (with achiral reagent) in the starting material to a chiral product with unequal amount of enantiomers is also enantioselective synthesis, if chiral part is detouched at the end of the reaction e.g. Prelog's synthesis.

Estrogen A female sex-hormone.

Fused (condensed) ring When two rings joined through two adjacent atoms which are common to both the rings is the fused ring or condensed ring.

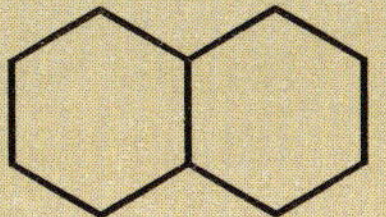

Fat A triester of glycerol that exists as a solid at room temperature.

First order asymmetric transformation It is defined as a spontaneous epimerization in solution involved single phase e.g. during mutarotation of glucose α-and β-D-glucopyranose.

FMO Frontier molecular orbital. That HOMO or LUMO which interact in the transition state in a pericyclic reaction.

Fullerene Fullerene has a definite formula C_{60}. Fullerene is a polygon with 60 vertices and 32 faces, 12 of which are pentagonal and 20 hexagonal is commonly encountered as the football. Each carbon in fullerene is bonded to three other and is SP^2 hybridized. It has a many resonance structures and appears to be aromatic. The fused ring aromatic system bends around and close to form a soccer ball-shaped molecule.

Fullerene is the third allotropic form of pure carbon known to exist after the network solids of diamond and graphite.

Gene A segment of DNA which contains the information necessary to control the synthesis of a specific protein via RNA. Gene is the unit of heredity. It is the fundamental unit responsible for he transmission of the potential characteristics of parents to their offspring.

Gabriel synthesis Conversion of an alkyl halide into a primary amine using phthalimide as a starting mterial.

Green chemistry A majority of organic reactions required the use of volatile solvents which may damage human health and environment and produce a number of hazardous substances or by-products which are harmful to the environment.

Green chemistry or Green reactions deals with the environmentally benign synthesis with a view to devise a pathway for the prevention of pollution, whether carried out in the laboratory or in the industry. In the Green reaction there is no need for separation or purification and energy required is kept to a minimum.

With the advancement of knowledge and development of new and better techniques a large number of organic reactions can be carried out in ECO-FRIENDLY conditions—these are Green reactions of Green chemistry.

Hydrophilic Literally 'water loving'; hydrophilic molecules are often charged and are readily soluble in water.

Hydrophilic bond : Polar attraction.

Hydrophobic Literally 'water hating'; hydrophobic molecules are non-polar and dissolve in organic solvents more easily than in water.

Hydrophobic bond : Van der waals' attraction (London forces).

HOMO Highest occupied molecular orbital; a type of frontier orbital. The HOMO contains the electrons of highest energy in a molecule, these are the electrons that take part in reactions as nucleophile.

Homo chiral and heterochiral Homo chiral will mean enantiomerically pure substance i.e. either all (+) or all (−) forms of chiral molecules. A pair enantiomers is a hetereochiral pair but all molecules of a particular enantiomer are homochiral.

If we consider a left hand and a right hand then they are heterochiral to each other but all left hands are homochiral and so also all right hands.

Heredity Heredity is the term applied to the transmission of the potential characteristics of parents to their offspring.

High dilution synthesis Synthesis of medium and large ring compounds by intramolecular cyclization at very low concentration of reactant avoiding intermolecular reaction.

Intramolecular rearrangement It is the migration of an atom or a group (e.g. Claisen or Pincol-pinacolone) or a bond (e.g. Sigmatropic) from one atom to another within a same molecule (in a concerted process) so that the product is an isomer of the starting molecule.

During the intramolecular migration, migrating group if asymmetric, migrates from the migrating origin with retention of configuration to the migrating terminus.

Intermolecular rearrangement It is the migration of an atom or a group from one atom to another between at least two separate molecules (in a stepwise process) and the product may or may not be an isomer of the starting material e.g. Fries rearrangement.

During the intermolecular migration, migrating group if asymmetric, migrates from the migrating origin with loss of asymmetry (i.e. racemised product) to the migrating terminus.

Intermediate

Transient species formed during a chemical reaction and it may take place in more than one step in a multistep reaction. Typically an intermediate is not stable under the conditions of its formations and proceeds further, to form product. Intermediates are genuine chemical entities with finite life times, although if they are high in energy their life-times may be very short. Unlike a transition state, which corresponds to a maximum along a potential energy surface, an intermediate lies at a potential energy minimum. Intermediates may be positively charged, negatively charged, radical and may be radical anion or even neutral molecule (e.g. R – N = C = O, benzyne) for a given reaction.

Inverse kinetic isotope effect

Inverse kinetic isotope effect is the one where K_H/K_D <1, as for example, if ElcB reaction is performed in D_2O in place of H_2O then the ratio of K_H/K_D is less than unity.

Iterative synthesis

A synthesis in which a reaction sequence is carried out more than once.

IHD

Index of hydrogen deficiency. The IHD is the number of 'pairs' of hydrogen atoms that must be removed from the saturated alkane formula (C_nH_{2n+2}) to give the molecular formula of the compound at hand.

The 'index' is therefore a number equal to the sum of the rings and multiple bonds (triple bonds require two index units) in the molecule.

As for example, the compound C_4H_8 (index = 1) has one of the following five combinations

Compound = C_4H_8
saturated alkane = C_4H_{10}

$\Delta H = 10 - 8 = 2$
Index = 2/2 = 1

1 double bond

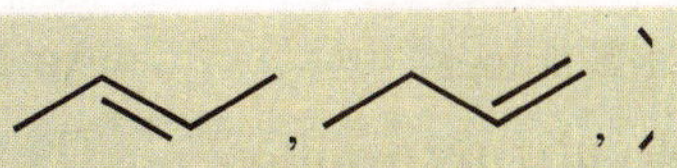

1 ring

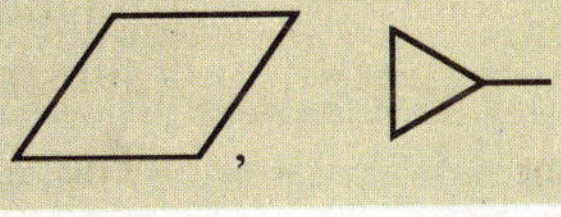

Latent polarity

The imaginary pattern of alternating positive and negative charges used to assist in the identification of suitable disconnections and synthons when using retrosynthetic analysis.

Latent polarities are arranged such that a positive charge is placed on the carbon adjacent to the heteroatom.

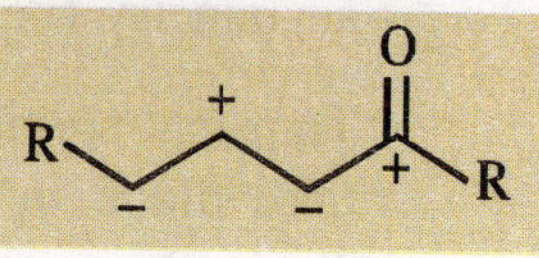

Levorotatory The enantiomer that rotates polarized light in a counter clockwise direction.

LUMO Lowest unoccupied molecular orbital. A type of frontier, the LUMO is the lowest in energy of the empty orbitals of a neutral molecule, cation, anion or radical. This is the orbital that takes part in reactions as electrophiles.

Lipids Lipids are naturally occuring organic molecules found in cells and tissues which are soluble in organic solvents but insoluble in water. Lipids can be subdivided into fats, waxes and steroids.

Linear synthesis A synthesis that builds a molecule step by step from starting materials.

Michael addition Conjugate addition of a stabilised carbanion to the β-position of an α, β-unsaturated double bond with a − M group.

Microscopic reversibility For a given reaction if forward and its reverse reaction follow exactly the same path but in opposite direction in the energy profile diagram. They have the same intermediate. As for example, sulphonation and desulphonation of benzene with conc. H_2SO_4 and dil H_2SO_4 respectively.

Material science The science of creating new materials such as metal, glass, wood, cardboard and paper.

MRI Magnetic resonance imaging. NMR used in medicine. The difference in the way water is bound in different tissues produces the signal vibration between organs as well as between healthy and diseased states.

MRI Scanner An NMR spectrometer used in medicine for whole body NMR.

Molecular recognition The recognition of one molecule by another as a result of specific interactions, for example, the specificity of an enzyme for its substrate.

Mechanism The sequence of steps (step-by-step process) that describes how a chemical reaction occurs, a description of the intermediates and transition states that are involved during the transformation of reactants to products. Mechanisms are 'theories' and not 'facts' of the subject.

Multistep synthesis Synthesis of a compound by a route that requires several steps.

Nucleophile Basic and electron rich centre either with a negative charge or with a lone-pair capable of donating electron to the carbon (and not to the proton) — a nucleophile is an electron source.

Nitrene A monovalent nitrogen species, analogous to a carbene. Nitrenes have only six electrons in their valence shell instead of normal eight. Nitrenes usually remains as triplet.

$$R-\ddot{\underset{\cdot\cdot}{N}}$$

NBMO Non bonding molecular orbital. The orbitals are too far apart to overlap significantly, so the molecular orbitals that results neither favours nor disfavours bonding.

Node That part of an orbital in which there is zero probability of finding an electron.

Non-covalent bindings Non-covalent bindings are H-bonding, Van-der-waals', electrostatic, hydrophilic, hydrophobic and π-stacking interactions, steric fit [receptor must have a cavity (pocket) that matches the size and shape of the guest], stereoelectronic fit (host-guest molecule must be properly aligned to maintain complementary between different subunits.

All these are chemistry of non-covalent bindings which brings order in the system—decrease in entropy.

Oxyanion A compound with a negatively charged oxygen ($R-\overset{\ominus}{O}$).

Oxonium ion A compound with a positively charged oxygen ($Me_3\overset{\oplus}{O}\,\overset{\ominus}{BF_4}$).

Oil A triester of glycerol that exists as a liquid at room temperature.

Organic synthesis Preparation of organic compound from other organic compound.

Photochemical reaction A reaction that takes place when a reactant absorbs light.

Pericyclic reaction Pericyclic reactions are those reactions in which a reactant is converted to a product through cyclic conjugated transition state without giving any intermediate either thermally or photochemically.

In the pericyclic reaction, the changes in bonding take place through reorganisation of electrons pairwise within a closed loop of interacting orbitals—hence the name pericyclic and they do not follow ionic or radical pathway.

Pro-chiral Molecules with enantiotopic ligands or faces are capable of giving two enantiomers and are known as pro-chiral. The ligands are thus known as pro-R and pro-S and the faces are Re-face and Si-face and the C-centre is known as pro-chiral centre.

Phase transfer catalysis Catalysis of a reaction by providing a way to bring a polar reagent into a non-polar phase so that the reaction between a polar and a non-polar compounds can occur. For example, oxidation of toluene to benzoic acid with $KMnO_4$ in presence of 18-crown-6.

Primary kinetic isotope effect (PKIE) The slowing of the rate of a reaction due to replacement of an atom by a heavier isotope when a bond to this atom is broken in the rate determining step; this is most commonly observed when hydrogen is replaced by deuterium. In the case of PKIE, the ratio of K_H/K_D is always greater than unity and the value usually ranges from 6 to 8. For example, oxidation of isopropyl alcohol with chromic acid.

$$Me_2CH-OH \xrightarrow{K_H} Me_2C{=}O$$

$$Me_2CD-OH \xrightarrow{K_D} Me_2C{=}O \qquad K_H/K_D > 6$$

Polymer A polymer is a large molecule made up of a repeating sequence of smaller units called monomer. Polymers which are synthesised from only one kind of monomer are called homopolymers and polymers which are prepared from more than one kind of monomers are called co-polymers.

Natural polymers are starch, cellulose, protein, natural rubber, DNA and RNA. Synthetic polymers are classified as chain-growth polymers (addition polymers) e.g. polythene, PVC, Teflon and step-growth polymers (condensation polymer) e.g. Terylene, Nylon and thermosetting polymers (resins) e.g. Bakelite which is a phenol-formaldehyde resin.

Petrochemicals The chemicals which are derived from petroleum or natural gas constituents, are called petrochemicals. The modern petrochemical industry produces plastics, fabrics, insectisides, herbicides, fertilizers, dyes, drugs, detergent and food preservatives.

Pharmaceuticals The study of the conversion of drug into a medicine (the design of dosage forms or formulations) and its manufacture, stability and the effectiveness of the drug dosage form is termed as pharmaceuticals.

There are three important stages in the use of a drug as a medicine (i) the discovery of the drug (ii) the manufacture of the drug in bulk-form, (iii) the formation of a drug into different dosage form like tablets, capsule, injection, syrup, cream, etc.

Phase transfer synthesis When organic synthesis is performed in presence of either insoluble inorganic supports or polymer bound reagents. Phase transfer synthesis is done under heterogeneous condition using at least one solid phase to which molecules are attached by covalent bond or ionic support. The reactions in the presence of supports are usually carried out in common organic solvents such as benzene, ethanol. For example, aldol and Michael addition occuring on a chromatographic column, Merrifield automated peptide synthesis.

Peptide synthesis Synthesis of amino acids-polymer where amino acids linked together by substituated amide bond.

Partial synthesis In many cases, the starting material may be a natural compound that can be efficiently converted into the desired product. In such cases, the synthesis of the desired product is referred to as a partial synthesis.

Photosynthesis The synthesis of glucose and oxygen from CO_2 and water in presence of visible light and chlorophyll.

Peterson synthesis Peterson synthesis is a silicon version of the wittig reaction for the synthesis of C—C double bond with desired stereochemistry.

Ring-strain The energy disadvantage in small rings due to the deviation of bond angles from their ideal values.

Rearrangement A reaction in which the arrangement of atoms or group of atoms within a molecule changes (change of C-skeleton or σ-frame-work) so that the product is an isomer of the starting material.

RNA A polymer of ribonucleotides.

Regioselective reaction A reaction that leads to the preferential formation of one constitutional (positional) isomer over another. A reaction that is 100% regioselective is the regiospecific.

Robinson annulation A Michael reaction followed by an intramolecular aldol condensation for cyclization.

RDX Research development explosive. It is a cyclic symmetrical trimethylene trinitramine $(CH_2)_3N_3(NO_2)_3$. RDX is prepared by the treatment of nitric acid with urotropine.

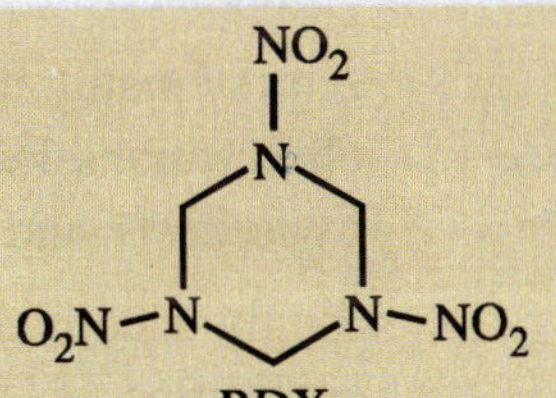

NO_2, O_2N–N, N–NO_2 (structure)

RDX

Ring-expansion rearrangement Rearrangement of a carbocation in which the positively charged carbon is bonded to a cyclic compound and as a result of rearrangement the size of the ring increases by one carbon.

RDS Rate determining step. Slowest step of a multistep reaction mechanism. The overall rate of a reaction can be no faster than its slowest step. RDS is the step in a multistep reaction that has the transition state with the highest energy.

Rate is an experimentally determined kinetic property.

Retrosynthesis Working backward (on paper) from the target molecule to the available starting materials by disconnection approach.

Simple molecules A simple molecule may be described as one which is small and whose total synthesis requires a small number of steps.

Spirocyclic compound A bicyclic compound joined through a single common atom (for a pair of rings) to give spirocyclic compond.

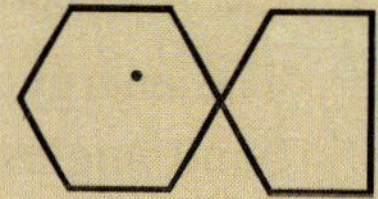

Soap Soap is a mixture of Na and K-salts of fatty acids containing 12 to 18 carbons.

Steroids Steroids can be defined as compounds which produce Diels hydrocarbon (one cyclopentane and three cyclohexane rings linked together) upon dehydrogenation with selenium.

The term steroid (sterol like) is derived from sterol (solid alcohol) as most of these compounds contain alcoholic group. Steroids exert a variety of physiological activities and may act as hormones i.e. chemical messengers secreted by glands in the body.

Stork enamine A reaction that uses an enamine as a Michael donor in a Michael condensation.

Scientific method A systematic approach to establishing new knowledge in which observations leads to laws, laws to theories, theories to testable hypothesis and hypotheses to experiments.

SOMO Singly occupied molecular orbital; the frontier orbital of a radical containing the unpaired electron.

Radical with high energy SOMO (when unpaired electron interact with the filled orbital) is the nucleophilic free radical e.g. $CH_2{=}CH-CH_2\cdot$ and $\cdot CH_3$ and radical with low energy SOMO (when unpaired electron interact with the vacant orbital π^*) is the electrophilic free radical e.g. $CH_3-\dot{C}H-CN$ and $CH_3-\underset{\underset{O}{\|}}{C}-CH_2\cdot$

Synthetic tree — An outline of the available routes to get the desired product (target molecule) from available starting materials is the synthetic tree.

Stereoelectronic factor — A combination of steric factor and electronic factor.

Stereoselectivity — The formation of one enantiomer, one diastereomer or one double bond isomer in preference to others.

Stereospecific reaction — A reaction in which the reactant can exist as a different stereoisomers and each stereoisomeric reactant leads to different stereoisomeric product. For example, cis and trans-2-butene reacts with singlet methylene individually to give exclusively cis and trans-1, 2-dimethyl cyclopropane respectively.

Stereoselective reaction — Stereoselective reactions are those reactions in which a given structure (may or may not be stereoisomeric) produce diastereomeric products in unequal amounts. For example, cis and trans-2-butene reacts with triplet methylene individually to give cis and trans-1, 2-dimethyl cyclopropane and in both the cases trans diastereomer is the major product.

SKIE — Secondary kinetic isotope effect. An isotope effect that arises for reasons other than in the primary kinetic isotope effect (PKIE) i.e. for any reaction if isotopic bond breaking does not take place in the rate determining step (RDS) but this isotopic bond breaking has the effect or contribution to the RDS is the SKIE. For example, rate of solvolysis of $(CD_3)_3CCl$ in water. For every deuterium incorporation for hydrogen in $(CH_3)_3CCl$, the rate is 10% slowed down.

Solid-phase synthesis — A technique in which one end of the compound being synthesized is covalently attached to a solid support. For example, peptide synthesis with Merrifield resin.

Second order asymmetric transformation — If one of the enantiomers or diastereomers crosses the phase-boundary i.e. comes out of the solution as crystals is the second order asymmetric transformation. For example, crystallisation of glucose from ethanol gives exclusively α-D-glucose and from pyridine β-D-glucose.

Statistical synthesis The technique used in the synthesis of catenanes and Rotaxanes is the statistical synthesis.

Supramolecule and Suprachemistry Supramolecule or supermolecule is a highly structured assembly of 'host and guest' molecules complementing each other both sterically and stereoelectronically through multiple contact points on common surfaces.

Suprachemistry or supermolecular chemistry is the chemistry of non-covalent intermolecular forces that bind two or more molecules leading to 'supermolecules'.

The host molecule usually very large with a cavity [e.g. Macro (mono or poly) cyclic polyethers] forms a stable complex with one or more guest molecules which are usually small (e.g. urea) through weak directional forces working within different subunits of the component molecules.

Tosylation A reaction which introduces the toluene-4-sulphonyl group into a molecule, generally by the reaction of an alcohol with tosyl chloride in presence of a weak base to give the tosylate ester.

Terpene A compound, isolated from a plant that contains carbon atoms in multiple of five (isoprene unit).

Limonene

Terpenoid A terpene that contains oxygen.

Camphor

Torsion angle (TA) Torsion angle (possible for 3D molecule and with at least four atoms and capable of giving rotamers) is the angle subtended by two ligands X and Y across the C—C bond axis for a type of molecule $X - C_1 - C_2 - Y$ which always are in a row.

TA = + 60° (+SC)

	Torsion angle has a magnitude and direction and therefore it is vector quantity. Torsion angle ranges from 0± 180° and it is looked through C—C bond axis.
Transcription	The synthesis of mRNA from a DNA blue-print.
Transition state	The point of maximum free energy in a reaction step in the conversion of starting material into product is the transition state (TS). Transition state is not a discrete, molecular species but it is a unique arrangements of reacting entities in which existing bonds may partly broken and new bonds partly formed.

$$A-B+C \longrightarrow \underset{\substack{\text{TS} \\ \text{(activated complex)}}}{[A\text{-----}B\text{-----}C]^{\neq}} \longrightarrow A + B-C$$

TS does not have a finite life time. In the TS, the chemical composition of the reacting substances are such in which the reacting molecules in a proper orientation or conformation for favourable (successful) collision.

Lastly, TS is a hypothetical description of the activated complex at the point of highest energy.

Testosterone	Male sex hormone.
Umpoled	Reversing the normal polarity of a functional group.
Umpolung synthesis	Synthesis involving reverse polarity of the substrate.
Van der waals' force (London force)	Induced-dipole induced-dipole interactions.
Van der waals' radius	A measure of the effective size of an atom or a group.
Van der waals' repulsion	A repulsive force occurs if two atoms or group of atoms approach each other at a distance less than the sum of their van der waals' radii.
Visible light	Electromagnetic radiation with wavelength ranging from 400-800 nm.
Vitamin	A substance needed in small amounts for normal body function that the body cannot synthesize or cannot synthesize in adequate amount.
Virus	These are very small micro-organisms which are parasitic within living-cells. These differ from bacteria in having only one kind of nucleic acid either DNA or RNA, in lacking the apparatus necessary for energy production and protein synthesis.

Wax An ester formed from a long chain carboxylic acid and a long chain alcohol. The acid and alcohol residues are usually saturated and contain an even number of carbon atoms.

$$CH_3-(CH_2)_x-\underset{\underset{O}{\|}}{C}-O-(CH_2)_y-CH_3 \qquad x = 24, 26 \qquad y = 29, 31$$

Williamson's synthesis Synthesis of ether by treatment of the halide with alkoxide.

Work up A term used to describe operations done when the main reactions is complete and before the product can be isolated such as addition of acid or base or extraction between water and organic solvents.

Ylide A molecule with positive and negative charges on adjacent covalently bonded atoms with complete octet.

$$Ph_3\overset{\oplus}{P}-\overset{\ominus}{C}H_2 \quad \text{and} \quad R_3\overset{\oplus}{N}-\overset{\ominus}{C}H_2$$

Zwitter ion A molecule that contains both positive and negative charged groups on non-adjacent atoms such as

$$H_3\overset{\oplus}{N}-CH_2-CO_2^{\ominus}, \quad H_3\overset{\oplus}{N}-C_6H_4-SO_3^{\ominus}.$$

Bibliography

Acheson R.M. *An Introduction to the Chemistry of Heterocyclic Compounds*, Second edition, Wiley Eastern Limited.

Bansal R.K. *Organic Reaction Mechanism*, Third edition, Tata McGraw Hill Publishing Company Limited.

Bansal R.K. *Synthetic Approaches in Organic Chemistry*, Narosa Publishing House, 1996.

Carey F.A., Sundberg R.J. *Advanced Organic Chemistry*, Part B: *Reactions and Synthesis*, Third edition, Plenum Press.

Carruthers W. *Modern Methods of Organic Synthesis*, Third edition, Cambridge University Press.

Chatwal G.R. *Synthetic Drugs*, Fifth edition, 1997, Himalaya Publishing House.

Finar I.L. *Organic Chemistry* (Vols. I and II), Elbs, Longman.

Hendrickson J.B., Donald J.C., Hammond G.S. *Organic Chemistry*, McGraw Hill, Kogakusha Ltd.

Jacobs A. *Understanding Organic Reaction Mechanisms*, Cambridge University Press.

Joule J.A., Mills K., Smith G.F. *Heterocyclic Chemistry*, Third edition, Stanley Thornes (Publisher) Ltd.

Kar A. — *Medicinal Chemistry*, Second edition, New Age International (P) Limited.

Krupadanan G.L.D., Prasad D.V., Rao K.V.P., Reddy K.L.N., Sudhakar C. — *Drugs*, University Press (India) Limited, 2001.

Laudon G.M. — *Organic Chemistry*, Fourth edition, Oxford University Press.

Mackie R.K., Smith D.M. — *Guidebook to Organic Synthesis* 1982, Longman, Scientific & Technical.

Mukherjee S.M., Singh S.P. — *Reaction Mechanism in Organic Chemistry*, Third edition, 1984, Macmillan India Limited.

Mukherjee S.M., Singh S.P., Kapoor R.P. — *Organic Chemistry* (Vols. I, II, and III) New Age International (P) Ltd., Publisher.

Nasipuri D. — *Stereochemistry of Organic Compounds*, Second edition, New Age International Publisher Limited.

Norman R.O.C. — *Principles of Organic Synthesis*, Methuen, 1968.

Norman R.O.C., Coxon J.M. — *Principles of Organic Synthesis*, Third edition, 1993, Chapman & Hall.

Shukla S.P., Trivedi G.L. — *Modern Organic Chemistry* 1997, S. Chand & Company Ltd.

Singh J., Singh J. — *Photochemistry and Pericyclic Reaction*, New Age International (P) Limited, Publisher.

Singh M.S. — *Advanced Organic Chemistry*, Reactions and Mechanisms, Pearson Education.

Sykes P. — *A Guidebook to Mechanism in Organic Chemistry*, Fourth edition, 1975, Orient Longman Ltd.

Warren S. — *Organic Synthesis: The Disconnection Approach*, John Wiley & Sons.

Warren S. — *Designing Organic Synthesis*, John Wiley & Sons, 1978.